Methods in Enzymology

Volume 173
BIOMEMBRANES
Part T
Cellular and Subcellular Transport:
Eukaryotic (Nonepithelial) Cells

METHODS IN ENZYMOLOGY

EDITORS-IN-CHIEF

John N. Abelson   Melvin I. Simon

DIVISION OF BIOLOGY
CALIFORNIA INSTITUTE OF TECHNOLOGY
PASADENA, CALIFORNIA

FOUNDING EDITORS

Sidney P. Colowick and Nathan O. Kaplan

Methods in Enzymology

Volume 173

Biomembranes

Part T

Cellular and Subcellular Transport:
Eukaryotic (Nonepithelial) Cells

EDITED BY

Sidney Fleischer
Becca Fleischer

DEPARTMENT OF MOLECULAR BIOLOGY
VANDERBILT UNIVERSITY
NASHVILLE, TENNESSEE

Editorial Advisory Board

ACADEMIC PRESS, INC.
Harcourt Brace Jovanovich, Publishers
San Diego   New York   Berkeley   Boston
London   Sydney   Tokyo   Toronto

ACADEMIC PRESS, INC.
San Diego, California 92101

United Kingdom Edition published by
ACADEMIC PRESS LIMITED
24-28 Oval Road, London NW1 7DX

LIBRARY OF CONGRESS CATALOG CARD NUMBER: 54-9110

ISBN   0-12-182074-2   (alk. paper)

PRINTED IN THE UNITED STATES OF AMERICA
89 90 91 92   9 8 7 6 5 4 3 2 1

Table of Contents

Section I. Red Blood Cells

A. Intact System

B. Derived Red Cell Preparations

C. Anion Transporter

Section II. Other Mammalian Cells: Intact Cells

Contributors to Volume 173

Article numbers are in parentheses following the names of contributors.
Affiliations listed are current.

THEO P. M. AKERBOOM (34), *Institut für Physiologische Chemie I, Universität Düsseldorf, D-4000 Düsseldorf, Federal Republic of Germany*

JAVIER ALVAREZ (22), *Departamento de Fisiología y Bioquímica, Facultad de Medicina, Universidad de Valladolid, 47005 Valladolid, Spain*

RICHARD A. ANDERSON (24), *Department of Pharmacology, University of Wisconsin, Madison, Wisconsin 53706*

POUL J. BJERRUM (31), *Department of Clinical Chemistry, Rigshospitalet, State University Hospital, DK-2100 Copenhagen O, Denmark*

P. F. BLACKMORE (35), *Department of Pharmacology, Eastern Virginia Medical School, Norfolk, Virginia 23501*

RHODA BLOSTEIN (23), *Departments of Medicine and Biochemistry, McGill University, Montreal General Hospital Research Institute, Montreal, Quebec H3G 1A4, Canada*

JESPER BRAHM (9, 29), *Department of General Physiology and Biophysics, The Panum Institute, University of Copenhagen, DK-2200 Copenhagen N, Denmark*

Z. I. CABANTCHIK (14, 27), *Department of Biological Chemistry, Institute of Life Sciences, Hebrew University, Jerusalem, Israel 91904*

PETER M. CALA (19), *Department of Human Physiology, School of Medicine, University of California, Davis, California 95616*

MITZY CANESSA (10), *Endocrine-Hypertension Division, Brigham and Women's Hospital, Harvard Medical School, Boston, Massachusetts 02115*

JOSEPH R. CASEY (32), *MRC Group in Membrane Biology, Departments of Medicine and Biochemistry, University of Toronto, Toronto, Ontario M5S 1A8, Canada*

HALVOR N. CHRISTENSEN (38), *Department of Pediatrics (Pediatric Genetics), University of California at San Diego, La Jolla, California 92093*

S. COHEN[1] (48), *Division of Cell Biology, The Hospital for Sick Children, Toronto, Ontario M5G 1X8, Canada*

JOHN CUPPOLETTI (25), *Department of Physiology and Biophysics, University of Cincinnati, College of Medicine, Cincinnati, Ohio 45267*

MARY BETH DE YOUNG (41), *Department of Physiology and Biophysics, Case Western Reserve University, Cleveland, Ohio 44106*

BERNHARD DEUTICKE (18), *Institut für Physiologie, Rheinisch-Westfälische Technische Hochschule Aachen, D-5100 Aachen, Federal Republic of Germany*

S. J. DIXON (48), *Department of Physiology, University of Western Ontario, Health Science Center, London, Ontario N6A 5C1, Canada*

A. ALAN EDDY (47), *Department of Biochemistry and Applied Molecular Biology, University of Manchester Institute of Science and Technology, Manchester M60 1QD, England*

J. CLIVE ELLORY (8), *University Laboratory of Physiology, University of Oxford, Oxford OX1 3PT, England*

LEIGH H. ENGLISH (42), *Ecogen Inc., Langhorne, Pennsylvania 19047*

ERLAND ERDMANN, (40), *Medizinische Klinik I der Universität München, Klini-*

[1] Deceased.

kum Grosshadern, D-8000 München 70, Federal Republic of Germany

EVAN A. EVANS (1), Departments of Pathology and Physics, University of British Columbia, Vancouver, British Columbia V6T 1W5, Canada

J. H. EXTON (35), Department of Molecular Physiology and Biophysics, Vanderbilt University School of Medicine, Howard Hughes Medical Institute, Nashville, Tennessee 37232

JEFFREY C. FREEDMAN (5), Department of Physiology, State University of New York, Health Science Center at Syracuse, Syracuse, New York 13210

OTTO FRÖHLICH (3), Department of Physiology, Emory University School of Medicine, Atlanta, Georgia 30322

GÜNTER FRED FUHRMANN (15), Department of Pharmacology and Toxicology, Philipps-Universität Marburg, D-3550 Marburg, Federal Republic of Germany

JAVIER GARCÍA-SANCHO (6, 22), Departamento de Fisiología y Bioquímica, Facultad de Medicina, Universidad de Valladolid, 47005 Valladolid, Spain

BARTOLOMEO GIANNATTASIO (41), Department of Physiology and Biophysics, Case Western Reserve University, Cleveland, Ohio 44106

JØRGEN GLIEMANN (39), Biomembrane Research Center, Institute of Physiology, University of Aarhus, DK-8000 Aarhus C, Denmark

J. D. GOETZ-SMITH (48), Division of Cell Biology, Research Institute, The Hospital for Sick Children, Toronto, Ontario M5G 1X8, Canada

S. GRINSTEIN (48), Division of Cell Biology, Research Institute, The Hospital for Sick Children, Toronto, Ontario M5G 1X8, Canada

ROBERT D. GRUBBS (36), Department of Pharmacology and Toxicology, Wright State University School of Medicine, Dayton, Ohio 45435

R. GRYGORCZYK (7, 30), Department of Physiology, University of Toronto, Toronto, Ontario M5S 1A8, Canada

ROBERT B. GUNN (3), Department of Physiology, Emory University School of Medicine, Atlanta, Georgia 30322

MARK HAAS (16), Departments of Pathology and Physiology, Yale University School of Medicine, New Haven, Connecticut 06510

P. HANKE-BAIER (30), Procter & Gamble GmbH, 6231 Schwalbach, Federal Republic of Germany

CATHERINE M. HARVEY (8), University Laboratory of Physiology, University of Oxford, Oxford OX1 3PT, England

WILLIAM J. HARVEY (23), Angelini Pharmaceutical Canada Inc., Montreal, Quebec, Canada H3F 1Z1

S. PAUL HMIEL (36), Department of Pharmacology, School of Medicine, Case Western Reserve University, Cleveland, Ohio 44106

D. HOF (7), I. Physiologisches Institut, Universität des Saarlandes, D-6650 Homburg/Saar, Federal Republic of Germany

KAREN S. HOFFMANN (19), College of Veterinary Medicine, University of Minnesota, St. Paul, Minnesota 55108

WILLIAM C. HORNE (24), Child Study Center, Yale University School of Medicine, New Haven, Connecticut 06510

E. R. JOHNSON (47), ConvaTec Biological Research Laboratory, New Tech Clwyd Ltd., Clwyd CH5 2NU, Wales

ROBERT M. JOHNSON (2), Department of Biochemistry, Wayne State Medical School, Detroit, Michigan 48201

CHAN Y. JUNG (25), Biophysics Laboratory, Veterans Administration Medical Center, Buffalo, New York 14214

RONALD S. KAPLAN (45), Department of Pharmacology, College of Medicine, University of South Alabama, Mobile, Alabama 36688

MICHAEL S. KILBERG (37), Department of Biochemistry and Molecular Biology, J. Hillis Miller Health Center, University of Florida, Gainesville, Florida 32610

PHILIP A. KNAUF (29), *Department of Biophysics, University of Rochester, School of Medicine and Dentistry, Rochester, New York 14642*

THOMAS L. LETO (24), *Section of Bacterial and Infectious Disease, National Institutes of Health/NIAID, Bethesda, Maryland 20892*

VIRGILIO L. LEW (6), *Physiological Laboratory, Cambridge University, Cambridge, England*

MICHAEL R. LIEBER (21), *Department of Pathology, Stanford University School of Medicine, Stanford, California 94305*

DEBRA M. LIEBERMAN (32), *MRC Group in Membrane Biology, Departments of Medicine and Biochemistry, University of Toronto, Toronto, Ontario M5S 1A8, Canada*

MICHAEL E. MAGUIRE (36), *Department of Pharmacology, School of Medicine, Case Western Reserve University, Cleveland, Ohio 44106*

FREDERICK R. MAXFIELD (46), *Department of Pathology, Columbia University, New York, New York 10032*

THOMAS J. MCMANUS (16), *Department of Cell Biology, Division of Physiology, Duke University Medical Center, Durham, North Carolina 27710*

TERRI S. NOVAK (5), *Department of Physiology, State University of New York, Health Science Center at Syracuse, Syracuse, New York 13210*

JOS A. F. OP DEN KAMP (12), *Centre for Biomembranes and Lipid Enzymology, State University of Utrecht, 3508 TB Utrecht, The Netherlands*

JOHN C. PARKER (17), *Department of Medicine, University of North Carolina at Chapel Hill, Chapel Hill, North Carolina 27599*

H. PASSOW (30), *Max-Planck-Institut für Biophysik, Abteilung für Zellphysiologie, D-6000 Frankfurt am Main 71, Federal Republic of Germany*

PETER L. PEDERSEN (45), *Laboratory for Molecular and Cellular Bioenergetics,* *Department of Biological Chemistry, The Johns Hopkins University School of Medicine, Baltimore, Maryland 21205*

PETER G. W. PLAGEMANN (44), *Department of Microbiology, University of Minnesota, Minneapolis, Minnesota 55455*

RAYMOND D. PRATT (45), *Laboratory for Molecular and Cellular Bioenergetics, Department of Biological Chemistry, The Johns Hopkins University School of Medicine, Baltimore, Maryland 21205*

REINHART A. F. REITHMEIER (32), *MRC Group in Membrane Biology, Departments of Medicine and Biochemistry, University of Toronto, Toronto, Ontario M5S 1A8, Canada*

BEN ROELOFSEN (12), *Centre for Biomembranes and Lipid Enzymology, State University of Utrecht, 3508 TB Utrecht, The Netherlands*

ASER ROTHSTEIN (26), *Research Institute, The Hospital for Sick Children, Toronto, Ontario M5G 1X8, Canada*

JOHN R. SACHS (4), *Department of Medicine, State University of New York at Stony Brook, Stony Brook, New York 11794*

ANTONIO SCARPA (41), *Department of Physiology and Biophysics, Case Western Reserve University, Cleveland, Ohio 44106*

JOHN T. SCHULZ (42), *Tufts University, School of Medicine, Boston, Massachusetts 02111*

W. SCHWARZ (7, 30), *Max-Planck-Institut für Biophysik, Abteilung für Zellphysiologie, D-6000 Frankfurt am Main 71, Federal Republic of Germany*

HELMUT SIES (34), *Institut für Physiologische Chemie I, Universität Düsseldorf, D-4000 Düsseldorf, Federal Republic of Germany*

RAYMOND A. SJODIN (43), *Department of Biophysics, University of Maryland, School of Medicine, Baltimore, Maryland 21201*

MARSHALL D. SNAVELY (36), *Department of Pharmacology, School of Medicine, Case Western Reserve University, Cleveland, Ohio 44106*

ARTHUR K. SOLOMON (11), *Biophysical Laboratory, Harvard Medical School, Boston, Massachusetts 02115*

THEODORE L. STECK (21, 33), *Department of Biochemistry and Molecular Biology, The University of Chicago, Chicago, Illinois 60637*

MICHAEL J. A. TANNER (28), *Department of Biochemistry, School of Medical Sciences, University of Bristol, Bristol BS8 1TD, England*

AHMAD WASEEM (33), *Clare Hall Laboratories, Imperial Cancer Research Fund, Herts EN6 3LD, England*

KARL WERDAN (40), *Medizinische Klinik I der Universität München, Klinikum Grosshadern, D-8000 München 70, Federal Republic of Germany*

W. F. WIDDAS (13), *Department of Biology, Royal Holloway and Bedford New College, University of London, Egham, Surrey TW20 0EX, England*

ROBERT M. WOHLHUETER (44), *Institute of Human Genetics, University of Minnesota, Minneapolis, Minnesota 55455*

PHILLIP G. WOOD (20), *Max-Planck-Institut für Biophysik, Abteilung für Zellphysiologie, 6000 Frankfurt am Main 71, Federal Republic of Germany*

Preface

Biological transport is part of the Biomembranes series of *Methods in Enzymology*. It is a continuation of methodology concerned with membrane function. This is a particularly good time to cover the topic of biological membrane transport because there is now a strong conceptual basis for its understanding. The field of transport has been subdivided into five topics.

1. Transport in Bacteria, Mitochondria, and Chloroplasts
2. ATP-Driven Pumps and Related Transport
3. General Methodology of Cellular and Subcellular Transport
4. Cellular and Subcellular Transport: Eukaryotic (Nonepithelial) Cells
5. Cellular and Subcellular Transport: Epithelial Cells

Topic 1 covered in Volumes 125 and 126 initiated the series. Topic 2 is covered in Volumes 156 and 157, Topic 3 in Volumes 171 and 172, and Topic 4 in Volumes 173 and 174. The remaining topic will be covered in subsequent volumes of the Biomembranes series.

Topic 4 is divided into two parts: this volume (Part T) which deals mainly with intact cells; major emphasis is on the red cell, derived red cell preparations, and the anion transporter, and Volume 174 (Part U) which covers transport by isolated subcellular organelle fractions, purified components as well as transport in plants, plant organelles, and single cell eukaryotes.

We are fortunate to have the good counsel of our Advisory Board. Their input ensures the quality of these volumes. The same Advisory Board has served for the complete transport series. Valuable input on the outlines of the five topics was also provided by Qais Al-Awqati, Ernesto Carafoli, Halvor Christensen, Isadore Edelman, Joseph Hoffman, Phil Knauf, and Hermann Passow. Additional valuable input for Volumes 173 and 174 was obtained from Ioav Cabanchik, John Exton, Arnot Kotyk, and Aser Rothstein.

The names of our advisory board members were inadvertently omitted in Volumes 125 and 126. When we noted the omission, it was too late to rectify the problem. For Volumes 125 and 126, we are pleased to acknowledge the advice of Angelo Azzi, Youssef Hatefi, Dieter Oesterhelt, and Peter Pederson.

The enthusiasm and cooperation of the participants have enriched and made these volumes possible. The friendly cooperation of the staff of Academic Press is gratefully acknowledged.

These volumes are dedicated to Professor Sidney Colowick, a dear friend and colleague, who died in 1985. We shall miss his wise counsel, encouragement, and friendship.

SIDNEY FLEISCHER
BECCA FLEISCHER

METHODS IN ENZYMOLOGY

VOLUME 157. Biomembranes (Part Q: ATP-Driven Pumps and Related Transport: Calcium, Proton, and Potassium Pumps)
Edited by SIDNEY FLEISCHER AND BECCA FLEISCHER

VOLUME 158. Metalloproteins (Part A)
Edited by JAMES F. RIORDAN AND BERT L. VALLEE

VOLUME 159. Initiation and Termination of Cyclic Nucleotide Action
Edited by JACKIE D. CORBIN AND ROGER A. JOHNSON

VOLUME 160. Biomass (Part A: Cellulose and Hemicellulose)
Edited by WILLIS A. WOOD AND SCOTT T. KELLOGG

VOLUME 161. Biomass (Part B: Lignin, Pectin, and Chitin)
Edited by WILLIS A. WOOD AND SCOTT T. KELLOGG

VOLUME 162. Immunochemical Techniques (Part L: Chemotaxis and Inflammation)
Edited by GIOVANNI DI SABATO

VOLUME 163. Immunochemical Techniques (Part M: Chemotaxis and Inflammation)
Edited by GIOVANNI DI SABATO

VOLUME 164. Ribosomes
Edited by HARRY F. NOLLER, JR., AND KIVIE MOLDAVE

VOLUME 165. Microbial Toxins: Tools for Enzymology
Edited by SIDNEY HARSHMAN

VOLUME 166. Branched-Chain Amino Acids
Edited by ROBERT HARRIS AND JOHN R. SOKATCH

VOLUME 167. Cyanobacteria
Edited by LESTER PACKER AND ALEXANDER N. GLAZER

VOLUME 168. Hormone Action (Part K: Neuroendocrine Peptides)
Edited by P. MICHAEL CONN

VOLUME 169. Platelets: Receptors, Adhesion, Secretion (Part A)
Edited by JACEK HAWIGER

Section I

Red Blood Cells

A. Intact System
Articles 1 through 19

B. Derived Red Cell Preparations
Articles 20 through 25

C. Anion Transporter
Articles 26 through 33

[1] Structure and Deformation Properties of Red Blood Cells: Concepts and Quantitative Methods

By EVAN A. EVANS

Introduction

The mature circulating red blood cell is a membrane-encapsulated liquid compartment which contains a hemoglobin solution for oxygen delivery to—and carbon monoxide removal from—the metabolizing tissues of an animal. These liquid-filled "bags" are not cells in the conventional eukaryotic sense. They cannot reproduce by cell division; instead they are "dead" remnants left over from several generations of nucleated cell division.[1] At the last stage of development, the genetic apparatus is disabled and the cytoplasmic reticular structure is broken down to leave a simple solution of hemoglobin. In mammals, the final stage is accompanied by expulsion of the nucleus prior to release into the circulation from the site of production (e.g., the bone marrow). On the other hand, for nonmammals (fish, birds, reptiles, amphibians), the nucleus remains attached on two faces by the cell membrane and is surrounded by a toroidal region of hemoglobin solution. When extracted from the circulation, the final products appear with smooth membrane contours as shown in Fig. 1. However, it is important to emphasize that the early-stage reticulocytes which just precede these circulating cells do not exhibit smooth or stationary membrane shapes.[1] Also, when enucleated *in vitro*, reticulocytes do not spontaneously form smooth surfaces. Consequently, it is very likely that the beautiful symmetric contours observed for circulating red cells are derived in part from the effects of high shear forces in the circulation which the reticulocyte experiences during its initial maleable phase. Subsequent to appearance in the circulation, these liquid-filled membrane capsules survive in the circulation for periods of time ranging from days (for birds) to years (for giant land tortoises).[2] In healthy humans, the circulating red cell has an average life span of about 120 days. Even with this 4-month longevity, red cells are replaced in the circulation at a phenomenal rate of millions per second! When we consider the mechanical trauma that red cells face in the circulation (e.g., high shear stresses in

[1] M. Bessis, "Living Blood Cells and Their Ultrastructure." Springer-Verlag, New York, 1972.

[2] O. W. Schalm, N. C. Jain, and E. J. Carroll, "Veterinary Hematology." Lea & Febiger, Philadelphia, 1975.

METHODS IN ENZYMOLOGY, VOL. 173

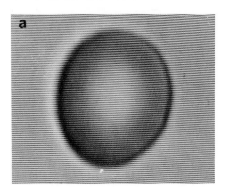

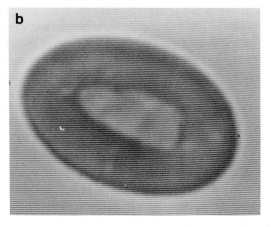

FIG. 1. Video micrographs of single red blood cells: (a) A *nonnucleated* human erythrocyte and (b) a *nucleated* avian erythrocyte. Both cell images were taken at the same magnification characterized by the human erythrocyte dimension of about 8×10^{-4} cm.

arteries, rapid extension and folding in the microcirculation, extrusion through small apertures or fenestrations as in the spleen), we are impressed by the resilience of the red cell membrane envelope which provides the structural integrity for the cell over the 120-day life span.

Red cell membranes are a molecularly thin, trilamellar composite made up of a superficial (liquid) double layer supported by a subsurface (rigid) network (Fig. 2). Additional components are associated with the red cell membrane, e.g., surface coats of carbohydrates and adsorbed gels of hemoglobin. The unifying feature of all membrane structures is the preferential assembly of molecular amphiphiles into two-dimensional con-

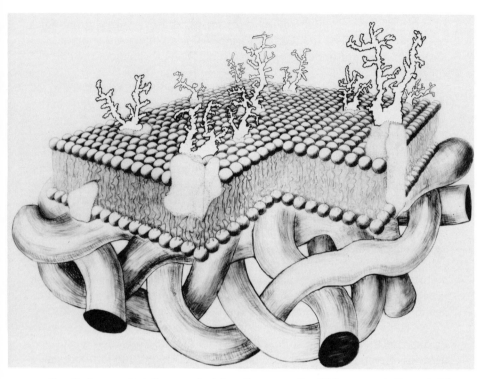

FIG. 2. A conceptual diagram of a trilamellar composite which represents the material structure of the red cell membrane: a superficial (liquid) double layer supported by a subsurface (rigid) network.

densed liquid bilayers which form tight cohesive (chemical) insulators to provide for a distinct cellular chemistry. Of equal importance, the subsurface network provides the "scaffolding" and support for the liquid bilayer; this rigid support creates the possibility of nonspherical cell geometries and provides the capability of large cell deformation without creating holes or leaks in the bilayer "sealer." The common constituents of the outer membrane double layer are lipid molecules which have their polar head groups anchored at the water interfaces and their hydrophobic acyl chains internalized toward the center of the bilayer. The lipid layers act as "a solvent" into which many amphiphilic "solutes" are incorporated (e.g., proteins and cholesterol).[3] The adjacent subsurface "scaffolding" is

[3] T. L. Steck, *J. Supramol. Struct.* **8**, 311 (1978).

made up of several proteins to form a filamentous network-like structure.[4] Because of the negligible solubility of membrane molecules in the adjacent aqueous phases, the red cell membrane is a condensed state of matter which does not undergo extensive remodeling throughout the cell life span.

Because it is the major cellular constituent in blood, deformability of the red cell greatly influences the flow properties of blood especially in the microcirculation. For high-shear situations in large vessels, red cells exhibit two types of kinematic behavior: tumbling or rotation of the cell as a rigid body; stable orientation and extension of the cell as the cellular envelope is forced to convect around the interior cytoplasm in a "tank tread" motion.[5] The transition from the former rigid body motion to the latter behavior characteristic of liquid drops is determined by the dynamic rigidity of the cell, which depends on both membrane and cytoplasmic factors. In the microcirculation, by comparison, individual red cells are extended and folded as they enter the small capillaries (Fig. 3). Here again, cell rigidity modulates the rate of entry into a capillary and influences the pressure drop across the small vessel, especially when the caliber is less than the cell diameter. The terms "cell rigidity" and "cell deformability" are descriptive representations of the cell response to external forces. These terms depend on several extrinsic and intrinsic factors. Extrinsic factors are those related to features of cell architecture such as surface area-to-volume ratio, membrane contour (smooth or wrinkled), presence of a nucleus, and cross-bridging of cell membrane surfaces by the nucleus. On the other hand, intrinsic factors are the elastic and viscous properties of the cell membrane and cytoplasmic and nuclear materials. The relative importance of extrinsic factors in cellular deformability also depends on the intrinsic material properties of the constituent materials. For instance, the surface-to-volume ratio of mammalian red blood cells is the most important extrinsic factor in deformability because the highly cohesive membrane double layer greatly resists area dilation and as such establishes a geometric constraint for cell deformation.[6] Within this constraint, the cell can be easily deformed as long as surface area is not required to increase or the cytoplasmic volume to decrease. Similarly, the effect of membrane wrinkles or corrugations on cellular rigidity depends on the intrinsic ratio of membrane bending to extensional stiffness. Finally, another obvious extrinsic factor is the presence of a nucleus in nonmammalian cells which is cross-bridged between opposite membrane faces. Be-

[4] V. Bennett, *Annu. Rev. Biochem.* **54,** 273 (1985).

[5] H. L. Goldsmith and J. Marlo, *Proc. R. Soc. London, Ser. B* **182,** 351 (1972).

[6] E. A. Evans and R. Skalak, "Mechanics and Thermodynamics of Biomembranes." CRC Press, Boca Raton, Florida, 1980.

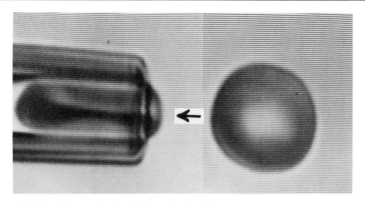

FIG. 3. Video micrograph of a single red cell before and after entry into a small capillary which demonstrates the extension and folding of the cell. Red cells experience this type of deformation 10^5 times during their 120-day life span.

cause of the rigidity of the nuclear material and the attachment to the cell membrane, deformation of the cell is restricted to regions of the toroidal membrane envelope. Because of such effects, the study of red cell deformability requires careful separation of intrinsic material properties from extrinsic cell geometry and structural assembly. This consideration is especially important when investigators attempt to relate abnormal cell deformability to pathophysiology of the cell.

For red blood cells, the material properties important in cell deformation reduce to those of the membrane and cytoplasm. The residual nucleus in nonmammalian cells appears to be an undeformable, rigid body. So for all cells, the portion of the membrane capsule which surrounds a cytoplasmic space forms the deformable region of the cell. Except in states of extreme dehydration or pathological situations like deoxygenated sickle cells, the hemoglobin solution inside red cells behaves as an ideal Newtonian liquid with a viscosity that depends strongly on concentration (i.e., from 6 dyn · sec/cm^2 at 31 g/dl to 40 dyn · sec/cm^2 at 40 g/dl hemoglobin concentration).[7] Because of the liquid character of the cytoplasm and its relatively low viscosity, most of the recognized deformation behavior of red cells is attributable to the mechanical properties of the membrane (e.g., shape recovery after deformation, dynamic rigidity of cells in the microcirculation).[6,8]

Because of the small size of red blood cells, measurement of deformation forces and precise definition of cell geometry are extremely difficult.

[7] S. Chien, *Blood Cells* **3,** 283 (1977).
[8] E. Evans, N. Mohandas, and A. Leung, *J. Clin. Invest.* **73,** 477 (1984).

Consequently, progress in red blood cell rheology has not come easily. However, developments over the past 15 years have culminated in a rheological approximation which can be successfully used to quantitate the intrinsic deformation properties of red blood cells. Mechanical experiments on single cells offer distinct advantages over the study of these capsules in suspension because deformation of individual cells can be directly controlled and measured. However, two significant difficulties are encountered in the design of mechanical tests for single cells: (1) because of cell size, optical diffraction limits the accuracy of measurement of cell geometry and deformation; (2) because the membrane is a thin structure supported by a soft cytoskeletal matrix, the forces required to deform the membrane (or even to fragment the surface!) may be 10^{-6} dyn (10^{-9} g) or less when applied to regions on the scale of 10^{-4} cm. To circumvent these difficulties, amplification methods are required to facilitate control and measurement at the laboratory level. The state-of-the-art equipment, methods, and analyses which have been developed to perform micromechanical tests on small cells will be described in this chapter. First, important principles will be introduced which form the basis for determination of intrinsic material properties of membranes.

Membrane Material Concepts and Properties

Experimentally, the determination of material properties of membranes (e.g., elastic moduli and coefficients of viscosity) involves the well-designed application of forces and observation of the *change* in membrane conformation over time. Because of its molecular thinness, the cell membrane can be treated as a continuous medium only in the two dimensions that characterize its surface plane. Usually, the curvatures of the cellular envelope are much smaller than the reciprocal of the membrane thickness; hence, we can model the mechanical behavior of the membrane as that of a thin shell. Conceptually, the surface is viewed as a smooth mosaic of very small rectangular elements, sufficiently small to be approximately "flat" when compared to the overall contour of the surface. Global deformation of the cell envelope can be reduced to local deformations of these small "imaginary" surface elements. Within this concept, *all* deformations can be reduced to three independent modes of action: dilation or condensation of each element area; simple extension (shear) of the element with its area held constant; and bending or alteration of the curvature of the element without changing its rectangular proportions.[6] These unique modes of deformation are illustrated schematically in Fig. 4. Similarly, the actions of external forces applied to the cell envelope reduce to a distribution of local force and moment "resultants"

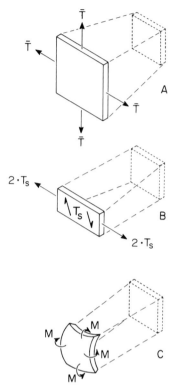

FIG. 4. Schematic illustration of the unique modes of deformation of local surface regions: (a) Area dilation (or condensation), (b) extension without area dilation, and (c) bending or curvature change. Also shown are the principal force and moment resultants which produce these types of deformation: isotropic tension, $\bar{T}$; surface shear, T_s (equivalent to the uniaxial tension $2 \cdot T_s$); and a membrane moment or torque, M.

also shown in Fig. 4; these include forces distributed per unit length along the edge of the element (tensions) which act tangent to the plane of the membrane surface and the bending moment or membrane torque per unit length which acts around the element edges as illustrated in Fig. 4.[6]

Mechanical equilibrium of the cell body is established by the balance of external forces applied to each membrane element (e.g., the transmembrane pressure normal to the surface and fluid shear stresses tangential to the surface) *opposed* by the actions of the membrane tensions and bending moments. The transmembrane pressure normal to the surface is opposed by the small inward projections of the membrane tension components (given by the tension times local curvature) plus a subtle contribution from the membrane bending moment (given by the second derivative of the bending moment with respect to curvilinear distance along the membrane contour).[6] This normal force equilibrium is a modern-day version of the "law of Laplace." The action of fluid shear stress tangent to the membrane surface is opposed primarily by a surface gradient of a membrane tension plus another subtle contribution of the bending moment (given by the surface gradient of the bending moment times the local membrane curvature).[6] Thus, for thin membrane shells, pressurization of the capsule produces smooth convex surfaces under tension and fluid shear stress acts to increase the membrane tension along the membrane contour.

Taking a more detailed approach, deformation of membrane elements can be explicitly represented by quantitative parameters for each independent mode of deformation. These include the area strain, α, which is the fractional increase in area, $\Delta A/A_0$; the shear strain, e_s, which is a measure of the in-plane extension through the local extension ratio, $\bar{\lambda}$, $[e_s = \frac{1}{4}(\bar{\lambda}^2 - \bar{\lambda}^{-2})]$; and the membrane "bending" strain which is proportional to the change in local membrane curvature, ΔC.[6] Since many cell deformations are dynamic and take place over small time intervals, it is also necessary to consider the corresponding local rates of deformation: the rate of area dilation or condensation $\partial \ln(1 + \alpha)/\partial t$; the surface shear rate, $\partial \ln\bar{\lambda}/\partial t$; and the rate of change of membrane curvature, $\partial \Delta C/\partial t$. These expressions quantitate the deformation and rate of deformation of local regions of the membrane surface and, because of surface continuity, are directly related to changes in global membrane conformation.

Similarly, the forces and moments that act on a membrane element can be quantitated by independent parts. This decomposition is arrived at by examination of the forces that act along the edge of the element tangent to the plane of the surface. There is a mean or isotropic part of the tension, $\bar{T}$, which is the average value obtained from rotation of the element through all angles about the surface normal.[6] *The mean tension acts to dilate or increase the element area!* By comparison, the other independent component is the maximum level of shear, T_s (which acts as a "cutting" force along the edge of the element), obtained by rotation of the element through 45° about the normal with respect to the simple extension illustrated in Fig. 4. For uniaxial extension, the membrane tension can be shown to be exactly equal to twice the maximum level of surface shear force.[6] *In-plane shear forces act to extend elements in the membrane plane!*

The material behavior of the membrane is explicitly represented by algebraic relations between these independent resultants of force (supported by the membrane) and the independent modes of deformation and rate of deformation. The most ideal behavior, for example, is that of either a perfectly elastic solid or liquid material. For the perfect elastic solid, the membrane forces and moments are proportional only to the modes of deformation.

$$\bar{T} = \bar{T}_0 + K\alpha$$
$$T_s = (\mu/2)(\tilde{\lambda}^2 - \tilde{\lambda}^{-2})$$
$$M = B\Delta C = B\Delta[(1/R_1) + (1/R_2)]$$

where K, μ, B are elastic coefficients or moduli of rigidity. These relations are obviously "static" in that deformation of an element creates forces which do not change with time. Also, elastic relations are conservative and represent "memory" in the material since removal of applied force leads directly to recovery of the initial shape.

For simple liquid behavior, on the other hand, membrane forces and moments are proportional to the *rate* of deformation; static changes in shape do not create forces in a liquid structure.

$$\bar{T} = \kappa \, [\partial\ln(1 + \alpha)/\partial t]$$
$$T_s = 2\eta(\partial\ln\tilde{\lambda}/\partial t)$$
$$M = \nu(\partial\Delta C/\partial t)$$

where κ, η, ν are coefficients of viscosity. Liquid behavior is totally nonconservative: removal of the external forces leaves the body permanently deformed with no restoration of shape.

Materials can only be approximately considered as elastic solids or viscous liquids and usually exhibit more embellished rheology which is often very complicated and must be defined empirically by experiment. There is a reasonable description of the spectrum of most material behavior: dissipative (viscoelastic) solid, elastic liquid (semisolid), and plastic or simple liquid. Dissipative solids are approximated by the superposition of elastic and viscous processes, e.g.,

$$T_s = (\mu/2)(\tilde{\lambda}^2 - \tilde{\lambda}^{-2}) + 2\eta(\partial\ln\tilde{\lambda}/\partial t)$$

The viscoelastic solid recovers its shape after deformation (i.e., has "memory") but the *rate* of recovery is limited by viscous dissipation.

Semisolid behavior is essentially that of an elastic liquid which is characterized by creep and relaxation processes. When forces applied to the material are held constant, the material "creeps" with a rate of deformation proportional to the force level. Likewise, when the shape is held constant, the material forces "relax" over time. On the other hand, rapid application of force produces instantaneous deformation similar to an elastic material. "Creep" behavior (which can be inverted to give "relaxation" response when the length is held constant) is represented by a serial coupling of elastic and viscous responses, e.g.,

$$\partial \ln \tilde{\lambda}/\partial t = \{1/[2(T_s^2 + \mu^2)^{1/2}]\}[(\partial T_s/\partial t) + (T_s/2\eta)]$$

The semisolid or elastic liquid has a "fading memory"; the capability for shape recovery attenuates with time to leave increasing residual deformation.

The simplest model for a material in transition to liquid behavior is that of an ideal plastic. Here, there is a yield threshold below which the material behaves as a solid and above which it flows as a simple liquid. The constitutive behavior is given by

$$\partial \ln \tilde{\lambda}/\partial t = (T_s - \hat{T}_s)/2\eta$$

when $T_s > \hat{T}_s$. The plastic or liquid has no "memory" or recovery of shape after deformation.

Even though such rheological complexity is disconcerting and even intimidating, red cell membranes exhibit all of these types of material response in appropriate situations.[6] Fortunately, for deformations like those experienced in the circulation, the normal red cell membrane can be represented as a highly extensible, dissipative solid (viscoelastic) shell. However, semisolid and plastic-like properties of the membrane are clearly important determinants in membrane failure or alteration processes (e.g., cell fragmentation and irreversible shape change as in sickle cell disease).

Membrane Chemical Equilibrium and Cell Conformation

Although the biconcave disk shape of mammalian red blood cells is considered to be the normal or "natural" state, cell conformation can vary significantly (from the disk shape); cell shapes range from cup forms (stomatocytes) to tightly crenated spherical forms (echinocytes) as a result of changes in the chemical environment. The multiplicity of agents and conditions which induce these transformations is well documented.[1,9,10] The exquisite sensitivity of the cell shape to changes in chemical environment is the result of the thin lamellar structure of the membrane. Specifically, large changes in membrane curvature are produced by small differential area changes between the constituent layers. From simple geometric considerations, it can be shown that the change in membrane curvature is simply proportional to the change in area of one

[9] E. Ponder, "Hemolysis and Related Phenomena." Grune & Stratton, New York, 1971.
[10] B. Deuticke, *Biochim. Biophys. Acta* **163**, 494 (1968).

layer relative to the next divided by the distance between layers.[6,11] For instance, a bump on the red cell surface with a radius of curvature of about 10^{-4} cm can be readily produced by a 1% increase in outer membrane surface area or less. Referred to as either "chemically induced moments"[11] or changes in "spontaneous curvature,"[12] these effects alter the reference geometry for the cell. Quantitatively, chemical effects can be incorporated in membrane mechanical behavior simply by the addition of a constant term to the membrane bending elasticity relation,

$$M = B\Delta C - M_c$$

Here, the parameter M_c directly represents the chemical effect and is given by the product of the free energy change per unit area, $\Delta\gamma$, between constituent layers times the separation distance between them, h. The change in equilibrium curvature (i.e., "spontaneous curvature") is approximately given by the induced moment, M_c, divided by the membrane bending elastic modulus,[6,11]

$$\Delta C_c \approx (\Delta\gamma \cdot h)/B$$

This relation can be used to estimate the changes in chemical potential responsible for morphological changes of red cells. *However,* it is emphasized that morphological changes are due to small relative changes between layers and cannot be used to determine changes in state of the membrane as a whole. For instance, uptake of free fatty acids from the environment can expand the red cell membrane by several percent although the spicules formed on the cell surface are due only to a *small* fractional difference in fatty acid composition between bilayer leaflets. Investigators should be extremely cautious when they attempt to relate red cell shape to membrane compositional changes!

From the viewpoint of mechanics, the force-free cell shape is merely a geometric reference from which to measure deformation or rate of deformation. The key question is whether or not the membrane can be treated as a "stress-free" material in this state. Micromechanical tests have shown that the biconcave geometry of the mammalian red cell *is* stress free within the limits of experimental resolution.[6] On the other hand, cupped and spiculated cells in general possess some level of membrane stress (tension) which increases as the geometric state becomes more aberrant. In the discussion to follow, subtle shape effects will be neglected in order to focus on the intrinsic material properties of the membrane.

[11] E. Evans, *Biophys J.* **14**, 923 (1974).
[12] W. Helfrich, *Z. Naturforsch. C* **28**, 693 (1973).

Micromechanical Test System

There have been many methods developed to test the deformability of red blood cells: micropipet aspiration, filtration through small-caliber sieves,[13] the "ektacytometer,"[14] the "rheoscope,"[15] extension by high-frequency electric fields,[16] etc. Of these approaches, micropipet aspiration methods are the most easily analyzed to derive the intrinsic mechanical properties of the red cell membrane. The other measurement techniques involve complicated kinematics of cell deformations so as to "mix up" the rheological properties; however, these other methods are very useful for qualitative evaluation of the dynamic deformability of red cells which is extremely important in clinical studies.

The micropipet aspiration system is centered around an inverted microscope which has several micromanipulators mounted directly on the microscope stage. The small pneumatically controlled manipulators are operated by "joy sticks" to provide smooth pipet-tip displacements of a fraction of a micron or more. Small glass suction micropipets are attached to the manipulators. The cells are placed in very small concentrations into one or more separate chambers on the microscope stage for observation. The time course of each experiment including the important data (e.g., temperature, pipet suction pressure, time) are simultaneously recorded on videotape with the use of video multiplexing. A mercury vapor lamp with narrow-band interference filters is used for the microscope illumination because monochromatic light provides the best image quality for video recording. It is essential that a high numerical aperture (greater than 0.6) objective be used with subsequent "empty" magnification to establish as large an image of the cell on the video monitor is possible. This large image ensures that there will be good video resolution since several video scan lines will lie within the diffraction border surrounding the cell. The high numerical aperture objective and monochromatic illumination minimize the optical diffraction at cell borders (also, interference contrast optics can be used to provide additional edge detection capability or image expression of structures inside the cell).

As mentioned above, it is important to optimize the optical video image; but even with optimization, diffraction at edges of the cell limits the accuracy of measurement of the whole cell geometry. Such limitations can lead to measurement errors of 10–20% for parameters such as surface area and volume of cells. Consequently it is expected that it would be

[13] W. H. Reinhart and S. Chien, *Am. J. Physiol.* **248**, C473 (1985).
[14] N. Mohandas, M. Clark, and S. B. Shohet, *J. Clin. Invest.* **66**, 563 (1980).
[15] T. Fischer and H. Schmid-Schönbein, *Blood Cells* **3**, 351 (1977).
[16] H. Engelhardt, H. Gaub, and E. Sackmann, *Nature (London)* **307**, 378 (1984).

extremely difficult to determine small deformations of the cell because of the limited optical resolution. Fortunately, it is possible to accurately detect displacements of positions on a cell even though the absolute location is not precisely defined. Hence, it is useful to employ various electronic devices (video edge enhancers and position analyzers) to facilitate measurement of displacements in the video image. With these add-on devices, it is possible to evaluate displacements of the cell surface with an accuracy on the order of 5% of the wavelength of light. The result, for example, is that we can detect changes in red cell membrane area on the order of 0.1% even though we cannot measure the total area more accurately than about 5–10%!

It is often desirable to compare properties of an individual cell in different solution environments (e.g., sickle red cells at various oxygen partial pressures and exposure of cells to membrane cross-linking agents). When properties can be compared for the same cell, cell-to-cell variations are eliminated to provide much better discrimination. A simple procedure has been developed for single cell comparative studies which involves the use of adjacent, separate chambers on the microscope stage. A large (on the order of 50×10^{-4} cm) pipet traverses one chamber, spans the air gap between chambers, and enters the second chamber through a small port in the side. Individual cells are inserted into the lumen of the large pipet; the stage is then translated to displace the large pipet from one chamber to the second chamber through the air gap interface. In the second chamber, the cell is withdrawn from the shelter of the transfer pipet with the use of a second (smaller) suction micropipet (sequence shown in Fig. 5). The trapped volume inside the transfer pipet is minimized by backfilling with oil such that only a small cavity remains near the tip of the pipet. The presence of the oil also restricts motion of fluid in the transfer pipet.

A significant component of the system is the apparatus used to control suction pressure which is applied by the pipet to deform the cell surface. Because of the small levels of force required to deform cells, small micropipets are used (internal diameters on the order of 10^{-4} cm) with suction pressures of 10^{-6} atm (dyn/cm^2) or greater. This approach provides a simple method for producing forces as low as 10^{-8} dyn (i.e., 10^{-11} g). Normally, in red cell deformation tests, suction pressures range from 10^2 to 10^5 dyn/cm^2. It is essential that the micropipet be coupled by a continuous water system (free of air bubble compliance) to the external pressure control system (usually a micrometer-positioned water manometer). Because of the small pressures involved, the system must be zeroed accurately before every cell test since alteration in curvature of the air–water interface through which the pipet enters the chamber can change the suction pressure. Fortunately, zero pressure adjustment is easily accom-

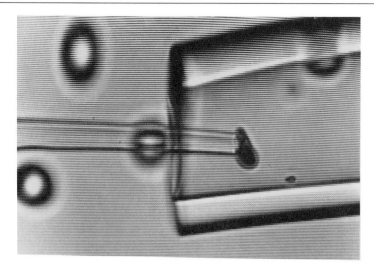

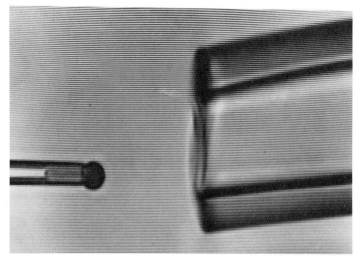

FIG. 5. Video micrographs which demonstrate manipulation and transfer of cells between adjacent chambers. A large pipet is used to span the air gap between chambers. Cells are inserted into the lumen of the large pipet; the microscope stage is then translated to displace the large pipet from one chamber to the other. Cells are then withdrawn from the shelter of the transfer pipet with the use of another smaller micropipet.

plished by observation of small particles within the lumen of the micropipet. When motion of these particles ceases then the pressure difference is negligible. Sensitive pressure transducers are available that can provide accurate measurement of differential pressures down to the level of 10^{-6}

atm. However, the resolution in pipet suction force is limited by the accuracy of the measurement of the pipet caliber. Because the internal diameter of the pipet can be as low as one wavelength of light, optical diffraction precludes accurate observation of the lumen size. To circumvent this problem, a gradually tapered microneedle is used as a "feeler gage" to determine the lumen dimension by insertion into the pipet. The microneedle is calibrated by scanning electron microscopy to provide a scale for cross-section as a function of length from the tip of the needle. With this method, the insertion depth inside the pipet provides a direct measure of the lumen size. Other methods have been tried (e.g., direct observation of the pipet entrance by scanning electron microscopy) but are not reliable because of a number of subtle artifacts.[17]

Direct Measurement of Red Blood Cell Membrane
Mechanical Properties

As was discussed in the section on concepts, there are elastic and viscous properties for three independent modes of membrane deformation: area dilation, in-plane extension (shear) without area change, and bending or curvature change (without area dilation or extension). Next, experimental procedures and computational algorithms for determination of membrane mechanical properties will be given for the normal behavior of the cell membrane as a "viscoelastic" solid material. This will be followed by more complicated rheology of the membrane in lysis and in shear failure ("creep" and plastic flow).

Red Cell Area Dilation

Because of the osmotic strength of the cytoplasmic contents, red cell area can be easily dilated by application of large micropipet suction pressures to osmotically preswollen red cells as illustrated in Fig. 6.[18,19] The red cell easily deforms and enters the micropipet until the portion of the cell outside the pipet becomes a rigid spherical surface; further displacement requires area dilation since the cell volume is restricted. As the pipet suction pressure is increased, the membrane isotropic tension is increased

---

[17] Although pipets are easily made by extrusion of locally heated glass tubes (about 1 mm in diameter) with commercially available devices, control of the final pipet dimension and production of a smooth, clean entrance require additional procedures. A simple method is to use a "microforge," with which the pipet after initial formation is fused to a molten glass bead, quenched by an air jet, and then broken by quick fracture to leave a smooth tip at the desired dimension.

[18] E. Evans, R. Waugh, and L. Melnik, *Biophys. J.* **16,** 585 (1976).

[19] E. Evans and R. Waugh, *Biophys. J.* **20,** 307 (1977).

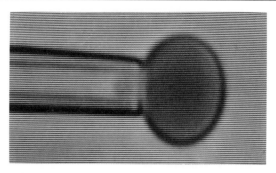

FIG. 6. Video micrograph of red cell membrane area dilation. The red cell was osmotically preswollen before aspiration by the pipet. Because the cell volume is essentially fixed, displacement of the cell into the micropipet requires surface area dilation. Aspiration pressures are in the range of 10^4–10^5 dyn/cm^2 for the pipet caliber of 2×10^{-4} cm.

proportionally,

$$\bar{T} = \mathrm{PR_p}/2(1 - R_p/R_o)$$

Here, P is the pipet suction pressure; R_p is the pipet inner radius; and R_o is the radius of the outer spherical segment of the red cell. For red cells in salt solutions of 0.1 M, less than 20% of the displacement of the aspirated length into the pipet is due to filtration of water from the cell interior. There is a simple correction for this small volume effect[19]; thus, area changes are approximately given by the simple expression,

$$\Delta A \cong 2\pi R_p(1 - R_p/R_0)\Delta L$$

where L is the aspirated length observed inside the pipet. Measurement of the aspirated length versus suction pressure can be simply transformed into membrane tension versus fractional change in surface area as shown in Fig. 7. For small fractional increases in membrane area up to lysis (area changes of 2 or 3% or less), large tensions are required in linear proportion to area dilation. The coefficient of proportionality is the elastic modulus of area dilation (area rigidity) of the red cell membrane which is in the range of 300 to 600 dyn/cm.[19] This large resistance to area dilation is provided by the superficial bilayer of amphiphilic molecules (lipids, proteins, cholesterol). As will be shown in the next section, aspiration of flaccid red cells requires suction pressures that are 10^3 times smaller, which shows that the area remains essentially constant for flaccid cell deformation; the low pressures represent membrane extensional and bending rigidities.

In principle, application of an instantaneous change in pipet suction pressure should produce a viscoelastic response which would appear as a

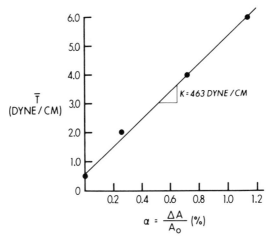

FIG. 7. From the aspiration pressure versus displacement of the cell projection inside the pipet shown in Fig. 6, the membrane tension is derived as a function of surface area dilation. This behavior is reversible and provides the elastic modulus of area compressibility for the red cell membrane.

rapid increase in area of the cell followed by exponential approach to the static limit observed in the above experiments. However, because of the lower limit on observation time (a minimum of 10^{-2} sec for the video system), it is not possible to evaluate the immediate viscous response to the application of isotropic tension. Indeed, estimates of the time constants for dilation of membrane area are extremely low, i.e., on the order of 10^{-6} sec.[6] Hence, area dilations subsequent to red cell sphering can be considered as "instantaneous" but limited to the order of about 3% before lysis. Consequently, the time course of rapid swelling of red cells is not limited by membrane viscoelasticity but rather by diffusional retardation through unstirred layers in the vicinity of the red cell.

Red Cell Membrane Extension without Area Dilation

Surface extensions of the membrane (as illustrated in Fig. 4) are produced by membrane shear forces. As a surface liquid, the superficial lipid bilayer component of the red cell membrane does not resist in-plane shear forces except by viscous flow; in other words, the modulus of extensional rigidity of the lipid bilayer is *zero*. By comparison, the red cell membrane as a whole possesses a small but significant extensional rigidity such that when the flaccid cell is extended (without surface area change), there is an increase in static level of force in proportion to extension of the cell. Aspiration of flaccid red cells by small micropipets (Fig. 8) shows a pro-

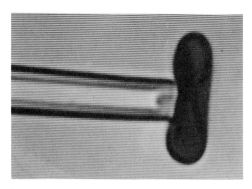

FIG. 8. Video micrograph of red cell membrane extension and bending without area dilation. Here, a small micropipet (caliber of 10^{-4} cm) aspirates a flaccid red cell with low suction pressures in the range 10^2–10^3 dyn/cm^2.

portional increase in projection length inside the pipet with increase in suction pressure. The stepwise increase in length versus aspiration pressure is reversible (time independent) and reduction of the aspiration pressure results in a proportional decrease in length. The pipet suction pressures are in the range of 100–900 dyn/cm^2 for small-caliber pipets. Analysis of this experiment has shown that the membrane extensional rigidity can be determined directly from the derivative of the suction pressure with respect to length[6,20]:

$$\mu \approx R_p^2(dP/dL)$$

(see Fig. 9). Here, it is necessary to apply the small-caliber pipet (10^{-4} cm in diameter) to the central concave portion of the red cell discocyte to establish an axisymmetric geometry and to minimize membrane buckling or folding complications. When the membrane folds or buckles, the membrane avoids extension and, thus, the aspiration pressure does not reflect the membrane extensional stiffness. For normal red cells, experiments yield values in the range of 5–7 × 10^{-3} dyn/cm for the modulus of shear rigidity at 25°.[21,22] This small extensional stiffness is provided by the subsurface protein (spectrin and others) meshwork or cytoskeleton.

Rapid application of pipet suction pressure to flaccid red cells (as in Fig. 8) should invoke a viscoelastic response where an initial rapid increase in aspiration length is followed by an exponential approach to a

[20] E. Evans, *Biophys. J.* **30**, 265 (1980).
[21] R. Waugh and E. Evans, *Biophys. J.* **26**, 115 (1979).
[22] G. B. Nash and H. J. Meiselman, *Ann. N.Y. Acad. Sci.* **416**, 255 (1983).

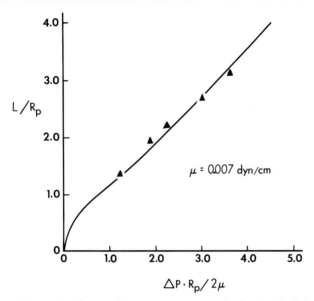

FIG. 9. The slope of the pipet suction pressure versus projection length of the cell shown in Fig. 8 is used to derive the membrane extensional rigidity. The pressure "loading" and "unloading" phases are reversible, which demonstrates the elastic memory of the cell membrane. This experiment is only valid over the range of aspiration lengths where the membrane surface remains smooth and does not fold or buckle. P, Pipet suction pressure; R_p, pipet inner radius; L, aspirated length inside pipet.

static length.[6,23] Such response has been observed[23]; however, careful examination of the cell surface contour (with the use of interference contrast optics) has shown that the initial rapid response is compromised by membrane buckling, which gives a more rapid increase in aspiration length than would occur if the membrane was forced to extend. An alternative approach to measurement of the viscous retardation of elastic extension is to observe the time course of recovery following an extensional deformation. For example, release of red cells which have been extended end to end by diametrically opposed pipets demonstrates the exponential-like behavior characteristic of a viscoelastic material (Figs. 10 and 11). Here, the characteristic time constant for membrane extensional deformation is given by the following equation,

$$\tau_e \approx (\eta_m + \tilde{\eta}_{Hb}\delta)/\mu$$

[23] S. Chien, K.-M. Jan, R. Skalak, and A. Tozeren, *Biophys. J.* **24**, 463 (1978).

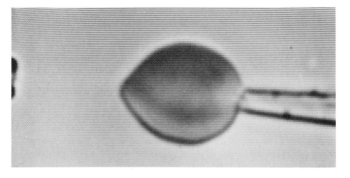

FIG. 10. Video micrographs of the elastic recovery of the red cell membrane after extensional deformation. A single red cell is extended end to end by diametrically opposed pipets; the cell is then released and the time course of the viscoelastic recovery is recorded.

where η_m is the membrane surface viscosity for extensional deformation; $\bar{\eta}_{Hb}$ represents dissipation in adjacent aqueous phases; δ is the characteristic dimension or thickness of the cell.[8,24,25] The recovery is driven by the elastic energy stored in the membrane but is opposed by the viscous dissipation in the membrane and adjacent aqueous phases (cytoplasm and external solution). Examination of the viscous dissipation in adjacent aqueous phases (as represented by the product of $\bar{\eta}_{Hb}\delta$) shows that cell response would be on the order of 10^{-3} sec if there were *no* membrane dissipation. On the contrary, as is shown in Fig. 11, the time for viscoelastic recovery is nearly 100-fold *slower* and, thus, the dissipation must be dominated by membrane viscous processes. Hence, with the observation

[24] R. M. Hochmuth, P. R. Worthy, and E. Evans, *Biophys. J.* **26,** 101 (1979).
[25] E. Evans and R. M. Hochmuth, *Biophys. J.* **16,** 1 (1976).

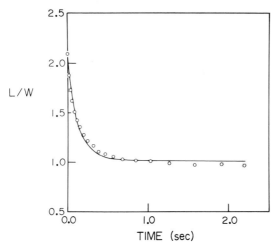

FIG. 11. The ratio of cell length (L) to width (W) is plotted as a function of time after release from end-to-end extension of the cell shown in Fig. 10. The time constant for recovery after extensional deformation is on the order of 0.1 sec, which is dominated by viscous dissipation in the cell membrane and is, thus given by the ratio of the membrane viscosity to the elastic modulus for extension.

of the time constant for viscoelastic recovery after extensional deformation *and* the previous measurement of membrane extensional stiffness, the coefficient of membrane surface viscosity can be easily determined from

$$\eta_m = \mu \tau_e$$

Again, it is important that cell surface buckling be avoided as much as possible in order to ensure that the recovery is dominated by extensional forces. The viscoelastic time constant also characterizes the response of the cell membrane to instantaneous changes in extensional forces and, thus, determines the dynamic (extensional) rigidity of the cell which is a significant factor in distribution and flow of red cells through the small microvessels. Dynamic rigidity is established by the product of the static rigidity (μ) times the characteristic time constant for the deformation response (τ_e) *divided* by the time interval over which the deformation is required to occur (i.e., the capillary entrance time, Δt_e). Because of the rapid entry into a microvessel, the dynamic rigidity can be at least an order of magnitude greater than the static rigidity of the cell. It is important to note that viscous dissipation in red cell extension is two orders of magnitude greater than dissipation in a lipid bilayer membrane (above the acyl chain crystallization temperature). Hence, it is the cytoskeletal meshwork that offers the primary viscous (dynamic) resistance to extension of the cell.

Red Cell Membrane Bending (Folding) without Area Dilation or Extension

In the previous discussion of membrane extensional rigidity, it was emphasized that planar deformations of the cell surface only occur if the surface contour remains smooth without buckling or folding. Indeed, it is observed that the red cell easily buckles or folds, as is shown in Fig. 12. Buckling (folding) of the membrane is an instability regulated by the ratio of membrane bending to extensional rigidity. In other words, when the bending stiffness is large, the membrane is able to sustain higher levels of extensional force (which implies a commensurate orthogonal compressional force) without folding. The observation of membrane buckling instability (Fig. 12) is direct evidence for the existence of *both* extensional and bending stiffness. When the suction pressures applied to flaccid red cells are sufficient to produce large aspiration lengths, the cell surface begins to wrinkle (buckle) and eventually large creases occur which enable the cell to fold and move up the pipet until limited by the surface area and volume restrictions. Analysis of the buckling instability shows that the bending modulus can be easily derived from the pressure, $\hat{P}$, at which the cell buckles or folds[26]:

$$B \approx cR_p^3\, \hat{P}$$

where the coefficient, c, depends on the ratio of the pipet inner radius to the cell outer radius and is in the range of 0.005–0.012. The best procedure for evaluation of buckling instability is to use a curved pipet so that the cell can be viewed from above.[26] Since this method is somewhat complicated, an alternative approach is simply to observe the pressure at which the cell folds and moves up the tube; this approach results in about a 60% overestimate in the pressure for buckling. Study of normal cells shows that pressures sufficient to initiate buckling are a strong function of pipet radius as predicted theoretically[26] and shown in Fig. 13. From these measurements, the bending modulus is determined to be on the order of 10^{-12} dyn · cm (erg). Similar levels of bending stiffness have been either measured or deduced for lipid bilayer membranes where the bilayer is in the liquid state above the crystallization temperature.[6,27,28] Analysis of red

[26] E. Evans, *Biophys. J.* **43**, 27 (1982).
[27] R. M. Servuss, W. Harbich, and W. Helfrich, *Biochim. Biophys. Acta* **436**, 900 (1976).
[28] M. B. Schneider, J. T. Jenkins, and W. W. Webb, *Biophys. J.* **45**, 891 (1984).

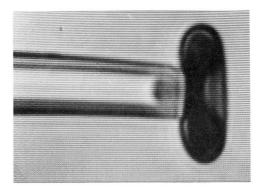

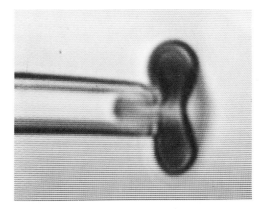

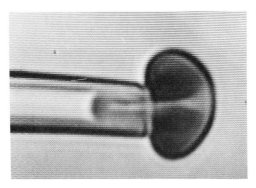

FIG. 12. Video micrographs of the buckling or folding instability caused by excessive aspiration in the pipet suction experiment. The critical pressure at which the cell buckles or folds is a direct measure of the membrane bending stiffness or curvature elastic modulus.

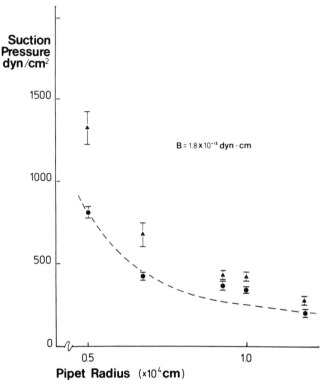

FIG. 13. The pressures necessary to initiate buckling and subsequent folding in pipet aspiration of flaccid red cells are shown as functions of pipet radius. Correlation of the theoretical prediction of buckling instability with experiment yields the upper bound on the bending stiffness or curvature elastic modulus (B) of the red cell membrane.

cell "flicker" (surface undulations) yields similar but somewhat lower values for bending rigidity.[29,30]

When cells enter small capillaries, they prefer to fold with little extension, as shown in Fig. 3. In the microcirculation, entry occurs over a very small time interval and, thus, gives rise to large viscous forces which oppose the deformation. Unlike the dynamic extensional rigidity discussed previously, the dynamic resistance to folding of the whole cell is primarily dominated by the viscous dissipation in the cytoplasm and exterior fluid phase. The characteristic folding time is approximated by the

[29] F. Brochard and J. F. Lennon, *J. Physiol.* (*London*) **36,** 1035 (1975).
[30] K. Fricke and E. Sackmann, *Biochim. Biophys. Acta* **803,** 145 (1984).

equation

$$\tau_f \approx (\nu + \bar{\eta}_{Hb}\delta/\Delta C_m^2)/B$$

where membrane dissipation in bending is represented by the coefficient, ν; dissipation in interior *and* exterior fluid phases is represented by the viscous coefficient, $\bar{\eta}_{Hb}$; δ is the characteristic dimension of the cell; and ΔC_m is the curvature of the fold. In contrast to measurement of static rigidities of the cell, dynamic response times for folding and extension are not easily separated. As shown before, end-to-end extension of red cells can be used to observe the rapid elastic recovery from extensional deformation provided that surface buckling is avoided. Similarly, the rapid elastic recovery from *folding* deformation can be used to establish the characteristic time constant τ_f if extensional deformations are minimized.

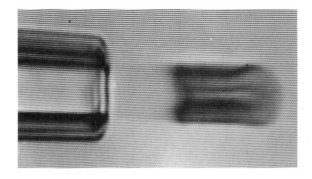

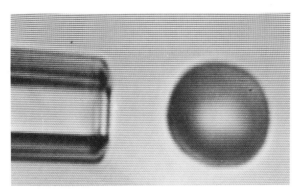

FIG. 14. Video micrographs of the elastic recovery from folding or bending deformations. A large-caliber pipet is used to aspirate the cell; then the cell is rapidly expelled from the end with a pressure pulse and the width recovery is recorded.

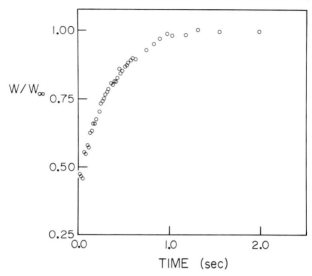

FIG. 15. The cell width (W) is plotted as a function of time after expulsion of the cell in Fig. 14. The characteristic time for cell unfolding to 60% of its final width is taken as the measure of the time constant, i.e., 0.3 sec for normal cells. The unfolding appears to be dominated by viscous dissipation in the adjacent aqueous phases (cytoplasm and exterior fluid).

Thus, large-caliber pipets are used to aspirate the cell. The cell simply folds upon entrance into the pipet and then is rapidly expelled from the end with a pressure pulse (Fig. 14). Because of the "creeping flow" properties in the local solution, the cell remains proximal to the pipet entrance for observation throughout the width recovery. The representative time for the cell width to recover 60% of its final width is taken as the e-fold measure of time constant for elastic recovery from folding. As shown in Fig. 15, the values are on the order of 0.3 sec for normal cells.[31] In principle, the membrane also contributes to the viscous retardation of the unfolding process; the magnitude of the coefficient for viscosity of membrane bending can be estimated from the surface shear viscosity by the following equation,

$$\nu \approx 3\eta_m h^2$$

which appears to be much less than the value characteristic of dissipation in the adjacent aqueous phases. From a functional viewpoint, measurements of τ_f and the bending stiffness establish the *dynamic* rigidity of red

[31] E. Evans, N. Mohandas, and A. Leung, *J. Clin. Invest.* **73**, 477 (1984).

cells in opposition to folding, i.e., $B(\tau_f/\Delta t_e)$, which again is a prominent factor in capillary entrance dynamics.

Red Cell Membrane Failure: Lysis and Fragmentation

As with its normal properties, the red cell membrane can fail by either area dilation (to give lysis) or by excess extensional deformation (to give plastic yield and fragmentation). Also, the membrane stresses necessary to produce these failure mechanisms differ widely in magnitude, i.e., the isotropic tension for lysis greatly exceeds the membrane shear force for fragmentation (by two orders of magnitude). Likewise, each failure mechanism has a drastically different effect on the cell. Lysis violates the chemical integrity of the cell interior, rendering the cell useless for oxygen transport, whereas fragmentation leaves the cell body (and its vesicular offsprings) with oxygen transport capability but with a reduced life span in the circulation.

As was shown many years ago,[32] red cell lysis due to mechanical dilation of surface area (by osmotic swelling or pipet pressurization) is a time-dependent (stochastic) process. A high level of membrane isotropic tension is required to cause rapid or immediate lysis whereas a much smaller level of stress is necessary to cause lysis after long times. The levels of tension range from 6 to 15 dyn/cm to produce lysis at room temperature. What has been observed recently for both red cells and phospholipid vesicles is that the level of tension required to lyse the capsule correlates with the membrane area compressibility modulus, i.e., higher levels of membrane stress at lysis are associated with larger moduli of area compressibility.[19,33] It appears that essentially a critical fractional increase in area is required to produce lysis (on the order of about 2–3%). Thus, red cells have a greater resistance to lysis at low temperature than at high temperature. It is important to note that measurements of lysis produced by mechanical stress are *unrelated* to classical measurements of "osmotic fragility." Osmotic fragility tests only evaluate the surface area-to-volume ratio of a cell population since red cells swell to spherical shapes with little or no stress build-up. When spherical cells are forced to swell, only a small amount of area increase beyond the critical spherical shape is possible before lysis. Hence, the cells swell easily until they reach spheres followed by essentially immediate lysis. Amphiphilic or hydrophobic compounds, which partition preferentially into the membrane bilayer, alter the cohesion of the membrane surface and its opposition to lysis. For instance, alcohols and detergent-like molecules greatly

[32] R. T. Rand, *Biophys. J.* **4**, 303 (1964).
[33] E. Evans and D. Needham, *J. Phys. Chem.* **91**, 4219 (1987).

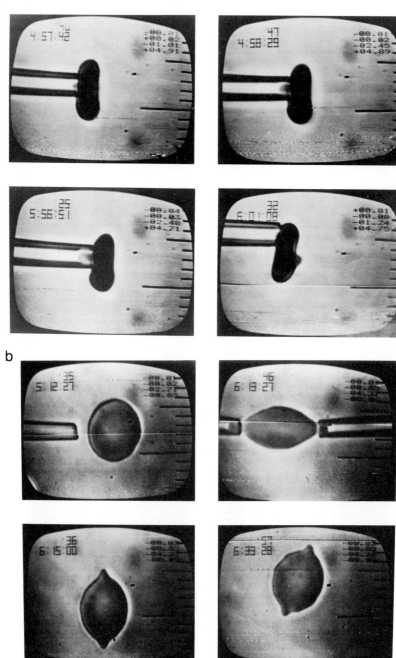

reduce the resistance to lysis whereas the presence of cholesterol greatly *increases* membrane cohesion.[34]

In contrast to the large membrane stresses required to rupture cells by area dilation, the red cell membrane structure can be permanently deformed and even fragmented by application of relatively small extensional forces. Structural rearrangement of the membrane cytoskeleton takes place when extensional forces applied to the membrane are either of sufficient magnitude or duration. For example, membrane "creep" or slow relaxation behavior has been observed when red cells are subjected to pipet aspirations or end-to-end extensions for long periods of time with levels of membrane extensional force used normally to study membrane elasticity (Fig. 16).[6,35,36] The residual or persistant deformations can be measured to establish a characteristic time constant for creep and relaxation which yields a coefficient of viscosity that represents structural relaxation processes in the cytoskeleton. Observations and analysis (as shown in Fig. 17) yield a characteristic time constant on the order of 10^3 sec with a viscous coefficient on the order of 6 dyn · sec/cm.[36] This viscous coefficient is four orders of magnitude larger than the surface viscosity determined for the normal viscoelastic response of the red cell as a solid material. Creep and relaxation are likely to be the crucial mechanisms in the formation of aberrant morphological forms such as irreversible sickle cells (ISC).

A more immediate type of material alteration is the continuous plastic extrusion of membrane from the cell produced by excessive extensional forces. For example, membrane microfilaments or "tethers" can be pulled from red cells which have point attachments to a substrate by simple extension in fluid shear fields[37] (Fig. 18) or by direct extrusion with

---

[34] Lysis tension is simply the maximum level of stress at which the cell ruptures and disappears up the tube in the area dilation experiment. Although simple in concept, care must be taken to determine the exact history of stress application (i.e., the duration and level). Also, pipets should be firepolished and the solution should contain 10% blood plasma to provide a surface coat of protein on the glass. Even with these precautions, many environmental factors influence the level of tension required to lyse red cells (e.g., static electrical charge on the microscope stage and chamber).

[35] E. Evans and P. L. LaCelle, *Blood* **45**, 29 (1975).

[36] D. R. Markle, E. Evans, and R. M. Hochmuth, *Biophys. J.* **42**, 91 (1983).

[37] R. M. Hochmuth, N. Mohandas, and P. L. Blackshear, Jr., *Biophys. J.* **13**, 747 (1973).

FIG. 16. Video micrographs of red cell membrane "creep" and relaxation behavior. Here, red cells have been subjected to pipet aspirations (a) or end-to-end extension (b) for long periods of time with levels of membrane extensional force used normally to study membrane elasticity. The residual or persistent deformations increase with time (where the cells are held for *several minutes to an hour*). (From Markle *et al.*[36] by permission.)

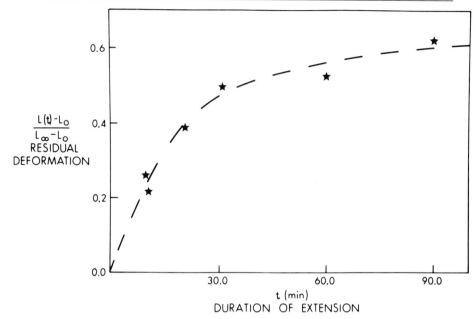

FIG. 17. Membrane "creep" and relaxation of the normal red cell membrane are shown as a function of time. The characteristic time constant for this process yields a coefficient of viscosity for structural relaxation processes in the membrane cytoskeleton. This viscous coefficient is 10^4 greater than the surface viscosity characteristic of normal membrane viscoelastic response. (From Markle et al.[36] by permission.)

micropipets.[38] The experimental evidence shows that the cell membrane can elastically support an initial extensional force applied to the small local process when the force is less than 10^{-6} dyn. Above this level, however, the membrane cylinder is continuously lengthened by extensional flow, the rate of which increases in proportion to the excess force above the threshold level. These observations can be modeled by a membrane yield above which the membrane flows limited by a coefficient of surface viscosity that represents dissipation concentrated in the annular region at the cell–filament junction.[6,39,40,41] The membrane yield-shear and coefficient of surface viscosity for flow onto the tether are given approxi-

[38] R. M. Hochmuth, H. C. Wiles, E. Evans, and J. T. McCown, Biophys. J. **39**, 83 (1982).
[39] E. Evans and R. M. Hochmuth, Biophys. J. **16**, 13 (1976).
[40] R. M. Hochmuth and E. Evans, Biophys. J. **39**, 71 (1982).
[41] R. Waugh, Biophys. J. **39**, 273 (1982).

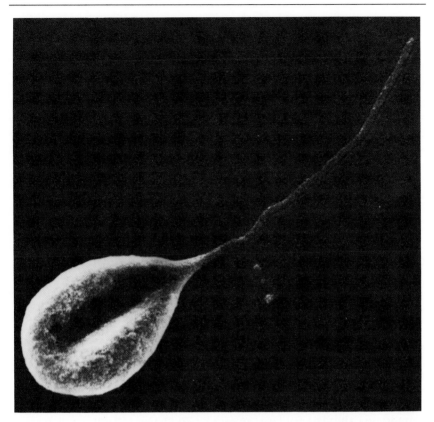

FIG. 18. A scanning electron micrograph of the plastic extrusion of membrane microfilaments pulled from red cells by application of an external fluid shear field. The yield and growth of the filaments can be modelled by simple plasticity theory based on a membrane yield shear above which the membrane flows limited by surface dissipation concentrated in the annular region at the cell–filament intersection. These material properties (along with those characteristic of "creep" and relaxation) represent the intrinsic factors in red cell membrane fragmentation. [From E. Evans and R. M. Hochmuth, *J. Membr. Biol.* **30**, 351 (1977) by permission.]

mately by

$$\hat{T}_s \approx \hat{F}/4\pi r_t$$
$$\eta \approx (F - \hat{F})/8\pi\dot{L}$$

where the threshold force to initiate flow is $\hat{F}$ and the level of force on the tether as it flows is F; r_t is the tether radius (10^{-6} cm or less); $\dot{L}$ is the rate of tether extrusion from the cell body. Experimental results show that the coefficient of surface viscosity for the extensional flow process is comparable to—or slightly greater than—that derived from the viscoelastic response of the cell membrane as a solid material, i.e., 10^{-3} dyn · sec/cm. Furthermore, the membrane yield threshold for plastic flow is comparable to the membrane shear force level deduced from relaxation experiments; these tests show that the red cell membrane appears to relax immediately when membrane extension exceeds ratios of $3:1$.[35,36] As yet, it is not certain that the extrusion of small membrane microcylinders from the cell surface includes the cytoskeletal material; it may be that the cytoskeleton is actually stripped off the membrane bilayer as it is extruded onto the small-caliber filament. In any case, such mechanical properties are clearly important factors in cell fragmentation and vesiculation. Again, membrane shear forces on the order of 10^{-1} dyn/cm produce fragmentation which is a level of stress that is two orders of magnitude less than the levels of isotropic tension necessary to cause lysis. It is apparent that red cells are vulnerable to fragmentation when forced to undergo large extensions or when held under moderate extensions for long periods of time. The material properties that represent creep, relaxation, and plastic flow establish a rational basis for study of membrane fragmentary disorders.

Summary

The lamellar configuration of the red cell membrane includes a (liquid) superficial bilayer of amphiphilic molecules supported by a (rigid) subsurface protein meshwork. Because of this composite structure, the red cell membrane exhibits very large resistance to changes in surface density or area with very low resistance to in-plane extension and bending deformations. The primary extrinsic factor in cell deformability is the surface area-to-volume ratio which establishes the minimum-caliber vessel into which a cell can deform (without rupture). Within the restriction provided by surface area and volume, the intrinsic properties of the membrane and cytoplasm determine the deformability characteristics of the red cell. Since the cytoplasm is liquid, the static rigidity of the cell is determined by membrane elastic constants. These include an elastic modulus for area compressibility in the range of 300–600 dyn/cm, an elastic modulus for in-plane extension or shear (at constant area) of $5-7 \times 10^{-3}$ dyn/cm, and a curvature or bending elastic modulus on the order of 10^{-12} dyn · cm. Even though small, the surface rigidity of the cell membrane is sufficient to return the membrane capsule to a discoid shape after deformation by external forces. Viscous dissipation in the peripheral protein structure (cytoskeleton) dominates the dynamic response of the cell to extensional forces. Based on a time constant for recovery after extensional deformation on the order of 0.1 sec, the coefficient of surface viscosity is on the order of 10^{-3} dyn · sec/cm. On the other hand, the dynamic resistance to folding of the cell appears to be limited by viscous dissipation in the cytoplasmic and external fluid phases. Dynamic rigidities for both extensional and folding deformations are important factors in the distribution of flow in the small microvessels. Although the red cell membrane normally behaves as a resilient viscoelastic shell, which recovers its conformation after deformation, structural relaxation and failure lead to break-up and fragmentation of the red cell. The levels of membrane extensional force required to extrude material from the cell are on the order of 0.1 dyn/cm, which is two orders of magnitude less than the level of tension necessary to lyse vesicles by rapid area dilation. Each of the material properties ascribed to the red cell membrane plays an important role in the deformability and survivability of the red cell in the circulation over its several-month life span.

Acknowledgment

The author gratefully acknowledges the support in part by the National Institutes of Health through Grant GM38331.

[2] Ektacytometry of Red Blood Cells

By ROBERT M. JOHNSON

The ektacytometer is an instrument for the measurement of the response of a population of red cells or red cell membranes to shear. The basic configuration developed by Bessis and Mohandas[1] is composed of a Couette viscometer, with two concentric cylinders constructed of an optically clear material (Fig. 1). The cells are suspended in a viscous medium and introduced into the gap between the cylinders (typically 0.5 mm). Cell elongation under shear after cylinder rotation begins is monitored by the diffraction of a laser beam directed normal to the cylinder axis. The two main factors affecting the viscosity of the blood are rouleaux formation, mediated primarily by fibrinogen, and the deformability of the erythrocyte. In the ektacytometer, rouleaux formation is suppressed by dilution and shearing forces. Moreover, the flow in the ektacytometer is laminar and the dilution is sufficient to eliminate cell–cell collision. As a result, the deforming forces acting on the cells are well defined, depending only on the shear rate and the viscosity of the suspending medium. The ektacy-

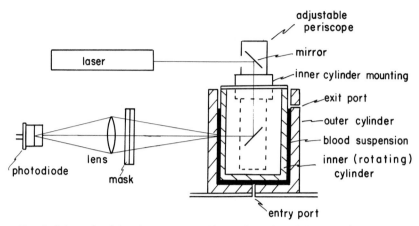

FIG. 1. Schematic of the ektacytometer. The configuration of the Technicon instrument is shown. The two cylinders are made of polished Lucite, and the inner cylinder is mounted on a bearing on the central shaft which also contains a periscope to direct the 1-mW helium–neon laser beam through the blood suspension.

tometer, therefore, measures the mechanical properties of the red cell in isolation. Because monitoring of cell shape is continuous, it is also possible to determine the time required to fragment cells or ghosts under high shear, and this quantitative measurement of mechanical stability, which is difficult to obtain by other methods, is one of the most valuable features of the ektacytometer.

Instrument

Until recently, it was likely that most of the instruments in use were homemade, following the original designs of Bessis's group at the Institut de Pathologie Cellulaire,[1,2] but ektacytometers are now commercially available from Technicon Instruments (Tarrytown, NY). Here, the use of the commercial instrument will be described. In addition to the basic configuration (Fig. 1), the Technicon instrument includes two pumps to generate buffer gradients, a microprocessor which controls the viscometer motor, the pumps and the image analyzer, and a keyboard with a display that will show the ektacytometric index or calibration information on demand. The concentric cylinder viscometer diameter is 50.7 mm and the cylinder gap is 0.5 ± 0.02 mm. The available rotation speeds are 0–256 rpm (inner cylinder), settable to 1 rpm. The instrument also has a relaxation mode, in which the time decay of the cell deformation is determined.

[1] M. Bessis and N. Mohandas, *Blood Cells* **1,** 307 (1975).
[2] B. Cavadini, Diplome de l'Ecole Pratique des Hautes Etudes, Paris, 1975.

The cylinder starts and stops 25 times in succession at 1-sec intervals, and the resulting decay curves of the index are averaged. In principle, these curves will characterize the dynamic rigidity of the erythrocyte, but this capability has not received much application to date. There are two outputs: a y axis output, which produces a voltage equal to the ektacytometric index, and an x axis output, which can produce a voltage proportional to rpm, osmolality, or time, depending on the setup of the experiment. These outputs are attached to an x–y recorder (Hewlett-Packard 7035B, or Sefram X1P, for example). Detailed manuals are available from the Technicon company.

Diluents

The two essential characteristics of the diluent used in ektacytometry are (1) a refractive index different from that of the cells or ghosts, so that a diffraction image is generated, and (2) a viscosity greater than the internal viscosity of the particles. Red cells in shear flow behave as liquid droplets, and will not deform unless external viscosity exceeds internal (if this condition is not met, the cells simply tumble). Tank-treading of the membrane transmits shear stress from the medium to the interior of the cell.[3,4] The internal viscosity of the erythrocyte is determined by the hemoglobin concentration, and is about 11 cP (centipoise) for an "average" value of 35 g% hemoglobin. The viscosity of protein solutions becomes highly nonlinear with concentration at the approximately 5 mM levels of hemoglobin found in normal cells. The usual range of mean corpuscular hemoglobin concentration (MCHC) of 29 to 37 g% implies an internal viscosity range of 5 to 17 cP (cf. Fig. 8 of Ref. 5). In practice, a medium viscosity of 20 cP is used, which is obtained by the addition of 20% dextran or 3% polyvinylpyrrolidone (PVP) to the suspending medium. This is sufficient to deform essentially all the cells in a normal population.[6] The buffers used are listed in Table I.

The shear rate (sec^{-1}) acting on the red cell is

$$r = R\Omega/\varepsilon \tag{1}$$

where R is the cylinder radius (cm); ε, the cylinder gap (cm); and Ω, the angular velocity (radians), which is equal to $(2\pi/60) \times$ rotations per minute.

$$\text{Shear stress (dyn/cm}^2) = r\eta \tag{2}$$

where η is the viscosity of the suspending medium (poise).

[3] T. M. Fischer, M. Stohr, and H. Schmid-Schönbein, *AIChE Symp. Ser.* **182**, 38 (1978).
[4] L. Dintenfass, *Acta Hematol.* **32**, 299 (1964).
[5] S. Chien, S. Usami, and J. F. Bertles, *J. Clin. Invest.* **49**, 623 (1970).
[6] N. Mohandas, M. R. Clark, M. S. Jacobs, and S. B. Shohet, *J. Clin. Invest.* **66**, 563 (1980).

TABLE I
EKTACYTOMETER SOLUTIONS

Base PVP buffer

Component	Grams per liter
3.1% PVP K90	31.0
6.34 mM Na_2HPO_4	0.90
2.00 mM NaH_2PO_4	0.24
0.04% NaN_3	0.4
1.00 mM EDTA	0.33

The PVP (MW = 360,000) requires overnight stirring to dissolve. Appropriate additions of HCl are made to achieve the desired pH (7.35 ± 0.05). The viscosity is 20 cP at 25° and 12 cP at 37°. This can be checked by a standard Cannon–Fenske viscometer, size #150. Once the viscosity of a given solution is known, the polymer concentration can be reproduced using refractometry to determine concentration. In practice, however, it is not inconvenient to determine viscosity directly on stock solutions which can be kept for a number of months. Alternatively, 20% dextran T40 or 30% Stractan[15] can be used to increase the viscosity of the solutions

Operating solutions

NaCl is used to adjust osmolality. The activity coefficient for 0.1 M NaCl is approximately 0.93, so that 1 mOsm = 0.54 mM. The osmolality should be checked after NaCl addition and adjusted if necessary. Either freezing point depression or vapor pressure (Wescor) measurements can be used

A. *Deformation vs shear curves:* The basic solution is made isotonic by the addition of 7.938 g/liter NaCl

B. *Osmoscan solutions:*
Isotonic solution: 7.938 g/liter NaCl (290 mOsm)
High tonicity solution: 22.428 g/liter NaCl (775 ± 15 mOsm)
For the low osmolality solution, the basic PVP solution is used. Its osmolality is 38 ± 2 mOsm

C. *Ghost fragmentation buffer:* This is much more viscous than the basic PVP solution.

Component	Grams per liter
6.34 mM Na_2HPO_4	0.90
2.00 mM NaH_2PO_4	0.24
0.04% NaN_3	0.4
1.00 mM EDTA	0.33
35% Dextran T40	350.0
NaCl	7.94

The dextran will dissolve without stirring if left to stand a few days. Alternatively, stir gently after adding a small amount of the dextran. Allow each addition to dissolve before adding the next. Osmolality can be determined by freezing point depression, but a vapor pressure instrument is more generally useful, since concentrated dextran

TABLE I (*continued*)

solutions often will not freeze in the cryoscopic measurement. The osmolality will be quite high, but this does not affect the fragmentation time. Test the viscosity of new solutions, using a size 300 Cannon–Fenske viscometer, to ensure that it is at least 120 cP. For reproducibility, it is convenient to make up enough solution for a number of months of experiments. The concentration of dextran can be obtained from the refractive index at 20° (RI) using the relationship:

$$RI = 1.3330 + 0.00153 \text{ (g/100 ml of solution)}$$

D. *Oxygen scan buffer:*

Component	Per liter
40 mM NaP$_i$	5.52 g NaH$_2$PO$_4 \cdot$ H$_2$O
0.04% NaN$_3$	0.4
3.1% PVP, MW 360,000	31 g

Appropriate additions of HCl and NaCl are made to achieve the desired pH and osmolality. The viscosity is 12 cP at 37°

Polymers

PVP: Polyvinylpyrrolidone K90, average MW 360,000 (GAF Corporation, Wayne, NJ)
dextran: Dextran T40, average MW 40,000 (Pharmacia Chemicals, Piscataway, NJ)
Stractan: Arabinogalactan (St. Regis Paper Co., Tacoma, WA, or Sigma Chemical Company, St. Louis, MO)

Image Analysis

At the rotation speeds used, the flow between the cylinders is laminar. The cells elongate and orient themselves with their long axis normal to the rotational axis. Because of its high hemoglobin (Hb) concentration, the interior of the red cell has a refractive index greater than that of the medium, and the erythrocyte will scatter light. The basic information about cell shape is derived from the diffraction pattern, which is circular when the cells are at rest, but becomes elliptical as they are stressed and elongate. As predicted by diffraction theory, the image is rotated 90°, and appears as an upright ellipse (Fig. 2). For the most direct measurement of cell deformation, the length L and width W of the first diffraction ring are measured, and the ellipticity ($L - W/L + W$) is calculated. This can be done by direct measurement of photographs of the pattern projected on a screen, but it is far more convenient to use the automated signal analyzer which is part of the Technicon instrument. The principle is shown in Fig. 3. A four-quadrant photodiode is placed behind a mask with four holes, two (a_0, a_1) on the major axis of the ellipse produced by stressed cells,

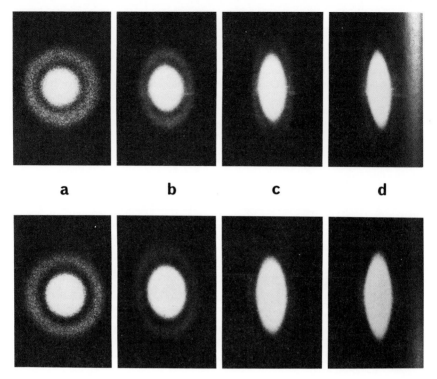

a	b	c	d

FIG. 2. Diffraction patterns from stressed normal human erythrocytes (top row) and normal rat cells (bottom row). (a) No stress, (b) 12.5 dyn/cm^2, (c) 50 dyn/cm^2, (d) 125 dyn/cm^2. (From Ref. 1.)

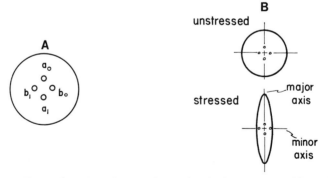

FIG. 3. (A) The configuration of the mask covering the four-quadrant silicon photodiode of the image analyzer. There are four holes in the mask, denoted a_0, a_1, b_0, and b_1. (B) Principle of the image analyzer. When cells are unstressed and the pattern is circular, the light intensity in all holes is equal. Under stress, the elongated pattern increases the light intensity in a_0 and a_1, relative to b_0 and b_1. The microprocessor calculates $A = (a_0 + a_1)/2$, $B = (b_0 + b_1)/2$, and the index $= (A - B)/(A + B)$. (C) Optical density scans of the major and minor axis of the diffraction patterns of normal human red cells under different applied stress values. The mask apertures are placed at the midpoint of the slope of the curve for unstressed cells. The x axes are not to scale. (From Ref. 1.)

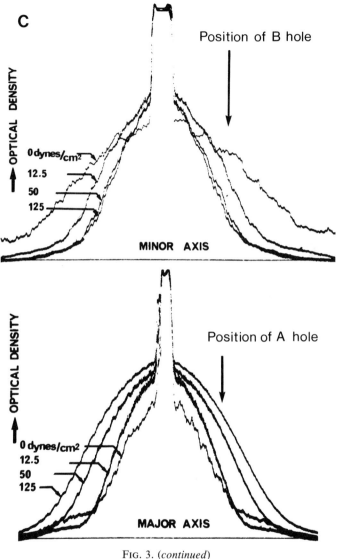

C

Position of B hole

OPTICAL DENSITY

0 dynes/cm²
12.5
50
125

MINOR AXIS

OPTICAL DENSITY

Position of A hole

0 dynes/cm²
12.5
50
125

MAJOR AXIS

FIG. 3. (*continued*)

and two (b_0, b_1) on the minor axis, equidistant from the center of the diffraction pattern. Each quadrant has an independent amplifier. The light intensity is passed to the microprocessor which calculates the averages $A = (a_0 + a_1)/2$ and $B = (b_0 + b_1)/2$, and uses these values to calculate the index $(A - B)/(A + B)$. An output voltage of 0 to 10 V proportional to the index appears at the output labeled y axis at the rear of the apparatus, which can be routed to an $x–y$ recorder. For unstressed cells, the pattern is circular and the light intensity in each hole is equal. Therefore, $(A − B/A + B) = 0$. As the cells are stressed, the pattern becomes ellipsoidal, and the light intensity in A increases as the intensity in B declines, and $(A − B/A + B)$ becomes positive. The instrument will also present the values of A, B, and B/A upon command. A video camera interfaced with a digitizer to a microcomputer has also been used for image analysis.[7]

The use of four holes also aids in centering the image analyzer. The intensity is measured at a_0, a_1, b_0, and b_1 with no cells in the viscometer. This is taken as the background value. Then a cell suspension is loaded and the intensities are again measured. After subtraction of background, a_0 should equal a_1 and b_0 equal b_1. There are adjusting screws to align the detector if required.

It can be shown that this image analyzer produces a quantitatively accurate measure of the red cell ellipticity. The argument proceeds in two steps. It must be demonstrated (1) that the ellipticity of the diffracted light is a true measure of the ellipticity of the stressed cells, i.e., $(L − W/L + W)_{cells} = (L − W/L + W)_{diffraction\ pattern}$; and (2) that use of the light intensity at the A and B holes is equivalent to the measurement of ellipse length and width, i.e., $(A − B/A + B) = (L − W/L + W)$. This was done by Groner et al.[8] They demonstrated first that the ellipticity of the diffraction image produced by the ektacytometer is equal to the ellipticity of stressed cells, by comparing ruler measurements of the diffraction pattern with the dimensions of red cells directly observed under the same stress in the rheoscope. For the second point, using an earlier version of the image analyzer, they empirically determined the correct placement of the detector and the mask holes to ensure that the index generated by the instrument, $(A − B/A + B)$, equals pattern ellipticity. A theoretical development based on the light scattering of prolate ellipsoids predicted a similar critical placement of the openings. The index calculated by the instrument is therefore a quantitative measure of the red cell ellipticity. The index has been given various names: DI (deformation index) or EI (ektacytometric index, or elongation index).

The image analyzer, although extremely convenient, has certain limitations that must be kept in mind. First, the size of the diffraction image is inversely proportional to the size of the cell (cf. Fig. 2). As a result, the image analyzer can produce an EI which differs from the ellipticity of the diffraction image.[8] For example, red cells that are smaller than human red cells will give an erroneously small value of the EI. This can be under-

[7] F. F. Vincenzi and J. J. Cambareri, in "Cellular and Molecular Aspects of Aging" (J. W. Eaton, D. K. Konzen, and J. G. White, eds.), pp. 213–222. Liss, New York, 1985.

[8] W. Groner, N. Mohandas, and M. Bessis, Clin. Chem. 26, 1435 (1980).

stood and corrected if the light intensity across the image is considered (Fig. 3C). The analyzer holes are placed on the center of the linear region of the intensity vs distance curve of human red cells. In effect, as the image widens and narrows, the aperture moves up and down the intensity curve. The placement of the holes has been chosen so that for the maximal observed widening or narrowing of the image of normal human erythrocytes, the holes remain on the linear portion of the intensity curve. For smaller cells with a larger diffraction image, this condition can be obtained by moving the image analyzer closer to the viscometer, so that it subtends a smaller solid angle of the diffracted light. This adjustment is not necessary in work with abnormal human erythrocytes, since the interpretation of the curves is largely empirical. If nonhuman cells that differ markedly in diameter are to be studied, however, the adjustment is necessary.

The second factor that affects the value of the EI is the shape of any nondeformable cells in the population. The EI for symmetrical rigid cells is zero, so that the presence of nondeformable discocytes simply reduces the signal in proportion to their concentration.[6] Asymmetric rigid cells (deoxygenated sickle cells or markedly dehydrated cells, for example) do not elongate in the shear field, but they do rotate and orient themselves with their long axes parallel to the viscometer axis.[9] Since the diffraction pattern is rotated by 90°, asymmetric rigid cells generate a horizontal component in the light pattern, and they therefore contribute a negative EI signal (Fig. 4). The magnitude will depend on the detailed cell shape and its orientation, but fortunately it is usually small. This negative EI term does, however, need to be considered in work with sickle cells.

Cell Preparation Procedures

Whole blood can be used as is. In general, it is not necessary to wash red cells or to remove leukocytes, since they are an insignificant fraction of the total cell number, although this can be done without affecting the results. (But see Fragmentation, below.) In the viscometer, the cell concentration can be anywhere between 0.5×10^8/ml and 2×10^8/ml without altering the signal. Higher or lower concentrations degrade the signal/ noise ratio excessively.

Instrument Operation

Most of the measurements that have been made in the ektacytometer are of four basic types, as illustrated in Fig. 5: (1) cell and mem-

[9] M. Bessis and N. Mohandas, *Blood Cells* **3,** 229 (1977).

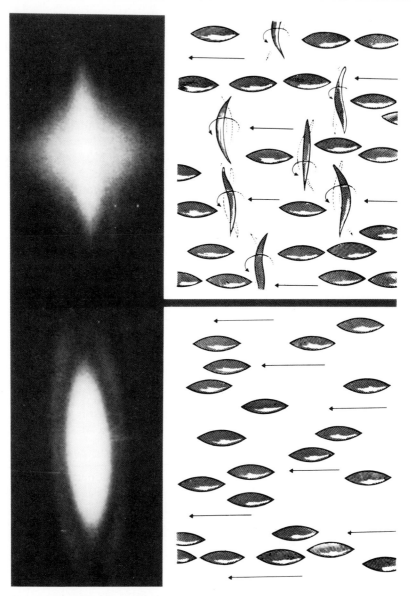

FIG. 4. The diffraction image of asymmetric rigid erythrocytes. A population of sickle cells containing numerous indeformable dehydrated cells (irreversibly sickled cells) during shear at 125 dyn/cm². (From Ref. 9.)

brane deformability as a function of applied stress, (2) osmotic gradient ektacytometry, (3) fragmentation of red cells and ghosts under high shear, and (4) oxygen effects on deformability, primarily with sickle cells.

Index vs Stress

A suspension of cells is made up in isotonic dextran or PVP buffer with a viscosity of about 20 cP (100 μl whole blood in 8 ml isotonic diluent). Five milliliters is introduced into the nonrotating viscometer through the lower port, using a syringe and tubing. The "ramp-up" command is keyed into the microprocessor, which increases the rotation speed to 256 rpm at a rate of 1 rpm/sec, and produces a voltage equal to (rpm/256)(10 V) at the x axis output terminal on the back of the instrument. A voltage proportional to the index is generated at the y axis output, so that a recording of the index as a function of rpm can be obtained. The shear rate for a given rpm can be readily calculated and entered on the x axis using Eq. (2) (Fig. 5A).

Data Analysis. Two types of information are obtained, the deformability of the population of erythrocytes, and an estimate of the membrane rigidity. The ektacytometric index (DI or EI) is an average of the deformability of all the cells in the population under the given experimental conditions, and for normal cells, the index reaches a maximal value of about 0.7. It is diminished in direct proportion to the number of undeformable cells in the population.[6] For an individual erythrocyte, deformability depends on three factors: the surface-to-volume ratio, the internal viscosity, and membrane rigidity.[10,11] Dehydrated cells, for example, will not deform, since their internal viscosity exceeds that of the medium. Intracellular dehydration leads to a sharp reduction in cell deformability in the ektacytometer with a decline in the maximal EI. The deformability index could in principle be written as an explicit function of these three factors and the shear rate, but insufficient information is available to do so at the present time. Of these factors, membrane rigidity is likely to be the least important for the whole cell, since the membrane bending modulus is normally small. Nevertheless, membrane rigidity can be estimated from the slope of the initial rise of these graphs.[6] Consequently, the measurement of cell deformation as a function of applied shear stress supplies an estimate of population cell deformability and of membrane stiffness.

[10] P. L. La Celle and R. I. Weed, *Prog. Hematol.* **7,** 1 (1971).
[11] N. Mohandas, W. M. Phillips, and M. Bessis, *Semin. Hematol.* **16,** 95 (1979).

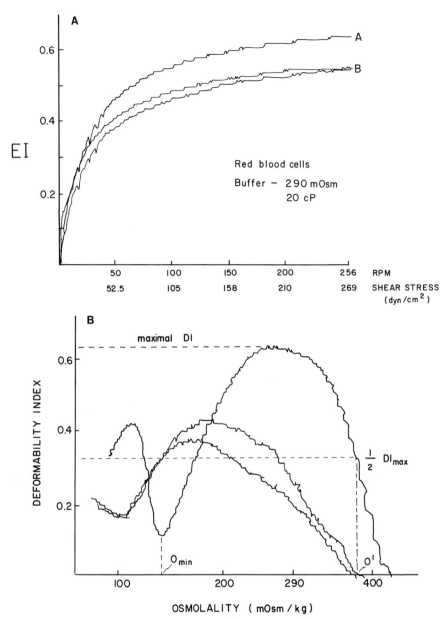

FIG. 5. Typical ektacytometric results. (A) EI as a function of applied stress for normal and two examples of slightly indeformable pathological red cells. (B) Osmoscans for a normal and two dehydrated cell samples. The parameters O' and O_{min} are indicated for the normal curve. (Modified from Ref. 13.) (C) Ghost fragmentation. Two curves for normal membranes are shown, together with the fragmentation curve of membranes with a defect in the self-association site of spectrin. These red cells are fragile and are associated with hemolytic anemia. (D) EI as a function of oxygen tension for sickle cells at two different pH values.

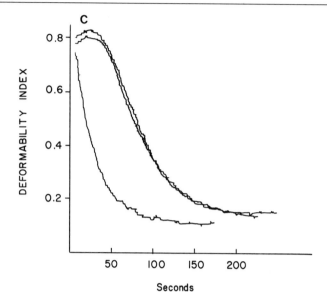

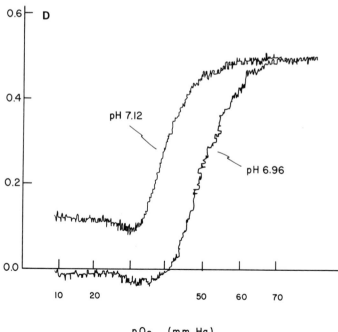

FIG. 5. (*continued*)

Osmotic Gradient Ektacytometry

In this mode, cells are mixed with a buffer–PVP stream with an increasing linear gradient of osmolality.[6,12,13] In the Technicon apparatus, the gradient is mixed and introduced into the viscometer by a peristaltic pump which is part of the instrument. The gradient-maker vessels contain 20 ml of low tonicity (38 mOsm) solution and 8 ml of high tonicity (775 mOsm) PVP. A suspension of red cells in isotonic buffer (100 μl of blood in 3 ml isotonic PVP) is added via a second peristaltic pump to the mixed stream at a T joint near the chamber entry port. An entire scan requires about 10 min. As there is no convenient way to monitor osmolality continuously, the desired solution osmolality is obtained by the addition of electrolytes and conductivity is measured instead. Two platinum wire electrodes are built into the wall of the viscometer, and the output of the instrument's conductivity circuit appears as a voltage at the y-axis output when the command C8E, start osmoscan, is entered. The output voltage is calibrated by manually filling the viscometer with PVP solutions with NaCl whose osmolalities have been independently determined. Changes in ambient temperature are compensated using a thermistor set in the wall of the viscometer. Unfortunately, the voltage output of the Technicon instrument is not linear with salt concentration and the resulting x axis (osmolality) is also not linear. However, for many purposes the nonlinearity is not a significant difficulty, as the interpretation of the curves is largely empirical. The resulting plot of the index (DI or EI) vs osmolality is called an osmoscan (Fig. 5B).

Data Analysis. The major numerical parameters obtained are O_{min} and O', but the general shape of the curve is also highly informative for an experienced operator. A comprehensive review can be found in Ref. 13. O_{min} is linearly related to the osmotic fragility curve and is an accurate measure of the surface-to-volume ratio of the erythrocyte population. O' is linear with 1/MCHC, a measure of cell dehydration. The osmoscan therefore provides information about water content and cell surface area, as well as cell deformability in isotonic solutions. Deviations of the curves from the normal shape are indicative of cell heterogeneity and are often diagnostic for various anemias.[12,13]

Transport Studies. Because the osmoscan can distinguish cell changes due to dehydration from changes brought about by membrane loss or membrane rigidity, it can be used to distinguish alterations in ion transport from the other possible actions of various agents. For example, Clark

[12] C. J. Feo, M. Nossal, E. Jones, and M. Bessis, *C.R. Hebd. Seances Acad. Sci.* **295,** 687 (1982).
[13] M. R. Clark, N. Mohandas, and S. B. Shohet, *Blood* **61,** 899 (1983).

et al.,[14] using an earlier version of the ektacytometer in the manual mode, were able to demonstrate that Ca^{2+} causes indeformability in red cells primarily by activating the Gardos channel for potassium, rather than by altering the mechanical properties of the membrane. Clark, Mohandas, and co-workers were also able to study the abnormal ion transport in desiccytes,[15] to determine that the indeformability of oxygenated sickle cells is related to abnormalities in ion transport,[16] and to characterize cation transport in ionophore-treated red cells.[13,17]

Fragmentation

The rate at which cells or ghosts fragment in the shear field applied by the ektacytometer is readily observed. The determination of the intrinsic mechanical stability of the membrane is one of the most valuable capabilities of the instrument, since this measurement is difficult or impossible to obtain by other methods.

Erythrocytes. If EI is followed as shear is increased, the red cells will be observed to reach a maximal elongation (cf. Fig. 5A). The index vs shear curve has a long plateau beginning at a shear of approximately 200 dyn/cm^2, but as shear is increased, the red cells can be made to fragment. In order to apply the fragmentation conditions at a defined time point, the cells are suspended in a highly viscous (124 cP) medium and introduced into the nonrotating viscometer. The rotation speed is brought to 150 rpm and simultaneously a time scan is begun. The index rises to a maximum which is related to the cell deformability, and then declines as the cells fragment, because the resulting small spherical vesicles do not deform and also because the size of the diffraction pattern, which is inversely proportional to the vesicle size, expands beyond the size of the image analyzer.

Data Analysis. The half-time is defined as the time when $EI = EI_0/2$ (Fig. 5C). The fragmentation times of intact erythrocytes are highly variable because of the variation in cytoplasmic viscosity found in a population of erythrocytes. As noted earlier, cell elongation does not occur unless the internal viscosity of the cell is lower than the viscosity of the suspending medium. At a given shear rate, those cells in a population with a relatively higher internal viscosity will be relatively less elongated and

[14] M. R. Clark, N. Mohandas, C. J. Feo, M. S. Jacobs, and S. B. Shohet, *J. Clin. Invest.* **67**, 531 (1981).
[15] M. R. Clark, N. Mohandas, V. Caggiano, and S. B. Shohet, *J. Supramol. Struct.* **8**, 521 (1978).
[16] M. R. Clark, N. Mohandas, and S. B. Shohet, *J. Clin. Invest.* **65**, 189 (1980).
[17] M. R. Clark, N. Mohandas, and S. B. Shohet, *J. Clin. Invest.* **70**, 1074 (1982).

deformed. The membranes of these cells suffer less mechanical stress due to deformation, and are slower to fragment. Erythrocyte fragmentation depends therefore on two factors: membrane strength and internal viscosity.

Ghost Fragmentation. The variability in internal viscosity can be eliminated by using resealed ghosts. Of the three factors that determine erythrocyte flexibility, two are reduced to negligible levels by the process of ghosting. Because ghosts have low hemoglobin content, their internal viscosity is very low, and their surface-to-volume ratio is quite high, eliminating the effect of this factor on deformability.[18] Therefore, the flexibility of the membrane alone determines the EI of ghosts. In order to generate a refractive index difference, ghosts are resealed and then placed in a medium of high refractive index. The refractive index depends on the weight percentage of polymer in solution, independent of molecular weight. While the intrinsic viscosity increases markedly with molecular weight. The choice of polymer is then determined by the requirement for a high refractive index with an appropriate viscosity. Dextran T40 (average MW 40,000) can be prepared at 35–40% concentrations to fulfill both requirements. In contrast, PVP (MW 360,000) at a concentration high enough to have an adequate refractive index would have an unworkably high viscosity. Stractan[18] can also be used in place of dextran.

Resealed Ghost Preparation. Careful attention to the conditions of lysis and resealing are necessary to obtain reproducible results. In addition, the viscosity of all solutions used in the ektacytometer must be measured before use, and carefully adjusted to the values given in the procedures below, which are based on those described by Heath *et al.*[18] with some modification. The leukocytes are removed by passing the blood through a cellulose column.[19] Erythrocytes are washed twice in 30 times the original blood volume of phosphate-buffered saline at room temperature. After each wash, the cells are collected by centrifugation (15,000 rpm, 1 min). They are cooled to 4°, and lysed in 30 vol of ice-cold 5 mM NaP$_i$, pH 8.0, 1 mM EDTA, 20 μg/ml phenylmethylsulfonyl fluoride (PMSF). After 5 min at 0°, the ghosts are collected by centrifugation at 15,000 rpm for 10 min. The supernatant and the tightly packed pink button at the bottom of the centrifuge tube are removed by aspiration, yielding pink membranes. The membranes are suspended in 15 vol of PBS and incubated for 1 hr at 37°. The resealed membranes are pelleted at 18,000 rpm, 10 min, and the supernatant removed.

[18] B. P. Heath, N. Mohandas, J. L. Wyatt, and S. B. Shohet, *Biochim. Biophys. Acta* **69,** 211 (1982).
[19] E. Beutler, "Red Cell Metabolism," 2nd Ed., p. 10. Grune and Stratton, New York, 1975.

Mechanical Fragmentation of Ghosts. One hundred and fifty microliters of resealed membranes is resuspended in 8 ml of fragmentation buffer (Table I), which contains 35% dextran T40 (Pharmacia) to make the viscosity 124 cP at 23°. The exact centipoise value is not critical, but it must be consistent between experiments if fragmentation times are to be compared. Viscosity becomes a power function of concentration at high concentrations, and small variations in the percentage of dextran significantly alter the viscosity. For this reason, the viscosity of every new solution of dextran buffer is checked with an appropriate Cannon–Fenske viscometer (size 300) in a thermostatted bath. The concentration of ghosts must be between 70×10^6 and 250×10^6 for an adequate signal-to-noise ratio. The ghost suspension is loaded into the ektacytometer with the rotational speed at zero. The cylinder rotational speed is brought to 150 rpm and recording begun simultaneously. The index rises initially and then declines as the red cell ghosts begin to fragment.

Data Analysis. The maximal initial index is denoted as $EI°$. This is a measure of membrane rigidity. The half-time, defined as the time in seconds at which $EI°/2$ is reached, is a measure of ghost stability. This number is quite reproducible, with values of 133 ± 40 sec for 40 normals in 1 series,[20] and 131 ± 32 sec in our laboratory ($n = 38$). This technique has been frequently used to identify carrier states for genetic variants of the membrane skeleton proteins that are associated with membrane fragility and hemolytic anemia.[20,21] It is also useful[18] for determining the effect of chemical treatments on the mechanical properties of the membrane. A recent application is found in the work of Chasis and Mohandas,[22] who combined measurements of ghost stability with the fragmentation assay and of deformability by the stress assay (see Index vs Stress, above) to show that these properties are independently regulated in the membrane.

Oxygen Scan

The technique cannot be performed on the Technicon instrument as purchased. A number of physical modifications[23,24] are necessary to generate and to monitor a continuous O_2 gradient in the viscometer (Fig. 6). Two bottles of buffer in a 37° bath are gassed with either N_2 or air. The

[20] N. Mohandas, M. R. Clark, B. P. Heath, M. Rossi, L. C. Wolfe, S. E. Lux, and S. B. Shohet, *Blood* **59**, 768 (1982).
[21] W. C. Mentzer, T. Turetsky, N. Mohandas, S. Schrier, C.-S. C. Wu, and H. Koenig, *Blood* **63**, 1439 (1984).
[22] J. A. Chasis and N. Mohandas, *J. Cell Biol.* **103**, 343 (1986).
[23] C. J. Feo, E. Jones, and M. Bessis, *C.R. Hebd. Seances Acad. Sci.* **293**, 57 (1981).
[24] M. Bessis, C. J. Feo, and E. Jones, *Cytometry* **3**, 296 (1983).

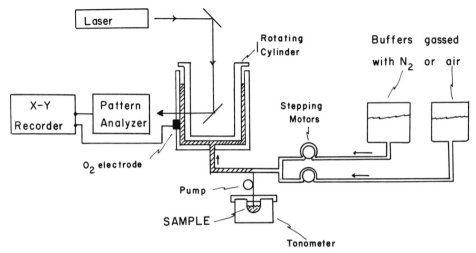

FIG. 6. Schematic of the ektacytometer modified for oxygen scans.

gradient is made by two peristaltic pumps driven by stepping motors which gradually increase the proportion of air-gassed buffer in the stream of buffer entering the viscometer. The blood sample, appropriately diluted and deoxygenated in a tonometer (IL model 237) at 37°, is added to the mixed buffer stream by a third peristaltic pump, the sampling pump. The cell suspension passes through a heating coil at 37° whose length is adjusted to give a transit time of 1 min. This permits complete mixing and equilibration of the cells and the O_2 gradient. The cells then enter the viscometer where O_2 concentration is measured with a Clark electrode installed in the outer cylinder. The outputs of the electrode and the image analyzer are fed into an $x–y$ recorder to give a continuous record of deformability as a function of O_2. The oxygen electrode is installed in a heating block to maintain its temperature at 37°, and the viscometer is wrapped in heating coils. The temperature in the viscometer can be monitored using the thermistor which is supplied as part of the conductivity circuit. Diagrams for the heating and monitoring circuits of the Clark electrode are available from the author. Note that cells are initially deoxygenated and are then mixed with the progressively oxygenated buffer (curves are recorded from low to high oxygen tension). The concentration of cells in the viscometer is approximately 25×10^6 ml^{-1}. The buffer was 40 mM NaP$_i$, 0.04% NaN$_3$ with 3.1% PVP, MW 360,000. Because of the elevated temperature in the experiment (37°) used to promote hemoglobin S polymerization, the solution viscosity is lowered to 12 cP.

Data Analysis. The deformability of sickle cells, as measured by the ektacytometer, increases during oxygenation, with a sigmoid dependence on O_2 concentration. Three features[25] of the ektacytometric curves have been used for quantitation of the effects of antisickling agents:

1. pEI_{50}, the pO_2 at which 50% of the total observed change in EI occurred. This is not necessarily the same as $P_{50}O_2$ the midpoint of the oxygen affinity curve, but is rather a measure of the relation between O_2 and cell rigidity.

2. The EI_{max}, the EI of completely oxygenated cells. This has been used as a minimum criterion for effective antisickling agents, which should not alter the flexibility of cells with depolymerized hemoglobin.

3. The EI_{min}, the index under completely deoxygenated conditions. This value is determined primarily by the fraction of cells that have become rigid, but is also influenced by shape factors. In general, EI_{max} differs somewhat between blood samples and depends on the exact experimental conditions. In order to compare results with different samples, the data is expressed as ΔEI:

$$\Delta EI = \left[\frac{(EI_{max} - EI_{min}) \text{ for treated cells}}{(EI_{max} - EI_{min}) \text{ for untreated cells}} \right] (100) \qquad (3)$$

Compounds that inhibit sickling by increasing oxygen affinity shift pEI_{50}, while compounds that increase deoxy-HbS solubility decrease ΔEI.[25] This distinguishes the two classes of agent.

Recently, a new experimental system was described by Sorette *et al.*,[26] in which sickle cells are deoxygenated in a gas-porous fiber gas exchange system and oxygen tension is measured externally. Although it requires careful calibration, this method appears to be more convenient than the original deoxygenation system described above.

Conclusion

The advantages of the ektacytometer lie in its speed and versatility. An osmoscan, for example, can be obtained in 10 min, and supplies information about population deformability, hydration state, and S/V ratio. Because of the possibility of gradient flow through the viscometer, combined with continuous recording, the response of erythrocytes or ghosts to a wide range of variables can be readily obtained. Although the instrument does not supply the precise quantitative data on individual cells

[25] R. M. Johnson, C. J. Feo, M. Nossal, and I. Dobo, *Blood* **66,** 432 (1985).
[26] M. P. Sorette, M. G. Lavenant, and M. R. Clark, *Blood* **69,** 316 (1987).

provided by micropipet aspiration, it yields population data that would be extremely tedious to acquire by micropipet techniques. The rheoscope has the advantage that cell shape is directly observed, but as for micropipet work, it is tedious to collect data under a range of conditions and there are no existing methods of automated image analysis. In addition, for all these techniques, the response to external variables can only be done point by point. The ektacytometer is therefore the method of choice for many investigations of the factors that affect the mechanical properties of red cells. Finally, the observation of the rate of fragmentation under shear stress supplies a valuable quantitation of the mechanical stability of the membrane.

[3] Methods and Analysis of Erythrocyte Anion Fluxes

By ROBERT B. GUNN and OTTO FRÖHLICH

Introduction

The purpose of this chapter is to provide as complete as possible a description of methods used to measure and analyze anion transport in red cells. The principles are the same for other transport systems, but because red cells behave as a single population of cells with a well-mixed cytoplasm, the extension to other cell types may need to be examined in detail. It is hoped that with this guide a novice scientist in this area can quickly become proficient. We are no exception to the rule that specialization leads to new words, or, worse, jargon, where old words have new specific meanings. These definitions are given when the term is first introduced.

Solutions

Human red blood cells behave as nearly perfect osmometers: they swell in hypotonic media and shrink in hypertonic media to make the internal osmolarity equal to that of the bathing medium. Human red cells have a normal mean cell volume of 87 μm^3 with a biconcave surface area of 142 μm^2. Since a sphere with this surface area has a volume of 164 μm^3, red cells can be expanded to about twice their volume before lysis. Some cells in the population of normal cells, of course, have lower surface areas or larger initial volumes and thus lyse with smaller relative volume changes or at higher osmotic concentrations than the average cell. This

METHODS IN ENZYMOLOGY, VOL. 173

heterogeneity of cell surface area and initial cell volume is the basis for the dispersion in the normal osmotic fragility curve. Since normal plasma osmolarity is 300 mOsm and normally less than two-thirds of the cell volume is water, more than doubling the water content is needed to double the cell volume. Consequently, the average cell can withstand a solution with as little as 100 mOsm of salts although 150 mOsm should swell it to just about double its water content.

Red cell experiments are only as good as the media used in preparing the cells and in the flux measurements. It is as important to know the actual composition of the solutions as to design them properly. Ideally, solutions should be analyzed after they are made or used to certify the concentration, osmolality, and ionic strength. It is better to measure the chloride concentration, for example, and find that it is 148 mM than it is to think it is 155 mM as the design protocol said it should be. With surprising frequency we find that results that appear strange can be traced to a solution whose concentration "changed," whose pH "shifted," or whose osmolality was higher than it "should be." As a precaution, all solutions should have their osmolality measured with a freezing point depression osmometer or a vapor pressure osmometer after they have been prepared *and* titrated to the desired pH value.

Since one often wishes to vary the concentration of one of the ions at constant osmolality, one of two strategies must be adopted: constant ionic strength using an indifferent or "spectator" ion; or varying ionic strength and using an impermeant nonelectrolyte such as sucrose to contribute the needed osmoles. One small cautionary note should be made regarding laboratory sucrose. A 1 M sucrose solution can have 700 μM Ca^{2+}. However, if this amount of Ca^{2+} is a problem, one should use grocery store sugar which is usually lower in contaminating salts but may contain sugars other than sucrose.

The expected osmolalities can be calculated from standard tables.[1] In nearly all cases the sum of osmolarities of the component salts is close to that actually measured.

Buffers which function to constrain fluctuations in solution pH are necessary, but often misleading. Red cells have a buffer capacity of 50 mmol/(liter packed erythrocytes · pH unit) over a wide range of pH values (~6–9). Consequently the cells themselves are the major buffer of many systems. If one has a 10% hematocrit and 5.5 mM buffer at its pK in the suspension medium, the buffer capacities of the cells and solution are equal. If the internal pH is 8.1 and the solution is 7.1, the final extracellu-

[1] "CRC Handbook of Chemistry and Physics" (R. C. Weast, ed.), pp. D229–D266. CRC Press, Boca Raton, Florida, 1983.

lar pH (pH_o) at proton equilibrium will be around 7.6. Red cells can be rather impermeant to protons so the rate of equilibration could be slow. Chloride–bicarbonate exchange is the most rapid way to equilibrate internal and external proton activities, so that in the presence of 22 mM HCO_3^-, equilibration takes 1–2 sec at room temperature. However, when cells are washed in HCO_3^--free media, as is often the case, the rate is 10^4 times slower. And if the anion exchanger is inhibited or one uses solutions free of exchangeable anions, the rate may be reduced another 100-fold. The reason for this additional reduction is that the anion exchanger also seems to cotransport protons and exchangeable anions (Cl^-, Br^-, NO_3^-, $H_2PO_4^-$, SO_4^{2-} [2]; Gunn, unpublished observations). In their absence (and strict absence of HCO_3^-) the red cells' proton permeability falls to that of phospholipid bilayers ($P = 10^{-4}$ cm/sec [3]).

The most common error is to wash cells with a buffer solution and then assume that the proton activities across the membrane are in equilibrium. This assumption can fail if one cools the cells and solutions and at the same time does not retain some HCO_3^- in the system. Fresh normal red cells in plasma at pH 7.4 and 37°, when washed free of plasma with ice-cold unbuffered 165 mM KCl or 165 mM NaCl solution, have an intracellular proton concentration in equilibrium with an extracellular pH_o = 8.1–8.2 at 0°. The pK values of hemoglobin depend on temperature much like those of phosphate and Tris [tris(hydroxymethyl)aminomethane] buffers: they increase as temperature decreases. Thus when cells are the only buffer in the system, the pH increases as the suspension is cooled. If one wishes to end up with cells at pH 7.4 at 37°, but one wants to wash them in an ice-cold medium, then one should either use an extracellular buffer with the same temperature dependence as hemoglobin, titrated to pH 7.4 at 37° and then cooled, or use no buffer at all. If the buffer has a different temperature dependence from hemoglobin, one will unknowingly be titrating the cell interior as one washes the cells at the low temperature. When the cells are rewarmed, their proton activity will not be immediately at equilibrium with that of the medium.

Preparation of Cells

Drawing Blood and Washing the Red Cells

We prepare a centrifuge tube or graduated cylinder by rinsing it with heparin (1000 U/ml) diluted 10- to 100-fold in 100 mM $MgCl_2$ or any other

[2] M. L. Jennings, *J. Membr. Biol.* **40**, 365 (1978).
[3] W. R. Nichols and D. W. Deamer, *Proc. Natl. Acad. Sci. U.S.A.* **77**, 2038 (1980).

isotonic salt solution. The residual droplets of heparin are more than adequate for the brief time the blood plasma remains with the cells. We usually bleed directly through the needle into this container, swirl the blood once during and at the end of collection, and place the blood temporarily on ice or immediately centrifuge it for 5 min at 20,000 g without any diluting fluid. The buffy coat is then formed as a sticky mass at the plasma–red cell interface and is easier to remove. Less centrifugation or diluting the plasma seems to inhibit this white cell clumping and thus results in more white cells contaminating the red cells. The plasma and buffy coat are removed by aspiration through a needle of #13 bore. The cells are diluted with appropriate medium and stirred with a glass rod to resuspend the cells, including the most dense cells that can be tightly packed on the bottom of the tube. They are then washed by repeatedly centrifuging them, aspirating off the supernatant fluid, and resuspending them. Assuming that the centrifuged cells have 10% extracellular space, the required number of wash cycles and volumes of each can be calculated to dilute any plasma factors not adherent to the cells.

Preparation of Cells with Different Intracellular pH

If one desires cells with a different internal pH, this can be achieved by titrating the cells with CO_2 or 165 mM HCO_3^- solutions and then washing away the HCO_3^- or CO_2 produced, respectively. Both CO_2 and HCO_3^- cross the membrane, the former by passive diffusion, the latter by band 3-mediated exchange with chloride. In the cytoplasm, CO_2 and H_2CO_3 rapidly interconvert through the mediation of carbonate dehydratase. The carbonic acid is also in very rapid equilibrium with H^+ and HCO_3^-, and proton titrates the hemoglobin. As a means for equilibrating transmembrane proton gradients, Cl^-–HCO_3^- exchange is therefore functionally equivalent to H^+–Cl^- cotransport. It is therefore important to have a functional anion exchanger and another exchangeable external anion (not HCO_3^-).

In order to increase pH_i, the cell suspension is titrated with a 165 mM $KHCO_3$ solution or 100 mM K_2CO_3 solution until the desired pH_i is achieved. Analogously, to lower pH_i the cells are suspended in an unbuffered salt solution (usually 165 mM KCl) to which is added some of the same solution but which has been bubbled with CO_2 for 5–10 min. The pH of this solution will be ~4, but it has little buffering capacity. By repeated washing of the cells with CO_2-bubbled saline, one can lower the pH to about 5. After the titration, the cells are washed in CO_2-free, unbuffered saline to remove CO_2 and HCO_3^- from the system. If a $pH_i < 5$ is desired, the volumes of CO_2-bubbled saline become prohibitive and a strong acid such as 300 mM HCl can be used (300 mM HCl is close to isotonic because the protons are rapidly buffered and never exceed 1 mM if the pH is above 3, whereas the Cl^- contributes about 270 mOsm to the osmotic strength of the solutions). Below pH 5, proton influx occurs increasingly by HCl cotransport on band 3 which has a unidirectional flux of 400 mmol (kg cells solids · min)$^{-1}$ at 20° and pH_o = 3, but a much smaller net influx as pH_i decreases (Gunn, unpublished observations). One can only feel confident that the proton activities are at equilibrium if the pH_o is above 5 and CO_2 is present in the system and the pH_o of a nominally unbuffered suspension medium is stable for 10 min. Below pH 5, equilibration of protons is always problematical and requires longer periods of observation of pH_o to be confident that protons are not continuing to enter or leave the cells.

Cells swell when they are acidified. When H^+ and Cl^- enter the cell, water follows for osmotic reasons. Since protons are not osmotically active because they bind to hemoglobin (Hb) and the other intracellular buffers, the stoichiometry between Cl^- and water is about 300 mM. The buffer capacity determines the amount of entering protons and thus the volume change. Theoretically one therefore expects about 50 mmol of Cl^- to be added to each liter of packed cells for each pH unit change. This is equivalent to 150 mmol/kg Hb or 135 mmol/kg cell solids. What has been observed is that as Cl^- content increased, water content increased by 0.24 kg/kg cell solids, or together that about 126 mmol Cl^-/kg dry cell solids is added for each unitary decrease in extracellular pH.[4] This value is lower when it is corrected for the fact that intracellular pH changes more than 1 unit when the extracellular pH decreases by 1 unit, due to the change in membrane potential. However, considering that these data are from experiments designed for another purpose, this experimental value of 126 mmol/(kg solids · pH unit)$^{-1}$ is probably not significantly different from the theoretical value of 135 estimated from the buffer capacity.

Preparation of Cells for Tracer Flux

Cells are resuspended to 30–50% hematocrit, briefly centrifuged, and radioactive anion solution added to the top of the supernatant. For chloride, we use 0.6 μCi/ml of suspension which results in 1300 cpm in 100 μl of a 100-fold dilution of the packed cells.[5] The isotope is mixed with the supernatant first and then with the cells using a glass rod. In this way the

[4] R. B. Gunn, M. Dalmark, D. C. Tosteson, and J. O. Wieth, *J. Gen. Physiol.* **61**, 185 (1973).
[5] "Packed cells" refers to the cell pellet after centrifugation of the cell suspension in the nylon tubes.

cells never are exposed directly to the tracer stock solution which in the case of Na³⁶Cl can be quite concentrated.

The isotope is allowed to equilibrate across the cell membranes at the same temperature as the cells are packed and subsequently the efflux measured. This assures that temperature changes do not alter the proton equilibration via temperature-dependent pK values, though these effects would usually be very small for anion fluxes.

The radioactive cell suspension at 30–50% hematocrit is transferred to custom-made plastic centrifuge tubes ("nylon tubes") that were fashioned by melting shut one end of a section of straight plastic tubing of 10-cm length and about 3-mm i.d. Each tube holds about 0.7 ml. They fit into holes in an otherwise solid holder the size of a 40-ml centrifuge tube. Alternatively, microcentrifuge tubes can be used although their characteristics in terms of extracellular trapped space will have to be determined. We pack seven tubes for each flux measurement. The packed cells from two tubes are used for duplicate efflux measurements, the cells from two more tubes are used for duplicate dry weight measurements, the cells from another two are used for duplicate acid precipitates and one packed cell sample is a spare in case the duplicates do not agree or one of the efflux measurements failed.

The compartments of the packed cell tube must be known for a quantitative understanding of what is happening and for calculations in these and other red cell experiments. The centrifuged cell suspension has three compartments: the supernatant (often called "plasma"[6] even if it is an artificial salt solution), the packed cell volume, and the trapped extracellular space within the packed cell column. The hematocrit is the packed cell volume as a percentage of the total volume of the suspension. Usually it is loosely used to mean the volume of the cell column (cells and extracellular trapped volume) as a percentage of the total volume.

We pack cells by centrifuging for 10 min at 12,000 rpm in a Sorvall SS-34 rotor which gives a mean effective force of 20,000 g. For cells under most conditions this results in a 3% trapped extracellular space by weight; i.e., 30 mg of trapped extracellular fluid per gram of wet packed cell column.

The distribution of chloride between the compartments of extracellular fluid and the centrifuged cells can be determined (see calculations). The ionic composition of the extracellular fluid in the supernatant and in

---

[6] "Plasma" refers to the cell-free supernatant (extracellular medium) after centrifugation of the tracer-loaded cell suspension in the nylon tubes; the plasma is used experimentally to determine the extracellular tracer and total Cl⁻ concentration; here the term no longer implies blood plasma.

the trapped space between cells is the same. Any anion exclusion from the trapped space due to the slight negative surface charge on the erythrocytes will be negligible in normal ionic strength media.

The plastic tubes containing the centrifuged cells with their radioactive label are immediately removed from the centrifuge and put on ice. We record the hematocrit by measuring the length of the cell column and the total fluid column. Each tube is cut with a razor blade such that the cut is made through the packed cell column parallel to the solution–cell interface and about 1–2 mm below the interface. The open end of the packed cell column is stroked 2–4 cm on paper towel to remove the cells and fluid just at the top, and then the tube is immediately capped. Small caps used for sealing the ends of blood–gas syringes are useful. The packed cells are then saved on ice. The remaining top part of the plastic tube that contains the supernatant and some red cells (at one end) is cut again with a razor blade to separate the red cells from the clear supernatant. For each cell type, the radioactive supernatants from several tubes are pooled in a small plastic test tube. The supernatant is needed to determine the specific activity of the radioactive tracer. The packed cells will be used to measure wet and dry weights and thus water contents, to make acid precipitates[7] for cell content (chloride) determinations, and for the efflux measurements. The next experimental step is to perform the actual tracer efflux experiments, but this will be described in a subsequent section.

Determination of Cell Contents

Cell Water Content ("Dry Weight")

Open glass vials (such as scintillation minivials) are marked for later identification and dried in an oven at 90–95° for 24 hr or to constant weight. They are moved to a desiccator jar to cool, and weighed after the scale is set to zero. Then the packed cells from a single sleeve are pushed with a plunger into the vial which is immediately reweighed. These wet red cells are returned to the oven and dried to constant weight (usually 24 hr), then the vials are cooled again in a desiccator jar and reweighed. These three weights are used to calculate (1) the wet packed cell mass and (2) the dry packed cell mass. Their difference divided by the wet packed cell mass is the uncorrected or apparent water fraction, F_w^{app}, uncorrected for trapped extracellular medium.

---

[7] "Precipitates" refers to the process of acid-denaturing the cell proteins to prepare protein-free cell extracts; supernatant of perchloric acid (PCA)- or trichloracetic acid (TCA)-treated and centrifuged cell pellet.

Trapped Space Fraction

Based on previous control experiments, we assume in our calculations that when intact cells are centrifuged under the above described conditions, the cell pellet typically contains an average of 3% extracellular volume that is trapped between the packed cells. This value of the trapped space fraction, F_{ts}, may vary somewhat with shrunken or swollen cells and is considerably larger with resealed cells ("ghosts").[8] If it is critical to know the trapped space for a given experimental condition, it can be determined with an impermeant radioactive marker such as ^{14}C-labeled sucrose, ^{14}C-labeled polyethylene glycol, or ^{22}Na. The tracer is added to an approximately 50% cell suspension which is subsequently centrifuged under the same conditions as for a flux experiment. The supernatant is saved to determine the activity of tracer in the extracellular space, A_{sup}. A sleeve of packed cells is then weighed and lysed in a known volume of distilled water. The dilution factor, K_{dil}, due to the water is calculated after converting the cell weight into cell volume assuming a specific density of 1.2 g/ml under isotonic conditions. After determining the tracer activity in the lysate, A_{lys}, one can calculate the trapped fraction as:

$$F_{ts} = A_{lys} K_{dil}/A_{sup} \tag{1}$$

Perchloric Acid Extracts ("Precipitates")

Two 5-ml disposable plastic test tubes are weighed. To each a sleeve of packed cells is added and weighed. Then 1 ml of 7% perchloric acid (PCA) in water is added and weighed (the weight should increase by 1.070 g). A glass rod is used to grind the cells in the acid. The rod is not removed until all the cells are mixed, so that the fluid removed with the rod has the same concentration of ^{36}Cl and ^{35}Cl as the remaining suspension. The tubes are capped and centrifuged in a desktop centrifuge. The supernatant fluid will be scintillation counted for radioactivity and titrated for total chloride in a chloridometer.

To each of two other plastic test tubes whose weight has been determined, 100–200 μl of supernatant fluid from the packed cells (saved and pooled earlier from the cut packing tubes) is added and the tubes weighed again. Then 1 ml of 7% PCA is added to each, with reweighing to assure that the proper amount was added. These tubes are vortexed and centrifuged. In a way similar to packed cells and supernatants, precipitates of

---

[8] Resealed ghosts are red cells with 1/10–1/1000 of normal dissolved intracellular constituents, with the low intrinsic permeability to cations restored after the hypotonic lysis procedure.

the whole blood suspension can be prepared if needed to check the measurements and calculations.

In this context, it might be valuable to know that the potassium-salt of perchloric acid has a low solubility. A PCA extract will therefore not yield the correct cellular K^+ content. If the K^+ content is to be determined, one should replace PCA by trichloroacetic acid (TCA).

Samples for Liquid Scintillation Counting

The design of the efflux experiments is such that up to this time very little care had to be exercised in terms of precision of the amounts used so long as all necessary samples were collected. For example, the hematocrit of the packed cells or amount of radioactivity did not have to be a precise value, the radioactive cells didn't have to be quantitatively injected into the efflux solution, or the filtered samples didn't have to be of a particular volume. From now on, quantitative measurements are needed.

The principle of sample preparation is to make all of the samples as similar as possible so that the variations in color, volume, and thus in scintillation counting efficiency are minimized. The efflux filtrates (see Flux Experiments) are counted in 2–3 ml scintillation cocktail in mini-vials, as duplicate 100- or 200-μl samples from each syringe and from a centrifuged infinity sample.[9] A blank using efflux medium, water, or 7% PCA is used to measure background. Thus there will be $2n + 3$ vials from the efflux measurements if n timed samples are taken. The radioactivity in the PCA extracts of packed cells and plasma is counted from duplicate 100-μl samples. We have found that a good electric diluter/pipetter is invaluable in rapidly handling the different samples. There will be eight vials from the precipitates plus one blank. Each is counted for 10 min or until 10,000 counts are accumulated. Vials with less than 500 cpm may be counted longer if greater counting accuracy is needed (which it rarely is).

Coulometric Chloride Titration

The efflux solutions and the supernatants should be measured for total chloride concentration. Direct measurement of the extracellular anion composition during the efflux is quite valuable both to check that solutions have the ionic composition they were designed to have and as a precaution that titration with HCl has not inadvertently altered the chloride concentration. It is essential to measure the chloride concentration in the supernatant of the packed cells or the PCA extract of this fluid. The

[9] "Infinity sample" refers to the efflux time point sample taken at long times, sufficiently long for complete tracer equilibration.

calculated flux value is directly proportional to this value, even though the actual flux may not be. To titrate the PCA extracts of the packed cells is a useful check on the methods and calculations but one should be aware that the coulometric titration includes not only Cl^-, Br^-, and I^- but also reduced glutathione (GSH) which is present in red cells (1–2 mM) and the PCA extract. Since red cell chloride concentration is in the range of 100 mM in the cell water, this amounts to a 1–2% error which is readily detectable.[4] The titration of standard chloride solutions is so dependable that it becomes a check on the delivery of the pipette and the accuracy of the experimenter and less a calibration of the machine.

Flux Experiments

Tracer Efflux in Exchange for Nontracer

The efflux of chloride isotope from red cells can either be in exchange for the influx of another anion (including nonradioactive chloride) or accompanied by a cation efflux, thus causing net salt loss and cell shrinkage. The latter is some 1000- to 10,000-fold smaller than exchange in the absence of cationophores (valinomycin, gramicidin) or cation efflux in the absence of low external ionic strength (<1 mM).[10] For the moment consider only the very rapid exchange of internal $^{36}Cl^-$ for external $^{35}Cl^-$.

The equipment includes a thermostatted chamber over a magnetic stirplate. Individual beakers with magnetic fleas sitting in an ice (and salt) bath are adequate but must be physically stable. Swinnex filter holders (Millipore Corp.; New Bedford, MA) have been modified by pressing out the porous disk that was designed to support the filter, milling with a lathe a seat for this disk in the other half of the holder, and placing the disk into it. This inverts the direction of fluid flow during filtration, permitting one to pull fluid through the filter into a syringe. The filter holders are reused the next day after being washed in tap water and air dried. Prior to each experiment sufficient filter holders are packed with one or two glass fiber prefilters (without a nitrocellulose filter or rubber gasket). They are tightened by hand and frozen at −10° to −20°. Before use, each filter holder is mated to a 10-ml plastic syringe on the Luer-lock end and to an extra-long (~10 cm) #13 hollow needle. Four to 10 of these filter assemblies are needed for each efflux measurement. They are put into a freezer for precooling. Also, 5-ml plastic syringes with 2–3 cm of plastic tubing mounted on their tips are frozen. This tubing should fit tightly over the

[10] G. S. Jones and P. A. Knauf, *J. Gen. Physiol.* **86**, 721 (1985).

end of the cut centrifuge tube which contains the packed radioactive cells. A timing device such as a foot pedal-operated clock-printer is especially convenient for rapid fluxes with half-times less than 20 sec. Some pre-knowledge of the half-time for tracer exchange under the given conditions is useful and the literature or exploratory experiments can be consulted as a guide. As a rule of thumb for chloride exchange in 150 mM Cl$^-$ solutions the rate coefficient (in seconds) times 10^4 is about the flux [in mmol (kg dry cell solids · min)$^{-1}$] or 7000 divided by the flux in these units gives the half-time in seconds. Given an approximate half-time and the number of data points one plans to take along the efflux time course, the best plan is to take points at equal intervals, with most data points taken before half-time.

Each efflux measurement involves the following steps in order. The nonradioactive efflux solution (20–35 ml) is placed in the thermostatted chamber and stirred continuously throughout. When the solution is at the desired temperature (usually 0°), several filter setups (5–10) and a 5-ml syringe with its plastic tubing are removed from the freezer and taken to the bench. (It is assumed that during the brief period until the start of the flux experiment they warm to near 0°. This precooling is deemed neces-sary to ensure that the efflux temperature is not raised above the intended value, due to the high temperature dependence of anion exchange: up to 2%/0.1°.[11]) About 2–3 ml of efflux solution is removed from the chamber with the 5-ml syringe and the fluid pulled back from the tip of the plastic tubing. A capped tube of packed cells is taken from the ice, the cap removed, and the packed cells mounted onto the tubing of the 5-ml syringe. Next the tip of the packed cell tube is cut off with a razor blade so that there is now a sleeve of packed cells open at one end and mounted on the other end to the air-filled tubing on the fluid-filled syringe. The efflux is initiated by injecting the cells and syringe content into the efflux chamber. Each sample at a known time after initiation of the efflux is taken by placing the needle of one filter assembly into the mixing cell suspension and drawing up the plunger sharply until fluid is seen in the syringe. The needle is quickly taken out of the efflux medium and the syringe is imme-diately separated from filter holder and needle. The syringes can be set in order on the bench until the last one is taken. One can wait a long time before taking an "infinity" sample from the efflux chamber, but as a means of economy of time and of saving the use of one filter holder we remove a 1- to 3-ml sample from the chamber and place it into two small marked test tubes, one for tracer counting and one for hemoglobin deter-mination. If the sample sits 1 min at room temperature, the specific activi-

[11] J. Brahm, *J. Gen. Physiol.* **70**, 283 (1977).

ties of the isotope will have equilibrated (5–10 min if maximal doses of inhibitors such as the stilbene disulfonates are present). The syringes from each time point are capped (again with blood–gas syringe caps) and placed in order in a large test tube rack. The remaining fluid in the efflux chamber is aspirated into a radioactive waste bottle and the chamber repeatedly filled with double-distilled water and aspirated before fresh nonradioactive efflux solution is added for the next efflux measurement.

Rapid Tracer Flux Techniques

The filter method described above can comfortably resolve a rate constant of 0.2 sec^{-1} by collecting four to five time points within 10–15 sec. At times it might be desirable to improve on this time resolution. The methods of Ku et al.[12] and Fröhlich and Jones[13] (Fröhlich, unpublished observations) permit an improvement of up to 5-fold. If rate constants as rapidly as 100 sec^{-1} need to be resolved, a flow-tube setup such as the one used by Brahm[14] is necessary.

The method of Ku et al.[12] is a quenching method, relying on a rapidly acting and fully inhibiting transport inhibitor to slow down the rate of tracer efflux. In this method, the cell suspension is pumped from the efflux chamber by means of a syringe to which a one-way ball valve is attached. Each piston stroke delivers an aliquot into a test tube containing a known volume of stopping medium with the transport inhibitor. The radioactivity in the extracellular space can then be determined either after centrifugation or filtration. When paced by a metronome, these investigators have sampled time points in 2-sec intervals.

The method of Jones and Fröhlich[13] uses the same filters and filter holders as in the conventional, hand-held method. The filter holders are, however, fitted into a thermostatted metal block. Instead of the #13 needle, each filter holder is connected to tubing that leads down to the efflux chamber. To ensure that samples are collected only after the cells have been suspended, each tubing ends in a large-gauge hollow needle that is held in place just above the waterline. The interior of the syringes is kept under negative pressure by connecting them to the vacuum line. As soon as the cells are dispersed (about 1 sec), the assembly of needles is pushed into the flux chamber. Immediately after that, the valves on the syringes are, in one motion, turned from closed to open and again to closed, permitting sampling for a fraction of a second. This technique permits one to take five time point samples in less than 4.5 sec after injection. As a test

[12] C.-P. Ku, M. L. Jennings, and H. Passow, *Biochim. Biophys. Acta* **553**, 132 (1979).
[13] O. Fröhlich and S. C. Jones, *Biochim. Biophys. Acta* **943**, 531 (1988).
[14] J. Brahm, *J. Gen. Physiol.* **81**, 283 (1983).

of the resolution of this technique, we have measured the temperature dependence of Cl^- exchange up to 12°, resolving rate constants as fast as 1 sec^{-1} (Fröhlich and Morris, unpublished results). This method has the advantage over the quenching method in that it can be used under conditions where no efficient inhibitor is available or a parallel, noninhibitable transport pathway exists.[13]

When still faster rates of transport need to be measured, the flow-tube technique of Brahm[11] is the one of choice. In this setup, packed cells from one syringe and the efflux medium from another syringe are driven, at a constant rate, into a small mixing chamber and then into a long tubing. This tubing contains, at predetermined intervals, filtration ports where extracellular medium is collected. The time a cell takes to travel from the point of mixing until it reaches the filter port is constant as long as the flow rate is constant. The distance between the filtration ports is converted into this time interval using the flow rate inside the tubing. This method has a time resolution of down to 5 msec.[14] In a modification of Brahm's flow-tube setup, Mayrand and Levitt[15] have devised a mixing mechanism that injects air into the flow tube. Since the air bubbles form segments of the cell suspension that move separately down the tubing, this approach obviates concerns about a possible lack of turbulence and the existence of inhomogeneous laminar flow velocity patterns at slow flow rates which could blur the time resolution. This method may therefore be most appropriate for processes of intermediate speed. However, the air-injection method requires estimates of the rate of expansion of the air bubbles as they travel down the tube since this rate influences the cell velocity between the different filtration ports.

Anion Net Flux

Net movements of substrate are the characteristics of a non-steady-state distribution of this substrate across the membrane, resulting in a net increase or decrease of the cellular content. It will be discussed later that under these conditions the unidirectional (tracer) influx and efflux are not the same, but that their difference can contain valuable kinetic information. There are several methods by which one can determine the rates of net flux of a substrate. In the simplest case, when the substrate concentration on the trans side is zero, the rate of net flux in that direction equals the rate of unidirectional flux, thus permitting one to use tracer flux experiments to measure net fluxes. If the substrate concentration is nonzero on both sides, the net flux rate is smaller than the rate of unidirectional (tracer) flux under the same conditions. One can then determine net flux

[15] R. R. Mayrand and D. G. Levitt, *J. Gen. Physiol.* **81,** 221 (1983).

as the difference between the two unidirectional fluxes, but this method loses its applicability when the unidirectional fluxes are significantly larger than the net flux. In the case of a tightly coupling exchanger such as the anion transporter, the difference between the unidirectional fluxes is so miniscule that it cannot be detected at all. It is then that one has to abandon the use of radioactive markers and follow the chemical changes in cellular content with time.

The most direct method is to determine the cellular contents as a function of time, for example, by collecting and washing the cells in the desired time intervals and assaying for the substrate contents after acid precipitation as described above. In terms of anion net transport, this means that one determines the cellular anion content which in the case of the halides can be done coulometrically in a chloride titrator. Alternatively, one can also measure the rates of cation loss or gain since during anion net transport cations have to move in parallel for electrostatic reasons, with a stoichiometry according to their relative charges. These parallel pathways are introduced into the membrane by adding the ionophores valinomycin or gramicidin[16–19]; the ionophores move the membrane potential close to the cation equilibrium potential so that in fact an applied cation gradient can be used to provide the thermodynamic driving force for anion net transport. Since now anion movements are rate limiting for the movements of salt, the net salt fluxes have the characteristics of the anion net transport pathway. The cation contents are conveniently analyzed by flame spectrophotometry.

A typical anion net transport experiment begins with preequilibrating the cells at 0° with gramicidin prior to packing. Gramicidin is insoluble in water and has to be added from ethanolic stock solution (10 mg/ml). The strategy is not to achieve a certain nominal gramicidin concentration in the medium but a certain stoichiometry of gramicidin to cells, assuming that all gramicidin will partition into the lipid bilayer of the red cells. Control experiments have demonstrated that 100–200 μg/ml packed cells yielded a sufficiently high cation permeability to equilibrate intracellular K^+ with extracellular Rb^+ in less than 10 sec. To initiate the net efflux experiment, the packed cells (typically 0.2–0.4 ml) are injected into the appropriate medium (typically 20–30 ml) that is thermostatted and vigorously stirred, as described before. At the appropriate time points, about 2

[16] J. H. Kaplan and H. Passow, *J. Membr. Biol.* **19,** 179 (1974).
[17] M. L. Hunter, *in* "Drugs and Transport Processes" (B. A. Callingham, ed.), p. 227. Macmillan, London, 1974.
[18] P. A. Knauf, G. F. Fuhrmann, S. Rothstein, and A. J. Rothstein, *J. Gen. Physiol.* **69,** 363 (1977).
[19] O. Fröhlich, C. Leibson, and R. B. Gunn, *J. Gen. Physiol.* **81,** 127 (1983).

ml of the suspension is removed and added to a test tube that contains 7 ml ice-cold stopping medium and a separating oil. The stopping medium serves to slow down the rate of net flux as well as to dilute extracellular K^+ to minimize its trapping in the cell pellet. It typically contains isotonic Tris–Cl or N-methylglucamine (NMG)-Cl plus 50–100 μM of the anion-exchange inhibitor DNDS (4,4′-dinitrostilbene-4,4′-disulfonate).[19,20] It is preferable to use the potassium or NMG salt instead of the sodium salt of DNDS to avoid K_i–Na_o exchange through gramicidin while the cells are in the stopping medium since this will reduce the cellular K^+ content in the pellet in the absence of Cl^- net efflux. Na^+-free DNDS is obtained by recrystallization; that is, by dissolving the commercially available Na-DNDS in distilled water and adding KCl or NMG-Cl whereupon the DNDS promptly precipitates as the appropriate salt. The oil (dibutyl phthalate) has a density between that of the cells and the stopping medium, which permits one to separate the cells from the stopping mixture by centrifugation in a clinical centrifuge.

After centrifugation, the aqueous supernatant and most of the oil are removed by aspiration and residual droplets of the stopping medium are collected after another centrifugation step, along with removal of the remaining oil. It is advantageous to use plastic (polystyrene) test tubes for these experiments instead of glass test tubes since the adhesive properties of the oil and water on a glass wall can make it difficult to remove the oil without also aspirating much of the cell pellet. Breakage of the plastic tubes at the top speed of the centrifuge can be minimized by keeping water in the metal carriers or shields that hold the centrifuge tubes. This reduces the contact pressure at the tip of tube where it rests on the rubber cushion.

The cell pellet is lysed in distilled water (1–2 ml). An aliquot of lysate is then diluted into modified Drabkin's solution (commercially available as solid salt mixture) to determine its hemoglobin concentration (an OD = 1 at 540 nm corresponds to 1.465 g hemoglobin/liter), and another aliquot is diluted into distilled water for the measurement of K^+ by flame photometry. Alternately, we have also employed a single-dilution step into distilled water and determined K^+ and Hb on the same dilution, reading the optical density of Hb at 412 nm. From the K^+ and Hb readings, using the appropriate dilution factors, the cellular K^+ content as mmol/kg Hb is calculated.

There are also two indirect methods by which net fluxes can be measured. They rely on the fact that substrate net flux is equivalent to the movement of osmotically active substances which in turn causes osmotic water movement and cell volume changes. Such volume changes are

[20] O. Fröhlich, *J. Gen. Physiol.* **84,** 877 (1984).

detectable as changes in light scattering of a cell suspension. This approach has been used in one study on anion net transport.[18] It is by this method that Knauf et al.[18] have demonstrated the correlation between the inhibition of anion net and exchange transport by DIDS and thus established that both transport modes are mediated by the band 3 protein. This method relies on careful calibration experiments since depending on the cell concentration an increasing cell volume can result in an increasing or decreasing light scattering intensity.

The second method of following cell volume changes is electronic sizing of the cells in suspension (also referred to as resistive pulse spectroscopy, or RPS[21]). This method has been used for measuring the relatively slow cell volume changes in other cells such as lymphocytes,[22] and we have adapted it to measurements of the more rapid erythrocyte net fluxes. It uses the principle employed in cell counters such as a Coulter counter, where a cell suspension is pumped through a narrow aperture that separates two compartments (a hollow glass finger placed in a vessel). Via two electrodes in these compartments, the electric resistance through the aperture is continuously monitored. The aperture resistance increases slightly when a cell passes through it, and the amplitude of this resistance pulse can be used as a measure of the cell size.

To employ this method for anion net flux experiments, modifications are necessary to permit one to determine cell size distributions as a function of time, in time windows as narrow as 1 sec (the commercially available sizers are too slow for most experimental conditions). We have achieved this by interfacing the output of a Coulter counter model ZBI (or a similar instrument) to a microcomputer such as an IBM-AT.[23,24] The interface consists of a peak detector and an A/D converter which converts the resistance signal into digital format at the time it reaches its maximum. The digital data are then processed in the microcomputer and converted into an amplitude histogram, that is, the microcomputer acts as a multichannel analyzer. We have used the top 40% of this distribution to determine the channel that contains the maximum and call this the peak cell size. The true cell volume or cell water content is determined by calibration experiments with cells of known water contents. Such a calibration curve is best established by treating cells with nystatin[25] in the presence of 150 mM KCl and 15–30 mM sucrose. It is by varying the sucrose concentration that one can vary the cell water content since K^+

[21] H. C. Mel and J. P. Yee, *Blood Cells* **1,** 391 (1975).

[22] S. Grinstein, A. DuPre, and A. Rothstein, *J. Gen. Physiol.* **79,** 849 (1982).

[23] V. Gottipaty, Ph.D. thesis. 1988.

[24] V. Gottipaty and O. Fröhlich, *Biophys. J.* **53,** 531a (Abstr.) (1988).

[25] A. Cass and M. Dalmark, *Nature (London), New Biol.* **244,** 47 (1973).

and Cl^- ions are permeant through the nystatin channels and the sucrose is the only extracellular impermeant osmolyte that balances the intracellular impermeant osmolytes, mainly hemoglobin. Anion net fluxes are calculated from the calculated rates of cell volume and water changes and from the osmolarity of the extracellular medium: the medium osmolality is equal to the ratio of the amount of cation and anion moved during net flux, to the amount of water that followed osmotically.

The primary source of artifactual volume changes in this experimental approach is exchange of chloride with bicarbonate. We mentioned before that $Cl^- - HCO_3^-$ exchange is equivalent to $Cl^- - H^+$ cotransport and is accompanied by osmotic water shifts. Indeed, measuring $Cl^- - HCO_3^-$ exchange by the concomitant cell volume changes can be used as an assay for anion-exchange transport activity. This is achieved by measuring the rates of volume decrease when cells are placed into zero-Cl^- (isotonic citrate) media containing known amounts of bicarbonate in the range of 20–200 μM (Gottipaty and Fröhlich, unpublished observations). Under these conditions, net efflux as measured by K^+ net efflux is only secondarily affected by the changes of pH_i that occur as a consequence of $Cl^- - HCO_3^-$ exchange. $Cl^- - HCO_3^-$ exchange is also visible as a volume change when cells have been stored at $0°$ and are immediately used at a different temperature. Since the pK of hemoglobin depends on temperature, a temperature change shifts pH_i and perturbs the transmembrane Cl^-–proton equilibrium. If bicarbonate exists in the system, its exchange with Cl^- toward a new equilibrium will cause a volume change. In most net flux experiments it is therefore very important to minimize the amount of bicarbonate present by extensively bubbling the solutions with N_2, and to use carbonate dehydratase inhibitors such as acetazolamide at all times.

Finally, the amplitude of the pulse originating from the counter is not a linear function of the cell size. With decreasing cell size the signal becomes a less sensitive measure of cell volume, with deviations from a linear relationship becoming visible near isotonic volume. This nonlinearity can be corrected by the calibration curve, but the experimental errors are increased at lower volumes because of the lesser sensitivity (lesser slope of the calibration curve). It is therefore advisable to preswell the cells, for example by the nystatin method described above, which brings them into a range of higher sensitivity. In control K^+ and tracer-Cl^- net flux experiments which measure directly the net salt loss we have established that anion net transport does not depend on cell volume (Fröhlich and Gottipaty, unpublished observations).

The advantage of the cell sizing method over the K^+ net flux method is severalfold. First, this method allows one to measure fluxes over as brief periods as 10–15 sec (with up to 10 time points within this period), compared to at least 30–40 sec in the K^+ flux method if one desires the same number of data points. Second, the availability of many time points where each measurement can have an accuracy of about 1% yields a better sensitivity compared to the K^+ net flux method where fewer time points with an accuracy of 1–2% are obtained. Third, there is no time-consuming analysis of cellular contents after the actual flux experiment which permits a considerably larger number of experiments in a given time period. Fourth, the number of cells needed is drastically smaller. Fifth, this method permits net influx experiments where the K^+ gradient is inwardly directed and the extracellular K^+ concentration is high. In K^+ net flux experiments, this experimental protocol would lead to unacceptable carryover of extracellular K^+ into the low-K^+ cell pellet.

Analysis of Anion Fluxes

Initial Rates and the Exponential Washout of a Single Compartment

Red cells behave as if they had a single, well-mixed, internal compartment separated from the medium by a single rate-limiting membrane. The rate coefficient for the transport of a molecule between two well-stirred compartments (cytoplasm and medium) can be measured by following the transport of tracer in exchange for nonradioactive (cold or traced) molecules. Ideally there is no isotope effect; that is, the tracer and the traced molecules have the same chemical properties. We will only consider this case since no isotope effect has been reported and any effect is likely to be small since ^{36}Cl has nearly the same mass as the naturally occurring mixture of ^{35}Cl (75.4%) and ^{37}Cl (24.6%), namely 35.5 Da. If the system is in steady state, both the appearance of the tracer into the trans compartment and its disappearance from the cis compartment follow a single exponential time course with the same rate coefficient. The exponential time course has nothing to do with mechanisms of transport.

There is a rigorous proof in the literature that influx and efflux have the same rate coefficient in a two-compartment system.[26] Here is a more intuitive argument. First, the steady state means that the concentration and content of the traced substance are constant in each compartment. Since in a two-compartment case the net movement of traced substance or water flow out of one compartment will change the content and/or concentration in the other compartment, the steady-state assumption for one compartment automatically assumes it for the other compartment.

[26] R. B. Gunn, C. S. Patlack, and J. Z. Hearon, *Math. Biosci.* **4,** 1 (1969).

Second, the traced substance consists of tracer, G^*, and nontracer, G', molecules. The steady-state assumption means that $G_{total} = G^* + G'$ is constant within each compartment, but G^* and G' are not constant. During the experiment net G^* moves from one compartment to the other and net G' moves equally in the opposite direction. If the G^* content of one compartment as a function of time is the function $G_1^*(t)$ then the G^* content of the other compartment is the function $G_2^*(t) = G_T^* - G_1^*(t)$, where G_T^* is a constant equal to the total amount of G^* in the two compartments. If $G_1^*(t)$ is a single exponential increasing from zero at the initial time ($t = 0$) to A at infinite time, i.e., $G_1^*(t) = A(1 - e^{-bt})$, then $G_2^*(t)$ in the other compartment decreases along the same exponential from G_T^* to B, i.e., $G_2^* = B + Ae^{-bt}$; $A + B = G_T^*$. Therefore, in the steady-state condition the information in the influx rate coefficient is the same as the information in the efflux rate coefficient since they must be exactly equal. However, the powerful assumption of the steady state must be proved otherwise grave errors may be made.

Since in the steady state the influx and efflux are equal, why prefer one measurement over the other? There are no theoretical reasons, only practical ones. On one hand, there must be a sufficient time resolution of the data points. One can usually isolate the medium uncontaminated with the cells faster than one can isolate cells uncontaminated with the medium. Thus if the rate coefficient is large, efflux is usually a better measurement. On the other hand, the composition of the medium is often better defined than the steady-state composition of cells in that medium. Consequently, the initial external specific activity, $G_1^*(0)/[G_1^*(0) + G_1'(0)]$, is more exactly defined, as is needed for influx measurements, than is the initial specific activity of the cells for efflux. The latter parameter, however, is not needed if one can wait until specific activity equilibrium is achieved after the efflux determination ("infinity" sample, see below). Often measurements are not made at steady state. In this case, the initial rate flux measurement requires determination of the initial specific activity on the cis side and thus tracer influx measurements are easier and frequently made. However, in the nonsteady state the unidirectional tracer influx and efflux are not equal and there is additional information to be obtained from measuring the initial net flux or the initial tracer efflux as well as the initial tracer influx.

The initial portion of an exponential process is well approximated by a straight line. The exponential uptake function $G_1^*(t) = A(1 - e^{-bt})$ can be rewritten as $G_1^*(t) = A[bt - (bt)^2/2! + (bt)^3/3! - \ldots]$ which if bt is small can be approximated by the line $G_1^*(t) = A(bt + \delta)$ where δ is the deviation of the straight line from the exponential (Table I). Two options are avail-

TABLE I
DEVIATION OF EXPONENTIAL PROCESS FROM LINEARITY

bt (dimensionless)	Error, δ (dimensionless)	Percentage of $t_{1/2}$ elapsed	Percentage of exchange completed
0.01	−0.00005	1.4	0.995
0.05	−0.001229	7.2	4.8
0.1	−0.004837	14.4	9.5
0.2	−0.018731	28.9	18.1
0.3	−0.040818	43.3	25.9
0.4	−0.070320	57.7	33.0
0.5	−0.106531	72.1	39.3
1.0	−0.367879	144.3	63.2
1.5	−0.723130	216.4	77.7
5.0	−4.006738	721.3	99.3

able to the experimenter: one may take samples over times comparable to the half-time and graph the results on semilogarithmic paper to make sure the plot forms a straight line and then fit the data with an exponential expression to extract the slope, b; or one can take samples at short times, graph the results on linear paper to be sure that this plot is along a straight line, and then fit the data with a linear regression to extract the slope, b.

How short is "short times" will depend on the rate coefficient, b. For slow processes when b is in the range of $(week)^{-1}$ (such as sulfate flux at $0°$) then samples can be taken leisurely; for fast processes when b is in the range of sec^{-1} (such as chloride flux at $>10°$) then samples must be taken before 0.1 sec to remain in the linear range. If t is measured in the same units as b^{-1}, then one can determine how good the linear approximation is and how fast one must sample. Remember that $bt_{1/2} = \ln 2 = 0.693$ and $bt_{1/e} = 1$. This means that in the course of one time unit (i.e., the unit used to express b^{-1}), $G^*(t)$ decreases to $1/e$ ($= 0.37$) of its value at the beginning of that time interval. Table I shows that data values should be collected before bt exceeds 0.2 or before the exchange has proceeded 18.1% of the way if the error between the straight line and the true exponential is to be less than 2%. Since the error is negative, the line of the linear initial slope lies above the rising exponential curve (both lines having the same value of b). If accurate data points are assumed to be on the straight line, when they are in fact below the straight line, the calculated slope and estimated value of b will be less than its true value. There is thus a tendency to underestimate b if points are taken late in the exponential curve and the linear approximation is used. The same is true in evaluating the declining exponential content of tracer of the other compartment: there is a tendency to underestimate b using the linear approximation.

If the steady-state self-exchange of tracer and traced molecules is a single exponential process and the initial portion of an exponential curve is linear, why might one want to collect data values beyond the first 18% of the exchange where perfect data will deviate more than 2% from the straight line? The reason is that one often cannot take data fast enough to determine the initial flux rate from the linear portion of the flux curve. If the data are graphed on semilogarithmic paper, the later points still are on the straight line and thus contribute to a proper estimate of b. We have found that many people do not appreciate that the slope of the semilogarithmic plot gives a true initial rate even though the data are collected beyond the half-time of the process. This is in fact one of the greatest strengths of steady-state self-exchange measurements.

Calculation of Cell Contents

The cellular substrate contents are calculated from the dry weight and precipitate measurements. They will be used below for the calculation of the flux rate. This involves, as first step, the calculation of the corrected cellular water content. For this purpose, the packed cell column can be envisioned to consist of three fractions: fraction a is the cell water content that we want to calculate, b is the fraction made up by the cell solids, F_{sol}, and c is the water in the trapped extracellular space, F_{ts}, that we want to subtract. The experimentally accessible uncorrected water fraction, F_w^{app}, can be expressed in terms of a, b, and c:

$$F_w^{app} = (a + c)/(a + b + c) \qquad (2)$$

It is converted into the corrected water fraction, F_w^{corr}, i.e., the water fraction of the packed cells after subtraction of the fraction of water due to extracellular trapped medium, by

$$F_w^{corr} = (F_w^{app} = F_{ts})/(1 - F_{ts}) = [(a + c) - c]/[(a + b + c) - c] \qquad (3)$$

Another often used quantity is the corrected water content, d_{corr}, which corresponds to the fraction a/b and typically has the units of kg water/kg cell solids. It is calculated by:

$$d_{corr} = \frac{F_w^{corr}}{F_{sol}} = \frac{F_w^{app} - F_{ts}}{1 - F_w^{app}} = \frac{(a + c) - c}{(a + b + c) - (a + c)} \qquad (4)$$

The cellular anion content is calculated from the extracellular anion concentration and the tracer distribution ratio, r_{Cl}, between the packed cells and their supernatant. To calculate r_{Cl}, one needs the radioactive counts of the cells and the extracellular medium:

$$\frac{cpm}{sample\ (g)} = \left(\frac{cpm}{pipetted\ volume}\right)$$

$$\left\{\frac{1\ ml\ PCA + [F_w^{app} \times sample\ weight\ (g)]}{weight\ of\ sample\ added\ to\ PCA\ (g)}\right\} \quad (5)$$

"Sample" means either the PCA extract of the packed cells (pc) or or their supernatant after cell packing ("plasma"). In the latter case, which only needs to be treated with PCA if it contains too much hemoglobin from lysis and could quench the scintillation counts, F_w^{app} has the value of unity. In the case of the packed cells, F_w^{app} has the value obtained from drying the cells. When the cellular content is to be determined by coulometric titration instead of tracer, one uses the titration reading instead of the sample counts. The following sequence of equations is used to correct for the radioactivity contributed by the extracellular trapped space, and to calculate r_{Cl} from the corrected sample radioactivities:

$$\left(\frac{cpm}{g\ pc}\right)_{corr} = \frac{cpm/g\ pc - (F_{ts})(cpm/g\ plasma)}{1 - F_{ts}} \quad (6)$$

$$\left(\frac{cpm}{g\ cell\ H_2O}\right)_{corr} = \frac{(cpm/g\ pc)_{corr}}{F_w^{corr}} =$$

$$\frac{cpm/g\ pc - (F_{ts})(cpm/g\ plasma)}{F_w^{app} - F_{ts}} \quad (7)$$

so that

$$r_{Cl} = (cpm/g\ cell\ water)/(cpm/g\ extracellular\ "plasma") \quad (8)$$

Note that r_{Cl} determined by the tracer method is usually slightly less than when determined by chloridometry, due to the contribution of other reducing compounds (especially glutathione) in the PCA supernatants.

Another useful quantity is the tracer content normalized to the cell solid content and corrected for trapped tracer:

$$\left(\frac{cpm}{g\ cell\ solids}\right)_{corr} = \frac{(cpm/g\ pc)_{corr}}{1 - F_w^{corr}} = d_{corr} \times \left(\frac{cpm}{g\ cell\ water}\right)_{corr} \quad (9)$$

The uncorrected value of cpm/g cell solids also can serve as an internal control measurement of the cells content since it can also be obtained from the samples of the cell suspension left over from the tracer efflux experiment: as the ratio of "infinity" counts to hemoglobin content, after conversion of Hb content to cell solids content (see Table II).

TABLE II
CONVERSION AMONG DIFFERENT CELL CONTENT
AND FLUX UNITS[a]

A. 1 kg cell solids corresponds to:
3.1×10^{13} cells
4.4×10^7 cm^2 membrane area
0.92 kg hemoglobin
1.8 kg cell water (pH 7.4, 0°)
2.5 liters of cells (pH 7.4, 0°)
B. 1 mmol (kg cell solids · min)$^{-1}$ corresponds to:
0.38 pmol cm^{-2} sec^{-1}
1.08 mmol (kg Hb · min)$^{-1}$

[a] Data from Refs. 5 and 27.

Calculation of Flux Rate

Calculating the rate of tracer or net flux during the initial, linear phase of the process is straightforward. The flux is the slope of a plot of extracellular or cellular content as a function of time, normalized by the measure of the number of cells in the system. Ideally, when studying the kinetics of a particular transport system, it would be useful to normalize the flux by the number of mediating transporter molecules. This is usually not done, mainly since the number of transporter molecules are not known, often also for reasons of tradition. In the case of erythrocyte transport in general and of anion transport in particular, one expresses the flux in terms of volume of cells, cellular dry solids content, or membrane surface area. Normally, it is not advisable to express the flux in terms of cell volume since the volume can change under different experimental conditions such as different pH values even though the number of cells and transporter molecules remains the same. If one plans to use these units, one should take care to convert the measured cell volume into an equivalent volume corresponding to a standard condition of tonicity, pH, and temperature. It is generally less confusing to normalize by the number of cells (determined by cell counting), by the cellular hemoglobin content (determined by hemoglobinometry), or by the total cell solids content (determined as "dry weights"). For comparison purposes, these normalizations can be rapidly interconverted with the values given in Table II.[5,27]

To calculate the flux rate from the exponential time course is only slightly more involved. The two-compartment analysis gives the following

[27] J. Brahm, *J. Gen. Physiol.* **82,** 1 (1983).

equation (see Appendix) to calculate flux rate from rate constant:

$$M = bS_1S_2/(S_1 + S_2) \tag{10}$$

where b is the slope obtained from the semilogarithmic fit and S_1 and S_2 are the substrate contents of the intracellular and extracellular compartment, respectively. Under the experimental conditions described above where the hematocrit in the tracer efflux medium is typically 1%, S_2 is usually much larger than S_1 so that Eq. (10) is reduced to Eq. (11):

$$M = -bS_1 \tag{11}$$

The fluxes calculated from Eq. (10) and (11) are identical only if the hematocrit approaches zero, but in our typical case they differ only by 1% (for an exception see below) which is within our experimental scatter. When S_1 is expressed as mmol (kg cell solids)$^{-1}$ and b is converted from sec^{-1} to min^{-1}, the calculated flux has the units mmol (kg cell solids $\times$ min)$^{-1}$.

It is clear from Eq. (11) that at constant turnover rate of the transporter protein and thus at constant flux rate, the rate constant is influenced by the cellular substrate content so that b is not necessarily proportional to M at different substrate concentrations or different cell volumes. For example, b is constant with increasing intracellular substrate concentration (at constant cell volume) when this concentration is far below the half-saturation constant for transport, since the increased cell content is matched by an increased activation of the transport mechanism. On the other hand, above saturating substrate concentrations, the value of b decreases even though M is constant (at V_{max}). It is for this reason that in many cases information is lost if the data are reported only as rate constants instead of flux rates.

Another factor influencing the value of b has been mentioned briefly before: the relative size of the two compartments. At low hematocrit and high extracellular anion concentrations, S_2 is always much larger than S_1. However, this is not the case at very low extracellular substrate concentrations when the extracellular compartment size (in terms of amount of substrate) becomes comparable or possibly even smaller than the intracellular compartment. In this case, using the exponential flux method and applying Eq. (11) for the flux calculation can lead to serious errors. There is, however, a compensating process that can cancel this error. The fraction of intracellular tracer that leaves the cell during isotopic equilibration, $S_2/(S_1 + S_2)$, decreases away from near unity as S_2 decreases. This means that b becomes faster since less substrate needs to be transported with a constant flux rate. The acceleration of b occurs by the factor $S_2/$

$(S_1 + S_2)$, as one can see from Eq. (10). The calculated value of b depends, however, on the way the "infinity" sample is taken. In order to obtain the exponential time course discussed above, this sample has to be collected in a time window during which the tracer has reached isotopic equilibrium through exchange but where chemically the anion concentration gradient has not yet been dissipated through net transport. The radioactivity in this sample differs from that in a sample that is taken at a much longer time when also the anion concentration gradient has disappeared, by the above-mentioned factor $S_2/(S_1 + S_2)$. If one chooses the latter, larger and "inappropriate" infinity value obtained after complete equilibration of tracer and concentration, the tracer efflux curve obviously is no longer exponential. However, within certain experimental limits one obtains the correct value of b: if the data are collected in the early phase of tracer equilibration, they will fit on an exponential curve for either infinity value. The calculated values of b in these two cases again differ by the factor $S_2/(S_1 + S_2)$. Thus, the error caused by not correcting for the comparable compartment size is cancelled by the error caused by the "inappropriate" choice of infinity samples. It is important to know, however, that this fortuitous righting of two wrongs only works when tracer equilibration has not proceeded too far, in extreme cases to no more than 10% completion.[28]

For the calculation of the flux rate using Eq. (11), the chloride content of the cells, S_1, is obtained from the product of three measured quantities: r_{Cl} [Eq. (8)]; the extracellular chloride concentration, Cl_o; and the water content, d_{corr} [Eq. (4)]. Because direct titration of the cell extracts for chloride includes cellular glutathione, the product of the extracellular concentration times the tracer chloride ratio, r_{Cl}, can be more accurate. Thus, the flux is given by

$$M = b r_{Cl} Cl_o d_{corr} \qquad (12)$$

The negative sign reverses the negative sign of b (as calculated from the semilogarithmic plot, see also Appendix) so that M is a positive number. The physical units of b are $\min^{-1}$, r_{Cl} is dimensionless, that is, (mol/kg cell water)/(mol/kg extracellular medium), Cl_o is in mmol/kg medium, and d_{corr} is in kg cell water/kg dry cell solids, so that the flux is calculated in mmol/(kg cell solids · min).

Appendix

The tracer flux between two compartments results in the equilibration of the specific activities in the two compartments and not necessarily the

---

[28] R. B. Gunn and O. Fröhlich, *J. Gen. Physiol.* **74**, 351 (1979).

equalization of nonradioactive concentrations or tracer concentrations. The net tracer flux between two compartments ends when the counts per minute per mole are equal in the two compartments. The fundamental equation for tracer fluxes is derived here.

Consider two connected compartments with R_1 and R_2 total amounts of tracer in them. R_1 and R_2 have the units of counts per minute (cpm). These compartments have S_1 and S_2 moles of the traced molecules (radioactive plus nonradioactive). The specific activities a_1 and a_2 are defined by $R_1 = a_1 S_1$ and $R_2 = a_2 S_2$. The instantaneous rate of gain of radioactivity by one compartment is equal to that lost by the other compartment: $dR_1/dt = -dR_2/dt$. This rate of gain by one compartment and rate of loss by the other is equal to $m_{12}a_1 - m_{21}a_2$, where m_{12} is the unidirectional flux (mol sec^{-1}) of traced molecules from compartment "1" to compartment "2" and m_{21} is the unidirectional flux (mol sec^{-1}) from compartment "2" to compartment "1." The net flux of traced molecules (not the tracer) is $dS_1/dt = m_{21} - m_{12}$. Then $dR_1/dt = d(a_1 S_1)/dt = a_1(dS_1/dt) + S_1(da_1/dt)$ by the definition of R_1 and by the derivative of a product. Substituting $dR_1/dt = m_{21}a_2 - m_{12}a_1$ and $dS_1/dt = m_{21} - m_{12}$ into this last equation we obtain

$$m_{21}a_2 - m_{12}a_1 = a_1(m_{21} - m_{12}) + S_1(da_1/dt) \qquad \text{(A1)}$$

or after canceling terms,

$$m_{21}(a_2 - a_1) = S_1(da_1/dt) \qquad \text{(A2)}$$

This equation says that the rate of change of specific activity in compartment one (da_1/dt) is inversely proportional to the moles of traced material in compartment "1" (S_1) and is proportional to the difference in the specific activities $(a_1 - a_2)$ between the two compartments and proportional to the unidirectional flux *into* compartment "1" from compartment "2." A symmetrical equation is obtained if dR_2/dt is calculated: $m_{21}(a_1 - a_2) = S_2(da_2/dt)$. The solution of these two differential equations under different boundary conditions allows the calculation of m_{12} or m_{21} from the time course of the specific activity in compartment one or two. In the steady state $m_{12} = m_{21}$ and $dS_1/dt = dS_2/dt = 0$ and the equations can be solved to give

$$\ln\{1 - [a_2(t)/a_\infty]\} = [-m(S_1 + S_2)/(S_1 S_2)]t \qquad \text{(A3)}$$

where a_∞ is the specific activity in both compartments at long times ("infinity"). If the left-hand side of this equation is graphed versus time (t) the line has a slope $b = -m(S_1 + S_2)/S_1 S_2$, which is Eq. (10). If S_2 (say the moles of traced molecules outside the cells) is much greater than S_1 (say the moles inside the cells; very low cell volume fraction or hematocrit) then $S_1 + S_2 \approx S_2$ which cancels with S_2 in the denominator and obtains $b \approx -m_{12}/S_1$ or $m_{12} = -bS_1$.

Acknowledgment

We thank Adrienne McLean for her excellent help in the preparation of this manuscript.

[4] Cation Fluxes in the Red Blood Cell: Na$^+$,K$^+$ Pump

By JOHN R. SACHS

Red blood cell (RBC) membranes are museums of transport processes. Some of these processes have functions which are physiologically important to the survival of the cell. For instance, the sodium pump of the red cells of many species, including human cells, combined with the cation permeability characteristics of the cell membrane, maintains constant cell volume[1]; dog red cells, however, have negligible Na$^+$ pump activity and cell volume regulation in this case seems to depend in large part on the presence of a Na$^+$–Ca^{2+} exchanger which exchanges cell Na$^+$ for external Ca^{2+}, and a Ca^{2+} pump which moves the Ca^{2+} back out of the cell so that the net result is the movement of Na$^+$ out of the cell against an electrochemical gradient to balance the passive leak of Na$^+$ into the cell.[2] In some cases, no apparent physiological function of the transport process is apparent; ferrett red cells have a highly developed Na$^+$,K$^+$,Cl$^-$ cotransporter,[3] but whether it has a function in maintaining cell viability is not clear. Ox red cells have high Na$^+$–Na$^+$ and Na$^+$–Li$^+$ exchange activity,[4] but the function of this exchanger is completely unclear. The transport systems present in red cell membranes seem to have properties very similar, if not identical, to the same systems present in other tissues. Because of the tractability and robustness of the red cells of most species, one can frequently obtain sophisticated information about the physiological characteristics of the transporter more readily than is possible with preparations derived from other tissues. Red cell volumes are very much larger than the volumes of the phospholipid vesicles in which transport proteins are frequently incorporated for transport studies, and the larger volume may make studies in which transport rate is related to intracellular

[1] D. C. Tosteson and J. F. Hoffman, *J. Gen. Physiol.* **44**, 169 (1960).
[2] J. C. Parker, *J. Gen. Physiol.* **71**, 1 (1978).
[3] P. W. Flatman, *J. Physiol.* (*London*) **341**, 545 (1983).
[4] A. L. Sorensen, L. B. Kirschner, and J. Barker, *J. Gen. Physiol.* **45**, 1031 (1962).

ionic concentrations more reliable. A great deal of what is known about the transport characteristics of several transport systems, such as the sodium pump, has come from studies using red blood cells or preparations derived from them. However, in order to study the characteristics of a particular cation transport mode, it must be possible to isolate it from the other transport systems resident within the cell membrane; the existence of a specific inhibitor, such as ouabain and other cardiotonic steroids[5] for the sodium pump, and furosemide[6] and bumetanide[7] for the Na^+,K^+,Cl^- cotransporter, greatly facilitates the isolation.

This chapter describes methods useful in the study of cation transport systems in intact red blood cells. The methods are described for use with human red cells, but they can usually be applied to cells from other species, either without change or with some modification, and I have usually used them to study the transport characteristics of the sodium pump, but they are readily applicable to the study of other cation transport systems. First, some methods for manipulating the intracellular composition of intact red blood cells will be described, and then methods for measuring unidirectional cation fluxes will be discussed.

Alteration of the Intracellular Cation Content of Intact Red Blood Cells

Blood is readily obtained by venepuncture and collected into heparin or citrate–phosphate–dextrose solution to prevent coagulation; standard blood bank methods are suitable. We usually use the cells within 3 days of collection; the anticoagulated blood is stored at 4°. Red cells can be sedimented by centrifugation for 10 min at 2000 rpm or, using a high-speed centrifuge, in the time it takes the rotor to reach 10,000 rpm and then stop. Before use the cells are separated from the plasma and the cells are washed by suspension in 5 or 10 vol of an isosmotic (295 mOsm/kg) solution of $MgCl_2$ (107 mM), choline chloride (160 mM), NaCl (160 mM), or of another salt, sedimented by centrifugation, and the supernatant aspirated. Washing solutions are frequently buffered to pH 7.4, but hemoglobin is such a strong buffer that this probably is not necessary.

Modification of Intracellular Monovalent Cation Content

Two methods are generally used for the modification of intracellular cation composition.

The first procedure is based on an old observation that red blood cells

[5] H. J. Schatzmann, Helv. Physiol. Pharmacol. Acta 11, 346 (1953).
[6] J. S. Wiley and R. A. Cooper, J. Clin. Invest. 53, 745 (1974).
[7] J. C. Ellory and G. W. Stewart, Br. J. Pharmacol. 75, 183 (1982).

become permeable to extracellular cations when they are treated with organic mercurials; the increased permeability correlates with the binding of mercury to the cell membrane, probably to sulfhydryl groups.[8] One of these compounds, p-chloromercuribenzenesulfonic acid (PCMBS), was found to interact with sulfhydryl groups and to permeabilize the cells to cations, and the permeability change was found to be reversible; treatment of the cells with cysteine removed the bound PCMBS and restored cation permeability to near normal.[9] Based on this observation, Garrahan and Rega[10] devised a method for preparing intact red cells with altered cation composition. The procedure consists of exposing cells for prolonged periods in the cold to solutions containing PCMBS and with the monovalent cation concentrations desired in the final cell preparation. The cells are then removed from the PCMBS solution and resuspended in solutions containing compounds which react with sulfhydryl groups; in the original report[10] cysteine was used. This suspension is incubated at 37°, and during this time the PCMBS is removed from the cell membrane and the cells reseal. We have modified the method[11] so that cells which contain less than 1 mmol/liter RBC Na^+ and less than 1 mmol/liter RBC K^+ can be obtained, the bulk of the cell cation having been replaced by choline. In our hands, the cells are of normal size, the passive permeability of the cell membranes to cations is no greater than the permeability of fresh cells, and we have found no detectable changes in any of the transport parameters we have measured. However, there have been reports that some transport phenomena are altered by the PCMBS procedure,[12] but the method used differed in significant respects from the procedure described here.

To use the method, red cells washed as described above are suspended at about 5% hematocrit (volume of cells/volume of suspension × 100) in a solution (pH 7.4) made up of (mM): PO_4 6.7, Cl^- 151, Mg 1.0, PCMBS 0.1 or 0.2, glucose 10, adenine 0.5, inosine 2.0, and varying quantities of Na^+, K^+, and choline to make up a total of 160 mM; the concentrations of Na^+ and K^+ are chosen to approximate the required final concentrations in the cells (in mmol/liter cell water; red cells are about 70% water by volume). The higher concentration of PCMBS should be used if the concentration of choline is high (greater than 120 mM). Sucrose is added at concentrations ranging from 0 in solutions in which

[8] R. I. Weed, J. Eber, and A. Rothstein, *J. Gen. Physiol.* **45,** 395 (1962).
[9] R. M. Sutherland, A. Rothstein, and R. I. Weed, *J. Cell. Physiol.* **69,** 185 (1967).
[10] P. J. Garrahan and A. F. Rega, *J. Physiol. (London)* **301,** 25 (1967).
[11] J. R. Sachs, *J. Physiol (London)* **264,** 449 (1977).
[12] M. Canessa, I. Bize, N. Adragna, and D. C. Tosteson, *J. Gen. Physiol.* **80,** 149 (1982).
[13] J. R. Sachs and L. G. Welt, *J. Clin Invest.* **46,** 65 (1967).

the only monovalent cation is choline to 69 mM (15 g/liter) in solutions in which only Na$^+$ and K$^+$ are present (PCMBS is more effective in making the cells permeable to Na$^+$ and K$^+$ than to choline, and the sucrose counteracts the colloid osmotic pressure of hemoglobin and maintains constant cell volume). The sucrose concentration is varied linearly with the fraction of the monovalent cation which is Na$^+$ + K$^+$. The suspension is incubated at 4° for 36 hr and the PCMBS solution is replaced twice during the incubation period. It is important to keep the cells in suspension during the incubation, either by shaking or by gentle magnetic mixing. If the cells are to be loaded with radioactive Na$^+$ or K$^+$, they are concentrated at the end of the incubation to about 85% hematocrit, isotope is added to the suspension, and the suspension is incubated at 4° for another hour or two. Care should be taken that the isotope solution does not markedly alter the pH or osmolality of the suspension, and it should be recognized that the cells will approach cation equilibrium with the final supernatant solution after the addition of isotope; if a significant amount of Na$^+$ or K$^+$ is added with the isotope, the final cell concentration will be greater than expected from the concentration of the ions in the PCMBS solution. At the end of the incubation in PCMBS solution, the cells are separated from the solution and resuspended in a solution similar to the PCMBS solution except that PCMBS is absent, the concentration of adenine is 3 mM, and the solution contains 4 mM dithiothreitol (DTT). We have found that DTT is much more effective than cysteine in restoring the impermeability of the cells to cations. If the cells are to contain Na$^+$, the DTT solution should be K$^+$ free (K$^+$ in the PCMBS solution can be replaced by choline) since operation of the Na$^+$,K$^+$ pump will lower the intracellular Na$^+$ concentration. The suspension of cells in DTT solution is incubated for 1 hr at 37°. At the end of the incubation, the cells are separated from the suspension and washed, and are then ready for use. If the cells are washed in MgCl$_2$ or choline chloride solution, the Na$^+$ and K$^+$ concentration can be measured by flame photometry.[13] The concentrations of adenine, inosine, and glucose present during the incubations result in cells with normal concentrations of ATP (1.0–1.5 mmol/liter red blood cells).

The second method for modifying intracellular Na$^+$ and K$^+$ concentrations is based on the observation[14] that the polyene antibiotic nystatin can be reversibly incorporated into red cell membranes. Since the antibiotic forms pores which are large enough to permit the passage of monovalent (but not divalent) cations, Na$^+$ and K$^+$ in the intracellular water of cells exposed to the antibiotic rapidly come to thermodynamic equilibrium

[14] A. Cass and M. Dalmark, *Nature (London) New Biol.* **244,** 47 (1973).

with the ions in the extracellular solution. To alter intracellular Na^+ and K^+ concentration by this method,[15] cells are first washed three or four times in a solution which contains PO_4^- (6.7 mM), Cl^- (151 mM), and Na^+, K^+, and choline to a total of 160 mM in the proportions required in the final cell water. The pH is adjusted to 6.7 since at that pH the net charge of hemoglobin is zero, there is no transmembrane potential difference, and therefore at thermodynamic equilibrium the concentration of cations in the cell water will equal the concentration in the loading solution[16]; alternatively, the procedure can be performed at pH 7.4 and the final intracellular cation concentration estimated from the fact that, at pH 7.4, the ratio of intracellular monovalent cation concentration (expressed in millimoles/liter cell water) to extracellular concentration is 1.4 at thermodynamic equilibrium. After washing, the cells are resuspended in the same solution to which 27 mM sucrose has been added, and which also contains nystatin. Nystatin can be dissolved at 5 mg/ml in methanol or dimethyl sulfoxide, and 1 ml of the nystatin solution is added to 100 ml of solution. At 5% hematocrit, this results in a ratio of 1.0 mg of nystatin/ml of cells, which permits equilibration of intracellular and extracellular Na^+ and K^+ in less than 1 min at room temperature; the cells are usually allowed to remain in suspension for 30 min at room temperature. Less nystatin per cell volume can be used; adequate equilibration can be achieved in 20 min at 0° (nystatin partitions better into cell membranes at low temperatures) at ratios as low as 0.2 mg nystatin/ml cells when the hematocrit is 17%; the lower concentrations of antibiotic facilitate its removal from the cells. If very low concentrations of Na^+ or K^+ are required, the supernatant solution should be changed one or more times during the incubation, and if the cells are to be loaded with radioactive Na^+ or K^+, they should be concentrated to 70 or 80% hematocrit, the isotope solution added (with care that the pH or osmolality is not drastically altered, and with attention to resulting alteration in the supernatant cation concentration), and the suspension incubated for an additional 10 or 20 min. After the cells are loaded, they are separated from the nystatin solution and resuspended at about 5% hematocrit in a solution of the same composition as the original loading solution but lacking nystatin and containing 100 mg/100 ml albumin and with the pH adjusted to 7.4; the suspension is allowed to incubate at room temperature for 10 min. The cells are then separated from the suspension and the procedure is repeated four or five times to completely remove the nystatin. The cells can then be washed with isosmotic $MgCl_2$ or choline chloride solution to remove ex-

[15] J. D. Cavieres and J. C. Ellory, *J. Physiol.* (*London*) **243**, 243 (1974).
[16] M. Canessa, C. Brugnara, D. Curi, and D. C. Tosteson, *J. Gen. Physiol.* **87**, 113 (1986).

tracellular Na^+ and K^+, and the intracellular Na^+ and K^+ content determined by flame photometry.

The nystatin method permits preparation of cells in a shorter period of time than the PCMBS procedure, although the amount of manipulation required is little different for the two methods. If the cells are to contain significant amounts of choline, the PCMBS method described above is much to be preferred since its use can result in cells in which Na^+ and K^+ are completely replaced by choline and yet are of normal size. Loading cells with choline by the nystatin method is much less successful, and the cells wind up badly shrunken since they lose intracellular K^+ without replacing it ion for ion with choline.

Modification of Intracellular Divalent Cation Concentration

Although PCMBS may increase membrane permeability to Mg^{2+}, the effect has not been characterized, and nystatin does not increase permeability to Mg^{2+} or Ca^{2+}. If it is required to alter the intracellular concentration of Mg^{2+} or Ca^{2+}, some other method must be used. One such method makes use of the divalent cation ionophore A23187,[17] which accomplishes the transmembrane exchange of a divalent cation (Mg^{2+} or Ca^{2+}) for two hydrogen ions, and which has been shown to modify intracellular Mg^{2+} and Ca^{2+} content of intact red cells in a reproducible fashion.[18,19] To alter intracellular Mg^{2+} concentration, cells are incubated at 10% hematocrit or less in a solution which, in addition to other substances, contains the concentration of ionic Mg^{2+} required in the cells, and which also contains 0.5 mM EGTA (ethylene glycol bis(β-aminoethyl ether)-N,N,N',N'-tetraacetic acid) in order to prevent the entry of Ca^{2+} which at low concentrations activates a K^+ channel in human red cells.[20] The ionophore is poorly soluble in water, but can be dissolved at 1 mg ionophore/ml in ethanol (1.9 mM). This solution is then added after the cells are in suspension in order to suppress the formation and precipitation of ionophore–divalent cation complexes. With a final concentration of 10 μM A23187, equilibrium distribution of Mg^{2+} across the cell membrane occurs within 5 to 10 min. Modification of intracellular Mg^{2+} concentration can be combined with modification of intracellular monovalent cation concentrations; if the PCMBS method is used, A23187 and ionic magnesium at the appropriate concentrations can be included in the dithiothreitol solution, or A23187 can be combined with nystatin if cations are altered by that

[17] P. W. Reed and H. A. Lardy, *J. Biol. Chem.* **247**, 6970 (1972).
[18] H. G. Ferreira and V. L. Lew, *Nature (London)* **259**, 47 (1976).
[19] P. Flatman and V. L. Lew, *Nature (London)* **267**, 360 (1977).
[20] G. Gardos, *Acta Physiol. Hung.* **15**, 121 (1959).

method. The suspension should be allowed to incubate for 20 min or more at room temperature or 37° (cation exchange mediated by the ionophore decreases with temperature and almost ceases at 0°). The cells are then separated from the suspension and washed four or five times at 4° with a solution identical to the loading solution (with the same concentration of Mg^{2+} and EGTA) but lacking ionophore. If necessary, the cells can then be washed with a Mg^{2+}-free solution and the intracellular Mg^{2+} concentration estimated by atomic absorption spectroscopy which will measure total magnesium concentration (including that bound to intracellular chelators) rather than ionic magnesium.

The ionophore can also be used to alter intracellular Ca^{2+} concentrations, but the situation is considerably more complicated than in the case with Mg^{2+}. The human red cell (but not the red cells of some other species) possesses a Ca^{2+}-activated K^+ channel[20] so that, as intracellular Ca^{2+} rises, K^+ is lost to the solution and the cell shrinks; since the channel is bidirectional, shrinkage can be prevented if the cells are suspended in an all-K^+ solution with 2.7 mM sucrose. The red cell membrane possesses a powerful Ca^{2+} pump[21] for which the substrate is ATP; increase in intracellular Ca^{2+} is impeded by the pump, and, in the process, intracellular ATP is depleted. Lew and Brown[22] have published an extensive discussion of the use of the ionophore for increasing red cell Ca^{2+} concentrations.

Modification of Intracellular ATP Concentration

The extent to which the intracellular ATP concentration of intact cells can be altered is limited; experiments which require such alterations are better performed with other preparations. However, cell ATP can be reduced to very low levels (10–30 μmol/liter cells) by incubating cells for 3 hr at 37° with 5 mM inosine and 5 mM iodoacetamide[23]; the process is somewhat faster in phosphate-loaded cells. Inosine is converted to hypoxanthine and ribose 5-phosphate by nucleoside phosphorylase; formation of triose phosphate from ribose phosphate utilizes ATP, but formation of ATP is prevented because glyceraldehyde-3-phosphate dehydrogenase is inhibited by iodoacetamide. High concentrations of ATP (at least 4.3 mmol/liter cells) can be obtained[24] by incubating cells at 10% hematocrit for 2 hr in a solution which contains 27 mM phosphate, 0.5 mM adenine, 1

[21] H. J. Schatzmann and F. F. Vincenzi J. Physiol. (London) 201, 369 (1969).
[22] V. L. Lew and A. M. Brown, in "Detection and Measurement of Free Ca in Cells" (C. C. Ashley and A. K. Campbell, eds.), p. 423. Elsevier/North-Holland, Amsterdam, 1979.
[23] V. L. Lew, Biochim. Biophys. Acta 233, 827 (1971).
[24] J. R. Sachs, J. Physiol. (London) 316, 263 (1981).

mM inosine, 10 mM glucose, 2 mM Mg^{2+}, and 2 μM A23187. In the absence of inonophore, ATP levels rarely exceed 2 mmol/liter cells; the ionophore probably keeps intracellular free Mg^{2+} constant even though Mg^{2+} is being complexed by the newly formed ATP. Perhaps phosphofructokinase is inhibited by ATP, but not by MgATP.

Measurement of Cation Movements

With either fresh red blood cells or cells modified by the manipulations described above, it is relatively simple to measure either unidirectional cation fluxes using labeled cations, or net cation movements by chemical methods.

Unidirectional Cation Influx

If a red cell suspension is considered as a two-compartment system, then

$$dX_c^*/dt = {}^ikX_s^* - {}^okX_c^* \tag{1}$$

in which X_s^* is the extracellular concentration (in some convenient unit such as cpm/liter) of the labeled species whose movement is to be measured, X_c^* its intracellular concentration, ^ik a constant which, when multiplied by the concentration of X in the external solution, gives the influx, and ^ok a similar constant which gives the efflux when multiplied by the concentration of X within the cells. Integration between $t = o$ and $t = t$ yields

$$^ik = (X_{c,t=t}^*/X_s^*)\{^ok/[1 - \exp(-^okt)]\} \tag{2}$$

if X_c at $t = o$ is taken as zero. Then the influx is

$$J^{o\rightarrow i} = {}^ikX_s = [X_{c,t=t}^*/(X_s^*/X_s)]\{^ok/[1 - \exp(-^okt)]\} \tag{3}$$

where X_s is the concentration of X in the external solution.

Influx experiments are therefore designed to measure the amount of the labeled species within the cells as a function of time; the concentration of X^* within the cells divided by the specific activity of X in the solution $[X_c^*/X_s^*/X_s]$ yields the uptake of X. A plot of uptake versus time yields, in general, a curved line with its concavity directed downward. In order to correct for efflux of tracer from the cell, the uptake must be multiplied by $^ok/[1 - \exp(-^okt)]$ to yield the influx. The magnitude of the correction depends on the values of ^ok and t; the smaller these values, the smaller the correction.

Cells, preferably washed in the solution in which the influx will be measured but without isotope, are added to the influx solution incubated at the appropriate temperature. The final hematocrit should be low (5% or less) so that changes in the composition of the solution over the course of the measurement are minimized. The composition of the solution should be appropriate for the measurement to be made and contain radioactive Na^+, K^+, or Rb^+. (By taking advantage of the difference in the half-lives of ^{42}K and ^{22}Na, it is possible to measure the influx of both ions simultaneously.) When setting the concentration of the ion whose influx is to be measured, it should be remembered that the measurement is most accurate when the ion concentration is low relative to the $K_{1/2}$ for the process; the proportion of the total influx contributed by the saturable process relative to that contributed by fluxes which increase linearly with ionic concentration is greater at relatively low concentrations. Samples of the suspension are taken at appropriate intervals, the cells separated from the suspension, and the cells and supernatant saved. The rapidity with which the cells must be separated from the solution depends on the rapidity of the process which is being measured; cation transport processes in human cells are slow enough to permit relatively leisurely methods such as centrifugation, but some processes (e.g., Na^+ and K^+ cotransport in ferrett red cells[3]) are fast enough to require more rapid procedures, such as centrifugation through n-butyl phthalate or silicone oil. Most transport processes are markedly slowed at low temperature, so that immersion in an ice bath slows ion movements while the cells are being separated. Although, in principle, it is possible to calculate isotope uptake by the cells by following the decrease in the supernatant isotope concentration, it is much more accurate to directly measure the isotope content of the cells. If the cells have been separated by centrifugation, the supernatant should be removed and the cells washed three times with 20 vol of ice-cold isosmotic $MgCl_2$ or choline chloride solution, which will reduce contamination with extracellular fluid to negligible amounts. If the cells have been centrifuged through n-butyl phthalate or silicone oil, the supernatant fluid and oil layer are carefully removed. The cell pellet will contain a variable amount of extracellular fluid (2–5%) depending on the time and force of centrifugation. If necessary, this can be measured by including an extracellular fluid marker such as [^3H]inulin, [^{14}C]sucrose, or $^{51}CrEDTA$; the latter substance has the advantage of being a γ emitter and therefore easily measured. Samples of the cells and supernatant are counted, the volume of cells which have been counted estimated by measurement of hemoglobin concentration, and the concentration of the cation in the influx solution estimated by flame photometry or atomic absorption spectroscopy. If it is believed that the efflux correction will be sizable even at short time intervals, ^ok should be determined as described below. The value of the influx can then be calculated from Eq. (3).

Unidirectional Cation Efflux

When cation efflux is measured, back diffusion of isotope lost from the cells can be ignored if the measurement is made at very low hematocrit so that the concentration of isotope in the extracellular solution remains negligible. Under this condition, Eq. (1) can be written

$$dX_c^*/dt = -{}^okX_c^* \tag{4}$$

This can be integrated and rearranged to yield

$$\ln[1 - (X_{s,t=t}^*/X_{s,t=\infty}^*)] = -{}^okt \tag{5}$$

where $X_{s,t=t}^*$ is the concentration of labeled cation in the supernatant at time t and $X_{s,t=\infty}^*$ its concentration at infinite time. The value of X_s^* at infinity can be estimated by measuring the concentration of the labeled cation in a sample of the whole suspension. If, as is usual, the experiment is terminated before more than 50% of the cell counts has been lost, an error in this estimate does not result in a serious error. If however, the measurement is extended longer, and there is reason to believe that the concentration of X^* in the whole suspension is not a good estimate of $X_{s,t=\infty}^*$, the Guggenheim procedure can be used.[25]

The efflux can be calculated from

$$J^{i \to o} = {}^okX_c \tag{6}$$

provided X_c (the concentration of cation within the cell) does not change over the course of the measurement. If the extracellular solution is free of X, X_c is not constant, but

$$dX_c/dt = -{}^okX_c \tag{7}$$

this can be integrated between $t = 0$ and $t = t$ to yield

$$X_{c,t=t} = X_{c,t=0} \exp(-{}^okt) \tag{8}$$

from which

$$J^{i \to o} = X_{c,t=0} - X_{c,t=t} = X_{c,t=0}[1 - \exp(-{}^okt)] \tag{9}$$

where $X_{c,t=0}$ is the measured intracellular concentration of cation X before the measurement is made. If neither the steady-state assumption nor the assumption of zero external cation concentration is valid, then

$$dX_c/dt = -{}^okX_c + {}^ikX_s \tag{10}$$

[25] H. Gutfreund, "Enzymes: Physical Principles," p. 118. Wiley, New York, 1975.

in which X_s is the concentration of cation X in the extracellular solution. Integration between $t = 0$ and $t = t$ yields

$$X_{c,t=t} = [X_{c,t=0} - (^ik/^ok)X_s]\exp(-^okt) + (^ik/^ok)X_s \qquad (11)$$

If ^ik is known or if a good estimate can be made, X_c can be calculated at the midpoint of the efflux period. Multiplying this value by ^ok will give a good estimate of $J^{i\to o}$.

To perform the experiment, cells must first be loaded with the isotopic species whose efflux is to be measured. If the intracellular concentration of ions is altered by the PCMBS or nystatin method, radioisotope can be added at that time as described above. If fresh cells are to be used, the cells should be incubated 3 hr or so at 37° in the presence of the radioisotope. The specific activity should be as high as reasonable and therefore the chemical concentration of the ion in the loading solution should be relatively low since influx by saturable processes will more than compensate for the low concentration; 16 mM K^+ is more than enough to saturate the Na^+ pump even in the presence of high concentrations of Na^+. The hematocrit should be high (>50%) to minimize the amount of radioisotope needed. It should be kept in mind that intracellular ionic composition may change during the loading incubations and conditions chosen to result in an appropriate final composition. After the cells are loaded, they are washed three or more times in 20 vol of an appropriate ice-cold isotonic solution, the last time in a solution similar in composition to that in which the efflux measurements will be made. At this time a sample of the cells is saved for later determination of cell ionic composition. The cells are then distributed to the efflux solutions which have been brought to the appropriate temperature; the final hematocrit should be less than 1%. Timed samples are taken and the cells separated from the suspension as described above for influx measurements; in this case, however, the cells are discarded since only the supernatants are needed. In addition, at some point, a sample of the well-mixed suspension is taken and hemolysed by detergent; one drop of Triton X-100 will hemolyse all the cells in 30 ml of suspension at 1% hematocrit. This sample serves as the infinite time sample; the cells are hemolysed so they will not settle during counting. The samples are counted, and after counting the hemoglobin concentration of the samples estimated from the optical density at 420 or 540 nm; the observed counts can then be corrected for the contribution made by hemolysis of the cells during the measurement. ^ok and $J^{i\to o}$ can then be calculated as described above.

Caution should be observed in equating the unidirectional efflux determined as described above with the net chemical efflux. Although plots of the data according to Eq. (4) are usually linear, such linearity may in part result from the short time intervals over which the measurements are made. It would not be surprising if the red cell population were heterogeneous both with regard to the maximal velocity of the transport process and the intracellular cation concentration; therefore, ^ok, X_c, and the specific activity of X_c are likely to be population values. ^ok may be a complicated function of X_c; e.g., for the Na^+,K^+ pump the efflux of Na^+ is approximately related to the cell Na^+ concentration by the expression[26]

$$J^{i\to o} = J^{i\to o}_{max}/(1 + K_N/Na_c)^3$$

in which $J^{i\to o}_{max}$ is the maximal Na^+ efflux, Na_c the intracellular Na^+ concentration, and K_N a constant. Since $^ok = J^{i\to o}/Na_c$, it is clear that ^ok will be a complicated function of Na_c, and therefore the value of $J^{i\to o}$ obtained by multiplying the population value of ^ok by the population value of Na_c may not be the net transport rate measured chemically; for the evaluation of net effluxes it is better to use chemically determined values when possible.

Net Cation Movements. In addition to the uncertainty about the fidelity with which unidirectional efflux measurements reflect net effluxes because of the heterogeneity of the cell population, the possibility of radioisotope exchanges raises uncertainty about the identity of both unidirectional influx and efflux measurements and the corresponding net movements when the measured cation is present on both sides of the membrane. There are many transport systems which, under appropriate circumstances, move a cation from one side of the membrane to the other where it is released, but cannot return the cation-binding site to the original surface unless a cation of the same species is transported. The process will be measured as a unidirectional flux, but it will not contribute to the net flux. If it is necessary to measure net fluxes, measurements should be made with the cation absent on the side of the membrane to which the movement is being measured, or the fluxes should be estimated by chemical means.

To make such estimates cells are prepared as described above and incubated in appropriate solutions. Samples of the suspension are taken at the beginning and end of the experimental period, the cells are separated from the solution, and washed three times with 20 vol of an ice-cold isosmotic $MgCl_2$ or choline chloride solution. Cation fluxes in red cells are slow enough, especially at 0°, that there is little change in intracellular cation content during the washing procedure. Change in intracellular cation content can be calculated from the change in the concentration of the extracellular solution or, preferably, from the change in intracellular cat-

[26] R. P. Garay and P. J. Garrahan, *J. Physiol.* (*London*) **231**, 297 (1973).

ion concentration. The cation concentration of the supernatant or of the cells can be determined by flame photometry or atomic absorption spectroscopy. In the case of the supernatant, this can be related to the change in the cation content of the cells by knowing the hematocrit of the suspension; in the case of the cell sample, the cation content of the cells can be calculated by comparing the hemoglobin concentration of the sample used for measuring cation concentration with the hemoglobin concentration of a sample of red cell suspension of known hematocrit.

Na$^+$ and K$^+$ Fluxes through the Sodium Pump

Since ouabain and other cardiotonic steroids are powerful and specific inhibitors of the sodium pump,[5] it is easy to isolate the fluxes carried out by the pump from those carried out by the other transport systems in the red cell membrane; fluxes of Na$^+$ or K$^+$ are measured under identical conditions in the presence and absence of 10^{-4} M ouabain, and the difference is the flux attributable to the sodium pump. It is important to avoid contamination of the ouabain-free samples by the samples containing ouabain; 10^{-8} M ouabain causes significant inhibition of the pump so that not much contamination is necessary to confound the results. Vanadate is also a powerful inhibitor of the pump,[27] but its specificity is uncertain and, since it inhibits at the inside surface, cells must be incubated with the inhibitor for a half-hour or so to permit equilibration of the vanadate across the membrane.

Fluxes Carried Out by the Sodium Pump of the Human Red Cell

The sodium pump is capable of mediating a variety of fluxes depending on the circumstances under which the measurements are made.[28] If the cells contain Na$^+$, with or without K$^+$, and the solution contains K$^+$, with or without Na$^+$, the pump exchanges intracellular Na$^+$ for extracellular K$^+$ with a stoichiometry of three Na$^+$ exchanged for two K$^+$. The velocity of the exchange is a saturable function both of internal Na$^+$ and external K$^+$, and in each case the curve relating velocity to cation concentration is sigmoid; in the case of K$^+$, the velocity is a function of the square of the cation concentration, and in the case of Na$^+$ it is a function of the cube of the concentration. External Na$^+$ competes with external K$^+$, and internal K$^+$ competes with internal Na$^+$.

[27] L. C. Cantley, Jr., L. Josephson, R. Warner, M. Yanagisawa, C. Lachene, and G. Guidotti, *J. Biol. Chem.* **252,** 7421 (1977).
[28] I. M. Glynn and S. J. D. Karlish, *Annu. Rev. Physiol.* **37,** 13 (1975).

If the external solution is K^+ free but contains high concentrations of Na^+, the pump exchanges internal for external Na^+. The apparent affinity for Na^+ at the external site is much lower than at the internal site. If the K^+concentration of the external solution is increased, the Na^+-Na^+ exchange is inhibited at the same concentration of K^+ at which the Na^+-K^+ exchange is activated. If cells are prepared to contain a high concentration of phosphate (by incubation for a short period in a solution in which phosphate replaces chloride), and if the cells contain K^+ but little Na^+ and the solution K^+ with or without Na^+, the pump carries out an exchange of internal K^+ for external K^+. The stoichiometry of both of the exchanges appears to be approximately $1:1$.

If cells are incubated in solutions free of both Na^+ and K^+, it is possible to observe both an Na^+ efflux and a K^+ efflux, neither of which are coupled to a cation influx. In order to demonstrate the uncoupled Na^+ efflux, the cells should be free of Na^+ since, in the absence of external Na^+, even the small amount of K^+ which leaks from the cells can activate a significant Na^+-K^+ exchange. The uncoupled K^+ efflux is best demonstrated with cells which are free of K^+, but contain phosphate and ATP.

Finally, if the cells contain low Na^+ but high K^+, and the solution high Na^+ but low K^+, the pump reverses and exchanges internal K^+ for external Na^+.

Steady-State Kinetic Experiments with the Sodium Pump

Values obtained for the Na^+ pump rate, whether measured as Na^+ efflux or K^+ influx, can be treated in the same way as values obtained in kinetic experiments with soluble enzymes.[29] By use of the powerful methods developed in the past 20 years for probing complex kinetic mechanisms by the analysis of the effects of substrate concentrations and inhibitors on apparent kinetic constants from steady-state experiments it is possible to critically evaluate proposed reaction mechanisms which arise from examination of the partial reactions of the Na^+,K^+-ATPase purified from tissues rich in the enzyme. Such experiments are facilitated with whole cells since it is possible to independently vary conditions at the two pump surfaces.

[29] I. H. Segal, "Enzyme Kinetics Behavior and Analysis of Rapid Equilibrium and Steady State Enzyme Systems." Wiley, New York, 1975.

[5] Use of Triphenylmethylphosphonium to Measure Membrane Potentials in Red Blood Cells

By Jeffrey C. Freedman *and* Terri S. Novak

The use of lipophilic ion distribution to monitor the transmembrane electrical potential, or E_m, was first developed by Skulachev and co-workers in order to determine the electrical polarity of mitochondria and submitochondrial membranes.[1] The time resolution of the technique was improved by the finding that the rate of accumulation of penetrating cations, including triphenylmethylphosphonium (TPMP), is increased by the presence of small amounts of lipophilic anions such as tetraphenylboron (TPB).[1] Subsequently, TPMP and other lipophilic cations were used to study electrically coupled cation transport in intact bacteria[2] and bacterial membrane vesicles,[3] and to characterize E_m in diverse preparations including Ehrlich ascites tumor cells,[4] cultured thyroid cells,[5] neuronal cells,[6] synaptosomes,[7] adipocytes,[8,9] fibroblasts,[10] and red blood

[1] L. E. Bakeeva, L. L. Grinius, A. A. Jasaitis, V. V. Kuliene, D. O. Levitsky, E. A. Liberman, I. I. Severina, and V. P. Skulachev, *Biochim. Biophys. Acta* **216,** 13 (1970); V. P. Skulachev, *Curr. Top. Bioenerg.* **4,** 127 (1971).

[2] F. M. Harold and D. Papineau, *J. Membr. Biol.* **8,** 27 (1972).

[3] H. Hirata, K. Altendorf, and F. M. Harold, *Proc. Natl. Acad. Sci. U.S.A.* **70,** 1804 (1973); K. Altendorf, H. Hirata, and F. M. Harold, *J. Biol. Chem.* **250,** 1405 (1975); S. Schuldiner and H. R. Kaback, *Biochemistry* **14,** 5451 (1975); S. Ramos, S. Schuldiner, and H. R. Kaback, *Proc. Natl. Acad. Sci. U.S.A.* **73,** 1892 (1976).

[4] E. Heinz, P. Geck, and C. Pietryzk, *Ann. N.Y. Acad. Sci.* **264,** 428 (1975); P. Geck, C. Pietrzyk, B.-C. Burckhardt, B. Pfeiffer, and E. Heinz, *Biochim. Biophys. Acta* **600,** 432 (1980).

[5] E. F. Grollman, G. Lee, F. S. Ambesi-Impiombato, M. F. Meldolesi, S. M. Aloj, H. G. Coon, H. R. Kaback, and L. D. Kohn, *Proc. Natl. Acad. Sci. U.S.A.* **74,** 2352 (1977).

[6] D. Lichtshtein, H. R. Kaback, and A. J. Blume, *Proc. Natl. Acad. Sci. U.S.A.* **76,** 650 (1979); D. Lichtshtein, K. Dunlop, H. R. Kaback, and A. J. Blume, *Proc. Natl. Acad. Sci. U.S.A.* **76,** 2580 (1979).

[7] S. Ramos, E. F. Grollman, P. S. Lazo, S. A. Dyer, W. H. Habig, M. C. Hardegree, H. R. Kaback, and L. D. Kohn, *Proc. Natl. Acad. Sci. U.S.A.* **76,** 4783 (1979); K. E. O. Åkerman and D. G. Nicholls, *Eur. J. Biochem.* **115,** 67 (1981).

[8] K. Cheng, J. C. Haspel, M. L. Vallano, B. Osotimehin, and M. Sonenberg, *J. Membr. Biol.* **56,** 191 (1980).

[9] M. L. Vallano and M. Sonnenberg, *J. Membr. Biol.* **68,** 57 (1982).

[10] O. Bussolati, P. C. Laris, N. Longo, V. Dall'Asta, R. Franchi-Gazzola, G. G. Guidotti, and G. C. Gazzola, *Biochim. Bipohys. Acta* **854,** 240 (1986).

METHODS IN ENZYMOLOGY, VOL. 173

cells.[8,11,12] Lipophilic ion distribution has also been used to check membrane potentials measured with optical potentiometric indicators[13] in suspensions of macrophages,[14] neutrophils,[15] lymphocytes,[16] platelets,[17] and adrenal chromaffin cells,[18] and in epithelial tissue.[19] The principles for using lipophilic ions to measure and monitor E_m have been discussed previously by others.[20]

The results of many studies indicate that the equilibrium distribution of TPMP, and its analog tetraphenylphosphonium (TPP), yields at least a qualitative or semiquantitative approximation of E_m. As with optical indicators, controls must be performed to ensure that the lipophilic ions being used do not perturb the transport systems under study, and do not interact adversely with other reagents in the experiments. Rigorous quantitative measurement of the plasma membrane potential of nonexcitable cells has been compromised by the problem of intracellular compartments and by binding of lipophilic ions to membranes and to intracellular proteins. Thus, in a study of E_m associated with catecholamine secretion from bovine adrenal chromaffin cells, it was necessary to analyze the subcellular distribution of TPP among mitochondrial, secretory granule, and bound and free cytoplasmic compartments.[18] With *Streptococcus faecalis,* binding of lipophilic cations was below saturation, implying that binding corrections determined at zero voltage may not generally be applicable to other voltages.[21] Upon hyperpolarization, for example, the amount of bound probe increases as the free intracellular cationic probe concentration increases. In attempting to correct for binding in red cells, it was assumed[8,11] that treatment with valinomycin at an external K^+ concentra-

[11] C. J. Deutsch, A. Holian, S. K. Holian, R. P. Daniele, and D. F. Wilson, *J. Cell. Physiol.* **99,** 79 (1979).

[12] A. Hunziger, F. W. Orme, and R. I. Macey, *J. Membr. Biol.* **84,** 147 (1985).

[13] See J. C. Freedman and T. S. Novak, this series, Vol. 172, p. 102 (1989); J. C. Freedman and P. C. Laris, *in* "Spectroscopic Membrane Probes" (L. Loew, ed.), Vol. 3, p. 1. CRC Press, Boca Raton, Florida, 1988.

[14] V. Castranova, L. Bowman, and P. R. Miles, *J. Cell. Physiol.* **101,** 471 (1979).

[15] B. E. Seligmann and J. I. Gallin, *J. Cell. Physiol.* **115,** 105 (1983).

[16] S. M. Felber and M. D. Brand, *Biochem. J.* **210,** 885 (1983).

[17] W. C. Horne, N. E. Norman, D. B. Schwartz, and E. R. Simons, *Eur. J. Biochem.* **120,** 295 (1981); L. T. Friedhoff and M. Sonnenberg, *Blood* **61,** 180 (1983).

[18] J. E. Friedman, P. I. Lelkes, E. Lavie, K. Rosenheck, F. Schneeweiss, and A. S. Schneider, *J. Neurochem.* **44,** 1391 (1985).

[19] J. P. Leader and A. D. MacKnight, *Fed. Proc., Fed. Am. Soc. Exp. Biol.* **41,** 54 (1982).

[20] H. Rottenberg, *Bioenergetics* **7,** 61 (1975); H. Rottenberg, this series, Vol. 55, p. 547 (1979); N. Kamo and Y. Kobatake, this series, Vol. 125, p. 46 (1986); J. B. Jackson and D. G. Nicholls, this series, Vol. 127, p. 557 (1986); H. Rottenberg, this series, Vol. 172, p. 63 (1989).

[21] A. Zaritsky, M. Kihara, and R. M. Macnab, *J. Membr. Biol.* **63,** 215 (1981).

tion, or $[K^+]_o$, that is equal to the internal K^+ concentration, or $[K^+]_c$, yields zero membrane potential. This assumption is only valid at the isoelectric point, or pI, of the cell contents; at any intracellular pH away from pI, a Gibbs–Donnan potential will still exist at equilibrium.[22] After adding valinomycin, the diffusion potential E_m is not precisely equal to the potassium equilibrium potential (E_K) because there is a net KCl flux described by $M_K = g_{K \cdot VAL}(E_m - E_K)/\mathscr{F}$, where $g_{K \cdot VAL}$ is the conductance of the $K \cdot$ valinomycin complex and $\mathscr{F}$ is the Faraday constant. According to the constant field equation,[23] E_m is determined by the value of the permeability ratio $P_{K \cdot VAL}/P_{Cl}$, thus necessitating other measurements to determine the difference between E_m and E_K. For a single-compartment system such as human red blood cells, the procedure described below avoids the need to correct for intracellular binding and gives a reproducible and reliable measure of E_m.

Theory

In order to illustrate the method, consider the case of red cells hyperpolarized by adding gramicidin at low $[K^+]_o$ in isotonic choline medium. Before addition of gramicidin, the initial membrane potential, E_m^1, is related to the equilibrium distribution of lipophilic cations by the Nernst equation:

$$E_m^1 = -(RT/F)\ln(a_c^1/a_o^1) \tag{1}$$

where a_c^1 and a_o^1 are the initial intracellular and extracellular concentrations of TPMP, respectively. After addition of gramicidin at low $[K^+]_o$, the lipophilic ions redistribute according to the new potential, E_m^2:

$$E_m^2 = -(RT/F)\ln(a_c^2/a_o^2), \tag{2}$$

where a_c^2 and a_o^2 are the new intracellular and extracellular concentrations. The change in membrane potential, $\Delta E_m = E_m^2 - E_m^1$, is given by

$$\Delta E_m = -(RT/F)\ln(a_c^2 a_o^1/a_o^2 a_c^1) \tag{3}$$

The external lipophilic ion concentration is varied at different $[K^+]_o$, and the total cell-associated TPMP is plotted vs $[TPMP]_o$ for each $[K^+]_o$ (see Fig. 2). Then, external concentrations are examined such that $a_c^2 = a_c^1$, in which case

$$\Delta E_m = -(RT/F)\ln(a_o^1/a_o^2) \tag{4}$$

Thus the change in potential induced by gramicidin at a particular $[K^+]_o$ is determined purely from the ratio of extracellular TPMP or TPP

[22] J. C. Freedman and J. F. Hoffman, *J. Gen. Physiol.* **74**, 157 (1979).
[23] D. E. Goldman, *J. Gen. Physiol.* **27**, 37 (1943); A. L. Hodgkin and B. Katz, *J. Physiol.* (*London*) **108**, 3777 (1949).

concentrations that correspond to equal levels of total intracellular amounts. These in turn are in equilibrium with equal free intracellular concentrations.

This method of determining ΔE_m is independent of the specific functional relationship between bound and free lipophilic ions, and thus circumvents the need for binding corrections. Other applications of this theoretical approach were developed previously by Halfman et al.[24]; a similar approach was also used by Hladky and Rink to analyze diS-C$_3$(5) distribution in red cells.[25]

Procedure

The analytical procedure described below is adapted from Cheng et al.[8] and has the advantage of using small samples that conserve isotope.

Reagents

Cell wash medium: 150 mM choline chloride, 5 mM HEPES, adjusted to pH 7.4 at 25° with 1 M Tris base (Sigma)

Variable potassium media: X mM KCl, $(150 - X)$ mM choline chloride, 0.3, 1, 3, or 10 μM TPMP, 0.3–1 μCi/ml [^{3}H]TPMP, and 5 mM HEPES, pH 7.4 at 25°

Gramicidin D: 0.33 mg/ml in ethanol (Sigma Chemical Co., St. Louis, MO)

TPMP: 5 mM in ethanol (Aldrich Chemicals Co., Milwaukee, WI), prediluted so as to add the same amount of ethanol to each flask with differing [TPMP]$_o$

TPB: 1 mM in ethanol (Aldrich); must be added after cells

[^{3}H]TPMP: 250 μCi/ml in ethanol (Amersham or New England Nuclear)

Protocol

1. Blood from human donors is drawn by venipuncture into heparanized tubes (green top Vacutainers, Becton-Dickenson, Rutherford, NJ) and immediately centrifuged at 13,800 g for 3 min at 4°. The plasma and buffy coat are aspirated and discarded and the packed cells are then washed three or four times by centrifugation, each time resuspending in about 5 vol of chilled isotonic cell wash medium. The cells are then adjusted to 50% hematocrit (HCT) in the cold cell wash medium and kept on ice for use on the same day.

---

[24] C. J. Halfman and J. Steinhardt, *Biochemistry* **10**, 3564 (1971); C. J. Halfman and T. Nishida, *Biochemistry* **11**, 3493 (1972).

[25] S. B. Hladky and T. J. Rink, *J. Physiol.* (*London*) **263**, 287 (1976); diS-C$_3$(5) is the optical potentiometric indicator 3,3'-dipropylthiodicarbocyanine.

2. At time zero, 50% HCT red cells is added to a final HCT of 1.2% to test tubes containing 3.0 ml of variable potassium media. For the experiments shown (Fig. 2), $[K^+]_o$ was 1 or 90 mM. The tubes are held in a custom-designed holder at a 45° angle in a waterbath shaker (Eberbach 6250) set at 150 oscillations per minute and thermostatted at 25°.

3. After about 1 min, TPB is added to 2 μM. The TPB should be added after the cells; if added to the medium before the cells, then the rate of uptake of TPMP in a repetitive series of identical experiments declines systematically with time, yielding scattered results.

4. Five minutes after adding cells, initial samples are taken in triplicate.

5. Three minutes later, gramicidin is added to 0.06 μg/ml.

6. After another 5 min, triplicate samples are again taken.

7. The samples (0.4 ml) of suspension are layered over the same amount of silicone oil (GE F50, General Electric Corp.), contained in 1.5-ml microcentrifuge tubes, and are spun immediately for 15 sec in a microcentrifuge.

8. Supernatant (10 μl) is sampled into a 20-ml scintillation vial, and the remaining supernatant and oil are removed by aspiration through a Pasteur pipet connected through a flask to house vacuum.

9. The cell pellet is lysed by addition of 0.3 ml deionized water and the hemolysate is transferred to a scintillation vial. The microcentrifuge tube is rinsed with 30% hydrogen peroxide, which is also emptied into the scintillation vial, such that a total of 0.6 ml peroxide is used. The peroxide bleaches the hemolysate, yielding a colorless extract suitable for scintillation counting.

10. To determine the specific activity of the medium, 10 μl is sampled in triplicate before addition of cells and processed as for supernatants.

11. To each vial is added 10 ml ACS scintillation fluid (Amersham), the vials are shaken vigorously, and the samples are counted in a β-counter at 10 min/sample or to 1% CV. All ^{3}H samples are corrected for quenching by using quench correction standards (Beckman).

12. The hemoglobin content of each suspension is also determined by hemolysing 0.4 ml in Drabkin's reagent.

Calculations

The cellular TPMP contents, or $[TPMP]_c$ (nmol/g Hb), are calculated as follows:

$$[TPMP]_c = (dpm_c[TPMP]_o^0)/(100 \ dpm_o^0[Hb]V_s)$$

where dpm_c is the disintegrations per minute, corrected for quenching, from the bleached hemolysates; dpm_o^0 is the corrected disintegrations per minute for the initial supernatant taken before adding cells (the factor 100 refers these counts to 1 ml); $[TPMP]_o^0$ is the initial supernatant concentration of TPMP (nmol/ml); [Hb] is the hemoglobin concentration (g/ml) in the suspension; and V_s is the volume of suspension sampled (0.4 ml).

The final extracellular concentration of TPMP, or $[TPMP]_o$ (micromolar) is

$$[TPMP]_o = (dpm_o^f[TPMP]_o^0)/dpm_o^0$$

where dpm_o^f is the corrected disintegrations per minute in the final supernatant.

Illustrative Results

The time course of uptake of TPMP with TPB indicates that equilibrium is reached within 5 min after hyperpolarization with gramicidin at 1 mM $[K]_o$ (Fig. 1). A similar time course is evident before adding the ionophore (not shown). When measuring diffusion potentials, it is important to check the time course of uptake and the stability of E_m under each set of experimental conditions. Controls established that 10 μM TPMP with 2 μM TPB had no effect on gramicidin-induced net K$^+$ efflux for 0–0.1 μg/ml gramicidin. The equilibrium distribution of TPMP before and after hyperpolarization of human red blood cells by gramicidin is shown in Fig. 2. The dashed line drawn between equal levels of $[TPMP]_c$ yields two values of $[TPMP]_o$ that enable calculation of ΔE_m by Eq. (4). In separate

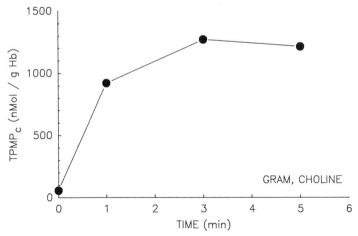

FIG. 1. Kinetics of uptake of [³H]triphenylmethylphosphonium (TPMP) into human red blood cells in the presence of tetraphenylborate (TPB).

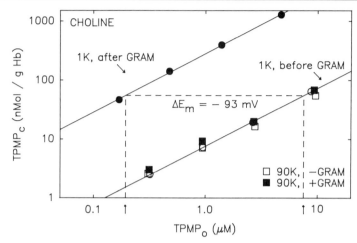

FIG. 2. Equilibrium distribution of [³H]triphenylmethylphosphonium in human red blood cells before (right line) and after (left line) hyperpolarization with gramicidin.

experiments with blood from four donors, the values determined for ΔE_m were -93, -93, -95, and -98 mV, or an average of -95 ± 2 mV (SD).

Acknowledgments

The silicone oil for cell separation was generously provided by the Silicone Products Division of General Electric Company. We also gratefully acknowledge the support of a grant from the National Institutes of Health (GM28839).

[6] Measurement and Control of Intracellular Calcium in Intact Red Cells

By VIRGILIO L. LEW and JAVIER GARCÍA-SANCHO

Introduction

Control of $[Ca^{2+}]_i$ in intact red cells is required for a variety of studies on Ca^{2+} transport and on the effects of intracellular Ca^{2+} on cell shape, metabolism, transport, volume, pH, cytoskeletal structure, or membrane–cytoskeleton interactions. There is no single established procedure that can be applied to control $[Ca^{2+}]_i$ over the vast concentration range that may be required for such studies. There are also fundamental limitations in the way metabolism and active Ca^{2+} transport interact in red cells to prevent the maintenance of steady cytoplasmic Ca^{2+} levels over critical concentration ranges. Because of this it seemed best to provide here broad methodological guidelines rather than set prescriptions. First we describe the relevant properties of red cells. We then provide a critical outline of the tools and experimental procedure available to control cell Ca^{2+}. Finally we analyze the pitfalls in the available methods and the fundamental limitations in the control of $[Ca^{2+}]_i$ in red cells. Some typical experimental protocols are provided at the end to illustrate specific applications.

Calcium Content and Transport in Human Red Cells

Fresh human red cells, washed in Ca^{2+}-free saline, have about 10–30 μmol Ca^{2+}/liter.[1,2] If the washing medium contained more than 0.1 mM EGTA the measured total cell Ca^{2+} is reduced to about 5 μmol/liter cells, indicating that most of the cell-associated Ca^{2+} is loosely bound to the external surface.[2] Of this, only about 1 μmol/liter cells can be equilibrated with external ^{45}Ca or mobilized out of the cells after permeabilization with the divalent cation ionophore A23187.[2] Red cells possess a powerful ATP-fueled Ca^{2+} pump.[3] When saturated with internal Ca^{2+}, and at nonlimiting ATP concentrations, it can extrude Ca^{2+} at rates of about 10–25 mmol/liter cells · hr.[4] The kinetics of the pump and of the passive inward leak result in cytoplasmic Ca^{2+} levels in the 10–30 nM range, with a pump-leak turnover of about 40 μmol/liter cells · hr under physiological conditions.[5] Passive Ca^{2+} influx saturates with external Ca^{2+}; the apparent $K_{1/2}$ is about 0.8–0.9 mM.[6,7] This means that Ca^{2+} influx cannot be effectively increased by raising external Ca^{2+} above its physiological level of 1.0–1.2 mM. Cytoplasmic Ca^{2+} buffering was shown to be adequately described by the expression $[Ca^{2+}]_i = \alpha[Ca^T]_i$, where $[Ca^{2+}]_i$ is the concentration of ionized Ca in the cytoplasm (micromolar); α, the fraction of ionized cell Ca, is about 0.15–0.40, and $[Ca^T]_i$ represents the total cell Ca^{2+} (in micromoles per liter of normal packed cells or per 340 g of hemoglobin).[8,9]

[1] D. G. Harrison and C. Long, *J. Physiol.* (*London*) **199**, 367 (1968).
[2] R. M. Bookchin and V. L. Lew, *Nature* (*London*) **284**, 561 (1980).
[3] H. J. Schatzmann, *in* "Membrane Calcium Transport" (E. Carafoli, ed.). Academic Press, London, 1982.
[4] G. Dagher and V. L. Lew, *J. Physiol.* (*London*) **407**, 569 (1988).
[5] V. L. Lew, R. Y. Tsien, C. Miner, and R. M. Bookchin, *Nature* (*London*) **298**, 478 (1982).
[6] T. Tiffert, J. Garcia-Sancho, and V. L. Lew, *Biochim. Biophys. Acta* **773**, 143 (1984).
[7] M. K. McNamara and J. S. Wiley, *Am. J. Physiol.* **250**, C1 (1985).
[8] H. G. Ferreira and V. L. Lew, *Nature* (*London*) **259**, 47 (1976).
[9] V. L. Lew and J. Garcia-Sancho, *Cell Calcium* **6**, 15 (1985); J. Garcia-Sancho and V. L. Lew, *J. Physiol.* **407**, 505 (1988); J. Garcia-Sancho and V. L. Lew, *J. Physiol.* **407**, 523 (1988); J. Garcia-Sancho and V. L. Lew, *J. Physiol.* **407**, 541 (1988); L. Almaraz, J. Garcia-Sancho, and V. L. Lew, *J. Physiol.* **407**, 557 (1988).

The negligible endogenous Ca^{2+} pool of normal human red cells allows accurate estimates of total cell Ca^{2+} using ^{45}Ca. If the Ca^{2+} gained or exchanged by the cells is from a large extracellular ^{45}Ca pool of known specific activity, measurement of the cell-associated radioactivity directly reports the total Ca^{2+} content of the cells.

Red cells from normal and anemic individuals have recently been shown to possess endocytic vesicles capable of ATP-dependent Ca^{2+} accumulation; vesicle accumulation causing elevated total cell Ca^{2+} has so far been documented only in sickle cell anemia red cells.[10] A similar origin for the elevated Ca^{2+} content of other abnormal red cells is likely but remains to be demonstrated. The fact that normal red cells do not accumulate Ca^{2+} throughout their life despite the presence of endocytic vesicles suggests either that their cytoplasmic Ca^{2+} levels do not increase in the circulation or that vesicles with accumulated Ca^{2+} are selectively pinched off in spleen sinusoids. In experiments with Ca^{2+}-permeabilized cells, on the other hand, when $[Ca^{2+}]_i$ is increased, the presence and possible effects of such vesicles on the total Ca^{2+} content of the cells has to be taken into consideration.[8]

Tools and Main Procedures Used to Control Red Cell Calcium Content

There are two main tools to measure and control total and ionized Ca^{2+} levels in red cells: divalent cation ionophores, like A23187,[11] and Ca^{2+} chelators that can be incorporated nondisruptively into red cells, like benz2 or quin2.[4,5,12,13] There are also two main metabolic states of intact red cells in which Ca^{2+}-related functions are usually investigated: fed or ATP depleted; these determine whether the pump is active or not. Ca^{2+} pump inhibition may be required to build up a measurable pool of total Ca^{2+} within the cells in the absence of incorporated chelators, or to allow passive distribution of Ca^{2+} in Ca^{2+}-permeabilized cells. Since there are no specific Ca^{2+} pump blockers, effective inhibition of the pump can only be obtained by reducing its fuel, ATP, its transported substrate, Ca^{2+}, or by cooling the cells to $0°$. There is a sharp increase in the Q_{10} of the pump below $5°$, from about 4 in the temperature range above $5°$ to $7–9$ below it.[9] The combination of tools and metabolic states enables all the manipulations that are required in order to measure and control intracellu-

[10] V. L. Lew, A Hockaday, M.-I. Sepulveda, A. P. Somlyo, and R. M. Bookchin, *Nature* (*London*) **315,** 586 (1985).
[11] B. C. Pressman, *Annu. Rev. Biochem.* **45,** 501 (1976).
[12] R. Y. Tsien, *Nature* (*London*) **290,** 527 (1981).
[13] R. L. Waller, L. R. Johnson, W. J. Brattin, and D. G. Dearborn, *Cell Calcium* **6,** 245 (1985).

lar Ca^{2+} over the restricted range of concentrations and conditions in which such control has become feasible.

The methods to measure and control total ionized cell Ca vary depending on whether the pump is active or not and on the level of Ca^{2+} within the cells. Before describing these methods in detail, it is necessary to explain the properties of the tools, the procedures to manipulate the cell ATP levels, and some fundamental limitations in the control of cell Ca^{2+} in fed and partially depleted cells.

We will consider here only the ionophore A23187 and the calcium chelators benz2 and quin2, which are, so far, the best characterized and more easily available agents.

A23187

A23187 is usually added to cell suspensions from concentrated stock solutions in ethanol (up to 2 mM) or in DMSO (usually up to 10 mM) to give less than 0.5% by volume of solvent. On addition to protein-free cell suspensions, A23187 immediately partitions into the cells, and at least half of it remains in the membrane, where it mediates a strict $M^{2+}:2H^+$ exchange with a turnover, at 37°, of about two to five Ca ions per second and per pair of ionophore molecules.[14] Net fluxes through the ionophore become zero when the equilibrium distribution given by $[M^{2+}]_i/[M^{2+}]_o = ([H^+]_i/[H^+]_o)^2$ is attained.[11] In red cells, the parallel operation of the anion transporter, as anion exchanger and electrodiffusional anion channel, secures that at equilibrium $[H^+]_i/[H^+]_o = [A^-]_o/[A^-]_i = \exp(-FV/RT)$, where A^- represents Cl^-, HCO_3^-, or any other highly permeant anion present, and $F, R, T,$ and V represent, respectively, the Faraday and ideal gas constants, the absolute temperature, and the membrane potential. These relations will apply under most experimental conditions, unless anion transport is inhibited, or special care is taken to remove dissolved CO_2. This means that in red cells the ionophore-induced distribution of divalent cations and protons, when unopposed by significant pump fluxes, generally coincides with the distribution expected from electrochemical equilibrium.

Addition of the ionophore to cells in the presence or absence of Ca^{2+} may profoundly affect their volume, composition, metabolism, and shape. In experiments designed to study the effect of $[Ca^{2+}]_i$ on a specific target, rather than on a global cell response, side effects of $[Ca^{2+}]_i$ should be reduced as far as possible. A few simple precautions will secure reasonable stability of volume, Na^+, K^+, and Mg^{2+} contents in most experi-

[14] V. L. Lew and L. O. Simonsen, *J. Physiol.* (*London*) **308,** 60 (1980).

mental conditions. The idea is to set the external concentrations of those ions, which become permeable during the Ca^{2+}-permeabilizing procedure, at electrochemical equilibrium across the membrane. This applies mainly to Mg^{2+}, which permeates through the ionophore, and to K^+, since internal Ca^{2+} can activate Ca^{2+}-sensitive, K^+-selective channels. At around 10% hematocrit, cells with normal membrane potentials of about -10 mV will maintain constant cell volume, Mg^{2+}, and K^+ contents in media with about 80–90 mM K^+ and 0.15 mM Mg^{2+}. Metabolically "souped up" cells, whose organic phosphate pool had been increased by incubation with inosine or adenosine, pyruvate, and phosphate, have a lower diffusible anion pool and higher membrane potentials. They will therefore require lower K^+ and Mg^{2+} concentrations in the medium in order to retain their volume and ionic composition when permeabilized to Ca^{2+} with the use of the ionophore. The principle to follow is to set the external K^+ or Mg^{2+} levels so that the concentration ratio across the membrane equals the proton ratio or the proton ratio squared, respectively.[15]

The ionophore-induced Ca^{2+} permeability can be easily controlled over a huge range simply by varying the ionophore concentration.[16] The pH and the presence of other divalent cations will affect Ca^{2+} transport by the ionophore. At external pH values within the 7.4–7.8 range and in the high-K^+, 0.15 mM Mg^{2+} medium described above, the relation between the rate constant for Ca^{2+} equilibration, k (in hr^{-1}), and cell ionophore concentration $[I]_c$ (in units of micromoles per liter cells) was found to be adequately represented by $k \approx ([I]_c)^{1.45}$. This gives approximate half-times for Ca^{2+} equilibration in ATP-depleted cells of about 42, 1.5, and 0.05 min for ionophore contents of 1, 10, and 100 μmol/liter cells, respectively. With 0.1 mM Ca^{2+} in the medium, for instance, this last ionophore concentration would cause a Ca^{2+} influx of about 50–100 mmol/liter cells · hr, enough to swamp even the highest reported Ca^{2+} pump fluxes of about 10–25 mmol/liter cells · hr. The ionophore-induced Ca^{2+} permeability was also shown to be uniform in all cells, whether ATP depleted or not.[9,17] The ionophore in the cells is in rapid dynamic equilibrium with free ionophore in the medium and can be easily extracted by washing the cells with albumin-containing media. Albumin binds ionophore in 1 : 1 molar ratio.[18] Ionophore washout fully restores the normal low divalent cation permeability of the cells. Since the ionophore-induced permeability has a relatively high temperature coefficient ($Q_{10} > 3$), ionophore washout in the

[15] P. Flatman and V. L. Lew, *Nature (London)* **267**, 360 (1977).
[16] L. O. Simonsen and V. L. Lew, *in* "Membrane Transport in Erythrocytes" (U. V. Lassen, H. H. Ussing, and J. O. Wieth, eds.). Munksgaard, Copenhagen, 1980.
[17] L. O. Simonsen, J. Gomme, and V. L. Lew, *Biochim. Biophys. Acta* **692**, 431 (1982).
[18] L. O. Simonsen, *J. Physiol. (London)* **313**, 34 (1981).

cold can be implemented with minimal change in the Ca^{2+} content the cells had in the presence of the ionophore. Cobalt, added to the medium at concentration exceeding those of Ca^{2+}, has recently been shown to block, instantaneously and completely, all Ca^{2+} transport by the ionophore.[4,6] Cobalt is transported by the ionophore and does not interfere with Ca^{2+} transport by the pump. It can therefore be used to expose uphill Ca^{2+} fluxes through the pump in Ca^{2+}-permeabilized cells without having to remove the ionophore.[4]

Benz2 and Quin2

The acetoxymethyl tetraester form of these chelators (benz2-AM and quin2-AM) is highly hydrophobic. Dissolved in dimethyl sulfoxide (DMSO) and added from concentrated stock solutions to well-stirred cell suspensions, the sterified chelators readily partition into the red cell membrane. During subsequent incubation at 37°, at least part of the ester links are hydrolyzed by cytoplasmic or inner membrane esterases and free chelator is released inside the cells. Ester hydrolysis releases 4 mol of formaldehyde/mol of free chelator formed. Formaldehyde depletes NAD^+ by enzymatic reduction and causes ATP depletion.[6] Depletion is accelerated by the presence of substrates. The metabolic block can be bypassed by the inclusion of 5–10 mM pyruvate in the incubation medium.[19] The Ca^{2+}-binding properties of benz2, determined *in situ* inside the cells, were shown to be the same as in free solution.[4] The apparent Ca^{2+} dissociation constant within the normal intracellular environment was found to be about 50 nM. Unlike benz2, quin2 can act as a fluorescent Ca^{2+} indicator, and it has been widely used as such in a variety of cells. The quenching effect of hemoglobin precludes measurement of quin2 fluorescence in red cells. Fluorescent Ca^{2+} indicators of higher quantum yield, like fura2,[20] may eventually help overcome this problem. The reported yield of benz2 or quin2 incorporation varies from 10 to 100%, depending on the origin of product and experimental conditions.

Control of Cell ATP Levels

Glycolysis is the only source of ATP in mammalian red cells. The physiological substrate of human red cells is glucose. The normal ATP content is about 1.0–1.5 mmol/liter cells and its turnover is in the 3–5 mmol/liter cells · hr range. Since the Ca^{2+} pump operates with a Ca^{2+} : ATP stoichiometry near 1 : 1, a Ca^{2+}-saturated pump consuming

[19] J. Garcia-Sancho, *Biochim. Biophys. Acta* **813**, 148 (1985).
[20] R. Y. Tsien, T. J. Rink, and M. Poenie, *Cell Calcium* **6**, 145 (1985).

ATP at rates over 10 mmol/liter cells · hr may rapidly deplete the cells of ATP, even in the presence of metabolic substrates. Internal Ca^{2+} was shown to stimulate glycolysis[21]; Ca^{2+}-stimulated ATP production may therefore match pump consumption and prevent ATP depletion. Whether Ca^{2+}-permeabilized cells become depleted of ATP or not depends therefore on the factors which control the $[Ca^{2+}]_i$-altered balance between ATP consumption and production. These include experimental variables such as nature of glycolytic substrate and extent of Ca^{2+} permeabilization, and intrinsic factors such as the kinetics of the Ca^{2+} pump and of the enzymes which control the nucleotide metabolism of the cell. $[Ca^{2+}]_i$-Induced ATP depletion is irreversible. The reason for this is that internal Ca^{2+} stimulates the AMP deaminase activity of the cell. Most of the cell's adenine nucleotide is then irreversibly converted to IMP.[9] Inosine is more efficient than glucose in sustaining normal ATP levels in Ca^{2+}-permeabilized human red cells. An additional experimental advantage of inosine is that it reduces the inorganic phosphate concentration of the cells to the micromolar level. This minimizes errors in the measurement of total cell Ca^{2+} due to the formation of local precipitates. Because of these properties, inosine has been the substrate of choice in most studies with Ca^{2+}-permeabilized cells.

Reversible ATP depletion is obtained by prolonged incubation (24–48 hr) of red cells at 37° in substrate-free media. Irreversible ATP depletion is usually induced by incubating cells in the presence of 4–6 mM iodoacetamide, which seems to have fewer side effects than iodoacetate.[22] Both agents are powerful inhibitors of the glyceraldehyde-3-phosphate dehydrogenase. Addition of glycolytic substrates in the presence of iodoacetamide speeds up ATP depletion because the substrate consumes ATP while glycolytic ATP production is blocked.[23] Inosine is also here more efficient than glucose. With 10 mM inosine and 5–6 mM iodoacetamide in the medium, cell ATP falls to the 1–5 μM level in less than 1 hr of incubation at 37°. With glucose, 2-deoxyglucose, or without substrates, iodoacetamide-poisoned cells would deplete to similar levels in 4 to 15 hr.[24] The low ATP level in inosine + iodoacetamide-depleted cells is maintained in rapid turnover for many hours (at about 200 μmol/liter cells · hr), probably sustained by the large reserves of 2,3-DPG within the cells.[25] This provides a supply of ATP that can fuel residual Ca^{2+} pumping at rates of about 100 μmol/liter cells · hr.[17]

[21] A. M. Brown and M. J. Johnston, *J. Physiol* (*London*) **341**, 63P (1983).
[22] G. A. Plishker, *Am. J. Physiol.* **248**, C419 (1985).
[23] V. L. Lew and H. G. Ferreira, *Curr. Top. Membr. Transp.* **10**, 217 (1978).
[24] V. L. Lew, *Biochim. Biophys. Acta* **233**, 827 (1971).
[25] I. M. Glynn and V. L. Lew, *J. Physiol.* (*London*) **207**, 393 (1970).

Control of Cell Calcium: Shortcomings and Limitations

The ideal methods to control cell Ca^{2+} in red cells should provide easy ways to set cell Ca^{2+} at any desirable level, uniformly in all cells, for the length of time and in the conditions required to perform whichever measurements are needed, and without any undesirable side effects like hemolysis, or unintended changes in cell volume, composition, or metabolic state. In practice, there are only two procedures which, within restricted ranges of cell Ca^{2+} concentrations, approach the main objectives of the ideal methods. The first is A23187-induced Ca^{2+} permeabilization of ATP-depleted red cells. The second is nondisruptive chelator incorporation.

Ca^{2+} Permeabilization Using the Ionophore A23187

Ionophore addition to ATP-depleted cells suspended at 10% hematocrit in a medium containing 80 mM K$^+$, 0.15 mM Mg^{2+}, and different Ca^{2+} levels, will redistribute Ca^{2+} to equilibrium at easily controllable rates. Total cell Ca^{2+} is measured from the cell-associated ^{45}Ca tracer, and cell Ca^{2+} is estimated from the external Ca^{2+} level and the proton ratio squared, as shown below. One problem common to this and all methods which use the ionophore is that ionophore addition always causes some hemolysis. This increases with ionophore content of the cells. Only part of the ionophore-induced lysis is Ca^{2+} dependent. Lysis may affect 10% of the cells at an ionophore concentration of 100 μmol/liter cells.[6] Lysis is higher in ATP-depleted cells than in substrate-fed cells. The origin of ionophore or ionophore + Ca^{2+}-induced lysis in unknown. Ionophore + Ca^{2+}-treated rabbit red cells had near normal survival *in vivo* despite substantial hemolysis during tracer labeling and Ca^{2+} loading *in vitro*.[26] This suggests that the cells which did not lyse had not been irreversibly harmed. An experimental problem created by ionophore-induced lysis, which becomes relevant when Ca^{2+} distribution across the membrane has to be accurately measured, is that released hemoglobin will bind some of the external Ca^{2+} and reduce the ionized Ca level in the medium. At 10% hematocrit and using ionophore concentrations in the 10–100 μmol/liter cells range, external Ca^{2+} is only about 75–80% of the total Ca^{2+} concentration, as determined from comparison of tracer and Ca^{2+} electrode measurements (V. L. Lew, R. Y. Tsien, and C. Miner, unpublished observations). Because of the lytic effects of ionophore (with and without Ca^{2+}) it would seem prudent not to exceed ionophore concentrations of about 100 μmol/liter cells. This condition restricts the use of Ca^{2+} permeabilization

---

[26] R. M. Bookchin, E. F. Roth, and V. L. Lew, *Blood* **66**, 220 (1985).

at low cell Ca^{2+} levels because residual Ca^{2+} pumping will effectively oppose even the highest ionophore-induced Ca^{2+} influx. For instance, with the highest ionophore concentration we ever used, of 200 μmol/liter cells, and in the presence of 0.1 μM externally buffered Ca^{2+}, the expected Ca^{2+} influx would be around 200 μmol/liter cells · hr, only about twice the residual pumping rate of rapidly depleted cells. Residual pumping could therefore prevent ionophore-induced Ca^{2+} equilibration. In addition, uncertainties concerning the uniformity of depletion and residual pumping among the cells restrict the use of this method in the submicromolar range of cell Ca^{2+} concentrations.

Fundamental Limitation in Control of Cell Calcium in Fed Cells

Ionophore-induced Ca^{2+} permeabilization can also be applied to fed cells. Here, however, it is impossible to obtain uniform control of cell Ca^{2+} levels in all the cells when the mean total cell Ca^{2+} is less than about 200 μmol/liter cells, and then only at the expense of unavoidable Ca^{2+}-induced ATP depletion. This is a fundamental limitation imposed by the intrinsic properties of Ca^{2+} pumps and red cell metabolism, and not by the Ca^{2+} permeabilization method. Each cell behaves as if it has a defined Ca^{2+} influx threshold. If Ca^{2+} influx is below this threshold, the cell can sustain steady states with balanced Ca^{2+} extrusion through the pump, normal ATP levels, and increased $[Ca^{2+}]_i$, in the submicromolar range. If it is above threshold, metabolic ATP production fails to match pump-mediated hydrolysis, ATP falls, adenine nucleotides become irreversibly converted to IMP, and cell Ca^{2+} approaches equilibrium with external Ca^{2+}. Thresholds vary from cell to cell and also with the nature of the substrate. This means that it is impossible to obtain a uniform Ca^{2+} distribution among uniformly permeabilized cells when the Ca^{2+} influx is set within the range of thresholds. In inosine-fed cells the Ca^{2+} influx thresholds vary within the 1–2 to 30 mmol/liter cells · hr range. Measurements of the mean total Ca^{2+} content of the cells 10–30 min after Ca^{2+} permeabilization show steady-state Ca^{2+} levels below equilibrium, but these represent mainly the Ca^{2+} contained within the fraction of Ca^{2+}-equilibrated cells plus some additional Ca^{2+} that may have accumulated within endocytic vesicles. The total Ca^{2+} content of the fraction of cells in dynamic pump–leak balance is a minute proportion of the total in the population and can only be exposed by special techniques.[9] Preliminary observations suggest that Ca^{2+} equilibration occurs at lower Ca^{2+} influx values in cells fed with glucose than in those fed with inosine, in line with the observation that glucose is a poorer substrate than inosine in human red cells.[23,27]

---

[27] O. Scharf, B. Foder, and U. Skibsted, *Biochim. Biophys. Acta* **730**, 295 (1983).

Calcium Loading of Chelator-Containing Cells

Control of cell Ca^{2+} levels in the physiological and subphysiological concentration range, i.e., below 10–30 nM, may be obtained by loading the cells nondisruptively with Ca^{2+} chelators. The detailed protocol is given below. The presence of an internal chelator allows the build-up of a sizeable total Ca^{2+} pool within the cells by Ca^{2+} influx through the normally small passive inward leak. Pump activation eventually establishes pump–leak Ca^{2+} steady states, but the total Ca^{2+} gained by the cells has now become easily measurable. The Ca^{2+} permeability may be increased with the use of the ionophore. This may be used to explore cytoplasmic Ca^{2+} levels above the physiological range in fed cells, or to investigate the submicromolar Ca^{2+} concentration range in ATP-depleted cells, by equilibration with externally buffered Ca^{2+}. However, if internal Ca^{2+} approaches or exceeds the optimal buffering concentrations of the incorporated chelator, the same limitations discussed above for Ca^{2+}-permeabilized fed or ATP-depleted cells will apply. It is therefore important to adjust ionophore and external Ca^{2+} levels so that Ca^{2+} influx does not cause internal Ca^{2+} to exceed those limits. Between 2 and 10% of the cells tend to lyse during the chelator-loading incubation after benz2-AM addition. Similar lysis was observed by addition of comparable amounts of DMSO and formaldehyde. Although free chelator seems to be present in all the cells, there is some evidence that it may not be uniformly distributed.[6] Formaldehyde binding to cell proteins may have effects other than inhibition of glycolysis, and which have not yet been identified.

Experimental Protocols to Measure and Control Red Cell Calcium

Solutions

All concentrations are given in millimolar concentration unless otherwise specified.

Solution A: NaCl, 145; KCl, 5; NaOH-neutralized HEPES buffer, pH 7.5, 10; $MgCl_2$, 0.15

Solution B: NaCl, 70; KCl, 80; otherwise the same as solution A.

Solutions C and D: Same as solutions A and B, respectively, but with neutralized EGTA, 0.1.

When required, the following agents may be added to the various incubation media to give (final concentrations): inosine, 10; glucose, 10; iodoacetamide, 5–6; pyruvate, 5–10. Usual concentrated stock solutions of A23187 and benz2 are as follow: A23187, 0.1 to 2 in ethanol, or 0.1 to 10 in DMSO; benz2-AM, 10 to 250 in DMSO. For A23187 or benz2-AM additions a practical rule is to deliver the desired amounts from the more concentrated stock solutions in volumes compatible with the required accuracy of delivery (5–10 μl, for instance), and in any case not to exceed 0.5% of solvent in the cell suspension. Indifferent use of solutions A or B, C or D is indicated by AB or CD, respectively. The specific activity of ^{45}Ca may vary in the range 10^6–10^8 cpm/μmol.

A23187-Induced Calcium Permeabilization of ATP-Depleted Red Cells

1. Use fresh blood obtained with anticoagulant, or bank blood.

2. Wash red cells by centrifugation and resuspension in 5–10 vol of solution CD. Spin at about 2000 g at room temperature. After each spin, suck off supernatant, buffy coat, and topmost cell layers for better white cell and reticulocyte removal. At least three washes are required for efficient medium exchange and removal of Ca^{2+} loosely bound to the external red cell surface. Bank cells may have become acid during storage in acid–citrate–dextrose; check pH of suspension in between washes and, if necessary, titrate with isotonic Tris base or sodium bicarbonate.

3. For ATP depletion, resuspend the cells at 10–20% hematocrit in solution CD with inosine and iodoacetamide.

4. Incubate at 37° for about 90 min. Use gentle stirring, continuous or intermittent.

5. Wash cells two or three times with 5–10 vol of solution B. Resuspend at about 10% hematocrit in solution B and distribute aliquots of this suspension into magnetically stirred vials containing the various external ^{45}Ca concentrations required.

6. Start incubation at 37° and add A23187 to a final concentration of 1–10 μM in the suspension (equivalent to about 10–100 μmol/liter cells), for rapid Ca^{2+} equilibration.

7. To measure the total Ca^{2+} content of the cells, sample 0.05- to 0.1-ml aliquots of the suspension into Eppendorf-type microcentrifuge tubes of 1.5-ml nominal capacity containing 0.4 ml of dibutyl phthalate oil, of relative density 1.042, and 0.9–1.0 ml of solution D, all maintained at ~0° in an ice bath. Delivery of the sample is followed by a rapid shake, to cool the cells and to dilute the extracellular medium uniformly, and by a 5- to 10-sec spin at 12,000–14,000 g in the microfuge. This is enough to separate the cells in a tight pellet at the bottom of the oil layer, with less than 1% contamination from the original suspending medium. The separated cells may wait indefinitely for further processing of the samples. After sucking the supernatant off and drying the tube walls with cotton swabs, the cells are lysed by addition of 6% trichloroacetic acid under vigorous vortex stirring, the denatured proteins pelleted by centifugation, and aliquots of the clear supernatant used for ^{45}Ca radioactivity measurement. All the cell-associated Ca^{2+} is recovered in the supernatant.

8. The specific activity of ^{45}Ca may be determined in 10-μl aliquots of the suspension, for instance, after removing the proteins as above. The measured radioactivity is divided by the amount of Ca^{2+} in the sample used for counting.

9. The total Ca^{2+} content of the cells is computed by dividing the activity in the cell sample by the volume of packed cells or the grams of hemoglobin represented in that sample and by the Ca^{2+} specific activity. The total Ca^{2+} content of cells is usually expressed in micromoles per liter cells or per 340 g of hemoglobin.

Calcium Control in Chelator-Loaded Cells

1. Red cells are washed and prepared as in steps 1 and 2 above.

2. For chelator loading, the washed cells are suspended at 10–30% hematocrit in medium CD, with inosine and pyruvate. Benz2-AM is added from concentrated stock to give the desired concentration of free chelator in the cells. This is calculated assuming complete partition and full hydrolysis. During addition, and for the subsequent minute, the cells are stirred magnetically or over a vortex. Incorporation of free chelator is fastest at 37° and requires about 90 min of incubation.

3. After this incubation the cells are washed two or three times with solution AB and resuspended in the same solution at about 10–20% hematocrit.

4. ^{45}Ca is added to this suspension in the final incubation bath at 37°. This initiates Ca^{2+} uptake by the cells.

5. Measurement of total cell Ca^{2+} here requires washing away the extracellular Ca^{2+} since, unlike with Ca^{2+}-permeabilized cells, the Ca^{2+} content of the cells is only a minute fraction of the Ca^{2+} concentration in the medium. Samples (0.1–0.2 ml) are delivered into 50–150 vol of ice-cold solution CD in a centrifuge tube, the cells are spun and washed once or twice more with the same ice-cold medium, and the final cell pellet is deproteinized and processed as described above for measurement of total cell Ca^{2+}.

6. Of the various procedures now available to estimate the chelator content of red cells,[5,6,13] the simplest and perhaps most accurate is the one based on the measurement of the ionophore-induced distribution of Ca^{2+} in the presence of nanomolar external Ca^{2+} levels.[6] To do this, an aliquot of the chelator-loaded cells is washed, suspended at about 10% hematocrit in medium B containing about 100 μM ^{45}Ca of the highest specific activity, and incubated at 37°. Ionophore A23187 is added to this suspension to give 50–100 μmol/liter cells. Ten to 15 min later EGTA is added to a final concentration of about 3 mM in the medium. Ca^{2+} redistributes to equilibrium between medium and chelator-containing cells, since $[Ca^{2+}]_i$

remains too low to activate the pump in these conditions. At higher $[Ca^{2+}]_i$, Ca^{2+} equilibration requires predepletion of the cells. Measurement of the total Ca^{2+} content of the cells and of the proton distribution ratio after the cells have reached equilibrium in excess EGTA allows assessment of the total chelator content of the cells, B, from the following equation[6]: $B = [Ca^T]_i \{K_B[(E-[Ca^T]_o/r^2K_E[Ca^T]_o] + 1\}$, where B is in micromoles/liter cells; E is the total extracellular concentration of EGTA (mM); K_B and K_E are the apparent dissociation constants of benz2 ($K_B \approx$ 50 nM) and of EGTA ($K_E \approx$ 40 nM at an external pH of 7.55 and in the presence of 0.15 mM Mg); $[Ca^T]_o$ is the total external Ca^{2+} concentration (mM); $r^2 = ([H^+]_i/[H^+]_o)^2$, and $[Ca^T]_i$ is the total cell Ca^{2+} content (in micromoles/liter cells).

These protocols offer only general guidelines and admit much variation within the restrictions and limitations outlined above.

Acknowledgments

We wish to thank the Wellcome Trust, the Medical Research Council, and EMBO for funding much of the research on which this chapter was based. We are also grateful to the British Council for travel grants.

[7] Recording Single-Channel Currents from Human Red Cells

By W. SCHWARZ, R. GRYGORCZYK, and D. HOF

Introduction

With the development of the improved patch-clamp technique[1] voltage-clamp experiments can now be performed on human erythrocytes.[2,3] This technique allows the measurement of net ion movements across a patch of the cell membrane under defined electrochemical conditions and the determination of parameters of conductance and gating at the level of the single transport molecule. The aim of this contribution is to point out the special features of the adaption of the patch-clamp technique to hu-

[1] O. P. Hamill, A. Marty, E. Neher, B. Sakmann, and F. J. Sigworth, *Pfluegers Arch.* **391**, 85 (1981).

[2] O. P. Hamill, in "Single Channel Recording" (B. Sakmann and E. Neher, eds.), pp. 451–471. Plenum, New York, 1983.

[3] R. Grygorczyk and W. Schwarz, *Cell Calcium* **4**, 499 (1983).

METHODS IN ENZYMOLOGY, VOL. 173

man erythrocytes. For a detailed description of the general features of the patch-clamp technique the reader is referred to the papers by Hamill *et al.*[1] and Sakmann and Neher,[4] and the book edited by Sakmann and Neher.[5] In this chapter we will also give a brief description of a computer system that supports performance and analysis of patch-clamp experiments, and we will present some results on ionic channels in the human erythrocytes which may help to illustrate how to optimize conditions for investigating single-channel events in these cells.

Methods and Materials

Solutions and Cell Preparation

The standard bath solution contains 150 mM KCl, 10 μM CaCl$_2$, and 5 mM of either 4-(2-hydroxyethyl)-1-piperazineethanesulfonic acid buffer (HEPES) or 4-morpholinopropanesulfonic acid buffer (MOPS) adjusted to pH 7.4 (solution I). For the investigation of anion conductances an NO$_3$ medium containing 150 mM NaNO$_3$, 10 mM KCl, 2.5 mM BaCl$_2$, and 5 mM HEPES adjusted to pH 6.9 (solution II) is usually used.

For the preparation of red blood cells a few drops of fresh blood are obtained by puncture and are washed three times in solution I or II and stored at about 10°. Immediately before an experiment a few cells are transferred to the test chamber (tissue culture dish) containing about 1 ml of bath solution.

Method of Patch Clamp

For the patch-clamp set-up any commercially available system may be used. In our laboratory we use the extracellular-patch-clamp system L/M EPC-5 (List-Electronic, Darmstadt, FRG). The suction pipet is brought into contact with an erythrocyte by means of a hydraulic micromanipulator (Narishige Sci. Inst. Lab., Tokyo, Japan) and an invertoscope (Zeiss, Oberkochen, FRG) at 400× magnification (Fig. 1).

Human erythrocytes are extremely small cells with a pronounced curvature. Therefore, for successful gigaseal formation suitably shaped micropipet are necessary. The best results are obtained with pipets made from thick-walled borosilicate glass (GC150, Clark Electromed. Instr., Reading, England). The pipets are pulled to give a resistance of 20–70 MΩ if filled with salt solution of 90% tonicity of solution I or II. Because of the low channel density in the erythrocytes membrane (see below), a large

[4] B. Sakmann and E. Neher, *Annu. Rev. Physiol.* **46,** 455 (1984).

[5] B. Sakmann and E. Neher (eds.), "Single Channel Recording." Plenum, New York, 1983.

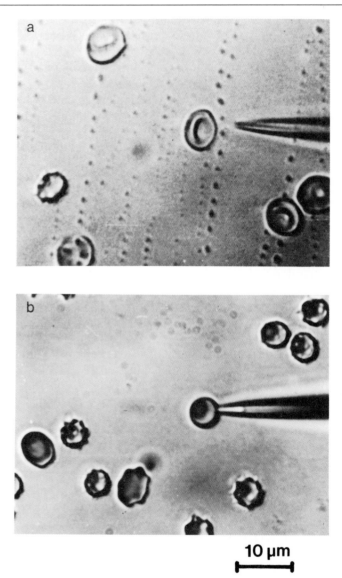

10 µm

FIG. 1. Light micrograph of erythrocytes and the suction pipet. (a) Erythrocyte before attachment to the pipet. (b) Erythrocytes after gigaseal formation; the light area in the tip of the pipet represents the portion of the erythrocyte drawn into the pipet by suction.

portion of the membrane should be sucked into the pipet (see Fig. 1b); in addition, for gigaseal formation a large contact area between the membrane and the glass surface should be established. This is guaranteed by using pipets which have large openings and long narrow shanks. A typical shape of the tip of the pipet is shown in Fig. 2. The tips are usually heat polished, but not coated with Sylgard. No enzymatic treatment of the cell surface is necessary to obtain seals of 5 to 100 GΩ.

Formation of Excised Membrane Patches

With human erythrocytes it is not possible to disrupt the patched membrane area to voltage-clamp the whole cell or to obtain excised membrane patches by conventional techniques. Nevertheless, excised membrane patches with the internal surface of the membrane in contact with the bath solution can be obtained routinely if after gigaseal formation the erythrocyte is brought into gentle contact with the bottom of the chamber. Usually after a few seconds the membrane touching the bottom disrupts and the cell disappears under the microscope due to leakage of the hemoglobin. The excised membrane configuration is verified by studying the effect of changing the bath solution on membrane currents. To allow fast changes of the solution in contact with the internal membrane surface of the excised patch the tip of the pipet is moved to the outlet of a fine tube that supplies fresh solution to the test chamber.

Data Recording

The currents from the voltage-clamped membrane can be recorded on analog magnetic tape Store-DS4 (Racal Recorders, Southampton, England). For the determination of current–voltage dependences, we preferred to perform experiments on-line by means of a PDP 11/23 computer (Digital Equipment, Maynard, MA) as described below. For further analysis of the records stored on tape, data were either transferred to a brush recorder (Gould Electronics, Cleveland, OH) and evaluated by hand, or were digitized and analyzed by computer.

Description of the On-Line Computer System

A data acquisition system based on the microcomputer PDP 11/23 was developed that is capable to generate voltage pulses and to record membrane signals synchronously.[6] The minimal configuration is shown in Fig. 3. For the terminal, a VT 100-compatible terminal with graphics compati-

---

[6] D. Hof, *Comput. Methods Program. Biomed.* **23**, 309 (1986).

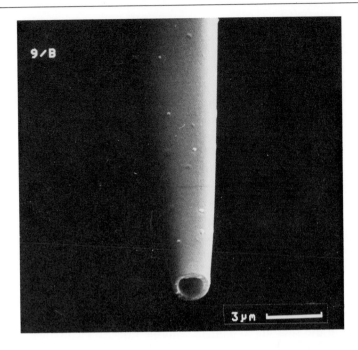

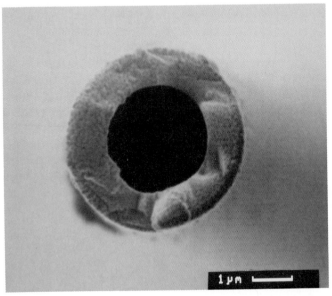

FIG. 2. Shape of suction pipet viewed under a scanning electron microscope demonstrating the shape recommended for gigaseal formation with human erythrocytes. *Top:* Side view of an unpolished pipet. *Bottom:* View on the pipet opening.

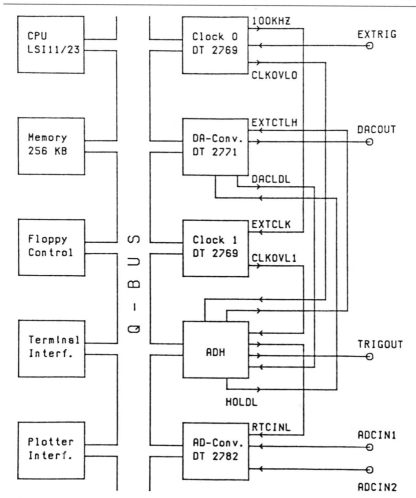

FIG. 3. Block diagram of the pulse-generating and data-recording system based on the microcomputer PDP 11/23. (From Hof[6] by permission). The 100-kHz signal generated by clock "0" is used as a common time base and is therefore fed to clock "1" (100 kHz → EXTCLK). The timing of pulse generation and data recording is controlled by CLKOVLO, which drives the DA converter as signal EXTCTLH (in DMA mode). The sampling clock CLKOVL1 is not directly connected to the clock input of the AD converter (RTCINL), but via a gating circuit on the ADH module. If the address-decoding circuit of the DA converter recognizes that the next word from the DA buffer has to be loaded into its data register, the module generates the DACLDL signal of the DA converter. In our concept, this signal is connected to the ADH module, where it serves to load bit 12 to 14 from the Q-bus to a register provided there. These bits are used in the following manner: BIT 12 (SLOWREC) decreases the selected sampling rate from CLKOVL1 by a factor selectable by a dip switch (standard factor 10), BIT 13 (ENREC) gates the sampling clock (RTCINL) to the AD converter, BIT 14 (TRIG) is connected to a BNC connector on the front panel and can be used to trigger the oscilloscope.

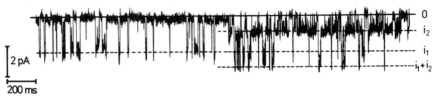

FIG. 4. Typical recording with two types of channels. The pipet and bath contain solution I with 50 mM KCl replaced by 50 mM NaCl. The holding potential is set to -150 mV. In addition to the closed state (0), the lines indicate the open states of the larger (i_1) and the smaller channel (i_2), and of the superposition of a smaller and a larger channel ($i_1 + i_2$).

ble with Tektronix 4010 is recommended. A floppy disk unit with two drives would be the minimal storage device; one drive is to supply the operating system and user programs, the other one is used for storage of data.

To use the computer as a pulse-generating and data-recording system, five Q-bus compatible modules are necessary (right components in Fig. 3), four of which were obtained from Data Translation (Marlboro, Massachusetts) and only slight modifications were necessary (see legend to Fig. 3). An additional hardware board (ADH) was designed for synchronization of the AD and DA conversions (for circuit diagram, see Hof[6]).

Analysis of Data

The computer was operated under the RT11 monitor, and all user programs were written in FORTRAN. For the hardware components from Data Translation the company supplies FORTRAN-callable subroutines that were used in the on-line programs. The off-line analysis of the data primarily consists of the interpretation of changes in the single-channel currents; for this, histograms of the current amplitudes or of the open and closed times are often useful. Appropriate programs were also written in FORTRAN.

Patch-Clamp Experiments on Red Cells

In our laboratory we have analyzed and identified two types of channels in the membrane of human erythrocytes.[3,7,8] An example of a record from an excised membrane patch with identical solution on both sides of the membrane is shown in Fig. 4. The two types of channels have a

[7] R. Grygorczyk and W. Schwarz, *Eur. Biophys. J.* **12,** 57 (1985).

[8] R. Grygorczyk, W. Schwarz, and H. Passow, *Biophys. J.* **45,** 693 (1984).

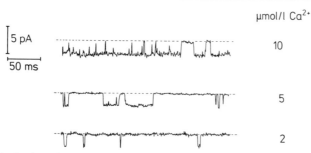

FIG. 5. Activation of the K^+ channels by Ca^{2+}. The pipet and bath contain solution I. The holding potential is set to -100 mV. The dotted lines indicate the closed state. (Redrawn from Grygorczyk and Schwarz.[7])

conductance of about 20 and 6 pS, respectively. Parallel to the patch-clamp experiments flux measurements with cell suspensions were usually performed with nearly identical composition of the solutions. The combination of the flux and current measurements is particularly useful since only direct comparison of these measurements allows conclusions about assignment of the currents to fluxes. In the following we will discuss the two types of channels and we will describe optimum conditions under which the channels can be investigated.

Ca^{2+}-Activated K^+ Channel

The first channel we discovered in the human erythrocyte was a channel that is activated if the intracellular Ca^{2+} activity increases to micromolar concentrations (see Fig. 5). The Ca^{2+} dependence of the probability of an open channel is similar to the Ca^{2+} dependence of K^+ permeability of the red cells determined by flux measurements (known as the Gardos phenomenon[9]). This dependence suggests that binding of two Ca^{2+} ions is necessary to open one K^+ channel. In addition, these channels exhibit a high selectivity for K^+.[7] These characteristics suggested that the Ca^{2+}-activated K^+ channels are the structures that are responsible for the so-called Gardos effect. Further essentially similar responses of fluxes and single-channel currents to the action of activators and inhibitors have confirmed this conclusion.[7,8,10,11]

[9] G. Gardos, *Biochim. Biophys. Acta* **30,** 653 (1958).
[10] M. Shields, R. Grygorczyk, G. F. Fuhrmann, W. Schwarz, and H. Passow, *Biochim. Biophys. Acta* **815,** 233 (1985).
[11] G. F. Fuhrmann, W. Schwarz, R. Kersten, and H. Sdun, *Biochim. Biophys. Acta* **820,** 223 (1985).

For the investigations of the Ca^{2+}-activated K^+ channels the selection of membrane potential and solutions may be optimized to achieve high single-channel activity and current. Maximum channel activity is obtained at Ca^{2+} concentrations exceeding 10 μM. Higher concentrations do not inhibit the activity; they have the welcome property of supporting gigaseal formation. Like other K^+ channels, single-channel conductance increases with increasing K^+; hence, if possible, experiments should be performed with 150 mM K^+ on both sides of the membrane. Under these conditions, a membrane potential of -100 mV gives single-channel currents of about 3 pA that can easily be analyzed. Reduction of the external K^+ to 10 mM reduces the single-channel current at a driving potential of -100 mV to 2 pA. The single-channel conductance as well as the single-channel activity shows inward rectification; therefore, alterations of these parameters can be analyzed best at negative membrane potentials. Channel activity can be investigated not only during voltage pulses but also during long-lasting membrane polarizations, because the K^+ channels do not exhibit voltage-dependent inactivation.

Anion-Selective Channels

In addition to the events related to the channels described above smaller events (compare Fig. 4) are occasionally observed. These pertain to a second type of channel which can be analyzed under conditions where the Ca^{2+}-activated K^+ channels are inhibited. This is easily achieved by adding 2.5 mM $BaCl_2$ to the pipet and bath solutions. Ba^{2+} blocks the K^+ currents and in addition improves gigaseal formation. In contrast to the Ca^{2+}-activated K^+ channels, single-channel conductance for anions does not show rectification. Nevertheless, if the experimental protocol permits, negative test potentials are recommended; at positive membrane potentials channel openings become a very rare event. Under the described conditions single-channel currents can be recorded that are most likely carried by anions. Figure 6 shows that replacement of Cl^- by NO_3^- increases the single-channel current by about 50%. Comparison of the effects of several inhibitors of the single-channel events and of anion fluxes suggests that these currents may be responsible for an H_2DIDS(4,4'-diisothiocyanostilbene-2,2'-disulfonate)-sensitive net flux component.

Number of Channels per Red Cell

From the rate of K^+ efflux the permeability of a single erythrocyte can be estimated. The single-channel conductance can be translated into a single-channel permeability, and the probability of a channel to be in the

(NO$_3^-$ medium, V$_H$ = -80 mV)

a

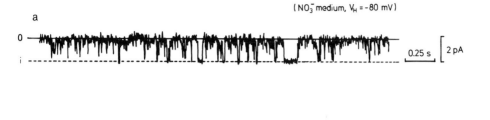

0.25 s   2 pA

(Cl$^-$ medium, V$_H$ = -120 mV)

b

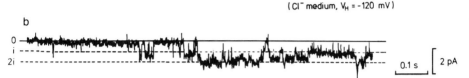

0.1 s   2 pA

FIG. 6. Anion-selective channels. The pipet and bath contain solution I in (a) and solution II in (b). The solid lines indicate the closed state, dotted lines the open states of one (*i*) or two (2*i*) channels.

conducting state can be estimated. From these factors, we can determine the number of K$^+$ channels that can be activated by Ca^{2+} in the intact erythrocyte. The channel density turns out to be only of the order of about 10 channels per cell.[8]

The density of the anion channels is also very low, but it is more difficult to give an estimate of the number of anion channels per red cell than of the number of the K$^+$ channels. This is due to the fact that within the patched membrane area several channels are usually simultaneously active. Hence, only a rough estimate of an upper limit can be made for the probability of a channel to be open. Based on such estimations, the number of anion channels per cell should be in the order of about 100.

Acknowledgments

We thank Drs. G. F. Fuhrmann and H. Passow for valuable comments on the manuscript, and Dr. K. Rascher and Dipl. Biol. B. Schweigert for the scanning micrographs.

[8] Identification of Amino Acid Transporters in the Red Blood Cell

By CATHERINE M. HARVEY and J. CLIVE ELLORY

Introduction to Red Cell Amino Acid Transport

As early as 1913 Van Slyke and Meyer[1] pointed out that animal tissues, specifically the liver, could contain amino acids at higher concentrations than surrounding tissue fluid, and could accumulate amino acids against concentration gradients. Subsequently Christensen *et al.*[2] found that the presence of various amino acids could affect the distribution of other amino acids between cells and extracellular fluid. It was from the detailed study of these effects, and the application of enzyme kinetics, that the definition of distinct transport systems for amino acid types began to emerge.[3] It became apparent that the kinetic properties of amino acid transport systems and stereoselectivity could not be accounted for by passive permeability.

Amino acid requirements of cells and tissues depend on many factors including differentiation and growth (with an obvious role for hormones and growth factors), protein synthesis, osmoregulation, neurotransmission, gluconeogenesis, GSH biosynthesis, and other specific metabolic pathways, to name a few. The ways in which these needs might modify transport of amino acids into and out of cells are clearly complex, and are the subject of recent interest in other cell systems.[4] How do red cells fit into the picture?

Red cells provide a unique experimental preparation in that they are easily obtainable as a homogeneous isolated cell preparation. It is not surprising therefore that the red cell is the obvious choice as an experimental paradigm to provide detailed kinetic information on the amino acid transport systems. Compared with most tissues the red cell has a simple metabolism, especially in the case of mammals, since the machinery for protein synthesis is often absent; there are clear advantages in using a tissue where the transported substance is not the subject of rapid metabolism or sequestration.

---

[1] D. D. Van Slyke and G. M. Meyer, *J. Biol. Chem.* **16**, 197 (1913–1914).
[2] H. N. Christensen, J. A. Streicher, and R. L. Elbinger, *J. Biol. Chem.* **172**, 515 (1948).
[3] D. L. Oxender and H. N. Christensen, *J. Biol. Chem.* **238**, 3686 (1963).
[4] M. H. Saier, G. A. Daniels, P. Boerner, and J. Lin, *J. Membr. Biol.* **104**, 1 (1988).

Several reasons have been proposed[5,6] for the existence of amino acid transport systems in red cells and it is useful to consider them here: (1) they might be functional relics of the large requirements for amino acids during reticulocyte development; (2) as proposed by Elwyn[7] and discussed further by Christensen,[8] the red cell could play a role in interorgan amino acid transport; (3) amino acids might be "accidental" substrates of systems designed to transport other substances such as band 3 protein[9]; (4) red cells have a requirement for amino acids for glutathione biosynthesis, a vital part of red cell metabolism. Cysteine, glycine, and glutamate are the relevant amino acids, and transport systems for cysteine and glycine have been characterized in some detail. It is interesting to note though that, at present, the transport of glutamate remains unresolved, at least for species other than carnivores.

As essentially dead-end cells which are highly adapted to carrying oxygen and carbon dioxide, to the exclusion of many other normal cell functions, red cells can tolerate changes in intracellular levels of certain amino acids which would be lethal for other tissues. The expression of amino acid transport systems and their evolutionary survival in red cells should not lead us to extrapolate carelessly to other tissues. However, the information we can obtain from experiments on red cells provides high-quality kinetic data which allow us to model the behavior of carriers in biological membranes.[10] This emphasizes an important role for red cells for studying transport systems at the fundamental level. Information from these experiments can give predictions and kinetic information for other transport systems and in other tissues.

Nomenclature and Types of Amino Acid Transporters

Amino acid transporters can be classified broadly into four types, in terms of kinetic mechanism, driving force, and complexity. The simplest are carriers of the facilitated diffusion type, capable of net or exchange fluxes. Band 3 protein is one of the best studied examples of this kind of transporter and can be considered an "honorary" amino acid transporter

[5] J. D. Young, and J. C. Ellory, in "Membrane Transport in Red Cells (J. C. Ellory and V. L. Lew, eds.), p. 301. Academic Press, New York, 1977.
[6] J. C. Ellory, in "Amino Acid Transport in Animal Cells" (D. L. Yudilevich and C. A. R. Boyd, eds.), Physiological Society Study Guide Vol. 2, p. 106. Manchester Univ. Press, 1987.
[7] D. H. Elwyn, Fed. Proc., Fed. Am. Soc. Exp. Biol. **25,** 854 (1966).
[8] H. N. Christensen, Physiol. Rev. **62,** 1193 (1982).
[9] J. C. Ellory, S. E. M. Jones, and J. D. Young, J. Physiol. (London) **320,** 403 (1981).
[10] R. Rosenberg, Biochim. Biophys. Acta **649,** 262 (1981).

since it transports glycine and sarcosine. The exchange mode of operation can be useful biologically since it can exchange a common substrate, present at high concentration, for a relatively rare substrate. Human L and T systems, and several members of the sheep, horse, and hagfish C/asc family of transporters fall into this category. More complex systems represent secondary active transport and are Na^+ dependent. The simplest case is for a stoichiometry of 1 Na^+ : 1 amino acid (aa) molecule such as the human ASC system. These systems can still operate in an exchange mode, even though they are predominantly driven by the Na^+ gradient. The next level of complexity concerns Na^+-dependent systems with a stoichiometry >1; the Na^+ : Glu system in dog and cat red cells is of this kind. The transport of >1 Na^+ : Glu gives a larger gradient to drive Glu accumulation. Finally some systems (Gly and β) have a requirement for Cl^- as well as Na^+. The most likely stoichiometry for binding is 1 Cl^- : 1 aa : 2 Na^+, although the situation is complex (see below). Figure 1 represents schematically the substrate specifications and Na^+ dependence of the known red cell transport systems.

The nomenclature of amino acid transport systems evolved as they were characterized in terms of affinities and Na^+ dependence. Most were named acronymically, based on substrate specificity, e.g., ASC (alanine,

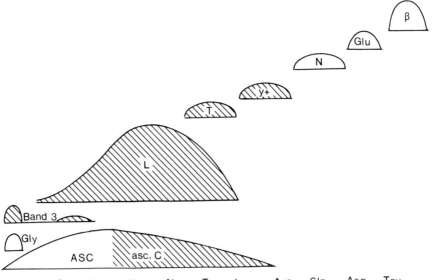

Pro   Ser   Val   Met   Ile   Tyr   Lys   Arg   Gln   Asp   Tau
Gly   Ala   Cys   Thr   Leu   Phe   Trp   Orn   His   Asn   Glu   ß-Ala

Fig. 1. Diagrammatic representation of amino acid transport systems in red cells. Hatched areas indicate Na^+ independent. (Modified after Ellory.[6])

cysteine, and serine), Ly$^+$ (lysine). A formal declaration was published in 1984, which set out criteria for naming amino acid transporters.[11]

Systems for basic amino acids, in the cationic form, are designated y$^+$ and in fact this is now the accepted label for the system previously known as Ly$^+$. For anionic amino acids x$^-$ is to be used. For both y$^+$ and x$^-$ systems a superscript is added, if desired, to denote preferred substrates, e.g., X^{-Glu}. Sodium dependence is denoted by capital letters, e.g., ASC versus asc, although certain earlier notations which do not fit with the present rules remain, e.g., L and T. Otherwise it is recommended by Bannai *et al.*[11] that transporters are characterized fully before being coded.

The Red Cell Amino Acid Transport Systems

One of the greatest problems confounding the separation and description of amino acid transport systems is the absence of specific inhibitors; investigators have depended almost entirely on functional criteria, and the fact that transporters have wide specificity, which might vary with species or tissue, complicates the task. It has been shown for example that glycine is transported by five separate routes in human red cells (see Fig. 10).[9] The problems of separating systems tend to be approached solely by kinetic analysis and interpretation, which are not always rigorous, as will be seen later.

System L

This Na$^+$-independent transport system has the widest tissue and species distribution, and was identified in red cells by Winter and Christensen.[12] Almost all neutral amino acids are transported by system L, even if they are preferred substrates of other systems. It has the highest capacity of human red cell amino acid transporters, which allows significant fluxes for even poor substrates. The system operates principally in the exchange mode under physiological conditions.[3] The human red cell L system has been the subject of detailed kinetic analyses, which will be discussed later.[13-16]

[11] S. Bannai, H. N. Christensen, J. V. Vadgama, J. C. Ellory, E. Englesberg, G. G. Guidotti, G. C. Gazzola, M. S. Kilberg, A. Lajtha, B. Sacktor, F. V. Sepulveda, J. D. Young, D. L. Yudilevich, and G. E. Mann, *Nature (London)* **311**, 308 (1984).
[12] C. G. Winter and H. N. Christensen, *J. Biol. Chem.* **239**, 872 (1964).
[13] D. G. Hoare, *J. Physiol. (London)* **221**, 311 (1972).
[14] D. G. Hoare, *J. Physiol. (London)* **221**, 331 (1972).
[15] R. Rosenberg, *J. Membr. Biol.* **62**, 79 (1981).
[16] A. R. Walmsley and A. G. Lowe, *Biochim. Biophys. Acta* **901**, 229 (1987).

System T

This Na^+-independent system in human red cells shows significant transport of the aromatic amino acids tryptophan, tyrosine, and phenylalanine. It was identified as a component of L-tryptophan uptake which cannot be due to the L system,[17] and was subsequently characterized further.[18] It is the major route for tryptophan transport at physiological substrate concentrations, and shows a weak response to trans stimulation. Tryptophan is important for serotonin synthesis in the central nervous system, and the transfer of tryptophan across the blood brain barrier has been a subject of great interest. This amino acid is one candidate for interorgan transport by red cells. The presence of the T system, in addition to the L system, in hepatocytes has been shown recently.[19] Because of differences in the affinities of the two systems for tryptophan, the contribution of the L system to transport becomes proportionally larger as the concentration of tryptophan increases. This means that, in the liver at least, the L system amino acids are more effective inhibitors of aromatic amino acid uptake at higher than at lower concentrations of the latter.[19] It may be that system L has a high enough capacity for tryptophan in some cell types. The L system in human red cells is somewhat more selective than in other tissues and the failure of tryptophan to inhibit the L system might reflect the need for a specific aromatic amino acid transporter in human red cells.

System y^+

Gardner and Levy[20] suggested the presence of a saturable Na^+-independent uptake route for dibasic amino acids in human red cells, but no attempt was made to relate such uptake to that of other amino acids. Antonioli and Christensen[21,22] found dibasic amino acid transport in rabbit erythrocytes and at an increased level in reticulocytes. The properties of the system were investigated in human red cells and compared with the C system in sheep.[23] The y^+ system (previously known as Ly^+) is highly stereoselective. Unusually for a Na^+-independent system, it has high affinity for substrate and is specific for the basic amino acids lysine, ornithine, and arginine. Recent investigations show that the system can

[17] J. D. Young and J. C. Ellory, *J. Neural Transm. Suppl.* **15**, 139 (1979).

[18] R. Rosenberg, J. D. Young, and J. C. Ellory, *Biochim. Biophys. Acta* **598**, 375 (1980).

[19] M. Salter, D. A. Bender, and C. I. Pogson, *Biochem. J.* **233**, 499 (1986).

[20] J. D. Gardner and A. G. Levy, *Metabolism* **21**, 413 (1972).

[21] J. A. Antonioli and H. N. Christensen, *J. Biol. Chem.* **244**, 1505 (1969).

[22] H. N. Christensen and J. A. Antonioli, *J. Biol. Chem.* **244**, 1497 (1969).

[23] J. D. Young, S. E. M. Jones, and J. C. Ellory, *Proc. R. Soc. London, Ser. B.* **209**, 355 (1980).

operate in an exchange mode for these substrates and *in vivo* this is probably dominant.[24] Neutral amino acids with Na^+ can mimic basic amino acids in their reactivity with this transporter.[25]

System ASC

Formally identified in the Ehrlich ascites tumor cell, the system was shown to be a third route for alanine transport which could not be accounted for by systems A and L.[26] A similar system was shown in rabbit reticulocytes[27] and pigeon erythrocytes.[28] The ASC system was distinguished from the A system by the following: exceptional stereoselectivity, less pH sensitivity, and the inability of MeAIB to inhibit substrate uptake. It is present in erythrocytes in the absence of the A system.

In human red cells the Na^+-dependent ASC system is the principal route for the entry of cysteine at physiological levels extracellularly ($K_m = 15 \ \mu M$). As for this system in other tissues, there is also a significant affinity for alanine and serine (giving rise to the acronym). In red cells, except human, the system operates predominantly in the exchange mode.[29] A coupling ratio of 1 : 1 for Na^+ : alanine has been demonstrated but the system does not appear to be electrogenic since it is not affected by changes in membrane potential.[30] ASC prefers amino acids with three to five carbon atoms in the chain (i.e., alanine, serine, cysteine) but does not tolerate an N-methylation (in contrast to system A). System ASC appears to be ubiquitous in tissues of higher animals, can operate uphill, and responds to trans stimulation. It has an absolute requirement for sodium, being intolerant to Li^+ substitution for Na^+.[31]

Systems C and asc

System C is the only amino acid transport system present in sheep red cells, apart from band 3 protein.[32] Substrate specificities are similar to those of ASC but affinities are typically lower: K_m values for cysteine and

[24] J. C. Ellory, F. C. Fervenza, C. M. Harvey, and B. M. Hendry, *J. Physiol. (London)*, **407**, 11P (1988).
[25] H. N. Christensen, M. E. Handlogten, and E. L. Thomas, *Proc. Natl. Acad. Sci. U.S.A.* **69**, 948 (1969).
[26] H. N. Christensen, D. L. Oxender, M. Liang, and K. A. Vatz, *J. Biol. Chem.* **240**, 3609 (1965).
[27] C. G. Winter and H. N. Christensen, *J. Biol. Chem.* **240**, 3594 (1965).
[28] E. Eavenson and H. N. Christensen, *J. Biol. Chem.* **242**, 5386 (1967).
[29] E. A. Al-Saleh and K. P. Wheeler, *Biochim. Biophys. Acta.* **684**, 157 (1982).
[30] J. C. Ellory, S. E. M. Jones, T. J. Rink, M. W. Wolowyk, and J. D. Young, *J. Physiol. (London)* **308**, 52 (1980).
[31] E. L. Thomas and H. N. Christensen, *J. Biol. Chem.* **246**, 1682 (1971).
[32] J. D. Young and J. C. Ellory, *Biochem. J.* **162**, 33 (1977).

alanine on system C are 12 and 16 mM, respectively, but on ASC they are 15 and 150 μM. System C also has a significant but low affinity for ornithine and lysine but not for arginine. It has been suggested that C might be a modified ASC system without the Na^+ dependence and the ability to transport the basic amino acids might reflect the presence of a cationic binding site representing a redundant Na^+ site.[6] This view is further supported by the identification of the Na^+-independent asc in horse red cells, with the same substrate specificities as ASC and C.[33] System asc also operates preferentially in the exchange mode like ASC (except in human erythrocytes[29]). Genetic variants of sheep and horses have been identified. In the case of sheep it is a diallelic system, and transport-negative and heterozygous phenotypes have been identified.[34,35] The horse system is more complex, but has been identified as triallelic.[33,36] Another similarity of asc and C is the capacity for dibasic amino acid transport. In this case, the cells which lack these transport systems (revealed in transport-negative phenotypes) accumulate high intracellular levels of dibasic amino acids. There seem to be many similarities between systems ASC, C, and asc; perhaps these represent expression of a family of transport systems in different species.

System N

The existence of this system was suggested by Joseph *et al.*[37] and identified formally in rat hepatocytes as a Na^+-dependent uptake of glutamine, histidine, and asparagine.[38] Subsequently, the presence of this system was shown in human red cells[39] and skeletal muscle.[40] Originally it was thought system N could be a candidate for the source of glutamate for red cell glutathione biosynthesis but in fact this cannot be the case, since there is no detectable intracellular glutaminase activity. Glutamine is a limiting substrate for intracellular parasite growth in malaria-infected red cells, and the parasite induces a massive increase in glutamine transport into the erythrocyte.[41]

[33] D. A. Fincham, D. K. Mason, and J. D. Young, *Biochem. J.* **227,** 13 (1985).
[34] E. M. Tucker, L. Kilgour, and J. D. Young, *J. Agric. Sci.* **87,** 315 (1976).
[35] J. D. Young, E. M. Tucker, and L. Kilgour, *Biochem. Genet.* **20,** 723 (1982).
[36] D. A. Fincham, D. K. Mason, and J. D. Young, *Biochim. Biophys. Acta* **937,** 184 (1988).
[37] S. K. Joseph, N. M. Bradford, and J. D. McGivan, *Biochem. J.* **176,** 827 (1978).
[38] M. S. Kilberg, M. E. Handlogten, and H. N. Christensen, *J. Biol. Chem.* **255,** 4011 (1980).
[39] J. C. Ellory and B. Osotimehin, *J. Physiol.* (*London*) **348,** 44P (1984).
[40] H. S. Hundal, P. W. Watt, and M. J. Rennie, *Biochem. Soc. Trans.* **14,** 1070 (1986).
[41] B. C. Elford, J. D. Haynes, J. D. Chulay, and R. J. M. Wilson, *Mol. Biochem. Parasitol.* **16,** 43 (1985).

System Gly

This was characterized formally in avian erythrocytes by Vidaver.[42] Its substrates are glycine and sarcosine, and it has a requirement for Na^+. The curve for Na^+ dependence of uptake is not a simple hyperbola but is sigmoid, which is consistent with more than one sodium binding for each amino acid. The characterization of this system in mammalian erythrocytes[9] reveals the further complexity of an absolute requirement for Cl^- ions. The system operates to perform net transport depending on the Na^+ gradient, and does not perform exchange fluxes. As the smallest neutral amino acid, glycine is a potential substrate for most of the mammalian systems, and discrimination could then prove difficult. Ellory used both Na^+ dependence and anion sensitivity as well as selected competitive inhibitors to identify the contribution of system Gly to uptake in human red cells.

System β

This Na^+-dependent system for β-amino acids has been identified in avian and fish red cells.[28,43] The substrates are taurine and β-alanine, the former of which, at least in fish, plays a role in red cell volume regulation. The system has high affinity for taurine ($K_m < 60 \ \mu M$) and although the sodium dependence is sigmoid, in a recent study a coupling ratio of 1 : 1 was measured for transport.[44] Exchange has not been detected.

System X⁻

Although most species' red cells are essentially impermeable to L-glutamate, carnivore red cells possess a high-affinity Na^+-dependent system for the anionic amino acids aspartate and glutamate.[45,46] An interesting study of an increase in glutamate transport in dogs has shown raised intracellular GSH and changes in activity of the Na^+,K^+-ATPase in these cells.

Malaria Induced

Although strictly outside the scope of this chapter, it is worth mentioning the large induced transport pathway found in human red cells infected

[42] G. A. Vidaver, *Biochemistry* **3**, 662 (1964).
[43] D. A. Fincham, M. W. Wolowyk, and J. D. Young, *J. Membr. Biol.* **96**, 45 (1987).
[44] C. M. Harvey and J. C. Ellory, *Biochem. Soc. Trans.* **16**, 555 (1988).
[45] J. C. Ellory, S. E. M. Jones, R. L. Preston, and J. D. Young, *J. Physiol. (London)* **320**, 79P (1982).
[46] M. Inaba and Y. Maede, *J. Biol. Chem.* **259**, 312 (1984).

with malaria (*Plasmodium falciparum*). The parasite is limited for certain amino acid substrates, particularly glutamine, and induces a transporter with up to a 100-fold increase in glutamine uptake. This system shows specificity for size and charge but has not been shown to exhibit any carrier kinetics (saturation, exchange, or stereospecificity). It may be best considered as a selective pore.[47]

Flux Measurements

Preparation of Red Cells

Blood samples can be taken into heparinized tubes, EDTA, or citrate. Care should be taken when using heparin at high concentrations since it usually contains formalin and has a low pH, which will acidify blood or substrates. A dose of 1000–5000 units/0.5 liters blood is sufficient. It is usually desirable to wash red cells out of plasma using isosmotic saline. It is also important to remove other cells (platelets, white cells, and reticulocytes), since a number of biochemical activities ascribed to red cells (e.g., glutaminase) are due to contamination with other cell types. Therefore, after centrifugation the buffy coat and upper layer of red cells should be removed by aspiration. It is also possible to remove contamination of these cells by filtration through synthetic wool. A column of washed cotton wool is prepared in a 5-ml syringe and the cells passed slowly through it by gravity.[48] If incubation in plasma is required, e.g., as in recent studies on the effects of plasma factors accumulating in renal failure on amino acid transport, then it is recommended to wash the cells and resuspend in plasma having removed the buffy coat.[49]

Washing red cells at room temperature also has the effect of removing intracellular amino acids differentially, and it cannot be assumed that such washing procedures will completely deplete the cells of amino acids.[24] If the transport system investigated operates in an exchange mode this could result in spurious estimates of kinetic parameters. (An example is given in Fig. 2.) If the intracellular contents are to be maintained, it is preferable to wash at 4° with ice-cold saline, and to keep cells on ice until required. For transport experiments it is important to use freshly drawn

[47] H. Ginsberg, S. Kutner, M. Krugliak, and Z. I. Cabantchik, *Mol. Biochem. Parasitol.* **14,** 313 (1985).

[48] J. Stuart, P. C. W. Stone, D. Bareford, N. M. Caldwell, J. E. Davies, and S. Baar, *Clin. Hemorrh.* **5,** 137 (1965).

[49] F. C. Fervenza, C. M. Harvey, B. M. Hendry, and J. C. Ellory, *Clin. Sci.* **76,** in press (1989).

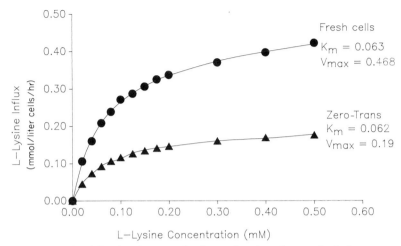

FIG. 2. L-Lysine influx into human red cells as a function of external L-lysine concentration. Zero-trans conditions were achieved after a 4-hr incubation of washed cells in saline.

blood, since there are several examples of changes in transport properties for stored or blood-bank blood. If storage is essential, blood samples can be stored for longer in Alsever's solution (see Solutions) but this is to be avoided if possible.

Routinely, three washes in 20 vol saline : 1 vol cells are sufficient and cells sedimented (10 min at 2000 g) using a low-speed centrifuge (refrigerated for preference). Although superficially a homogeneous population, in fact blood contains red cells of varying age. For both intracellular enzyme activity and membrane transport, there are examples of variations in populations of different age. Further, there are species differences in terms of circulating reticulocytes and young red cells. For example in the rabbit there are normally 3% but in the monkey there are less than 0.1%.[50] In humans, the red cell population distribution can be different for various clinical conditions, e.g., hemochromatosis. In fact there is an extensive literature on the changes in properties of transport systems with erythrocyte maturation.[21,22,27,51,52] However, having pointed out that this phenomenon occurs for many transport systems investigated so far, it should be emphasized that the fluxes for whole blood red cell populations are ex-

[50] E. M. Tucker and J. D. Young, in "Red Cell Membranes: A Methodological Approach" (J. C. Ellory and J. D. Young, eds.), p. 31. Academic Press, London, 1982.
[51] E. M. Tucker and J. D. Young, Biochem. J. 192, 33 (1980).
[52] R. M. Johnstone, M. Adam, and C. Turbide, J. Biol. Chem. 262, 9412 (1987).

tremely consistent both within and between individuals. All kinetic work to date on all red cell transport systems (including the sodium pump and glucose transporter) has been performed on unfractionated blood.

The density of red cells increases with age and so most separation techniques involve centrifugation and separation according to density. A relatively quick and simple method involves the use of a continuous density gradient of a colloidal silica sol, Percoll.[53] The self-forming gradient (stable at room temperature) is generated by centrifuging the Percoll solution (25,000 g, 15 min, 20°), and modifications of conditions will give different gradients. Washed blood in buffered saline is layered on the gradient and centrifuged (800 g, 10 min, room temperature). Collection of cells is achieved by upward displacement with 60% sucrose solution, and the fractions washed in saline to remove Percoll solution.

Cells should be resuspended finally in incubation medium and kept at 4° until required, and for the purposes of flux measurements a final hematocrit of 5–10% is desirable.

Solutions and Incubation Media

Preliminary washing and incubation of red cells can be performed in isosmotic saline buffered to pH 7.4. Phosphate buffer (10 mM), or one of the "Good" buffers (see Ref. 56) is preferred because of its low temperature coefficient. Fluxes in mammalian red cells are usually performed at 37°, although the body temperature of some species can be higher (in birds up to 45°), and with solutions whose pH was set at room temperature.

Incubation salines resemble plasma, and usually contain either 150 mM NaCl, 5 mM KCl, 5 mM glucose, 10 mM MOPS or 140 mM NaCl, 15 mM Na$_2$HPO$_4$; NaH$_2$PO$_4$, buffered to pH 7.4 at 37°. If cells are to be incubated at 37° for a significant period (>1 hr), it is desirable to change the medium periodically to preserve metabolism. Alsever's solution is an excellent anticoagulant for use in the long-term storage of blood, up to 3 weeks at 4°.[54] This comprises trisodium citrate (31 mM), sodium chloride (72 mM), glucose (114 mM), citric acid (3 mM), to pH 6.1. This is used in 1:4 ratio with blood.

Since Na$^+$ dependence is an important criterion for identifying certain amino acid transporters, it is necessary to define an appropriate Na$^+$ substitute for characterizing fluxes. Substitution of sodium is often achieved with choline, though this is not ideal unless the choline is puri-

[53] M. W. Wolowyk, in "Red Cell Membranes: A Methodological Approach" (J. C. Ellory and J. D. Young, eds.), p. 1. Academic Press, London, 1982.

[54] E. M. Tucker and J. C. Ellory, in "Red Cell Membranes: A Methodological Approach" (J. C. Ellory and J. D. Young, eds.), p. 95. Academic Press, London, 1982.

fied by recrystallization from hot ethanol, and checked by osmometry for deliquescence. Simpler alternatives include potassium salts, or N-methyl-D-glucamine (NMDG), a pure crystalline base. When varying the sodium concentration, care must be taken to ensure that changes in chloride concentrations or osmolarity do not occur. A useful example to consult is the sodium substitution for flux measurements on a Na^+–Cl^--dependent system.[43] Chloride replacement should be undertaken with similar caution with regard to osmolarity and maintenance of sodium concentration. Replacement of chloride can be achieved using methyl sulfate, which is available as the free acid.[55] Fine adjustments of osmolarity can be made using sucrose. Similar considerations with regard to osmolarity and pH apply to kinetic studies where changes in concentrations of substrate or competitor amino acids greater than 10 mM occur. In this case, it is possible to adjust osmolarity by reducing the concentration of NaCl, but fluxes might also be reduced accordingly. In preference, then, sucrose should be added to maintain a constant tonicity.

Finally, at the end of incubations, when using the rapid washing technique, cells are washed free of extracellular label with ice-cold wash medium. This consists of 107 mM $MgCl_2$ and 10 mM Tris–HCl (pH 7.4).

Use of Isotopes

The movement of amino acids across red cells is usually followed by using radioactive isotopes; most amino acids can be purchased with ^3H, ^{14}C, or ^{35}S labels. It is important to store these carefully since degradation could result in spurious results from the formation of more permeable substances, e.g., ^3H_2O. It is suggested that sterility be maintained and aliquots kept frozen separately until required. If isotopes are kept for a long time it is desirable to freeze-dry them before use. The exact amount of tracer to be used depends upon the hematocrit and the magnitude of the flux to be measured, but as a general guide 0.1 μCi for ^{14}C and 1 μCi for ^3H per milliliter will be sufficient. Certain amino acids, e.g., tryptophan and histidine, are more labile than others. Similarly, there can be differences in stability between ^3H- and ^{14}C-labeled products. The former are usually cheaper and less stable, but present a higher specific activity. Some uniformly ^{14}C-labeled amino acids have a significant increase in molecular weight due to the number of ^{14}C atoms per molecule. Figure 3 shows an example of spurious data for degraded glycine.

For Na^+-dependent systems parallel measures of Na^+ uptake and labeled amino acid uptake can be made using ^{22}Na or ^{24}Na at 0.2 μCi/ml.

[55] P. B. Dunham, G. W. Stewart, and J. C. Ellory, *Proc. Natl. Acad. Sci. U.S.A.* **77,** 1711 (1980).

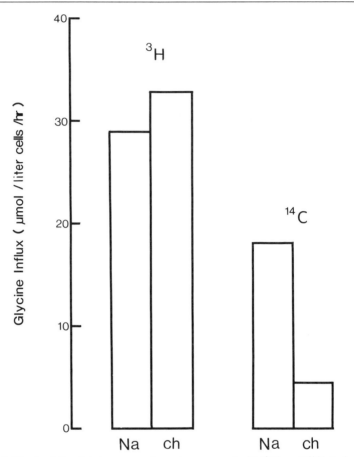

FIG. 3. Glycine influx into human erythrocytes measured with [³H]- and [¹⁴C]glycine. The [³H]glycine had been stored for 6 months at 4°, and was contaminated. Fluxes were measured in Na⁺ and choline (ch) media.

For rapid flux measurements, an extracellular space marker is used, usually [³H]inulin with the simultaneous ¹⁴C-labeled substrate amino acid. The two isotopes can be separated by standard β-scintillation spectrometry. The exact tracer ratio will vary but normally [³H]inulin at 1.0 μCi/ml and ¹⁴C-labeled amino acid at 0.2 μCi/ml is satisfactory.

Experimental Protocol

The principle behind flux measurements is the rapid separation of cells from medium. For influx measurements this can involve either a repeated washing procedure, or centrifugation through a distinct phase, of intermediate density between cells and incubation medium. Flux experiments can be conveniently performed in standard 1.5-ml polypropylene conical (Eppendorf) centrifuge tubes. It is usual to work to a final volume of 1 ml, with 0.05–0.1 ml cells/tube (i.e., a hematocrit of 5–10%).

Influx. For the fluxes with a relatively long time course it is possible to start by adding aliquots of prewarmed cell suspension to aliquots of incubation medium containing the isotope. Incubation at required temperature proceeds for an appropriate time, with gentle mixing by inversion every 10 min to prevent settling of cells. Incubation can then be stopped, by rapid centrifugation (12,000 *g*, 15 sec), aspiration of supernatant, and addition of an aliquot of ice-cold washing medium. For shorter times (5–15 min) or large batches of tubes it is more convenient to mix ice-cold cell suspension and isotope in tubes in an ice bath and to start the flux by transferring them to 37°; fluxes are stopped by returning the tubes to an ice bath. The cooling curves are symmetric and equilibrate within 2 min (see Fig. 4).

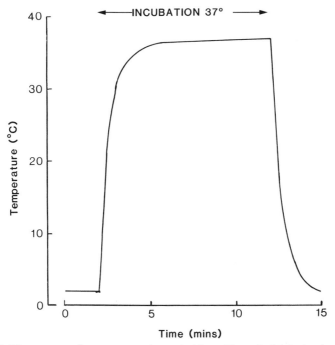

FIG. 4. Time course of temperature change inside an Eppendorf 1.5-ml polypropylene tube. Measured on transfer from an ice bath to a 37° water bath and back again. Equilibration is within 2 min. (From Young and Ellory.[56])

At the end of the incubation cells are washed by addition of 1 ml of ice-cold $MgCl_2$ medium and resuspension using a vortex mixer. Sedimentation between washes can be achieved by centrifugation at 12,000 g for 10 sec using bench microcentrifuges (e.g., Eppendorf 5415). The procedure can be used easily for batches of 12 or 18 tubes; washing this number of tubes 4 times with 5 spins should take less than 10 min. Extracellular space is negligible and so this technique is simple in that no extracellular space marker is required. The use of the washing procedure is limited to flux times of 5 min or more. Rapid fluxes, where incubation times are less than 5 min, should be stopped by spinning a sample of the suspension through wash solution layered on 0.5 ml n-dibutyl phthalate (12,000 g for 10–20 sec). A cohort of these separation tubes can be prepared in advance and placed in centrifuges, allowing the separation of cells from isotope after only a few seconds. The net trapped extracellular space is 3–5%, which can be estimated using an extracellular space marker such as [^{3}H]inulin. This allows simultaneous estimates of uptake and space if ^{14}C-labeled substrate is used. After separating the cells through oil, the upper layers can be removed by careful aspiration and the sides of the tubes dried with, for example, dental tampons, and the tubes inverted. Although dibutyl phthalate is appropriate in terms of density for human red cells, other species, e.g., fish, may differ. It is possible to use other phthalate esters, mixtures of dibutyl phthalate and corn oil, or various silicone oils as other separation media of different density.

In all cases the final cell pellets are lysed with 0.5 ml 0.1% (w/v) Triton X-100 in water and protein precipitated by the addition of 0.5 ml 5% (w/v) trichloroacetic acid (TCA), and pelleted by centrifugation (12,000 g, 2 min). The supernatant can then be transferred to scintillation fluid for counting.[56]

Results are usually expressed in terms of micromoles per milliliter cells per minute. The calculation of micromoles comes directly from specific activity of the isotope and amino acid concentration used, determined by counting an aliquot of either the tube supernatant at the end of the initial centrifugation or an aliquot of the concentrated, labeled medium originally added. Under most circumstances, the contribution of tracer to the total concentration can be ignored, but at low concentrations it can be significant. The volume of cells per tube can be determined either from Coulter counter measurements of an appropriately diluted sample of the cell suspension used for flux measurements, or a microhematocrit

[56] J. D. Young and J. C. Ellory, *in* "Red Cell Membranes: A Methodological Approach" (J. C. Ellory and J. D. Young, eds.), p. 119. Academic Press, London, 1982.

determination, or from hemoglobin measurements. In fact it is usually convenient to use hemoglobin, measured in Drabkin's solution as cyanmethemoglobin, which can be converted to packed cell equivalent by the relationship $OD_{540 \; nm}$ packed human red cells = 247, i.e., cells diluted 1 : 1000 (0.1% hct) into Drabkin's solution will have an OD_{540} of 0.247. Further refinements can include determination of cell water by wet weight/dry weight measurements with an appropriate extracellular marker. It is only in particular instances, such as cells from certain clinical disorders or comparisons of cells at different densities (young versus old cells) and experiments under anisotonic conditions, where cell water may vary. In these cases it is appropriate to express fluxes per gram of hemoglobin determined as above.

Efflux. In order to obtain detailed kinetic information, it is also necessary to study amino acid efflux. Cells are preloaded with appropriate concentrations of labeled amino acid. It is necessary to use a much higher specific activity of isotope in order to load with sufficient radioactivity to follow efflux. After loading, cells must be kept and washed on ice to minimize efflux.

Efflux is started by the addition of preloaded ice-cold cells to prewarmed incubation medium. Aliquots can be removed at various times and centrifuged through *n*-dibutyl phthalate, allowing subsequent sampling of cell-free supernatant for processing for scintillation counting at leisure.

It is important in efflux experiments to determine, from the radioactivity in cells and medium after loading, that concentrations of substrate have reached equilibrium. Care must be taken to ensure that the substrate is not metabolized during the preloading period. It is also necessary to measure the radioactivity in the extracellular medium at time zero or efflux may be overestimated.

Nuclear Magnetic Resonance (NMR)

Nuclear magnetic resonance (NMR) spectroscopy is a useful technique for measuring the concentrations of free metabolites in biological samples. Red cell amino acid analysis and flux measurements are obvious areas for the application of this technique.

The use of NMR has been applied recently to studies of taurine levels in pigeon erythrocytes and plasma (C. M. Harvey, unpublished observations). The method has the advantage of real time flux analysis, and the determination of true intracellular levels of substrate. The fact that radioisotopes are not necessary is an advantage for studying exotic amino acids, e.g., D-penicillamine, taurine, or di- and tripeptides. Kinetic measurements have been made for a variety of amino acids and the results are generally in good agreement with conventional tracer methods. However, a high hematocrit is required to see an adequate NMR signal and this poses problems associated with viscosity, accurate volume measurement, and substrate concentration changes. Also, relatively long measurement times are needed (>20 sec), limiting the usefulness for rapid transport systems.

Experimental Strategies

Initial Rate

In order to ensure that the initial rate of amino acid entry is being measured, preliminary investigations of the time course of uptake must be performed. Different transporters show very different rates, but it is essential that the plot of amino acid uptake with time is linear for the particular molecule studied (see Fig. 5).

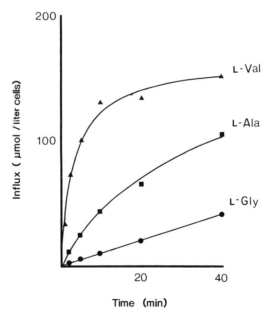

FIG. 5. Time course of influx of three amino acids into human red cells. (From Young et al.[23])

Concentration Dependence

Classical Michaelis–Menten enzyme kinetics can be applied to the substrate dependence of amino acid transport via specific transport systems. The initial rate of flux is proportional to [S] for low values of [S], but with higher values the rate asymptotically approaches a maximum, giving a hyperbolic relationship. It is possible, and appropriate therefore, to think in terms of kinetic parameters V_{max} and K_m when describing transport properties.

$$S + C \underset{k_{-1}}{\overset{k_1}{\rightleftharpoons}} CS \underset{k_{-2}}{\overset{k_2}{\rightleftharpoons}} C + P$$

In this case S and P are actually the same chemical species, differing only in their position with regard to the membrane. C represents the catalytic site of the transporter.

Estimation of Kinetic Parameters

Atkins and Nimmo[57] have reviewed some of the difficulties in the current trends in the estimation of kinetic parameters. Steady-state kinetic experiments involve then the measurement of P at one time (t) or at several times. The data can be treated in two ways. The most common method is to estimate the initial rate (v_i) of uptake at several different concentrations of substrate, and examine the relationship between v_i and [S]. An alternative is to fit an integrated Michaelis–Menten equation to the time course of amino acid uptake. However, this is fraught with difficulties[58,59] and it is sensible to limit measurements to initial rates; it is from these that kinetic parameters should then be determined.

Even so, fitting the Michaelis–Menten equation to amino acid data is not an easy task. The form of the equation describing amino acid uptake in red cells is usually represented as two components given by

$$v = [V_{max}S/(S + K_m)] + K_dS \tag{1}$$

but sometimes as one component given by

$$v = V_{max}S/(S + K_m) \tag{2}$$

Equation (1) is often appropriate[60] since amino acid transport appears to be described by a combined saturable and nonsaturable flux. The second

[57] G. L. Atkins and I. A. Nimmo, *Anal. Biochem.* **104**, 1 (1980).
[58] P. F. J. Newman, G. L. Atkins, and I. A. Nimmo, *Biochem. J.* **143**, 779 (1974).
[59] A. Cornish-Bowden, *Biochem. J.* **149**, 305 (1975).
[60] J. D. Young and J. C. Ellory in "Membrane Transport in Red Cells" (J. C. Ellory and V. L. Lew, eds.), p. 301. Academic Press, New York, 1977.

component, linear in terms of its concentration dependence, is sometimes ascribed to diffusion. However, this is nearly always erroneous; the flux in fact represents transport by another transporter with very low affinity for this substrate. The analysis of uptake data with this Eq. (1) is still useful, since the identification and extraction of a linear component of uptake simply shows that within the concentration range used, saturation cannot be detected and multiple transporters are involved.

After subtraction of a linear component, traditionally the curve [that described in Eq. (2)] has been transformed into a linear form: Lineweaver–Burk ($1/v : 1/s$), Hanes ($s/v : s$), or Eadie–Hofstee ($v : v/s$). These are attractive because of their simplicity and are widely used. However, there are statistical problems with simplistic analysis of this kind, as pointed out by Atkins and Nimmo.[57]

The advent of microcomputers means that Eq. (2) can be fitted by nonlinear least-squares regression where estimates of K_m are made and tried in repetitive cycles until a minimum error occurs. Then the value for V_{max} is derived. There are several software packages available, designed for enzyme kinetics and drug binding, which are directly applicable for amino acid transport and allow the three-parameter fit of Eq. (1). This method provides more accurate estimates of K_m and V_{max}, including nonlinear weighting procedures.

For red cells, the uptake of most amino acids present in the physiological substrate range conforms to Eq. (1), giving transport by a saturable and a nonsaturable route. Sometimes, if the system of interest is Na^+ dependent, the linear component can be determined experimentally by performing parallel fluxes in Na^+-free medium and subtracted directly. This will only be the case, though, if the nonsaturable component is caused by the substrate being carried additionally by Na^+-independent systems or diffusion. If all components of transport are Na^+ independent, one has to rely on computer fitting to a three-parameter model or selective inhibition studies to obtain kinetic parameters. However, fits to multicomponent Michaelis–Menten-type equations are also notoriously artifactual. Equations containing a number of transport routes can be built into the more sophisticated computer models and the best way of obtaining the most accurate estimates of kinetic parameters is to fit the data with several different components of transport and select the most appropriate, using competitive inhibition studies.

Equations (1) and (2) above can be used for this and, if necessary, an additional saturable component can be added to elucidate whether two saturable components can be extracted in the concentration range used, i.e.,

$$v = [V_1 S/(K_{m1} + S)] + [V_2 S/(K_{m2} + S)] + K_d S \qquad (3)$$

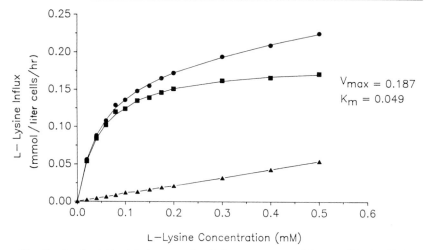

FIG. 6. L-Lysine influx into human red cells in zero-trans conditions, at 37°. Total influx, ●; saturable component, ■; linear component, ▲.

This approach has already been used by Rosenberg[15] in the detailed analysis of the leucine data of Hoare.[13,14] The concentration range selected from which to estimate kinetic parameters will only be effective if the range is large enough to extract the linear component accurately (Fig. 6).

Experimental Procedures

The Zero-Trans Procedure. If cells are depleted completely of intracellular amino acids (the trans side of the membrane) and amino acid uptake is measured at varying external (the cis side) substrate concentrations, by fitting the data for a saturable component it is possible to estimate the zero-trans (zt) parameters. An alternative zero-trans procedure involves preloading cells and following efflux into the zero-trans (substrate-free) extracellular medium. The limiting velocity for transport is given by $V_{12}(zt)$, where side 1 is the cis side. The amino acid concentration at which half of this maximum velocity is reached is $K_{12}(zt)$. When side 2 is the cis face, these numbers are simply reversed. It is necessary for detailed kinetic analysis of a transporter that both zero-trans procedures be carried out.

Equilibrium Exchange Procedure. Cells are first loaded with substrate by allowing preincubation with nonradioactive amino acid until equilibrium is reached. (The problems associated with charged substrates and the existence of transmembrane potentials should be recognized here.) Tracer amino acid is then added externally (such that it does not alter the concentration) and uptake is then measured. A range of concentrations is used, and in this way the kinetic parameters V^{ee} and K^{ee} can be obtained, as for $V(zt)$ above. The choice of direction (1 to 2 or 2 to 1) is irrelevant here since the parameters V^{ee} and K^{ee} are identical for both directions.

Infinite-Trans Procedure. From the results of zero-trans experiments, it is possible to define substrate concentrations which are saturating for transport. Using unlabeled amino acid at concentrations above this level, it is possible to measure the flux of labeled substrate in varying concentrations from the other compartment—the cis face. As the cis substrate concentration rises, uptake will approach V^{ee}. The constant for half-saturation is given by $K_{\frac{1}{2}}^{it}$.

Infinite-Cis Procedure. Again selecting an appropriate saturation concentration, one side of the membrane is set at a limitingly high concentration. This will be the cis side, from which uptake of labeled substrate is measured into the trans side, which has varying concentrations of unlabeled substrate. In the absence of trans substrate the flux of substrate will be that of the corresponding zero-trans procedure. As the trans substrate is increased the uptake from the cis face will be reduced. K^{ic} is the concentration for half-saturation at the trans face, which reduces uptake from the cis face to half of the maximal value.

Experiments at infinite substrate concentrations are sometimes technically difficult to perform. For example Rosenberg was not able to measure infinite-trans fluxes of leucine in human red cells reliably, because the transport rate was too high and immediately elevated the concentration of leucine in the cells.

Role of Ions

Primary active transport of Na^+ via the Na^+/K^+ pump is responsible for generating and maintaining an inwardly directed electrochemical gradient of Na^+ in nearly all cells. This is then used to drive an enormous variety of organic solutes by secondary active transport (see Fig. 7).

The coion can influence translocation in two ways: (1) it might affect the affinity of the carrier for the substrate or (2) it might increase the velocity of translocation. For amino acid transport systems both effects appear.[30,61] A crude distinction is often attempted on the basis of the two Michaelis–Menten parameters K_m and V_{max}: if the K_m for the amino acid is raised with the removal or reduction of the coion then this is said to reduce the affinity of the binding site for the substrate. It can also be used as evidence of sequential binding.[62,63]

[61] R. K. Crane, *Rev. Physiol. Biochem. Pharmacol.* **78**, 99 (1977).
[62] A. R. Chipperfield, *J. Physiol. (London)* **386**, 78P (1987).
[63] P. A. King and R. B. Gunn, *J. Gen. Physiol.* **90**, 23a (Abstr.) (1987).

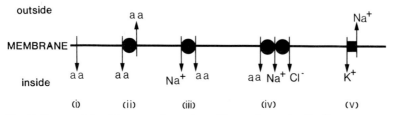

FIG. 7. Four models of the routes for amino acid entry into red cells. (i) Uncoupled; (ii) exchange; (iii) symport with Na^+; (iv) symport with Na^+ and Cl^-; (v) Na^+,K^+-ATPase establishes and maintains the Na^+ gradient.

A change in the V_{max}, without a change in K_m, would indicate that the principal role of the ion is to facilitate translocation without enhancing binding. As indicated at the start of this chapter, there are transport systems which require more than one sodium ion. The Hill plot analysis is frequently used in this situation. The Hill equation can be found in standard texts on enzyme kinetics[64] and is given by

$$Y_s = s^H/(K_s + s^H) \qquad (4)$$

where Y_s is flux as a proportion of V_{max}; s is the concentration of ligand. Rearranging and taking logs we obtain

$$\log[Y_s/(1 - Y_2)] = H \log s - \log K_s \qquad (5)$$

and so a plot of $\log[Y_s/(1 - Y_s)]$ against $\log s$ will give a slope of H and an intercept of $-\log K_s$. This analysis assumes that there are no stable intermediates in the transport sequence, and the value of H corresponds to the number of binding sites. Thus in the case of a single binding site or a number of independent binding sites the value of H would be 1, and the curve of velocity versus s would correspond to a rectangular hyperbola.

Data on the ion dependence of amino acid transport are often treated in this way and the number of binding sites, thus indicated, extrapolated to give coupling ratios of Na^+ to amino acid. Figure 8 shows data for taurine uptake treated in this way.

Na^+ and the Driving Force

In a recent and comprehensive consideration of Na^+-dependent amino acid transport,[65] Eddy concludes that in general, the evidence strongly suggests that amino acid transport is driven by the sodium electrochemical gradient. This in turn is established with the metabolic energy from

[64] M. Dixon and E. C. Webb, "Enzymes." Longman, London, 1979.
[65] A. A. Eddy, in "Amino Acid Transport in Animal Cells" (D. L. Yudilevich and C. A. R. Boyd, eds.), Physiological Study Guide Vol. 2, p. 47. Manchester Univ. Press, 1987.

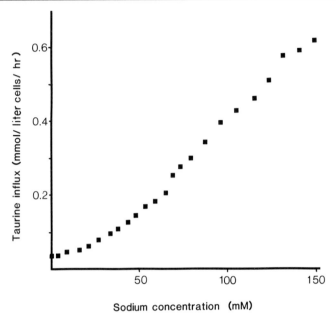

FIG. 8. Taurine influx into pigeon red cells as a function of sodium concentration. The nonhyperbolic relationship is typical for two Na^+ binding per molecule of taurine transported. (Modified from Harvey and Ellory.[44])

ATP provided by the sodium pump. In ascites tumor cells it seems that this is the sole and sufficient source of energy for amino acid transport by system A.

In Na^+-independent systems, it is likely that accumulation of some amino acids will drive the uptake of others by heteroexchange. Such systems are not dependent upon Na^+ gradients, K^+ gradients, or metabolic energy.[20,24] However, it has been pointed out that for the pigeon erythrocyte, the Na^+-independent system does not show countertransport while the Na^+-dependent system does. This might discourage the generalization that a function for Na^+ as a cosubstrate necessarily implies energization.[28]

Red cells represent an ideal system in which to test the Na^+ gradient hypothesis, since it is possible to vary intracellular Na^+ by treatment with p-chloromercuribenzene sulfate (PCMBS) or nystatin.[66]

[66] J. D. Cavieres, in "Red Cell Membranes: A Methodological Approach" (J. C. Ellory and J. D. Young, eds.), p. 179. Academic Press, London, 1982.

Coupling Ratios and Stoichiometry

Vidaver[67] examined the uptake of sodium and glycine in the pigeon erythrocyte, and concluded that the transporter complex is quarternary; two Na^+ accompany the glycine–carrier complex into the cell. He also found that glycine uptake was accelerated by the presence of chloride in a way which implied that translocation of Cl^- occurred, too. The net direction of mediated glycine transport depends strictly upon the direction of the Na^+ gradient across the cell.[68] He also found that the rate of glycine entry increased with the Na^+ concentration according to the Michaelis–Menten equation when $[Na]^2$ was introduced rather than $[Na]$. These observations have been confirmed but they do not hold at Na^+ concentrations below 20 mM.[69] In contrast Wheeler and co-workers found that alanine uptake in the same cell and also in the rabbit reticulocyte depends on first-order kinetics.

Further investigations revealed that the Na^+-dependent uptake of eight neutral amino acids by pigeon erythrocytes was associated with increases in both influx and efflux of Na^+, but no net movement of Na^+ was detected.[70] The Na^+-dependent fluxes of amino acids and the amino acid-dependent fluxes of sodium were determined directly, allowing the comparison of stoichiometry and coupling ratios.[71]

The sigmoid form of the curves describing the uptake of glycine as a function of the Na^+ concentration could arise if the rate-limiting step is the reorientation of the complex $(Na)_2C(Gly)$ as suggested by Vidaver. In contrast, the hyperbolic form for alanine influx could correspond to the complex $(Na)C(Ala)$. Similarly, the nonhyperbolic form for β-alanine corresponds to $(Na)_nC(\beta\text{-Ala})$, where $n > 1$. Traditionally, and even after the warnings implied in this chapter, the Na^+-dependence curves have been used to prescribe coupling as well as stoichiometry (e.g., see Ref. 43). From this, it would follow that 2 Na^+, 1 Na^+, and more than 1 Na^+ are transported with glycine, alanine, and β-alanine, respectively. However, Wheeler and Christensen found coupling ratios of 1.5, 2.5, and 1.0 for these amino acids.[71] In recent investigations in pigeon erythrocytes Harvey and Ellory[44] have shown that there is a discrepancy in the stoichiometry and coupling between 1 taurine and Na^+. The coupling ratio agrees

[67] G. A. Vidaver, *Biochemistry* **3**, 799 (1964).

[68] G. A. Vidaver, *Biochemistry* **3**, 795 (1964).

[69] K. P. Wheeler, Y. Inui, P. F. Hollenberg, E. Eavenson, and H. N. Christensen, *Biochim. Biophys. Acta* **109**, 620 (1965).

[70] K. P. Wheeler and H. N. Christensen, *Fed. Proc., Fed. Am. Soc. Exp. Biol.* **26**, 394 (1967).

[71] K. P. Wheeler, and H. N. Christensen, *J. Biol. Chem.* **242**, 3782 (1967).

with those for β-alanine above the 1 Na^+ : 1 taurine, and the stoichiometry agrees with that found in fish red cells of 2 Na^+ : 1 taurine.[43] It is possible that some ions are required for amino acid binding to the carrier without being translocated, but the reverse discrepancy of Wheeler and Christensen (amino acid $>$ Na^+) cannot be explained in this way. The "delta" analysis of coupling and stoichiometry[72] might be useful in some circumstances, where the transport of the amino acid by the route in question is overestimated due to the contribution of other systems.

Sodium at the Binding Site

The subject of the role of sodium at the binding sites of amino acids has been studied extensively by Christensen and interpreted at various stages in the light of new information.[73,74] While clarifying the transport of lysine by systems for neutral amino acids, he discovered that neutral amino acids could also be carried by cationic systems. The reactivity of the latter is dependent on the presence of Na^+ and is strongly enhanced by the presence of a terminal hydroxyl group on the amino acid, particularly if there are four or five carbon atoms in a chain. The findings suggest that the Na^+ ion serves as a surrogate for the cationic structure.[31,75]

Anion Effects

The absolute requirement for Cl^- by the glycine transporter in pigeon erythrocytes was identified by Vidaver[42]; the requirement also exists in human erythrocytes[9] and in the β system for taurine in fish[76] and pigeon erythrocytes.[44] In the Gly system in pigeons anion specificity is rather low: F^-, NO_3^-, HCO_2^-, SCN^-, and I^- all replace Cl^- to a greater or lesser extent. Acetate and larger anions are excluded. Only the K_m and not V_{max} for glycine entry depends on Cl^- concentrations, so it seems that Cl^- binds before glycine.[77] In human erythrocytes Br^- but not I^- can substitute for Cl^-.[9] Recently, King and Gunn[63] investigated the sequence of binding for ions and glycine in the human Gly system. Glycine influx increases sigmoidally with Na^+ concentration, giving a Hill plot slope of 1.98, and together with direct measures of glycine and Na^+ uptake this indicates coupling of 2 Na^+ : 1 glycine. The Cl^- dependence followed

[72] J. Y. F. Paterson, F. V. Sepulveda, and M. W. Smith, *Biochim. Biophys. Acta* **603**, 288 (1980).
[73] H. N. Christensen, *Adv. Enzymol.* **32**, 1 (1969).
[74] H. N. Christensen, *Adv. Enzymol.* **49**, 41 (1969).
[75] E. L. Thomas and H. N. Christensen, *Biochem. Biophys. Res. Commun.* **40**, 277 (1970).
[76] D. A. Fincham, M. W. Wolowyk, and J. D. Young, *Biochem. Soc. Trans.* **13**, 686 (1985).
[77] J. R. Imler and G. A. Vidaver, *Biochim. Biophys. Acta* **288**, 153 (1972).

Michaelis–Menten kinetics with a K_m of 9.5 mM. At low Cl$^-$ concentrations (5 mM) the Na$^+$ dependence became linear, suggesting that the Cl$^-$ facilitates binding of the second Na$^+$.

Identification and Discrimination of Systems and Use of Inhibitors

A brief survey of the distinct amino acid transporters found in erythrocytes was presented initially. Some of these systems are general and widely distributed. Others are specific to red cells and one or two other tissues.

The identification and discrimination of amino acid transporters relies on several criteria which involve the determination of (1) the initial rate of uptake of a particular amino acid under clearly defined (preferably zero-trans) conditions as a function of substrate concentration, leading to (2) the estimation of the kinetic constants K_m and V_{max} under these conditions for specific amino acids, (3) the ionic (Na$^+$, Cl$^-$) requirements for transport, (4) cross-inhibition studies using selected amino acid substrates or inhibitors, and (5) trans stimulation studies for both homo- (by the same amino acid) or hetero- (by others) exchange.

The first three of these are considered in other sections, and here the subject of inhibitor studies is addressed.

Use of Inhibitors

Although we lack specific inhibitors for amino acid transport systems, some attempts have been made to use SH-reactive reagents such as N-ethylmaleimide (NEM). The mercurial PCMBS, and MK-196 as differential inhibitors. In some cases substrate protection with amino acid has allowed [^{14}C]NEM to be used to label the amino acid transport system. This approach is of limited usefulness unless the transporter is relatively abundant in the membrane. Nevertheless, the use of NEM and MK-196 has allowed the separation of systems L and ASC activity.[6]

Competitive Inhibition

The inhibitor will probably be a substrate analog, reacting at the same site on the carrier as the substrate. In theory a compound dissimilar to the substrate might bind to the carrier elsewhere and act allosterically. The classical kinetic treatment of inhibitor action can be found in standard texts. In the presence of a competitive inhibitor, the flux of substrate will be given by

$$v = V_{max}[S]/\{[S] + K_m(1 + [I]/K_i)\} \tag{6}$$

where [I] is the concentration of inhibitor and K_i is the dissociation constant for the carrier–inhibitor complex.

The value of K_i can thus be determined, and will serve as a quantifier for a particular inhibitor. It follows that if the inhibitor is present at the K_i concentration, then the K_m for the amino acid will be twice that without inhibitor. The V_{max} for the substrate remains unchanged.

Uncompetitive Inhibition

The inhibitor will bind to the substrate–carrier complex but not to the carrier alone.

Here V_{max} is reduced, becoming

$$V_{max} = V/(1 + [I]/K_i) \tag{7}$$

K_m is also reduced, becoming

$$K_m = K/(1 + [I]/K_i) \tag{8}$$

Noncompetitive Inhibition

Here the inhibitor can bind to the substrate–carrier complex or the carrier alone. If $K_{ics} = K_{ic}$ we have a special case of simple noncompetitive inhibition in which the K_m is unaltered. If $K_{ic} > K_{ics}$ then K_m will decrease; if $K_{ics} > K_{ic}$ then K_m will increase; V_{max} is always reduced.

The usual case for amino acid transport inhibition will be competitive inhibition.

Determination of K_i

The initial rates of flux for a particular amino acid should be measured for a series of substrate concentrations in the absence of inhibitor and at two to four fixed inhibitor concentrations. K_i can be found by substituting several values for [I] and solving Eqs. (7) and (8) above.

The K_i of an inhibitor is the concentration of the inhibitor which reduces the rate of transport to half of its maximum, assuming that the observations are made at a negligibly small concentration of substrate. Hence, formally K_i is equivalent to the K_m for transport of the inhibitor alone in the absence of another substrate.

Homogeneity Tests

Inhibition of transport does not, in itself, prove that an analog shares the transport system. This requires testing the equality of K_m and K_i when A is substrate and B is inhibitor. The test is repeated, reversing the roles of substrate and inhibitor. Christensen suggests that the range of analogs tested should be as diverse as possible among the structures of the system.[78] This is called the A,B,C test and is limited by the precision with which K_m and K_i can be determined (see above). If an amino acid is carried by more than one route, these two parameters must be determined for each component of transport separately or their use will be invalid.

Such heterogeneity was overlooked in the determination of K_m and K_i values for methionine in Ehrlich cells. Consequently the wrong conclusion regarding sharing a carrier with glycine was reached.[73] A sensitive test of heterogeneity in the interaction of two analogs for transport is the Inui constant-ratio test. If the saturable fluxes of the two share a common transport system (and if the inspected flux does not contain a hidden route), then when they are present at concentrations of constant ratio the velocity of transport of one will be in a constant ratio to that of the other.[79]

Ideally, a *model* substrate with absolute specificity for a particular route should be used for its identification. For example, N-methylation of an amino acid restricts its transport to system A; then *N*-methyl-AIB is used to discriminate between the A and L system in Ehrlich cells.[80]

The problem of an amino acid being carried by more than one system can be resolved by inhibition studies. Thus, glycine entry into human red cells can be resolved into five separate components: (1) a high-affinity Na^+/Cl^--dependent component, specific for glycine, sarcosine, and proline. Defined by Cl^- substitution, the properties of this route resemble those of the Gly system found in pigeon red cells; (2) a Na^+-dependent, but Cl^--independent route with further specificity for alanine, cysteine, and serine, (ASC), defined by Na^+ replacement and cross-inhibition studies with ASC substrates; (3) a Na^+-independent route (L), inhibitable by leucine; (4) a Na^+-independent, SITS-sensitive route by the anion-exchange system (band 3 protein). By this route uptake is raised when measured in isotonic SO_4^{2-} medium or when pH is increased; (5) a residual Na^+-independent route. Such detailed dissection has allowed the comparative contributions of these routes to be assessed (Fig. 9).

Kinetics of the Simple Carrier

Membrane carriers have a feature not usually present in other treatments of enzyme kinetics; they display reorientation and conformational

[78] H. N. Christensen, "Biological Transport," 2nd Ed. Benjamin, Reading, Massachusetts, 1975.
[79] Y. Inui, and H. N. Christensen, *J. Gen. Physiol.* **50**, 203 (1966).
[80] H. N. Christensen, *J. Membr. Biol.* **84**, 97 (1985).

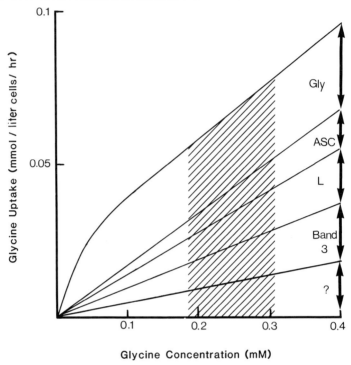

Glycine Concentration (mM)

FIG. 9. Routes for glycine influx in human erythrocytes as a function of glycine concentration. The hatched area represents physiological plasma glycine concentration range. (From Ellory *et al.*[30])

change during transmembrane movement of their substrates. While doing so they are themselves in effect, though not in reality, transferred across the membrane. Such carriers exist in two conformational states: their binding sites are alternately exposed at one and then at the other face of the membrane.

The conventional carrier model can be described with four sequential reversible steps as shown in Fig. 10: (1) binding of the substrate on the cis side, (2) translocation of the substrate and carrier, (3) release of substrate at the trans side, and (4) relocation of the transporter to its original state. Only steps 2 and 4 are vectorial. For most transporters the carrier can reorient loaded (exchange) or empty. It is possible to define extreme experimental conditions under which the carrier performs in a particular transport mode, which can be useful in the kinetic analysis of the carrier. For example the phenomenon of trans stimulation, that is the increase in amino acid uptake into human red cells seen when intracellular amino

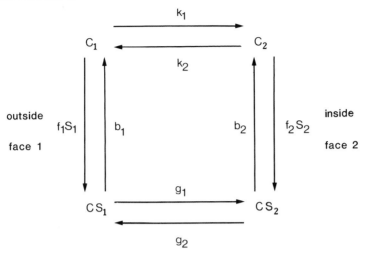

FIG. 10. The four-state model of the simple carrier. C represents the carrier; CS represents the carrier–substrate complex. Rate constants are shown. (Modified from Stein.[83])

acids are present, is an important criterion for distinguishing between carriers and channels.[81] Coupling to the "driving" reaction can be at any of the four stages. In secondary active transport a ternary complex is involved: between the substrate, the carrier, and the coion.[82] At least in Na^+-dependent amino acid transport systems there is some evidence that this latter system operates.

Detailed Kinetic Analyses

It is appropriate and necessary to introduce the descriptive model for a simple carrier of Lieb and Stein at this stage. The reader is referred to the comprehensive treatment of carrier-mediated transport.[81,83] It is with the increasing importance of this treatment in mind that we present some of the basic ideas underlying the selection of experimental strategies.

The model differs from the simple pore in that there are two states of the free carrier C_1 and C_2 interacting with faces 1 and 2 of the membrane.

The equation describing the steady-state unidirectional flux of sub-

[81] W. R. Lieb, in "Red Cell Membranes: A Methodological Approach" (J. C. Ellory and J. D. Lew, eds.) p. 135. Academic Press, London, 1982.
[82] P. Geck and E. Heinz, in "Membrane Transport" (S. L. Bonting and J. J. H. H. M. De Pont, eds.), p. 285. Elsevier/North-Holland, Amsterdam, 1981.
[83] W. D. Stein, "Transport and Diffusion across Cell Membranes." Academic Press, London, 1986.

strate v_{12} as a function of the substrate concentrations S_1 and S_2 at the two faces of the membrane is

$$v_{12} = (KS_1 + S_1 S_2)/(K^2 R_\infty + KR_{12} S_1 + KR_{21} S_2 + R_{ee} S_1 S_2) \quad (9)$$

[from Stein (1986), see Ref. 83] where

$$K = (k_1/f_1) + (k_2/f_2) + (b_1 k_1/f_1 g_1) \quad (10)$$

The reverse flux can be obtained by exchanging 1 for 2.

To find out what Eq. (9) predicts for kinetic parameters in zt and ee one sets S_1 or S_2 to 0 for the two cases of zt experiments and $S_1 = S_2$ for ee. Results for all parameters are presented by Stein; e.g., for zero-trans (zt_{12}) $V_{max} = 1/R_{12}$ and $K_m = KR_\infty/R_{12}$.

Also, for infinite cis: it is necessary to find the net flow of substrate from v_{12} and v_{21}. Side 1 is the cis side so that S_1 is limitingly high:

$$\text{Net } v_{12} = v_{12} - v_{21} = K/(KR_{12} + R_{ee} S_2) \quad (11)$$

This has a maximum value $1/R_{12}$ when S_2 is set to 0, i.e., at V_{max} of a zt situation. To show how the simple carrier model can account for transacceleration Eq. (9) is differentiated to give

$$dv_{12}/dS_2 = (R_{12} - R_{ee}) = K(K + S_1)(v_{12})^2/(K + S_2)^2 S_1 \quad (12)$$

Terms in K and S and all squared terms are positive. Therefore the whole expression is positive.

Hence when $R_{12} > R_{ee}$ transstimulation will be found.

Carrier or Pore

The difference between the simple pore and the simple carrier is the basis on which the carrier kinetics can be tested to determine whether an amino acid transporter functions as a simple carrier. Details of kinetic and mathematical derivation can be found elsewhere.[83]

Only the carrier has a term involving the product $S_1 S_2$ allowing for mutual interaction of the substrate species at the two membrane faces. This is not the case for the simple pore.

If in the equation for the flux via a simple carrier rate constants k_1 and k_2 become very large, only the terms containing these will remain in the equation and it becomes that of the simple pore.

Thus, if the rate constants for the interconversion of the two forms of the carrier (C_1 and C_2) become very large, there is no detectable activation energy barrier between these two forms, and the simple carrier reverts to the simple pore.

Conversely, if terms in k are not infinitely great there is a detectable energy barrier (or if infinite-cis or infinite-trans experiments give measurable values for half-saturation parameters). The carrier need not move bodily across the membrane during transition from C_1 to C_2; what is required is the transition of accessibility of the binding site. If in a real system this transition occurs at a finite rate, terms that do not contain k (i.e., those with S_1 and S_2) cannot be ignored and the predictions of a simple pore must be rejected.

The fact that infinite-cis and -trans experiments yield finite values of the respective half-saturation concentration leads to the rejection of the simple pore model, but the actual values of these experimental parameters may or may not be consistent with the simple carrier model. In fact one can develop rejection criteria for the simple carrier in terms of the experimentally measured parameters.

Symmetry

Kinetic parameters V_{max} and K_m using zero-trans, equilibrium exchange, and infinite-cis procedures allow the calculation of resistance parameters in R:

$$R_\infty = R_{12} + R_{21} - R_{ee} \tag{13}$$

For nucleoside transport, Lieb[81] calculated five estimates for half-saturation parameter K using every available experimental half-saturation parameter and the relevant resistance terms in R. These were not significantly different from one another, meaning that the simple carrier holds for uridine transport.

If, as in the case for uridine transport, infinite-cis uptake and efflux parameters are far from equal, by this criterion also the system is asymmetric and to the same extent. The ee parameters are necessarily the same for efflux as influx since the system is everywhere in equilibrium.

Of the five parameters used to describe the simple carrier K, R_{ee} and R_∞ remain the same when one interchanges the subscripts 1 and 2. They are independent of direction and thus symmetric.

R_{12} and R_{21}, however, do change and hence these parameters exhibit "sidedness" of the membrane system.

R_{12}/R_{21} is used as a measure of a symmetry or asymmetry of the membrane. All half-saturation parameters and maximum velocities (except those of ee) contain either R_{12} or R_{21}. Hence all these parameters will be asymmetric if the system is asymmetric and in each case the ratio of the relevant parameters for the 1 to 2 direction will be given by this same ratio.

There is nothing in the definition of the simple carrier that makes it necessary for the system to be symmetric. The only thermodynamic requirement is that a carrier operating without an outside energy source cannot perform a net transport of substrate when the concentrations at the two faces are equal.

From Eq. (9) net transport in the 1 to 2 direction is given by

$$V_{net_{12}} =$$
$$v_{12} - v_{21} = K(S_1 - S_2)/(K^2 R_\infty + KR_{12}S_1 + KR_{21}S_2 + R_{ee}S_1 S_2) \quad (14)$$

This is zero when $S_1 = S_2$, as required for a transport system in the absence of added energy imput.

In the uridine carrier of the red cell, the system actually seems to behave as a symmetric carrier with fresh red cells but storage at low temperature converts the system to an asymmetric one.

The resistance term describing the conformational change of the unloaded carrier (R_∞) is 17-fold larger than the corresponding parameter for the loaded carrier R_{ee}. This is a very clear example of how the substrate facilitates the transmembrane movement of the carrier just as the carrier facilitates the transport of substrate. Also, R_{ee} is smaller than both R_{12} and R_{21}, which means that transacceleration operates in both directions.

Using Parameters to Characterize Carrier

The parameter K and the three values of the resistance parameters in R are sufficient to characterize a system which behaves as a simple carrier. Further details of application are to be found in Stein.[83] Terms in R are best considered as resistances and are defined as sums of the reciprocals of the rate constant (see above). A rate constant is a measure of the ease with which a "reaction" takes place along a certain path. Hence, the reciprocal is the resistance (per unit site) for the reaction. Resistances add algebraically if they are in series, e.g., R_{12} is the resistance for the zt experiment in the 1 to 2 direction. In this experiment the concentration of S_1 is by definition limitingly high. Formation of CS at side 1 cannot be rate limiting. Thus what determines the value of the maximum velocity is the overall resistance for the return journey, given by the resistance experienced by the complex on breaking down at side 2, plus that for the conformational change leading to the carrier's becoming available again at side 1.[83]

K is the intrinsic dissociation constant, and so for fluxes zt_{in} and zt_{out} and ee:

$$K_{12_{zt}} R_{12} = K_{21_{zt}} R_{21} = K_{ee} R_{ee} = KR_\infty \quad (15)$$

KR_∞ contains the resistance when substrate is present at neither side of the membrane. The term K therefore expresses the half-saturation concentration appropriate to the hypothetical procedure when substrate is absent from both sides of the membrane, i.e., the free carrier distribution across the membrane is unperturbed by the substrate. It represents the reciprocal of the intrinsic affinity of the substrate for the system, being determined by the relative tendency of the free carrier to combine with substrate and move across the membrane compared with its tendency to move as a free carrier.[81,83,84]

Analysis of Amino Acid Transporters

From the use of this detailed kinetic analysis on collected data by Rosenberg, for both the L and the T system, it is clear that much information can be gained with regard to the functioning of amino acid carriers.[10,18]

For the L system, Rosenberg[10,18] found that the simple pore could be rejected for leucine transport across the red cell membrane. The L system behaves as a simple carrier, since the three estimates of K are not different while R_{ee} is lower than R_{12} and R_{21}. The system shows transacceleration, but behaves as a symmetric carrier, since R_{12} is not different from R_{21}. The ratio of the rate constants k_1 and k_2, for the interconversion of the two forms of the carrier, is not far from unity—consistent with the systems being symmetric. All the rate constants, describing the interconversion rates for the loaded carrier, consistent with the phenomenon of transacceleration, and with the value of R_{ee} being less than R_{12} or R_{21}, are greater than corresponding rates for the unloaded carrier. He concluded that both the dissociation and translocation of the carrier complex is faster than the translocation of the empty carrier; no translocation step is rate determining and the carrier complex is equally distributed across the membrane at equilibrium.

Again, analysis of tryptophan transport via the T system has given no reason to reject the simple carrier, and the estimated value for R_∞ for tryptophan differs from that for leucine via the L system by an order of magnitude, confirming that the systems are independent. (The term R_∞ is the total resistance experienced by the empty carrier as it undergoes a complete cycle of conformational change leading to the substrate-binding site being first at one face, then at the opposite face, and finally back to the original face.)

[84] W. D. Stein, *in* "Membrane Transport" (S. L. Bonting, and J. J. H. H. M. De Pont, eds.), p. 123. Elsevier/North-Holland, Amsterdam, 1981.

Changes in Transport with Cell Maturation

One reason for the existence of amino acid transport systems in red cells might be as relics of the requirements for amino acids during reticulocyte metabolism.

There are species differences, and also differences for various clinical conditions, in the number of circulating reticulocytes.[50] Separation of reticulocytes will be necessary if the blood sample has been taken with a view to examining the changes in properties of transport systems with maturation. The separation procedures were introduced briefly in a previous section.

It appears then that development from blast cell to erythrocyte involves the loss of transport function, if not the loss of transporters, for some amino acids. Which of these is the case remains to be elucidated and the discussion of the role for amino acid transport retention remains. However, in recent studies on sheep reticulocyte development, it has been shown that portions of membrane containing amino acid transporters are lost. Vesicles are released from the developing reticulocytes, containing a number of activities characteristic of the reticulocyte membrane. These activities are known to disappear, or at least diminish, during development, the implication being a "shedding" of membrane functions.[52]

Thermodynamic Effects on Transport

Temperature and Pressure

Pure phospholipid bilayers undergo phase transitions at temperatures which depend on the chain length and structure of the lipid head group. Cholesterol orders the chains above their transition temperature but disorders them below this, making simple physical and chemical interpretation of temperature effects difficult. Since the current perception of transporters follows the view that carriers are discrete protein entities embedded in the membrane, then temperature effects on membranes should be reflected in changes in transport function.[85]

In intact red cells, there are at least three ways in which temperature can affect transport. The first is by direct action on the carrier molecules themselves, the second is the referred effect through lipid molecules and

[85] D. C. Lee and D. Chapman, *in* "Temperature and Animal Cells" (K. Bowler and B. J. Fuller, eds.) p. 35. Soc. Exp. Biol., Cambridge, England, 1987.

their physical environment, and the third is via chemical changes to charged groups.[86]

The effects of temperature on transport can be analyzed in terms of changes in the kinetic parameters K_m and V_{max}, and can provide information on the mechanisms of transport. For red cells these kinetic changes have been studied in some detail for the L system[14-16] and the glucose carrier.[87] In all of these studies the rate of transport was shown to be highly sensitive to temperature. The analysis of Lowe and Walmsley can be applied to the kinetic constants derived from zero-trans entry and exit, and equilibrium exchange.

Temperature and Initial Rate

The extent to which the rate of a transport is affected by a change in temperature is determined by the magnitude of its energy of activation.

The Arrhenius equation

$$k = Ae^{-E/RT} \tag{16}$$

explains the sensitivity of the rate of transport to changes in temperature, since the relationship is exponential.

$$\ln k = \ln A - (E/RT) \tag{17}$$

[k, rate constant; T, temperature (in K); R, gas constant]. So

$$\log k = \log A - [(E/2.303)(1/T)] \tag{18}$$

The activation energy E can be determined exprimentally by plotting $\log k$ against $1/T$. The slope of the straight line obtained will be $-E/19.14$.

At two temperatures T_1 and T_2 the rate constants are k' and k'':

$$\log k'' - \log k' = \log A - \left(\frac{E}{2.303R}\right)\left(\frac{1}{T_2}\right) - \left[\log A - \left(\frac{E}{2.303R}\right)\left(\frac{1}{T_1}\right)\right] \tag{19}$$

$$\log\left(\frac{k''}{k'}\right) = -\left(\frac{E}{2.303R}\right)\left[\left(\frac{1}{T_2}\right) - \left(\frac{1}{T_1}\right)\right] = \left(\frac{E}{2.303R}\right)\left[\frac{(T_2 - T_1)}{T_1T_2}\right]$$

so

$$\log(k''/k') = E(T_2 - T_1)/19.14T_1T_2 \tag{20}$$

$$C + S \underset{k_{-1}}{\overset{k_1}{\rightleftharpoons}} CS \underset{k_{-2}}{\overset{k_2}{\rightleftharpoons}} C + P$$

[86] J. C. Ellory, and A. C. Hall, in "Temperature and Animal Cells" (K. Bowler and B. J. Fuller, eds.) p. 53. Soc. Exp. Biol., Cambridge, England, 1987.
[87] A. G. Lowe and A. R. Walmsley, Biochim. Biophys. Acta 857, 146 (1986).

A change in temperature will affect the value of k_1 ($K_s = k_{-1}/k_1$) and thus alters the steady-state concentration of CS at a nonsaturating concentration of substrate. Also k_2 changes: this can be studied specifically at saturating concentrations of substrate

$$[CS] = c_0, \qquad v_0 = V_{max} = k_2 c_0$$

where c_0 is the carrier concentration.

At temperature T_1, the maximum velocity V'_{max} and rate constant for rate-limiting step (CS $\rightarrow$ C + P) is k'_2. At temperature T_2 these parameters are V''_{max} and k''_2. Since $V_{max} = k_2 c_0$,

$$\log V''_{max}/\log V'_{max} =$$
$$k''_2 c_0 / k'_2 c_0 = \log k_2^{11}/\log k_2^1 = E(T_2 - T_1)/19.14 T_1 T_2 \quad (21)$$

by plotting $\log V_{max}$ against $1/T$ and finding the slope of $E/19.14$ the value of E can be found. Also $E = H + RT$ where H is the enthalpy of activation of the transport and is the difference between the enthalpy of the "reactants" and that of the transition state.

Lowe and Walmsley Analysis

This was used originally for the glucose carrier, but has also been applied to the L system for leucine transport.[16] In this study, kinetic data (including those of Hoare) are fitted to the conventional four-state carrier model and the thermodynamics of the two systems (L and glucose) compared. The only assumption in the analysis is that the rates of association with, and dissociation from, the carrier are much faster than the reorientation steps.

The exact procedures are given in Lowe and Walmsley[87] (see also Walmsley[88]) but the key findings, as outlined in their subsequent comparison of systems, are as follow:

Near 0° the rates for transport in equilibrium exchange of both glucose and leucine are much faster than both types of zero-trans transport. Thus these measures of zero-trans fluxes give good approximations to the rate constants g_1, g_2 (those for empty carrier reorientation), and their activation energies. The rates for equilibrium exchange can be used to set lower limits to rate constants k_1 and k_2.

Enthalpy and entropy changes accompanying reorientation were calculated for both the glucose and leucine systems. Calculation of Gibbs free energies allowed the determination of the distribution of carriers between the four possible states. Such detailed thermodynamic descrip-

[88] A. R. Walmsley, *Trends Biochem. Sci.* **13**, 226 (1988).

tion led Walmsley and Lowe to conclude that (1) there are marked differences in the binding of the two substrates; unlike glucose, leucine binding is strongly exothermic on both sides of the membrane; (2) at low temperatures the carriers and their substrate complexes are mainly inward facing; (3) increasing temperature, and to some extent binding of substrate, promotes outward-facing conformations; (4) at physiological temperatures a significant fraction of the glucose carrier is found in all four possible forms but the leucine carrier predominantly in the outward-facing form.

Separation and Reconstitution Strategies

The characterization of amino acid transport systems is proceeding with greater speed in recent years and the detailed analysis of carrier transport can yield important information. However, there are other unanswered questions. Some work on the pigeon erythrocyte membrane and its transporters gives an indication of possibilities.

Watts and Wheeler[89] isolated the pigeon red cell membrane, to yield whole-cell "ghosts." They also used other methods giving membrane fragments, and demonstrated possible disadvantages. The protein components were analyzed using gel electrophoresis, revealing similar composition to human erythrocytes. Two major bands were identified which do not appear in human cells, and bands 4.2 and 6 of the human cells are absent from pigeon. Watts and Wheeler[90] further describe the partial separation of an Na^+-dependent amino acid transport system, in which the transport activity seems to be associated with the band 3 region of polypeptides. The characterization of two Na^+-dependent systems, Gly and ASC, distinguishable by anion sensitivity, was achieved in vesicles prepared from pigeon red cell membranes. The only major integral protein retained in these vesicles was band 3. While it was not proved that this was the transporter, the evidence appears to be strong.[91]

The methods involved in the preparation of these vesicles differ for human and pigeon red cells, probably because of the differences in interactions of peripheral proteins, notably spectrin, with the membrane.

Concluding Comment

The ultimate aims of membrane transport studies include the establishing of *in vivo* functioning of transporters and the physiological meth-

---

[89] C. Watts and K. P. Wheeler, *Biochem. J.* **173,** 899 (1978).
[90] C. Watts and K. P. Wheeler, *FEBS. Lett.* **94,** 241 (1978).
[91] C. Watts and K. P. Wheeler, *Biochim. Biophys. Acta* **602,** 446 (1980).

ods of control[92]; the ability to isolate transporters and perform reconstitution studies in liposomes; the examination of the importance of the lipid environment on carrier function by lipid substitution; the chemical identification of transporters and the derivation of their amino acid sequences; the raising of antibodies to them and, finally, their cloning and sequencing. These rest on the continuous refining of techniques to identify, and discriminate between, amino acid transporters.[93] The human red cell glucose transporter has already been taken through this route. Amino acid transporters differ in their abundance and the inhibitors available for biochemical studies. Nevertheless, there is every possibility that we shall soon have similar information for the L system and then for others.

[92] M. A. Shotwell, M. S. Kilberg, and D. L. Oxender, *Biochim. Biophys. Acta* **737**, 267 (1983).
[93] J. V. Vadgama, and H. N. Christensen, *J. Biol. Chem.* **260**, 2912 (1985).

[9] Transport Measurement of Anions, Nonelectrolytes, and Water in Red Blood Cell and Ghost Systems

By JESPER BRAHM

Introduction

The fact that several chemical reactions, e.g., enzymatic reactions, and biological transport processes proceed very rapidly is one of the greatest challenges Nature presents for scientists in the biological fields. Several elaborate methods have been used for solving rather specific and limited problems.

In transport physiology the red blood cell has gained importance as a test object. Red blood cells are generally easy to obtain in sufficient quantities, they are uniform in size, and their membrane structure is simple compared to other cell types. It is also possible to prepare resealed ghosts with intact membrane transport properties and deliberate solute content. The advantageous conditions given above for using red cells in transport studies may be overshadowed by the fact that under physiological conditions several transport processes are so rapid that more sophisticated methods are needed.

It is obvious that the determination of the time dependence of transmembrane solute movement requires a reliable method to reveal the interrelation between the transport event and time. Until the early 1930s such transport studies were restricted to experimental conditions which imply a net flux of solute across the membrane(s). With the introduction of radioactively labeled solutes the experimental range was expanded to include solute transport under steady-state conditions (self-exchange diffusion), i.e., the intra- and extracellular chemical concentrations kept constant during the experiment. Such tracer methods may also be useful under non-steady-state conditions. More recently nuclear magnetic resonance techniques have also proved suitable for investigations of rapid transport processes in isolated cells (for a review, see Macey and Brahm[1]).

The goal of this chapter is to describe the flow tube technique, which is well suited for studying rapid transport of radioactively labeled solute across membranes from red cells and ghosts. The technique has a time resolution of a few milliseconds and may be used in the determination of transport processes with half-times in the range from about 3 msec to 1–2 sec. Slower transport processes may be determined by the Millipore–Swinnex filtering method,[2] which has a time resolution of a few seconds, and by centrifugation techniques whose time resolution is even lower.

Continuous Flow Tube Method

The principle of this method is to convert time to distance,[3] as demonstrated in 1897 by Rutherford,[4] who blew ionized gases through long glass pipes with a constant velocity to determine the decay constant of the deionization process. Given a certain measure, e.g., a decay constant k with the unit of reciprocal time $(1/t)$ for a process with a monoexponential time course, the process is halfway completed after a half-time $t_{1/2} = (\ln 2)/k$. Let $k = 0.693$ sec^{-1} for the process to be determined, then $t_{1/2} = 1$ sec. If the reaction is initiated at the inlet of the tube, and the average linear velocity is 1 m/sec, the reaction is halfway completed 1 m down the tube as it takes 1 sec to pass that distance. Thus, the time from the start of the reaction to the observation depends on the distance and the flow velocity, both of which can be adjusted appropriately.

The method was introduced into biology by Hartridge and Roughton,[5] and can be used in different versions (stopped flow, continuous flow, accelerated flow) for studies of rapid reactions such as enzymatic reac-

[1] R. I. Macey and J. Brahm, in "Water Transport in Biological Membranes" (G. Benga, ed.), in press. CRC Press, Boca Raton, Florida.
[2] M. Dalmark and J. O. Wieth, J. Physiol. (London) 224, 583 (1972).
[3] D. C. Tosteson, Acta Physiol. Scand. 46, 19 (1959).
[4] E. Rutherford, Philos. Mag. 44, 422 (1897).
[5] H. Hartridge and F. J. W. Roughton, Proc. R. Soc. London, Ser. A104, 376 (1923).

tions. Stopped flow and continuous flow techniques have also been used for determinations of solute and water permeabilities in red cell membranes. The present chapter is confined to the continuous flow tube technique; a recent review deals with the stopped flow technique.[1]

Figure 1 shows schematically the principles of the method developed in our laboratory for the determination of efflux of radioactive solutes from loaded cells in a dilute cell suspension. The original design was made by Piiper.[6] The syringe (1) contains 1–2 ml of packed red cells or ghosts loaded with the radioactive isotope under study. The pump driving the syringe injects cells continuously into the mixing chamber (2). These cells are mixed with the nonradioactive medium, which is injected from the container (3) by means of air pressure regulated with a precision valve. The cell suspension prepared in the mixing chamber flows down the observation tube made of constant-bore tubing and is directed either to a waste bucket (9) or a cylinder glass (8) during the measuring period by an electrically controlled valve (7). At predetermined distances, which can be varied in the experiments, the tube is replaced by Perspex filtering units (6) whose walls in part consist of micropore filters. We have used Nucleopore filters (Shandon Southern Instruments, Inc., Sewickley, PA) with pore diameters of 0.4–0.8 μm and Millipore prefilters (Millipore Corp., Bedford, MA) with pore diameters of 1.2 μm. With regard to internal measurements, the diameters of the inlets, the tube, and the outlets are 2 mm as described in the setup by Piiper.[6] We have reduced the external measurements so that the minimum length of a tube with six filtering units is 72 mm. With flow rates of about 475 cm/sec the time resolution of the flow tube, i.e., the time for the suspension to pass from one filtering unit to the neighboring unit, is about 2–3 msec.

Though extremely simple in principle, the use of the method raises a series of methodological questions which we will consider next.

Is the Flow Rate Constant?

It is critical for the calculation of the sampling times corresponding to the distances from the mixing chamber that the flow of the cell suspension along the observation pipe be linear. The average velocity of flow can be calculated by measuring the time required for collection of a certain volume of cell suspension (and in practice ignoring the volume of the filtrates that here amounts to 1–2% of the collected volume). However, the calculation does not reveal any fluctuations in the flow rate during the experiments. One way, which in simplicity by far outdistances complicated techniques such as the Doppler method, is to record using a strain gauge

[6] J. Piiper, *Pfluegers Arch. Gesamte Physiol. Menschen Tiere* **278**, 500 (1964).

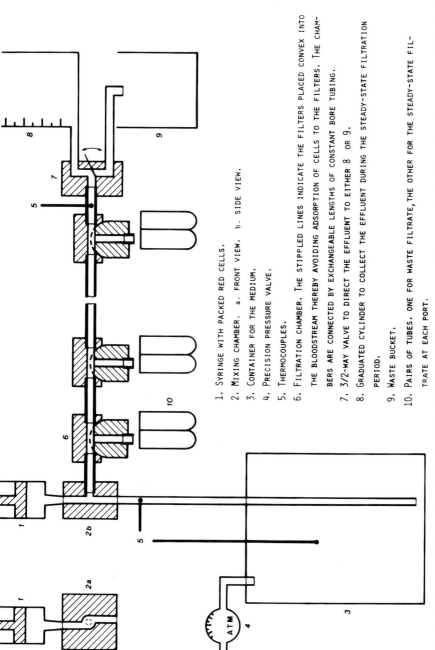

1. SYRINGE WITH PACKED RED CELLS.
2. MIXING CHAMBER. a. FRONT VIEW. b. SIDE VIEW.
3. CONTAINER FOR THE MEDIUM.
4. PRECISION PRESSURE VALVE.
5. THERMOCOUPLES.
6. FILTRATION CHAMBER. THE STIPPLED LINES INDICATE THE FILTERS PLACED CONVEX INTO
   THE BLOODSTREAM THEREBY AVOIDING ADSORPTION OF CELLS TO THE FILTERS. THE CHAM-
   BERS ARE CONNECTED BY EXCHANGEABLE LENGTHS OF CONSTANT BORE TUBING.
7. 3/2-WAY VALVE TO DIRECT THE EFFLUENT TO EITHER 8 OR 9.
8. GRADUATED CYLINDER TO COLLECT THE EFFLUENT DURING THE STEADY-STATE FILTRATION
   PERIOD.
9. WASTE BUCKET.
10. PAIRS OF TUBES, ONE FOR WASTE FILTRATE, THE OTHER FOR THE STEADY-STATE FIL-
    TRATE AT EACH PORT.

FIG. 1. A diagram of the version of the flow tube technique used to measure rapid transport of radioactive solutes across isolated cell membranes. (From Brahm[18] by permission.)

the buoyancy of a Perspex bar dipped in the cell suspension that accumulates in the cylinder [(8) in Fig. 1]. A constant flow rate implies a constant increase in buoyancy. In our studies the flow rate varies <2% in the measuring period, and is independent of the flow rate.

Is Turbulence Present?

Turbulence ensures that the mixing of the cell suspension initiated in the mixing chamber is maintained in the pipe whereby the diffusion resistance in "unstirred layers" of the medium adjacent to the outside of the membrane is minimized. Furthermore, the velocity profile of the suspension in the pipe becomes flattened by turbulence compared to the parabolic profile when the flow is laminar. The mathematical description of the profile under turbulence is very complicated (see Lamb[7]), and it suffices to emphasize that turbulence is a *sine qua non* of the experimental conditions.

The fundamental work on flow of fluids along narrow pipes by Reynolds[8] shows that a laminar flow becomes turbulent as a critical flow rate is exceeded. The transition to turbulence appears at a definite value of the dimensionless Reynolds number (Re), which is defined by the linear velocity of flow (v, cm/sec), the pipe radius (r, cm), the viscosity of the fluid (η, poise), and the fluid density (ρ, g/cm^3) according to

$$Re = vr\rho/\eta \tag{1}$$

Generally, the Reynolds number indicating the appearance of turbulence has a value of 800–1000. However, turbulence may very well appear at a lower value, if the pipe has constrictions, irregularities, or bendings, which promote turbulence. The Reynolds number, therefore, represents but a crude estimate of whether the flow is turbulent. Turbulence can be judged more precisely from a comparison of the relations between flow rate and the pressure gradient for laminar flow:

$$(P_1 - P_2)/L = 8\eta v/r^2 \quad \text{(dyn/cm}^3) \tag{2}$$

and turbulent flow, which has the empirically determined relation[9]:

$$(P_1 - P_2)/L = 0.079\,\eta^{0.25}v^{1.75}/r^{1.25} \quad \text{(dyn/cm}^3) \tag{3}$$

where P_1 and P_2 are pressures at the inlet and the outlet of the flow tube, respectively, and L is the length of the tube. It appears from Eqs. (2) and (3) that the pressure gradient under laminar flow varies proportionally

[7] H. Lamb, "Hydrodynamics." Cambridge Univ. Press, Cambridge, England, 1932.
[8] O. Reynolds, *Trans. R. Soc. London* **1974,** 935 (1883).
[9] R. L. Whitmore, "Rheology of the Circulation." Pergamon, Oxford, England, 1968.

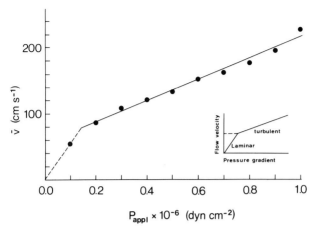

FIG. 2. The linear velocity of flow against the applied pressure which drives the dilute cell suspension through the tube in the flow tube apparatus. (From Brahm[13] by permission.)

with the mean velocity of flow, whereas the pressure gradient under turbulent flow by approximation varies proportionally with the square of the mean velocity of flow. It is, however, difficult to measure the pressure difference in the narrow tube, so we depict the linear velocity of flow against the applied pressure, which can be considered to be proportional to the pressure gradient. Figure 2 shows such a diagram obtained from a series of flow measurements. This depiction is more sensitive in revealing whether the flow is laminar or turbulent than the diagram of flow velocity vs the square root of the applied pressure used by others.[10] Figure 2 shows that velocities <80–90 cm/sec cannot be included in the linear relationship obtained for the measured velocities above the critical flow rate. Furthermore, the value of the slope of the upper part of the line is 1.79, as expected from Eq. (3) if the flow is turbulent.[11]

The crude Reynolds number for the critical flow rate is about 100 cm/sec for a dilute cell suspension (assuming that $\eta = 0.01$ P and $\rho = 1$ g/cm³) moving in the pipe depicted in the present version of the flow tube. The fact that turbulence appears at a lower velocity of flow agrees with the intended construction of the pipe with the filters in the sampling units bulging into the bloodstream so that the tendency of the fluid to adapt a laminar flow is further reduced. The conclusion is that laminar flow as a methodological error can be excluded at velocities of flow above 90–100 cm/sec.

[10] C. V. Paganelli and A. K. Solomon, *J. Gen. Physiol.* **41**, 259 (1957).
[11] J. Brahm, Ph.D. thesis. Univ. of Copenhagen, Copenhagen, Denmark, 1978.

Is the Mixing Chamber Efficient?

Mixing of the packed cells and the medium must be over before the cell suspension passes the first sampling unit, i.e., in <2–3 msec at the highest flow velocities of 450–500 cm/sec. The mixing chamber used has a tangential introduction of both the cells and the medium (cf. Fig. 1) as was concluded to be the best by Hartridge and Roughton,[5] who evaluated the efficacy of mixing chambers with different geometry.

The efficiency of the mixing chamber in the present version was tested by two different approaches: (1) An extracellular marker trapped between the cells in the syringe is released by the mixing of cells with the medium. (2) The cells are loaded with a solute to which the membrane is highly permeable, so that the efflux depends on an instantaneous exposure of the total membrane area for the transport process and on the barriers created on both sides of the membrane due to incomplete stirring of the medium adjacent to the membrane surfaces ("unstirred layers effect").

Experiments with $Na_2{}^{51}CrO_4$ and ^{51}Cr-labeled EDTA show that the flow tube versions have a complete mixing of the cell suspension before it reaches the first sampling unit.[10,12,13]

Is the "Unstirred Layers" Determination of Membrane
*   Permeability Invalid?*

Efficient mixing in the mixing chamber and turbulent flow in the pipe do not rule out that solutes, which are transported between two water phases separated by the membrane, experience other diffusion barriers than that represented by the membrane itself. An ideal mixing of the extracellular fluid is hardly possible, and the cytosol is considered to be unstirred. Increasingly imperfect mixing of layers adjacent to the membrane implies complicated concentration profiles. For practical purposes one operates with an equivalent totally unstirred layer separating the cell membrane and a perfectly stirred medium. The thickness of the layer has been estimated from red cell experiments with the lipophilic alcohol *n*-butanol,[13] which was previously used to determine unstirred layers in different methods applied to measure solute transport in planar and spherical bilayers and liposomes.[14,15] Figure 3 shows the results of two experiments with *n*-butanol. In one experiment (Fig. 3A) the *n*-butanol concentration was 1 m*M*, and the washout of *n*-[^{14}C]butanol from the cells

[12] V. W. Sidel and A. K. Solomon, *J. Gen. Physiol.* **41,** 243 (1957).
[13] J. Brahm, *J. Gen. Physiol.* **81,** 283 (1983).
[14] R. Holz and A. Finkelstein, *J. Gen. Physiol.* **56,** 125 (1970).
[15] M. Poznansky, S. Tong, P. C. White, J. M. Milgram, and A. K. Solomon, *J. Gen. Physiol.* **67,** 45 (1976).

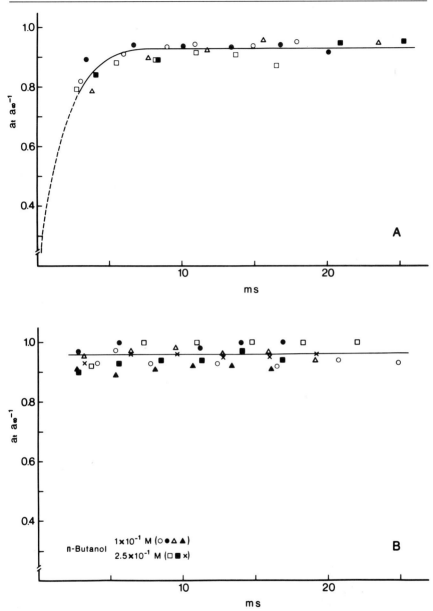

FIG. 3. The efflux of n-[^{14}C]butanol from labeled red cells at butanol concentrations of 1 mM (A), and 100 and 250 mM (B). For explanation, see text. (From Brahm[13] by permission.)

was completed after about 6 msec. By increasing the n-butanol concentration to 250 mM (Fig. 3B), equilibrium was reached before the first sampling about 3 msec after mixing was initiated in the mixing chamber. Apparently the red cell membrane permeability to butanol increases by increasing butanol concentration. Under such conditions the membrane does not constitute a measurable diffusion barrier to butanol. If one assumes that (1) an unstirred layer represents the only diffusion barrier to butanol, (2) a Fickian diffusion regime applies to butanol in the experimental situation, (3) the diffusion coefficient of butanol (D_B) is 10^{-5} cm^2/sec, and (4) the process is completed after six half-times (99% completion), i.e., $t_{1/2}$ is <0.5 msec, then the permeability of the unstirred layer to butanol (P_B) is >6 × 10^{-2} cm/sec, and the equivalent layer is $D_B/P_B \leq 1.7$ μm.[13] The result agrees with conclusions made in a hydrodynamic study of turbulence in dilute cell suspensions and a theoretical work on unstirred layers' thickness.[16,17]

Diffusion through unstirred layers sets an upper limit for which transport processes are determined by means of the flow tube. As shown in Table I membrane permeabilities up to 10^{-2} cm/sec can be determined if one accepts an error of 10–20% caused by the effect of unstirred layers.

Interception of Efflux Curve with Ordinate

In our version of the flow tube the cytocrit of the dilute cell suspension is kept low (<1%) intentionally to reduce the degree of (hemo)lysis of intact cells and ghosts.[18] The cost of this advantage is that for most transport processes one is restricted to measuring radioactive tracer efflux only. Under these conditions and at steady state a monoexponential decrease of intracellular radioactivity with time follows:

$$a_i(t) = a_i(0)e^{-kt} \qquad \text{(mol or Ci/cm}^3) \qquad (4)$$

where $a_i(t)$ and $a_i(0)$ are the intracellular concentrations of the radioactive isotope at time t and $t = 0$, respectively, and k (1/sec) is the rate coefficient for the transport process. Since the experimental conditions of a low cytocrit imply that the extracellular space is finite, though large, and since it is more convenient to determine the extracellular increase of radioactivity than the intracellular decrease, Eq. (4) is transformed to Eq. (5), which expresses the increase of extracellular radioactivity with time:

$$a_0(t) = a_0(\infty)(1 - e^{-kt}) \qquad (5)$$

[16] M. Gad-el-Hak, J. B. Morton, and H. Kutchai, *Biophys. J.* **18**, 289 (1977).
[17] S. A. Rice, *Biophys. J.* **29**, 65 (1980).
[18] J. Brahm, *J. Gen. Physiol.* **70**, 283 (1977).

TABLE I

ERROR CAUSED BY 1.7-μm UNSTIRRED LAYER IN
DETERMINATION OF MEMBRANE PERMEABILITY[a]

P_m (10^3 cm/sec)	P (10^3 cm/sec)	$1 - P/P_m$ (%)	k (1/sec)	$t_{1/2}$ (msec)
10.0	8.5	15	198.8	3.5
5.0	4.6	9	107.2	6.5
3.7	3.5	6	81.0	8.6
2.4	32.3	4	53.6	12.9
1.0	0.98	2	22.9	30.3

[a] A diffusion barrier composed of an erythrocyte membrane
with the permeability P_m and an unstirred layer with the
thickness u and the permeability P_u has the compounded
permeability, P, according to

$$1/P = 1/P_m + 1/P_u = 1/P_m + u/D_u = 1/k(A/V)$$

where k is the compounded rate coefficient, and A/V is the
ratio of the cell area to the cell volume. D_u, which is the
diffusion coefficient of the solute in a dilute suspension, is
assumed to be 10^{-5} cm^2/sec. A/V is set to 2.3×10^4 ($A =$
1.42×10^{-6} cm^2/cell; $V = 6.3 \times 10^{-11}$ cm^3 H$_2$O/cell).

where $a_0(\infty)$ is the extracellular tracer concentration at equilibrium. Rearranging Eq. (5) to

$$\ln\{1 - [a_0(t)/a_0(\infty)]\} = -kt \qquad (6)$$

shows that the intracellular radioactivity depicted against time is linear in
a semilogarithmic plot. Ideally all radioactivity is in the cellular phase at
time $t = 0$ as indicated by the value of the intercept (I) of 1.0. However, it
is a consistent finding that $I < 1.0$, indicating that a fraction of the radioactivity is extracellular at the depicted $t = 0$. $I \geq 0.85$, which is characteristic for plots of chloride, urea, and glucose efflux,[18-20] can be explained in
terms of extracellular trapping of radioactivity between the cells in the
syringe, and a delay in the mixing chamber. Lower values of I are found
for efflux of water and alcohols[13,21] and, in addition to the above-mentioned causes, may be determined by properties of the cell membrane.

Extracellular Trapping of Radioactivity. Usually the preparation and
incubation procedures end by centrifuging the cell suspension so that the

[19] J. Brahm, *J. Gen. Physiol.* **82,** 1 (1983).
[20] J. Brahm, *J. Physiol.* (*London*) **339,** 1 (1983).
[21] J. Brahm, *J. Gen. Physiol.* **79,** 791 (1982).

hematocrit of the cell sample in the syringe (Fig. 1) is about 85%. In accordance with the control experiments using ^{51}Cr-labeled EDTA the isotope trapped between the cells is distributed instantaneously in the extracellular phase of the dilute suspension. Less centrifugation increases the trapped volume and the fraction of extracellular isotope which lowers the intercept. Figure 4 plots four efflux curves of experiments in which the hematocrit in the syringe was varied. Under the experimental conditions the delay in the mixing chamber (see below) is negligible (<1%), and the value of I represents the hematocrit because the distribution of ^{36}Cl$^-$ between the cytosol and the medium is 1 at the chosen pH (6.3 at 38°).[18] The linearity and similar slope of the plots indicate that the same rate coefficient determines the transport process, independent of a change of the hematocrit from 85 to 50%.

Delay in Mixing Chamber. For practical purposes the distances from the chamber to the sampling units are measured from the back wall of the chamber. Because the diameter of the mixing chamber is twice that of the tube, the transit time for the cell suspension is longer in the mixing chamber than in a section of the tube of the same length (cf. Table II). If the plot is not corrected for the delay the depicted time values are all too small by the same magnitude, and the parallel line is shifted to the left toward the zero-time point, thereby reducing the value of I. As illustrated in Table III

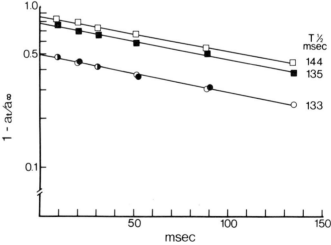

FIG. 4. Plots of ^{36}Cl$^-$ efflux from labeled human red blood cells, which were packed to a varying degree. The ordinate expresses the fraction of tracer that remains intracellularly at the sampling time. The similar slopes of the lines indicate that the hematocrit in the syringe has no effect on the efflux process. (pH 6.3; temperature, 38°.)

TABLE II
DELAY IN MIXING CHAMBER OF FLOW TUBE APPARATUS

Parameter	Mixing chamber	Section of tube
Length (cm)	0.2	0.2
Diameter (cm)	0.4	0.2
Cross-sectional area (cm^2)	0.126	0.0314
Volume (cm^3)	0.025	0.0063

	Velocity of flow		Transit time (msec)/0.2 cm		Delay in mixing chamber (msec)
	cm/sec				
cm^3/sec	Mixing chamber	Tube	Mixing chamber	Tube	
4.71	37.4	150	5.4	1.3	4.1
7.85	62.3	250	3.2	0.8	2.4
14.13	112.1	450	1.8	0.4	1.4

this error increases in magnitude the faster the transport process. It should be emphasized that the error has no effect on the calculated rate coefficient, if only the experimentally determined points are used for the linear regression analysis.[21]

Properties of Membranes. Plots of alcohol and water efflux show lower values of I than can be accounted for by trapped plasma radioactivity and a delay in the mixing chamber. It can, however, be ruled out that

TABLE III
SHIFT OF INTERCEPTION OF EFFLUX CURVE WITH
ORDINATE CAUSED BY DELAY IN MIXING CHAMBER[a]

Transport process		Interception by delay in mixing chamber (msec)		
Half-time (msec)	Rate coefficient (1/sec)	0	1.4	2.4
100	6.9	0.85	0.84	0.84
10	69.3	0.85	0.77	0.72

[a] The interception of 0.85 of the efflux curve with the ordinate by a delay of 0 msec is caused by extracellular radioactivity trapped between the cells in the syringe that amounts to 15% of the total radioactivity in the syringe.

TABLE IV
INTERCEPTION OF ^3H_2O EFFLUX CURVES WITH
THE ORDINATE IN EXPERIMENTS WITH CHICKEN
AND HUMAN RED BLOOD CELLS

Red blood cell	T (°C)	$t_{1/2}$ (msec)	Interception with ordinate
Chicken	38	13.5	0.87
Human	25	13.1	0.50

the low I values are methodological artifacts. Table IV shows a comparison of ^3H_2O efflux in human and chicken red cells performed under identical experimental conditions with the exception of temperature so that the two transport processes proceed at similar rates. With chicken red cells a

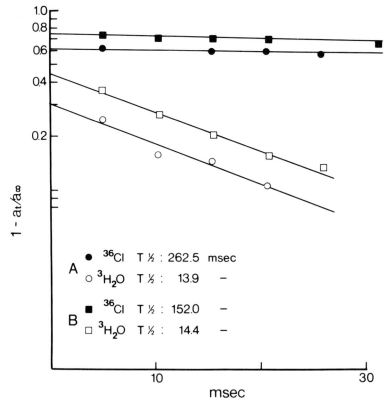

FIG. 5. Plot of two experiments with simultaneous efflux of ^3H_2O and $^{36}Cl^-$ from red cells. For explanation, see text. (pH 7.40; temperature, 25°.)

high I value is obtained in agreement with a trapped volume of 10–15%. The reduction of the I value to 0.5 for human red cells cannot be due to trapped volume. Figure 5 illustrates the results of two experiments in which the human red cell samples were loaded with both $^{36}Cl^-$ and ^3H_2O resulting in high and low I values, respectively. Cell sample A was spun less than sample B so that the efflux curves are displaced downward. The difference in the I values in each experiment is 0.3, independent of the degree of packing of the cell sample. The extrapolated efflux curves of each experiment intersect in abscissa (time) values of −10 to −12 msec. Since the flow rate in the experiments was about 250 cm/sec it can be concluded that the depicted difference of 0.3 in the paired I values is caused by the delay in the mixing chamber, which only amounts to 2–3 msec at the applied velocity of flow (cf. Table II).

Figures 6 and 7 depicting ^3H_2O and [^3H]methanol efflux, respectively, illustrate that the apparent value of the extracellular trapped tracer fraction at $t = 0$ increases with temperature. The pronounced temperature dependence of the I value is observed only for solutes that permeate the lipid phase of the membrane to a significant degree. I suggested[21] that the cells experience a transitory shear-stress caused by the acceleration of

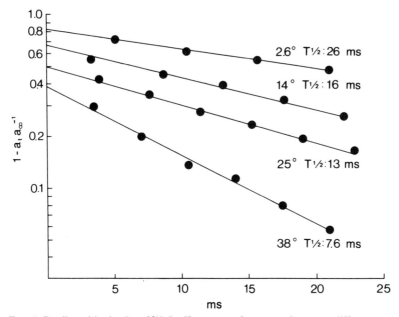

FIG. 6. Semilogarithmic plot of ^3H_2O efflux curves from experiments at different temperatures, to illustrate the temperature dependence of the interception with the ordinate. (From Brahm[21] by permission.)

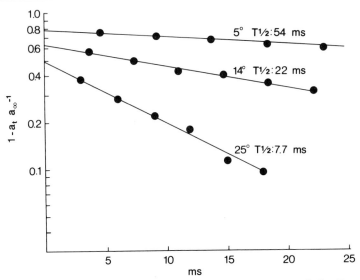

Fig. 7. The rate of methanol efflux from human erythrocytes (pH 7.2). Semilogarithmic plot of [³H]methanol efflux curves from experiments at different temperatures, to illustrate the temperature dependence of the interception with the ordinate.

flow rate as the cell enter the narrower tube after mixing in the chamber (cf. Table II). During this acceleration period the membrane lipids are transitorily rearranged ("membrane thinning")[22]; this may induce a brief increase in the permeability of the lipid phase of the membrane. The permeability increase is only detected for solutes, such as water and alcohols, which predominantly permeate the lipid phase, and is almost absent at low temperatures where the stiffening of the lipids may prevent membrane thinning.

If the temperature-induced displacement of the intercept for water at 38° (Fig. 6) is caused by an initial rapid exchange of ³H₂O, this process must have ended before the first sampling unit was reached by the cell suspension because the linearity of the plot indicates that only one rate coefficient determined the efflux at $t > 3$–4 msec. Thus, the permeability of the "rapid exchangeable fraction" must be $>10^{-2}$ cm/sec $[P > (\ln 2)/(t_{1/2})(V/A) = 0.69/0.003 \times 4.3 \times 10^{-5}$ cm/sec]. The high I value of the experiment performed at 2.6° (Fig. 6) indicates that all the intracellular ³H₂O exchanges with a rate that equals the experimentally determined permeability of 10^{-3} cm/sec. By raising the temperature to 38° the permeability briefly increases by a factor of 10. If one assumes that the increased

[22] S. Chien, K. P. Sung, R. Skalak, S. Usami, and A. Tözeren, *Biophys. J.* **24**, 463 (1978).

permeability is related to a diffusion coefficient in the membrane (D_m, cm^2/sec) that depends on the membrane viscosity (cf. the Einstein–Stokes relation: $D_m = RT/(6\pi rnN_A$; R is the gas constant; T is the absolute temperature; r is the radius of the spherical molecule, N_A is Avogadro's number) the transitory permeability increase is caused by a 10-fold decrease of the membrane viscosity due to the rise in temperature. Studies of the temperature dependence of the membrane viscosity of human red cells[23,24] show that the viscosity decreases 6–10 times between 5 and 38°, in support of this hypothesis.

That a similar increase of membrane permeability to chloride, urea, and glucose is not observed is in agreement with the proposed hypothesis. The solutes are predominantly transported by systems localized to integral membrane proteins. For each solute the fraction which is transported across the lipid membrane phase is so small that a brief increase in permeability is not detectable as a decrease of the value of the intercept with the ordinate.

In transport studies of water and some nonelectrolytes it is important to realize that the theoretical value of the intercept with the ordinate does not agree with the intercept obtained from a linear regression analysis of the experimental data points. By including the theoretical intercept value in the analysis,[10] the slope of the efflux curve, and thus the permeability coefficient, is overestimated.

Concluding Remarks

The continuous flow tube technique is an elaborate and reliable method to determine rapid transport processes. At the lower limit of its performance it overlaps with the upper limit of the Millipore–Swinnex filtering technique, and similar results are obtained with the two methods at the limit of their performance.[18] The upper limit of performance is set by the thickness of unstirred layers, which in our studies is reduced to a thickness close to the theoretical one. Thus, membrane permeabilities up to 10^{-2} cm/sec are within measurable range. More recent results of solute transport across red cell membranes obtained by means of more sophisticated methods, such as nuclear magnetic resonance and fluorescent probe techniques, have confirmed the results obtained by means of the flow tube technique, and thereby reject the recently raised criticism of the continuous flow tube method.[25]

[23] M. Dembo, V. Glushko, M. E. Aberlin, and M. Sonenberg, *Biochim. Biophys. Acta* **522**, 201 (1979).
[24] R. M. Hochmuth, K. L. Buxbaum, and E. A. Evans, *Biophys. J.* **29**, 177 (1980).
[25] N. P. Illsley and A. S. Verkman, *Biochemistry* **26**, 1215 (1987).

[10] Kinetic Properties of Na^+/H^+ Exchange and Li^+/Na^+, Na^+/Na^+, and Na^+/Li^+ Exchanges of Human Red Cells

By MITZY CANESSA

Kinetic Properties of Human Red Cell Na^+/H^+ Exchange

Na^+/H^+ exchange is a transport system widely distributed in eukaryotic cells and implicated in many cellular functions, such as control of cell pH and cell volume, response to mitogens, growth factors, and hormones, and in Na^+ reabsorption in the kidney. The expression of this transporter in the mature anucleated red cell might be a remnant of its activity in the precursor stem cells.[1] The maximal rate of proton gradient-driven Na^+ influx has a mean value in mature human red cells of 30 mmol/liter cells × hr, an activity 5–10 times higher than of the Na^+ pump which was believed to be the most active Na^+ transport.[2] However, the exchanger might not be operative in mature cells because the high activity of the Cl^-/HCO_3^- exchanger rapidly equilibrates outward H^+ gradients. The method described in this chapter can be used also to study the activity of the exchanger in rat and rabbit red cells.

Kinetics studies of the human red cell Na^+/H^+ exchanger indicate that it has a similar K_m for internal H^+ to activate Na^+ influx; the dependence on internal H^+ is very sigmoidal and has a high Hill coefficient ($n = 2$–3), revealing the presence of proton activatory site.[2] This kinetic property of Na^+/H^+ exchange has important physiological implications for the regulation of cell pH because it determines how steeply the transporter will respond to a slight fall in cell pH below pH_i 7.0.

External Na^+ activates Na^+ influx into acid-loaded cells by binding to a single class of low-affinity Na^+ sites (K_m 60 mM). Human red cell Na^+/H^+ exchanger is only partially amiloride-sensitive.[1,2] Amiloride and its high-affinity analogs inhibit 60 to 80% of the proton gradient Na^+ influx. Because the stoichiometric ratio between H^+ efflux driven by an inward Na^+ gradient and Na^+ influx driven by an outward H^+ gradient is $1:1$, the system can be assayed measuring the inhibitory effect of external acid pH on the net movement of Na^+ into acid-loaded cells incubated in Na^+ media with pH 8.0 and 6.0.[2]

The V_{max} of Na^+/H^+ exchange displays large interindividual differences in red cells of normal subjects (10–40 mmol/liter cells/hr); however,

[1] N. Escobales and M. Canessa, *J. Membr. Biol.* **90**, 21 (1986).
[2] A. Semplicini and M. Canessa, *J. Membr. Biol.*, in press (1989).

METHODS IN ENZYMOLOGY, VOL. 173

the factors involved in the expression of the transporter have not yet been investigated.

Red Cell Preparation

Blood is drawn into heparinized or EDTA tubes and centrifuged at room temperature at 1850 g for 10 min. Plasma and buffy coat are removed by aspiration and the red cells washed four times with cold choline washing solution (CWS). A 50% cell suspension is made in CWS to measure hematocrit (Ht) by centrifugation in capillary tubes in a microhematocrit centrifuge for 10 min. The initial red cell volume is estimated (optical density/liter cells) by measuring the hematocrit and the optical density (OD) of hemoglobin (Hb) at 540 nm. A 50-μl sample of the cell suspension in CWS is lysed (1/50) with Acationox (American Scientific Products, McGraw Park, IL) solution for cell Na$^+$ (1/50 dilution) and subsequently (1/500) for hemoglobin and cell K$^+$ determinations.

The Na$^+$ concentration of the cell lysate is measured in an atomic absorption spectrophotometer or flame photometer using appropriate Na$^+$ standards. The cell Na$^+$ content (mmol/liter cell) is calculated using the hematocrit of the cell suspension.

Modification of Cell Sodium Content

Cell Na$^+$ appears not to influence significantly Na$^+$/H$^+$ exchange activity in human red cells.[2] However, it is convenient to remove cell Na$^+$ to improve the determinations of cell Na$^+$ content. This step is not necessary for rat red cells which contain very low Na$^+$ content (3 mmol/liter cells).

A modified nystatin method is used to bring cell Na$^+$ to the desired concentration. The ionophore nystatin has a high partition coefficient (membrane/medium) for membrane cholesterol. The insertion of nystatin into the membrane is favored by low temperature (4°). Under these conditions, the membrane becomes permeable to cations and the nystatin holes will equilibrate internal and external cation concentrations. The expected red cell cation content is calculated in millimoles per liter of red blood cells dividing the cation concentration (mM) of the nystatin loading solution by the cell water content (65%). To avoid cell swelling during the insertion of nystatin, all loading solutions contain 50 mM sucrose, a nonpenetrating molecule, to balance the fixed cell anions (Hb, phosphate).

One milliliter of washed packed red cells is slowly added, while mixing, to 5 ml cold loading solution (NLS) containing 40 μg/ml of nystatin. Do not reverse this order of addition, otherwise cell lysis will be produced. The cell suspension is kept at 4°, in the dark, for 20 min, with periodic vortexing. Subsequently, the cell suspension is centrifuged, the supernatant discarded, the loading solution (without nystatin) renewed, and the cells incubated for an additional 10 min in the cold.

Nystatin is removed by adding 5 ml of warm (37°) nystatin washing solution (NWS) to the cell pellet and incubating for 10 min at 37°; NWS contains the same K^+ concentration as NLS. Nystatin removal is completed by four washes with NWS.

Modification of Cell pH

In red cells, pH_i and pH_o rapidly equilibrate through the actions of carbonate dehydratase and the Cl^-/HCO_3^- exchange system so that $[H^+]_i/[H^+]_o = [Cl^-]_o/[Cl^-]_i$. Thus, to modify the cell pH, the erythrocytes are incubated in acid-loading solutions made hypertonic with sucrose to avoid the cell swelling produced by the increase in Cl^- content. Afterward, the cell pH is clamped by inhibition of the anion exchanger with maximal doses of DIDS (4,4′-diisothiocyanatostilbene-2,2′-disulfonic acid) and inhibition of carbonate dehydratase by methazolamide (Neptazane, Lederle Co., NJ).

The cell pH of fresh or nystatin-treated cells was modified by incubating the cells at 10% Ht in acid-loading solutions for 10 min in a shaking water bath at 37° (Table I). The hypertonic media (around 350 mOsm) should not increase red cell volume less than 5% of that of the original cells. After a 10-min incubation, the pH_i is clamped by addition of 1.0 ml

TABLE I
COMPOSITION OF ACID-LOADING SOLUTIONS TO
MODIFY CELL pH OF HUMAN RED CELLS[a]

Cell pH	pH_o	KCl (mM)	Sucrose (mM)	mOsm
5.85	5.6	170	40	360
6.2	5.8	170	40	360
6.5	6.0	170	40	360
6.7	6.2	170	40	360
6.8	6.5	150	40	340
7.2	7.0	150	—	315
7.4	7.4	150	—	315

[a] All loading solutions contain 0.1 mM ouabain, 10 mM glucose, 20 mM Tris–MES for pH 5.6–7, or Tris–MOPS for pH 7.2 and 7.4, 0.15 mM MgCl₂, and 0.01 mM bumetanide (add fresh daily).

of acid-loading solution (ALS) containing 20 μl of 200 mM DIDS and 10 μl of 0.85 M methazolamide in dimethyl sulfoxide (DMSO) to inhibit the anion exchanger and carbonate dehydratase, respectively. The final concentration of these drugs is 0.2 mM (1.75 mg DIDS/ml RBC) and 0.5 mM methazolamide. The DIDS–methazolamide solution is prepared by adding 20 μl of 200 mM DIDS stock in DMSO and 10 μl of 0.84 M methazolamide stock to 1 ml of ALS. The cell suspensions are incubated in a shaker bath in their respective ALS for 30 more minutes at 37°. Table I shows the relationship between the pH of the loading solutions and the final cell pH. Because we found that there is interindividual variability, the pH of the incubating media can be measured prior to the addition of DIDS to ensure that the desired cell pH is achieved. If the pH of ALS is too far away from the desired pH, the ALS can be replaced with fresh solution or the pH of incubated media adjusted to the desired pH. The acid-loaded cells (pH$_i$ 5.9–6.9) were washed four times with 5 vol of cold (4°) unbuffered washing solution (AWS) and kept cold until flux measurement. For cells with pH$_i$ over 6.9, AWS was made isotonic by reducing KCl to 140 mM and eliminating sucrose.

The pH of the loaded cells is measured by lysing 0.2 ml of packed red cells in 2 ml of Acationox lysing solution and the pH of the lysate measured with a pH meter at room temperature. Use a small pH electrode and allow to equilibrate 3–5 min.

The cell pellet of acid-loaded cells is kept on ice ready for flux measurements. An aliquot of the cell suspension (50% Ht) is used for the measurements of intracellular Na$^+$ and hemoglobin as previously described. The cell volume after the acid loading (optical density of hemoglobin/liter cells) should be lower than 2.5% the values of the fresh cells.

Net Sodium Influx Measurements

To determine net Na$^+$ influx driven by an outward H$^+$ gradient, red cells loaded with different cell pH values are incubated in Na$^+$ media of pH 8.0 and 6.0; the difference between pH 8.0 and pH$_o$ 6.0 is a measure of the H$^+$ gradient-driven Na$^+$ influx. For V_{max} measurements cells loaded to pH$_i$ 6.0 should be incubated in a 150 mM NaCl medium. To determine the K_m for external Na$^+$, the Na$^+$ concentration of the medium can be varied between 0 and 100 mM. Sixty to 80% of net Na$^+$ influx driven by a proton gradient is inhibited by 1 mM amiloride and its high-affinity analogs. Four milliliters of Na$^+$ influx media, pH 8.0 and 6.0, are preincubated at 37° in a shaking water bath and the transport reaction is started by addition of 0.1 ml of packed acid-loaded cells (1% final Ht). This protocol avoids significant dissipation of the H$^+$ gradient, which takes place even at 4°; similarly, to separate cells from the Na^+ incubation media, cell washing with cold solutions is avoided. At timed intervals, the transport reaction is stopped by pipetting 0.25-ml duplicate aliquots of the warm cell suspension into previously chilled Eppendorf tubes (1.5 ml) containing 0.7 ml of cold Na^+-free solution layered over 0.4 ml dibutyl phthalate ($d = 1.04$). Duplicate samples from pH 8.0 Na^+ media are taken at 1, 5, and 11 min; for pH 6.0 Na^+ media duplicate samples are taken at 1 and 21 min. The Na^+-free solution (See Solutions for the Assay of Na^+/H^+ Exchange) over the oil reduces the trapping of Na^+ in the cell pellet. The tubes are immediately centrifuged in an Eppendorf centrifuge at 12,000 g for 15 sec; a longer time makes the subsequent lysing step more difficult. The supernatants are removed by aspiration, and from there on, the tubes wiped to remove external Na^+ contamination and placed in plastic racks washed with distilled water and covered with wippete paper.

Thereafter, the bottoms of the tubes are cut with a needle cutter (Destruclip from Fisher, Fairlawn, NJ) into 3-ml plastic tubes containing 1 ml Acationox lysing solution. Vortex the lysates periodically until all cells have lysed and keep overnight at 4°. Spin the lysates for 10 min at 4000 rpm for measurements of the optical density of hemoglobin. Fifty microliters of the lysate is diluted 1/10 to measure hemoglobin. The Na^+ concentration of the lysate is measured in an atomic absorption spectrophotometer or a flame photometer using Na^+ standards (20 to 150 μM) in 3 mM KCl. The Na^+ trapping is less than 1.0 μM Na^+. The cell Na^+ content (millimoles per liter of cells) is computed according to the following equation:

$$\text{Cell } Na^+ \text{ (mmol/liter cells)} = Na_{lys}OD_{cs}/Ht_{cs}OD_1 1000$$

where Na_{lys} is the Na^+ concentration (μM) of the lysate; OD_1, the optical density of Hb from the lysate of the flux media sample; OD_{cs}, the optical density of Hb from the lysate of the fresh cell suspension; and Ht_{cs}, the hematocrit of the fresh cell suspension.

The slope of the regression line of cell Na^+ content vs time is calculated with the least-squares method and net Na^+ influx expressed in millimoles per liter cells × hour. The coefficient of variation of the triplicate influx samples is 14% at pH_o 6.0 and 4% at pH_o 8.0, indicating that the external Na^+ contamination of the tubes is negligible if they are appropriately manipulated.

Net Na^+ influx driven by an outward proton gradient ($\Delta pH_o Na^+$ influx $= v$) is calculated from the difference between pH 8.0 minus pH 6.0. A Hill plot [$\log(V_{max} - v/v)$ as a function of $\log H_i^+$ (nM)] permits one to determine the K_m for internal protons from the $x = 0$ intercept and the Hill coefficient (n) from the slope.

Measurements of Unidirectional Sodium Influx

The experimental design to measure the unidirectional Na$^+$ influx is the same as for the net Na$^+$ influx except that the acid-loaded cells are incubated in 3 ml of medium containing 3 μCi/ml of ^{22}Na at a final hematocrit of 10%. ^{22}Na in the lysate of the influx media is counted in a well-type gamma counter up to 15,000 cpm. The cell Na$^+$ content is calculated according to

$$\text{Cell Na}^+ \text{ (mmol/liter cells)} = \text{cpm}_l \text{OD}_{cs}/\text{OD}_l \text{Ht}_{cs} \text{SA}$$

where cpm$_l$ is the counts per minute per liter of cells of the influx sample; OD$_l$, the optical density at 540 nm of the influx media sample; OD$_{cs}$, the optical density of the fresh cell lysate; Ht$_{cs}$, the hematocrit of the fresh cell suspension; and SA, the specific activity of the Na$^+$ influx media prior to flux measurement in counts per minute per millimole.

The unidirectional Na$^+$ influx (mmol/liter cells $\times$ hr) is computed from the slope of the regression line of cell Na$^+$ content versus time using the least-squares method.

Measurements of Proton Efflux

Na$^+$/H$^+$ exchange can also be estimated as the external Na$^+$-driven H$^+$ efflux. For this purpose, proton extrusion rates are measured into K$^+$ and Na$^+$ media. The flux media are unbuffered, and kept at constant pH$_o$ 8 (or 7.4, as needed) in a pH-stat apparatus (Brinkman 632) using KOH. The unbuffered K$^+$ or Na$^+$ media (4 ml) are first equilibrated at 37° in a water-jacketed chamber. Acid-loaded cells (0.3–0.4 ml of 50% cell suspension in AWS) are added to efflux media with constant stirring. The external pH is quickly (<30 sec) adjusted by the addition of 0.1 M KOH by the titrator to the set end point (pH$_o$ = 8.0) using the manual command. Thereafter the pH stat is set up in the automatic mode; the amount of 0.1 M KOH (μl) added to keep the pH$_o$ at the set end point is recorded for 10 min. A sample of efflux medium (50 μl) is diluted 1/50 in Acationox for Hb determination. Under these experimental conditions, the volume of KOH added is linear for 3–5 min and then declines due to the dissipation of the outward H$^+$ gradient. The initial rate (IR) of KOH solution added to the efflux media is calculated from the slope of the recording of volume of KOH as a function of time (μl vs time) if it is linear for more than 5 min; when the KOH volume, as a function of time, is linear for a shorter time a nonlinear regression program is used to calculate IR from the whole recording period according to

$$\text{IR} = (V - V_0)/\ln t$$

where IR is the initial rate of KOH addition (μl/min); V, the volume of KOH (μl) added at time t; V_0, the volume of KOH added to bring the pH of the efflux media to 8.0 (μl); and t, the time (min).

H^+ efflux is computed from the initial rate of KOH addition, i.e., of H^+ efflux, the concentration of the titrating solution, and the hematocrit of the titrated cell suspension, according to

H^+ efflux (mmol/liter cells $\times$ hr) =
$$(IR \times 60 \times C \times OD_{cs})(V_{cs} \times Ht_{cs} \times OD_1)$$

where IR is the initial rate of KOH addition (μl/min); C, the concentration of the titrating solution (M); OD_{cs}, the optical density of Hb from the lysate of the acid-loaded cell suspension prior to flux measurement; V_{cs}, the volume of the efflux media (ml); OD_1, the optical density of Hb from the lysate of the flux media sample; and Ht_{cs}, the hematocrit of the acid-loaded cell suspension.

Solutions for the Assay of Na^+/H^+ Exchange

Choline washing solution (CWS): Combine choline chloride, 149 mM; $MgCl_2$, 1 mM; Tris–MOPS, 10 mM; pH 7.4 at 4°. Measure osmolarity (280–300 mOsm) and pH

Acationox lysing solution: 0.02% in double-distilled water

Nystatin stock solution: Preweigh the complete nystatin sample into 5-mg portions and place in tubes. Cap the tubes, protect them from light with aluminum foil, and keep them in the refrigerator. This is to avoid inactivation of nystatin by humidity. The day of the experiment add 1.3 ml of DMSO to the tube

Nystatin loading solution (NLS): Combine KCl, 150 mM, and sucrose, 40 mM. Keep frozen in 50-ml aliquots. To change cell Na^+ content replace KCl for NaCl, keeping the sum at 150 mM. The day of the experiment add 45 μl of nystatin stock solution to 5 ml of NLS to be used to load 1.0 ml of packed red cells. Check the osmolarity of this solution

Nystatin washing solution (NWS): Combine KCl, 150 mM; sucrose, 40 mM; glucose, 10 mM; KPO_4 buffer, 10 mM, pH 7.4; albumin, 1 g/liter

Na^+-free diluting solution: Combine choline chloride, 80 mM; $MgCl_2$, 0.25 mM; Tris–MOPS, 10 mM, pH 7.4 at 4°; 82 mM KCl

Acid-loading solutions: Solutions (as given in Table 1) with decreasing pH (pH$_o$) are used to obtain red cells with different cell pH (pH$_i$)

Acid-washing solutions (AWS): To wash acid-loaded cells prepare two types of solution to be used for the indicated pH$_i$ (see Table II)

TABLE II
COMPOSITION OF ACID-WASHING SOLUTIONS[a]

pH_i	KCl	Sucrose	$MgCl_2$
5.9–6.8	170	40	0.15
7.2–7.4	150	—	0.15

[a] All washing solutions are unbuffered.

Bumetanide stock solution (10 mM): M_r 364.4. Dissolve 3.6 mg in 1 ml of DMSO

Methazolamide stock solution (0.84 M): Neptazane (from Lederle Laboratory Division of American Cyanamide Co., Pearl River, NJ). M_r 236.0. Dissolve 20 mg in 100 μl DMSO

DIDS stock solution (200 mM): From Sigma Chemical Company. M_r 498.5. Dissolve 10 mg in 100 μl of DMSO; make fresh daily. Keep reagent bottle away from light and humidity

Sodium influx media (mM) (see Table III)

Proton efflux media: NaCl or KCl, 150 mM, glucose, 10 mM; sucrose, 40 mM, $MgCl_2$, 0.15 mM; ouabain, 0.1 mM; bumetanide, 0.01 mM; methazolamide, 0.5 mM; phloretin, 0.1 mM

Amiloride stock solution: Prepare 1 M solution in DMSO. Add to the Na⁺ influx media (final concentration, 1 mM)

TABLE III
COMPOSITION OF Na⁺ INFLUX MEDIUM TO BE
USED FOR MEASUREMENTS OF Na⁺/H⁺
EXCHANGE IN HUMAN RED CELLS[a]

pH_i	NaCl (mM)	KCl (mM)	Sucrose (mM)	mOsm
5.85	150	20	40	360
6.0	150	20	40	360
6.2	150	20	40	360
6.5	150	20	40	360
6.8	150	—	40	340
7.2	150	—	—	315
7.4	150	—	—	315

[a] All flux media contain (mM): 0.1 ouabain, 10 glucose, 10 Tris–MES for pH 6 or Tris–MOPS for pH 8 media, 0.15 $MgCl_2$, and 0.01 bumetanide (add fresh daily).

Chemicals: Dibutyl phthalate from Fisher Scientific Company (Fairlawn, NJ). Amiloride, ouabain, Tris, MES, MOPS, DIDS, nystatin, and albumin (bovine fraction V) from Sigma Chemical Company (St. Louis, MO). Phloretin is from Nutritional Biochemicals Corporation (Cleveland, OH) and methazolamide is from Lederle Laboratories Division of the American Cyanamide Company (Pearl River, NY). Choline chloride is from Calbiochem, Behring Diagnostic (San Diego, CA); this brand was tested against recrystallized reagent and was shown to be ammonia free

Li^+/Na^+, Na^+/Na^+, and Na^+/Li^+ Exchanges of Human Red Cells

In human red cells, lithium is extruded against its own electrochemical gradient driven by an inward Na^+ gradient. The Na^+-dependent uphill extrusion is saturable, ouabain-insensitive, inhibited by phloretin, and does not require the presence of cellular ATP.[3-5] Unidirectional Li^+ efflux is stimulated by Na^+ on the trans side and inhibited by Na^+ on the cis side of the membrane. The K_m for internal Li^+ to stimulate Na^+-dependent Li^+ efflux is 0.5 mM and the K_m for external Na^+ is 20–25 mM.[5]

The maximum rate of Na^+-dependent Li^+ efflux (Li^+/Na^+ countertransport, Li^+/Na^+ exchange) varies markedly between individuals and it is under exchange genetic control. Genetic studies carried in families of Salt Lake City[6] and Rochester[7] indicate that 70 to 80% of the interindividual variance can be acccunted for by heritability factors. A model of polygenic inheritance and/or a recessive major gene can explain 80% of the variability of Li^+/Na^+ exchange in the human population.[6] A subset of patients with essential hypertension[8] and hyperlipidemia shows elevated V_{max} of Li^+/Na^+ exchange.

The Li^+/Na^+ countertransport appears to be a mode of operation of the Na^+/H^+ exchanger in human red cells.[9] It is sigmoidally stimulated by acid cell pH and external Na^+ and competitively inhibited by external acid pH; internal lithium also stimulate Na^+/H^+ exchange. However, this

[3] M. Haas, J. Schooler, and D. C. Tosteson, *Nature (London)* **258,** 425 (1975).
[4] G. N. Pandey, B. Sarkadi, M. Haas, R. B. Gunn, J. M. Davis, and D. C. Tosteson, *J. Gen. Physiol.* **72,** 233 (1978).
[5] B. Sarkadi, J. K. Alifimof, R. B. Gunn, and D. C. Tosteson, *J. Gen. Physiol.* **72,** 249 (1978).
[6] M. M. Dadone, S. J. Hasstedt, S. C. Hunt, J. B. Smith, O. Ash, and R. R. Williams, *Am. J. Med. Genet.* **1117,** 565 (1984).
[7] E. Boerwinkle, S. T. Turner, R. Weinshilboum, M. Johnson, E. Richelson, and C. F. Sing, *Genet. Epidemiol.* **3,** 365 (1986).
[8] M. Canessa, N. Adragna, H. Solomon, T. M. Connolly, and D. C. Tosteson, *N. Engl. J. Med.* **302,** 772 (1980).

pathway, as well as the Na^+/Na^+ exchange mode, is not inhibited by amiloride and its analogs.[9] Because there is a significant fraction of 1 : 1 Na^+/H^+ exchange insensitive to amiloride, the lack of effect of this inhibitor does not exclude that this exchange is a transport mode of the pH regulation system.

The Li^+/Na^+ exchange is insensitive to inhibitors of the anion exchanger such as DIDS, furosemide, and dipyridamole.[10] However, a fraction of Na^+/Na^+ and Na^+/Li^+ exchanges instead are blocked by all these inhibitors. Thus, measurements of external Li^+-stimulated Na^+ efflux are not equivalent to Na^+-stimulated Li^+ efflux because of the asymmetry of the Li^+ and Na^+ sites and their sensitivity to inhibitors; assays of Li^+-stimulated Na^+ efflux may comprise DIDS-sensitive cation movement coupled to the anion exchanger.[10]

Li^+/Na^+ exchange is best measured as external Na^+-stimulated Li^+ efflux. Because external Mg^{2+} concentration is inhibitory of Li^+ efflux into Na^+-free media, the assay is performed maintaining constant Mg^{2+} concentration (1 mM) in the washing and efflux media. To perform V_{max} measurements red cells are loaded with 6 to 10 mmol/liter cells of Li^+ to saturate internal sites. Li^+ efflux is measured in the absence of Na^+ (choline media) and with 150 mM NaCl; the difference between both effluxes measures Li^+/Na^+ countertransport. Under the condition of the assay, Li^+ efflux is linear up to 75 min in normal subjects.

The principal sources of error in this technique are (1) incomplete washing of the Li^+-loaded cells, which may lead to high Li^+ concentration at time-zero efflux; (2) irregular pipetting of the cell suspension into Na^+-free and Na^+ media; (3) low values of Li^+ efflux into choline media (it is recommended that a double amount of cells of Na^+ media be used); and (4) lithium measurements by atomic absorption. The use of standards in the same media is especially critical in certain atomic absorption spectrophotometers such as IL instruments.

Na^+/Na^+ exchange is best measured by the component of Na^+ efflux stimulated by external Na^+ in the presence of ouabain to inhibit the Na^+ pump, of bumetanide to inhibit $Na^+/K^+/Cl^-$ cotransport and DIDS to inhibit the anion exchanger.

Na^+/Li^+ exchange can be measured by the component of Na^+ efflux stimulated by external Li^+ (10 mM)[6,10] or as phloretin-sensitive Li^+ influx into Na^+-loaded cells.[5,11] Both procedures give rates three times lower than Li^+/Na^+ exchange.

[9] M. Canessa, K. Morgan, and A. Semplicini, *J. Cardiovasc. Pharmacol.*, in press (1988).
[10] P. A. Hannaert and R. P. Garay, *J. Gen. Physiol.* **72**, 353 (1986).
[11] J. Duhm, B. O. Gobel, R. Lorenz, and P. C. Weber, *Hypertension* **4**, 477 (1982).

Blood Sampling

Draw 5 ml of blood into a vacutainer tube containing Na^+–heparin or EDTA. Transfer the blood into a Sorvall centrifuge tube and spin 15 min at 3000 rpm at $4°$. Aspirate and discard the plasma and buffy coat. Wash the cells four times with 10 ml of choline washing solution (CWS), keeping the tubes in ice and spinning 2 min between washes at 7000 rpm at $4°$.

Dilute an aliquot of cells with an equal volume of CWS. Label as fresh cell sample (FC). Use the remaining packed cells for lithium loading by nystatin or diffusional method. Measure the hematocrit of the suspension in duplicate. Dilute (1/50) 50 μl of FC 50% suspension to 2.5 ml Acationox solution for cell Na^+ determination. Dilute $10\times$ the 1/50 dilution with double-distilled water for cell K^+ determination and OD at 540 nm for hemoglobin.

Red Cell Preservation

Washed red cells are suspended at 20% hematocrit in the preservation solution. Keep the suspension at $4°$. Transport assays should be performed within 3 days.

Red Cell Lithium-Loading Procedures

Nystatin Loading. This method permits one to prepare Li^+-loaded cells in a shorter time than the diffusional method. It is convenient when a large number of samples are processed. Pipette 5 ml of cold Li^+-loading solution and 40 μl of nystatin stock solution into a cold 15-ml Sorvall centrifuge tube. While vortexing the above mixture, slowly add 1 ml of packed red blood cells to the tube. Do not change this sequence because it is designed to avoid hemolysis. Keep the tube at $4°$ in ice water for 20 min, periodically vortexing.

Spin for 2 min at 3000 g in a refrigerated centrifuge, remove the supernatant, and add 5 ml of cold $(4°)$ loading solution without nystatin. Place the tubes in the ice-water bath for an additional 10 min. Preincubate four Sorvall tubes (15 ml) with 10 ml of lithium nystatin washing solution (LiNWS) at $37°$ in a water bath. Spin the loaded cells at room temperature, remove the supernatant, and incubate the cells suspension in a $37°$ water bath for 5 min with one of the aliquots of warm LiNWS; spin and wash three times with 10 ml of warm $(37°)$ LiNWS.

Wash the cells five times with cold CWS, keeping the tubes on ice. After the last wash, make a 50% cell suspension with CWS. Keep the suspension at $4°$. Measure hematocrit, electrolytes, and Li^+ and K^+ content, diluting as described for fresh cell sample.

To ensure that nystatin removal is complete measure Na$^+$ efflux from fresh cells and nystatin-loaded cells (10 mmol/liter cell Na$^+$) into a choline medium in the presence of ouabain and furosemide. The rate constant of Na$^+$ efflux should not increase more than 2% upon nystatin loading. Inadequate loading can be caused by a nystatin batch with low activity; test three different nystatin concentrations (40, 45, 50 μg/ml) in the loading solution if inadequate loading is obtained. The rate constant of Li$^+$ efflux (flux/cell Li$^+$ content) into choline media should not be higher than 0.035.

Diffusional Lithium Loading. Red cells are incubated for 3 hr in 150 mM Li medium buffered at pH 7.4. The cells gain Li$^+$ by diffusion, by exchange with internal Na$^+$, and by transport by the Na$^+$ pump. Wash the cells four times with choline washing solution. Make a 50% suspension and measure hematocrit, Li$^+$, Na$^+$, and K$^+$ cell content (1/50 dilution) and optical density of hemoglobin (1/500 dilution). This method is very safe for Li$^+$ loading but too long to process large number of samples.

Sodium and Radioactive Sodium Loading of Red Cells

For determinations of Na$^+$/Na$^+$ and Na$^+$/Li$^+$ exchanges, sodium content can be varied using the ionophore nystatin. Washed packed cells are incubated at 15% hematocrit in cold nystatin loading solution (NLS) containing 40 μg/ml nystatin for 20 min at 4° in the dark. To vary the cellular Na$^+$, the NLS should contain different ratios of NaCl : KCl according to the desired Na$^+$ content. For V_{max} measurements of DIDS-insensitive Na$^+$/Na$^+$ exchange load red cells with 10 mmol/liter cell Na$^+$. For V_{max} measurements of Na$^+$/Li$^+$ exchange load cells to 20 mmol/liter cell Na$^+$. To prepare cells with low internal Na$^+$ (depleted or less than 10 mmole/liter cells) the supernatant was removed and the cells incubated for a further 20 min in NLS without nystatin. The cells are loaded with ^{22}Na, after the first incubation at 4°, by resuspending the cells at 50% hematocrit in the same loading solution, without nystatin, containing 5 μCi ^{22}Na/ml packed cells. After a 20-min incubation at 4°, with adequate shaking, the suspension is warmed to 37° for 5 min, subsequently spun for 2 min at 5000 g, and washed four times with warm (37°) nystatin wash solution (NWS). After the final wash the supernatant is removed and the cells washed with CWS to remove external cations.

Lithium Efflux Measurements

Prepare two plastic tubes (20–30 ml) with 7.0 ml of choline and Na$^+$ media and keep them in ice. Use class A volumetric pipets to measure this volume, because flux calculations are based on the precision of this volume. Label six tubes (4-ml plastic tubes) for each medium for duplicate samples at 0, 30, and 60 min. Keep the tubes at 4° in an ice-water bath. Color code the tubes for the choline and Na^+ media. Vortex the cell suspension (50%) of the Li^+-loaded cells and then add 0.7 ml of suspension to the choline and Na^+ media. Vortex the flasks and using plastic Pasteur pipets divide the contents (1.5 ml) of each flask into the six appropriately labeled tubes and cap them. The tubes are incubated for 10, 30, and 60 min at 37° in a shaking water bath and vortex every 15 min. Remove the tubes at the appropriate time, cool in ice water for 1 min, and spin the tubes at 5000 rpm for 5 min in a refrigerated (4°) centrifuge. Carefully aspirate the supernatant with a plastic pipet, avoid putting the tip of the pipet close to the cells. Transfer the supernatants into prelabeled 3-ml plastic tubes. Cap the tube if Li^+ concentration is not measured immediately. These supernatants are used for Li^+ determination by atomic absorption spectrophotometry.

Li^+ efflux is calculated from the change in Li^+ concentration in the media per unit time ($\mu mol/min$) divided per volume of cells used to measure the flux from

Li^+ efflux (mmol/liter/cells/hr) =

$$\frac{V_{media}(Li_{60\,min} - Li_{0\,min})/1000 + (1 - Ht_0 \times V_c)1000}{V_c Ht_0}$$

where $Li_{60\,min}$ is the Li^+ concentration (μM) of the 60-min sample; $Li_{0\,min}$, the Li^+ concentration (μM) of the zero-time sample; Ht_0, the hematocrit of the 50% cell suspension Li^+-loaded cells; V_c, the volume, in milliliters, of cell suspension of Li^+-loaded cells added to the flux media (0.7 ml); and V_{media}, the volume of the efflux medium (7.0 ml).

The change in concentration between 0 and 60 min can be used only if the rate is linear. The linear regression for the six determinations (slope) can be used for a more accurate determination. In subjects with very high Li^+/Na^+ exchange the efflux is linear only up to 30 min. The zero-time concentration of Li^+ should not be higher than 3 μM.

Sodium Efflux Measurements

$^{22}Na^+$ efflux from cells loaded with different Na^+ contents is measured by incubating the ^{22}Na-loaded cells at 2% Ht at 37°. Na^+/Na^+ exchange is measured as the difference in ^{22}Na efflux between Na^+ medium (150 mM) and Na^+-free K^+ medium (150 mM), i.e., as external Na^+-stimulated Na^+ efflux. ^{22}Na-loaded cells, with the required internal Na^+, are suspended at 2% (Na^+ medium) and 4% hematocrit (K^+ medium) in 7 ml of ice-cold flux medium. Following distribution of the cell suspension into 4-ml tubes on ice, Na$^+$ efflux is started by warming up the cell suspension to 37°. Initial rates were determined by taking duplicate samples at 0, 30, and 60 min. The samples are centrifuged for 2 min, 2000 g at 4°, and the radioactivity in the supernatant counted in a 0.8-ml aliquot. The Na$^+$ efflux is calculated (mmol/liter cells × hr, flux unit = FU) from the specific activity of the loaded cells and the cpm of ^{22}Na appearing in the efflux media as a function of time.

The Na$^+$ content of the efflux media was calculated according to

Na$^+$ content of the efflux medium (mmol/liter cells) =
$$\text{cpm}_{sp}/(SA_a 1 \times Ht_f)$$

where cpm$_{sp}$ is the radioactivity in the supernatant (cpm/liter); $SA_a 1$, the specific activity of the acid-loaded cells prior to flux measurement (cpm/mmol Na$^+$); and Ht$_f$, the hematocrit of the efflux medium.

^{22}Na efflux was calculated from the slope of the regression line of the Na$^+$ content of the efflux medium as a function of time with the least-squares method divided per volume of cells used in the flux media.

Na$^+$/Li$^+$ exchange is determined measuring Na$^+$ flux into 150 mM K$^+$ medium (or choline) and 140 mM KCl plus 10 mM LiCl.

Solutions

Stock solutions: It is convenient to prepare the following stock solutions in double-distilled water for rapid preparation of several solutions used in these assays. Prepare 1 M glucose (180.6 g/liter), 0.5 M sucrose (171.15 g/liter), 10 mM ouabain (7.286 g/liter). Tris–MOPS, 100 mM, pH 7.4 (20.93 g/liter) is prepared by adding Tris base to a MOPS solution up to the pH at room temperature which corresponds to pH 7.4 at 37°. Use a temperature chart for Tris buffers. NaCl (1 M; 58.44 g/liter), 1 M MgCl$_2$ (203.31 g/liter), 1 M LiCl (42.39 g/liter), and 1 M choline chloride (139.6 g/liter) should be kept cold; also maintain reagent bottle in the refrigerator to avoid bacterial degradation

Potassium phosphate buffer, pH 7.4: Prepare 100 mM K$_2$HPO$_4$ (8.709 g/500 ml) and KH$_2$PO$_4$ (6.8045 g/500 ml) and mix them to the correct pH

Nystatin stock solution: Preweigh all nystatin out of a big bottle into tubes containing 5 mg. Cap the tubes and protect them from the light with aluminum foil and keep them in the refrigerator. This is to avoid inactivation of nystatin by humidity

Choline washing solution (CWS): Combine choline chloride, 149 mM; MgCl$_2$, 1.0 mM; Tris–MOPS, 10 mM; pH 7.4 at 4°. Measure osmolarity (280–300 mOsm) and pH. Used choline chloride from Calbiochem or otherwise should be recrystallized to eliminate ammonia

Acationox lysing solution: Prepare a 0.02% solution in double-distilled water

Nystatin loading solutions (NLS): For lithium loading use KCl, 140 mM; LiCl, 10 mM; sucrose, 40 mM. For Na$^+$ loading use 10 mM NaCl and 140 mM KCl. Keep frozen in 50-ml aliquots. On the day of the experiment add 45 μl of nystatin stock solution to 5 ml of NLS to be used to load 1.0 ml of packed red cells. Check the osmolarity of this solution

Nystatin washing solution (NWS): To wash lithium-loaded cells use KCl, 154 mM; LiCl, 10 mM; sucrose, 40 mM; glucose, 10 mM; KPO$_4$ buffer, pH 7.4, 10 mM; albumin, 1 g/liter. For sodium-loaded cells replace lithium for 10 mM NaCl. Nystatin washing solutions should be aliquoted in 50-ml capped tubes and stored frozen. Do not reuse them for more than 1 day after warming up because it is a good culture medium

Choline efflux medium: Prepare using stock solutions of 148 mM choline chloride, 1 mM MgCl$_2$, 10 mM Tris–MOPS, pH 7.4 at 37°, 10 mM glucose, and 0.1 mM ouabain. The Na$^+$ contamination of this medium should not be higher than 6 μM

Na$^+$ efflux medium: Combine 150 mM NaCl, 1.0 mM MgCl$_2$, 10 mM Tris–MOPS, pH 7.4 at 37°, 10 mM glucose, and 0.1 mM ouabain

Preservation solution: Combine 140 mM KCl, 10 mM NaCl, 2.5 mM phosphate buffer, and 1.0 mM MgCl$_2$

Na$^+$ efflux medium: 150 mM NaCl and KCl to give an external concentration of Na$^+$ of 0 to 150 mM, 0.15 mM MgCl$_2$, 0.1 mM ouabain, 10 mM glucose, 10 mM Tris–MOPS, pH 7.4 at 37°

Atomic absorption standards: Dry LiCl, NaCl, or KCl in a large weighing dish in a 110°F oven overnight. Place the weighing dish into a clean desiccator to cool. When cool, check the weight and note it. Place the dish back into the desiccator overnight. Check the weight again the next day to see if there is any change. Repeat this until there is no change in the weight for 2 days. Weigh out exactly 584.4 mg to make 1 liter of 10 mM stock, 423.9 mg of LiCl (very hygroscopic), or 745 mg of KCl

Rinse a 1-liter volumetric flask (class A) with a small amount of double-distilled water three times and pour the weighed salt. Rinse the weighing dish and bring the flask slowly almost to volume with double-distilled water. Allow a few minutes for the residual water on the sides of the flask to settle, then with a plastic transfer pipet bring it to volume slowly, drop by drop. Rinse a 1-liter Nalgene bottle with a small amount of the stock solution three times, and later transfer the remaining stock solution into the bottle. Cap the bottle tightly and wrap some paraffin film around it to avoid any evaporation.

To prepare standards, never pipet directly from the bottle; always pour needed amount into a plastic tube and then recap the bottle. To make standards from the stock solutions always use 500-ml volumetric flasks (class A) and volumetric pipets (TD, class A). TD means to deliver, do not blow out. Fill the pipet, deliver the volume mark, wipe off outer excess with a clean Kimwipe, slowly bring volume to the mark, then allow remainder to drain into flask.

The composition of the standards should have a background similar to the samples to be read. If there is a background of 150 mM NaCl in the sample, then prepare Li$^+$ standards in 150 mM NaCl. These standards should be checked against commercially available atomic absorption stock standards. For measurements of Li$^+$ concentrations of Li$^+$ efflux media prepare standards with 10, 20, 50, and 100 μM LiCl in choline and Na$^+$ media. For measurements of Li$^+$ concentration in cell lysate prepare Li$^+$ standards in 3 mM KCl. For Na$^+$ measurements in cell lysate prepare standards in 3 mM KCl.

Measure the absorbance of the standards with an atomic absorption spectrophotometer and make a graph of concentration vs optical density. A linear relationship between concentration and optical density should be seen up to 120 μM with a slight bending between 120 and 240 μM. Keep the graphs as a record, and compare each new batch of standards to the previous ones to be sure there are no differences.

Chemicals. Dibutyl phthalate is from Fisher Scientific Company (Fairlawn, NJ). Amiloride, ouabain, Tris, MES, MOPS, DIDS, nystatin, and albumin (bovine fraction V) are from Sigma Chemical Company (St. Louis, MO). Phloretin is from Nutritional Biochemicals Corporation (Cleveland, OH) and methazolamide is from Lederle Laboratories Division of the American Cyanamide Company (Pearl River, NY). Choline chloride was obtained from Calbiochem-Behring Diagnostic (San Diego, CA); this brand was tested against recrystallized reagent and shown to be ammonia free.

[11] Water Channels across the Red Blood Cell and Other Biological Membranes

By ARTHUR K. SOLOMON

Water may cross the red cell or other biological membranes either by diffusion or filtration, two fundamentally different processes. Diffusion is normally measured by tracer flux, usually of ^{3}HHO (THO), in the absence of any chemical potential gradient for bulk water, so that the only driving force is the concentration gradient of the THO, which is negligible in a macroscopic sense. In contrast, filtration, or hydraulic conductivity, is measured by the net water flux in response to a hydrostatic or osmotic pressure gradient. In aqueous channels through a membrane the hydraulic conductivity is greater than the diffusional conductivity and in human red cells the ratio of these permeability coefficients[1] is 6.3. This chapter has been written specifically about the red cell membrane, but the principles and many of the methods are also applicable to transport across biological membranes in vesicles or other single-cell preparations.

Water Diffusion across the Red Cell Membrane

For diffusion, the unidirectional flux of the ith solute, J_i (in mol cm^{-2} sec $^{-1}$) is given by

$$J_i = c_i v_i \tag{1}$$

in which the concentration, c_i, is given in mol cm^{-3} and the velocity, v_i, is given in cm sec^{-1} (see Schultz[2]). The velocity is directly related to the force, f_i, by the proportionality coefficient, the mobility, u_i (the velocity per unit force), so that $v_i = u_i f_i$. For diffusion in the x direction, $f_i = -d\bar{\mu}_i/dx$ in which $\bar{\mu}_i$ is the electrochemical potential of i, so that

$$J_i = -c_i u_i (d\bar{\mu}_i/dx) \tag{2}$$

This generalized form of the diffusion equation is the Nernst–Planck equation in one dimension, which can easily be simplified to the case of water diffusion by substituting the subscript w for i and the water potential gradient, $d\mu_w/dx$, for the electrochemical potential gradient, so that

$$J_w = -c_w u_w (d\mu_w/dx) \tag{3}$$

[1] A. K. Solomon, *J. Membr. Biol.* **94,** 227 (1986).
[2] S. G. Schultz, "Basic Principles of Membrane Transport." Cambridge Univ. Press, Cambridge, England, 1980.

If the solutions are considered to be ideal and temperature and pressure are constant, $d\mu_w/dx$ may be replaced by $RT \, d \ln c_w/dx$ so that

$$J_w = -u_w RT (dc_w/dx) \tag{4}$$

For diffusion across a membrane in the steady state, assuming u_w to be constant across the membrane, Eq. (4) can be integrated. Using the Einstein relation, $D_w = RTu_w$, we obtain the diffusion equation in its usual form:

$$J_w = -D_w(\Delta c_w/\Delta x) = -P_d \Delta c_w \tag{5}$$

in which Δc_w is the water concentration difference across a membrane of thickness Δx. Δc_w is taken to represent the gradient only in the direction in which the tracer moves, so that Δc_w in pure H_2O would equal 55.56×10^3 mol cm^{-3}. In a homogeneous aqueous solution, D_w is the diffusion coefficient of water; when water passes through channels in a membrane, the hindered diffusion coefficient, D'_w, is $<D_w$. The water diffusional coefficient through the membrane, P_d, is $D'_w/\Delta x$ with the dimensions, cm sec^{-1}. P_d is related to the permeability coefficient, ω (mol dyn^{-1} sec^{-1}) used in the irreversible thermodynamic treatment of membrane processes[3] by $P_d = \omega RT$. The characteristic exchange time for efflux from the cell, $t_e = (V'/A)/P_d$. The cell water volume, V', to area, A, ratio used by Dix and Solomon[4] is $V'/A = 5.33 \times 10^{-5}$ cm (using $V = 100 \times 10^{-12}$ cm^3; $A = 1.35 \times 10^{-6}$ cm^2; cell H_2O content $= 0.72$, following Canham and Burton,[5] Jay,[6] and Sidel and Solomon[7]).

To measure water diffusion in a two-compartment system, tracer is added to one compartment at $t = 0$, usually in the absence of any gradient in water activity. The THO tracer flux is first determined in cpm cm^{-2} sec^{-1} [in which cpm is counts (or disintegrations, dpm) min^{-1}] and then divided by the specific activity in cpm (or dpm) (mol H_2O)$^{-1}$ to determine the water flux in (mol H_2O) cm^{-2} sec^{-1}. D'_w is then obtained from Eq. (5). Δc_w is obtained from the bulk water concentration on the tracer side.

Continuous Flow Method Using Radioactive Tracers

Water diffuses into the human red cell very quickly with an efflux exchange time, t_e, of 12.7 msec at 20°,[4] so that rapid reaction methods are

[3] A. Katchalsky and P. F. Curran, "Nonequilibrium Thermodynamics in Biophysics," pp. 113–132. Harvard Univ. Press, Cambridge, Massachusetts, 1965.
[4] J. A. Dix and A. K. Solomon, Biochim. Biophys. Acta 773, 219 (1984).
[5] P. B. Canham and A. C. Burton, Circ. Res. 22, 405 (1968).
[6] A. W. L. Jay, Biophys. J. 15, 205 (1975).
[7] V. W. Sidel and A. K. Solomon, J. Gen. Physiol. 41, 243 (1957).

necessary. In 1957, Paganelli and Solomon[8] used a continuous flow method to measure the diffusion of THO out of the human red cell, based on the designs of Hartridge and Roughton,[9] Dirken and Mook,[10] and Chance.[11] When Hartridge and Roughton devised the continuous flow method in 1923 to extend the measurement of reaction kinetics down to the millisecond range, they used optical methods to detect the advancement of the reaction by changes in the hemoglobin (Hb) spectrum. Dirken and Mook[10] extended the continuous flow method in 1931 by separating samples of plasma from red cells by filtration at known times after mixing (with a resolution of about 20 msec) so that these timed samples of filtrate could later by analyzed for pH, total CO_2, and Cl^- by chemical methods. Paganelli and Solomon extended the Dirken and Mook filtration technique and devised an apparatus (Fig. 1A and B) that could filter four plasma samples from cells over a time span of 8 msec with a time resolution of about 0.2 msec. In 1964, Barton and Brown[12] modified the method to require smaller samples of blood and to provide 10 collecting ports spaced over a time span of about 17 msec with the same time resolution and an exponential time course, as shown in Fig. 1C. The continuous flow method has also been used to determine red cell permeability to other solutes, particularly Cl^-, which was measured by Tosteson[13] in 1959 and Brahm[14] in 1977. Brahm's apparatus, which he later used to measure red cell water diffusion,[15] operated with a much lower hematocrit and provided six samples over a 20- to 25-msec time span.

There are a number of artifacts of the continuous flow method which, as is usually the case, are discussed most carefully in the early papers describing the method, particularly by Roughton and Millikan[16] in 1936, in Chance's 1940 Ph.D. thesis,[11] and subsequently in 1963 by Roughton[17] again. One question is the performance of the mixing chamber. It is necessary to take the time spent in the mixing chamber into account in computing the reaction times at the four filtration ports; Paganelli and Solomon[8]

[8] C. V. Paganelli and A. K. Solomon, *J. Gen. Physiol.* **41**, 259 (1957).
[9] H. Hartridge and F. J. W. Roughton, *Proc. R. Soc. London, Ser. A* **104**, 376 (1923).
[10] M. N. J. Dirken and H. W. Mook, *J. Physiol. (London)* **73**, 349 (1931).
[11] B. Chance, Ph.D. thesis. Univ. of Pennsylvania, Philadelphia, 1940.
[12] T. C. Barton and D. A. J. Brown, *J. Gen. Physiol.* **47**, 839 (1964).
[13] D. C. Tosteson, *Acta Physiol. Scand.* **46**, 19 (1959).
[14] J. Brahm, *J. Gen. Physiol.* **70**, 283 (1977).
[15] J. Brahm, *J. Gen. Physiol.* **79**, 791 (1982).
[16] F. J. W. Roughton and G. A. Millikan, *Proc. R. Soc. London, Ser. A* **155**, 258 (1936).
[17] F. J. W. Roughton, *in* "Rates and Mechanisms of Reactions: II. Techniques of Organic Chemistry" (S. L. Friess, E. S. Lewis, and A. Weissberger, eds.), Vol. 8, p. 704. Wiley (Interscience), New York, 1963.

A

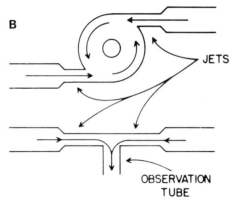

TOGGLE VALVES

|←— d —→|

FILTRATION CHAMBER

TO PRESSURE TANK

TRITIATED BUFFER

MIXING CHAMBER

EFFLUENT RED CELL SUSPENSION

RED CELL SUSPENSION

$$ t = \frac{d}{u} \quad \text{WHERE} \quad \begin{aligned} d &= \text{DISTANCE} \\ u &= \text{LINEAR FLOW} \\ t &= \text{REACTION TIME} \end{aligned} $$

B

JETS

OBSERVATION TUBE

FIG. 1. (A) Continuous flow apparatus of Paganelli and Solomon.[8] Flow was initiated by opening two spring-loaded toggle valves simultaneously, forcing the solutions through the tangential mixing chamber (B) and the filtration chamber, whose ports are closed by Millipore filters which filter the suspension medium from the cells. The driving pressure which propels the suspension also provides the pressure gradient for the filtration. (C) Data of Barton and Brown[12] showing exponential time course of THO loss from human red cells in a continuous flow apparatus with 10 ports. p_n is the THO concentration in the suspension medium and p_∞ is its equilibrium value.

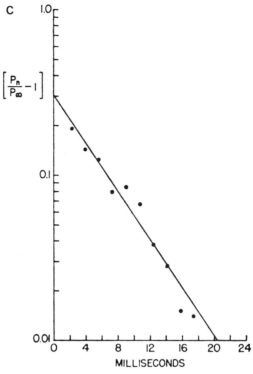

FIG. 1. (*continued*)

used the formula of Hartridge and Roughton[9]:

$$L_0 = r_m^2 L_m / r_0^2 \qquad (6)$$

in which L_0 is the equivalent length of the mixing chamber (cm), r is the radius in cm (of the mixing chamber, r_m, and the observation tube, r_0) and L_m is the length of the mixing chamber. For our chamber, L_0 was 0.54 cm, = 0.7 msec at our flow velocity. In our tangenitial two-jet mixing chamber, we showed that by the time the reaction mixture had passed 1.6 mm beyond the exit orifice of the mixing chamber, 0.9 msec had elapsed and mixing was at least 97% complete. When other apparatuses are built with different dimensions of the jets and mixing chamber, it is necessary to confirm that the mixing is complete and to calculate the equivalent length.

For red cells, it is particularly important to make sure that the flow down the observation tube is turbulent because, as Coulter and Pappenheimer[18] have shown, in Poiseuille flow the red cells are concentrated

---

[18] N. A. Coulter, Jr. and J. R. Pappenheimer, *Am. J. Physiol.* **159**, 401 (1949).

along the axis of the observation tube. If the concentration of the cells in the suspending medium is not radially isotropic, samples filtered at the periphery, as in our apparatus, are not representative of the mean hematocrit. Coulter and Pappenheimer have shown that the Reynolds formula

$$\mu_c = 1000\eta/\rho r_0 \tag{7}$$

in which μ_c is the critical velocity (cm sec^{-1}), η is the viscosity of the suspension (poise), and ρ is the density (g cm^{-3}) applies to bovine blood at many hematocrits. In our apparatus, the critical velocity of 640 cm sec^{-1} was always exceeded by the velocity, which was usually 814 cm sec^{-1}. Paganelli and Solomon[8] also showed that the flow velocity was linearly proportional to the (applied pressure)$^{0.5}$, which is typical of turbulent flow. When experiments are done in experimental systems with different observation tubes or flow velocities, or different cell or vesicle suspensions, it is necessary to establish that the flow is turbulent.

The tracer permeability coefficient was computed by Paganelli and Solomon on the basis of conventional steady-state two-compartment kinetics, but there are certain problems that need to be addressed in the continuous flow method, particularly those of the unstirred layer and diffusion within the red cell. Dainty[19] pointed out the importance of the unstirred layer of fluid that may still adhere to the external surface of the red cells after their environment has been suddenly changed in a jet mixing chamber. Dainty gave the following equation to relate the measured permeability coefficient, $P_{d,\,obs}$ to the corrected coefficient, $P_{d,true}$:

$$P_{d,\,obs}^{-1} = P_{d,\,true}^{-1} + \delta/D_w \tag{8}$$

in which δ is the thickness in centimeters, of the unstirred layer. Sha'afi *et al.*[20] showed that the thickness of the unstirred layer in a mixing chamber similar in general design to that of Paganelli and Solomon[8] (but operating at a much lower hematocrit, a lower solution velocity, and with larger jets) was 5.5 μm. They pointed out that this thickness of unstirred layer caused only a small correction in P_d. Recently Williams and Kutchai[21] used the diffusional quenching of a fluorescent probe, tetramethylrhodamine isothiocyanate, adhering to the red cell membrane in a stopped-flow apparatus, to measure an unstirred layer of 4.7 μm (or 6.9 μm using a slightly different model), in good agreement with the earlier value.

It is also necessary for water diffusion within the cell to be fast compared with the exchange time, $t_e = 12.7$ msec. Paganelli and Solomon[8]

[19] J. Dainty, *Adv. Bot. Res.* **1**, 279 (1963).
[20] R. I. Sha'afi, G. T. Rich, V. W. Sidel, W. Bossert, and A. K. Solomon, *J. Gen. Physiol.* **50**, 1377 (1967).
[21] J. B. Williams and H. Kutchai, *Biophys. J.* **49**, 453 (1986).

computed that the time required to attain 90% of diffusion equilibrium within the red cell was 0.2 msec so the interior of the cell acts as a well-mixed water compartment. As will be seen below, the use of proton nuclear magnetic resonance (NMR) makes it possible to follow the reaction for many half-times with no sign of departure from an exponential curve, which confirms the two-compartment assumption. When the continuous flow method is used to measure water diffusion across other cell membranes, particularly in large cells containing many organelles, it will be necessary to confirm that diffusion within the cell is fast compared to diffusion across the plasma membrane.

THO is a larger molecule than H_2O and Wang et al.[22] found its diffusion coefficient to be 14% smaller than that for $H_2^{18}O$, which is considered to be the most nearly ideal tracer for H_2O, so that it is necessary to increase the measured THO fluxes by 14% to compensate (the same correction is required[22] for ^2HHO). The measured parameter in the two-compartment system is the rate, k, of water diffusion across the red cell membrane, with the dimensions of sec^{-1} $[k = 1/t_e = P_d/(V'/A)]$.

Tracer Diffusion in Packed Cells

Redwood et al.[23] developed an ingenious method to obtain P_d from measurements of diffusion through a tightly packed column of red cells and derived the required set of equations. The advantage of the method is that the time required to measure diffusion through a packed red cell column is measured in hours rather than milliseconds as required for diffusion across a single cell membrane. The disadvantage is that diffusion across such a column includes contributions from diffusion through the cell and through the interstices between the tightly packed cells, as discussed carefully by Redwood et al.[23] Some of the required measurements can be made directly, such as the fraction of extracellular space, which is measured conventionally by an extracellular fluid marker, and the diffusion through concentrated Hb solutions and through plasma which are measured at the same time. There are other assumptions which depend directly on the geometry of the system. Redwood et al.[23] assume that the packing is similar to that in a brick wall with unidirectional diffusion longitudinally along each course of bricks; except that in the construct of Redwood et al., the bricks are replaced with cylinders, each cylinder representing one red cell, a close packing arrangement which appears to be consistent with the electron micrographs Redwood et al.[23] present. The cylinders are uniform with length, L, and cross-sectional area, A_2,

[22] J. H. Wang, C. V. Robinson, and I. S. Edelman, J. Am. Chem. Soc. **75**, 466 (1953).
[23] W. R. Redwood, E. Rall, and W. Perl, J. Gen. Physiol. **64**, 706 (1974).

with an exchange area, S, on the side. It seems reasonable to accept the arguments given in the paper that the path length for diffusion around the cells is approximated by that for diffusion through the cells and that the permeability of the red cell is isotropic. In the final fitting process, Redwood et al.[23] set $S/A_2 = 8$, which characterizes a cylinder whose length is twice the end face diameter, and then treat L as an adjustable parameter to fit their data to the permeability coefficients given by Sha'afi et al.[24] for THO, formamide, and propionamide in dog red cells. This procedure leads to $L = 8.5$ μm, in reasonable agreement with the 7-μm diameter for the dog red cell, which is given in the literature.[25]

Osberghaus et al.[26] have improved the method by adapting it to Wang's[27] procedure used to measure the self-diffusion coefficient of D_2O, which offers a considerably simplified method of measurement. Osberghaus et al. set $L = 4.5$ μm, calculated from the cube root of the human red cell volume and set $S/A_2 = 4$, which is characteristic for a cube. This leads to $P_d = 3.7 \pm 0.1 \times 10^{-3}$ cm sec^{-1}, after conversion to a standard red cell[4] in good agreement with the values obtained by the THO method, as can be seen in Table I.

Bulk Diffusion of D_2O across Membranes

The bulk properties of D_2O differ significantly from those of H_2O and Lawaczeck[28,29] has used two of these differences as a basis for methods of measuring water diffusion across vesicle and red cell membranes (see Additional Notes). The fluorescence quantum yield of a chromophore, such as tryptamine-HCl, is greater in D_2O than in H_2O. When lipid vesicles, prepared with tryptamine HCl sealed inside in an H_2O milieu, are mixed with a deuterated solution, the observed fluorescence intensity will increase as D_2O diffuses into the vesicle and increases the quantum yield. The permeability coefficient can be determined from the rate of D_2O diffusion just as it can from THO diffusion. Subsequently, Lawaczeck[29] found that the light scattering from these vesicles also depended on the vesicular D_2O concentration, which he attributed to the difference in refractive index between D_2O and H_2O, and he showed that the light scattering from DPPC (dipalmitoylphosphatidylcholine) vesicles followed the same time course as the fluorescence intensity. Subsequently, Pit-

[24] R. I. Sha'afi, C. M. Gary-Bobo, and A. K. Solomon, J. Gen. Physiol. **58**, 238 (1971).
[25] P. L. Altman and D. S. Dittmer, "Respiration and Circulation," p. 151. Fed. Am. Soc. Exp. Biol., Bethesda, Maryland, 1971.
[26] U. Osberghaus, H. Schonert, and B. Deuticke, J. Membr. Biol. **68**, 29 (1982).
[27] J. H. Wang, J. Am. Chem. Soc. **73**, 1182 (1951).
[28] R. Lawaczeck, J. Membr. Biol. **51**, 229 (1979).
[29] R. Lawaczeck, Biophys. J. **45**, 491 (1984).

TABLE I
CORRECTED PERMEABILITIES FOR DIFFUSIONAL PERMEABILITY IN THE
HUMAN RED CELL[a]

Method	P_d (cm sec^{-1} × 10^3)	Ref.
^{3}HHO	4.2	Paganelli and Solomon[b]
^{3}HHO	4.5	Barton and Brown[c]
^{3}HHO	3.8	Vieira et al.[d]
^{3}HHO	3.7	Osberghaus et al.[e]
T_2	4.2	Conlon and Outhred[f]
T_2	4.0	Conlon and Outhred[g]
T_2	4.2	Chien and Macey[h]
T_2	5.4	Morariu and Benga[i]
T_1, T_2, T_{12}	3.8	Fabry and Eisenstadt[j]
T_1	4.0	Dix and Solomon[k]

Mean   4.2 ± 0.5

Values that diverge from mean above

^{3}HHO	2.4	Brahm[l]
^{17}O T_1	2.7	Shporer and Civan[m]
T_2	2.3	Pirkle et al.[n]
T_1	2.4	Brooks et al.[o]
T_1	2.6	Chien and Macey[h]

[a] Corrected to 20°, V/A = 5.33 × 10^{-5} cm, cell water content of 0.72 and 14% difference in diffusion coefficient between THO and H$_2$^{18}O, as described in the text.
[b] C. V. Paganelli and A. K. Solomon, *J. Gen. Physiol.* **41**, 259 (1957).
[c] T. C. Barton and D. A. J. Brown, *J. Gen. Physiol.* **47**, 839 (1964).
[d] F. L. Vieira, R. I. Sha'afi, and A. K. Solomon, *J. Gen. Physiol.* **55**, 451 (1970).
[e] U. Osberghaus, H. Schonert, and B. Deuticke, *J. Membr. Biol.* **68**, 29 (1982).
[f] T. Conlon and R. Outhred, *Biochim. Biophys. Acta* **288**, 354 (1972).
[g] T. Conlon and R. Outhred, *Biochim. Biophys. Acta* **511**, 408 (1978).
[h] D. Y. Chien and R. I. Macey, *Biochim. Biophys. Acta* **464**, 45 (1977).
[i] V. V. Morariu and G. Benga, *Biochim. Biophys. Acta* **469**, 301 (1977).
[j] M. E. Fabry and M. Eisenstadt, *J. Membr. Biol.* **42**, 375 (1978).
[k] J. A. Dix and A. K. Solomon, *Biochim. Biophys. Acta* **773**, 219 (1984).
[l] J. Brahm, *J. Gen. Physiol.* **79**, 791 (1982).
[m] M. Shporer and M. M. Civan, *Biochim. Biophys. Acta* **385**, 81 (1975).
[n] J. L. Pirkle, D. L. Ashley, and J. H. Goldstein, *Biophys. J.* **25**, 389 (1979).
[o] R. A. Brooks, J. H. Battocletti, A. Sances, S. J. Larson, R. L. Bowman, and V. Kudravcev, *IEEE Trans. Biomed. Eng.* **BME-22**, 12 (1975).

terich and Lawaczeck[30] applied the D_2O light-scattering technique to re-sealed red cell ghosts and found $P_d = 1.2 \times 10^{-3}$ cm sec^{-1}, smaller than the value of 2.9×10^{-3} cm sec^{-1} given by Brahm.[15]

Nuclear Magnetic Resonance Methods

Conlon and Outhred[31] developed a nuclear magnetic resonance method of measuring water diffusion by the spin–spin relaxation time, T_2, of water protons. When a brief radio-frequency pulse is applied to a sample of protons placed in a suitable magnetic field, their spins become oriented. If these protons are contained in water inside the red cell, the spins lose their orientation with a spontaneous relaxation time, T_2, of about 140 msec at 9 MHz. In a system with suitable time constants, the spin can be used as if it were a radioactive tracer with a decay time of 140 msec, which is much longer than the transmembrane exchange time of 12.7 msec. The decay time for the protons in the plasma water can be reduced to a very few milliseconds by adding the very slowly permeable paramagnetic ion Mn^{2+}, to the plasma. As soon as the labeled cell protons from the cell water cross the membrane they decay rapidly with the characteristic time of the plasma decay. There is a negligible contribution from plasma protons entering the cell so that the decay process is domi-nated by two characteristic time constants, ≈ 12.7 msec, and the decay time of the extracellular protons, which is measured independently in an experiment on supernatant alone. The intrinsic time constant inside the cell is also measured independently with packed red cells in the absence of supernatant. The total decay of magnetization in a T_2 experiment under these conditions is described by a set of coupled linear differential equa-tions:

$$dM_i/dt = -([1/T_{2(i)}] + k_{io})M_i + k_{oi}M_o \qquad (9a)$$
$$dM_o/dt = k_{io}M_i - ([1/T_{2(o)}] + k_{oi})M_o \qquad (9b)$$

in which M is the magnetization with subscripts i, inside the cell, and o, outside. The rate constants, which are related to one another by the relative i and o volumes, are efflux, k_{io}, and influx, k_{oi}. The solution for the observed sum of the two magnetizations $(M_i + M_o)$ is a double expo-nential equation whose coefficients, given by Fabry and Eisenstadt,[32] (their Eqs. 4 and 5), have two time constants, each constant containing contributions from the intrinsic decay of M_i and M_o as well as the ex-change rate.[31,32] It is advisable to deconvolute this equation by nonlinear least squares to obtain the time constants accurately. If conditions can be

[30] H. Pitterich and R. Lawaczeck, *Biochim. Biophys. Acta* **821**, 233 (1985).
[31] T. Conlon and R. Outhred, *Biochim. Biophys. Acta* **288**, 354 (1972).
[32] M. E. Fabry and M. Eisenstadt, *Biophys. J.* **15**, 1101 (1975).

chosen in which the decay of the extracellular magnetization is very fast, the equations can be considerably simplified, as Conlon and Outhred[31] have shown (their Eqs. 9 and 10).

Fabry and Eisenstadt[33] also measured red cell water diffusion by the proton spin-lattice relaxation time, T_1. In this method, the interpretation of the data is complicated by the need to take account of a third spin system, the hemoglobin protons, which can gain magnetization from the intracellular water protons by spin diffusion. The three coupled linear differential equations required for the T_1 method and their solution are given in the appendix to Dix and Solomon[4] and by Fabry and Eisenstadt.[33] To extract the exchange time from their data, Dix and Solomon simulated the total magnetization decay, using constants which were either extracted from the literature or determined in independent measurements, except for the transmembrane exchange rate, which could be set arbitrarily. The data were fit by nonlinear least squares to a single exponential over the experimental time range and the transmembrane exchange rate was varied until the simulated data time constant matched that obtained from the fit to the experimental data.

Other NMR methods have also been used, such as the T_{12} method used by Fabry and Eisenstadt[33] and the measurements with $H_2^{17}O$ by Shporer and Civan,[34] a method which has the advantage of not requiring the addition of Mn^{2+} to the suspending medium. Using both the T_2 and the T_{12} methods, Fabry and Eisenstadt[33] made the important observation that the transmembrane exchange process was characterized by over two decades of linear decay (see Fig. 2). This means that, to 1% or better, the red cell behaves as a well-mixed compartment and that intracellular diffusion through the tightly packed Hb is fast compared to the flux across the membrane. Similarly, exchange with any form of bound water within the cell, if significant, must be fast compared with the transmembrane diffusion.

There are several special precautions that need to be taken when using the NMR method. When the red cell sample is placed in the NMR magnet, the spinner should be turned off to avoid packing the cells on the side of the sample tube. Other precautions arise from the presence of the extracellular Mn^{2+} and the high hematocrits which are routinely used to maximize the signal/noise ratio. It is not necessary to suspend the cells in plasma, though it has been commonly used. However, as Conlon and Outhred[35] pointed out, red cells form rouleaux in albumin-containing solu-

[33] M. E. Fabry and M. Eisenstadt, *Biophys. J.* **42**, 375 (1978).
[34] M. Shporer and M. M. Civan, *Biochim. Biophys. Acta* **385**, 81 (1975).
[35] T. Conlon and R. Outhred, *Biochim. Biophys. Acta* **511**, 408 (1978).

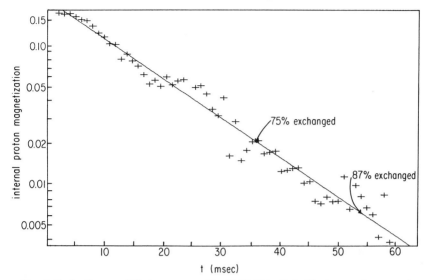

FIG. 2. Red cell water diffusion measured by T_1 NMR by J. A. Dix (personal communication). This portion of the decay is dominated by proton exchange between cell water and extracellular water; the corrections for spin diffusion described by Dix and Solomon[4] must be applied to compute the exchange time. These data show that the exchange rate remains exponential for at least three half-times.

tions and rouleaux formation is reversed at high Mn^{2+} concentrations. Rouleaux formation provides a larger effective cell volume, leading to overestimates of t_e. In practice, it is wise to use albumin in the extracellular solution to enhance the Mn^{2+} paramagnetic effect, but to take care that the albumin concentration is low enough not to cause rouleaux formation. Although Mn^{2+} diffuses into the red cell slowly, Dix and Solomon[4] found that significant quantities of Mn^{2+} had entered by 30 min, so that it is necessary to add the Mn^{2+} immediately prior to the NMR experiment. There has been considerable concern about the effect of high extracellular Mn^{2+} but these concerns have been put to rest by Brahm,[15] who used the THO method to show that high concentrations of Mn^{2+} did not affect red cell water diffusion.

Dix and Solomon[4] took, as a uniform standard, a red cell with the V'/A and water content given below Eq. (5) and corrected the results of a number of red cell water diffusion measurements to this standard cell at 20°, using the activation energy of 5.3 kcal mol^{-1} given by Fabry and Eisenstadt.[33] As shown in Table I, they found very good agreement among 10 measurements by 9 groups of investigators, using T_1, T_2, and two different THO methods, with an average P_d of 4.2 ± 0.5 cm sec^{-1}.

Five groups of investigators reported results with a P_d lower by a factor of about two. The results of three of these groups are probably in error due to neglect of spin diffusion or rouleaux formation and the reason for the divergence of the other two results is not known.

Osmotic Water Flow across the Red Cell Membrane

When an osmotic pressure gradient is applied across a semipermeable membrane, there is a net volume flow of solvent, which is driven by the difference in the chemical potential of water across the membrane. The presence of a bulk driving force sets osmotic flow apart from diffusion, which is normally measured in the absence of any gradient of bulk (as opposed to tracer) water activity. In net volume flow, also called viscous flow, there is a bulk transfer of momentum among the molecules comprising the flowing stream, which is not observed in diffusion. In Onsager's words,[36] ". . . viscous flow is a relative motion of adjacent portions of a liquid. Diffusion is a relative motion of its constituents."

Consider the pair of rigid chambers (denoted by ' and "), separated by an ideal semipermeable membrane, which initially contain distilled water on both sides, as illustrated in Fig. 3. If an impermeable solute, say a protein (denoted by the subscript, i), is introduced into the ' chamber, at a concentration, c_i', and the solution is ideal, the level of fluid on that side of the system will rise and the system will come to equilibrium when there is a pressure difference $(P' - P'')$. At equilibrium, the chemical potential of water in the " chamber, μ_w'', must equal μ_w' so that, at constant T:

$$\mu_w^0 + \bar{V}_w P'' = \mu_w^0 + \bar{V}_w P' + RT \ln x_w' \tag{10}$$

in which μ_w^0 is the standard chemical potential of water, $\bar{V}_w$ is the partial molar volume of water, and x_w' is the mole fraction of water in '. $x_w' = n_w'/(n_w' + n_i')$ in which n is the number of moles of the species, w and i, in the ' compartment so that

$$\bar{V}_w \Delta\pi = RT \ln[1 + (n_i'/n_w')] \tag{11}$$

in which the osmotic pressure $\Delta\pi = P' - P''$. Since the solution in ' is dilute, $\ln[1 + (n_i'/n_w')] \approx n_i'/n_w'$ and the volume of the solution in ' $\approx n_w \bar{V}_w$, so that, using $c_i = n_i'/n_w'$:

$$\Delta\pi = RTc_i \tag{12}$$

The water flux, J_v, is measured as a volume flux with units of $cm^3\ cm^{-2}\ sec^{-1}$ and is given by $J_w \bar{V}_w$. In a system such as Fig. 3, in which the

[36] L. Onsager, *Ann. N.Y. Acad. Sci.* **46**, 241 (1945).

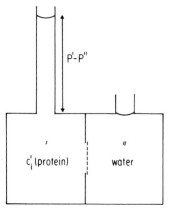

FIG. 3. Schematic diagram of osmotic pressure across semipermeable membrane.

compartments are large enough for the concentration of the driving solute to remain constant, the steady-state J_v is related to the driving force, $\Delta\pi$, by the hydraulic conductivity, L_p, a proportionality coefficient with units of cm^3 dyn^{-1} sec^{-1}:

$$J_v = L_p\Delta\pi \tag{13}$$

The relationship between the osmotic permeability coefficient and the diffusional coefficient has been used to determine the equivalent pore radius in lipid membranes and the red cell (see Solomon[37]). For this purpose it is necessary to express both the fluxes and the coefficients in the same units, so that the flux is taken as J_w (mol H$_2$O cm^{-2} sec^{-1}) and the driving force is the concentration gradient of impermeable solute, Δc_i (mol cm^{-3}). To take account of the nonidealities of the solute driving the flow, it is also necessary to introduce the osmotic coefficient of the solute, ϕ_i, leading to

$$J_w = \phi_i P_f \Delta c_i \tag{14}$$

in which the filtration coefficient, $P_f = RTL_p/\bar{V}_w$ with units of cm sec^{-1}. The conversion factor, $RT/\bar{V}_w = 1.34 \times 10^9$ dyn cm^{-2} at 20°.

If the channels in the membrane separating the two chambers are large enough to permit passage of the solute as well as the solvent, the osmotic pressure becomes smaller than that of an impermeable solute, as pointed out by Staverman[38] and discussed in detail in the next section. He introduced the reflection coefficient, σ, which is the ratio of the observed

[37] A. K. Solomon, J. Gen. Physiol. **51**, 335s (1968).
[38] A. J. Staverman, Rec. Trav. Chim. **70**, 344 (1951).

osmotic pressure to that predicted for an impermeable solute. The operational equation for volume flow produced by a single permeable solute in the absence of a hydrostatic pressure gradient then becomes (the permeable solute is denoted by s)

$$J_v \approx \sigma_s \phi_s L_p RT \Delta c_s \tag{15}$$

Strictly speaking, the volume flux should include a small contribution from solute flux, but the solute contribution is negligible in experiments in which the osmotic pressure gradient is produced by solutes of very low permeability, so that J_v is usually taken as equal to $J_w \bar{V}_w$ in such experiments.

Stopped-Flow Method

The stopped-flow method of measuring red cell osmotic flow was introduced by Sha'afi et al.,[20] following the continuous flow method of Sidel and Solomon[7] and the hemolysis method introduced by Jacobs.[39] The method uses the intensity of light scattered by, or transmitted through, a suspension of red cells as a measure of red cell volume. Measurements have been made by a number of investigators (including Farmer and Macey,[40] Blum and Forster,[41] Colombe and Macey,[42] Galey,[43] Papanek,[44] and Levin et al.[45]). Our discussion will be based on the apparatus of Terwilliger and Solomon[46] shown in Fig. 4A, but will also draw on the careful discussion in Papanek's Ph.D. thesis.[44]

The stopped-flow apparatus illustrated in Fig. 4A has been designed from the apparatus of Gibson and Milnes.[47] The intensity of 90° scattered white light is used to measure the time course of red cell volume changes after a sudden osmotic shock. Two coupled syringes, one filled with a red cell suspension and the other with a test solution, are driven by a hydraulic ram so that the solution travels in turn through a mixing chamber and an observation tube and is driven into a collecting syringe. When the hydraulic ram is driven onto its rigid stop, flow ceases and observation

[39] M. H. Jacobs, *Biol. Bull.* **62**, 178 (1932).
[40] R. E. L. Farmer and R. I. Macey, *Biochim. Biophys. Acta* **196**, 53 (1970).
[41] R. M. Blum and R. E. Forster, *Biochim. Biophys. Acta* **203**, 410 (1970).
[42] B. W. Colombe and R. I. Macey, *Biochim. Biophys. Acta* **363**, 226 (1974).
[43] W. R. Galey, *J. Membr. Sci.* **4**, 41 (1978).
[44] T. H. Papanek, Ph.D. thesis. Massachusetts Inst. of Technol., Cambridge, Massachusetts, 1978.
[45] S. W. Levin, R. L. Levin, and A. K. Solomon, *J. Biochem. Biophys. Methods* **3**, 255 (1980).
[46] T. C. Terwilliger and A. K. Solomon, *J. Gen. Physiol.* **77**, 549 (1981).
[47] Q. H. Gibson and L. Milnes, *Biochem. J.* **91**, 161 (1964).

A

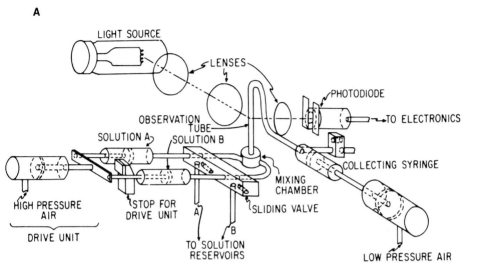

B

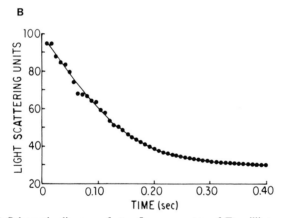

FIG. 4. (A) Schematic diagram of stop-flow apparatus of Terwilliger and Solomon.[46] When high-pressure air is forced into the drive unit, solutions A and B are driven from the syringes into the mixing chamber, through the observation tube, and into the collecting syringe where flow is halted when the hydraulic ram driving the syringes hits the stop. After flow is stopped, 90° scattered light intensity is measured by a photodiode and the data are transmitted to an on-line computer. Low-pressure air is used to apply a constant pressure to the fluid in the collecting syringe throughout the cycle. Valves are controlled pneumatically by electronically actuated solenoids. (B) Fit of volume–time curve to Eq. (17). Points are the average light-scattering units (lsu) from a series of 25 runs (after subtraction of 25 control runs) when bank blood is exposed to an osmotic pressure gradient of 440 mOsm NaCl (unpublished data of Toon and Solomon). The data were transformed to volume units and fit to Eq. (17) using a nonlinear least-squares numerical integration program written by J. A. Dix (personal communication).

begins. The scattered light intensity is detected by a photodiode, amplified and fed through an analog-to-digital converter for on-line data collection by a computer. Conventionally, the data collection is initiated when the collecting syringe plunger hits its stop, but in our system the movement of the ram is monitored by both position and velocity transducers and data collection begins when the velocity trace decreases below 10% of maximum.

The dead time was determined in our apparatus as 15.5 msec by Levin et al.,[45] who mixed 10 mM Fe(NO$_3$)$_3$ (in 0.1 N H$_2$SO$_4$) with 10 mM KSCN and determined the time course of the color change, using a method similar to that described by Sha'afi et al.[20] in 1967 and similar to the methods generally available in instruction manuals for commercial stopped-flow devices.

Determination of Osmotic Permeability Coefficient

The osmotic permeability coefficient, also called the hydraulic conductivity, L_p, has been defined in Eq. (13), which is equivalent to

$$dV/dt = L_p(\pi^o - \pi^i) \tag{16}$$

in which π is the osmotic pressure with the superscripts, o, outside, and i, inside the cell. This equation was integrated by Sidel and Solomon[7] and is given by Levin et al.[45] in the following form:

$$L_p = \frac{V_{c,iso}}{ART(\pi^o)^2 t} \pi^i_{iso}[1 - (V_b/V_{c,iso})]$$

$$\left[\ln \frac{\pi^i_{iso} - \pi^o}{\pi^i_{iso} - Q\pi^o} - \frac{\pi^o[(V_c/V_{c,iso}) - 1]}{\pi^i_{iso}[1 - (V_b/V_{c,iso})]} \right] \tag{17}$$

in which the subscript iso refers to isosmolal. A is the red cell area, which for the human red cell is taken to be independent of cell volume. When Eq. (17) is applied to other cell suspensions, or to vesicles, the dependence of A on cell volume must be included explicitly. V_b is the apparent nonosmotic water plus cell solutes and membrane volume, as determined from a plot of cell volume, V_c, as a function of medium osmolality, which follows the following equation[48]:

$$\hat{V}_c = (1 - \hat{V}_b)\pi^i_{iso}/\pi^i + \hat{V}_b \tag{18}$$

Volumes have been normalized so that $\hat{V}_c = V_c/V_{c,iso}$ and $\hat{V}_b = V_b/V_{c,iso}$. $Q = [(\hat{V}_c - \hat{V}_b)/(1 - \hat{V}_b)]$.

[48] D. Savitz, V. W. Sidel, and A. K. Solomon, J. Gen. Physiol. 48, 79 (1964).

Macey[49] has pointed out that after rearranging an equation analogous to Eq. (17) in the following form:

$$\frac{L_p ART(\pi^o)^2 t}{V_{c,iso}\pi^i_{iso}} = [1 - (V_b/V_{c,iso})]\left[\ln\frac{\pi^i_{iso} - \pi^o}{\pi^i_{iso} - Q\pi^o} - \frac{\pi^o[(V_c/V_{c,iso}) - 1]}{\pi^i_{iso}(1 - (V_b/V_{c,iso}))}\right] \quad (19)$$

a plot of the right-hand side of the equation as a function of t will give a straight line whose slope is proportional to L_p.

Stopped-flow measurements in heterogeneous solutions, such as red cell suspensions, pose numerous problems which are absent in conventional stopped-flow measurements of homogeneous solutions. When a flowing cell suspension is brought to a sudden stop in the apparatus, standing waves are generated and, for the anisotropic human red cell, the scattered light intensity depends on the relation of the red cell axis to the incident light axis, as discussed extensively by Papanek.[44] The scattered light contains a damped sinusoidal component with a period of 20–40 msec, which lasts for about 100–200 msec. When data from 5 to 10 runs are averaged, the effect on the data becomes manageable. Recently, we have averaged 25 runs and have been able to obtain good data over the first 8–160 msec. In order to accumulate this much data, we have automated the hydraulic drive, purge, and refill cycle. Data are collected online by computer, processed, and averaged. Since bubbles are occasionally produced in the system, it is necessary to examine each trace before letting the computer include it in the average. Blum and Forster[41] observed that swelling the red cells in hypoosmolal solutions to make them spherical reduced the stopping artifact, which means that the problem may not be important for experiments with other cell suspensions or vesicles. Mlekoday et al.[50] made their cells spherical by treatment with lecithin and carried out control experiments to show that L_p for lecithin-treated cells was within 2% of that in swollen cells. Though this comparison reduces the possibility of an artifactual difference in L_p, it does not entirely eliminate it.

In order to separate the signal produced by the osmotic gradient from other possible artifacts produced by the stopped-flow device, we subtracted control data. For control experiments, cells suspended in isosmolal buffer were put in one syringe in the stopped-flow device and the

[49] R. I. Macey, in "Membrane Transport in Biology" (G. Giebisch, D. C. Tosteson, and H. H. Ussing, eds.), Vol. 2, p. 1. Springer-Verlag, Berlin, 1979.
[50] H. J. Mlekoday, R. Moore, and D. G. Levitt, J. Gen. Physiol. 81, 213 (1983).

same isosmolal buffer was put in the other. There were approximately the same number of control runs as experimental runs and the difference curve was generated by the computer for further processing. Outside of an artifact in the first few milliseconds, the control signal was generally a straight line parallel to the time axis.

The stopped-flow device determines cell volume from the intensity of 90° scattered white light, measured in light-scattering units (lsu), which must be converted to red cell volume for further processing in the computer. Initially, we calibrated the system by carrying out a number of control runs at different osmolalities and measured both the light-scattering units and the hematocrit as a function of osmolality, to give a direct calibration. The linearity of Eq. (18) has been firmly established and we replaced the measurements of graded osmolalities with measurements at only two, interpolating linearly. This procedure would be exact if light scattering were also linearly related to red cell volume. Though both Levin et al.[45] and Terwilliger and Solomon[46] have shown this dependence to be curvilinear, the data of Levin et al. show that the points fit a linear approximation well over the range of $V_c/V_{c,iso} = 0.8–1.0$. The computation is very much simplified by the use of the linear approximation and the simplification appears to us to outweigh the relatively small inaccuracy over this restricted cell volume range. The value of the constant, $\hat{V}_b$, in Eq. (18) has been very well established as 0.43 for fresh red cells, initially by Savitz et al.,[48] and has more recently been confirmed by Solomon et al.[51] Since it is an important parameter in the equations it is necessary to verify the value of $\hat{V}_b$ under the exact experimental conditions (we find the same value for bank blood as for fresh cells) and for the cell or vesicle preparation used. Initial and final cell volumes are measured in light-scattering units during the experiment and converted to cell volumes using the isosmolal volume of 100×10^{-12} cm^3, following Jay.[6] Taking the red cell area[6] $A = 1.35 \times 10^{-6}$ cm^2, L_p is then computed by nonlinear least squares from Eq. (17) or by numerical integration of Eq. (16) to obtain the fit illustrated in Fig. 4B.

Terwilliger and Solomon[46] have made a particularly careful study of the value of L_p in fresh human red cells, taking special care to apply very small osmotic pressure gradients to minimize any nonlinearities which might introduce errors. They determined the possible contributions of all the other sources of error they could find and developed a perturbation method to make an exact calculation of L_p with minimal error, obtaining a value of $1.8 \pm 0.1 \times 10^{-11}$ cm^3 dyn^{-1} sec^{-1} at 25–26°, which they believe to be a particularly accurate measurement. The computational difficulty of

[51] A. K. Solomon, M. R. Toon, and J. A. Dix, *J. Membr. Biol.* **91**, 259 (1986).

the exact perturbation method is such that it is not used for routine measurements of L_p.

Farmer and Macey[40] have devised a simplified perturbation method to measure an approximate value of L_p, which permits considerable simplification of the computations. The method depends upon applying an osmotic pressure difference small enough that Eq. (16) can be integrated to a single exponential solution in which L_p can be determined directly from the time constant, τ_w, of the exponential. In order to use the method, the excursion between the initial cell volume and final cell volume, $V_{c,\infty}$ must be small enough that the ratio $M = (V - V_{c,\infty})/(V_{c,\infty} - V_b)$ be $\ll 1$. Under these conditions, $1/(1 + M) \cong (1 - M)$, and the solution to Eq. (16) is

$$V_c = V_{c,\infty} + V_{c,\text{init}}e^{(-t/\tau_w)} \tag{20}$$

in which $V_{c,\text{init}}$ is the initial cell volume and

$$\tau_w = (V_{c,\text{iso}} - V_b)\pi_{\text{iso}}^0/AL_p(\pi^o)^2 \tag{21}$$

Farmer and Macey[40] plotted the signal in light-scattering units as a function of time and determined τ_w graphically; now a computer would generate a three-parameter fit. Since the only parameter required to determine $L_{p,\text{perturb}}$ is τ_w there is no necessity to determine either $V_{c,\text{init}}$ or $V_{c,\infty}$. Consequently, as Farmer and Macey[40] pointed out, the light-scattering signal does not need to be transformed into volume units, though the method requires a linear dependence of light-scattering units upon cell volume. It is also not necessary to make measurements near zero time, since the only parameter to be extracted is the exponential time constant. For small excursions in osmotic pressure, the perturbation method is very good; for a 50 mOsm osmotic gradient, the ratio, $L_{p,\text{perturb}}/L_{p,\text{true}}$, is 0.97. However, as the gradients increase, the percentage difference between $L_{p,\text{perturb}}$ and $L_{p,\text{true}}$ increases linearly, reaching 8% for $\Delta\pi = 100$ mOsm and 17% for $\Delta\pi = 200$ mOsm.

The Farmer and Macey[40] perturbation method also permits a great simplification in the apparatus, since there is no need for rapid mixing and hence no stopping artifacts. Farmer and Macey simply injected 0.5–0.6 ml of the perturbing solution through a dispersing nozzle into the cell suspension in a Beckman-type cuvette, containing a stirring magnet rotating at 50–60 rps. The cuvette was illuminated by a small bulb and the transmitted light was amplified and recorded.

Table II gives a compilation of the values for L_p that have been obtained since 1957 for human blood that has been obtained fresh or stored for only a few days, after conversion to common units of V/A. It is interesting that the L_p has remained so nearly constant over more than 30 years, as methods have changed and techniques have improved. All of the

TABLE II

DETERMINATION OF HYDRAULIC CONDUCTIVITY, L_p, FOR THE HUMAN RED CELL[a]

Method	L_p $(cm^3\ dyn^{-1}\ sec^{-1} \times 10^{11})$	Ref.
Continuous flow	1.3 ± 0.2	Sidel and Solomon[c]
Stopped flow	1.25 ± 0.06	Sha'afi et al.[d]
Stopped flow	1.78	Rich et al.[e]
Stopped flow	1.2 ± 0.1	Sirs[f]
Stopped flow	1.85[b]	Blum and Forster[g]
Perturbation	1.97[b]	Farmer and Macey[h]
Perturbation	1.8 ± 0.1[b]	Colombe and Macey[i]
Stopped flow	1.79 ± 0.05	Galey[j]
Stopped flow	1.4 ± 0.3	Papanek[k]
Stopped flow	2.0 ± 0.4	Levin et al.[l]
Stopped flow	1.8 ± 0.1	Terwilliger and Solomon[m]
Stopped flow	1.52 ± 0.02	Mlekoday et al.[n]
Mean (since 1970)	1.8 ± 0.2	

[a] Blood either fresh or less than a few days old. All values have been converted to a cell volume of $100 \times 10^{-12}\ cm^3$ and area of $1.35 \times 10^{-6}\ cm^2$ following Jay[6] and Canham and Burton,[5] as used by Dix and Solomon[4] and in Table III. In many cases temperatures were not specified by the authors, which means that experiments were done at room temperature, which is generally in the range of 20–26°. Since the apparent activation energy [F. L. Vieira, R. I. Sha'afi, and A. K. Solomon, *J. Gen. Physiol.* **55**, 451 (1970)] is only 3.3 kcal mol^{-1}, we have not corrected these data to a common temperature.

[b] These investigators did not specify the cell area used in their computations and we have assumed that they used the conventional value of that time ($1.65 \times 10^{-6}\ cm^2$) since they did not refer to Canham and Burton[5] and the paper by Jay[6] was not yet published. We have converted these values to a cell area of $1.35 \times 10^{-6}\ cm^2$.

[c] V. W. Sidel and A. K. Solomon, *J. Gen. Physiol.* **41**, 243 (1957).

[d] R. I. Sha'afi, G. T. Rich, V. W. Sidel, W. Bossert, and A. K. Solomon, *J. Gen. Physiol.* **50**, 1377 (1967).

[e] G. T. Rich, R. I. Sha'afi, A. Romualdez, and A. K. Solomon, *J. Gen. Physiol.* **52**, 941 (1968).

[f] J. A. Sirs, *J. Physiol.* (*London*) **205**, 147 (1969).

[g] R. M. Blum and R. E. Forster, *Biochim. Biophys. Acta* **203**, 410 (1970).

[h] R. E. L. Farmer and R. I. Macey, *Biochim. Biophys. Acta* **196**, 53 (1970).

[i] B. W. Colombe and R. I. Macey, *Biochim. Biophys. Acta* **363**, 226 (1974).

[j] W. R. Galey, *J. Membr. Sci.* **4**, 41 (1978).

[k] T. H. Papanek, Ph.D. thesis. Massachusetts Inst. of Technol., Cambridge, Massachusetts, 1978.

[l] S. W. Levin, R. L. Levin, and A. K. Solomon, *J. Biochem. Biophys. Methods* **3**, 255 (1980).

[m] T. C. Terwilliger and A. K. Solomon, *J. Gen. Physiol.* **77**, 549 (1981).

[n] H. J. Mlekoday, R. Moore, and D. G. Levitt, *J. Gen. Physiol.* **81**, 213 (1983).

values obtained since 1970 cluster in a narrow band around the mean of $1.8 \pm 0.2 \text{ cm}^3 \text{ dyn}^{-1} \text{ sec}^{-1}$.

Measurement of the Reflection Coefficient in Red Cell Suspensions

When a semipermeable membrane separates a solution containing a single permeant solute from a compartment containing solvent alone, the resultant flows of solute and solvent in the steady state are described by a set of two equations developed by Kedem and Katchalsky[52] and discussed extensively in the text by Katchalsky and Curran.[3]

$$J_v = L_p \Delta P + L_{pd} \Delta \pi_s \qquad (22a)$$
$$J_D = L_{dp} \Delta P + L_d \Delta \pi_s \qquad (22b)$$

These equations relate the flows to the two driving forces, the hydrostatic pressure difference, ΔP, and the solute (subscript s) concentration difference, expressed in terms of the osmotic pressure difference, $\Delta \pi_s$ ($= RT \Delta c_s$). In cases in which there is no applied hydrostatic pressure difference, but an osmotic pressure difference due to an impermeable solute (subscript i), ΔP is replaced by $\Delta \pi_i$. J_v is total steady-state volume flux, which now includes both solute and solvent flow, and J_D is an unfamiliar measure of solute flux, the velocity of solute flow relative to solvent flow. The Onsager reciprocal relation requires that the cross-coefficients, $L_{pd} = L_{dp}$, so there are three phenomenological coefficients, L_p, L_{pd}, and L_d. Although the flows and phenomenological coefficients are given in unfamiliar terms, Eqs. (22a,b) illustrate a most important property because three coefficients are required to describe the system, rather than two, one apiece for solute and solvent flux. L_p is the straight coefficient relating volume flow to its conjugate force, ΔP; and L_d is the straight coefficient relating solute flow to its conjugate force, $\Delta \pi_s$. In addition there is the coupling coefficient, or cross-coefficient, L_{pd}, whose meaning will become clearer when the relations expressed in Eqs. (22a,b) are given in more conventional terms.

Kedem and Katchalsky[52] transformed Eqs. (22a,b) into the following form, in which the fluxes correspond to those normally measured in the laboratory.

$$J_v = L_p \Delta \pi_i - \sigma L_p \Delta \pi_s \qquad (23a)$$
$$J_s = \bar{c}_s (1 - \sigma) J_v + \omega \Delta \pi_s \qquad (23b)$$

J_s is the steady-state solute flux measured conventionally, as with radioactive tracers and $\bar{c}_s$ is ordinarily taken as the mean of the concentrations

52 O. Kedem and A. Katchalsky, *J. Gen. Physiol.* **45**, 143 (1961).

of the permeant solute in the two compartments on either side of the membrane.

The three coefficients required to describe the fluxes across the membrane may be defined using Eqs. (23a,b).

$$L_p = (J_v/\Delta\pi_i)_{\Delta\pi_s=0} \tag{24a}$$
$$\sigma = (-L_{pd}/L_p) = -(J_v/L_p\Delta\pi_s)_{\Delta\pi_i=0} \tag{24b}$$
$$\omega = (J_s/\Delta\pi_s)_{J_v=0} \tag{24c}$$

It will be seen that the definition of the hydraulic conductivity, L_p, is the same as in Eq. (13). As previously mentioned, the cross-coefficient, σ, called the reflection coefficient, was introduced by Staverman[38] to permit quantitative evaluation of the osmotic pressure of a permeant solute; it is a measure of the ability of a membrane to discriminate between solute and solvent. Staverman used the term reflection coefficient for σ because every time an impermeant molecule strikes a membrane it is reflected back into the solution from which it came; the fraction of solute molecules reflected is unity, so that $\sigma = 1$. For permeant solutes, a certain fraction of collisions result in transmission across the membrane, so that $\sigma < 1.0$. In the limit when the membrane cannot discriminate between solute and solvent, no net solute molecules are reflected and $\sigma = 0$. An example of a very permeant solute is D_2O, for which Durbin[53] determined that $\sigma = 0.002$ for passage across dialysis tubing. The third phenomenological coefficient, ω, is the solute permeability coefficient, whose measurement will be discussed in this volume by Brahm [9].

The symmetry of L_{pd} in Eqs. (22a,b) is reflected in the presence of σ in both cross-terms in Eqs. (23a,b). A practical example of the importance of these cross-terms is the phenomenon known as solvent drag, which is described quantitatively by the term containing σ in Eq. (23b) for J_s. Consider a situation in which the permeant solute concentration is the same in two compartments separated by a semipermeable membrane, so that $\Delta\pi_s = 0$. Then introduce an impermeant solute to produce a volume flow. The volume flow will entrain solute and carry it across the membrane, with a flux, J_s, given by

$$J_s = \bar{c}_s(1 - \sigma)L_p\Delta\pi_i \tag{25}$$

Equation (25) shows that the solvent drag is driven by $\Delta\pi_i$ and is proportional to $(1 - \sigma)$, falling to 0 when $\sigma = 1$, and rising smoothly as σ approaches 0 and the membrane becomes less able to discriminate be-

[53] R. P. Durbin, J. Gen. Physiol. 44, 315 (1960).

tween solute and solvent. Andersen and Ussing[54] used double-labeled thiourea to demonstrate solvent drag across isolated toad skin. When $J_v = 0$, they found that the thiourea flux in one direction was equal to that in the other. When an osmotic pressure was imposed on the system the flux ratio rose, linearly proportional to J_v.

Zero-Time Method for Measurement of σ

Although J_v can be determined relatively easily from the time course of red cell swelling, rapid analytical methods, specific to the permeable solute, are required to measure both J_s and $\Delta\pi_s$. Goldstein and Solomon[55] took advantage of the fact that $\Delta\pi_s$ is known exactly at $t = 0$ and pointed out that when $J_v = 0$ at $t = 0$, Eq. (23a) simplifies to

$$\sigma = (\Delta\pi_i/\Delta\pi_s)_{t=0, J_v=0} \tag{26}$$

To obtain $J_{v,t=0}$ Goldstein and Solomon measured J_v as a function of time in a continuous flow apparatus in which the sampling ports were equally spaced between 45 and 190 msec; they obtained the $t = 0$ point by extrapolation. In each experiment, they compared three solute concentrations, usually two shrinking and one swelling, and found the value for $(J_v = 0)_{t=0}$ by interpolation. This method has the inherent advantages of a null method, which does not require any assumptions of a linear dependence of light-scattering units on cell volume. As long as there are no singularities around zero, the dependence of the rate of change of the light-scattering signal on the applied osmotic gradient can have an arbitrary shape, provided enough points are taken around $J_v = 0$ to determine the interpolated value accurately.

The continuous flow method suffers from the difficulty of extrapolating to zero time since the first point is not obtained until 45 msec after mixing. The stopped-flow method, which permits observations beginning at 10–15 msec after mixing, was used by Sha'afi *et al.*,[56] to determine $\sigma_{urea} = 0.55 \pm 0.02$, which agreed reasonably well with Goldstein and Solomon's[55] earlier determination of $\sigma_{urea} = 0.62 \pm 0.02$. Under the conditions of Sha'afi *et al.*,[56] the dependence of $J_{v,t=0}$ on the osmolality of the medium was linear, as shown in Fig. 5. The problem with these measurements, as Owen and Eyring[57] pointed out, was the noise caused by the

[54] R. Andersen and H. H. Ussing, *Acta Physiol. Scand.* **39**, 228 (1957).
[55] D. A. Goldstein and A. K. Solomon, *J. Gen. Physiol.* **44**, 1 (1960).
[56] R. I. Sha'afi, G. T. Rich, D. C. Mikulecky, and A. K. Solomon, *J. Gen. Physiol.* **55**, 427 (1970).
[57] J. D. Owen and E. M. Eyring, *J. Gen. Physiol.* **66**, 251 (1975).

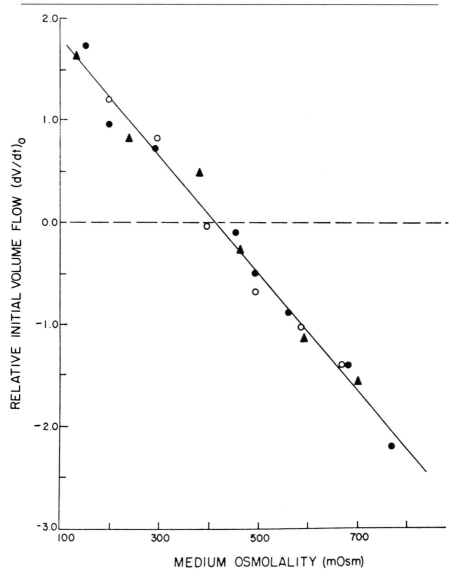

FIG. 5. Dependence of zero-time rate of volume change on external osmolality in human red cells in the presence of urea. Data from two (of three) experiments were linearly transformed to superpose the data. (From Sha'afi *et al.*[56])

stopping artifact, which was particularly troublesome at zero time. To avoid these problems, Owen and Eyring made measurements over the time frame of 20–130 msec and measured their "zero-time" slope at the midpoint of this time frame, obtaining a value of $\sigma_{urea} = 0.79 \pm 0.02$.

Chasan and Solomon[58] modified the zero-time method by measuring J_v as a function of $\Delta\pi$ in paired sets of experiments, one set when $\Delta\pi_s = 0$, and the other when $\Delta\pi_i = 0$. Under these conditions, σ is given by the ratio of the two slopes

$$\sigma = (J_v/\Delta\pi_s)_{\Delta\pi_i=0}/(J_v/\Delta\pi_i)_{\Delta\pi_s=0} \qquad (27)$$

Since this is a ratio method, it does not require that the rate of change of cell volume be measured in volume units; it suffices to measure volume in light-scattering units, provided that the light-scattering units are a linear function of cell volume. Figure 6A gives an example of the determination of σ_{urea} by this method; the observation that the data fit a straight line shows that both L_p and σ_{urea} are essentially independent of the solute osmolality over this concentration range.

In order to minimize the problem of the stopping artifact, Chasan and Solomon[58] averaged 10–15 runs, drew a smooth curve through the data over the range 20–200 msec (including the dead time), and obtained the tangent at 20 msec graphically. They also developed the following equation to correct the 20-msec value of σ to the zero-time value

$$\lambda = 0.47\omega - 1.8L_p \qquad (28a)$$
$$\sigma_{t=0} = \sigma_t(1 + \lambda t) \qquad (28b)$$

In Eq. (28a), ω is given in units of 10^{-15} mol dyn^{-1} sec^{-1} and L_p in units of 10^{-11} cm^3 dyn^{-1} sec^{-1}. For example, $\lambda t = 0.03$ at 10 msec, for urea with $\omega = 14 \times 10^{-15}$ mol dyn^{-1} sec^{-1} and red cell $L_p = 1.8 \times 10^{-11}$ cm^3 dyn^{-1} sec^{-1}. Chasan and Solomon[58] pointed out that the time required for the solute to penetrate the unstirred layer, ≈ 10 msec, should be subtracted from the 20-msec observation time, leading to a computed correction of $<5\%$, which was not applied since they felt the correction lay within the uncertainty of their measured $\sigma_{urea} = 0.70 \pm 0.02$.

Toon and Solomon[59] found that a further significant reduction in noise was attainable by averaging 25 runs (after subtraction of 25 control runs), which produced a time course that could be fit with a second order polynomial with high accuracy, as shown for urea in Fig. 6B. The zero-time method only requires determination of the slope at $t = 0$ and so it is not necessary to fit the curve to an accurate equation such as Eq. (17) or (19);

[58] B. Chasan and A. K. Solomon, *Biochim. Biophys. Acta* **821**, 56 (1985).
[59] M. R. Toon and A. K. Solomon, *Biochim. Biophys. Acta* **898**, 275 (1987).

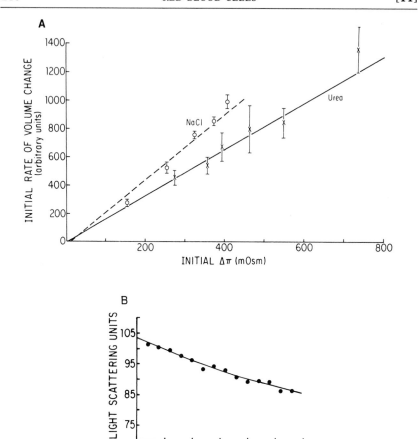

FIG. 6. (A) Determination of σ_{urea} in human red cells by the Chasan and Solomon method [Eq. (27)] (unpublished data of M. R. Toon and A. K. Solomon). The NaCl data have a slope of 2.3 ± 0.1 lsu sec^{-1} mOsm^{-1}, as compared to a slope of 1.62 ± 0.07 lsu sec^{-1} mOsm^{-1}, leading to $\sigma_{urea} = 0.71 \pm 0.07$. (B) Initial phase of cell-shrinking curve when human red cells (bank blood) are exposed to an osmotic pressure gradient produced by 400 mOsm urea (unpublished data of M. R. Toon and A. K. Solomon). The red cell volume, expressed in light-scattering units (●) has been fitted to a second degree polynominal ($a_1 + a_2t + a_3t^2$) with coefficients $a_1 = 103 \pm 1$ lsu, $a_2 = -200 \pm 30$ lsu sec^{-1}, $a_3 = 400 \pm 200$ lsu sec^{-2}.

empirically we have obtained better fits to a polynomial rather than to an exponential. The slope at the time of the first recorded observation, 8 msec, is given directly by the coefficient of the first order term, the contribution of the second order term at 8 msec being negligible. In view of the balance between the dead time and the unstirred layer traverse time, we assumed that the 8-msec observation was close enough to $t = 0$ to require no correction. The great advantage of this method is that the slope is printed out directly by the computer and there is very little judgmental input in fitting the data. The only decision required is the time period of the fit. We truncated our fits at 120 msec for urea, 160 msec for ethylene glycol, and 260 msec for NaCl, because the time course diverged from a second order polynomial at longer times. Using this method, Toon and Solomon[59] determined that $\sigma_{\text{urea}} = 0.65 \pm 0.03$, in good agreement with the prior determinations.

The zero-time method makes it possible to determine the effect of a permeant solute on L_p, provided the solute is at equilibrium across the membrane at zero time, that is, $(\Delta\pi_s = 0)_{t=0}$. Under these conditions, Eq. (23a) leads to

$$(dJ_v/d\Delta\pi_i)_{t=0,\Delta\pi_s=0} = L_p \tag{29a}$$
$$([d(\text{lsu})/dt]/d\Delta\pi_i)_{t=0,\Delta\pi_s=0} = L_{p,\text{lsu}} \tag{29b}$$

assuming a linear relation between light-scattering units and cell volume. This method permits exploration of possible interactions of permeant solutes with the red cell membrane that could affect the aqueous channel. Toon and Solomon[60] studied the effect of urea using Eq. (29b) to form the ratio, $L_{p,\text{lsu},+\text{urea}}/L_{p,\text{lsu,control}}$ and showed that high urea concentrations, such as the red cell encounters in its passage through the kidney, can inhibit red cell water flux by as much as 39%.

These high concentrations of urea cause changes in the refractive index both of the suspending buffer and the red cell. Since the cells normally comprise <2% of the suspension, the extracellular solute concentrations are essentially constant and no time-dependent artifact is introduced. However, as Levitt and Mlekoday[61] showed, the refractive index of the red cells changes as urea permeates during an experiment. Levitt and Mlekoday convincingly demonstrated the existence of the refractive index effect in their light-scattering apparatus by showing that the volume of individual red cells, exposed to a 500-mOsm urea gradient, did not return to the initial volume when all the gradients had been dissipated, using red cells that had been treated with lecithin to make them spherical and thus minimize the injection artifact. Levitt and Mlekoday added dextran, which changes the refractive index significantly with little effect on red cell volume, to a red cell suspension and found that the light-scattering signal depends linearly and independently both on the refractive index and on the cell volume, confirming the observations previously made by

[60] M. R. Toon and A. K. Solomon, *Biochim. Biophys. Acta*, **940**, 266 (1988).
[61] D. G. Levitt and H. J. Mlekoday, *J. Gen. Physiol.* **81**, 239 (1983).

Terwilliger and Solomon[46] with sucrose. Levitt and Mlekoday[61] found that the refractive index effect accounted quantitatively for the observed difference between initial and final cell volume, and incorporated a refractive index correction in their subsequent calculations.

The importance of the refractive index effect depends on the characteristics of the cell or vesicle suspension, the concentration of the solute, the exact conditions of measurement and the experimental apparatus. Toon and Solomon[59] used outdated bank blood and solved the stopping artifact problem by averaging repeated runs, rather than treating the cells to make them spherical. Under their conditions, the red cells returned almost to their initial volume after exposure to a 500-mOsm urea gradient, so that refractive index effects were expected to be small. Toon and Solomon determined that a refractive index difference, equal to that caused by 250-mOsm urea, would produce a 7% change in the conversion of light-scattering units to cell volume. Since their data were collected only over the period from 8 msec, when there was virtually no urea in the cell, to 120 msec, when the cell urea content was $\approx$50% of the equilibrium value, they concluded that no correction was required. For ethylene glycol, which permeates at about 25% of the urea rate, the effect is much smaller. In vesicle suspensions, the refractive index effect does not appear to have any influence on the determination of ω_{urea}[62] while it cannot be neglected in measurements of σ_{urea}.[63] These examples illustrate the necessity of carrying out control experiments whenever large concentrations of permeant solutes are accumulated in the cell or vesicle suspension.

Table III is a compilation of the measurements of σ_{urea} since the first determination in 1960. It is interesting that all of the experimental determinations, though not the computer simulations, fall in a relatively narrow band, similar to the results for P_d and L_p in Tables I and II.

Determination of L_p, ω, and σ from a Single Time Course

When a red cell is suddenly exposed to a solution containing impermeant solutes with a single permeant species, the resultant volume–time curve is determined by the three parameters, L_p, σ, and ω. In principle, all three parameters might be extracted by computer analysis of a single curve, rather than designing specific conditions and using selected parts of the curve to obtain these parameters in separate experiments, as de-

[62] A. S. Verkman, J. A. Dix, and J. L. Seifter, *Am. J. Physiol.* **248**, F650 (1985).
[63] E. G. Holmberg, M. B. Singer, and J. A. Dix, *Biophys. J.* **51**, 341a (Abstr.) (1987).

TABLE III
DETERMINATION OF σ_{urea} FOR THE HUMAN RED CELL

Method	σ_{urea}	Ref.
Experimental determination		
Zero-time continuous flow	0.62 ± 0.02	Goldstein and Solomon[a]
Zero-time stopped flow	0.55 ± 0.02	Sha'afi et al.[b]
Early-time stopped flow	0.79 ± 0.02	Owen and Eyring[c]
Total time course analysis	0.75–0.95	Levitt and Mlekoday[d]
Modified zero-time method	0.70 ± 0.02	Chasan and Solomon[e]
Modified zero-time method	0.65 ± 0.03	Toon and Solomon[f]
Computer simulation		
Fit to Sha'afi et al. data[b]	≈1	Macey and Wadzinski[g]
Fit to Owen and Eyring data[c]	0.75	Owen[h]
Fit to Sha'afi et al. data[b]	0.8	Owen and Galey[i]

[a] D. A. Goldstein and A. K. Solomon, *J. Gen. Physiol.* **44**, 1 (1960).
[b] R. I. Sha'afi, G. T. Rich, D. C. Mikulecky, and A. K. Solomon, *J. Gen. Physiol.* **55**, 427 (1970).
[c] J. D. Owen and E. M. Eyring, *J. Gen. Physiol.* **66**, 251 (1975).
[d] D. G. Levitt and H. J. Mlekoday, *J. Gen. Physiol.* **81**, 239 (1983).
[e] B. Chasan and A. K. Solomon, *Biochim. Biophys. Acta* **821**, 56 (1985).
[f] M. R. Toon and A. K. Solomon, *Biochim. Biophys. Acta* **898**, 275 (1987).
[g] R. I. Macey and L. T. Wadzinski, *Fed. Proc., Fed. Am. Soc. Exp. Biol.* **33**, 2323 (1974).
[h] J. D. Owen, *Biochim. Biophys. Acta* **443**, 306 (1976).
[i] J. D. Owen and W. R. Galey, *Biochim. Biophys. Acta* **466**, 517 (1977).

scribed above. In 1977, Milgram and Solomon[64] developed the necessary equations and described a method to extract the relevant parameters from a volume–time curve by direct numerical integration, but the method was so complex that it was never applied. A year or two later, R. L. Levin (personal communication) developed the software that made it possible for us to extract the parameters by nonlinear least-squares methods. We carried out an extended series of experiments to determine the effect of medium osmolality on L_p, ω_{urea}, and σ_{urea}. The data revealed a number of consistent but unexpected relationships, which went counter to results obtained by the more conventional analyses, made one parameter at a time. We concluded that, in our hands, this application of the nonlinear least-squares method had converged on values that were incorrect, and abandoned the attempt.

Levitt and Mlekoday[61] have used a variant of the total time-course analysis method to measure three parameters from fitting a shrink–swell

[64] J. H. Milgram and A. K. Solomon, *J. Membr. Biol.* **34**, 103 (1977).

curve. They first showed that urea permeation follows single-site binding kinetics which are described by a maximum permeability ω_{max} and a binding constant K_s. They then measured L_p in a separate experiment. Next, they used this value of L_p in determining σ_{urea}, $\omega_{max,urea}$, and $K_{s,urea}$ from the best least-squares fit to the shrink–swell curve. Levitt and Mlekoday's procedure leads to $\sigma_{urea} = 0.95$, higher than the values obtained by other methods, but Levitt and Mlekoday point out that the fit for $\sigma = 0.75$ is also acceptable and conclude ". . . that these methods do not provide an accurate measure of σ for the red cell." Smith et al.[65] have studied the total time course method and concluded that the three membrane parameters are so tightly coupled that independent measurements of both L_p and ω are required in order to obtain precise measurements of σ.

Additional Notes

1. Recently, Kuwahara and Verkman[66] have developed a method based on that of Lawaczeck[28,29] using the fluorescent chromophore, aminonaphthalene trisulfonic acid (ANTS) instead of the indole derivative, tryptamine HCl, used by Lawaczeck. Kuwahara and Verkman point out that ANTS is superior to the indoles with respect to quantum yield and cell impermeability; they have used the method in studies of diffusional and osmotic water permeability[67] of kidney collecting tubules.

2. Chen et al.[68] have developed another method using a fluorescent chromophore to measure water transport in cells and vesicles, based on self-quenching of the entrapped fluorophore, fluorescein sulfonate; they have applied this method to measurements of water transport in kidney vesicles.

Acknowledgment

The author would like to express his gratitude to Dr. James A. Dix for his critical comments and valuable suggestions.

[65] K. R. Smith, J. Myslik, and J. A. Dix, *Biophys. J.* **49**, 273a (Abstr.) (1986).
[66] M. Kuwahara and A. S. Verkman, *Biophys. J.* **54**, 587 (1988).
[67] M. Kuwahara, C. A. Berry, and A. S. Verkman, *Biophys. J.* **54**, 595 (1988).
[68] P.-Y. Chen, D. Pearce, and A. S. Verkman, *Biochemistry* **27**, 5713 (1988).

[12] Transbilayer Mobility of Phosphatidylcholine in the Red Blood Cell

By Jos A. F. Op den Kamp and Ben Roelofsen

The rate of transbilayer movement ("flip-flop") of phospholipids in biological membranes has not been studied in great detail, except for that in the red cell membrane. Devaux and co-workers[1] used spin-labeled analogs of phosphatidylcholine to determine its transbilayer mobility in the human erythrocyte. It may be argued, however, that these reporter molecules are not truly representative of the lecithin molecules naturally occurring in this membrane not only because of the nitroxide spin label they contained, but also because of their relatively short fatty acyl chains, which were an essential prerequisite to enable a spontaneous incorporation of those probe molecules into the membrane of the intact red cell. Mohandas *et al.*[2] and Bergmann *et al.*[3] developed a method using radiolabeled lysophospholipids as probe molecules. Although the simplicity of the latter method is an advantage, it should be noted that lysophospholipids may be subject to acylation to form diacylglycerophospholipids, a process taking place specifically in the inner membrane leaflet and therefore interfering in the transbilayer equilibration of the probe. Moreover, lysophospholipids are also structurally different from the naturally occurring diacylglycerophospholipids, and cannot therefore be considered as their genuine representatives. Renooij and colleagues labeled the endogenous erythrocyte phosphatidylcholine by incubating intact cells either in ^{32}P-labeled plasma[4-6] or in the presence of ^{14}C-labeled fatty acids.[6] This distribution of labeled phosphatidylcholine over both halves of the bilayer was assessed using phospholipases. The results of these studies indicated a slow, if any, transbilayer mobility of phosphatidylcholine in the erythrocyte membrane. Half-time values of 2–7 and 30 hr or more have been reported for the flip-flop of phosphatidylcholine in the membranes of rat

[1] A. Rousselet, C. Guthmann, J. Matricon, A. Bienvenue, and P. F. Devaux, *Biochim. Biophys. Acta* **426**, 357 (1976).
[2] N. Mohandas, J. Wyatt, S. F. Mel, M. E. Rossi, and S. B. Shohet, *J. Biol. Chem.* **257**, 6537 (1982).
[3] W. L. Bergmann, V. Dressler, C. W. M. Haest, and B. Deuticke, *Biochim. Biophys. Acta* **772**, 328 (1984).
[4] W. Renooij, L. M. G. van Golde, R. F. A. Zwaal, and L. L. M. van Deenen, *Eur. J. Biochem.* **61**, 53 (1976).
[5] W. Renooij and L. M. G. van Golde, *FEBS Lett.* **71**, 321 (1976).
[6] W. Renooij and L. M. G. van Golde, *Biochim. Biophys. Acta* **470**, 465 (1977).

METHODS IN ENZYMOLOGY, VOL. 173

and human erythrocytes, respectively. The availability of phospholipid transfer proteins opened the possibility of introducing intact long-chain diacyl glycerophospholipids into the membrane of the intact cell. The phosphatidylcholine-specific transfer protein (PC–TP) from beef liver has shown to be particularly suitable for such studies as it mediates a one-for-one exchange of those lipid molecules.[7] The first application of this protein in such studies concerned the human erythrocyte, using phospholipases for the localization of the inserted reporter molecules.[8] A special feature of the PC–TP-mediated exchange process is that it enables a determination of the flip-flop rate of individual molecular species of phosphatidylcholine.

Principle of the Assay

Trace amounts of chemically defined, radiolabeled phosphatidylcholine molecules of known specific radioactivity are introduced into the outer layer of the membrane of intact erythrocytes by using the phosphatidylcholine-specific transfer protein, purified from bovine liver. Under the conditions applied, total lipid content and overall phospholipid compositions remain unchanged, provided that the fatty acid composition of the newly introduced phosphatidylcholine molecules is similar to that of the native phosphatidylcholine.

Following the introduction of the phosphatidylcholine reporter molecules, cells are incubated to allow an equilibration of these molecules over both halves of the bilayer. At various times the amount of radiolabeled phosphatidylcholine still present in the outer membrane layer is assayed by treatment of the intact cells with phospholipase A_2 plus sphingomyelinase C (sphingomyelin phosphodiesterase) and determination of the specific radioactivity of the lysophosphatidylcholine thus formed.

Materials

Erythrocytes

Fresh human erythrocytes are obtained from healthy individuals by venepuncture, using acid–citrate–dextrose[9] as anticoagulant. Cells are collected by centrifugation for 10 min at 2500 g in a swing-out rotor, washed twice with at least 5 vol of 155 mM NaCl and once with a buffer

[7] B. Bloj and D. B. Zilversmit, *Mol. Cell. Biochem.* **40**, 163 (1981).
[8] G. van Meer and J. A. F. Op den Kamp, *J. Cell. Biochem.* **19**, 193 (1982).
[9] H. M. Anderson and J. C. Turner, *J. Clin. Invest.* **39**, 1 (1960).

containing 90 mM KCl, 45 mM NaCl, 44 mM sucrose, 10 mM glucose, 100 IU/ml penicillin, 100 μg/ml streptomycin, and 10 mM Tris–HCl, pH 7.4. After each wash, supernatant and buffy coat are carefully removed by aspiration.

Phosphatidylcholine Donor Vesicles

Unilamellar phosphatidylcholine donor vesicles are prepared as follows: equimolar amounts of cholesterol (Merck, Darmstadt, FRG) and phospholipid mixture are dissolved in chloroform/methanol (2 : 1, v/v). The phospholipid mixture consists of 97% egg phosphatidylcholine[10] (Sigma Chemical Company, St. Louis, MO), 3% phosphatidic acid[11] (derived from egg phosphatidylcholine, Sigma Chemical Company, St. Louis, MO), trace amounts of radiolabeled [^{14}C]phosphatidylcholine[12] (0.1 mCi/mmol vesicle phosphatidylcholine), and trace amounts of [^{3}H]glycerol trioleate (The Radiochemical Centre, Amersham, England; 0.5 mCi/mmol vesicle phosphatidylcholine), the latter serving as nonexchangeable marker.

The lipid solution is dried at 37° under a stream of nitrogen. Lipids are subsequently dispersed by vortexing in buffer, containing 90 mM KCl, 45 mM NaCl, 44 mM sucrose, 10 mM glucose, 100 IU/ml penicillin, 100 μg/ ml streptomycin, and 10 mM Tris–HCl, pH 7.4. Unilamellar vesicles are prepared by ultrasonication of the above dispersion for 15 min under a nitrogen atmosphere, using a Branson sonifier at 70 W. During sonication, the temperature of the mixture is kept below 20° by cooling in melting ice. Larger lipid aggregates and metal particles released from the sonifier probe are removed by centrifugation for 1 hr at 100,000 g. The final vesicle suspension is assayed for radioactivity (^{3}H and ^{14}C) as well as lipid phosphorus,[13] to determine the specific radioactivity of the phosphatidylcholine donor system. The final phosphatidylcholine concentration in the vesicle suspension should be approximately 10 mM.

---

[10] Egg phosphatidylcholine is used as the bulk phosphatidylcholine in the donor vesicles, because its fatty acid composition is similar to that of the phosphatidylcholine in the human erythrocyte and will not therefore cause a change in the phosphatidylcholine molecular species composition in the erythrocyte during the exchange procedure.

[11] Phosphatidic acid is included in the vesicles to render them a negative surface charge which minimizes a contamination of red cells with donor vesicles after the exchange procedure.

[12] The ^{14}C label should preferentially be present in the N-methyl groups of the choline moiety or alternatively either in both fatty acids or in the fatty acid at the 1-position of the glycerol backbone.

[13] G. Rouser, S. Fleischer, and A. Yamamoto, *Lipids* **5**, 494 (1979).

Phosphatidylcholine-Specific Transfer Protein

The phosphatidylcholine-specific transfer protein is purified from beef liver according to the procedure of Westerman et al.[14] Glycerol, routinely used at a 50% (v/v) concentration during storage of the transfer protein at $-20°$, is removed by overnight dialysis at $4°$ against 300 vol of buffer (90 mM KCl, 45 mM NaCl, 44 mM sucrose, 10 mM glucose, 100 IU/ml penicillin, 100 μg/ml streptomycin, and 10 mM Tris–HCl, pH 7.4). To avoid nonspecific protein adherence, the dialysis tubing is preequilibrated for 1 hr with a 0.1% solution of bovine serum albumin. After dialysis, the volume of the protein solution is reduced by covering the dialysis bag with flake polyethylene glycol (Aquacide III, Calbiochem-Behring), until a concentration of 100 mg protein/ml is reached.

Phospholipases

Phospholipases A_2 (EC 3.1.1.4) from *Naja naja* and bee venom (both from Sigma Chemical Company, St. Louis, MO; specific activities, respectively, 970 and 1500 IU/mg protein) are dissolved in 50 mM Tris–HCl, 5 mM CaCl$_2$, pH 7.4 at concentrations of approximately 0.5 IU/μl and used without further purification.

Sphingomyelinase C (EC 3.1.4.12, sphingomyelin phosphodiesterase) from *Staphylococcus aureus* is purified according to Zwaal et al.[15] and stored in 50 mM Tris–HCl, 5 mM CaCl$_2$, 50% (v/v) glycerol at $-20°$ at a concentration of approximately 0.1 IU/μl.

Incubation Procedures

Introduction of [^{14}C]Phosphatidylcholine into Intact Erythrocytes

Trace amounts of [^{14}C]phosphatidylcholine are introduced into the outer membrane leaflet of the intact erythrocyte by incubation of the cells together with the [^{14}C]phosphatidylcholine donor vesicles in the presence of the phosphatidylcholine-specific transfer protein. The incubation is carried out at $37°$ under gentle agitation, in buffer containing 90 mM KCl, 45 mM NaCl, 44 mM sucrose, 10 mM glucose, 100 IU/ml penicillin, 100 μg/ml streptomycin, and 10 mM Tris–HCl, pH 7.4. The mixture contains the red cells (30–40% suspension), 2 to 3 μM of the transfer protein (50–70 μg protein/ml), and [^{14}C]phosphatidylcholine donor vesicles. The ratio

[14] J. Westerman, H. H. Kamp, and K. W. A. Wirtz, this series, Vol. 98, p. 581.
[15] R. F. A. Zwaal, B. Roelofsen, P. Comfurius, and L. L. M. van Deenen, *Biochim. Biophys. Acta* **406**, 83 (1975).

of vesicle to erythrocyte phosphatidylcholine may vary from 1 to 4. After 1 hr, the exchange reaction is stopped by centrifugation for 5 min at 2500 g. Residual transfer protein and lipid vesicles are removed by two additional washes of the cells with the above buffer.

Equilibration of [¹⁴C]Phosphatidylcholine between Both Membrane Leaflets

The [¹⁴C]phosphatidylcholine is enabled to equilibrate between both halves of the membrane bilayer by subsequent incubation of the cells (33% suspension) in the above buffer (see the previous section) at 37°. At timed intervals, samples (comprising 310 μl of packed cells) are taken from this incubation mixture and the cells collected by centrifugation for 5 min at 2500 g. Sampling is usually continued up to a total incubation time of 8 hr.

Incubation of Cells with Phospholipases

To determine the transbilayer distribution of the [¹⁴C]phosphatidylcholine in the red cell membrane, the phosphatidylcholine in the outer monolayer is selectively converted into its 1-acyl lyso derivative by treatment of the intact cells with phospholipases. The cells, sampled during the above incubation (see the previous section), are suspended in 6 vol of the following buffer: 90 mM KCl, 45 mM NaCl, 10 mM CaCl$_2$, 0.25 mM MgCl$_2$, 22 mM sucrose, and 10 mM Tris–HCl, pH 7.4. A mixture of highly purified phospholipases A$_2$ from *N. naja* and bee venom is added, each at a concentration of 6 IU/100 μl packed cells. After 1 hr of incubation at 37°, highly purified sphingomyelin phosphodiesterase (sphingomyelinase C, 2 IU/100 μl packed cells) is added and the incubation continued for another 60 min in order to achieve a complete (phospholipase A$_2$-induced) hydrolysis of the phosphatidylcholine in the outer membrane layer of the intact erythrocytes.[15] The cells are subsequently collected by centrifugation (5 min at 2500 g), and the action of the phospholipases is terminated by resuspending the cells in 0.5 ml of 100 mM EDTA in 0.9% NaCl. The supernatant of the last centrifugation step can be used to determine the extent of hemolysis[16] by measuring its absorbance at 418 nm.

Lipid Analysis

The erythrocytes, suspended in 0.5 ml 0.9% NaCl, containing 100 mM EDTA, are lysed by the addition of at least 5 ml ice-cold water, previously

---

[16] Cell lysis should be <5%.

saturated with CO_2. After 1 hr at 4°, the ghost membranes are spun down at 2500 g for 20 min and their lipids extracted following the procedure of Rose and Oklander.[17] After extraction, the solvent is evaporated under reduced pressure and the residue taken up in 3.5 ml chloroform/methanol (2 : 1, v/v). From this solution, the following samples are taken: (1) 1.5 ml for one-dimensional thin-layer chromatography, (2) 1.5 ml for two-dimensional thin-layer chromatography, and (3) 0.3 ml for transferral to a scintillation vial to determine the extent of contamination of the cells with the [14C]phosphatidylcholine donor vesicles used in the initial exchange incubation.

1. One-dimensional thin-layer chromatography is carried out on precoated silica gel plates (Silicagel 60, Merck, Darmstadt, FRG), using chloroform/menthol/25% ammonia/water (90 : 54 : 5.5 : 5.5, v/v) as developing system. After drying, lipid spots are visualized by exposing the plate to I_2 vapor and those spots corresponding to, respectively, phosphatidylcholine and lysophosphatidylcholine indicated. After evaporation of the I_2, these two spots are scraped off and quantitatively transferred into scintillation vials. Toluene, containing 2,5-diphenyloxazole (5 g/liter), methyl-1,4-bis[2-(5-phenyloxazolyl)]benzene (0.25 g/liter), and Biosolv (2%, v/v) is added and radioactivity measured by the 14C single-channel rate method, using an appropriate liquid scintillation spectrometer.

2. Two-dimensional thin-layer chromatography is carried out on silica gel plates according to Broekhuyse.[18] Lipid spots are again visualized by iodine (I_2) staining and the areas corresponding to, respectively, phosphatidylcholine and lysophosphatidylcholine scraped off and ashed with 70% perchloric acid. Phosphate analysis is performed according to Rouser et al.[13]

3. Radioactivity measurements on the total lipid extract are performed after evaporation of the solvent and addition of scintillation solution as described above (2). An 3H/14C dual-label procedure is applied to assess the amount of the nonexchangeable [3H]glycerol trioleate in the lipid extract in order to determine the extent of contamination of erythrocyte [14C]phosphatidylcholine with [14C]phosphatidylcholine originating from donor vesicles adhering to the cells.[19]

[17] H. G. Rose and M. Oklander, J. Lipid Res. 6, 428 (1965).
[18] R. M. Broekhuyse, Clin. Chim. Acta 23, 457 (1965).
[19] Under the experimental conditions described here, this contamination is always <2% of the total erythrocyte phospholipid.

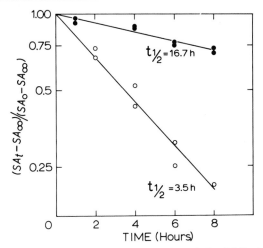

FIG. 1. Transbilayer movement of 1,2-dioleoyl- and 1-palmitoyl-2-linoleoylphosphatidyl-choline in the human erythrocyte at 37°. Transbilayer movements were determined by following the transbilayer equilibration of 1,2-[1-^{14}C]dioleoylphosphatidylcholine and 1-pal-mitoyl-2-linoleoylphosphatidyl[N-$methyl$-^{14}C]choline, previously introduced into the outer membrane leaflet of the intact cells by using the phosphatidylcholine-specific transfer pro-tein from beef liver. The redistribution of the probe molecules was followed by measuring the specific radioactivity of the phosphatidylcholine in the outer monolayer, modified to lysophosphatidylcholine by phospholipase treatment. The relative specific radioactivities of the lysophosphatidylcholines [$(SA_t - SA_x)/(SA_0 - SA_x)$], derived from typical experi-ments, are plotted semilogarithmically vs the time of incubation at 37°. Half-time values for transbilayer movement ($t_{1/2}$) can be calculated from the slope of these lines for 1,2-dioleoyl- (●) and 1-palmitoyl-2-linoleoylphosphatidylcholine (○), respectively. (From Middelkoop *et al.*[22])

Calculations

Corrections of the analytical data for contamination of erythrocytes with phosphatidylcholine donor vesicles are carried out as described by van Meer *et al.*[20] for similar contaminations using rat liver microsomes as phosphatidylcholine donor system. The mathematical background of the calculations, based on a two-pool closed system, is presented in detail elsewhere.[8] In summary: the half-time value for the equilibration of phos-phatidylcholine molecules over both halves of the membrane, the so-called flip-flop, is determined from a semilogarithmic plot of $(SA_t - SA_\infty)/(SA_0 - SA_\infty)$ versus time of incubation (compare Fig. 1), in which SA_t is

[20] G. van Meer, B. J. H. M. Poorthuis, K. W. A. Wirtz, J. A. F. Op den Kamp, and L. L. M. van Deenen, *Eur. J. Biochem.* **103**, 283 (1980).

the specific radioactivity of the lysophosphatidylcholine produced by treatment of the intact cells with phospholipase A_2 (plus sphingomyelin phosphodiesterase), which treatment started at the time t of incubation B. SA_∞ is the specific radioactivity of the lysophosphatidylcholine, representing the phosphatidylcholine in outer monolayer, at indefinite time of incubation when complete equilibration will be reached. This value is of course identical to the specific radioactivity of the entire pool of phosphatidylcholine in nonphospholipase-treated cells at any time. SA_0 is the specific radioactivity of the lysophosphatidylcholine at zero time. Since in normal human erythrocytes 76% of the phosphatidylcholine is in the outer monolayer,[15] it follows that $SA_0 = SA_\infty/0.76$.

Comments

By using ^{14}C-labeled egg phosphatidylcholine, which can be easily prepared from egg phosphatidylcholine by the method of Stoffel,[21] the procedure described above results in the determination of the average flip-flop rate experienced by the total phosphatidylcholine complement of the membrane which, similar to egg phosphatidylcholine, is composed of a great variety of molecular species differing in fatty acid composition.

The same method can also be used, however, to determine the transbilayer mobility of individual molecular species of phosphatidylcholine in the red cell membrane (compare Fig. 1). To that end, the phosphatidylcholine donor vesicles are supplemented with the ^{14}C-labeled species of choice.[22] It should be remembered, however, that in this case the bulk phosphatidylcholine of these vesicles should still be composed of (cold) egg phosphatidylcholine, in order to avoid considerable changes in the native molecular species composition of the red cell membrane. It has been shown that extensive replacement of the phosphatidylcholine in the intact human erythrocyte by either disaturated or highly unsaturated species has most dramatic consequences for a number of membrane parameters and the structural integrity of the cell.[23,24]

It is essential to mention that the procedure described in this chapter cannot be used in case phosphatidylcholine molecules experience fast transbilayer dynamics with half-times of flip-flop of only a few hours or

[21] W. Stoffel, this series, Vol. 35, p. 533.
[22] E. Middelkoop, B. H. Lubin, J. A. F. Op den Kamp, and B. Roelofsen, *Biochim. Biophys. Acta* **855**, 421 (1986).
[23] F. A. Kuypers, B. Roelofsen, J. A. F. Op den Kamp, and L. L. M. van Deenen, *Biochim. Biophys. Acta* **769**, 337 (1984).
[24] F. A. Kuypers, B. Roelofsen, W. Berendsen, J. A. F. Op den Kamp, and L. L. M. van Deenen, *J. Cell Biol.* **99**, 2260 (1984).

less. This is due to the relatively lengthy incubations that are required for the introduction of the probe molecules and subsequent phospholipase treatments of the cells.

The method has been successfully applied to determine phosphatidyl-choline flip-flop rates under a variety of conditions in both normal and pathologic erythrocytes, such as sickle cells[25,26] and hereditary pyropoiki-locytes.[27]

The method fulfills the requirement that the discrimination between the inner and outer pools of phosphatidylcholine is complete and that the initial distribution of the probe molecule is not disturbed by subsequent treatment of the cells with phospholipases. This is illustrated by the observation that, immediately after its insertion, 96% or more of the newly introduced [^{14}C]phosphatidylcholine is detected in the outer membrane layer.[8]

In principle, the method is not necessarily restricted to phosphatidyl-choline. Using a nonspecific phospholipid transfer protein, trace amounts of radiolabeled phosphatidylethanolamine, phosphatidylserine, and sphingomyelin can be also introduced into the outer membrane layer of the intact (human) erythrocyte.[28] It should be stressed again, however, that also in those cases the bulk phospholipid in the donor vesicles should consist of egg phosphatidylcholine.

[25] P. F. H. Franck, D. T.-Y. Chiu, J. A. F. Op den Kamp, B. Lubin, L. L. M. van Deenen, and B. Roelofsen, *J. Biol. Chem.* **258,** 8435 (1983).
[26] P. F. H. Franck, E. M. Bevers, B. H. Lubin, P. Comfurius, D. T.-Y. Chiu, J. A. F. Op den Kamp, R. F. A. Zwaal, L. L. M. van Deenen, and B. Roelofsen, *J. Clin. Invest.* **75,** 183 (1985).
[27] P. F. H. Franck, J. A. F. Op den Kamp, B. Lubin, W. Berendsen, P. Joosten, E. Briët, L. L. M. van Deenen, and B. Roelofsen, *Biochim. Biophys. Acta* **815,** 259 (1985).
[28] L. Tilley, S. Cribier, B. Roelofsen, J. A. F. Op den Kamp, and L. L. M. van Deenen, *FEBS Lett.* **194,** 21 (1986).

[13] Sugar Transport in Red Blood Cells

By W. F. WIDDAS

Introduction

Sugars are very hydrophilic molecules and as such are quite unsuitable to cross the bimolecular lipid layers which form the basis of cell membranes but which are readily permeated by hydrophobic substances. Nevertheless, it is essential to life in mammals that glucose should enter cells to serve as a substrate for metabolism. Evidence has accumulated that glucose does penetrate cell membranes, but only by a special mechanism involving components of the cell membrane with which glucose (and other monosaccharides) react and which in some way facilitate the transport across the membrane.

Work to study and characterize this special mechanism has chiefly been done using human red cells because they are a convenient experimental material; they also have the peculiarity of possessing a very rapid transfer rate for glucose (and closely related sugars such as 2-deoxyglucose, 3-O-methylglucose, mannose, and galactose) so that the sugar in the cells equilibrates with that in the medium.

By contrast, the red blood cells of common laboratory and farm animals which one might naturally turn to as an experimental source have rates of transfer only about one-thousandth that in human red cells, and due to metabolism of the glucose which does penetrate they fail to equilibrate with the outside medium. Although this is the situation with adult laboratory animals, Widdas[1] showed that the red cells from fetal and newborn animals of several species have rates of transfer comparable with human blood. Red cells from fetal guinea pigs have a faster rate than human red cells but have broadly similar properties (Aubby and Widdas[2]).

Source of Red Cells

Human Red Cells. Fresh human red cells obtained by sterile venepuncture of the antecubital vein are ideal if adequate numbers of volunteers are available.

Failing this, most laboratories have access to blood from blood banks, which are the best source when very large quantities are required for the experiments planned. The blood usually available for laboratory use is "time expired" and in such blood the rates of sugar transfer are reduced to a large and variable degree—it may be as low as half that in fresh blood. This variability is a drawback in kinetic experiments and attempts have been made to restore a more normal activity. A method used by Ginsburg[3] for a number of years consists of incubating a 10% erythrocyte suspension for 60–90 min at 37° in a medium consisting of NaCl (120 mM), KCl (5 mM), NaH_2PO_4 (10 mM), glucose (5 mM), and inosine (15 mM), pH 7.6. The present author has confirmed the rejuvenating effect of this treatment for time-expired blood and it is recommended where the red cells are to be used for kinetic studies.

[1] W. F. Widdas, *J. Physiol.* (*London*) **127**, 318 (1955).
[2] D. S. Aubby and W. F. Widdas, *J. Physiol.* (*London*) **309**, 317 (1980).
[3] H. Ginsburg, personal communication (1985).

Fetal Red Cells. Fetal and newborn guinea pigs have been extensively used in the author's laboratory. Newborn rabbits have been used by Augustin *et al.*[4] and newborn dog red cells by Lee *et al.*[5]

Other Red Cells. Apart from the above cells, which have very fast transport rates, some studies have been made with cells having very slow rates. Chief among these have been erythrocytes from adult rabbits[6,7] and avian erythrocytes as used by Wood and Morgan.[8]

Preparation of Red Cells

Anticoagulants. Fresh blood tends to clot and some form of anticoagulant is needed. Citrate,[9] heparin,[1] and EDTA (1 mg/ml)[10] are all suitable, as is the acid–citrate–dextrose (ACD) mixture used in the Blood Transfusion Service. Alternatively, the blood may be defibrinated by stirring with a glass loop[1] and then by filtering through glass wool.[7]

Heparinized unwashed blood can readily be kept up to 48 hr in a refrigerator at 4° and used for repeat experiments.

Suspending Media. Saline media buffered with either phosphate[1,9] or Tris[11] are used to dilute and wash the red cells prior to use. After the first centrifugation the buffy coat, which contains the white cells, is carefully removed and discarded. If the first dilution is about 5-fold this will reduce the normal (5 mM) blood glucose to 1 mM; two further dilutions of this order will reduce the inside concentration to under 0.1 mM, which is low enough for most experimental purposes, but some investigators have used a fourth wash to ensure that the residual glucose is insignificant.

A typical suspending medium for human red cells would be NaCl (119 mM), KCl (4.8 mM), CaCl$_2$ (2.6 mM), MgCl$_2$ (1.7 mM), Tris (35 mM), pH 7.4; tonicity, 300 mOsm.[11]

For rabbit red cells the medium used was NaCl (137 mM), KCl (5.9 mM), CaCl$_2$ (1.3 mM), MgCl$_2$ (2.4 mM), KH$_2$PO$_4$ (1.2 mM), imidazole (4.2 mM), glycylglycine (7.6 mM), bovine serum albumin (0.2%), pH 7.35.[7]

[4] H. W. Augustin, L. van Rohden, and M. R. Häcker, *Acta Biol. Med. Ger.* **19**, 723 (1967).
[5] P. Lee, J. Auvil, J. E. Grey, and M. Smith, *Fed. Proc., Fed. Am. Soc. Exp. Biol.* **35**, 780 (1976).
[6] H. E. Morgan, C. F. Kalman, R. L. Post, and C. R. Park, *Fed. Proc., Fed. Am. Soc. Exp. Biol.* **14**, 336 (1955).
[7] D. M. Regen and H. E. Morgan, *Biochim. Biophys. Acta* **79**, 151 (1964).
[8] R. E. Wood and H. E. Morgan, *J. Biol. Chem.* **224**, 1451 (1969).
[9] P. G. LeFevre, *J. Gen. Physiol.* **31**, 505 (1948).
[10] R. D. Taverna and R. G. Langdon, *Biochim. Biophys. Acta* **298**, 412 (1973).
[11] P. G. LeFevre, *Biochim. Biophys. Acta* **120**, 395 (1966).

Experimental Techniques

In general there are three ways of investigating sugar transport in red cells but these are not usually applicable to the same type of problem. Consideration must be given both to the object of the experiment and to the source of red cells it is proposed to use when planning the most appropriate experimental approach. The three main techniques are (1) standard or modified biochemical techniques, (2) optical methods involving osmotic swelling or shrinking of erythrocytes, and (3) radioactive techniques.

Standard or Modified Biochemical Techniques

The biochemical estimation of sugars involves preparing a protein-free extract and, if the cellular concentration is required, a centrifugation to separate the cells is also needed. Centrifugation takes a time which is long relative to the rates of transport of sugars by cells with fast transfer rates. Thus biochemical methods were unable to make useful contributions to the kinetics of glucose transport in human red cells. They have, however, been used in important work on sugar transport in rabbit red cells.

The estimation of sugars as reducing agents is not sufficiently specific but glucose determined enzymatically[12] is more useful. However, in competitive studies the need arises to estimate two sugars in the presence of each other. Park and Johnson[13] used paper chromatography to separate the sugars and by eluting the spots and developing with a suitable reagent (e.g., aniline hydrogen phthalate[14] for galactose) were able to estimate the sugars quantitatively.

Morgan et al.[6] created conditions in rabbit red cells whereby either influx or efflux could be measured independently. They used 10^{-2} M NaF to inhibit utilization so that glucose was accumulated to the required amount. The cells were then washed free of external glucose at $0°$ and the efflux measured into a medium containing yeast hexokinase.

Influx was measured into cells in the presence of 2.1×10^{-5} M Methylene Blue, which promoted phosphorylation of the glucose that penetrated, so that the internal free glucose concentration was kept less than 0.1% of the outside concentration. In this way experiments to determine the influx as a function of the outside concentration and efflux as a function of the internal concentration were carried out. These showed that in both directions a saturable transport mechanism was involved similar to

---

[12] M. W. Slein, G. T. Cori, and C. F. Cori, *J. Biol. Chem.* **186,** 763 (1950).
[13] C. R. Park and L. H. Johnson, *Am. J. Physiol.* **182,** 17 (1955).
[14] S. M. Partridge, *Nature (London)* **164,** 443 (1949).

that shown to exist in human red cells by optical experiments but about a thousand times slower than in human cells.

Using rabbit red cells Park et al.[15] were the first to demonstrate an important property of facilitated transport which has been termed "uphill transport induced by counterflow."[16] It was predicted by Widdas[17] that competition for a "carrier" transfer could result in one sugar being transferred against its concentration gradient if a competitive sugar was transferring down its own concentration gradient. Park et al.[15] set out to test this prediction using xylose and glucose as the competing sugars. The rabbit red cells were first equilibrated with xylose, and since this is a nonmetabolizing sugar the cells approached equilibrium after a 2 hour incubation. Glucose was then added to the outside medium, and since this sugar on entering the cells was removed by metabolism its concentration gradient, inward, was maintained. The concentration of xylose in the cells was found to be reduced although its outward movement was against the concentration gradient for that sugar.

The results of this now classic experiment are shown in Fig. 1. It is important since the phenomenon was a prediction from the kinetics of a mobile carrier and could not arise from a mechanism involving only simple diffusion.

Although the mobile carrier model may be an oversimplification of the sugar transport mechanism which is now attributed to a transmembrane protein, the phenomenon of uphill transfer by counterflow could only arise if there was some sort of conformational change within the protein which provided the equivalent to movement in the mobile carrier model.

Optical Methods Involving Osmotic Swelling or Shrinking

It is common experience that suspensions of intact red cells are opaque but the same suspension becomes a colored transparent liquid if the cells are hemolysed either chemically or osmotically. The opacity of intact cell suspensions is due to the higher refractive index of the inside medium containing hemoglobin at high concentration. On hemolysis the hemoglobin becomes uniformly distributed and light does not suffer refraction at the cell boundaries. The change in light transmission with hemolysis is large and can readily be detected with the naked eye and a convenient light source. In experiments the time for hemolysis can be noted with a stopwatch and if hemolysis is produced by the penetration of

[15] C. R. Park, R. L. Post, C. F. Kalman, J. H. Wright, Jr., L. H. Johnson, and H. E. Morgan, Ciba Found. Colloq. Endocrinol. [Proc.] 9, 240 (1956).
[16] T. Rosenberg and W. Wilbrandt, J. Gen. Physiol. 41, 289 (1957).
[17] W. F. Widdas, J. Physiol. (London) 118, 23 (1952).

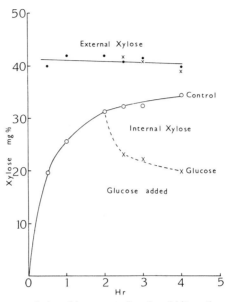

FIG. 1. Uphill transport induced by counterflow in rabbit erythrocytes. ○, Xylose concentration in control cells during a 4-hr incubation in a 40 mg/100 ml xylose medium. ×, Xylose concentration in the cells after addition of glucose. Reduction in the internal xylose concentration showed that xylose had moved out into the xylose-containing medium although this was against the concentration gradient. (After Park et al.[15])

a nonelectrolyte from an isosmotic solution the speed with which hemolysis occurs is a measure of the rate of penetration of the nonelectrolyte concerned. This technique was used extensively by Jacobs and colleagues in the 1920s and 1930s and the quantitative aspects were described by Jacobs and Stewart.[18]

However, it was observed by Masing[19] as long ago as 1914 that human red cells do not hemolyse in isosmotic glucose solutions, so the gross light changes produced by hemolysis are not available for following sugar penetration rates. The same light changes can be followed more accurately using photoelectric devices and Ørskov[20] observed that red cell volume changes which did not involve hemolysis could nevertheless be determined from the much smaller changes in light transmission which they caused.

[18] M. H. Jacobs and D. R. Stewart, *J. Cell. Comp. Physiol.* **1,** 71 (1932).
[19] E. Masing, *Pfluegers Arch. Gesamte Physiol. Menschen Tiere* **156,** 401 (1914).
[20] S. L. Ørskov, *Biochem. Z.* **279,** 241 (1935).

Parpart,[21] working in Jacob's laboratory, had also developed photo-electric techniques for following red cell volume changes and this apparatus was used by LeFevre[9] to follow the changes due to glycerol and glucose penetration.

Widdas[22] modified an apparatus to follow the osmotic hemolysis in fetal sheep red cells to make it sensitive enough to follow the small changes in light transmission obtained in glucose penetration experiments.

Wilbrandt and co-workers[23] had also used photoelectric techniques but the precision of their method was greatly increased by Fuhrmann's development of an apparatus for directly measuring the scattering of light at right angles to the incident beam through the suspension.[24]

The three techniques have differences which will be mentioned but full details of only the author's technique will be given. An account of the Fuhrmann technique is the subject of a later chapter in this volume.[25]

LeFevre's Technique. The apparatus used by LeFevre in his earliest experiments has not been published in detail but his modified apparatus as used later was fully described in 1966.[11] The cuvette, of about 15-ml capacity, was enclosed in a thermostatically controlled housing and illuminated by a specially regulated 6-V tungsten lamp. The light beam was split by a glass surface to provide a reference beam containing an adjustable glass wedge while the main beam went through the cuvette. The reference and cuvette beams were detected by two separate photomultipliers supplied from the same battery source. A cathode follower coupling circuit fed into a T-Y recorder.

Suspensions of cells were made up to 1/300–1/200 relative to the volume of the medium. Typically, 12 ml of cell suspension was allowed to reach a steady level of light transmission and then up to 1.5 ml of medium containing sugar at 1.7 M was added. When the cells shrank the scattering of light was increased and the forward transmission was reduced. As reswelling of the cells occurred the light transmission increased. Runs of 10 to 15 min were recorded; stirring was not specified.

Widdas's Technique. Widdas[22] developed a technique which differed from LeFevre's in certain respects. The cuvette was 2 cm wide (capacity 25 ml) and was especially constructed to be boot shaped. The light beam went through the "toe" of the boot as a 2.5-cm-diameter beam of parallel

[21] A. K. Parpart, *J. Cell. Comp. Physiol.* **7,** 153 (1935).

[22] W. F. Widdas, *J. Physiol. (London)* **120,** 20P (1953).

[23] W. Wilbrandt, E. Guensberg, and H. Lauener, *Helv. Physiol. Acta* **5,** C20 (1947).

[24] G. F. Fuhrmann, P. Liggenstorfer, and W. Wilbrandt, *Experientia* **27,** 1428 (1971).

[25] G. F. Fuhrmann, this volume [15].

light while in the "leg" of the boot a paddle stirrer was driven by a synchronous motor at about 375 rpm.

The light was divided into reference and main beams which passed through Polaroid filters set mutually at right angles. The two beams converged onto a single photocell in front of which was a third Polaroid filter held in a mounting which was rotated at 12.5 Hz. A shutter in the reference beam was used to get a null (as in LeFevre's technique) but the out-of-balance signal (a sine wave of 25 Hz) was amplified with a low-bandwidth ac amplifier prior to phase-sensitive rectification and display on a recording galvanometer.

The 21 ml of suspension containing only 3 μl of packed cells was first allowed to reach a steady state; then sugar dissolved in the same medium was added. For entry experiments 0.5 ml of 1.67 M sugar was added in four consecutive runs, allowing equilibration to occur before each new addition.

The double-beam arrangement, single photocell, and ac amplification minimized zero drift and gave good long-term stability. A second apparatus using a chopped light source in place of the Polaroid filters has also been used. In recent years a silicon photodiode incorporating a color correction filter (visible radiation) and solid-state amplifiers have replaced the original photocell and valve amplifier. A chart recorder is used in place of the recording galvanometer, but the basic optical principles have remained unchanged for over 30 years.

Exit Experiments. The very low hematocrit (the cells occupied only 1 part in 7000 of the medium) used by Widdas made the technique suitable for adaptation to follow glucose exits from cells.[26,27] For an exit it is not practicable to add 3 μl of packed cells to 21 ml saline in the cuvette as the packed cells take too long to be dispersed homogeneously. It was found that a preliminary suspension in 0.15–0.2 ml of the glucose-containing medium was, however, rapidly dispersed when added to the cuvette and gave satisfactory exit records. In practice the cells are preincubated in these same proportions, i.e., 10.3 ml buffered saline, 0.5 ml 1.67 M glucose in buffered saline, 0.2 ml packed cells which have been previously washed. After incubation, long enough to reach equilibrium, aliquots of 0.15–0.2 ml are tested for exits and a suitable value in this range chosen. Prior to taking an aliquot the suspension is stirred for a few seconds on a vortex mixer to ensure uniform mixing and the absence of rouleaux.

When the cells are incubated at 76 mM glucose the addition of 0.15 ml to the cuvette makes the outside concentration 0.6 mM as a minimum.

[26] F. Bowyer and W. F. Widdas, *J. Physiol. (London)* **141**, 219 (1958).
[27] A. K. Sen and W. F. Widdas, *J. Physiol. (London)* **160**, 392 (1962).

However, this proved to be no serious drawback as the exit times are linearly related to the outside glucose concentration (up to about 12 mM) and the time for exit into a glucose-free solution could be obtained by extrapolation (see Figs. 2 and 3). Note, however, that at the very low hematocrit (1 in 7000) the exit of even 76 mM from the cells could only have changed the outside concentration by 0.01 mM, which is negligible.

At the low hematocrit used in this technique the light transmission actually increased as the cells shrank and vice versa. The cells in the 2-cm light beam would not be quite sufficient to occupy the whole area if they could be brought into the same plane; and it may be presumed that cells which shrink allow more light to pass between them and this light is not subject to scattering.

Fuhrmann's Technique. As in the two methods described above the cuvette is surrounded by a water jacket except for where the light enters and leaves (in this case at right angles to the entry beam). The cuvette holds 15 ml of cell suspension which has a hematocrit of about 1 in 1000. Stirring is by means of a magnetic "flea." The light scattered at right angles increases if the cells shrink and their refractive index is increased. In this respect the changes are qualitatively similar to those in the Widdas technique although the basis is probably different. Both are in contrast to what is seen in the LeFevre technique.

For further details of the Fuhrmann technique, see Chapter [15] in this volume.[25]

Results from Optical Experiments

Although simple osmotic and biochemical techniques had highlighted anomalies in the apparent rates of uptake of glucose at different concentrations,[28] and Klinghoffer[29] had found that equilibration, which occurred readily at low glucose concentrations, was delayed indefinitely at isosmotic concentrations, it was the optical techniques which first allowed the rates of uptake to be displayed and measured.

The characteristics of the uptake curves for glucose led LeFevre[9] to conclude that simple diffusion could not be responsible and he took the important step in suggesting that transfer may involve the temporary formation of a complex with some constitutent of the cell membrane.

This idea led to the formulation of simple kinetics for the transfer process; the most successful of these have been based on the model suggested by Widdas,[17] who proposed that the rate-limiting step was a

---

[28] R. Ege, thesis. University of Copenhagen, Copenhagen, Denmark [cited by O. Bang and S. L. Ørskov, *J. Clin. Invest.* **16**, 279 (1937)].

[29] K. A. Klinghoffer, *Am. J. Physiol.* **111**, 231 (1935).

slow transfer through the membrane while the reactions with the membrane constituents at the interfaces were deemed to be rapid. In its simplest form transfer through the membrane could therefore be represented by

$$dS/dT = V_{\max}\{[(S/S + K_{\mathrm{s}})] - [S'/(S' + K_{\mathrm{s}})]\} \qquad (1)$$

where $V_{\max}$ represents the rate when all the components on one side are saturated and there is no backflux. S and S' represent the outside and inside sugar concentrations and K_{s} is the half-saturation concentration equivalent to the Michaelis constant for an enzyme-catalyzed reaction.

Widdas[17,30] showed that this equation had two approximations which could be applied when K_{s} was either very large or very small relative to the concentrations. Thus, if $S \ll K_{\mathrm{s}} \gg S'$

$$dS/dT = (V_{\max}/K_{\mathrm{s}})(S - S') \qquad (2)$$

but if $S \gg K_{\mathrm{s}} \ll S'$

$$dS/dT = V_{\max}K_{\mathrm{s}}[(1/S') - (1/S)] \qquad (3)$$

Equation (2) is equivalent to diffusion, but Eq. (3), which represents a near-saturation condition, suggests that the rate will be proportional to the difference of the reciprocals of the concentrations. The integrated forms of Eqs. (2) and (3) were applied to the swelling curves of red cells in glucose and sorbose solutions by Widdas.[31] The integrated form of Eq. (1) was used by LeFevre[32] to estimate the $V_{\max}$ and K_{s} for a number of different monosaccharides.

The use of optical techniques to follow exits, either of high-affinity sugars such as glucose[27] or of such sugars in the presence of inhibitors,[33] has many advantages over their use for following entries, chief among these being the greater speed of the light transmission change and the ease of quantification.

Figure 2 is a reproduction of the type of records obtained from glucose exits into a series of low outside concentrations. The traces on the recorder are mostly straight lines since the efflux may be considered to be maximal for a large part of the exit. There is, of course, a small constant influx due to the glucose in the outside medium and as this is increased the exit is slowed and takes longer. To measure the records the straight line part is produced until it intersects the equilibrium value and the exit time

[30] W. F. Widdas, *J. Physiol.* (*London*) **115**, 36P (1951).
[31] W. F. Widdas, *J. Physiol.* (*London*) **125**, 163 (1954).
[32] P. G. LeFevre, *Am. J. Physiol.* **203**, 286 (1962).
[33] A. K. Sen and W. F. Widdas, *J. Physiol.* (*London*) **160**, 404 (1962).

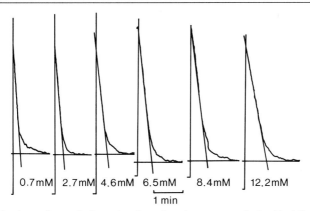

0.7mM 2.7mM 4.6mM 6.5mM 8.4mM 12.2mM

1 min

FIG. 2. Tracings of records from the photoelectric apparatus during "exit" experiments at 37°. Cells preequilibrated in 76 mM were losing glucose into medium at the concentrations shown. The linear part of each record was produced to the base line and the times of exit measured for analysis as in Fig. 3. (From Sen and Widdas.[27])

read off. This represents the time the exit would have taken if it had continued at the initial linear rate. It is found empirically, and there is theoretical justification from the integrated equations, that the exit times are a linear function of the outside concentration.

Lines drawn as in Fig. 3 have two intercepts. The one with the ordinate is the extrapolated time for exit into a glucose-free solution, and knowing the concentration at which the cells were preincubated it is possible to calculate the V_{max} for the cells at that concentration. The intercept on the abscissa is the outside concentration which would double the exit time compared with that into a glucose-free solution. On a simple view it can be interpreted as the half-saturation for the components on the outside of the membrane but recent work suggests the intercept is an estimate of the K_m for a modifier site on the glucose transporter.[34]

The use of exit experiments in the presence of inhibitors can distinguish between those which are competitive for the outside sites and those which are noncompetitive. However, care must be taken since an inhibitor which is only competitive at the inside sites would appear to be noncompetitive in exit experiments.[35,36]

Drawbacks to Optical Experiments. The need for specially constructed apparatus at one time limited the number of laboratories where optical experiments could be carried out. The minimum requirement now

[34] G. F. Baker and W. F. Widdas, *J. Physiol. (London)* **395,** 57 (1988).
[35] D. A. Basketter and W. F. Widdas, *J. Physiol. (London)* **278,** 389 (1978).
[36] W. F. Widdas, *Curr. Top. Membr. Transp.* **14,** 165 (1980).

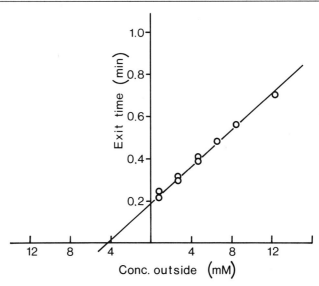

FIG. 3. "Exit" times obtained from records such as in Fig. 2 plotted against the outside concentration. The line (drawn by eye) has two intercepts: that on the ordinate represents the time for exit into a glucose-free medium (t_0) while that on the abscissa is the concentration which would have an exit time twice t_0. (From Sen and Widdas.[27])

is probably a specially adapted cuvette and water jacket, since Bloch[37] was able to use a commercially available spectrophotometer and recorder. Carruthers and Melchior[38] also carried out exit experiments with a modified commercial instrument.

The swelling and shrinking of red cells call for changes in the osmotic contents relative to the medium and changes in concentration of the order of 20–30 mM are probably required as a minimum. Since the normal blood glucose is only 5 mM, the criticism could be levelled that the concentrations used are unphysiological. Also, since the volume changes are needed to produce refractive index changes the methods are best used with intact red cells. Partially hemolysed red cells and resealed ghosts have been used in special cases (e.g., Carruthers and Melchior[38]).

Exchange fluxes, where there is neither a volume nor a concentration change, are not measurable by either biochemical or optical techniques and it is in this field that important discoveries have been made using the third technique to be described and which involves the use of radioactively labeled sugars.

[37] R. Bloch, *Biochemistry* **12,** 4799 (1973).
[38] A. Carruthers and D. L. Melchior, *Biochim. Biophys. Acta* **728,** 254 (1983).

Radioactive Techniques

[14]C-Labeled glucose first became generally available in the 1950s and since then a wide variety of labeled sugars and sugar derivatives have been produced. Labeling with tritium is also obtainable and since scintillation spectrometers permit the separate determination of the tritium and ^{14}C label in mixtures there are now many experimental possibilities, not all of which have so far been exploited.

To use radioactive sugars to measure rates of transport across the red cell membrane, however, requires that a number of difficulties be overcome. Most of these arise from the great rapidity of the transfer process, the need to separate the cells from the suspending medium, and the need to deproteinize the samples. These are the same limitations as were mentioned when considering the application of biochemical techniques to red cells which have rapid rates of sugar transfers; but the increased knowledge of the transfer mechanism, gained largely from the use of optical experiments, has provided us with ways of meeting these difficulties.

Temperature Effects. Although sugar transfers are very fast at 37° the process has a moderately high Q_{10} of about 2.5 and consequently it is advantageous to work at lower temperatures. Many laboratories use 20° but a lower temperature of 16° was used by Baker and Widdas[39] and became the temperature of choice in the author's laboratory.

Taking the temperature down to 0° caused sufficient slowing to permit the separation of rabbit red cells from their suspending media in the experiments of Morgan et al.,[6] but this would not be enough for human red cells.

Stopping Solutions. To prevent changes in the concentration in human red cells during centrifugation, the effect of temperature is usually combined with the use of a "stopping solution" containing powerful inhibitors of the glucose transfer mechanism. The stopping solutions are based on the inhibition produced by mercuric ions but Levine and Stein[40] found empirically that the effect was improved in the presence of 1.25 mM KI, whereas Karlish et al.[41] used a lower concentration of mercuric ions but included the inhibitor phloretin. Their stopping solution had the following composition: NaCl (1%, w/v), $HgCl_2$ (10^{-6} M), KI (1.25 mM), phloretin, dissolved in ethanol to give 10^{-4} M phloretin, and 1% ethanol.

The stopping solution used in the author's laboratory had the follow-

[39] G. F. Baker and W. F. Widdas, *J. Physiol.* (*London*) **231**, 143 (1973).
[40] M. Levine and W. D. Stein, *Biochim. Biophys. Acta* **127**, 179 (1966).
[41] S. J. D. Karlish, W. R. Lieb, D. Ram, and W. D. Stein, *Biochim. Biophys. Acta* **255**, 126 (1972).

ing composition: NaCl (0.34 M), KI (0.65 mM), HgCl$_2$ (2 mM), phloretin (0.1 mM).

Alternatives to the use of stopping solutions have been reported. Wilbrandt and Kotyk[42] centrifuged the suspension rapidly at 0° over silicone oil through which the cells passed. Mawe and Hempling[43] separated the cells from the medium by a rapid filtration through Millipore filters.

Sampling Procedures. The procedures for taking and processing samples used in the author's laboratory from 1971 were based on those described by Miller.[44] Typically, samples are taken at appropriate intervals from the experimental suspension in a 1-ml automatic syringe and injected into 10 ml of ice-cold stopping solution in a conical tube. The tubes are kept on ice until they are centrifuged, which is done as soon as possible. After centrifugation the supernatant is carefully withdrawn so as not to disturb the cell pellet. A further 2 ml of ice-cold stopping solution is introduced from an automatic burette to wash down the sides of the tube, which is again centrifuged. After removal of the supernatant the cell pellet is resuspended in 50 μl of saline buffer with vigorous stirring on a vortex mixer while 1.2 ml of absolute alcohol is added. The cell debris is centrifuged and 1 ml of the supernatant alcoholic solution (containing labeled sugar) is carefully pipetted into a counting vial containing 10 ml of scintillation fluid.

The scintillation fluid used was 0.4% (w/v) 2,5-diphenyloxazole and 0.02% (w/v) 1,4-D-[2-(5-phenyloxazolyl)]benzene in a 2 : 1 mixture of toluene and Triton X-100.

Some laboratories have preferred to use 0.1 ml 20% trichloroacetic acid solution in place of absolute alcohol (e.g., Karlish *et al.*[41]).

The procedures outlined above presuppose that the same quantities of red cells are taken up by the automatic syringe at each sampling time. To ensure this the suspension must be adequately stirred; a Perspex apparatus was made to surround a 30-ml conical flask with a water jacket which was mounted over a magnetic stirrer so that a magnetic flea in the flask was rotated at 10 Hz. A second Perspex apparatus was designed so that the tubes holding the stopping solution were held in a repeatable position relative to the automatic syringe when samples were injected but so that stopping solution did not splash up onto the needle of the syringe. With practice, samples at 5-sec intervals could be taken but longer intervals are preferable if the experimental design permits.

Treating the Results. Usually five or six samples are taken during the

[42] W. Wilbrandt and A. Kotyk, *Arch. Exp. Pathol. Pharmakol.* **249,** 279 (1964).
[43] R. C. Mawe and H. G. Hempling, *J. Cell. Comp. Physiol.* **66,** 95 (1965).
[44] D. M. Miller, *Biophys. J.* **8,** 1329 (1968).

first 60–90 sec and two further samples are taken when it is judged the cells will have come to equilibrium, which may be from 4 to 20 min, depending on the type of experiment and the presence or absence of inhibitors. The radioactivity initially used is adjusted so that over the counting period (say 20 min) about 10,000 counts are recorded. The counts for the final equilibrium samples may be less if in the experiment radioactively is effluxing from the cells. The standard error of the counts is equal to the square root of the total counted and at this level is about 1% and probably less than other experimental errors in the technique.

In an equilibrium exchange situation where the concentrations are the same inside and out but the radioactivity starts inside (and effluxes during the experiment), the conditions are ideal in that there are neither chemical concentration nor volume changes to be considered. The radioactively labeled sugar which is leaving the cells is being replaced by an equal amount of unlabeled sugar from the outside medium which is relatively large in volume. Under these circumstances the radioactivity in the samples should fall off exponentially to the final equilibrium value. This may be represented as

$$-2.303 \log(X - X_\infty) = kt + \text{constant} \qquad (4)$$

where X is the number of counts per minute in the sample at time t (seconds) and X_∞ is the number when equilibrium has been reached. Note that by subtracting these two terms the effect of any radioactivity outside the cells is also canceled out. Figure 4 taken from the results of Baker and

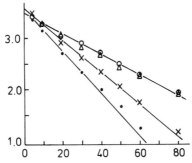

FIG. 4. Logarithmic fall-off of intracellular radioactivity during glucose exchange at 20 mM (16°) in a control experiment and in the presence of the nontransportable inhibitor ethylidene glucose. Samples taken over 60 sec in the control and 80 sec when inhibited. ●, Control; ×, with 50 mM inhibitor inside the cells; ○, with 200 mM inside; △, with 25 mM outside the cells. The slopes of such lines were used to derive the exchange flux rates as described in the text. (After Baker and Widdas.[39])

Widdas[39] illustrates the manner of plotting the results according to Eq. (4).

Once the slope (k) of the line has been determined and irrespective of the exact volume of the cells used in the samples (provided that each sample contained a like volume), the flux represented by J is obtained from the relation

$$J = 60kS \qquad (5)$$

where S is the concentration of sugar in the cells. If S is expressed in millimolar units then the units of J are millimoles (liters cell water)$^{-1}$ minute^{-1}.

Similar considerations apply to radioactive uptake experiments but in these the amount of radioactivity outside the cells will be much larger. The terms in parentheses in Eq. (4) will be reversed and the log will be positive. Although the outside radioactivity will tend to be canceled out, it is likely to give rise to greater variability between samples due to differences in washing down the centrifuge tubes and other preparative procedures. For this reason the use of radioactive efflux is to be preferred where possible, e.g., in exchanges since efflux and influx are then equal.

An experiment where this is not possible and which is illustrated in Fig. 5 is one designed to show uphill transport induced by counterflow in human red cells. In this experiment cells were preincubated with 76 mM of a nonradioactive sugar and then placed in a medium containing 4 mM ^{14}C-labeled 3-O-methylglucose at 16°. The nonradioactive sugars were glucose, 3-O-methylglucose, galactose, and sorbose. Samples were taken at the times shown and their radioactivity compared with that in the equilibrium samples when the inside concentration was the same as outside (4 mM). With glucose as the competitive sugar it will be seen that at 60 sec the inside 3-O-methylglucose had risen to over 9-fold its equilibrium value and must have been entering the cells against its concentration gradient. The other sugars were less effective in inducing the uphill transfer and the ketose sugar sorbose did not appear to induce any.

The experiments in Fig. 5 were carried out by final-year honor students between 1971 and 1981 as laboratory exercises and illustrate the power and the relative simplicity of the radioactive techniques when applied to sugar transport problems.

Another more difficult application of radioactive techniques has been to measure initial uptake rates under a variety of experimental conditions (e.g., Lacko et al.[45,46]). Here the uptake must be sampled at very short

[45] L. Lacko, B. Wittke, and P. Geck, *J. Cell. Physiol.* **80**, 73 (1972).
[46] L. Lacko, B. Wittke, and P. Geck, *J. Cell. Physiol.* **82**, 213 (1973).

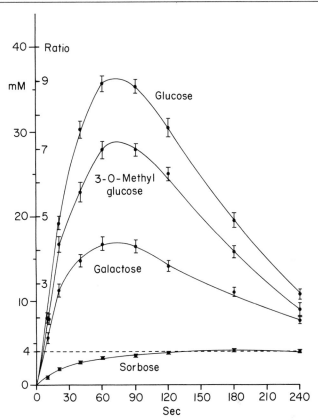

Fig. 5. Uphill transport induced by counterflow in human red cells at 16°. Cells were preincubated at 76 mM in the nonradioactive sugars shown and then rapidly introduced into a medium containing 4 mM ^{14}C-labeled 3-O-methylglucose. By comparing the intracellular radioactivity at times up to 4 min with that in the cells when equilibrium had been reached the ability to induce against the gradient uptake of 3-O-methylglucose was demonstrated (except for the case of sorbose). The means and standard errors are given for over 24 experiments for each sugar carried out by honors students in the author's laboratory between 1971 and 1981.

times since if any radioactively labeled sugar is allowed to accumulate in the cell it soon has an effect upon the influx rate.

With the availability of radioactively labeled sugars of high specific activity it has been possible to adapt fast-flow techniques to measure sugar fluxes.[47,48] Such techniques can be used to study rates of transport

[47] J. Brahm, *J. Physiol.* (*London*) **339**, 339 (1983).
[48] A. G. Lowe and A. R. Walmsley, *Biochim. Biophys. Acta* **857**, 146 (1986).

at body temperature, or above, without some of the difficulties described for other radioactive procedures. At present the experiments are limited to measurements of rates of efflux from cells loaded with radioactive sugars at various concentrations. However, the outside concentration of nonradioactive sugar can be varied. If it is the same as that in the cells, the exchange efflux is determined. At the other extreme, efflux rates into a sugar-free medium can be measured. The fast-flow technique does not lend itself to the measurement of influx rates.

Reference should be made to the original articles of these and other related studies before undertaking such experiments.

Results Using Radioactive Techniques

One of the earliest consequences of the use of radioactive techniques in red cell transport studies was the demonstration that the exchange of sugar can occur at a faster overall rate than is possible in either net entry or net exit experiments. The half-saturation for exchange[44] was also found to be different from that determined by glucose–sorbose competition or by the Sen–Widdas exit method. The shortcomings were attributed to the simple kinetics of the mobile carrier model of Eq. (1) and developments followed rapidly in which alternative models of transfer and kinetics were advanced. The kinetics which were closest to those in the original hypothesis were derived from a carrier model with asymmetric affinities at the two sides and in which the rate constants for attaining equilibrium at the two interfaces could be finite relative to the rate constants for movement through the membrane. In this model[49] the complex equation describing transport had five parameters for each sugar (and four of these were independent) and in addition there could be two parameters for diffusion inside and outside the cells.

It is not possible to review these kinetics in detail here, nor is it possible to discuss analyses based on other kinetics (e.g., that by Eilam and Stein,[50] in which they used the tetramer model of Lieb and Stein[51]). Suffice it to say that an experiment designed to measure an influx or an efflux under standardized conditions, e.g., as an equilibrium exchange or with defined conditions as to inside and outside concentrations, can still be analyzed on the basis that the measured flux will depend on the concentration in a simple Michaelis manner. Such a flux (J) can be represented by the following equation:

$$J = V_{max}[S/(S + K_s)] \tag{6}$$

[49] D. M. Regen and H. L. Tarpley, *Biochim. Biophys. Acta* **339,** 218 (1974).
[50] Y. Eilam and W. D. Stein, *Biochim. Biophys. Acta* **266,** 161 (1972).
[51] W. R. Lieb and W. D. Stein, *Nature (London), New Biol.* **230,** 108 (1971).

which is a well-recognized form in enzyme kinetics and either a double-reciprocal or preferably a Hanes[52] plot can be used to determine K_s and V_{max}. For the Hanes plot Eq. (6) is rewritten as

$$S/J = (1/V_{max})(S + K_s) \qquad (7)$$

and plotting S/J as ordinate and S along the abscissa K_s is obtained as the negative intercept on the abscissa. V_{max} is the reciprocal of the slope.

Such defined experiments can be used to study inhibitors of transport and will distinguish between competitive and noncompetitive substances. Other experiments which are valuable when studying inhibitors were outlined by Devés and Krupka.[53]

If, however, a more ambitious project is proposed in which fluxes are to be measured under widely different experimental conditions with a view to characterizing the overall transport process, then one of the more detailed kinetic analyses should be consulted and applied. Some of the ways in which the Regen and Tarpley[49] kinetics have been used for this purpose were reviewed by Widdas.[36] The usual approach involves measuring influx into sugar-free cells, influx into cells preloaded with fixed concentrations of nonradioactive sugar, and influx (or efflux) under equilibrium exchange conditions. The great difficulty of measuring the initial influx into sugar-free cells has made this more complete analysis hard to attain except in cells with low transport rates such as avian erythrocytes as used by Cheung et al.[54]

An Indirect Technique

A novel technique was developed by Taverna and Langdon[10] which, while not applicable to intact red cells, could be used with resealed erythrocyte ghosts. These authors found that with partial hemolysis the erythrocytes were permeable to protein molecules and they were able to introduce glucose oxidase and catalase into such cells. If the cells were resealed by incubation at 37° the glucose oxidase and catalase were trapped in the cells and all the external enzymes could be removed by repeated washings. The entry of glucose into such cells induced oxygen utilization and by measuring the rate of oxygen disappearance with an oxygen electrode they could indirectly determine the rate of glucose influx. This technique allowed uptake to be measured under conditions where free glucose was not accumulating inside the cells.

They were also able to use this technique with inside-out vesicles

[52] C. S. Hanes, *Biochem. J.* **26**, 1406 (1932).
[53] R. Devés and R. M. Krupka, *Biochim. Biophys. Acta* **510**, 186 (1978).
[54] J. Y. Cheung, D. M. Regen, M. E. Schworer, C. F. Whitfield, and H. E. Morgan, *Biochim. Biophys. Acta* **470**, 212 (1977).

prepared from hemolysed red cells[55] and these further studies suggested that the sugar transport system in the membrane was not so asymmetric as had been presumed from the results with intact cells.

Conclusion

Various ways of studying glucose transport in red blood cells have been described and some examples of the results obtained by use of the three main techniques are illustrated.

In general the biochemical techniques have been restricted in usefulness to cells with slow rates of transport and are not suitable for following the fast transport found in human and some fetal red cells. They might be exploited further in such cells if they could be combined with the use of "stopping solutions" as employed in the radioactive techniques.

Optical methods require either the construction of a special apparatus or the modification of a suitable commercial instrument. Probably their most useful role is in checking substances which are thought to be possible inhibitors of sugar transport.

Radioactive techniques are within the compass of most biological laboratories; they are easy to apply and honors students can quickly learn to use them to obtain meaningful results. They are the techniques of first choice for anyone entering this field of study.

Whichever of the above methods is chosen, practice at the laboratory bench will be required to gain experience in preparing and handling the red cell suspensions.

[55] R. D. Taverna and R. G. Langdon, *Biochim. Biophys. Acta* **298,** 422 (1973).

[14] Nucleoside Transport across Red Cell Membranes

By Z. I. CABANTCHIK

Introduction

Red blood cells from a variety of species are known to possess nucleoside transport system(s)[1-3] which subserve(s) manifold intracellular and

[1] S. M. Jarvis, J. R. Hammond, A. R. P. Paterson, and A. S. Clanachan, *Biochem. J.* **208,** 83 (1982).

[2] J. D. Young and S. M. Jarvis, *Biosci. Rep.* **3,** 309 (1983).

[3] P. G. W. Plagemann and R. M. Wohlhueter, *Curr. Top. Membr. Transp.* **14,** 225 (1980).

systemic functions. By transporting adenosine from the plasma and subsequently metabolizing it, the human red cell, on the one hand, acquires an important metabolite and, on the other, it removes a potent vasodilator,[4] thus contributing to circulatory homeostasis. In porcines, the red cells lack glucose transport capacity, and the transport of inosine is apparently essential for fueling the intracellular systems.[5] In other cells, a very similar transport system has been used as a route for delivering (into cells) and/or as a target for therapeutic agents against viral and neoplastic conditions.[6,7]

The nucleoside transporter of human red cells, with which this chapter is primarily concerned, is a system of nonconcentrative, facilitated diffusion[8] which has a demonstrably broad substrate specificity for purine and pyrimidine nucleosides.[2] The system has been studied in detail in terms of its kinetic properties, particularly with pyrimidine nucleoside substrates, as those have been shown to be free of metabolic conversion by the intracellular machinery.[6] Interestingly, the nucleoside transporter provided the first example of a membrane transport system which was fully compatible with the simple carrier model.[9,10] It was found to be directionally asymmetric at 25° in that the V_{max} of net influx was approximately four times higher than the corresponding V_{max} of net efflux.[9] However, that was apparently a property of cells isolated from outdated blood, since cells from relatively fresh blood had demonstrably symmetric properties.[11,12] The asymmetry was also highly dependent on temperature, even reaching values as high as 14 at 4°.[12]

The nucleoside transporter is also a recent target for intensive biochemical studies. Those have been facilitated by the development of a series of high-affinity binding probes which are members of two families of pentofuranosides of S^6-substituted 6-thiopurines and of N^6-substituted adenine derivatives.[6] The most studied of these probes, p-nitrobenzylthioinosine (NBMI), was instrumental in the development of methodo-

[4] R. M. Berne, R. M. Knabb, S. W. Ely, and R. Rubio, *Fed. Proc., Fed. Am. Soc. Exp. Biol.* **42**, 3136 (1983).
[5] S. M. Jarvis, J. D. Young, M. Ansay, A. L. Archibald, R. A. Harkness, and R. J. Simmonds, *Biochim. Biophys. Acta* **597**, 183 (1980).
[6] A. R. P. Paterson, N. Kolassa, and C. E. Cass, *Pharmacol. Ther.* **12**, 515 (1981).
[7] R. J. Suhadolnik, *in* "Nucleosides as Biological Probes." Wiley, New York, 1979.
[8] J. M. Oliver and A. R. P. Paterson, *Can. J. Biochem.* **49**, 262 (1971).
[9] Z. I. Cabantchik and H. Ginsburg, *J. Gen. Physiol.* **69**, 75 (1977).
[10] W. R. Lieb and W. D. Stein, *Biochim. Biophys. Acta* **373**, 178 (1976).
[11] S. M. Jarvis, J. R. Hammond, A. R. P. Paterson, and A. S. Clanachan, *Biochem. J.* **210**, 461 (1983).
[12] P. G. W. Plagemann and R. M. Wohlhueter, *Biochim. Biophys. Acta* **778**, 176 (1984).

logies for assaying nucleoside transport[13,14] in the determination of the number of transporters per cell (10^4/cell),[2] and in the identification of the putative functional protein of approximately 55 kDa by photoaffinity labeling,[15] by isolation purification,[16] and by reconstitution procedures.[17] The carrier copurifies with another major transporter of erythrocytes, the glucose carrier, which shows a similar mobility in SDS-PAGE in the so-called band 4.5 area.[18] However, the two proteins differ in their proteolytic profiles and exofacial susceptibility to trypsin administered in low ionic strength media.[19] The two classes of proteins reside in the membrane, apparently as dimers, as judged by the radiation inactivation profiles of lyophilized membranes, either before or after alkali-EDTA stripping of extrinsic polypeptides.[20,21] But judging from the NBMI binding-inhibition profiles and Hill plots, it appears as though the functional polypeptide unit is the monomer (M. Zangvill and Z. I. Cabantchik, unpublished observations).

Although inhibition by NBMI is of a competitive nature, it is not clear whether the probe acts on the transport site itself or on an allosteric site. Binding to the transporter and the elicited inhibitions are very fast,[9,22] the site of action probably being located at the outer membrane surface[23] but inaccessible to trypsin. Although hitherto the chemical nature of the binding site has not been elucidated, several lines of evidence point toward the involvement of a hydrophobic pocket with e^- capacity as a constituent of the binding site of the high-affinity probes.[24]

Red blood cells from species other than human have a nucleoside transport system with a K_m similar to that of the human species, but the V_m and the corresponding number of transporters/cell vary over almost

[13] P. G. W. Plagemann and R. M. Wohlhueter, in "Regulatory Function of Adenosine" (R. M. Berne, T. W. Rall, and R. Rubio, eds.), p. 179. Nijhoff, Boston, 1983.
[14] A. R. P. Paterson, E. R. Harley, and C. E. Cass, *Methods Pharmacol.* **6**, 165 (1985).
[15] J. D. Young, S. M. Jarvis, M. P. J. Robins, and A. R. P. Paterson, *J. Biol. Chem.* **258**, 2202 (1983).
[16] S. M. Jarvis and J. D. Young, *Biochem. J.* **213**, 331 (1984).
[17] C. M. Tse, J. A. Belt, S. M. Jarvis, A. R. P. Paterson, J. S. Wu, and J. D. Young, *J. Biol. Chem.* **260**, 3506 (1985).
[18] J. S. Wu, S. M. Jarvis, and J. D. Young, *Biochem. J.* **214**, 995 (1983).
[19] N. S. Janmohamed, J. D. Young, and S. M. Jarvis, *Biochem. J.* **230**, 777 (1985).
[20] S. M. Jarvis, D. A. Finchan, J. C. Ellory, A. R. P. Paterson, and J. D. Young, *Biochim. Biophys. Acta* **772**, 227 (1984).
[21] S. M. Jarvis, J. C. Ellory, and J. D. Young, *Biochim. Biophys. Acta*, in press.
[22] S. M. Jarvis, N. S. Janmohamed, and J. D. Young, *Biochem. J.* **216**, 661 (1983).
[23] S. M. Jarvis, D. McBride, and J. D. Young, *J. Physiol. (London)* **324**, 31 (1982).
[24] A. R. P. Paterson, E. S. Jakobs, E. R. Harley, C. E. Cass, and M. J. Robins, in "Development of Target Oriented Cancer Drugs" (Y. C. Cheng ed.), p. 41. Raven, New York, 1983.

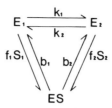

FIG. 1. Schematic representation of the two-E (carrier) model.

three orders of magnitude.[1] The transporter from porcine erythrocytes has an apparently higher molecular weight than that of human cells, as well as a different binding behavior toward an ion-exchange column.[25] In rat erythrocytes, there are apparently two distinctly different nucleoside transporters which differ in their substrate affinity, their susceptibility to inhibition by NBMI, and to the SH-reagent p-hydroxymercuribenzene sulfonate.[26] Sheep display two red cell variants: one (Nu^+) which transports nucleosides, and the other, which does not (Nu^-). In the fetal stage and also in reticulocytes, both red cell variants have demonstrable nucleoside transport activity; however, upon maturation, both lose most of those transporters, the Nu^+ remaining with 20/cell and the Nu^- remaining with virtually none.[27] Despite this change, the turnover rate for the transporter remains practically unchanged at the various stages of red cell differentiation.[28] Dogs have apparently no detectable nucleoside transporter in their red cell membranes.[1]

Characterization of the Nucleoside in Terms of the Simple
Carrier Model

Elementary Kinetic Analysis. The handling of substrates by membrane carriers (E)[10,29] is depicted in the model of the simple carrier (Fig. 1). S_1 and S_2 represent substrate concentrations at membrane faces 1 and 2, respectively; b_1 and b_2 represent the respective rate constants for *breakdown* of the carrier–substrate complex at faces 1 and 2; f_1 and f_2 are the respective rates of *formation* of the complex, and k_1 and k_2 are the rates of

[25] Y. P. Kwong, Y. M. Chong, S. M. Jarvis, and J. D. Young, *Biochem. Soc. Trans.* **13**, 708 (1985).

[26] S. M. Jarvis and J. D. Young, *J. Membr. Biol.* **93**, 7 (1986).

[27] S. M. Jarvis and J. D. Young, *J. Physiol. (London)* **324**, 47 (1982).

[28] S. M. Jarvis and J. D. Young, *Biochem. J.* **190**, 377 (1980).

[29] W. R. Lieb, *in* "Red Cell Membranes: A Methodological Approach" (J. C. Ellory and J. D. Young, eds.), p. 135. Academic Press, London, 1982.

interconversion of the free carrier forms. The unidirectional flow of substrate from one membrane face to the other is given by

$$v_{12} =$$
$$(K[S_1] + [S_1][S_2])/K^2R_{00} + KR_{12}[S_1] + KR_{21}[S_2] + R_{ee}[S_1][S_2]) \quad (1)$$

where the basic parameters R (resistances or reciprocals of rate constants) and K, the apparent affinity constant, are defined in Table I in terms of the rate constants shown in Fig. 1.

Experimental Procedures

In order to obtain a full kinetic description of the carrier with regard to basic measurable parameters, several experimental procedures must be applied.

In the *zero-trans* procedure (*zt*), the concentration of substrate at one face of the membrane is held at zero, while the other is varied. The unidirectional flux (here also a net flux) proceeds from the trans (nonzero) face into the face lacking substrate. For $S_2 = 0$, Eq. (1) becomes Michaelian:

$$v_{1,2}^{zt} = (1/R_{12}[S_1])/\{K(R_{00}/R_{12}) + [S_1]\} \quad (2)$$

where the Michaelis $K^{zt} = KR_{00}/R_{12}$ and $V_{1,2}^{zt} = 1/R_{12}$.

A similar equation with the interchanged subscripts identifying flows in the 2,1 direction yields $K_{2,1}^{zt} = KR_{00}/R_{21}$ and $V_{2,1}^{zt} = 1/R_{21}$.

In the *equilibrium exchange* procedure (ee), the concentration of substrate is set equally at both membrane faces, and the unidirectional flux of a labeled substrate is measured as a function of [S] (where $[S_1] = [S_2]$).

TABLE I
STEADY-STATE VALUES FOR SIMPLE CARRIER[a]

Basic parameters	Molecular rate constants
nR_{12}	$(1/b_2) + (1/k_2)$
nR_{21}	$(1/b_1) + (1/k_1)$
nR_{ee}	$(1/b_1) + (1/b_2)$
nR_{00}	$(1/k_1) + (1/k_2)$
K	$(k_1/f_1) + (k_2/f_2)$
Constraint	$b_1f_2k_1 = b_2f_1k_2$

[a] n is the total number of carriers per unit area of membrane, while nR terms are specific resistances of the membrane to particular carrier forms. (From Lieb and Stein.[10])

The resulting equation is

$$v^{ee}_{1,2} = v^{ee}_{2,1} = (1/R^{ee}[S])/\{K(R_{00}/R_{ee}) + [S]\} \qquad (3)$$

where $K^{ee} = KR_{00}/R_{ee}$ and $V^{ee} = 1/R^{ee}$.

Thus, by performing zt experiments in the 1,2 direction and in the 2,1 direction, and ee experiments in either direction, it is possible to obtain all the parameters required for defining the system in terms of a simple carrier. Those are R_{12}, R_{21}, and $1/KR_{00}$ from zt experiments; R_{ee} and KR_{00}/R_{ee} from ee experiments; and the equality-constraint $R_{ee} + R_{00} = R_{12} + R_{21}$, which is used for obtaining K.

A third procedure which can be used to test the simple carrier is that of *infinite-cis* (ic), where [S] is held at limitingly high values at one face of the membrane and the net flow of substrate from that cis face is determined as a function of [S] at the other (trans) surface. Such a flux is given by

$$v^{ic}_{1,2} \text{ (net)} = (1/R_{ee})/\{K(R/R_{ee}) + [S_2]\} \qquad (4)$$

and an analogous equation is applicable to the net flux in the opposite direction. This procedure can be used as a test for consistency of the model as applied to a particular carrier system, since the apparent Michaelis constants $K^{ic}_{1,2} = KR_{12}/R_{ee}$ and $K^{ic}_{2,1} = KR_{21}/R_{ee}$ can be predicted from data obtained by the two zt and ee experiments, and compared with the experimental values obtained by the two ic procedures.

An additional procedure used to study carrier systems is that of *infinite-trans* (it), in which [S] at one surface is held at a limitingly high value, defined as so high that a further increase would not change the unidirectional flux of labeled substrate into that surface. Thus, for example, while $[S_2]$ is held at infinity, $[S_1]$ is varied and the unidirectional flux from face 1 to face 2 is measured with labeled substrate. Equation (5) describes this procedure.

$$v^{it}_{1,2} = \{(1/R_{ee})[S_1]\}/\{K(R_{21}/R_{ee}) + [S_1]\} \qquad (5)$$

An analogous equation is applicable to the net flux in the opposite direction. The two Michaelis constants are given by $K^{it}_{1,2} = KR_{21}/R_{ee}$ and $K^{it}_{2,1} = KR_{12}/R_{ee}$, and the maximal velocities $V^{ic}_{1,2} = V^{ic}_{2,1} = V^{ee}$.

A compilation of the kinetic parameters obtained by the various experimental procedures is given in Table II. It can be readily noted that the K and R terms themselves number just four independent variables (the R values being interconnected), so that by using all the above listed procedures, the parameters are heavily overdetermined. However, this property can be used as a consistency test for assessing the simple carrier model to a particular transport system, an example of which is shown above for ic parameters.

TABLE II
INTERPRETATION OF EXPERIMENTAL DATA IN TERMS OF BASIC
MEASURABLE PARAMETERS[a]

Procedure	V_{max}	K_m
Zero trans	$v_{1 \to 2}^{zt} = 1/R_{12}$	$K_{1 \to 2}^{zt} = K(R_{00}/R_{12})$
	$v_{2 \to 1}^{zt} = 1/R_{21}$	$K_{2 \to 1}^{zt} = K(R_{00}/R_{21})$
Infinite trans	$v^{it} = v_{1 \to 2}^{it}$	$K_{1 \to 2}^{it} = K(R_{21}/R_{ee})$
	$= v_{2 \to 1}^{it} = 1/R_{ee}$	$K_{2 \to 1}^{it} = K(R_{12}/R_{ee})$
Infinite cis	$v_{1 \to 2}^{ic} = 1/R_{12}$	$K_{1 \to 2}^{ic} = K(R_{12}/R_{ee})$
	$v_{2 \to 1}^{ic} = 1/R_{21}$	$K_{2 \to 1}^{ic} = K(R_{21}/R_{ee})$
Equilibrium exchange	$v^{ee} = 1/R_{ee}$	$K^{ee} = K(R_{00}/R_{ee})$
Constraint	$R_{ee} + R_{00} = R_{12} + R_{21}$	

[a] From Lieb and Stein.[10]

Methodological Aspects of Flux Measurements

Because it is nonmetabolized in red cells,[2] uridine was the favorite substrate for measuring nucleoside transport and for testing the applicability of the simple carrier model.[9–11] Whether in cells from recently outdated[9,10] or freshly drawn blood,[10,11] uridine transport fulfills all the kinetic criteria of a simple carrier mechanism. These properties are the subject of the present chapter. Outside the scope of this chapter are recent developments of techniques for measuring transport of adenosine by quenched-flow procedures[30] applied to intact cells, or by rapid mixing of cells and substrate in ATP-depleted adenosine deaminase-blocked human red cells.[31] Because of the rapidity of uridine fluxes, it is necessary to have means for stopping the fluxes swiftly and efficiently at given periods of time, either by separating the compartments (i.e., cells and media) or by adding an inhibitor.[32,33]

Efflux Measurements

The filter technique is highly suitable for efflux studies.[9] Filters (0.45 μm) and prefilters are mounted on 25-mm-diameter holders (Sartorius or Millipore), the latter connected to 5-ml Luer-lock syringes (tested for air tightness). Erythrocytes are loaded with uridine (0.1–20 mM in pH 7.4

[30] A. R. P. Paterson, E. H. Harley, and C. E. Cass, *Methods Pharmacol.* **6**, 165 (1985).
[31] P. G. W. Plagemann, R. M. Wohlhueter, and M. Kraupp, *J. Cell. Physiol.* **125**, 330 (1985).
[32] R. M. Wohlhueter, R. Marz, J. C. Graff, and P. G. W. Plagemann, *Methods Cell Biol.* **20**, 211 (1978).
[33] A. R. P. Paterson, E. R. Harley, and C. E. Cass, *Biochem. J.* **224**, 1001 (1984).

buffered saline), usually at 5% hematocrit, for 1 hr at 37°, followed by centrifugation, resuspension to 50% hematocrit in the same uridine medium to which a small aliquot of [^{3}H]- or [^{14}C]uridine is added, and incubated for another 1-hr period at 37°. After packing by spinning on a Beckman microfuge (Beckman Instruments, CA) or equivalent for 5 min, the cells are aspirated into a 2-cm-long Teflon tubing mounted on the needle of a 2-ml syringe containing buffered isotonic medium (with or without uridine at the preset concentration—same as in the loading medium). To initiate fluxes, the cells are flushed instantaneously into a thermostatted beaker containing a vigorously stirred washout medium (magnetic spin bar, 300 rpm), so that homogeneous mixing is achieved within 1 sec. The final hematocrit was held lower than 1%. At the desired time intervals, samples of 0.5–1.0 ml were aspirated through the filters, and the filter holders immediately disconnected to avoid further filtration. Time resolution is 1–2 sec. Radioactivity is measured on 50–250 μl of the filtrate. The radioactivity at ∞ (equilibrium) time is obtained either by incubating the remaining suspension for 1 hr at 37° and sampling by filtration or by adding to the suspension TCA to a 5% final concentration, followed by centrifugation.

Efflux can also be measured by the cold-stopper technique. Aliquots of 0.1–0.5 ml of cell suspension are withdrawn from the flux medium and transferred to a 15-ml Corex tube containing 5 ml of ice-cold buffered medium with inhibitor [e.g., μM NBMI, 20 μM HNBMG, 50 μM dipyridamole (CIBA), or 1 mM dilazep]. After centrifugation (7000 rpm for 10 sec in a Sorvall RC2B), the pellet is washed once with cold stopper medium, the supernatant discarded, and the cells lysed with 1–3 ml H$_2$O. Alternatively, the stopper medium contains also 1 ml of a fluid denser than water ($\rho = 1.035$ g/ml), such as a mixture of silicone fluids (Dow-Croning) or dibutyl phthalate, so that after centrifugation, there is no need for washing the cells. The upper aqueous supernatant and the oil layer over the packed cells are aspirated, the walls of the tubes are carefully wiped, and the cells are lysed with 1–3 ml H$_2$O. For relatively fast fluxes, it is recommended to carry out single point experiments. This is accomplished by jetting the aliquot of uridine-loaded cells via flushing with 0.5 ml flux medium into 0.5 ml medium vortexed in a 15-ml Corex tube. Stopping is accomplished by adding 5 ml of ice-cold medium containing the inhibitor. Various manual, mechanical, and electromechanical devices can be applied either to administer solutions at preset times or rapidly mix cells and medium and repetively dispense them on centrifuge tubes containing oil mixtures, or rapidly mix cells and medium and stopper in a reaction line. Irrespective of the stopper method used, radioactivity is always checked in the cells. The number of cells in each individual sample can be obtained from the hemoglobin content measured by absorption at either 410 or 540 nm, depending on the dilution used.

For large-scale operations (i.e., dozens of samples), it is recommended to read the hemoglobin samples on a 96-well plate (flat bottom) with the aid of an ELISA reader (e.g., Biotek, Dynatech Instruments, or equivalent). For each individual blood sample it is necessary to establish the correspondence between hemoglobin content and the number of cells (measured either on an electronic counter or simply by checking hematocrits, where 1 ml packed cells corresponds to 10^{10} cells). The cell water content for a given number of cells or for the aliquot used in the particular flux experiment is obtained by equilibrating cells with ^3H_2O and an extracellular marker (e.g., [^{14}C]sucrose or [^{14}C]inulin) in analogous conditions to those of the experiment separating the suspension through an oil layer and checking the ^3H and ^{14}C. The extracellular contamination through various oil mixtures is between 5 and 10%, and the cell water is about $0.7 \times$ cell volume for isotonic buffered solutions (pH 7.4).

Zero-Trans Efflux (zt-out, or 12)

The range of the concentrations commonly used for this procedure is 0.2–10 mM. Cells are loaded with uridine (Ur), cold and labeled. With the filter technique, the first five points of the efflux profile are used to obtain the initial extracellular radioactivity dpm(o). This is done by linear regression of dpm(t) versus t. Subsequently, the rate constant of the initial velocity is obtained by linear regression of [dpm(t) $-$ dpm(o)]/[dpm(∞) $-$ dpm(o)] versus t. This is applicable to time courses for which $S_i(t)$ does not change more than 20% $S_i(o)$. Since the resulting slope is the rate constant which is essentially equivalent to S/v, a plot of the slope versus S (Hanes–Woolf) gives immediately K^{zt}_{12} (the $-x$ intercept) and V^{zt}_{12} ($-x$ intercept/slope), thus overcoming measurements of specific activities and other forms of data manipulation. Plagemann and Wohlhueter have used an integrated form of Eq. (2) and generated the kinetic constants by an iterative method based on nonlinear least squares.[32] So as to measure influx, they have used a rapid mixer which repetitively dispenses the reaction mixture on Eppendorf tubes containing oil, followed by centrifugation and processing of the cell pellet.

For the stopper method, one measures the radioactivity associated with cells, dpm, normalized for the number of cells (from the Hb determinations). The internal concentration of substrate $S_i(t)$ is given by

$$S_i(t) = H[(D - B)/(O_i S_a)]L$$

where $H = \frac{1}{2}(OD_{st}/HCT_{st})$, OD_{st} is the optical density (either at 410 or 540 nm) of a standard cell suspension (at hematocrit HCT_{st}) which is hemolysed 1 : 200. D is dpm(t), B is dpm background, O_i is the optical density of the sample (at either 410 o4 540 nm), S_a is the specific activity of UR (dpm/μl loaded medium/mM Ur). S_i is given in millimoles per liter cell water. L is a correction factor which takes into account the dilution due to addition of TCA, the aliquot taken for counting, and the cell water fraction (0.7 in the present experimental conditions).

The initial rate of efflux is obtained from linear regression of S_i versus t, which is used to generate either the S/v versus S or v/S versus v linearized plots analyzed by linear regression or to analyze the Michaelis form v versus t by nonlinear regression.

Zero-Trans Influx (zt-in, or 21)

This procedure was based on the stopper technique, essentially as described above for efflux, yielding K_{21}^{zt} and V_{21}^{zt} by both linearized and nonlinear forms of the Michaelis equation.

Equilibrium Exchange Efflux (ee)

Cells are preloaded with cold and labeled Ur as described for zt-out, and fluxes are performed in media containing unlabeled Ur at the same concentration as in the loading medium. This procedure uses the integrated rate equation which, in its simplest form, describes the rate of material appearing in the medium (filter method)

$$\ln[(C_\infty - C_t)/(C_\infty - C_0)] = -kt$$

where C_t and C_0 as well as C_∞ are the respective concentrations of radioactivity appearing in the supernatant at time t, 0, and at equilibrium (i.e., infinity). Again, k obtained by linear regression is plotted against S to yield the parameters K^{ee} and R^{ee}.

When using the stopper method, an equivalent equation is obtained, S_i replacing for C. The values of S_i are obtained as described for zt-out.

Equilibrium Exchange Influx

The procedure is analogous to that of zt-in, except that cells are preloaded with cold Ur (see zt-out) in the rate constants of influx and obtained in an essentially similar fashion as those of efflux, by plotting

$$\ln[(C_\infty - C_t)/C_\infty] = -kt$$

Whereas in the efflux mode the natural log term represents the fractional radioactivity within the cells, in the efflux mode it represents the equivalent fraction in the medium.

Infinite-Cis Efflux (ic-out, or 12)

Cells are loaded with cold Ur at an infinitely high concentration (≥ 4 mM), followed by loading with labeled substrate (as for zt-out). Fluxes are measured by either the filter or stopper technique as for zt-out. They are performed in media containing varying concentrations of Ur (S_2) containing the same specific radioactivity as the loaded medium. The net unidirectional flux is given by

$$v_{net} = V - [VS_2/(K^{ic}_{12} + S_1)] \quad \text{or} \quad 1/v_{net} = (1/V) + (S_0/K^{ic}_{12}V)$$

K^{ic}_{12} is obtained either by linear or nonlinear regression analysis of these equations applied to experimental data.

Infinite Cis Influx (ic-in, or 21)

Cells are loaded to different S_i with cold and labeled Ur, the same as for the zt-out procedure, and fluxes are performed into media containing 4 mM Ur at the same specific activity as in the loading media. Fluxes are done by the stopper technique. Data analysis is essentially analogous to that shown for ic-out, the calculated constant being K^{ic}_{21}.

The ic procedures are not only valuable for assessing the consistency of the carrier model but methodologically they are essential for testing possible unstirred layer effects on transport measurements. This may have a considerable retarding effect on relatively fast fluxes such as in the case of adenosine, thus obliterating flux profiles. As proposed by Lieb and Stein,[10] by plotting ic data as

$$1/v_{12} = C_1[S_2/(V_{12} - v^{ic}_{12})] - (C_1/P_2) \quad \text{or}$$
$$1/v_{21} = C_2[S_2/(V_{21} - v^{ic}_{21})] - (C_2/P_1)$$

(P_2 is the effective permeability coefficient of the unstirred layer at side 2, and C is a constant), one obtains a straight line for simple carrier kinetics and an x intercept which is not significantly different from zero for a system unperturbed by unstirred layers. The results indicate[10] that such layers at either membrane surface need not be considered for uridine fluxes.[9]

For all the procedures described above, it is extremely important to carry out the experiments in well-stirred and thermostatted conditions, since small variations in temperature are immediately reflected in the V values because of the relatively high energy of activation of the system (about 20 kcal/mol).

Kinetic Parameters of Uridine Transport

In order to obtain meaningful kinetic parameters, it is necessary to use at least eight different substrate concentrations. For each concentration, the initial velocity, v_i, or rate constant, k, is derived from averaged tripli-cates taken for each time point, plotted against time, and analyzed by linear regression (for v_i) or by nonlinear regression (for k). The kinetic parameters are computed both from S/v versus S and v/S versus v by linear regression and by nonlinear regression of the Michaeliis S versus v plot. This is done because each plot emphasizes data obtained from differ-ent ranges of substrate concentration. However, when the S values are properly spaced out, the computed parameters obtained by the different analytical procedures are statistically not different from each other (Table III).

The kinetic parameters of uridine transport obtained by the various experimental procedures at 25° with recently outdated human blood[9] were compiled from Ref. 9 and adapted. A salient feature of the experimental parameters is that the V/K values were essentially the same for all the procedures, an inherent property of the simple carrier model. Further-more, the R and K values obtained by the different procedures were fully consistent with each other, again as predicted for the simple carrier.

An interesting property observed in cells from outdated blood is a directional asymmetry manifested in V^{zt} values ($V_{12}^{zt} > V_{21}^{zt}$). The 4-fold asymmetry, which is absent in fresh cells,[11,12] appears within 7 weeks of blood storage, but not after ATP depletion of the fresh cells.[12] The kinetic data are further interpreted in terms of molecular rate constants (Table I) after the respective bounds are calculated[9,29] (Table IV). Certain of these bounds are close enough together to provide a good estimate of the actual values of the rate constants. Clearly, the translocation rates of the un-loaded carrier are rate-limiting steps, the one in the outbound direction being about four times slower than that in the inbound direction. The possibility that in ee conditions, the concentration of loaded carrier at either membrane surface is equal, but that of unloaded carrier is not, was

TABLE III

COMPARATIVE VALUES OF KINETIC PARAMETERS OF URIDINE TRANSPORT DERIVED BY DIFFERENT ANALYTICAL PROCEDURES[a]

Parameter	Method		
	v/S versus v	S/v versus S	NLSQ
K^{ee} (mM ± SE)	1.27 ± 0.11	1.31 ± 0.092	1.32 ± 0.14
V^{ee} (mM/min ± SE)	8.24 ± 0.83	8.23 ± 0.092	8.27 ± 0.24

[a] Equilibrium exchange efflux was performed by the filtration technique at 25° in the range of 0.025–20.0 mM Ur and analyzed by the indicated procedure, where NLSQ represents the analysis by nonlinear least-squares.

TABLE IV
KINETIC PARAMETERS OF URIDINE TRANSPORT[a]

Procedure	Direction	Experimental values ($\pm$ SE)		
		V (mM/min)	K_m (mM)	V/K
Zero trans	In $\rightarrow$ out	$v_{1\rightarrow2}^{zt} = 1.98\ (\pm0.31)$	$K_{1\rightarrow2}^{zt} = 0.40\ (\pm0.12)$	4.91 (±1.67)
	Out $\rightarrow$ in	$v_{2\rightarrow1}^{zt} = 0.53\ (\pm0.038)$	$K_{2\rightarrow1}^{zt} = 0.073\ (\pm0.069)$	7.27 (±6.29)
Infinite cis	In $\rightarrow$ out	$v_{1\rightarrow2}^{ic} \equiv v_{1\rightarrow2}^{zt}$	$K_{1\rightarrow2}^{ic} = 0.252\ (\pm0.096)$	—
	Out $\rightarrow$ in	$v_{2\rightarrow1}^{ic} \equiv v_{2\rightarrow1}^{zt}$	$K_{2\rightarrow1}^{ic} = 0.937\ (\pm0.226)$	—
Equilibrium exchange	Both averaged	$v^{ee} = 7.54\ (\pm0.45)$	$K^{ee} = 1.29\ (\pm0.11)$	5.70 (±0.8)

Simple carrier resistance parameters	R (min/mM) ($\pm$ SE)	nR (msec) ($\pm$ SE)
$R_{12} = (v_{1\rightarrow2}^{zt})^{-1}$	0.505 (±0.079)	$nR_{12} = 10.1\ (\pm1.6)$
$R_{21} = (v_{2\rightarrow1}^{zt})^{-1}$	1.887 (±0.135)	$nR_{21} = 37.7\ (\pm2.7)$
$R_{ee} = (v^{ee})^{-1}$	0.133 (±0.008)	$nR_{ee} = 2.7\ (\pm0.16)$
$R_{00} = R_{12} + R_{21} - R_{ee}$	2.259 (±0.157)	$nR_{00} = 45.1\ (\pm3.1)$

Independent estimates of simple carrier affinity parameter K estimated using:	Value (μM) ($\pm$ SE)
$K = K_{1\rightarrow2}^{zt}(R_{12}/R_{00})$	89 (±31)
$K = K_{2\rightarrow1}^{zt}(R_{21}/R_{00})$	61 (±58)
$K = K_{1\rightarrow2}^{ic}(R_{ee}/R_{12})$	66 (±28)
$K = K_{2\rightarrow1}^{ic}(R_{ee}/R_{21})$	66 (±17)
$K = K^{ee}(R_{ee}/R_{00})$	76 (±10)

Mean ($\pm$ standard deviation) = 72 (±11) μM

[a] Outdated human blood, 25°, pH 7.4. Data were obtained and analyzed as given in Ref. 9 and further elaborated[29] according to the model shown in Fig. 1. The carrier resistance parameters and affinity parameter K were calculated using the relationships given in Table II.[9] The millimolar units stand for millimoles Ur/liter isotonic cell water. The nR values were obtained by using $n = 3.3 \times 10^{-4}$ mmol uridine transport systems/liter isotonic cell water,[29] based on 1.25×10^4 carriers/cell, an average red cell volume of 87 μm^3, and average water content of 7.7% (v/v).[29]

given consideration on the basis of data shown in Table V.[34] Further description of kinetic and biochemical properties of nucleoside transport systems can be found in a recently published review.[35]

[34] H. Ginsburg and Z. I. Cabantchik, *J. Gen. Physiol.* **70**, 679 (1977).
[35] P. G. W. Plagemann, R. M. Wohlhueter, and C. Woffendin, *Biochim. Biophys. Acta* **947**, 405 (1988).

TABLE V
UPPER AND LOWER BOUNDS OF RATE
CONSTANTS IN TERMS OF BASIC MEASURABLE
PARAMETERS[a]

$$27\ (\pm 2)\text{sec}^{-1} < k_1 < 29\ (\pm 2)\ \text{sec}^{-1}$$
$$99\ (\pm 16)\text{sec}^{-1} < k_2 < 136\ (\pm 29)\ \text{sec}^{-1}$$
$$b_1, b_2 > 370\ (\pm 22)\ \text{sec}^{-1}$$
$$f_1 > 3.7\ (\pm 0.6) \times 10^5\ M^{-1}\ \text{sec}^{-1}$$
$$f_2 > 1.4\ (\pm 0.3) \times 10^6\ M^{-1}\ \text{sec}^{-1}$$

[a] Outdated blood, 25°, pH 7.4. The molecular
rate constants are defined in Table I. The values given in Table IV were used to compute
the bounds, as described elsewhere.[9,29]

Acknowledgments

This work was supported in part, by Grants R22 AI 20342 and R01 HL 40158 from the National Institutes of Health. The author gratefully acknowledges the collaboration of Professors W. D. Stein and H. Ginsburg, as well as Professors A. R. P. Paterson and R. G. W. Plagemann, and Dr. S. M. Jarvis for providing manuscripts of their studies prior to publication.

[15] Transport Studies in Red Blood Cells by Measuring Light Scattering

By GÜNTER FRED FUHRMANN

Introduction

Net changes of osmotically active contents inside the red blood cell area accompanied by an osmotic water flow and reflected by a proportional change in cell volume. Since movement of water itself is extremely fast with half-times of a few milliseconds in mammalian red blood cells, it does not become a limiting factor in transport studies of substrates, which usually permeate a few orders of magnitude slower than water.

Volume changes of red blood cells can be followed directly by recording the intensity of light scattered by the cell suspension. It has been found that a light beam passed through the cell suspension varies in light intensity in direct proportion to the change in cell volume. This so-called "light scattering" or "opacimetric" method has a long history as described by Ponder[1] in his famous book, "Hemolysis and Related Phenomena." The types of apparatus applied and modified especially for sugar transport studies are described in detail by Widdas[2] in this volume.

The Right-Angle Light-Scattering Apparatus

The apparatus modified by Fuhrmann et al.[3] is different from the ones described by Ponder[1] and Widdas,[2] mainly in the arrangement of the light beam and the detection system.

A light beam (LB) of parallel white light enters the measuring cuvette, the red blood cells suspended in the cuvette scatter the light, and a sensitive photomultiplier (PM) arranged at a right angle to the incoming light beam detects the scattered light (SL) by transformation of the light intensity into a voltage signal which is continuously recorded (Fig. 1).

The intensity of the incoming light beam is set at such a low level that without cells in the cuvette no significant light is detected by the photomultiplier.

The Signal Is Dependent on a Small Number of Cells. Since only the scattered light from the cells is detected, the signal of the photomultiplier is directly proportional to the number of cells in the cuvette. This is, however, only the case at low hematocrit values.

Usually we suspend 10 μl of packed red blood cells or equivalently a 100-μl cell suspension with 10% hematocrit in 15 ml (15,000 μl) of the solution in the cuvette. Thus, the dilution of the cells is 1500-fold and the final hematocrit only 0.067%.

A small number of cells suspended in a large volume has several advantages for kinetic analysis in membrane transport. At a given outside concentration of a permeable substance (S_o) the change in inside concentration (S_i) of the small intracellular compartment (in total approximately 7 μl) does not influence effectively the concentration of the large outside compartment (15,000 μl).

Since the concentration change of S_o is negligible, the undirectional influx, which is a function of S_o, remains nearly constant during the transport experiments. Linear equations similar to the Lineweaver–Burk linearization procedure satisfactorily describe the observed transport kinetics.[4–6]

[1] E. Ponder, "Hemolysis and Related Phenomena," pp. 50–109. Grune & Stratton, New York, 1971.

[2] W. F. Widdas, this volume [13].

[3] G. F. Fuhrmann, P. Liggenstorfer, and W. Wilbrandt, *Experientia* **27,** 1428 (1971).

[4] G. F. Fuhrmann, *in* "Erythrocytes, Thrombocytes, Leucocytes" (E. Gerlach, K. Moser, E. Deutsch, and W. Wilmanns, eds.), pp. 102–105. Thieme, Stuttgart, Federal Republic of Germany, 1973.

[5] W. Wilbrandt, *in* "Carrier Diffusion in Biomembranes" (F. Kreuzer and J. F. G. Seegers, eds.), Vol. 3, pp. 79–99. Plenum, New York, 1972.

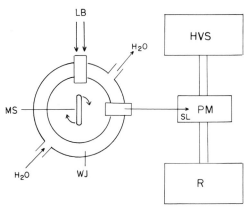

FIG. 1. Right-angle light-scattering apparatus. MS, Magnetic stirrer inside the measuring cuvette; WJ, water jacket; LB, light beam; SL, scattered light; PM, photomultiplier; HVS, high-voltage supply; R, recorder. (Modified from Fuhrmann et al.[3])

The Signal Is a Function of the Refractive Index of the Light. The scattered light intensity detected by the photomultiplier at a right angle to the incoming light beam is a function of the refractive index difference between the red blood cells and the surrounding medium.

The intracellular concentration of hemoglobin determines the refractive index of the red blood cells. The intensity of scattered light is directly proportional to the concentration of hemoglobin. In Fig. 2 a typical calibration curve for the signal is plotted against the cellular hemoglobin concentration. The calibration curve is obtained by adding a concentrated NaCl solution of about 5 mol/liter to the cell suspension and recording osmotic shrinkage of the cells by the increase in photomultiplier signal (ordinate, light scattering in arbitrary units). The hemoglobin concentration at the abscissa is obtained by dividing the hemoglobin content of an average human red blood cell by the relative cell volume (V) calculated by the equation given by Ponder[1]:

$$V = W[(1/T) - 1] + 100$$

This equation is one form of the van't Hoff Mariotte law for ideal osmotic behavior. At every calibration point the tonicity (T) of the cell suspension is calculated by the ratio of the freezing point depression of the solution in mOsm/liter (measured by an Advanced osmometer, Advanced Instruments, Inc., Needham Heights, MA) to that of an isoosmotic 300 mOsm/liter salt solution. The initial volume of the cells in the latter solution is denoted by 100 in Ponder (1) and in the graph in Fig. 2 by 1. The cell water

[6] W. Wilbrandt and G. F. Fuhrmann, *Experientia* **33**, 1472 (1977).

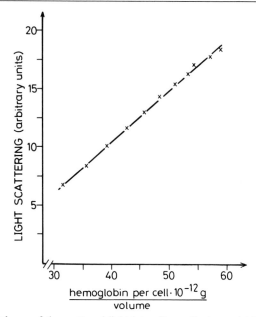

FIG. 2. Dependence of the scattered light intensity on the hemoglobin concentration in the cell. One hundred microliters of a 10% cell suspension was added to 14.9 ml buffer solution (145 mmol/liter NaCl, 1 mmol/liter KCl, and 20 mmol/liter Tris, pH 7.4) of 306 mOsm/liter in the measuring cuvette at 37°. The first cross at the left side shows the scattered light intensity for this cell suspension and the calculated hemoglobin concentration. The hemoglobin content of an average human red blood cell was taken to be 31.2×10^{-12} g and the volume of the cell was calculated as described in the text. The following values were obtained by addition of 100 μl 5.2 mol/liter NaCl solution to the cuvette and recording the scattered light. The osmolarity of the buffer solution and of the solution after NaCl addition was determined on an Advanced osmometer. Results represent the average of four determinations. The crosses are fitted by a straight line ($r = 0.998$).

(W) is assumed to be 70%. The results show a straight line relationship between the light-scattering signal and the hemoglobin concentration ($r = 0.998$). This is in agreement with ideal osmotic behavior of the red cells and also in accordance with the dependence of the scattered light intensity on the difference in refractive index between hemoglobin and the surrounding medium. Hemolysed cells do not scatter light.

Sealed ghosts with buffered salt solution inside also give no signal with this method. If, however, ghosts are sealed with at least a quarter of their original hemoglobin content, volume changes can be detected.

When the refractive index of the suspension solution is increased, the scattered light detected will be diminished in direct proportion to the reduction in refractive index difference between the cells and the surrounding medium. This can be observed, for example, if sucrose is used to replace the osmolarity of an NaCl solution. The higher refractive index of the sucrose in relation to the NaCl solution reduces the intensity of scattered light in the same proportion.

The Signal Is Influenced by the Shape of the Cells. In order to hold the cells in suspension and to keep them well mixed, the contents of the measuring cuvette are stirred using a magnetic stirrer (Fig. 1, MS). Freshly prepared, biconcave red blood cells in motion produce the so-called Schlieren phenomenon. This phenomenon is reflected by the photomultiplier signal with increased background noise. If the shape of the cells changes from biconcave to spheric, the noise is reduced in relation to the change in shape.

To avoid these fluctuations in the signal due to the Schlieren phenomenon a capacitor or an electronic filter (H. Müller, Department of Physics, University of Marburg, FRG) is inserted, so that the noise level in the base line is diminished, while the signal from the photomultiplier due to scattered light is not affected.

The remaining noise signal is a valuable tool in detecting shape changes of the cells as measured by an increase or decrease in noise which might be produced by several chemical agents.

Special Features of the Apparatus for Reproducibility. The temperature in the measuring cuvette is kept constant using a water jacket (Fig. 1, WJ). The heat produced by the light bulb for the illuminating beam is removed by a second water jacket for the bulb house. The second water jacket is connected to the same thermostatic system.

The photomultiplier tube is supplied by a stabilized constant high-voltage supply (Fig. 1, HVS). The source of the light beam entering the measuring cuvette is regulated by a stabilized constant power supply. These are the main features required for production of reproducible signals. Deviations during 1 day of experiments are insignificant.

The accuracy of this method depends on the exactness of pipetting cells and suspension solution into the measuring cuvette.

Additional Instrumentation for the Light-Scattering Method. In addition to measuring light scattering, the apparatus has also been equipped with electrode systems for analysis of potassium ions (W. Riemann, Department of Applied Physiology, University of Marburg, FRG). The electrode system is built into the cover. For potassium selectivity valinomycin has been incorporated into a PVC membrane.[7] The electrode produces

[7] D. Ammann, R. Bissig, Z. Cimerman, U. Fiedler, M. Güggi, W. E. Morf, H. Osswald, E. Pretsch, and W. Simon, *in* "Ion and Enzyme Electrodes in Biology and Medicine" (M. Kessler, L. C. Clark, Jr., D. W. Lübbers, J. A. Silver, and W. Simon, eds.) pp. 22–37. Urban & Schwarzenberg, Munich, Federal Republic of Germany, 1976.

nearly linear signals down to 10^{-5} mol/liter potassium concentration. This allows simultaneous and continuous measurement of cellular volume change and potassium loss.

A pH-electrode and a chloride electrode has also been added to the light-scattering apparatus to record the pH changes of the solution and to follow chloride concentration changes in the suspension medium.

A further addition in instrumentation consists of a feedback system that controls the volume of the cells, i.e., the light-scattering signal. With this device the swelling of the cells during the transport process can be prevented: the electrical signal of the photomultiplier is fed into an automatic titrator (pH-Stat, Radiometer Copenhagen, Denmark) and held constant by automatic titration of a concentrated salt solution into the measuring cuvette, causing feedback shrinkage of the cells. This system of osmotitration measures net transport of substances by a constant volume of cells. The microliters of concentrated salt titrated per unit time is equivalent to the flux.

Application of the Right-Angle Light-Scattering Method in Transport Studies

Transport of Nonelectrolytes. In general, net influx or net efflux experiments of nonelectrolytes can be accomplished. However, for methodological implications and kinetic reasons the net efflux experiments are of greater advantage.

Net efflux experiments: The right-angle light-scattering method has been used to measure net efflux of glucose in human red blood cells[3-5] and to follow net effluxes of glycerol and ethylene glycol in bovine red blood cells.[6]

The particular advantage of this method is revealed by a glucose efflux experiment at 37° in human red blood cells (Fig. 3). The efflux was followed in cells initially loaded with 300 mmol glucose/liter in three sets of experiments with 0, 5, and 10 mmol glucose/liter added to the suspension medium. Even within the brief period of 30 sec the efflux of glucose with no glucose added can be accurately determined. From a practical standpoint this experiment can be regarded as a true zero-trans efflux,[8] since the contaminating glucose from 10 μl of packed cells is negligible and transported glucose from the cell interior is very low (approximately 100 μmol glucose/liter). However, a special technique described below is needed for rapid suspension of packed red cells into the medium of the cuvette: After cells loaded with glucose are pelleted in a Sorval centrifuge at more than 14,000 g, they are sucked into a 1-ml tuberculin syringe

[8] Y. Eilam and W. D. Stein, *Methods Membr. Biol.* **2,** 283 (1978).

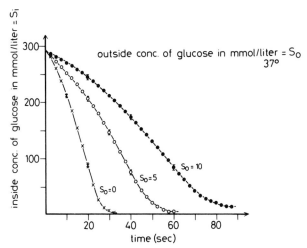

FIG. 3. Efflux of glucose from preloaded human red blood cells at 37° and pH 7.4. (×), No glucose added to the outside (mean of 16 experiments); (○), 5 mmol/liter glucose added (mean of 6 experiments); and (●), 10 mmol/liter glucose added (mean of 6 experiments). The bars at the points indicate the standard error.

inserted into a microsyringe holder. Using this device, exactly 10 μl packed cells is injected into a small tube attached to the tuberculin syringe needle. The tube containing the 10 μl of packed cells is removed from the needle and inserted at the tube's opposite end to another needle attached to a second tuberculin syringe. This second syringe is filled before with about 200 μl of exit solution from the cuvette. The experiment is started by flushing the exit solution in the second syringe through the tube into the measuring cuvette with the shutter of the photomultiplier open. This technique allows fast dispersion of red cells and measurement of glucose exit shortly thereafter.

Since in the net efflux experiments the change in inside concentration of glucose (S_i) has been analyzed per time (t), the conventional net flux of dm_i/dt (m_i is the amount of glucose inside the cell) was rearranged to dS_i/dt. Between dm_i and dS_i the following relationship exists.

$$dm_i/dt = (dm_i/dS_i)(dS_i/dt)$$

dm_i/dS_i can be solved by using the following equation for ideal osmometer behavior of the cell:

$$V = (m_i + n)/(S_o + N)$$

where V is the volume of the cell, n the osmotic content of the cell, S_o the outside concentration of glucose, and N the outside concentration of electrolyte in mOsm/liter. By solving the equation for m_i

$$m_i = nS_i/(S_o + N - S_i)$$

and differentiating,

$$dm_i/dS_i = n(S_o + N)/(S_o + N - S_i)^2$$

the rearranged form becomes

$$dS_i/dt = [(S_o + N - S_i)^2/n(S_o + N)](dm_i/dt)$$

The term $(S_o + N - S_i)^2/n(S_o + N)$ is the reciprocal of the cell volume at time t.

The analysis of changes in inside glucose concentration at constant outside glucose concentration allowed the kinetic analysis of the apparent half-saturation constants for the unidirectional influx and efflux at two different temperatures.[4,5] These values given in Table I reveal the asymmetrical properties of glucose transport in human red blood cells.

In order to compensate for the 300 mmol glucose/liter in the cell interior in addition to the normal osmotic content (300 mOsm/liter), the solution in the cuvette contained 600 mOsm/liter electrolyte solution. Thus, at the beginning of the glucose efflux experiment the osmotic pressures inside and outside the cell are about equal. The subsequent net efflux of glucose causes the cells to shrink.

Calibration of the photomultiplier signal in the efflux experiment for the glucose inside concentration S_i at time t has been performed by a separate experiment. To control cells suspended in 300 mOsm/liter elec-

TABLE I
DIFFERENCES IN APPARENT HALF-SATURATION CONCENTRATION FOR UNIDIRECTIONAL
EFFLUX AND INFLUX OF GLUCOSE IN HUMAN RED BLOOD CELLS[a]

Temperature (°C)	Method	Unidirectional efflux	Method	Unidirectional influx
20	a	74 (60)	a	2.4 (60)
	b	74.9 ± 5.8 (SE) (25)	c	2.3 (7)
37	a	59 (105)	a	5.3 (105)
	b	64.5 ± 5.6 (SE) (20)	c	5.8 (6)

[a] The half-saturation concentrations are given in mOsm/liter as mean values of the number of experiments as indicated in parentheses. For method (a), values were calculated by the general rate equation given by Wilbrandt[5] and as outlined by Fuhrmann.[4] In method (b), values were analyzed in experiments with practically no glucose outside by equations similar to Lineweaver–Burk and in (c) by the Sen–Widdas procedure as described by Widdas.[2]

trolyte solution portions of 5 mol/liter NaCl solution were added to the cuvette and the resulting changes in light scattering were recorded. The osmolarity at each calibration level has been analyzed by an Advanced osmometer described previously. As for the glucose efflux, ideal osmotic behavior of the cells is assumed also for the calibration experiment. In this experiment the signal is proportional to

$$n/V = N_1 \text{ (300 mOsm/liter)} + \text{added NaCl (mOsm/liter)}$$

and in the efflux experiment

$$n/V = N_2 \text{ (600 mOsm/liter)} + S_o - S_i$$

By combining the two equations a calibration curve can be constructed for each photomultiplier signal and the related inside glucose concentration S_i:

$$\text{Light-scattering signal} \approx S_i = S_o + N_2 - N_1 - \text{added NaCl}$$

From the continuous recording of the light scattering in the glucose efflux experiment each point after the first 5 sec can be taken. Five seconds is the time required for mixing the cells. As seen from the bars in Fig. 3 the experimental error is relatively small.

A limitation of this method, however, is the relative high concentration gradient of permeant between the inside and outside of the cells. If the concentration gradient is lower than about 50 mOsm/liter, volume changes are too small to be detected.

By means of suitable calibrations not only the inside concentration S_i but also the volume or the intracellular amount of substrate m_i can be obtained as a function of time t.

As an example of the application of the right-angle light-scattering method for a nonsaturable transport, glycerol and ethylene glycol diffusion in bovine red blood cells is offered to demonstrate the importance of osmotic volume changes in kinetic analysis.[6] If osmotic volume changes occur during an efflux of these diffusible substances and are not taken into account, the kinetic analysis yields apparent saturation constants, which are in error.

Net influx experiments: Using this method in nonelectrolyte influx experiments has several disadvantages which should be mentioned. The first is the necessary high outside concentration S_o, which reduces the difference in refractive index between the cells and the medium because of its usually high refractive index. This causes a substantial reduction in the light-scattering signal. The second disadvantage is the relatively long adaption time of the apparatus of about 5 sec, which is caused mainly by the time required for equal distribution of the cells in the cuvette. Because of this delay the important rapid initial phase of net influx cannot be measured. Furthermore, it usually takes a long time to reach equilibrium because the outside nonelectrolyte concentration of a saturable transport system greatly exceeds the half-saturation concentration. The third problem is that in the presence of a high concentration of nonelectrolyte (which usually penetrates slowly until equilibration is obtained), the light-scattering signal cannot be calibrated practically with the concentrated NaCl solution method. This difficulty can be overcome in nonelectrolyte influx experiments by the osmotitration technique. The swelling of the cells during net influx of glucose, for example, can be prevented by using the feedback system of the previously described automatic titrator to keep the light-scattering signal constant by titration of a concentrated NaCl solution into the cuvette. The volume of salt solution added per unit time is proportional to the net influx of glucose. This new osmotitration method provides a valuable tool in transport studies.

Transport of Electrolytes. The right-angle light-scattering method can also provide an accurate determination of net fluxes of electrolytes. In addition to net flux experiments, the new technique of osmotitration allows the determination of heteroanion exchange by an indirect method.

As mentioned earlier water permeability is very high and the red blood cells behave as perfect osmometers; therefore it can be deduced that the osmolarity of electrolytes entering or leaving the cell must equal that of the surrounding solution.[9] Thus, the net flux of electrolytes (J) can be calculated:

$$J = (dV_{H_2O}/dt) \times \text{osmolarity of the solution}$$

If t is given in minutes and V_{H_2O} is expressed as volume of cell water per liter of cell water originally present in cells in a 300 mOsm/liter electrolyte solution, the flux has the unit of milliosmoles electrolyte per liter cell water per minute. For reasons of electroneutrality the sum of anions and cations must be equivalent. To convert the milliosmoles per liter of electrolyte into millimoles per liter, the corresponding experimental values given for the electrolyte species in the "Handbook of Chemistry and Physics"[10] should be taken.

Calibration of the light-scattering signal for the cell volume V_{H_2O} was done at the beginning and the end of the experiment by a separate experiment similar to the nonelectrolyte studies. Portions of 100-μl samples of 5 mol/liter NaCl solution were added to the cuvette and the resulting

[9] P. A. Knauf, G. F. Fuhrmann, S. Rothstein, and A. Rothstein, *J. Gen. Physiol.* **69**, 363 (1977).
[10] R. C. Weast (ed.), "Handbook of Chemistry and Physics." CRC Press, Boca Raton, Florida, 1988.

changes in light scattering recorded. The intensity of scattered light from control cells in 300 mOsm/liter electrolyte solution was arbitrarily considered to indicate the cell water volume V_{H_2O} of 1 liter. After making a small correction for the dilution of the suspension by the addition of small samples of NaCl, a calibration curve similar to the nonelectrolyte calibration curve was constructed relating the light-scattering signal to the cell water volume.[9]

Net anion transport studies: The analysis of net anion transport in red blood cells is rather complex. Because of electrical neutrality, an equivalent net flux of anions and cations should be assumed. The net anion flux must be measured under conditions where known anion concentration gradients and membrane potentials are imposed. The electrolyte movements were induced in human red blood cells by addition of valinomycin.[9] The potassium efflux down its electrochemical gradient from high potassium cells into a low potassium–high sodium chloride solution induces an electrical field, which provides the driving force for the movement of the anions.

In Fig. 4a the change in light-scattering signal is shown after the addition of valinomycin to cells suspended in chloride solution. The light-scattering traces indicate that valinomycin induces a substantial shrinkage of the cells equilibrated with chloride. In Fig. 4b the cells have been pretreated with 4,4′-diisothiocyanostilbene 2,2′-disulfonate (DIDS), a potent inhibitor of anion exchange.[11] The traces show that DIDS inhibits only by about 60% the valinomycin-induced shrinkage of the cell. In order to ensure that the light-scattering signal was still sensitive to volume changes in the cells, samples of 5 mol/liter NaCl solution were added at the end of the experiment. These produced an increase in light-scattering signal as with control cells, thus demonstrating that DIDS exerts no direct effect on the light-scattering response of the cells.

By using the rate of shrinkage after addition of valinomycin and suitable calibration procedures, as described above, the rate of shrinkage can be converted into the net flux rate expressed as millimoles Cl^- per liter cell H_2O per minute. This is shown in Fig. 5a, where the net flux of Cl^- is depicted as a function of the valinomycin concentration with and without DIDS pretreatment. In Fig. 5b the net fluxes of SO_4^{2-} determined from volume loss in sulfate-equilibrated cells are plotted against the valinomycin concentration. Instead of DIDS, the anion-exchange inhibitor 4-acetamido-4′-isothiocyanostilbene 2,2′-disulfonate (SITS) has been used for pretreatment. In each case (Fig. 5a and b) the anion net flux tends to reach a limiting value at high concentrations of valinomycin. In the presence of

[11] Z. I. Cabantchik and A. Rothstein, *J. Membr. Biol.* **10**, 311 (1972).

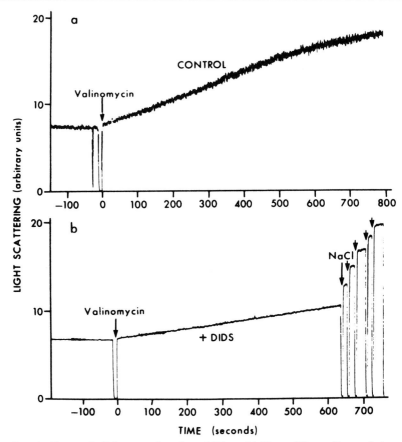

FIG. 4. Changes in light scattering after addition of 1.33 μmol/liter valinomycin to cells equilibrated in chloride medium without (a) and with 30 μmol/liter DIDS pretreatment (b). At the points labeled NaCl (0.1 ml of 5.2 mol/liter), NaCl solution samples were added. (From Knauf et al.[9])

inhibitors of anion transport, the limiting value is considerably lower. Comparing Cl⁻ and SO_4^{2-} net fluxes, it becomes apparent that the maximum flux for SO_4^{2-} is significantly lower than that for Cl⁻. The tendency of the net anion fluxes to reach limiting values would be expected if the potassium permeability increased by the carrier valinomycin becomes limited by a lower anion permeability. Under these circumstances the net electrolyte efflux should be sensitive to anion transport inhibitors and should be affected by the specific properties of the anion species. Both predictions are qualitatively fulfilled in these experiments.

By applying the Goldman equation, theoretical curves for Cl⁻ and SO_4^{2-} fluxes (Fig. 5, broken lines) were calculated using values for net

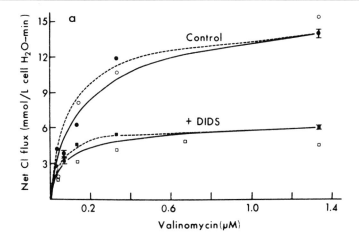

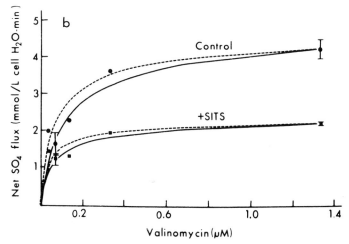

FIG. 5. Dependence of net anion flux on the valinomycin concentration in cells equili-brated in chloride medium (a) and in sulfate medium (b). The upper points (circles) are from control cells; the lower points (squares) for cells pretreated with 125 μmol/liter DIDS (a) and with 110 μmol/liter SITS (b). The bars indicate the standard error of the mean for five determinations. The open symbols (a) are results from a separate experiment on cells from the same donor. Broken and solid lines are calculated as described in the text. (From Knauf et al.[9])

anion permeability determined from the flux data at the highest valinomy-cin concentration (1.33 μmol/liter) and values for the potassium perme-ability determined as a function of the valinomycin concentration. There was no significant difference between these curves and those resulting from the assumption of a linear relationship between potassium permeability and valinomycin concentration (Fig. 5, solid lines). The experimental points are relatively well fitted by theoretical curves. An advantage of the net anion flux method described is that there exists an almost linear relationship between the net anion flux and the corresponding anion permeability. Thus, the net potassium efflux provides a useful empirical measurement of the net anion permeability.

The partial inhibition by DIDS and SITS of net anion flux is in contrast to the nearly complete inhibition of anion exchange.[11] The discussion has led to the conclusion that, at least for the inhibited part, a common element, the anion exchanger, is involved.[9] Recent measurements in red blood cell membranes by the Gigaseal patch-clamp technique,[12] however, have demonstrated the presence of anion channels partially inhibited by H_2DIDS and different in number than the anion exchanger.[13] These results favor the existence of completely separate and distinct membrane transport mechanisms.

Net cation transport studies: The right-angle light-scattering method has been used successfully to characterize the Ca^{2+} (or Pb^{2+})-activated potassium channel in human red blood cells.[14–16] These channels induce a rapid efflux of potassium from the red blood cells of humans and other species. The permeability change for potassium occurs without a significant increase in sodium permeability. The osmotic content decreases and the cells shrink, indicating that the potassium ions are accompanied by an equivalent amount of anions. The selective increase in potassium efflux can be produced by a variety of treatments of the red blood cells. All of these treatments involve the use of calcium. If the intracellular activity rises to micromolar levels, the potassium-selective channels become activated.

From the investigations of Ørskov[17] in 1935 lead has also been known to evoke a rapid efflux of potassium from red blood cells. One problem, however, with comparative studies on the effects of calcium and lead is the fact that most analytical grade chemicals used for preparing isoosmotic solutions contain contaminating calcium ions in the micromolar

[12] R. Grygorczyk, W. Schwarz, and H. Passow, *Biophys. J.* **45**, 693 (1984).
[13] H. Passow, W. Schwarz, M. Glibowicka, N. Aranibar, and M. Raida, *in* "Molecular Basis of Biomembrane Transport" (F. Palmieri and E. Quagliariello, eds.), pp. 121–139. Elsevier, New York, 1988.
[14] G. F. Fuhrmann, J. Hüttermann, and P. A. Knauf, *Biochim. Biophys. Acta* **769**, 130 (1984).
[15] M. Shields, R. Grygorczyk, G. F. Fuhrmann, W. Schwarz, and H. Passow, *Biochim. Biophys. Acta* **815**, 223 (1984).
[16] G. F. Fuhrmann, W. Schwarz, R. Kersten, and H. Sdun, *Biochim. Biophys. Acta* **820**, 223 (1985).
[17] S. L. Ørskov, *Biochem. J.* **297**, 250 (1935).

range. Therefore, it is unclear to what extent the potassium flux induced by lead is the consequence of an increase in Ca^{2+} influx into the cells or a liberation from intracellular Ca^{2+}-binding sites.

Light-scattering traces, in Fig. 6a show that analytical grade reagents contain sufficient contaminating calcium to activate the potassium-selective channels in human red blood cells after addition of the calcium ionophore A23187. Cell shrinkage is indicated by deflection of light scattering in this trace.

If the potassium permeability of the membrane is greatly increased by the addition of valinomycin, the rate of net KCl efflux is limited by the net permeability of the chloride anion as shown by net anion transport studies. In nitrate media, this limiting value is reached at much higher rates of net potassium efflux,[12] indicating that the nitrate anion has a much higher net permeability than the chloride anion, approximately four times as large. The net potassium efflux in the presence of calcium is within the range where the nitrate flux is not a limiting factor. Therefore, investigations of the potassium channel in red cells were performed in nitrate media. In chloride media, however, at such high potassium efflux, the net potassium efflux is limited by the chloride permeability.[14]

When analytical grade $NaNO_3$, which contains in isoosmotic concentration about 7–8 μmol calcium/liter, is replaced by $NaNO_3$ Suprapur (E. Merck, Darmstadt, FRG), the addition of A23187 induces no KNO_3 ef-

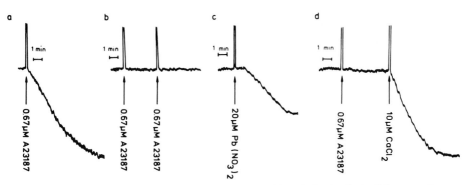

Fig. 6. Shrinkage of human red blood cells detected by a deflection in the light-scattering signal after addition of A23187 or lead. Red cells were suspended at a hematocrit of 0.067% in a medium containing 160 mmol/liter $NaNO_3$, 1 mmol/liter KNO_3 (adjusted to pH 7 at 37°) made from either reagent grade (a) or Suprapur (b, c, and d) $NaNO_3$. The calcium concentration in the reagent-grade solution was 7–8 μmol/liter and in the Suprapur solution 0.2 μmol/liter. At the times indicated, A23187, calcium, or lead were added. The sharp upward deflection in the traces indicates the opening and closing of the measuring cuvette (in Fig. 4 there are downward deflections because of reverse recording) during the various additions. (From Shields et al.[15])

flux, i.e., there is no shrinkage of the cells (Fig. 6b). This is expected from the analytically observed calcium concentration of about 0.2 μmol/liter, which is below the threshold for activation. Addition of 10 μmol calcium/liter to the A23187-containing medium immediately elicits shrinkage of the cells (Fig. 6d). When $Pb(NO_3)_2$ is added in place of calcium, an essentially similar response is observed (Fig. 6c). Therefore, the metal lead by itself can induce the potassium efflux. Also the possibility that calcium is released from intracellular calcium-binding sites could be ruled out.

These results on selective potassium permeability in the red cell demonstrate the value of the light-scattering technique even in qualitative screening experiments. As shown previously, the qualitative light-scattering signal can easily be converted into quantitative net fluxes by an appropriate calibration technique.

Heteroanion-exchange transport studies: When red blood cells are suspended in an isotonic solution of NH_4Cl equilibrated with the atmosphere, hemolysis of the cells occurs with time according to the cycle described by Jacobs and Stewart.[18] The red blood cell is relatively impermeable to NH_4^+ but highly permeable to NH_3. NH_3 diffuses quickly into the cell, in which a new equilibrium will then be established by hydrolysis:

$$NH_3 + H_2O \rightleftharpoons NH_4^+ + OH^-$$
$$NH_3 + H_2CO_3 \rightleftharpoons NH_4^+ + HCO_3^-$$

The new anions formed intracellularly exchange with extracellular Cl^- via the anion-exchanger membrane protein of band 3 to obtain anion equilibration. The sum of the reactions of the cycle is a net influx of NH_4Cl into the cell, leading to osmotic swelling and finally to hemolysis.

The rate-limiting step in this overall reactions is, however, the heteroanion exchange of OH^-/Cl^- and HCO_3^-/Cl^- by the anion exchanger. Therefore, SITS, a blocker of the anion exchanger, inhibits nearly completely hemolysis of bovine red blood cells incubated in NH_4Cl solution.[19] Swelling of cells in the NH_4Cl system and subsequent hemolysis is a measure of the heteroanion-exchange flux. Instead of determination hemolysis,[19] which is not sharply measurable, the newly developed feedback titration preventing cellular swelling by the osmotitration method was applied. The milliosmoles of NaCl titrated per unit time by the automatic titration device are directly proportional to the heteroanion-exchange flux.

In this new method the light-scattering signal obtained immediately

[18] M. H. Jacobs and D. R. Stewart, *J. Gen. Physiol.* **25,** 539 (1942).
[19] J. L. Cousin, R. Motais, and F. Sola, *J. Physiol. (London)* **253,** 385 (1975).

after suspension of cells in 150 mmol/liter NH_4Cl solution buffered with 20 mmol/liter HEPES adjusted to the desired pH was kept constant by automatic titration of a 4.5 mol/liter NaCl solution into the cuvette. At 20° the speed of the titration system is well adjusted to the biological reaction, so that after an adaptation time of about half a minute a reproducible titration curve of NaCl solution per unit time is obtained. By using an Advanced osmometer, the microliters of NaCl solution titrated can be converted into milliosmoles per unit time. At low hematocrit the titration curve is independent of the number of cells suspended in the cuvette. However, the titration curve exhibits increasing stability when twice the usual hematocrit is used (0.13 instead of 0.067%). The NH_4Cl solution was in equilibrium with the CO_2 of the atmosphere. By removing the CO_2 from the solution by flushing with nitrogen the titration rate is reduced in proportion to the diminished HCO_3^- concentration of the system.

The great advantage of this method is the fast and easy approach to determine inhibition of the heteroanion exchange by diverse chemical probes. If for simple analysis of the kinetics the Dixon plot[20] is used, by plotting the inverse rate of the nearly linear titration rate at the first 4 min versus the concentration of inhibitor (I), a straight line is obtained for many inhibitors. From the x-intercept, which corresponds to the apparent K_i, the concentration required for 50% inhibition can be calculated. For example, the dihydro derivative of DIDS, H_2DIDS, demonstrated in four experiments a mean value ± SD of 0.43 ± 0.13 μmol/liter for the apparent K_i value, which is in good agreement with that given in the literature.[21]

Not only cells equilibrated with chloride, but also those equilibrated with other anions like SO_4^{2-} can be used in the osmotitration method with ammonium. In this case the titrant was a 2 mol/liter Na_2SO_4 solution, to keep the volume constant.

Conclusions

The right-angle light-scattering method for transport studies in red blood cells has been shown to be a rapid and convenient tool to measure net transport rates of nonelectrolytes and electrolytes, which usually permeate a few orders of magnitude slower than H_2O. The advantages of fast screening of experiments as well as the ability to measure true flux rates by appropriate calibration techniques have been demonstrated. In addi-

[20] J. L. Webb, in "Enzyme and Metabolic Inhibitors," Vol. 1. Academic Press, New York, 1963.
[21] Y. Shami, A. Rothstein, and P. A. Knauf, *Biochim. Biophys. Acta* **508,** 357 (1978).

tion to net flux rates, anion-exchange rates also can be measured by applying the Jacobs–Stewart cycle in ammonium solutions and the osmotitration method at constant cell volume.

Acknowledgment

I would like to thank Drs. K. J. Netter, H. Passow, W. Schwarz, and E. S. Vesell for their valuable comments on the manuscript and Miss D. Wörsdörfer for her excellent secretarial assistance.

[16] Cation–Anion Cotransport

By Mark Haas and Thomas J. McManus

Introduction

Specific pathways for the coupled transport of cations and anions in the same direction across cell membranes have been described in a wide variety of cells, including reabsorptive and secretory epithelia,[1,2] excitable cells,[3,3a] cultured cell lines,[4–7] Ehrlich ascites tumor cells,[8,9] and red blood cells from mammalian and avian species.[10–13] The majority of these pathways involve the cotransport of the alkali cations sodium and/or potassium with chloride, although cotransport systems involving other ions (e.g., sodium plus phosphate[14]) have also been described. In general,

[1] R. Greger, E. Schlatter, and F. Lang, *Pfluegers Arch.* **396**, 308 (1983).

[2] H. C. Palfrey, P. Silva, and F. H. Epstein, *Am. J. Physiol.* **246**, C242 (1984).

[3] J. M. Russell, *J. Gen. Physiol.* **81**, 909 (1983).

[3a] S. Liu, R. Jacob, D. Piwnica-Worms, and M. Lieberman, *Am. J. Physiol.* **253**, C721 (1987)

[4] J. A. McRoberts, S. Erlinger, M. J. Rindler, and M. H. Saier, *J. Biol. Chem.* **257**, 2260 (1982).

[5] N. E. Owen and M. L. Prastein, *J. Biol. Chem.* **260**, 1445 (1985).

[6] C. D. A. Brown and H. Murer, *J. Membr. Biol.* **87**, 131 (1985).

[7] K. Amsler, J. J. Donahue, C. W. Slayman, and E. A. Adelberg, *J. Cell. Physiol.* **123**, 257 (1985).

[8] P. Geck, C. Pietrzyk, B. C. Burckhardt, B. Pfeiffer, and E. Heinz, *Biochim. Biophys. Acta* **600**, 432 (1980).

[9] E. K. Hoffmann, C. Sjoholm, and L. O. Simonsen, *J. Membr. Biol.* **76**, 269 (1983).

[10] M. Haas, W. F. Schmidt III, and T. J. McManus, *J. Gen. Physiol.* **80**, 125 (1982).

[11] P. W. Flatman, *J. Physiol.* (*London*) **341**, 545 (1983).

[12] P. K. Lauf, *J. Membr. Biol.* **77**, 57 (1983).

[13] M. Canessa, C. Brugnara, D. Cusi, and D. C. Tosteson, *J. Gen. Physiol.* **87**, 113 (1986).

[14] P. Gmaj and H. Murer, *Physiol. Rev.* **66**, 36 (1986).

the ion specificities of (Na^+ + Cl^-), (K^+ + Cl^-), and (Na^+ + K^+ + Cl^-) cotransport pathways are fairly stringent; rubidium can substitute for potassium, lithium for sodium (though in most cases with lower affinity), and bromide for chloride.[10,15,16] Other cations and anions are usually not transported by these pathways, though some may serve as inhibitors as discussed below. In this chapter, we describe in some detail two methods for the study of cation–chloride cotransport in intact red blood cells which are directed at establishing the stoichiometric relationship between cotransported ions, the effect of cell volume on this stoichiometry and on the magnitude of cotransport, and the electrogenicity or electroneutrality of the cotransport process. It should be noted that these methods, as described, were developed specifically for use in red blood cells in order to take advantage of (or circumvent) certain properties of red cell membranes including their low conductance to sodium and potassium, their relatively high chloride conductance, and the rapid anion-exchange system which also appears to mediate a substantial fraction of the chloride conductance. The experiments presented to illustrate these methods were performed with duck red cells, which exhibit an (Na^+ + K^+ + Cl^-) cotransport that is stimulated either by cell shrinkage[17,18] or β-adrenergic catecholamines,[10,15] and a (K^+ + Cl^-) cotransport system that is stimulated by cell swelling.[19–21] The methods described below have also been successfully employed in studies of red cells of other species that have lower cotransport rates than those of duck red cells. In addition, these methods may be adapted for use in the study of cation–chloride cotransport in other types of cells in suspension, such as cultured cells or Ehrlich ascites cells.

Before presenting the specific methods, it is important to include a brief discussion of how cation fluxes occurring via cation–chloride cotransport are defined. It is common practice to define cotransport as the ouabain-insensitive influx or efflux of sodium and/or potassium that is inhibited by "loop" diuretics such as furosemide or bumetanide. Although in most cases this is quite acceptable, there are several potential problems associated with using diuretic-sensitive fluxes as the sole criterion for cotransport. First, furosemide, at a concentration (1 mM) needed

[15] W. F. Schmidt III and T. J. McManus, *J. Gen. Physiol.* **70**, 81 (1977).

[16] M. Canessa, I. Bize, N. Adragna, and D. Tosteson, *J. Gen. Physiol.* **80**, 149 (1982).

[17] W. F. Schmidt III and T. J. McManus, *J. Gen. Physiol.* **70**, 59 (1977).

[18] F. M. Kregenow, *J. Gen. Physiol.* **58**, 396 (1971).

[19] F. M. Kregenow, *J. Gen. Physiol.* **58**, 372 (1971).

[20] M. Haas and T. J. McManus, *J. Gen. Physiol.* **85**, 649 (1985).

[21] T. J. McManus, M. Haas, L. C. Starke, and C. Y. Lytle, *Ann. N.Y. Acad. Sci.* **456**, 183 (1985).

to fully inhibit most $(Na^+ + K^+ + Cl^-)$, $(Na^+ + Cl^-)$, and $(K^+ + Cl^-)$ cotransport systems also partially inhibits other transport pathways in red cells[22] including sodium-dependent glycine transport and anion exchange (which in the presence of bicarbonate can mediate monovalent cation transport via cation–CO_3 ion pairs[23]). Second, while bumetanide inhibits $(Na^+ + K^+ + Cl^-)$ cotransport systems with high affinity, this is generally not true with $(K^+ + Cl^-)$ systems. Thus, $(K^+ + Cl^-)$ cotransport systems have been occasionally termed "bumetanide insensitive" (e.g., Ref. 24) on the basis of the finding that they are not significantly inhibited by 10^{-5} M bumetanide (a concentration which fully inhibits $(Na^+ + K^+ + Cl^-)$ cotransport under physiological ionic conditions[22]), when in fact higher concentrations of bumetanide (1 mM) do inhibit $(K^+ + Cl^-)$ cotransport (e.g., Refs. 25 and 25a). Third, the inhibitory potency of bumetanide and furosemide toward $(Na^+ + K^+ + Cl^-)$ and $(Na^+ + Cl^-)$ cotransport, and in at least one case $(K^+ + Cl^-)$ cotransport as well, is dependent on the ionic composition of the extracellular medium. Bumetanide inhibits $(Na^+ + K^+ + 2Cl^-)$ cotransport in duck red cells by apparent competition for a chloride site,[26] and furosemide likewise competes with chloride in inhibiting $(Na^+ + Cl^-)$ cotransport in toad cornea.[27] In contrast, increasing extracellular sodium, $[Na]_o$, and potassium, $[K]_o$, enhances inhibition of $(Na^+ + K^+ + 2Cl^-)$ cotransport by "loop" diuretics,[22] and binding of radiolabeled bumetanide to duck red cells[28] and membranes from dog kidney outer medulla[29] is likewise enhanced by increasing $[Na]_o$ and $[K]_o$. In the complete absence of $[Na]_o$ and $[K]_o$, 2×10^{-4} M bumetanide (or 1 mM furosemide) is required for complete inhibition of $(Na^+ + K^+ + 2Cl^-)$ cotransport in duck red cells. In the case of $(K^+ + Cl^-)$ cotransport in sheep red cells treated with the sulfhydryl reagent N-ethylamaleimide, furosemide inhibition is also enhanced by increasing $[K]_o$.[12]

An alternative to using diuretic-sensitive fluxes as a measure of cation–chloride cotransport is to measure chloride-dependent cation fluxes, i.e., the difference in ouabain-insensitive sodium or potassium flux in the presence of chloride and in the complete absence of chloride. Most, if not

[22] H. C. Palfrey, P. W. Feit, and P. Greengard, *Am. J. Physiol.* **238,** C139 (1980).
[23] J. Funder and J. O. Wieth, *Acta Physiol. Scand.* **71,** 168 (1967).
[24] C. Brugnara, A. S. Kopin, H. F. Bunn, and D. C. Tosteson, *J. Clin. Invest.* **75,** 1608 (1985).
[25] L. R. Berkowitz and E. P. Orringer, *Am. J. Physiol.* **252,** C300 (1987).
[25a] D. Kaji, *J. Gen. Physiol.* **88,** 719 (1986).
[26] M. Haas and T. J. McManus, *Am. J. Physiol.* **245,** C235 (1983).
[27] J. H. Ludens, *J. Pharmacol. Exp. Ther.* **223,** 25 (1982).
[28] M. Haas and B. Forbush III, *J. Biol. Chem.* **261,** 8434 (1986).
[29] B. Forbush III and H. C. Palfrey, *J. Biol. Chem.* **258,** 11787 (1983).

all cation-chloride cotransport systems are completely inhibited when chloride is totally replaced by a permeant anion other than bromide. Ideally, chloride-dependent fluxes should be equal to furosemide- or bumetanide-sensitive fluxes when a fully inhibitory concentration of diuretic is used, and in avian red cells we find this to be the case when the anions methyl sulfate or methane sulfonate are used to replace chloride. However, relatively lipophilic anions such as nitrate and thiocyanate increase passive, nonspecific fluxes of sodium and potassium across red cell membranes,[23] and thus ouabain-insensitive cation fluxes in the presence of these anions will be higher than in a chloride medium containing a dose of furosemide or bumetanide that fully inhibits cation–chloride cotransport. It is also advisable to avoid the use of nitrate in studies of cation-chloride cotransport for another reason: in several different cell types it appears that nitrate may compete with chloride for sites on $(Na^+ + K^+ + Cl^-)$ cotransport[5,6,28] and $(K^+ + Cl^-)$[30] cotransport pathways.

Stoichiometry of Cotransport: The Zero-Trans Efflux Method

Measurement of net sodium and potassium effluxes into media free of these cations has proved to be a useful way of determining stoichiometric relationships between cotransported cations in red blood cells. This method is particularly applicable to cells such as red cells in which the baseline permeability of the cell membrane to monovalent cations is low, and most of the transport of sodium and potassium occurs via specific mediated pathways. In duck red cells in particular, >50% of sodium and potassium effluxes into isotonic choline or tetramethylammonium (TMA) media are inhibited either by 1 mM furosemide or by replacement of intra- and extracellular chloride with methyl sulfate; this percentage increases when $(Na^+ + K^+ + 2Cl^-)$ cotransport or $(K^+ + Cl^-)$ cotransport is stimulated by shrinking or swelling the cells, respectively (see below). Choline and TMA do not appear to interact with either $(Na^+ + K^+ + 2Cl^-)$ or $(K^+ + Cl^-)$ cotransport systems in a variety of cell types, and are thus the cations of choice for use in the incubation medium in zero-trans efflux experiments. We have used TMA in most of our experiments simply because TMA salts of several monovalent anions other than chloride are available commercially or can be easily prepared from TMA hydroxide and the appropriate acid. Efflux determinations can be made by measuring either cell or medium sodium and potassium contents before and after given time periods of incubation in the zero-trans medium. When the effluxes of sodium and potassium are large, as in duck red cells, measure-

---

[30] P. K. Lauf, *J. Membr. Biol.* **82,** 167 (1984).

ment of cell cation contents gives good results. However, it is important to note that when substantial net salt efflux occurs, the cells will also lose water and shrink during the incubation period. Therefore, cell sodium and potassium levels (and values for net effluxes) should be determined and expressed as total contents (millimoles/kilogram cell solid or millimoles/ liter original cells) rather than as concentrations (millimoles/liter cell water). When the effluxes of sodium and potassium are small, as in human red cells, it is preferable to measure the accumulation of these cations in the extracellular medium, either directly or by isotopic tracer methods using cells preloaded with ^{22}Na, ^{24}Na, ^{42}K, or ^{86}Rb (the latter being a potassium congener). If the hematocrit and cell contents of sodium and potassium (and specific activities of their tracers, if applicable) are known, measurements of sodium and potassium accumulation in the medium will yield values for their effluxes.

An example of a zero-trans efflux experiment using duck red cells is shown in Fig. 1. In this experiment, the cells were loaded using the nystatin method[31] (see Fig. 1 legend) to contain a constant level of potassium and three different sodium contents. They were then incubated 10 min in sodium- and potassium-free TMA chloride media of different osmolalities, with or without 1 mM furosemide (the range of osmolalities employed was different for the three groups of cells in order to bring cells of three different sodium and total salt contents into the same range of cell volumes). In Fig. 1, furosemide-sensitive net sodium efflux ($-\Delta Na_c$) and potassium efflux ($-\Delta K_c$) are plotted as a function of the average cell water content over the 10-min incubation period; this average was used because the cells were progressively shrinking as they lost sodium, potassium, and chloride into the TMA chloride media. The isotonic water content of duck red cells is 1.50 kg H_2O/kg cell solid, or 60% (w/w). In swollen cells with $W_c > 1.5$, the cells lose potassium without a concomitant sodium loss, indicative of the ($K^+ + Cl^-$) cotransport seen in swollen duck red cells.[20,21] This swelling-induced ($K^+ + Cl^-$) cotransport is independent of [Na]$_c$, at least over the range of values tested. In contrast, cell shrinkage ($W_c < 1.5$) stimulates net sodium and potassium efflux with an $Na^+ : K^+$ stoichiometry of 1:1, indicative of the ($Na^+ + K^+ + 2Cl^-$) cotransport activated by shrinkage of duck red cells.[10,17] Increasing cell sodium from 41.1 to 73.6 mmol/kg cell solid increases the level of outward ($Na^+ + K^+ + 2Cl^-$) cotransport; the effect of increasing cell sodium on this process appears to saturate above the latter level.

The zero-trans experiment in Fig. 1 nicely displays the two volume-sensitive cation-chloride cotransport systems in duck red cells, and dem-

[31] A. Cass and M. Dalmark, *Nature (London), New Biol.* **244**, 47 (1973).

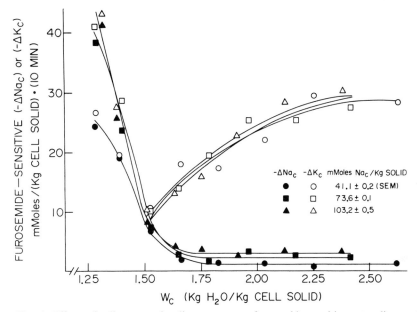

FIG. 1. Effects of cell water and sodium content on furosemide-sensitive net sodium and potassium effluxes from duck red cells into sodium- and potassium-free media. Cells were prepared to contain 202 mmol potassium/kg cell solid and the three listed sodium contents by the nystatin method as follows: Cells were incubated for 1 hr at 0°, 2% hematocrit in media containing 20 μg/ml nystatin, 50 mM sucrose, 100 mM KCl, and either 20, 35, or 50 mM NaCl. This procedure was then repeated by centrifuging and resuspending the cells in the same media without nystatin. Nystatin was eluted from the cells, restoring the membrane to its normal state of low cation permeability, by washing the cells six times at 25°, 2% hematocrit in the same solution as above (without nystatin), but containing 0.1 mM ouabain and 0.25% bovine serum albumin (fraction V). The albumin-containing solutions were titrated to pH 7.4 at 25° with KOH. Test incubations were performed for 10 min at 3% hematocrit, 41° (body temperature for ducks) in media containing varying concentrations of TMA chloride, with and without 1 mM furosemide. The ranges of osmolalities over which the cells were incubated were as follows: 140–350 mOsm/kg (●), 170–380 (■), and 200–410 (▲), all in increments of 30 mOsm/kg. Open symbols represent furosemide-sensitive net potassium efflux ($-\Delta K_c$), closed symbols represent furosemide-sensitive net sodium efflux ($-\Delta Na_c$).

onstrates the 1 : 1 Na$^+$: K$^+$ stoichiometry of the (Na$^+$ + K$^+$ + 2Cl$^-$) cotransport pathway. Whether experiments such as that in Fig. 1 can be routinely used to show cation : chloride stoichiometries of cotransport systems in red cells remains an open question; this is because a small but significant fraction of anion exchange remains even in cells treated with a maximal inhibitory concentration of DIDS, a stilbene derivative that is a potent inhibitor of the red cell anion exchanger. Because chloride fluxes via the anion exchanger are so rapid, even when this exchanger is 99.99% inhibited, the residual chloride fluxes (at 37°) would be expected to be of the same order of magnitude as those occurring via ($Na^+ + K^+ + 2Cl^-$) cotransport in maximally shrunken duck red cells. Lytle[32] has circumvented the anion-exchange problem by performing measurements of net Na^+, K^+, and Cl^- effluxes from duck red blood cells incubated in TMA-gluconate medium. The cells were first treated with DIDS. Under these conditions, bumetanide (1 mM)-sensitive efflux stoichiometries of approximately 1 Na^+:1 K^+:2 Cl^- and 1 K^+:1 Cl^- were obtained in shrunken and swollen cells, respectively. Substitution of impermeant, nonexchangeable anions such as gluconate for extracellular chloride does markedly alter the membrane potential of red blood cells, and is thus best applied in studies of transport processes known to be electroneutral (see below). Treatment of the cells with DIDS is also warranted to minimize changes in intracellular pH resulting from Cl^-/OH^- exchange or ($H^+ + Cl^-$) cotransport via the anion-exchange pathway.[32a] Hall and Ellory[33] have reported experiments in which the stoichiometry of bumetanide-sensitive tracer sodium, potassium, and chloride influxes in ferret red cells (which exhibit bumetanide-sensitive fluxes similar in magnitude to those in duck red cells[11]) was found to be 2 Na^+:1 K^+:3 Cl^-, similar to the cotransport stoichiometry determined in squid axon.[3] In these experiments of Hall and Ellory[33] anion exchange was minimized not only by treating the cells with two inhibitors of this exchange (DIDS plus SITS), but also by lowering the temperature to 21°. The rationale for lowering the temperature in these experiments is that the red cell anion-exchange pathway has a greater sensitivity to temperature (higher Arrhenius activation energy) than does ($Na^+ + K^+ + Cl^-$) cotransport.[33]

A major advantage of the zero-trans efflux method with regard to determination of stoichiometries of cotransported cations is that potential errors due to K^+/K^+ and Na^+/Na^+ exchanges are avoided. Both duck[10,20,32,34] and human[13,34a] red cells exhibit bumetanide-sensitive, chloride-dependent, 1:1 K^+/K^+ and Na^+/Na^+ exchanges, and kinetic analysis of these exchanges has provided strong evidence for their being partial reactions of ($Na^+ + K^+ + 2Cl^-$) cotransport.[32,34,34a] When bumetanide-sensitive tracer influxes of sodium and potassium are measured in either

[32] C. Lytle, Ph.D. thesis, Duke University (1988).
[32a] M. L. Jennings, *J. Membr. Biol.* **40,** 365 (1978).
[33] A. C. Hall and J. C. Ellory, *J. Membr. Biol.* **85,** 205 (1985).
[34] C. Lytle, M. Haas, and T. J. McManus, *Fed. Proc., Fed. Am. Soc. Exp. Biol.* **45,** 548 (1986).
[34a] J. Duhm, *J. Membr. Biol.* **98,** 15 (1987).

duck or human red cells, the potassium influx exceeds that of sodium because the high intracellular potassium concentration of these cells favors K^+/K^+ exchange relative to Na^+/Na^+ exchange. In duck red cells prepared by the nystatin method to contain 150 mM sodium and 10 mM potassium, bumetanide-sensitive sodium influx now exceeds potassium influx.[32,34] Indeed, it is quite possible that the stoichiometry of 2 Na^+ : 1 K^+ : 3 Cl^- reported for bumetanide-sensitive influxes in ferret red cells[33] reflects a component of Na^+/Na^+ exchange, since ferret red cells have very low ouabain-sensitive cation fluxes and thus contain sodium as their major intracellular cation.[11] In zero-trans efflux experiments, K^+/K^+ and Na^+/Na^+ exchanges cannot occur, and the true stoichiometry of (Na^+ + K^+ + Cl^-) cotransport processes is determined.

Electrogenicity vs Electroneutrality of Transport:
  The Valinomycin Method

A decade ago it was shown that partial replacement of extracellular chloride, $[Cl]_o$, by an impermeant anion such as gluconate markedly inhibited furosemide-sensitive net sodium and potassium influx into duck red cells which were shrunken or treated with catecholamines.[35] Because in red cells the membrane potential (E_m) is equal to the Nernst potential for chloride (E_{Cl}), replacement of $[Cl]_o$ by gluconate at constant intracellular chloride, $[Cl]_c$, causes E_m to become more positive. It was thus initially thought that the decrease in (Na^+ + K^+) uptake when $[Cl]_o$ was partially replaced by gluconate was the result of this depolarization.[35] In this model, (Na^+ + K^+) was thought to cross the membrane in an electrogenic manner, with chloride (or presumably, any other permeant anion) following passively to maintain electroneutrality. However, the finding that the coupled, furosemide-sensitive movements of sodium and potassium in duck red cells have a specific anion requirement for chloride (or bromide) and are completely inhibited when chloride is fully replaced by other permeant anions such as nitrate[10,36] led to the consideration that chloride might in fact move across the membrane in an electrically neutral complex with the cations, i.e., (Na^+ + K^+ + $2Cl^-$). In this latter case, the inhibitory effect of partial replacement of $[Cl]_o$ at constant $[Cl]_c$ can be attributed to a change in the direction of the chloride chemical potential gradient, independent of E_m. However, as long as the ratio of $[Cl]_c/[Cl]_o$ affects both E_m and the chloride chemical gradient, as is the case in red

---

[35] W. F. Schmidt III and T. J. McManus, *J. Gen. Physiol.* **70,** 99 (1977).
[36] F. M. Kregenow and T. Caryk, *Physiologist* **22,** 73 (1979).

cells, it is not possible to distinguish between these two possible mechanisms of transport [electrogenic (Na^+ + K^+) cotransport versus electroneutral (Na^+ + K^+ + $2Cl^-$) cotransport].

This problem was resolved by employing the potassium ionophore valinomycin to increase the potassium conductance (P_K) of the membrane sufficiently so that P_K exceeded P_{Cl}. Upon further addition of DIDS, which reduces P_{Cl} in red cells by ~50% (in addition to inhibiting electroneutral anion exchange[37]), P_K sufficiently exceeds P_{Cl} so that E_m can be held constant near the potassium equilibrium potential (E_K) while $[Cl]_o$ is varied at constant $[Cl]_c$.[10] It is actually possible to estimate P_K/P_{Cl} in valinomycin-treated cells by measuring 1- or 2-min potassium effluxes from these cells into NaCl and RbCl media. If sufficient valinomycin is present (10^{-7} M or more in human red cells[38]), the potassium efflux into NaCl gives an estimate of P_{Cl}, since the rate of this efflux is limited only by the conductance of the major permeable anion available to follow potassium out of the cell to maintain electroneutrality.[38] If, however, the external medium contains only RbCl, valinomycin promotes a K^+/Rb^+ exchange which is limited by P_K, since valinomycin preferentially selects rubidium over potassium.[39] Thus, potassium efflux into RbCl gives an estimate of P_K, and P_K/P_{Cl} in valinomycin-treated cells can be estimated by the ratio of K^+ efflux into RbCl divided by K^+ efflux into NaCl (see Table I of Ref. 10 for further details). Knowing P_K/P_{Cl} and assuming P_{Na} is small relative to P_K and P_{Cl} (a safe assumption in valinomycin-treated red cells), E_m can be estimated from the constant-field (Goldman–Hodgkin–Katz) equation.

With E_m held constant near E_K by the use of valinomycin and DIDS, it is possible to vary $[Cl]_o$ (at constant $[Cl]_c$) independently of E_m. Figure 2 shows an experiment in which net changes in cell sodium were measured in DIDS-treated duck red cells as a function of $[Cl]_o$, at constant $[Cl]_c$. Methyl sulfate, which has a conductance about one-half that of chloride in DIDS-treated cells,[10] was used to replace chloride; with DIDS present the potential exchange of internal chloride for external methyl sulfate was inhibited and $[Cl]_c$ remained essentially constant over the 5-min incubation period. The cells were incubated with norepinephrine (1 μM) to stimulate (Na^+ + K^+ + $2Cl^-$) cotransport, in the presence and absence of 2 μM valinomycin and 1 mM furosemide. In the presence of valinomycin, E_m was held constant near E_K (about -8 mV in this experiment), whereas E_m became progressively more negative in the absence of valinomycin as

[37] P. A. Knauf, G. F. Fuhrmann, S. Rothstein, and A. Rothstein, *J. Gen. Physiol.* **69**, 363 (1977).

[38] M. J. Hunter, *J. Physiol. (London)* **268**, 35 (1977).

[39] T. E. Andreoli, M. Tieffenberg, and D. C. Tosteson, *J. Gen. Physiol.* **50**, 2527 (1967).

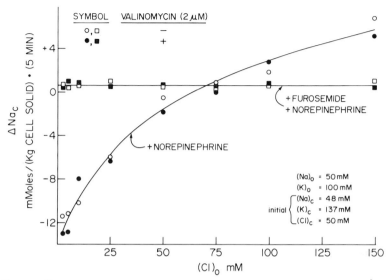

FIG. 2. Effect of the chloride gradient on norepinephrine-stimulated, furosemide-sensitive net sodium movements in duck red cells at constant and variable E_m. Cells were preincubated 8 hr at 4° in a medium containing 80 mM sodium methyl sulfate (MeSO$_4$), 60 mM NaCl, and 20 mM KCl to increase cell sodium and also replace ~50% of cell chloride with MeSO$_4$. Next, an additional preincubation was done for 90 min at 41° (body temperature for ducks) in media of the same ionic composition but also containing 0.1 mM ouabain (to further increase [Na]$_c$) and 10^{-5} M DIDS. Test incubations were done at 41°, 3% hematocrit in media containing 50 mM sodium, 100 mM potassium, 10^{-5} M DIDS, 10^{-6} M norepinephrine, and 0.1 mM ouabain, with and without 2×10^{-6} M valinomycin and 1 mM furosemide. Methyl sulfate substituted for external chloride, maintaining isotonicity. In the presence of valinomycin, P_K/P_{Cl} was estimated to be 7.5 (see text), and E_m (as estimated from the constant-field equation) was constant at -7.9 ± 0.5 mV (SEM, $n = 8$) over the entire range of [Cl]$_o$ from 2.5 to 150 mM. E_m in the absence of valinomycin varied from -0.4 mV at 2.5 mM [Cl]$_o$ to -18.3 mV at 150 mM [Cl]$_o$, as estimated from the constant-field equation with $P_{MeSO_4} = 0.5 P_{Cl}$ (as determined in DIDS-treated duck red cells[10]) and assuming $P_{Cl} \gg P_K, P_{Na}$. (Reproduced from *The Journal of General Physiology*, 1982, volume **80**, pp. 125–147, by copyright permission of the Rockefeller University Press.)

[Cl]$_o$ was increased (see legend for Fig. 2). Figure 2 shows that at low [Cl]$_o$ there was a furosemide-sensitive net loss of sodium, whereas at high [Cl]$_o$, when the chloride gradient was inwardly directed, the cells gained sodium. This effect of [Cl]$_o$ on furosemide-sensitive net sodium movements was identical in the presence and absence of valinomycin, showing that the parameter affecting this sodium transport is the chemical potential gradient of chloride and not E_m.

The experiment shown in Fig. 2 strongly supports the idea that catecholamine-stimulated, furosemide-sensitive sodium transport in duck red cells occurs via an electrically neutral cotransport of $(Na^+ + K^+ + 2Cl^-)$ rather than an electrogenic $(Na^+ + K^+)$ cotransport. Further evidence that this is the case has come from experiments similar to that in Fig. 2, only in which a significant sodium gradient was present and furosemide-sensitive net transport of sodium against this gradient was driven by a chloride gradient at constant E_m (see Fig. 7 of Ref. 10). Finally, two methods can and have been used to confirm that the cotransport pathway is indeed electrically neutral. First, in cells treated with DIDS, the effect of E_m on furosemide-sensitive net sodium transport can be studied by varying $[K]_o$ at constant $[K]_c$ in the presence of valinomycin (varying E_m; $E_m = E_K$) and in its absence (constant E_m; $E_m = E_{Cl}$). When this experiment was done in duck red cells, the same effect of varying $[K]_o$ was seen whether or not valinomycin was present, confirming a lack of effect of E_m on the cotransport pathway and strongly suggesting its electroneutrality.[10] In addition, changes in E_m upon activation or inhibition of transport pathways can be detected using potential-sensitive, fluorescent carbocyanine dyes such as diS-C$_3$[5] (for a detailed study of the use of such dyes in red cells, see Ref. 40). Briefly, these dyes are taken up by red cells upon hyperpolarization of the membrane, and released upon depolarization. Because the fluorescence of the dye is quenched when it is within (or associated with) the cells, changes in dye fluorescence can be used to monitor changes in E_m.[40] This dye method was used to demonstrate the electrogenicity of the Na^+/K^+ pump in red cells, as a depolarization of the membrane (increase in diS-C$_3$[5] fluorescence) was seen upon addition of ouabain.[41] This result is not surprising, considering that the Na^+/K^+ pump transports three sodium ions out of the cell in exchange for two potassium ions.[42] However, in duck red cells incubated in a TMA chloride medium containing 1 mM KCl and 1 mM NaCl [conditions designed to maximize outward net $(Na^+ + K^+ + 2Cl^-)$ cotransport], we found no effect of either adding norepinephrine (to stimulate cotransport) or subsequently increasing $[K]_o$ on diS-C$_3$[5] fluorescence, confirming the electroneutrality of $(Na^+ + K^+ + 2Cl^-)$ cotransport in these cells[21] (also see Ref. 43). Likewise, the $(K^+ + Cl^-)$ cotransport stimulated by cell swelling was found to be electrically neutral using the dye method in duck red cells,[21] as well as

[40] P. J. Sims, A. S. Waggoner, C. H. Wang, and J. F. Hoffman, *Biochemistry* **13**, 3315 (1974).

[41] J. F. Hoffman, J. H. Kaplan, and T. J. Callahan, *Fed. Proc., Fed. Am. Soc. Exp. Biol.* **38**, 2440 (1979).

[42] R. L. Post and P. C. Jolly, *Biochim. Biophys. Acta* **25**, 108 (1957).

[43] F. M. Kregenow, *in* "Membrane Transport in Red Cells" (J. C. Ellory and V. L. Lew, eds.), p. 383. Academic Press, New York, 1977.

in human red cells,[44] where E_m was determined from the ratio of $[H^+]_c/[H^+]_o$ in cells treated with the proton-specific ionophore CCCP and DIDS.[45]

Summary

Two methods have been described for the study of cation–chloride cotransport systems. The zero-trans efflux method is designed to determine stoichiometric relationships between cotransported ions under conditions where ion exchanges cannot occur. These exchanges (e.g., Na^+/Na^+, K^+/K^+) may occur as partial or incomplete reactions of a cotransport process and can lead to erroneous determinations of the stoichiometry of the cotransport process. The zero-trans efflux method can also be used to study the effects of cell volume, pH, and intracellular ion concentrations on cotransport processes. The valinomycin method is used to determine the electrogenicity or electroneutrality of transport, and in this regard can be used in conjunction with other methods such as those employing potential-sensitive dyes or microelectrodes.

Other, more recently developed ionophores with specificity for lithium rather than potassium[46] have now been used to study the effect of E_m on the ATP-dependent Na^+/K^+ pump.[47] It may be possible to use such ionophores to confirm the suspected electroneutrality of $(K^+ + Cl^-)$ cotransport systems, as well as for other studies of specific potassium transport processes in which valinomycin obviously cannot be used.

Both methods discussed in detail in this chapter, and particularly the valinomycin method, were originally devised for use in red blood cells in order to take advantage of (or circumvent) properties of the red cell membrane, such as its low permeability to sodium and potassium and relatively high permeability to chloride. However, valinomycin has been used successfully to demonstrate the electroneutrality of $(Na^+ + K^+ + 2Cl^-)$ cotransport in MDCK cells,[4] and the zero-trans efflux method should be applicable to the study of transport processes in other types of cells in suspension, so long as the transport system being studied can be accurately defined (e.g., as an inhibitor-sensitive or chloride-dependent cation flux) and comprises a significant fraction of the total salt efflux.

[44] C. Brugnara, T. Van Ha, and D. C. Tosteson, *J. Gen. Physiol.* **92,** 42a (1988).
[45] R. I. Macey, J. S. Adorante, and F. W. Orme, *Biochim. Biophys. Acta* **512,** 284 (1978).
[46] R. Margalit and A. Shanzer, *Biochim. Biophys. Acta* **649,** 441 (1981).
[47] R. Goldshlegger, S. J. D. Karlish, A. Rephaeli and W. D. Stein, *J. Physiol. (London)* **387,** 331 (1987).

[17] Sodium–Calcium and Sodium–Proton Exchangers in Red Blood Cells

By JOHN C. PARKER

Introduction

Na^+–Ca^{2+} and Na^+–H^+ exchange are found in many cell types, including muscle, nerve, and epithelia.[1,2] Some species of red cells have these transport pathways[3–11] (Table I).

TABLE I
Na$^+$–Ca^{2+} AND Na$^+$–H$^+$ EXCHANGE IN RED BLOOD CELLS[a]

Species and conditions	Na^+–Ca^{2+} exchange	Na^+–H^+ exchange
Species of red cell		
Human		(3, 4)
Dog	(5–7)	(9–11)
Ferret	(8)	
Rabbit		(12–14)
Amphiuma		(15, 16)
Trout		(17)
Mouse leukemia	(18)	
Activating stimuli		
Cell swelling	(5)	
Cell shrinkage		(10, 13, 15, 16)
Cytoplasmic acidification		(11, 13, 19)
Alkalinization	(5)	
Catecholamines		(17)
Hypoxia		(20)
Lithium in the cytoplasm		(19)
Dimethyl sulfoxide	(18)	
Diamide	(21)	
Nitrate and thiocyanate	(22)	
Lysis and resealing	(23)	
Inhibitors		
Amiloride and its analogs	(18)	(4, 9, 15, 16)
Quinidine and quinine	(5)	(9)
Extracellular acidification		(13, 19)
Extracellular lithium		(19)
Nitrate and thiocyanate		(9, 24)

[a] Reference numbers are given in parentheses.

In red cells, Na$^+$–Ca^{2+} and Na$^+$–H$^+$ exchange do not function at all times but must be activated. A list of stimuli is given in Table I. Because the two pathways act in physiological solutions to correct cell swelling or shrinkage, it is thought that they are part of a volume-regulatory system. In fibroblasts and lymphocytes Na$^+$–H$^+$ exchange is activated by phorbol esters and growth factors,[25] but in red cells these agents do not have straightforward effects on Na$^+$ or proton fluxes. Amiloride and quinidine inhibit both Na$^+$–Ca^{2+} and Na$^+$–H$^+$ exchange.

There are difficulties in the precise assay of both Na$^+$–Ca^{2+} and Na$^+$–H$^+$ exchange in red cells. Neither system can be measured unless it is activated by some stimulus, and the question therefore arises whether a given effect is exerted directly on the transporter or on the activation mechanism.[24] Because the operation of both pathways tends to change cell volume, and because cell volume influences the activity of each system, it is difficult to keep the transporters working at a constant rate while ion movements through them are being measured. Thus, "steady-state flux" data from which one can perform compartmental analysis, or from which kinetic models can be formulated, are difficult to obtain. A partial

[1] P. S. Aronson, *Annu. Rev. Physiol.* **47,** 545 (1985).
[2] K. D. Philipson, *Annu. Rev. Physiol.* **47,** 561 (1985).
[3] N. Escobales and M. Canessa, *J. Biol. Chem.* **260,** 11914 (1985).
[4] N. Escobales and M. Canessa, *J. Membr. Biol.* **90,** 21 (1986).
[5] J. C. Parker, *J. Gen. Physiol.* **71,** 1 (1978).
[6] O. E. Ortiz and R. A. Sjodin, *J. Physiol. (London)* **354,** 287 (1984).
[7] A. A. Altamirano and L. Beauge, *Cell Calcium* **6,** 503 (1985).
[8] M. A. Milanick and J. F. Hoffman, *Ann. N.Y. Acad. Sci.* **488,** 174 (1986).
[9] J. C. Parker, *Am. J. Physiol.* **237,** C324 (1983).
[10] J. C. Parker and V. Castranova, *J. Gen. Physiol.* **84,** 379 (1984).
[11] S. Grinstein and J. D. Smith, *J. Biol. Chem.* **262,** 9088 (1987).
[12] M. L. Jennings, M. Adams-Lackey, and K. W. Cook, *Am. J. Physiol.* **249,** C63 (1985).
[13] M. L. Jennings, S. M. Douglas, and P. E. McAndrew, *Am. J. Physiol.* **251,** C32 (1986).
[14] N. Escobales and M. Canessa, *J. Cell. Physiol.* **132,** 73 (1987).
[15] P. M. Cala, *J. Gen. Physiol.* **76,** 683 (1980).
[16] A. W. Siebens and F. M. Kregenow, *J. Gen. Physiol.* **86,** 527 (1985).
[17] A. Baroin, F. Garcia-Romeau, T. Lamarre, and R. Motais, *J. Physiol. (London)* **356,** 21 (1984).
[18] R. L. Smith, I. G. Macara, R. Levenson, D. Housman, and L. Cantley, *J. Biol. Chem.* **257,** 773 (1982).
[19] J. C. Parker, *J. Gen. Physiol.* **87,** 189 (1986).
[20] R. Motais, F. Garcia-Rameau, and F. Borgese, *J. Gen. Physiol.* **90,** 197 (1987).
[21] J. C. Parker, *Am. J. Physiol.* **253,** C580 (1987).
[22] J. C. Parker, *Am. J. Physiol.* **244,** C318 (1983).
[23] J. C. Parker, *Biochim. Biophys. Acta* **943,** 463 (1988).
[24] J. C. Parker, *J. Gen. Physiol.* **84,** 789 (1984).
[25] S. Grinstein, S. Cohen, J. D. Goetz, and A. Rothstein, *J. Cell. Biol.* **101,** 269 (1985).

solution to this problem may be afforded by the recent demonstration that both Na^+-Ca^{2+} and Na^+-H^+ systems can be locked in the "on" or "off" position by the use of low concentrations of fixatives or cross-linkers.[21,24,26]

Ca^{2+} transport in all species of red cells is dominated by the ATP-dependent Ca^{2+} extrusion pump for which there is no specific inhibitor. Thus, Ca^{2+} fluxes via the Na^+-Ca^{2+} exchanger are best measured in the inward direction. Even so, the action of an unopposed Ca^{2+} pump will confound attempts to judge unidirectional inward calcium movements and thus to obtain reliable estimates of the affinity or stoichiometry of the transporter.[27] A strategy for approaching this problem is suggested by a recently reported technique in which red cells are preloaded with an agent that binds calcium so tightly that the Ca^{2+} pump cannot function.[28] Alternatively, the Ca^{2+} pump can be inactivated by vanadate.[8] In such cells net and unidirectional calcium influx should be equivalent. Another way to study Na^+-Ca^{2+} exchange is to measure Na^+ efflux[5] (see below).

Reliable measurements of Na^+-H^+ exchange are complicated by the tendency of proton gradients across the cell membrane to be dissipated rapidly by the action of band 3-mediated chloride–bicarbonate exchange. This problem can be surmounted only partly by the use of band 3 inhibitors, the best of which leave enough residual activity to influence estimates of unidirectional proton flux.[10]

Because of these difficulties the procedures detailed below yield mainly qualitative data about the presence or absence of the pathways in question. Each transport system is given an "operational definition" based on inhibitor, trans stimulation, or cis competition effects. Thus, $Ca^{2+}-Na^+$ exchange is defined as (1) the portion of Ca^{2+} influx that is inhibited by quinidine or by Na^+ on the cis side, or (2) the portion of Na^+ efflux that is stimulated by external Ca^{2+}. Sodium–proton exchange is defined as (1) amiloride-sensitive Na^+ influx or (2) amiloride-sensitive proton influx. To confirm the presence or absence of either pathway it is important to design studies (using the methods below) showing the trans dependence of the flux in question. Thus Ca^{2+} influx by Na^+-Ca^{2+} exchange and proton influx by Na^+-H^+ exchange should be stimulated in high-Na^+ cells and inhibited in low-Na^+ cells. Dog red cells are naturally high in Na^+, and the red cells of most other species are low in this cation. Sodium in dog red cells can be easily manipulated via the reversible

[26] J. C. Parker and P. S. Glosson, *Am. J. Physiol.* **253**, C60 (1987).

[27] J. C. Parker, *Am. J. Physiol.* **237**, C10 (1979).

[28] V. L. Lew, R. Y. Tsien, C. Miner, and R. M. Bookchin, *Nature* (*London*) **298**, 478 (1982).

permeabilizing effect of external ATP.[29] In red cells of other species Na$^+$ content can be altered by preincubation with nystatin.[30]

In addition to establishing a trans requirement for the countertransported ion there is a further consideration that must be dealt with in the case of Na$^+$–H$^+$ exchange, namely, its electroneutrality. Inasmuch as epithelial tissues contain amiloride-sensitive Na$^+$ and proton channels that are conductive[31,32] neither amiloride-sensitivity nor ion selectivity are adequate criteria for establishing the presence of electroneutral Na$^+$–H$^+$ exchange. This latter can be done, however, by using ionophores to clamp the membrane potential.[10,15,33]

A K$^+$ conductance pathway opens up in red cells of many species when free Ca^{2+} levels in the cytoplasm increase. This may complicate many of the measurements described below, because the large K$^+$ fluxes that develop when the channel is activated may alter cell volume, and change membrane potential. Activation of the Na$^+$–Ca^{2+} exchanger in dog red cells results in large accumulations of Ca^{2+}; if the cells are in an all-K$^+$ medium, they promptly swell and lyse.[34] Thus, when K$^+$ is used as the principal external cation in the procedures below (see proton influx measurements), it is important to include EGTA in the medium to prevent traces of Ca^{2+} from entering the cell and triggering the Ca^{2+}-dependent K$^+$ channel.

Methods

The methods given below are for dog red blood cells. Suitable adaptations for other species can be made, keeping in mind the dependence of inward Ca^{2+} and proton movements on a high cytoplasmic Na$^+$ concentration (see introductory discussion).

Sodium–Calcium Exchange: Calcium Influx[5]

Principle

Calcium influx is measured in circumstances that maximize Na$^+$–Ca^{2+} exchange. Thus, internal Na$^+$ is high, external Na$^+$ is zero, medium pH is

---

[29] J. C. Parker, V. Castranova, and J. M. Goldinger, *J. Gen. Physiol.* **69**, 417 (1977).
[30] J. D. Cavieres, in "Red Cell Membranes: A Methodological Approach" (J. C. Ellory and J. D. Young, eds.), p. 179. Academic Press, New York, 1982.
[31] D. J. Benos, *Am. J. Physiol.* **242**, C131 (1982).
[32] L. G. Palmer, *J. Membr. Biol.* **67**, 91 (1982).
[33] P. M. Cala, *Mol. Physiol.* **4**, 33 (1983).
[34] J. C. Parker, *Am. J. Physiol.* **244**, C313 (1983).

7.40 or higher, and the cells are swollen so that the transporter will be activated. The "stopping solution" uses Li^+ as the principal cation to halt outward Ca^{2+} movements via $Ca^{2+}-Na^+$ exchange.

Media and Reagents

Wash medium: 100 mM LiCl, 10 mM HEPES, 5 mM glucose; pH adjusted to 7.40 at 37° with Tris base

Solution A: Wash medium plus $CaCl_2$ (0–1 mM)

Solution B: Solution A plus quinidine sulfate (0.4 mM)

Solution C: 100 mM NaCl, 10 mM HEPES, 5 mM glucose, 0–1 mM $CaCl_2$; pH 7.40 at 37°

Stopping solution: 150 mM LiCl, 5 mM HEPES, 5 mM EGTA, pH 7.5 at room temperature. Cool in an ice bath

Detergent: Acationox detergent (Scientific Products, McGaw Park, IL)

Procedure

Wash red cells from heparinized, freshly drawn dog blood in 10–20 vol of wash medium three times at room temperature by suspension and centrifugation. Suspend cells in 10–20 vol of solutions A–C, place at 37° in a shaker bath, and at time zero add ^{45}Ca (1 μCi/ml suspension) to each flask. At intervals thereafter remove aliquots of the suspension and mix with 2–3 vol of ice-cold stopping solution. At the end of the flux period (30–60 min) centrifuge all samples at 2–4°, remove some supernatant for counting, and discard the remainder. Resuspend cells in enough stopping solution to transfer them to a suitable centrifuge tube for cell packing, centrifuge the cells at 25,000 g at 2–4°, remove all of the supernatant, and weigh aliquots of packed cells (200–300 mg) into tared vials. Solubilize the cells with 5.0 ml distilled water plus a drop of detergent. To 4.0 ml of this hemolysate add 1.0 ml 7% perchloric acid. Add 1.0 ml of this to 9.0 ml scintillation fluid. Count the incubation medium by adding 0.25 ml to 4.0 ml distilled water plus one drop of detergent and 1 ml 7% perchloric acid. Add 1.0 ml of this to 9.0 ml scintillation fluid. The amount of calcium that enters the cells (mmol/kg wet weight) is equal to the number of counts in the cells divided by the specific activity of the medium. Sodium–calcium exchange is operationally defined as the calcium influx measured in solution A minus the calcium influx measured in solutions B or C. There is considerable variation in the amount of Na^+-Ca^{2+} exchange in red cells from different dogs.[23]

To use this assay for Na^+-Ca^{2+} exchange in red cells from most other species (nondog) the cytoplasmic Na^+ content must be raised by preincubation of the cells with ouabain at 4° or by reversible permeabilization, e.g., with nystatin.[30]

Sodium–Calcium Exchange: Sodium Efflux[5]

Principle

Sodium efflux is measured in circumstances that maximize Na–Ca exchange (see preceding method). Loading of dog red cells with ^{22}Na can be achieved rapidly by taking advantage of the reversible permeabilizing action of external ATP.[29] Red cells of other species can be loaded with ^{22}Na (10 μCi/ml suspension) by incubation for 2–4 hr in physiologic saline media at 37°. In red cells that have an Na$^+$–K$^+$ pump, baseline Na$^+$ efflux should be reduced by adding 0.1 mM ouabain to all flux media. Na–Ca exchange is operationally defined as the increment in Na$^+$ efflux seen when Ca^{2+} is added to the external medium.

Media and Reagents

Wash medium: 100 mM LiCl, 10 mM HEPES, 5 mM glucose; pH 7.40 at 37° with Tris base

Loading solution (for dog red cells): 140 mM NaCl or KCl, 10 mM HEPES, 5 mM glucose, 1.0 mM ATP; pH 7.5 at room temperature with Tris base·

Solution A: 100 mM LiCl, 10 mM HEPES, 5 mM glucose, 0.1 mM free EDTA, pH 7.40 at 37° with Tris base

Solution B: Solution A plus CaCl$_2$ (0–1 mM)

Procedure

Wash red cells from heparinized, freshly drawn dog blood in 10–20 vol of wash medium three times at room temperature by suspension and centrifugation. Resuspend the cells in 2–3 vol of loading solution, add ^{22}Na (0.5–1.0 μCi/ml suspension), and incubate 30–45 min at 37° in a shaker bath. Centrifuge the radioactive suspension at 2–4° and wash the cells four times with 20 vol of ice-cold wash medium. Remove sufficient supernatant after the last wash so the remaining suspension is 50–75% cells. Pipette aliquots of this well-mixed cell concentrate into 30–50 vol of prewarmed solutions A and B at 37°. At 10 to 15-minute intervals remove samples from the dilute suspension and rapidly separate cells from supernatant by centrifugation. Discard the cells. Prepare a hemolysate of suspensions A and B by adding a drop of detergent to them after the last sample has been taken. Measure radioactivity and hemoglobin (OD at 540 nm) in each supernatant sample and in the hemolysed whole suspension. Calculate percentage hemolysis (should be less than 2%) and subtract it from the percentage counts released. Plot $-\ln(1$ − fraction of counts released) versus time for first order efflux rate constant. To obtain flux multiply the rate constant by the internal Na^+ concentration. Sodium efflux through Na^+-Ca^{2+} exchange is operationally defined as the increment in Na^+ efflux that is seen when Ca^{2+} is added to the efflux medium.

Sodium–Proton Exchange: Sodium Influx[19]

Principle

Sodium influx is measured at several cell volumes to demonstrate the activation of the transporter by cell shrinkage. Sodium–proton exchange is operationally defined as the proportion of Na^+ influx that is inhibited by amiloride.

Media and Reagents

Wash medium: 150 mM NaCl, 10 mM HEPES, 5 mM glucose; pH 7.40 at 37° with Tris base

Incubation solutions	Wash medium plus:
Solution A	No additions
Solution B	NaCl or KCl (30 mM)
Solution C	NaCl or KCl (60 mM)

Solution D: Solution A plus 0.4 mM amiloride
Solution E: Solution B plus 0.4 mM amiloride
Solution F: Solution C plus 0.4 mM amiloride

Procedure

Wash red cells from heparinized, freshly drawn dog blood in 10–20 vol of wash medium three times at room temperature and divide into six equal portions labeled A–F. Wash each aliquot of cells once at room temperature with its corresponding incubation solution and resuspend each in the same solution at a cell/medium ratio of 1/10. Place the cell suspensions at 37° in a shaker bath and after 2–5 min add ^{22}Na (1 μCi/ml suspension) to each flask. At 15 to 30-min intervals thereafter centrifuge aliquots of each suspension at 25,000 g and 2–4°. Save a sample of each incubation medium for analysis and discard the rest, cleaning all supernatant from the top of the cell pellet. Weigh aliquots of the packed cell pellets into tared vials, dilute one aliquot of each to 5–10 ml with distilled water plus a drop of detergent. Determine radioactivity and Na⁺ concentration of the resulting hemolysates and supernatants. Sodium influx is equal to the rate of increase in cell counts divided by the specific activity of the medium. Sodium–proton exchange is defined operationally as the influx in each control flask (A, B, C) minus the influx in its amiloride-containing counterpart (D, E, F, respectively).

Sodium–Proton Exchange: Proton Influx[9,10]

Principle

Cells are introduced into an unbuffered medium under conditions (cell shrinkage) that will activate the Na⁺–H⁺ exchanger. A large driving force for outward Na⁺ movement is created by placing the (high Na⁺) cells in a zero-Na⁺ medium containing K⁺ as the principle cation. The tendency of the proton gradient developed by the Na⁺–H⁺ exchanger to be dissipated by chloride–bicarbonate countertransport is opposed by the inclusion in the flux medium of dipyridamole, an anion transport inhibitor. The change in medium pH can be recorded, or, alternatively, one can add protons or hydroxyl ions to the suspension so as to maintain the pH at 7.40, thus allowing a quantitative measurement of proton flux. Sodium–proton exchange is operationally defined as the proton influx that is inhibited by amiloride. A procedure very similar to this measures proton efflux from low-Na⁺ cells into a high-Na⁺ medium.[10]

Media and Reagents

Buffered wash medium: 150 mM NaCl, 10 mM HEPES, 0.1 mM K_2HCO_3, 5 mM glucose, pH 7.40 at 37°
Unbuffered wash medium: 150 mM NaCl
Solution A: 210 mM KCl, 0.1 mM EGTA, 0.05 mM dipyridamole (in DMSO, final DMSO concentration 0.4 vol%)
Solution B: Solution A plus amiloride (0.2 mM)

Procedure

Wash red cells from heparinized, freshly drawn dog blood at room temperature three times in 10–20 vol of buffered wash medium and then three times in unbuffered wash medium. One milliliter of freshly washed cells is introduced into 50 ml of 37° solution A or B in a thermostatted 37° beaker containing a pH electrode. Suspension pH is recorded as a function of time, or, alternatively, 0.020 ml of 40 mM HCl or KOH can be added as needed every 10 sec to keep the pH at 7.40–7.45. Proton influx is estimated qualitatively from the rise in pH as a function of time or quantitatively by the amount of HCl that has to be added to the suspension to maintain the pH within physiologic limits. Sodium–proton exchange is operationally defined as the difference in proton influx between solutions A and B.

Acknowledgment

This work was supported by NIH Grant AM 11357 from the United States Public Health Service.

[18] Monocarboxylate Transport in Red Blood Cells: Kinetics and Chemical Modification

By Bernhard Deuticke

Introduction

Conclusive evidence has been accumulated in recent years that anion transport across the red blood cell membrane is not exclusively mediated by the transport system located in the major intrinsic protein, band 3. At least one group of anions, namely the short-chain aliphatic monocarboxylates, penetrate by a system differing from this "classical" anion-exchange system[1–15] (for a review, see Deuticke[16]). The physiologically most

[1] A. P. Halestrap and R. M. Denton, *Biochem. J.* **138,** 313 (1974).
[2] A. P. Halestrap, *Biochem. J.* **156,** 193 (1976).
[3] B. Deuticke, I. Rickert, and E. Beyer, *Biochim. Biophys. Acta* **507,** 137 (1978).
[4] D. R. Leeks and A. P. Halestrap, *Biochem. Soc. Trans.* **6,** 1363 (1978).
[5] D. R. Leeks and A. P. Halestrap, *in* "Function and Molecular Aspects of Biomembrane Transport" (E. Quagliariello *et al.,* eds.), p. 427. Elsevier/North-Holland, Amsterdam, 1979.
[6] W. P. Dubinsky and E. Racker, *J. Membr. Biol.* **44,** 25 (1978).
[7] B. Deuticke, E. Beyer, and B. Forst, *Biochim. Biophys. Acta* **684,** 96 (1982).
[8] B. Deuticke, *Alfred Benzon Symp.* **14,** 539 (1980).
[9] M. L. Jennings and M. Adams-Lackey, *J. Biol. Chem.* **257,** 12866 (1982).
[10] J. A. Donovan and M. L. Jennings, *Biochemistry* **24,** 561 (1985).

relevant members of this group are lactate and pyruvate. Some of their homologs, such as β-hydroxybutyrate,[13-15] acetoacetate, and acetate, deserve interest because their translocation by similar systems in other membranes may bear functional relevance.[17-22]

Major evidence for an independent monocarboxylate system is outlined as follows:

1. Monocarboxylate transport differs in its kinetic properties from the transport of small inorganic anions such as Cl^- or HCO_3^-. The transport of these latter anions almost exclusively occurs by an obligatory anion exchange which accounts for electroneutrality.[23,24] The monocarboxylate system can also mediate anion net movements at a considerable rate,[7] electroneutrality being maintained by cotransported H^+.[4,6,9]

2. Monocarboxylate movements are inhibited by covalently bound and noncovalent inhibitors which do not affect, or have very little effect on, inorganic anion exchange.[3,4,10,25,25a] On the other hand, some of the most potent inhibitors of inorganic anion exchange have almost no effect on monocarboxylate transport.[2,3,7]

3. While the classical anion-exchange system also accepts and translocates monocarboxylates,[2,7] no transport of inorganic anions by the monocarboxylate carrier could yet be demonstrated.[7,16]

Detailed analysis of monocarboxylate transport is still not complete. Compared to the transport of inorganic anions, this analysis is complicated by problems arising from the fact that (1) the physiologically rele-

[11] A. W. De Bruijne, H. Vreeburg, and J. Van Steveninck, *Biochim. Biophys. Acta* **732**, 562 (1983).
[12] A. W. De Bruijne, H. Vreeburg, and J. Van Steveninck, *Biochim. Biophys. Acta.* **812**, 841 (1985).
[13] B. L. Andersen, H. L. Tarpley, and D. M. Regen, *Biochim. Biophys. Acta* **508**, 525 (1978).
[14] D. M. Regen and H. L. Tarpley, *Biochim. Biophys. Acta* **508**, 539 (1978).
[15] D. M. Regen and H. L. Tarpley, *Biochim. Biophys. Acta* **601**, 500 (1980).
[16] B. Deuticke, *J. Membr. Biol.* **70**, 89 (1982).
[17] A. P. Halestrap, *Biochem. J.* **148**, 85 (1975).
[18] A. P. Halestrap, *Biochem. J.* **172**, 377 (1978).
[19] T. L. Spencer and A. L. Lehninger, *Biochem. J.* **154**, 405 (1976).
[20] G. E. Mann, B. V. Zlokovic, and D. L. Yudilevich, *Biochim. Biophys. Acta* **819**, 241 (1985).
[21] J. P. Monson, J. A. Smith, R. D. Cohen, and R. A. Iles, *Clin. Sci.* **62**, 411 (1982).
[22] P. Fafournoux, C. Demigne, and C. Remesy, *J. Biol. Chem.* **260**, 292 (1985).
[23] P. A. Knauf, *Curr. Top. Membr. Transp.* **12**, 249 (1979).
[24] H. Passow, *Rev. Physiol. Biochem. Pharmacol.* **103**, 61 (1986).
[25] J. H. Johnson, J. A. Belt, W. P. Dubinsky, A. Zimniak, and E. Racker, *Biochemistry* **19**, 3836 (1980).
[25a] J. D. Donovan and M. L. Jennings, *Biochemistry* **25**, 2538 (1986).

vant substrates of the system undergo metabolic conversion in the cell, and (2) transport not only occurs by the two parallel transport systems, but also by nonionic diffusion of the undissociated acid.[7,26,27] These three pathways contribute to a varying extent depending on the experimental conditions.

On the other hand, certain aspects, e.g., the structure of the active (substrate-binding) site, can be assessed with greater ease due to the availability of a large number of substrate analogs and of possibilities to study their transport parameters by indirect means without the requirement of radioactive tracers.

The most relevant properties of monocarboxylate transport via the specific transport system can be categorized within the framework of a description of the experimental procedures by which they can be investigated. Transport of monocarboxylates via the classical inorganic anion-exchange system will not be treated in detail. For procedures that measure transport via this system the reader is referred to the relevant contributions elsewhere in this volume.[28,29] Available results have also been summarized[16] and approaches to the study of nonionic diffusion of monocarboxylic acids in red cells have been discussed in a review,[16] in which further references are compiled. A more general treatise has been published.[30]

Measurement of Tracer Fluxes

Measurements of tracer fluxes generally follow regimes originally developed for the study of anion homoexchange via the classical exchange system.[28,29,31] The following procedures have proved useful in our studies.

Substrates. L-Lactate is available as the sodium (purity $> 99\%$, Fluka) or lithium salt ($> 99\%$, Calbiochem) and D-lactate as the lithium salt (95–97%, Sigma). Sodium glycolate has to be prepared as a stock solution from commercial glycolic acid (99%, Fluka) by titration to neutrality with sodium hydroxide and is stored frozen at $-30°$.

Labeled lactates are available at a purity >98%. ^{14}C-Labeled sodium pyruvate poses problems due to decomposition and condensation reactions. Storage as the free acid at $-20°$ has been claimed to improve the stability.[32] If possible, the purity of the labeled compounds should be

[26] B. Deuticke, *Alfred Benzon Symp.* **4,** 307 (1972).
[27] L. Aubert and R. Motais, *J. Physiol. (London)* **246,** 159 (1975).
[28] R. B. Gunn and O. Fröhlich, this volume [3].
[29] J. Brahm, this volume [9].
[30] A. Walter and J. Gutknecht, *J. Membr. Biol.* **77,** 255 (1984).
[31] K. F. Schnell, S. Gerhardt, and A. Schöppe-Fredenburg, *J. Membr. Biol.* **30,** 319 (1977).
[32] R. W. von Korff, *Anal. Biochem.* **8,** 171 (1964).

checked by thin-layer chromatography (TLC), e.g., by the system devised by Hansen[33] and subjected to radiochromatogram scanning.

Incubation System. Human blood obtained by venipuncture is anticoagulated by heparin (0.2 USP units/ml). Erythrocytes are isolated by centrifugation at room temperature (6 min, 5500 g) in a clinical centrifuge with swing buckets. Plasma and "buffy coat," consisting of platelets, leukocytes, and chylomicrons, are removed by aspiration. The cells are washed three times by resuspension in 3 vol of 140 mM NaCl plus 7.5 mM Na$_2$HPO$_4$/NaH$_2$PO$_4$, pH 7.4, and centrifugation. If complete removal of endogeneous lactate and pyruvate is desired, an additional preincubation of the cells at elevated pH (pH 8–8.2) is helpful.[11]

The packed cells are resuspended in 20 vol of a medium of the following composition (concentrations in mM): NaCl, 140; Na$_2$HPO$_4$/NaH$_2$PO$_4$, 6.5; sodium L-lactate (or any other monocarboxylate to be investigated), 5.4; pH 7.4.

Addition of glucose is only necessary when long incubations (>60–90 min) are intended. When experimental membrane modifications are expected to make membranes leaky to alkali cations (and Cl$^-$) it is advisable to use media of the following composition (mM): KCl, 100; NaCl, 36; Na$_2$HPO$_4$/NaH$_2$PO$_4$, 6.5; L-lactate or any other monocarboxylate to be tested, 5.4; sucrose, 44; gramicidin D, 5 μg/ml. In such media cation leaks induced by membrane modification will not lead to uncontrollable cation shifts, since cations are already at equilibrium due to the choice of the medium and the high alkali cation permeability induced by the channel-forming antibiotic. Colloid-osmotic swelling is prevented by sucrose. The concentrations of sucrose required for any given situation of ion composition, pH, and temperature may in principle be estimated from the careful work of Freedman and Hoffman[34] or of Dalmark.[35]

Isotope Preloading. Equilibrium exchange is easily measured following the rationale and procedure outlined in detail by Gardos *et al.*[36] We use their procedure with some modifications: Erythrocyte suspensions (hematocrit 5%) are first incubated for time periods sufficient to adjust the pH and to bring all readily permeable components to equilibrium. The cells are then spun down (10 min, 5500 g) and part of the supernatant is removed to elevate the hematocrit to about 30–50%. The supernatant removed (preloading medium), which is at full equilibrium with the cell

[33] S. A. Hansen, *J. Chromatogr.* **124**, 123 (1976).
[34] J. C. Freedman and J. F. Hoffmann, *J. Gen. Physiol.* **74**, 157 (1979).
[35] M. Dalmark, *J. Physiol.* (*London*) **250**, 65 (1975).
[36] G. Gardos, J. F. Hoffman, and H. Passow, in "Laboratory Techniques in Membrane Biophysics" (H. Passow and R. Stämpfli, eds.), p. 9. Springer-Verlag, Heidelberg, Federal Republic of Germany, 1969.

content, is kept at 0° and later serves as the suspension medium during the tracer efflux period.

The concentrated cell suspension is incubated with labeled substrate (usually 0.1 μCi/ml of suspension) at 37°. After equilibration of the tracer the suspension is centrifuged (10 min, 5500 g) and the supernatant removed. When no appreciable loss of tracer is likely to occur, the cells are washed once in preloading medium at 0°.

Efflux Measurement. Tracer efflux is started by injecting 1 ml of packed cells into 19 ml of preloading medium. When no particular temperature is required the following temperatures will allow sampling at suitable intervals: L- and D-lactate, 10° and 20°; glycolate, 10°; acetate, 0°. The suspensions are incubated deeply immersed in a gently shaking water bath. Samples (1.2 ml) are removed at intervals and centrifuged (20 sec, 12,000 g). Supernatant (0.8–0.9 ml) is removed and added to 50 μl of concentrated $HClO_4$ in a 1.5-ml reaction vessel in order to remove traces of protein. When high rates of flux are to be expected, stop techniques may become necessary. Besides rapid cooling to 0° or dilution by ice-cold medium, the following inhibitors have proved useful for this purpose: *p*-chloromercuribenzene sulfonate (PCMBS) (0.1 mM), phloretin (0.3 mM) alone or in combination with isobutylmethyl xanthine (3–5 mM), or α-cyano-4-hydroxycinnamate (2.5 mM). Whether stilbene disulfonates should be used in addition depends on the fractional contribution of band 3 to the flux.

Because of the low hematocrit of the suspension, the extracellular content of tracer after attainment of tracer equilibrium will be essentially equal to the tracer content of the whole suspension. An aliquot of the whole suspension, treated in the same way with $HClO_4$ as the supernatants, therefore provides the equilibrium value.

Evaluation Procedure. The equilibrium exchange rate coefficient k^{ee} is obtained under the assumption of a closed two-compartment system by plotting $\ln(1 - cpm_t/cmp_\infty)$ against time, according to Eq. (1):

$$\ln(1 - cpm_t/cpm_\infty) = -k^{ee}t \tag{1}$$

where cpm_∞ and cpm_t denote the radioactivities in a standard volume of supernatant, or total suspension, at various times (t) and after attainment of equilibrium (∞).

The assumption of a closed two-compartment system is verified by demonstrating linearity of the resulting graph. In case of the equilibrium exchange of the important monocarboxylates (lactate, glycolate) linearity prevails up to at least 90% tracer equilibration. The rate coefficient is best obtained from a least-squares fit of the data.

The unidirectional flux J^{ee} is calculated by Eq. (2):

$$J^{ee} = k^{ee}[C_iV_iC_eV_e/(C_iV_i + C_eV_e)](1/A_c) \qquad (2)$$

C_i and C_e are the intra- and extracellular concentrations (moles/liter H_2O), V_i and V_e are the respective aqueous spaces (liters H_2O/standard volume of cells). A_c is the cell surface area. In case of a low hematocrit, where $C_iV_i \ll C_eV_e$, Eq. (2) simplifies to

$$J^{ee} = k^{ee}C_iV_i/A_c \qquad (3)$$

V_i/A_c can be equalled to the ratio $(V_i^n/A_c)(1 - DW_a^{fr}/1 - DW_n^{fr})$ where V_i^n is the mean aqueous space of the cell in native state, DW_a^{fr}, the actual fractional dry weight, and DW_n^{fr}, the fractional dry weight in the native state.

The actual dry weight is determined by drying a carefully weighed sample of the pelleted cells at the end of the flux measurement to constant dry weight, usually at 90–95°. Appropriate values for the other parameters are well established for the human erythrocyte: V_i^n may be calculated as mean cellular volume (MCV)(1 − fractional dry weight in the native state). This may be approximated as 90×10^{-12} cm$^3 \times 0.67 = 60.3 \times 10^{-12}$ cm^3 (at pH 7.4). Values of $130–140 \times 10^{-18}$ cm^2 for A_c seem adequate.[37,38] For other mammalian species approximate value for MCV and A_c have been listed,[37,38] and values for the actual and the native fractional H_2O content should be determined in each case.

The intracellular concentration of the substrate can usually be obtained from the extracellular concentration and the distribution ratio of the tracer. Under equilibrium conditions this ratio should be equal to the chloride distribution ratio, which depends on numerous variables, of which pH, temperature, and organic phosphate concentrations are the most important ones. In many cases it may therefore be sufficient to use published values,[35,39,40] or to derive the values from the equations given by Freedman and Hoffman.[34] If reliable estimates are impossible, the distribution ratio has to be measured. A very convenient double-labeling procedure has been described by Schnell and Besl.[41] Gross deviations of the distribution ratios from the chloride ratio may indicate adsorption of the tested anion to intracellular constituents which then must be determined.

[37] W. Gruber and B. Deuticke, *J. Membr. Biol.* **13**, 19 (1973).
[38] L. Sewchand and P. B. Canham, *Can. J. Physiol. Pharmacol.* **54**, 437 (1976).
[39] J. Funder and J. O. Wieth, *Acta Physiol. Scand.* **68**, 234 (1966).
[40] J. Duhm, *Pfluegers Arch. Gesamte Physiol. Menschen Tiere* **326**, 341 (1971).
[41] K. F. Schnell and E. Besl, *Pfluegers Arch.* **402**, 197 (1984).

Fluxes calculated by Eq. (3) have the dimension of mol $\times$ cm^{-2} $\times$ sec^{-1}. They can be transformed into mol/ml cells $\times$ min by considering that 1 ml of human erythrocytes consists of 1.1×10^{10} cells under normal conditions and has a total surface area of 1.51×10^4 cm^2.

Equilibrium exchange fluxes may also be derived from tracer influx measurements. In these cases initial rates are usually determined and stopping techniques (cooling[14] or inhibitor stop) or fast separation procedures[11] will be indispensable.

Interference of Metabolic Reactions

Problems in measuring the tracer efflux of monocarboxylates may arise from their metabolic (enzymatic) conversion, which will produce labeled species other than the added one. Since pyruvate and L-lactate are not further metabolized in glycolyzing nonnucleated erythrocytes, their enzymatic interconversion by lactate dehydrogenase (LDH) is the major metabolic complication. The enzyme will promote net reduction of α-oxo acids in dependence on the available reducing NADH equivalents, but also induce interconversion of the two acids by isotope exchange.[42] This obstacle might in principle be overcome by (1) using "normal"[43] or even white[44] resealed ghosts, (2) removing NADH and NAD from the cells, and (3) inhibiting LDH. While possibility (1) is feasible (although not yet tested thoroughly[3]), appropriate realizations of possibilities (2) and (3) are presently not at hand (see Holbrook et al.[42] for a review on LDH and its experimental modification).

Separation of the Parallel Pathways of Monocarboxylate Transfer

Equilibrium exchange fluxes obtained as described above cannot directly be used for characterizing the monocarboxylate transporter since they are composed of three components contributing to a varying extent.[3,7] Besides transport by the monocarboxylate transporter, transport via the anion-exchange system and by nonionic diffusion has to be considered. Separation of three parallel transport systems by kinetic means is almost impossible, but inhibitors are available for a rather clear-cut discrimination.

Selective Blockage of the Monocarboxylate Transporter. The component mediated by the monocarboxylate transporter can be distinguished by its sensitivity to certain SH reagents, particularly mercurials or,

[42] J. J. Holbrook, A. Liljas, S. J. Steindel, and M. G. Rossman, *in* "The Enzymes" (P. Boyer, ed.), Vol. 11, 3rd Ed., p. 191. Academic Press, New York, 1975.
[43] G. Schwoch and H. Passow, *Mol. Cell. Biochem.* **2**, 197 (1973).
[44] P. G. Wood, this volume [20].

TABLE I
COVALENT OR TIGHTLY BOUND INHIBITORS FOR SELECTIVE INHIBITION OF
MONOCARBOXYLATE TRANSPORT BY THE MONOCARBOXYLATE TRANSPORTER[a]

Inhibitor	Concentration	Conditions, comments
PCMBS,	10–100 μM, or	Present during efflux measurement (Hct 5%)
PCMB	50 nmol/ml cells	Pretreatment not required unless flux very fast (full inhibition within 10 sec)
HgCl$_2$	50–100 μM, or	Present during efflux measurement (Hct 5%)
	1 μmol/ml cells	At higher concentrations, and at lower hematocrit, unspecific flux acceleration by leak formation may occur
DTNB	2–4 mM	pH 8.2, 120 min, 37°, subsequent washing advisable, since DTNB is a noncovalent inhibitor of band 3-mediated anion transport[b]

[a] Details given in Deuticke et al.[3,7]
[b] From R. A. F. Reithmeier, Biochim. Biophys. Acta 732, 122 (1983).

slightly less reliable, impermeable dithiol reagents, e.g., 5,5′-dithiobis(2-nitrobenzoate) (DTNB).[7] Other reagents [N-ethylmaleimide, 4-chloro-7-nitrobenzene-2-oxa-1,3-diazole (NBD-chloride), 4,4′-dithiodipyridine (DTDP)] are not sufficiently specific.[7] As yet there is only one noncovalent inhibitor (isobutylmethylxanthine) suitable for this purpose.[16] α-Cyanocinnamates which are sometimes referred to as "specific" inhibitors of monocarboxylate transfer cannot be used in erythrocytes for this purpose, since they are unselective inhibitors of both anion-transport systems.[2,3,16] Lactate equilibrium exchange is inhibited with an I_{50} of 0.15 mM (10°), inorganic anion equilibrium exchange at an I_{50} of >5 mM.[45] In all cases of a discrimination between parallel pathways by inhibitors the appropriate concentration of inhibitor should be verified by dose–response curves, obtained for at least two inhibitors of the component to be studied.

The flux mediated by the specific system (MC) is obtained from $k_{total} - k_{inhibitor}$. Details are summarized in Table I. Whenever possible, PCMBS should be used. Under the conditions of full inhibition of monocarboxylate transport (0.04 μmol/ml cells) the enhancement of potassium leak flux by this mercurial is only negligible.[46,47] PCMBS should be prepared as a fresh stock solution daily. To obtain complete dissolution dissolve 10 mg PCMBS in 1 ml 0.04 N NaOH and add 1 ml 125 mM sodium phosphate

[45] B. Deuticke, unpublished results.
[46] R. M. Sutherland, A. Rothstein, and R. I. Weed, J. Cell. Physiol. 69, 185 (1967).
[47] B. Deuticke, Membr. Biochem. 6, 309 (1986).

buffer. Stock solutions of DTNB may be 1 M, dissolved in pure dimethyl-formamide, and should be freshly prepared.

Blockage of Monocarboxylate Transport via the Classical Pathway. The component of monocarboxylate transport mediated by the classical anion-exchange system is in principle accessible by using selective inhibitors of band 3-mediated transport. An ideal inhibitor with this specification is not yet available; most of the well-established[48] inhibitors of anion exchange (noncovalent as well as covalent) also inhibit[7] the monocarboxylate transporter (see below).

For most purposes, stilbene disulfonates are appropriate. They produce full inhibition of band 3-mediated transport at concentrations at which essentially no effect on the monocarboxylate transporter is observed.[8,9] When the component mediated by band 3 is small it may not be easy to obtain this component from the difference of the total and the stilbene disulfonate-inhibited flux. It is then more appropriate to calculate the difference between the PCMBS-insensitive component and the (PCMBS + stilbene disulfonate)-insensitive component of the monocarboxylate flux.

Stilbene disulfonate inhibitors which have been well tested are 4,4'-diisothiocyanostilbene 2,2'-disulfonate (DIDS) and 4,4'-dinitrostilbene 2,2'-disulfonate (DNDS) (Table II). As a further disulfonate inhibitor, tetrathionate has proved adequate.[7] At neutral pH this inorganic anion produces reversible specific inhibition of band 3-mediated transport. Only prolonged treatment of cells with tetrathionate at alkaline pH induces irreversible inhibition of the monocarboxylate *and* the inorganic anion-exchange transport system due to a modification of SH groups.[45]

Contribution of Nonionic Diffusion. The component of simple non-ionic diffusion of the undissociated acid (usually assumed to occur via the lipid domain of the membrane[48]), can be separated from the mediated components by combined treatment of the cells with the inhibitors specified above.[7] In addition, a considerable number of less specific noncovalent inhibitors may be used which suppress both systems simultaneously (cf. Table III). The residual flux in the presence of these inhibitors is independent of the type of inhibitor and behaves like a process involving the membrane lipid domain.[7]

Quantitative Aspects of the Three Pathways. Contributions of the three pathways have been determined in a few instances. Data for human erythrocytes and a few other species are summarized in Table IV. These data are valid only for the indicated conditions of temperature, pH, and

---

[48] B. Deuticke, *Rev. Physiol. Biochem. Pharmacol.* **78**, 1 (1977).

TABLE II
INHIBITION OF THE COMPONENT OF MONOCARBOXYLATE TRANSPORT MEDIATED BY
BAND 3 PROTEIN[a]

Inhibitor	Concentrations	Conditions, comments
DIDS	$1.5–2 \times 10^6$ molecules/cell or	Pretreated 30 min at 37°, pH 7.4
	25–35 nmol/ml cells	Lower temperatures may necessitate longer exposure and higher pH. Cell lysis will increase the amount of DIDS required. The doses suggested include a safety margin of about 50%
DNDS	150–200 μM	Present during efflux measurement. DNDS exists in a cis and a trans form, the trans isomer being the more efficient inhibitor.[b] The isomers can be distinguished by spectrophotometry[c]
SITS	100 μM	Present during flux measurement[d]
Tetrathionate ($Na_2S_4O_6$)	10 mM	Present during efflux measurement

[a] Details given in Deuticke et al.[7]
[b] From O. Fröhlich and R. B. Gunn, Fed. Proc., Fed. Am. Soc. Exp. Biol. **39**, 1714 (1980).
[c] From O. Fröhlich, J. Membr. Biol. **65**, 111 (1980).
[d] From Leeks and Halestrap.[4]

substrate concentration. Fractional contributions will vary with all three parameters.

As evident from Table IV, situations quite different from those in human erythrocytes may prevail in other mammalian species, since not only the activities of the transport systems but also the permeabilities of

TABLE III
COMPOUNDS APPROPRIATE FOR SIMULTANEOUS
COMPLETE INHIBITION OF THE TWO MEDIATED
PATHWAYS OF MONOCARBOXYLATE TRANSPORT[a]

Inhibitor	Concentration
Salicylate	20 mM
Phloretin	250 μM
2,4,6-Trichlorobenzene sulfonate	20 μM
Niflumic acid	20 μM

[a] From Deuticke et al.[3,7]

TABLE IV
FRACTIONAL CONTRIBUTIONS OF THREE PATHWAYS OF MONOCARBOXYLATE TRANSPORT
FOR VARIOUS SUBSTRATES AND TYPES OF RED CELL[a]

Species	Substrates	MC system	Band 3 system	Nonionic diffusion	Temperature (°C)
Man	L-Lactate	0.9	0.06	0.04	30
	D-Lactate	0.54	0.31	0.15	30
	Glycolate	0.34	0.65	<0.01	15
Ox	L-Lactate	0.15	0.53	0.32	37
Rabbit, rat,	L-Lactate	>0.9	<0.1	nd[b]	0
guinea pig	Acetate	0.8–0.9	nd	nd	0 (pH 8)

[a] From Deuticke et al.[3,7,45] MC, Monocarboxylate. Substrate concentration 5 mM, equilibrium exchange, pH 7.4.
[b] nd, Not determined.

the lipid domain vary. Selection of the appropriate species therefore greatly helps to study a particular component. Ox and sheep cells are suited for studying the transfer of monocarboxylate via band 3 or by nonionic diffusion, while rabbit, rat, and guinea pig are suitable objects for studying the monocarboxylate transport system. Details available are summarized in Ref. 3.

Characteristics of Monocarboxylate Transfer via the Specific Transport System

Equilibrium Exchange Fluxes. Evaluation of the PCMBS-sensitive component of monocarboxylate equilibrium exchange flux (measured as described above) in terms of the Michaelis–Menten formalism provides kinetic constants,[16] summarized in Table V, together with data determined without consideration of the different components of transport. Studies have hitherto been focused on lactate, glycolate, and β-hydroxybutyrate. Reliable data for pyruvate equilibrium exchange are not yet available due to the problems outlined in the previous section. Influx measurements have been carried out by Leeks and Halestrap[4] (see following paragraph). For an evaluation of these data in terms of kinetic models the reader is referred to the section, Kinetic Models of the Monocarboxylate Transporter.

Net Fluxes. Chloride efflux from erythrocytes is depressed dramatically when no readily exchangeable anion is available on the trans side of the membrane.[23,49,50] The fluxes of lactate and other aliphatic monocar-

[49] J. O. Wieth, *Alfred Benzon Symp.* **4**, 265 (1972).
[50] M. L. Jennings, *J. Gen. Physiol.* **79**, 169 (1982).

TABLE V
KINETIC PARAMETERS FOR MONOCARBOXYLATE TRANSPORT VIA THE SPECIFIC
MONOCARBOXYLATE TRANSFER SYSTEM

Species, substrate	Type of measurement	$T(°C)$	pH_{cis}	J_{max} (pmol × cm^{-2} × sec^{-1})	K_T (mM)	J_{max}/K_T	Ref.
Man	Equilibrium	20	7.4	4.3	46.3	0.093	a
L-Lactate	exchange	10		2.1	72.8	0.029	a
	Net efflux	20	7.15	0.97	10.5	0.095	a
	(zero-trans)	10		0.10	3.0	0.033	a
		5		0.02	1.7	0.012	a
	Net influx	25	7.4	0.82	13.1		b
	(zero-trans)			1.2	7.6		c
				1.2 ^{d}	11.3 ^{d}		c
Glycolate	Net efflux	15	7.25	0.16	14	0.011	e
	(zero-trans)						
Pyruvate	Net influx	25	7.4	1.27	1.26		c
				1.05 ^{d}	1.90 ^{d}		c
Rat							
3-Hydroxybutyrate	Net influx	10	7.4	1.57	16		f

^{a} From Deuticke.[16]
^{b} J_{max} is a mean value from numbers scattering between 0.45 and 1.1.[11]
^{c} From Leeks and Halestrap.[4,5]
^{d} Data from pH changes due to lactate/H^+ cotransport.
^{e} From Deuticke.[45]
^{f} From Regen and Tarpley.[14]

boxylates into media free of monocarboxylate are only partly reduced under these conditions,[7] suggesting considerable net movements of lactate by means of a cotransport with H^+ or an exchange against OH^-. While these two alternatives cannot be distinguished on a kinetic basis, a monocarboxylate/OH^- exchange seems unlikely, as outlined by De Bruijne et al.,[11] in view of the extremely high, selective affinity for OH^- as a substrate one would have to postulate for the purported carrier.

Thus, lactate–H^+ cotransport is more likely to be involved. This process can be analyzed quantitatively and qualitatively by following the increase of extracellular pH upon addition of sodium lactate to suspensions of erythrocytes in slightly buffered media. Details can be found in publications by Dubinsky and Racker,[6] Leeks and Halestrap,[5] or Jennings and Adams-Lackay.[9] Helpful information is also given in the work of Spencer and Lehninger, who studied ascites tumor cells.[19] In such experiments blockage of pH equilibration by OH^- (HCO_3^-)/Cl^- exchange via band 3 protein is indispensable in order to prevent secondary changes of

pH_e. Stilbene disulfonates are appropriate for this purpose. It has to be kept in mind that a mere alkalinization of the extracellular compartment upon addition of a monocarboxylate to a red cell suspension does not prove the involvement of the specific monocarboxylate transporter. Nonionic diffusion of the undissociated acid also produces this phenomenon, although usually much faster.[26,27] Inhibition of the pH shift by inhibitors of the specific system (e.g., PCMBS) has to be demonstrated. Finally, the particular kinetic situation of an H^+–anion cotransport system makes the rate of pH changes (and of lactate movements) susceptible to the pH in the trans compartment. Differences among published data are in part due to this feature.

Kinetic data for lactate net transport can be obtained more conveniently by following lactate movements. Flux measurements under zero-trans conditions, i.e., when no substrate is present on the transside of the membrane, are most easily interpretable. Most investigators have used tracers,[2–5,7–15] but enzymatic analysis[6,50a,50b] and even spin-echo nuclear magnetic resonance techniques[51] have also been used. Net efflux experiments can be executed like equilibrium exchange measurements using labeled substrates. In order to maintain the initial conditions, sufficient buffering should warrant constant pH at both sides of the membrane. Care should also be taken to avoid significant changes of substrate concentrations during the flux measurements in order to maintain true zero-trans conditions and defined substrate concentrations. Practically, this is achieved by working in the range of 100–90% of the original cis concentration. Rate coefficients may then be calculated as for equilibrium exchange (see the section, Efflux Measurement).

Net influx experiments will require stop techniques[4] or fast separation techniques[11,14] since the extracellular (cis) radioactivity has to be removed completely prior to analysis in order to determine the uptake of a very low percentage of the extracellular radioactivity with sufficient accuracy.

Kinetic data obtained for zero-trans flux conditions are summarized in Table V. The half-saturation constants are much lower than those for equilibrium exchange. This facilitates experiments, since relative fluxes (J/J_{max}) can be studied over a broader range of concentrations than for equilibrium exchange ($K_T = 50–70$ mM), where substrate concentrations exceeding $J/J_{max} = 0.5–0.7$ are hard to realize in intact cells due to the restrictions of isotonicity.

[50a] W. N. Fishbein, J. W. Foellmer, J. I. Davis, T. M. Fishbein, and P. Armbrustmacher, *Biochem. Metabol. Met. Biol.* **39**, 338 (1988).
[50b] W. N. Fishbein, J. I. Davis, J. W. Foellmer, and M. R. Casey, *Biochem. Med. Metabol. Biol.* **39**, 351 (1988).
[51] K. M. Brindle, F. F. Brown, I. D. Campbell, C. Grathwohl, and P. W. Kuchel, *Biochem. J.* **180**, 37 (1979).

Kinetic Models of the Monocarboxylate Transporter

The concept of a monocarboxylate (MC)/H^+ cotransport can be formulated by several alternative kinetic models, which are open to experimental testing under a number of aspects. Quite generally, these aspects concern the symmetry of the rate coefficients and binding constants, the involvement of mobile or nonmobile binding sites, and the problem of the rate-limiting step. More specifically, in cotransport systems the order of binding of substrates has to be clarified. One may distinguish random binding from ordered binding and in the latter case the two types of sequence "H^+ first, then MC," or "MC first, then H^+." In a further refinement of the models one would try to take into account the possibility that a substrate which is bound first on the cis side might either leave last (mirror type) but could also leave first (glide type) on the trans side.

Random and ordered models are summarized in Fig. 1. The molecular constants (K_m, α, P_C, P_{CA}) in this scheme can be used to describe the fluxes under various conditions. For a detailed derivation of the equations the reader is referred to the work of Heinz,[52] Hopfer and Groseclose,[53] Van den Broeck and Van Steveninck,[54,55] Devés and Krupka,[56] and Turner.[57]

Usually the molecular constants can be lumped together into the macroscopic parameters of a Michaelis–Menten-type equation ($J = J_{max} S/(S + K_T)$). The Michaelis–Menten parameters (J_{max}, K_T) can be derived from the experimental data by the usual linearization procedures or a nonlinear least-squares fit. In formulating the equations defining K_T or J_{max} one can either aim at the most general formulation, making no assumptions whatsoever concerning rate-limiting steps, mobility or immobility of the binding site, symmetry or asymmetry of rate and binding constants, or restrictions in mobility of the partially loaded forms of the carrier (CH^+, CS^-). Recent examples for this approach are found in the work of Turner, where relevant earlier work is also reviewed.[57]

For many purposes it may be sufficient to formulate the equations under simplifying assumptions and to be specific only in those parameters and coefficients that can actually be varied or determined.[16] This more specified approach has the evident advantage that the resulting equations are must more easy to evaluate, since parameters that cannot be separated are condensed.

[52] E. Heinz, "Mechanics and Energetics of Biological Transport." Springer-Verlag, Berlin, 1978.
[53] U. Hopfer and R. Groseclose, *J. Biol. Chem.* **255**, 4453 (1980).
[54] P. J. A. Van den Broek and J. Van Steveninck, *Biochim. Biophys. Acta* **602**, 419 (1980).
[55] P. J. A. Van den Broek and J. Van Steveninck, *Biochim. Biophys. Acta* **693**, 213 (1982).
[56] R. Devés and R. M. Krupka, *Biochim. Biophys. Acta* **556**, 533 (1979).
[57] R. J. Turner, *J. Membr. Biol.* **76**, 1 (1983).

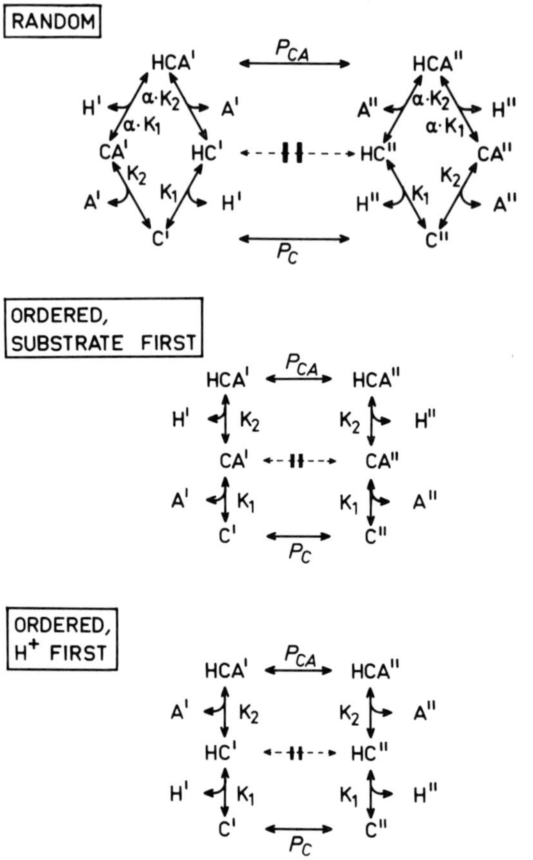

FIG. 1. Kinetic models of the monocarboxylate transport system in the erythrocyte membrane. The model is illustrated by the classical ferry-boat symbolism, but assumed to represent the alternating and rapid association and dissociation of substrate and H^+, at both sides of the membrane, to and from conformational isomers of an intrinsic protein spanning the barrier domain. Simplifying assumptions include (1) rapid equilibrium for binding of substrate and H^+, and (2) the symmetry of the equilibrium constants for substrate and H^+ binding, and of the isomerization constants in the forward and the backward direction. Moreover, immobility of the intermediate states of loading has been assumed.

In case of cotransport processes, assignment to a type of model is possible by determination of the Michaelis–Menten parameters for the fluxes of substrate (i.e., monocarboxylate) as a function of the concentration of cosubstrate (i.e., H^+). The measurements have to be carried out under various experimental conditions concerning substrate concentrations. The following conditions are usually preferred: (1) equilibrium exchange, i.e., $S_{cis} = S_{trans}$, (2) zero trans, i.e., $S_{trans} = 0$, and (3) infinite trans, i.e., $S_{trans} = \infty$.

Ideally, these conditions should also be applied to the cosubstrate, in order to simplify the mathematics. This requirement can be fulfilled under equilibrium exchange conditions but not for the two other experimental conditions unless one relaxes the definition of zero and infinite, e.g., to 0.1 or 0.01 times K_H and 10 or 100 times K_H, or to conditions where a further increase of the cosubstrate concentration will not affect the Michaelis–Menten parameters.[58]

General information may already be obtained by measuring the macroscopic parameters for conditions (1) to (3) defined above, but including measurements of net fluxes in both directions, at the same fixed concentrations of the cosubstrate. J_{max} and K_T values thus obtained are related to each other by a number of equations, as outlined by Lieb and Stein[59] or Devés and Krupka[56] for single-substrate carrier systems, and for cotransport systems by Turner.[60]

Testing the experimental data by these equations serves two purposes: Conceptually, the validity of the rapid equilibrium carrier model can be substantiated or rejected. Experimentally, once the rapid equilibrium model can be accepted, the equations form a test for the reliability and inner consistency of the data obtained under various experimental conditions.

In order to illustrate the foregoing considerations, a number of simplified equations are compiled in Table VI that have already allowed or will allow the assignment of certain molecular properties to the monocarboxylate–H^+ cotransport system in the erythrocyte.[11,12,16] As an example, Table VII demonstrates that these equations [and some of the interrelationships[56,59,60] between parameters as obtained under conditions (1)–(3)] provide means to discriminate among the possible models of the monocarboxylate transporter.

From Table VII it follows that in transporting lactate the system behaves as a mobile carrier transporting substrate and cosubstrate in a 1 : 1 coupled mode.[4-6] Equilibration of the substrate with the binding site is rapid as compared to the reorientation of the binding site.[16] Binding probably occurs in ordered sequence, H^+ before monocarboxylate, at least in human erythrocytes.[11,12] Measurements demonstrating a lack of influence of pH on J_{max}^{ee} are still required to prove finally the ordered binding model (cf. Table VI). The loaded form of the carrier reorients faster than the unloaded form, while the intermediate form (one substrate bound) is immobile. The symmetry problem is not yet solved. The data fit the mirror

[58] W. R. Lieb, in "Red Cell Membranes: A Methodological Approach" (J. C. Ellory and J. D. Young, eds.), p. 135. Academic Press, London, 1982.
[59] W. R. Lieb and W. D. Stein, Biochim. Biophys. Acta 373, 178 (1974).
[60] R. J. Turner, Biochim. Biophys. Acta 689, 444 (1982).

TABLE VI

RELATIONSHIP BETWEEN "MOLECULAR" KINETIC PARAMETERS DEFINED FOR THE VARIOUS MODELS GIVEN IN FIG. 1 AND THE MACROSCOPIC PARAMETERS OF THE MICHAELIS–MENTEN FORMALISM

$\rho = P_{CA}/P_C$ C_T = sum of all forms of the carrier

	Random	Ordered, H⁺ first	Ordered, A⁻ first
J_{max}			
J^{ee}	$\rho P_C \dfrac{C_T}{2\left(1 + \alpha \dfrac{K_1}{H}\right)}$	$\rho P_C \dfrac{C_T}{2}$	$\rho P_C \dfrac{C_T H}{2(K_2 + H)}$
J^{a}	$\rho P_C \dfrac{C_T}{1 + \rho\left(1 + \dfrac{H''}{K_1}\right) + \alpha \dfrac{K_1}{H'}}$	$\rho P_C \dfrac{C_T}{1 + \rho\left(1 + \dfrac{H''}{K_1}\right)}$	$\rho P_C \dfrac{C_T}{1 + \rho + \dfrac{K_2}{H'}}$
K_T			
J^{ee}	$\alpha K_2 \dfrac{1 + \dfrac{K_1}{H}}{1 + \alpha \dfrac{K_1}{H}}$	$K_2\left(1 + \dfrac{K_1}{H}\right)$	$K_1 \dfrac{K_2}{H + K_2}$
J^{a}	$\alpha K_2 \dfrac{1 + 2\dfrac{K_1}{H'} + \dfrac{H''}{H'}}{1 + \rho\left(1 + \dfrac{H''}{K_1}\right) + \alpha \dfrac{K_1}{H'}}$	$K_2 \dfrac{1 + 2\dfrac{K_1}{H'} + \dfrac{H''}{H'}}{1 + \rho\left(1 + \dfrac{H''}{K_1}\right)}$	$K_1 \dfrac{2\dfrac{K_2}{H'}}{1 + \rho + \dfrac{K_2}{H'}}$

model (first on, last off) but experiments aiming in particular at the discrimination from the glide model (first on, first off) may still be required. Table VIII summarizes sets of pH'/pH" combinations that will help to discriminate between the various models. The realization of such experiments is discussed in the next section.

A further treatment of the monocarboxylate–H⁺ cotransport system in terms of model parameters and their analysis by a suitable combination of observable parameters would exceed the scope of this chapter. For more detailed information the reader is referred to the work of Heinz,[52] Hopfer and Groseclose,[53] Turner,[57,60] or more generally, Lieb and Stein[59] or Devés and Krupka.[56]

Stable Transmembrane pH Gradients as a Tool in Monocarboxylate Transport Analysis

As mentioned above, the boundary conditions for the cosubstrate H⁺ cannot be changed to the same extent as for cosubstrates in other (e.g., Na⁺) cotransport systems. However, even qualitatively the changes of the macroscopic parameters with pH_{cis} and pH_{trans} will be helpful for the assignment of a model to the system. This became evident from Table VIII.

TABLE VII
Assignment of Molecular Kinetic Properties to the H^+–Monocarboxylate (MC$^-$) Cotransport System on the Basis of Michaelis–Menten Parameters Obtained under Various Flux Conditions

Question	Answer	Evidence (or way to obtain evidence)	Ref.
"Carrier" mobile or immobile?	Mobile	1. Transacceleration (see below)	Data: a, b
		2. Uphill counterflow (not definitely demonstrated)	Theory: c
Rate-limiting step? (1)	Reorientation of binding site, not association/dissociation of substrate	$J^{ee}_{max} \gg J^{zt}_{max}$ (equilibrium exchange $\gg$ zero-trans flux)	Data: a, d
Rate-limiting step? (2)	Reorientation of binding site in the unloaded state ($P_C < P_{CA}$)	Transacceleration by substrates	Data: a Theory: c, e
Stoichiometry MC$^-$: H$^+$?	Approximately 1 : 1	Simultaneous measurement of H$^+$/lactate movements	Data: f, g
Is the fully loaded or unloaded transport site charged? That is, is the transport electrogenic?	Probably not	Membrane potential does not affect transport rates	Data: h
Symmetric or asymmetric? i.e., $K'_n = K''_n$ $P'^{\rightarrow''}_{CA} = P''^{\rightarrow'}_{CA}$ $P'^{\rightarrow''}_{C} = P''^{\rightarrow'}_{C}$ (at $H' = H''$!)	Not known	Determination of K^{zt} and J^{zt}_{max} for influx and efflux at $H' = H''$	Theory: c
Sequence of MC$^-$ and H$^+$ binding?	Human Ordered, H$^+$ first Rat Possibly random	J^{zt}_{max} *not* dependent on pH$_{cis}$ K^{zt} dependent on pH$_{cis}$ J^{ee}_{max} pH dependent	Data: a, i Theory: i Data: j Theory: a
Intermediate form (CH$^+$) mobile?	Most likely not	(1) Stoichiometry MC$^-$: H$^-$ (2) Distribution ratios: $H^+_i/H^+_e = MC^-_e/MC^-_i$	f, g, i

^{a} Deuticke.[16]
^{b} Deuticke.[8]
^{c} Heinz.[52]
^{d} Hoare.[62]
^{e} Devés and Krupka.[61]
^{f} Dubinsky and Racker.[6]
^{g} Leeks and Halestrap.[5]
^{h} Deuticke.[45]
^{i} De Bruijne et al.[11]
^{j} Regen and Tarpley.[15]

TABLE VIII
pH (H$^+$) DEPENDENCE OF MICHAELIS–MENTEN PARAMETERS (K_T, J_{max})
FOR MONOCARBOXYLATE/H$^+$ COTRANSPORT AS A TOOL FOR THE
ASSIGNMENT OF TRANSPORT MODELS[a]

Model	K_T	J_{max}	J_{max}/K_T
Zero-trans flux			
Ordered binding (H, MC$^-$)	H', H''	H''	H', H''
Ordered binding (MC$^-$, H$^+$)	H'	H'	H'
Random binding	H', H''	H', H''	H', H''
Equilibrium exchange ($H' = H''$)			
Ordered binding (H$^+$, MC$^-$)	H', H''	No effect	H', H''
Ordered binding (MC$^-$, H$^+$)	H', H''	H', H''	H', H''
Random binding	H', H''	H', H''	H', H''

    [a] $'$, Cis; $''$, trans.

Steep and stable transmembrane pH gradients can be realized in studies of monocarboxylate–H$^+$ cotransport since the major high-capacity
pathway for the dissipation of pH gradients, i.e., the inorganic anion
exchanger, can be blocked without alteration of ion movements via the
monocarboxylate transporter. Experiments may be carried out under
equilibrium-exchange or zero-trans (for substrate) conditions, the latter
providing much more easily interpretable data.

Procedures. Cells are loaded with unlabeled and labeled substrate as
for the usual net efflux measurements. The intracellular substrate level is
defined by the extracellular level and the known Donnan ratio. If desired,
the extracellular pH is set to a value which produces a desired intracellular pH deviating from the normal. Data required for this procedure may be
extracted from the work of Funder and Wieth,[39] Duhm,[40] or Dalmark.[35] At
the end of the loading period with tracer, DIDS is added (hematocrit
45%) at a concentration of 25 μmol/liter medium or equivalent, of 30
nmol/ml cells. The suspension is incubated for 30 min at 37°, cooled to 0°,
centrifuged, and the supernatant removed. The cells are washed twice in
an excess of substrate-free medium at 0°. Efflux is started by injecting 1
ml of cells into 19 ml of an appropriate saline medium containing 20 mM of
an impermeable organic zwitterionic buffer (HEPES, etc.) having its pK
as near as possible to the desired extracellular pH. The sulfonic acid
derivatives (HEPES, etc.) are preferable to the carboxylates or amino
compounds (Bicine, tris, etc.) since the latter may permeate when their
cationic group is deprotonated. Flux measurements should be carried out
at 5 or 10° in order to keep residual fluxes via band 3 as low as possible.
On the other hand, conditions should be chosen to keep the flux period as

brief as possible (<3 min). Extracellular pH values should be measured in the medium before addition of cells and at the end of the flux period. In our hands essentially constant values were obtained.

A range of extracellular pH values between 6 and 10 is thus accessible. To safeguard the data against artifacts resulting from leak formation control experiments with added PCMBS are required. Above pH 10 inhibition by PCMBS is no longer complete, suggesting formation of membrane leaks. Within the range mentioned, the combination of blockage of pH equilibration and appropriate buffering allows the study of transport processes in erythrocytes at stable pH disequilibrium.

Substrate Specificity of the Monocarboxylate Transporter

Experimental Approach. To establish the substrates bound to and transported by the carrier, direct methods are practicable only to a limited extent. This is due to the lack of commercially available, labeled monocarboxylates and of simple methods to determine unlabeled monocarboxylates by chemical or enzymatic techniques. Direct flux measurements are thus only feasible with few substrates, marked in Table IX.

Among indirect methods, demonstration of competitive inhibition is very popular as a tool to study substrate specificity. This approach, however, suffers from the disadvantage that in the simple setup for measuring initial fluxes under zero-trans conditions it does not determine whether the competitor is bound *and* transported (i.e., is a true substrate) or only bound (i.e., is a mere competitive inhibitor).

Transport parameters of true substrates can be determined more conveniently by the "transacceleration" procedure. This approach is based on the effect of an unlabeled substrate analog on the rate of transport of a labeled substrate, present at a very low concentration (relative to its K_T value) on the opposite side of the membrane.[52,61] It can be used when, as usual,[52] the rate of reorientation of the unloaded binding site (P_C) is different from (usually lower than) the reorientation rate of the loaded binding site (P_{CA}). Details of the derivation are given by Devés and Krupka[61] for single-substrate systems. At invariant conditions of pH the approach can be applied without changes to cotransport systems.

The conceptual setup is explained in Fig. 2. Experimentally, the cells are preloaded at pH 7.4 with labeled monocarboxylate (A') as for net efflux studies at a concentration of substrate well below the K_T for zero-trans efflux. In case of L-lactate $C_e = 0.5$ mM ($C_i = A' = 0.35$ mM) was found to be appropriate (K_T for zero-trans efflux = 3 mM at 10°).[16] After

[61] R. Devés and R. M. Krupka, *Biochim. Biophys. Acta* **556,** 524 (1979).

TABLE IX

KINETIC PARAMETERS OF THE TRANSPORT OF VARIOUS MONOCARBOXYLATES BY THE SPECIFIC TRANSPORT SYSTEM, AS DERIVED FROM TRANSACCELERATION (TRANSINHIBITION) EXPERIMENTS[a]

T''	C_2 K_T (mM)	$J^{ta}_{max}/J_{T''=0}$	T''	C_3 K_T (mM)	$J^{ta}_{max}/J_{T''=0}$	T''	C_4 K_T (mM)	$J^{ta}_{max}/J_{T''=0}$
Acetate[b]	2.6	1.9	Propionate	No effect		Butyrate	3.9	0.3
CN-acetate	0.74	2.2	2-Cl-propionate	1.3	3.2	2-Cl-butyrate	0.3	1.5
F-acetate	1.16	2.6	2-Br-propionate	0.8	2.3	2-OH-butyrate	1.1	2.0
Cl-acetate[b]	0.80	2.7	2-Cl₂-propionate	2.7	2.2	2-Oxobutyrate	1.4	2.1
Br-acetate	0.51	3.0	L-Lactate[b]	3.8	3.8	3-OH-butyrate[b]	15.3	1.8
I-acetate[b]	0.96	3.3	D-Lactate	11.5	1.9	3-Oxobutyrate	5.8	2.0
Cl₂-acetate	0.4	3.7	Pyruvate	0.7	5.5	(=acetoacetate)		
Glycolate[b]	12.4	3.4	2-Thiolactate	0.6	1.8	Crotonate	3.8	0.6
			3-Cl-propionate	3.1	2.9			
			3-Nitropropionate	1.7	3.3			

[a] B. Deuticke, unpublished results. For the meaning of the kinetic constants see text.
[b] Also available in labeled form for direct measurements.

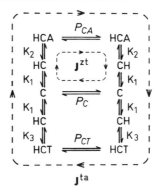

FIG. 2. Model for the transacceleration of the net efflux of a substrate A by a substrate analog T present on the trans side of the membrane at fixed protonation of the system. The model compares the events during the zero-trans efflux (J^{zt}) of A with its efflux transaccelerated by T (J^{ta}). For symbols and simplifying assumptions see Fig. 1.

preloading, the cells are centrifuged and washed once with lactate-free medium at 0°. Efflux is initiated by injecting aliquots of 1 ml of cells into 19 ml lactate-free media containing DIDS (5 μM) and the substrate analog T at various appropriate concentrations.

Aliquots are taken after time intervals at which the boundary conditions $A'' = 0$, $T' = 0$ are still essentially valid. Rate coefficients are proportional to fluxes, since A' is identical in all samples, and are calculated as usual. Fluxes of substrate A in the presence of the substrate analog T on the trans side of the membrane will usually be enhanced relative to analog-free controls.

Evaluation of Data. Rationale: From the general rate equations for a mobile carrier system it can be deduced[61] that the following equation [Eq. (4)] relates the transaccelerated flux of A' (normalized to the flux at $T'' = 0$) to the concentration of the transaccelerating substrate T:

$$J^{ta}/J_{T''=0} = J^{ta}_{T''\to\infty}/J_{T''=0} + K^{zt}_{T}[1 - (J^{ta}/J_{T''=0})]1/T'') \tag{4}$$

[Eq. (2) from Devés and Krupka][61], where J^{ta} is the transaccelerated flux of A'; $J_{T''=0}$, the control flux in the absence of T; and K^{zt}_{T}, the half-saturation constant for zero-trans flux of T from $''$ to $'$.

It can further be shown that

$$(J^{ta}_{T''\to\infty}/J_{T''=0})_{A'\to0} = J_{\max A} = f J^{zt \overset{''\to'}{}}_{\max T} \tag{5}$$

where $J^{zt \overset{''\to'}{}}_{\max T}$ is the maximum of zero-trans flux of T from $''$ to $'$. The proportionality factor f is the harmonic means ($P_{\bar{c}}$) of the rate coefficient for inward (P_C) and outward (P_{-C}) reorientation of the unloaded binding site[62]:

$$(1/P_{-C}) + (1/P_C) = 2/P_{\bar{c}} \qquad (6)$$

The proportionality factor f is thus independent of the nature of the substrate analog T. The ratio between J_{max_A} values in the presence of two analogs T_1 and T_2 is equal to the ratio of the maximum transport rates of these two analogs under zero-trans conditions.

If a reference value is available for one substrate, for which J_{max}^{zt} has actually been measured, the transacceleration approach even provides a means to obtain J_{max}^{zt} values for a large number of substrate analogs. Transinhibition by substrate analogs may also be observed in such trans effect studies, indicating that the carrier substrate–analog complex reorients slower than the free carrier (i.e., $P_{CT} < P_C$) or is even immobile.

Procedure and results: Evaluation of experimental data in terms of Eq. (4) is best carried out by plotting $J^{ta}/J_{T''=0}$ against the absolute value of $[(J^{ta}/J_{T''=0}) - 1](1/T'')$ (see Fig. 3). The slope of the resulting straight line is negative for transaccelerating and positive for transinhibitory anions.[61] The K_T values are obtained from the slope, relative maximal rates $(J_{max}^{ta}/J_{T''=0})$ from the intercept with the ordinate. The kinetic parameters for a number of important analogs of lactate are compiled in Table IX. A preliminary but more complete qualitative survey including anions not serving as substrates has been given elsewhere.[16] Reproducible measurements require rigorous control of extracellular pH, since the kinetic constants depend strongly on the extracellular pH.

From the data in Table IX it becomes evident that for the transported substrates studied in detail the affinities range from one-fourth to four times the K_T of L-lactate. The maximal transport rates are more similar, with the notable exception of pyruvate. As a further relevant feature it should be noted that the transporter discriminates between enantiomers (D- versus L-lactate), which was also observed in direct measurements.[3] The affinity data available have been used elsewhere to construct a hypothetical scheme of the substrate-binding site.[16]

Inhibitors of Monocarboxylate Transfer

Reversible Inhibitors. Some aspects of this problem have already been discussed above. A considerable number of compounds inhibit monocarboxylate transport reversibly. Most of them, however, are not specific for this system, but also inhibit anion exchange via band 3.[23,28] A list of inhibitors is given in Table X. A detailed analysis of the type of inhibition has been carried out in only a few cases. The inhibitor constants of competitive inhibitors (such as the cyanocinnamates[2]) exhibit a marked pH

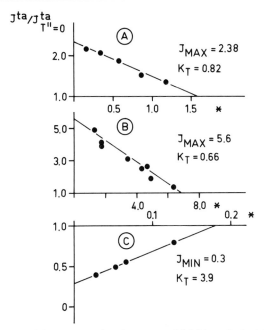

FIG. 3. Evaluation of the transacceleration or transinhibition of L-lactate efflux by other monocarboxylates. (A) 2-Bromopropionate; (B) pyruvate; and (C) butyrate. Efflux measurements were carried out as described in the text, at $pH_e = 7.4$ and $5°$. $J_{max} = (J^{ta}/J_{T''=0})_{max}$; * (abscissa) $= (J^{ta}/J_{T''=0} - 1)T''^{-1}$. K_T values in millimolar units. For further details, see Devés and Krupka.[61]

dependence.[45] This results from the fact that inhibitor binding, like substrate binding, is influenced by the protonation of the transport system.

Covalent Modifiers. Covalently or very tightly bound inhibitors are potential tools for the identification of the transporter. So far, mainly sulfhydryl and amino reagents have been studied (Table XI). Since many amino reagents are powerful inhibitors of the inorganic anion-exchange system,[23,24,48,61a] this class of compounds is only of limited value for labeling the putative monocarboxylate transporter. Stilbene disulfonates can be used, taking advantage of different concentration required for inhibition of the two anion-transport systems, as outlined below (see H_2DIDS Binding). Recently, N-hydroxysulfosuccinimido-activated esters of mono- and dicarboxylic acids have been shown to be much more potent inhibitors of monocarboxylate transport than of inorganic anion exchange.[25a]

61a M. L. Jennings, *J. Biol. Chem.* **257,** 7554 (1982).

TABLE X
Noncovalent Inhibitors of
Monocarboxylate Transport[a]

Compound	Concentration (mM)	Ref.
Phenopyrazon	7.5[b]	g
Phloretin	0.3[c]	g
Salicylate	0.5	g
Benzene 1,3-disulfonate	10	h
2,4,5-Trichlorobenzene sulfonate	0.6	h
8-Anilino-1-naphthalene sulfonate	0.05	h
Niflumic acid	0.002	h
Tetracaine	5	g
Dipyridamol	0.02	i
Pentoxifyllin	1.7	i
Methylisobutylxanthine	0.1	i
Lysolecithin	0.03	i
Oleic acid	0.25	i
Triton X-100	0.06	i
Tween 80	0.03	i
Brij 96	0.010[d]	i
Brij 76	0.027	i
Aromatic monocarboxylates:		
trans-Cinnamate	0.6[e]	i
4-Hydroxycinnamate	0.4[e]	i
α-CN-4-OH-cinnamate	0.05[e]	i
α-CN-3-OH-cinnamate	0.25[e]	i
α-CN-4-OH-cinnamate	0.06[f]	j
α-CN-cinnamate	0.09[f]	j

[a] Concentrations refer to half-maximal inhibition unless specified otherwise. Conditions: J_{Lac}^{ee}, pH 7.4, unless specified otherwise.
[b] Ninety percent inhibition.
[c] Ninety-six percent inhibition.
[d] Same I_{50} for J^{ee} and J^{zt}, inhibitor trans side.
[e] J^{zt}, inhibitor on trans side.
[f] J^{zt}, inhibitor on cis side.
[g] Deuticke et al.[3]
[h] Deuticke et al.[7]
[i] Deuticke.[45]
[j] Halestrap.[2]

TABLE XI

COVALENT INHIBITORS OF THE TRANSFER OF MONOCARBOXYLATES[a]

Agent	Concentration (mM)	Exposure time (min)	Inhibition (%)	Comments concerning the treatment with inhibitor	Ref.
Pyridoxalphosphate	0.6		50 (I_{50})	Present during flux measurement	b
Fluorodinitrobenzene	3.5	60	95		b
Trinitrobenzene sulfonate	1.0	60	35		b
DIDS, etc.	See Table II				
Isobutylcarbonyl lactyl anhydride	0.01	60	50 (I_{50})	Rabbit erythrocytes; treated at 0°, pH 7.4	c
Bis(sulfosuccinimido)-suberate	0.075	30	50 (I_{50})	Rabbit erythrocytes; treated at pH 7.4, room temperature	f
Formaldehyde/sodium borohydride (reductive methylation)	16/5	3 × 6	87	Rabbit erythrocytes; Hct 5%	f
Phenylglyoxal	10	5 × 1.5	67	Rabbit erythrocytes; Hct 20%, pH 10, 5 treatments with the agents for 90 sec at room temperature, interrupted by quenching at pH 6 and washing.	f
Iodoacetate	5	120	0		d
Iodoacetamide	20	90	30	Treatment at pH 8.0	d
N-Ethylmaleimide	10	90	86	Treatment at pH 8.0	d
Diamide	7	60	30	Treatment at pH 8.0	d
NBD-chloride	2	10	50 (I_{50})		e
DTNB	0.4	120	50 (I_{50})	Treatment at pH 8.2	b
DTDP	5	5	50 (I_{50})	Inhibition disappears on prolonged treatment with the agent	b
PCMBS	0.002		50 (I_{50})	Present during efflux, Hct 5%, pH 7.4	b
	0.012	90	50 (I_{50})	Cells pretreated, pH 7.4, 37°, Hct 20%	
HgCl$_2$	0.02		50	Present during flux measurement	b
Hydroquinone	12	75	64	Effective only at pH >6.5	d
o-Iodosobenzoate	12	75			
Pronase	2.5 mg/ml	60	47		b
Trypsin	2 mg/ml	60	0		b

[a] Data relate to fluxes of L-lactate measured under equilibrium exchange conditions. Cells were pretreated with the agents at pH 7.4 and 37° at a hematocrit of 10% unless specified otherwise.
[b] Deuticke et al.[3]
[c] Donovan and Jennings.[10]
[d] Deuticke.[45]
[e] Deuticke et al.[7]
[f] Donovan and Jennings.[25a] Related activated monocarboxylate ester have similar effects.

A low specificity is also true for the arginyl-modifying agent, phenylglyoxal, a well-established inhibitor of inorganic anion exchange,[61b] which inhibits monocarboxylate transport under appropriate conditions.[25a]

In the case of SH reagents the situation is different. Mercurials are selective inhibitors, among the anion-transport systems, of monocarboxylate transport (besides their effects on the transport of monosaccharides,[63] cations,[64] water, and urea,[64a,64b,64c] and their perturbing influence on cation barrier properties.[46,47] For this reason the mercurials may also be used for a further molecular identification of the monocarboxylate transporter (see below).

Membrane Labeling by Inhibitors of Monocarboxylate Transport

Determination of Numbers of Binding Sites. PCMBS and PCMB, for which the erythrocyte membrane has a rather low permeability,[65] inhibit monocarboxylate transport by the specific system at very low concentrations.[3,7] Maximal inhibition is reached within a few seconds.[3] From dose-response curves (Fig. 4) it becomes evident that maximal inhibition only requires the exposure of the cells to 30–40 nmol mercurial/ml cells. This allows binding studies with labeled PCMB. The following procedure has proved useful in our studies.

Washed cells are suspended in 8 vol of phosphate-buffered saline, pH 7.35, 20°. ^{14}C-Labeled PCMB is added at varying concentrations (e.g., 5–120 nmol/ml cells). After an exposure of 1 min the suspension is diluted by adding another 16 vol of ice-cold buffered saline containing 1 mM N-ethylmaleimide (NEM) in order to block readily reactive SH groups. The cells are centrifuged (5 min, 5000 g) and washed three times in 6 vol of cold phosphate-buffered isotonic saline. Subsequently, they are suspended in 1 vol of isotonic phosphate buffer (pH 7.4) and lysed by addition of 20 vol 5 mM phosphate buffer (0°, pH 7.4, containing 1 mM NEM). The ghost membranes are washed three times in lysing buffer containing 1 mM NEM. The washed ghosts are packed by centrifugation, solubilized by addition of 0.1 vol SDS (200 g/liter), and heated for 2 min in boiling water. An aliquot (50–100 μl) is used for liquid scintillation counting,

[61b] J. O. Wieth, P. J. Bjerrum, and C. L. Borders, Jr., *J. Gen. Physiol.* **79,** 283 (1982).
[62] D. G. Hoare, *J. Physiol. (London)* **221,** 311 (1972).
[63] J. Van Steveninck, R. I. Weed, and A. Rothstein, *J. Gen. Physiol.* **48,** 617 (1965).
[64] A. F. Rega, A. Rothstein, and R. I. Weed, *J. Cell. Physiol.* **70,** 45 (1967).
[64a] R. I. Macey, *Am. J. Physiol.* **246,** C195 (1984).
[64b] M. R. Toon and A. K. Solomon, *Biochim. Biophys. Acta* **860,** 361 (1986).
[64c] D. M. Ojcius and A. K. Solomon, *Biochim. Biophys. Acta* **942,** 73 (1988).
[65] P. A. Knauf and A. Rothstein, *J. Gen. Physiol.* **58,** 211 (1971).

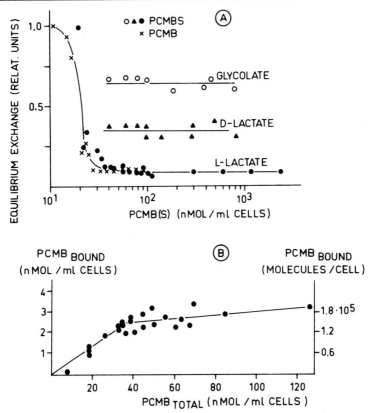

FIG. 4. Inhibition of monocarboxylate transfer by PCMBS or PCMB and their binding to the erythrocyte membrane. (A) Flux measurements in the presence of the inhibitor (Hct 5%, pH 7.4, 10°). (B) PCMB binding at 0° as described in the text.

another aliquot for protein determination by the Lowry procedure.[66] Since inhibition of monocarboxylate transport by PCMB is not reversed by washing with saline, even including NEM, all the procedures are not supposed to remove label from the relevant sites on the membrane.

From the binding isotherm shown in Fig. 4 it can be seen that about 2.5 nmol/ml cells, equivalent to about 1.5×10^5 molecules of PCMB per cell is firmly bound at the conditions of complete inhibition of transport. The duration of the exposure slightly affects firm binding: after 3 min about 25% more PCMB is firmly bound than after a 0.5-min exposure.[45]

Because of the high affinity of mercurials for SH groups elution of

[66] O. H. Lowry, N. J. Rosebrough, A. L. Farr, and R. J. Randall, Biol. Chem. 193, 265 (1951).

mercurial is not likely to be a problem. Migration between thiols will not pose problems as long as only total binding is investigated.

In view of the low permeability of PCMB the thiols to which the mercurials are bound during a brief exposure must be located on the outer membrane surface. According to studies with fully impermeable thiol labels the total number of exofacial thiols in red cells is about 9×10^5/ cell.[67,68] The PCMB-binding studies imply that a maximum of about 20% of these groups is involved in monocarboxylate transport.

Identification of Carrier Protein. Two preliminary attempts have been made to identify polypeptide fractions involved in monocarboxylate transport.

PCMB labeling: Membranes from human red cells labeled with [^{14}C]PCMB as described above were solubilized, omitting the heating at 100°, and subjected to SDS-gel electrophoresis[69] on rod gels containing 5% acrylamide, using buffers free of reducing thiols such as mercapto-ethanol. Duplicate gels were run at 7–8 mA/gel. One gel was stained by Coomassie Blue, the other one cut into 2-mm slices. The slices were treated with 0.7 ml of a tissue solubilizer (e.g., Soluene 350, Packard Instruments) at 50° overnight in scintillation vials, mixed with 10 ml of a commercial scintillation fluid, and counted. The integral recovery amounted to 85–90% of the total number of counts applied to the gels.

Most of the label (up to 85–90%) was found in the region of bands 4.5 and 7[16] (enumeration of the bands according to the nomenclature of Steck[70]). Labeling of band 3 protein, which was sometimes also observed, is probably artifactual, since according to recent observations[64] band 3 protein is likely to be labeled by PCMB only under more drastic conditions. Labeling of band 4.5 and 7 proteins was insensitive to the unavoidable manipulations of the membrane material before and after solubilization. In particular, there was no labeling of spectrin, which contains a high amount of SH groups[71] precluding a major migration of mercurial between membrane proteins during solubilization. Treatment of ghosts with PCMB led to extensive labeling of spectrin as well as band 3 protein.[45] Nevertheless, PCMB probably labels too many sites not involved in monocarboxylate transfer, since covalent, SH-reactive agents, which do not inhibit monocarboxylate transfer, reduce the labeling with PCMB at

[67] R. E. Abbott and D. Schachter, *J. Biol. Chem.* **251,** 7176 (1976).
[68] P. C. Chan and M. S. Rosenblum, *Proc. Soc. Exp. Biol. Med.* **130,** 143 (1969).
[69] C. W. M. Haest, G. Plasa, D. Kamp, and B. Deuticke, *Biochim. Biophys. Acta* **509,** 21 (1978).
[70] T. L. Steck, *J. Cell Biol.* **62,** 1 (1974).
[71] J. M. Anderson, *J. Biol. Chem.* **254,** 939 (1979).

least up to 50% without changing the inhibitory action of the mercurial.[45] Results of studies with DIDS binding support this contention.

H₂DIDS binding: Although the stilbene disulfonates are highly selective inhibitors of anion transport via band 3, they also inhibit monocarboxylate transport.[2,7,9] Jennings and collaborators[9,10] took advantage of this feature. They pretreated rabbit erythrocytes (known to have a high capacity for monocarboxylate transport[3]) with 5 μM H₂DIDS (80 min, 37°C, hematocrit 5%) which is sufficient to block completely anion transport via band 3 protein. Subsequently, cells were treated for 1 hr at 37° with 10 or 20 μM [³H]H₂DIDS. After two washings in at least 20 vol of phosphate-buffered saline medium containing 0.02% (w/v) bovine serum albumin the membranes were isolated, solubilized, and subjected to SDS-gel electrophoresis following the Laemmli procedure. Gel slabs were stained in Coomassie Blue G, scanned spectrophotometrically, sliced, and counted for radioactivity. By this procedure a protein band at M_r 40,000–45,000 was labeled with high preference. Labeling was proportional to inhibition and could be suppressed[10,25a] by other covalent inhibitors of lactate transport, isobutylcarbonyllactylanhyride (IBCLA), and bis(sulfosuccinimido-suberate).[25,25a] In human red cells labeling of the M_r 43,000 region was only minimal. In rabbits Jennings and collaborators[9,10] found 2.5 nmol/ml cells of [³H]H₂DIDS bound to the M_r 40,000–45,000 region. If one assumes the number of transport sites to be 100-fold lower in human than in rabbit erythrocytes,[3] the expected number of copies of the transporter in human red cells would be about 1.5×10^3/cell, about 1% of the number derived from mercurial labeling. The results of the labeling procedure described here are supported by recent studies using ³H-labeled *N*-hydroxysulfosuccinimido-activated esters as labels.[25a] Selective binding of these NH₂-modifying agents to an M_r 40,000–45,000 band was observed in rabbit erythrocytes pretreated with 4 μM H₂DIDS. To clearly identify the transport-active protein fraction in human erythrocytes more selective labels will be required.

Acknowledgments

Work from the author's laboratory was supported by the Deutsche Forschungsgemeinschaft (Sonderforschungsbereich 160, Projekt C 3). The secretarial help of H. Thomas is greatly appreciated.

[19] Alkali Metal/Proton Exchange

By PETER M. CALA and KAREN S. HOFFMANN

Introduction

Since the late 1970s, alkali metal/H^+ exchange pathways have been the focus of much scientific inquiry. Alkali metal/H^+ exchange (Am^+/H^+) has been found to play a central role in a variety of fundamental cellular processes which include cell volume and pH regulation, epithelial fluid and electrolyte transport, and cell growth and development. Because of the versatility of alkali metal/H^+ exchangers, the pathways have attracted the interest of a broad cross-section of the scientific community. The intent of this chapter is to detail the methods involved in measuring net alkali metal exchange flux and ultimately to explain criteria for establishing the existence or participation of alkali metal/H^+ exchange transport. The examples presented will be drawn from our ongoing studies of the *Amphiuma* red blood cell. The methods used are readily adaptable to other cell types assuming that cells are available in 6- to 15-mg (wet weight) quantities and that net fluxes are at least 10–15 mmol/kg of dry cell solid/hr.

Alkali metal/H^+ exchange in *Amphiuma*, as in other red blood cells, is functionally coupled to Cl^-/HCO_3^- exchange.[1] Consequently, Na^+ or K^+, H^+, Cl^-, and HCO_3^- are substrates for transport. The overall net transport of the above ions will be dictated by the carrier's kinetic behavior and the thermodynamic forces acting on all transported species. In light of the above it is necessary to be able to obtain accurate measurements of the concentrations of all transported ions as well as to be able to document time-dependent changes in cell ion contents and concentrations. Further, since the alkali metal/H^+ exchangers are volume sensitive, cell volume (H_2O content) must also be known. This chapter will detail methods which have been useful in the study of Am/H^+ exchange in the *Amphiuma* red blood cell. Part of the power of the method is that it permits determination of Na^+, K^+, Cl^-, H_2O, and even H^+, on the same sample. Since all of the experiments described involve measurements of cell ion and H_2O content, this "general" method will be presented first. Subsequently, we will describe modifications of this method necessitated by the particular experiment being described.

[1] P. M. Cala, *J. Gen. Physiol.* **76**, 683 (1980).

General Method for Net Flux Experiment on *Amphiuma* Erythrocytes

Procedure

Blood is obtained from live *Amphiuma tridactylum* by cardiac puncture using a heparinized syringe. The blood is centrifuged using a clinical centrifuge and red blood cells are separated from plasma. The cells are then resuspended and "washed" three times at a 10% hematocrit in isotonic Ringer's solution (IR : 110 mM NaCl, 3 mM KCl, 0.1 mM MgCl$_2$, 0.5 mM CaCl$_2$, 18 mM HEPES, 5 mM glucose). Ringer's solution was previously gassed with water-saturated room air and adjusted to pH 7.65 with NaOH. The cells are incubated for at least 90 min (at 20°, in the dark) to ensure that steady state is obtained with respect to volume and intracellular ion content. In order to initiate an experiment the cells are resuspended (at 10% hematocrit) in experimental media (see below) containing [^{14}C]inulin (1.25 μCi/ml Ringer's) as an extracellular space marker. Samples are obtained by removal of 400-μl aliquots of the cell suspension at predetermined intervals. The samples are then transferred to preweighed 400-μl PE tubes (Stockwell Scientific, Menlo Park, CA) and centrifuged for 4 min at 12,000 g (Eppendorf Micro Centrifuge model 5414, Brinkman Instruments Co.). The supernatant is separated from the cell pellet and both supernatant and cells are saved for analysis. Following separation any remaining supernatant in the tube containing the cell pellet (including that on the sides of the tube) is removed by aspiration. Typically the uppermost (top) layer of cells is also aspirated in order to minimize trapped supernatant.

Experimental Media. Ouabain (1 mM) can be used in the preincubation and experimental media to inhibit the Na$^+$/K$^+$ pump. Osmotic variations of media are usually made by changing NaCl concentration. Hypoosmotic media used to study K$^+$/H$^+$ exchange responsible for regulatory volume decrease (RVD) contain less than the 110 mM NaCl in isosmotic medium. In contrast the hyperosmotic media used to study Na$^+$/H$^+$ exchange responsible for regulatory volume increase (RVI) are more than 110 mM NaCl (See protocol A for typical experimental osmolarity variations). Table I contains a list of some possible media substitutions and their known effects on the transport.

Lysis of Cells

The "wet cell pellets" are weighed, then lysed by addition of 230 μl of 5 mM MgSO$_4$ and mechanically agitated with a thin kinked wire attached to a power rotary tool. (MgSO$_4$ is necessary for cellular nuclease activity which prevents gelation of cellular DNA.) The protein is precipitated by

TABLE I
POSSIBLE MEDIUM SUBSTITUTIONS FOR NET FLUX EXPERIMENTS

Substitute	Ion(s) replacing	Available forms	Effects/comments
1. TMA (tetra-methyl-ammonium)	Na^+ and/or K^+	TMA-Cl	Not a substrate for AM/H^+ Usually requires purification by recrystallization before using[a] May degrade into ammonia and methane
2. Choline	Na^+ and/or K^+	Choline chloride	Same as TMA
3. NMG (N-methylgluc-amine)	Na^+ and/or K^+	NMG	Not a substrate for Am/H^+
4. Lithium	Na^+ and/or K^+	LiCl	Will serve as substrate for AM/H^+ [b]
5. Propionate	Cl^-	Various salt forms	Permeable in undissociated form, dissociating at intracellular pH acidifying cell interior (see text)[c]
6. Gluconate	Cl^-	Various salt forms	Not a substrate for Cl^-/HCO_3^- Buffers Ca^{2+}—be careful
7. PAH (p-amino-hippurate)	Cl^-	Na-PAH	Poor substrate for Cl^-/HCO_3^-; very low rate of PAH transport on Cl^-/HCO_3^- exchanger[d]
8. SO_4, SCN, NO_3	Cl^-	Various salt forms	Serve as substrates for Cl^-/HCO_3^- with variable affinities; variable inhibitory effects on Na^+/H^+ [d]
9. Sucrose	Any osmotically active particle	Sucrose	Lowered ionic strength of media Be careful: exposure to low ionic strength for long periods can cause major changes in permeability and other membrane characteristics

[a] M. Haas, W. F. Schmitd III, and T. J. McManus, *J. Gen. Physiol.* **80**, 125 (1982).
[b] J. Funder, D. C. Tosteson, and J. O. Wieth, *J. Gen. Physiol.* **71**, 721 (1978).
[c] R. Motais, in "Membrane Transport in Red Cells" (J. C. Ellory and V. L. Lew, eds.), p. 197. Academic Press, New York, 1977.
[d] J. S. Adorante and P. M. Cala, *J. Gen. Physiol.* **90**, 209 (1987).

addition of 20 μl of 0.5 M ZnSO$_4$ and the pellet is packed by centrifugation for 8 min. Aliquots of the remaining "lysate" are removed for analysis of ions, H$_2$O, and isotope content. The insoluble pellet is dried in an oven to constant weight (at least 18 hr at 70°).

Analysis for Intracellular Water and Ion Content

Wet and dry pellet weights and cell H$_2$O are determined gravimetrically. Na$^+$ and K$^+$ are determined using a flame photometer (Instrumentation Laboratory, Inc., model 343). Chloride is measured by coulometric titration with silver ions (Buchler Chloridometer, model 4-2500). Aliquots for scintillation counting are diluted with 5 ml ACS (aqueous counting scintillant, Amersham Corp.) and counted on a liquid scintillation counter (Beckman Instruments, Inc., LS 6800). The analysis of lysate and supernatant is carried out in the same manner with the following exceptions: (1) larger volumes of lysate (about 10 times more) are used for all ion determinations since the intracellular concentrations are small and have been further lowered by dilution necessary for cell lysis. (2) The range of flame photometer is expanded 10-fold for the lysates (by adjusting the calibration) to improve the accuracy of the Na$^+$ and K$^+$ measurements.

The correction for trapped extracellular supernatant is determined by comparing the [^{14}C]inulin (counts) associated in the lysate with [^{14}C]inulin in a known volume of supernatant. Alternatively, if extracellular space is invariant for a particular treatment, a blanket correction can be made, eliminating the need for [^{14}C]inulin as an extracellular space marker. After examination of many past net flux experiments performed by this laboratory, we have found that a blanket correction based on a constant ratio of volume of trapped supernatant per weight of dry cell solid yields excellent results when compared with actual ^{14}C extracellular space measurements. However, this type of blanket correction must be established by each laboratory for each cell type as the results will be dependent on the technique employed as well as cell type.

Measuring Intracellular pH: Freeze–Thaw Method

As little as 400 μl of suspended cells at 10% hematocrit can be used for intracellular pH determination. These aliquots must be taken separately from those used for analysis of intracellular water and ion content. The samples are centrifuged for 2 min at 12,000 g. The supernatant is removed and the packed cell pellet is immediately frozen in liquid nitrogen. As the pellet thaws, the cells lyse and the intracellular pH can be measured directly using a radiometer Copenhagen capillary electrode (model G297/

G2) with a calomel reference electrode (model K497). The supernatant pH can also be measured with this electrode. This method will give reasonable values for cytosol pH in cells such as red blood cells which have few organelles. Cells where there is intracellular H^+ compartmentation are not suited to this method.

Studies of Net Alkali Metal/H^+ Exchange: Criteria Used to Identify Alkali Metal/H^+ Exchange

In some cells alkali metal/H^+ exchange flux pathways may be operative in the steady state, as has been suggested for cardiac muscle.[2,3] In other cell types like the *Amphiuma* red blood cell or the human lymphocyte alkali metal/H^+ exchange is inducible and appears to play a non-steady-state role in the regulation of cell volume and/or pH. In the case where alkali metal/H^+ exchangers are contributors to steady-state ion flux, it will not be possible to measure net ion fluxes (see General Method, above) by such pathways while the Na^+/K^+ pump is running since net fluxes by the dissipative alkali metal/H^+ exchange pathways are countered by conservative alkali metal ion fluxes through the pump.

In the case of regulatory (volume or pH) alkali metal/H^+ exchange it is necessary to activate the pathways. If, as is the case for volume-sensitive alkali metal ion fluxes by *Amphiuma* red blood cells, the *inducible* pathway is capable of much higher flux rates than the Na^+/K^+ pump (one to three orders of magnitude), net alkali metal exchange transport will not be significantly altered by a functional Na^+/K^+ pump. As a general rule it is advisable to perform net flux studies in the presence of ouabain (10^{-4} M) and thereby avoid the possibility that the pump may mask net alkali metal/H^+ exchange flux. Our studies have focused upon inducible alkali metal/H^+ exchange in the cell volume regulatory response of *Amphiuma* red blood cells. A protocol describing manipulations involved in evaluating volume-activated ion flux pathways (or identifying their existence) is included in the Appendix (Protocol A).

We have employed a variety of strategies in our attempts to understand alkali metal/H^+ exchange by *Amphiuma* red blood cells (see Fig. 1). Below we will present general details of those approaches that we have found to be most useful. Specific experimental examples may be found in the Appendix with explanations, modifications, and warnings where appropriate.

---

[2] S. Anderson, P. Cala, and E. Cragoe, *Physiologist* **29**, 126 (1986).

[3] D. Pinwica-Worms, R. Jacob, C. R. Horres, and M. Lieberman, *J. Gen. Physiol.* **85**, 43 (1985).

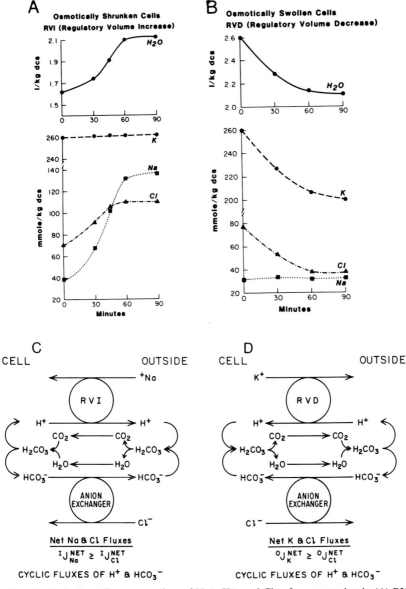

FIG. 1. Typical net fluxes over time of Na^+, K^+, and Cl^- after suspension in (A) RVI (hypertonic, 360 mOsm) or (B) RVD (hypotonic, 150 mOsm) media. Intracellular ion contents are expressed as mmol/kg dry cell solid (dcs). Chloride and osmotically obliged water follow Na^+ uptake in RVI or K^+ loss in RVD. Net Cl^- flux is often less than that of Na^+ or K^+ as mentioned above. Ouabain (1 mM) is used to inhibit the Na^+/K^+ pump. (From Cala.[11]) Models depicting (C) Na^+/H^+ exchange and (D) K^+/H^+ exchange both coupled through pH to Cl^-/HCO_3^- exchange. Net Cl^- flux is functionally coupled to Na^+ or K^+ flux in a manner inversely related to the buffer capacity of the system.

Establishing Electroneutrality

If net ion flux is via Am/H$^+$ exchange, the Am flux should neither contribute to nor respond to the membrane voltage. It is necessary to establish that the ion flux does not carry current. This can be done using a variety of approaches which will only be mentioned here. First conventional microelectrodes may be used to measure transport related changes in voltage, current, or resistance. It is also possible to establish electroneutrality using patch-clamp techniques and finally, if the above approaches are not possible, the membrane potential can be evaluated using voltage-sensitive dyes. The questions of relevance are (1) is the net Am flux associated with a current? (2) Is the net Am flux capable of alerting the membrane voltage? (3) Is the membrane voltage a driving force for net Am flux?

Question (3) can be evaluated electrically with microelectrodes by clamping the membrane voltage and measuring the effect of voltage on the pathway in question. Alternately, it is possible to use an ionophore as a current source.[1,4,5] Briefly, if the cells are exposed to an ionophore, which is able to increase the conductance to some ion (i), such that the membrane's conductance is now approximated by the ionophore-induced conductance of that ion, then the cell E_m will approach the equilibrium potential of that ion (E_i). It is important that the transmembrane concentration ratio of the ion (and therefore E_i and E_m) be constant (or nearly constant) during the period of flux measurement. If E_i and E_m are changing, the results of the flux experiment will be difficult if not impossible to interpret unless those changes are known with precision.

Pharmacological Tests: Studies Employing Amiloride and Its Analogs

It has become increasingly more common to use amiloride as a probe for Na$^+$/H$^+$ exchange. This may, however, be the weakest criterion for establishing the existence of Na$^+$/H$^+$ exchange. The diuretic amiloride has been shown to inhibit Na$^+$/H$^+$ exchange, Na$^+$ conductance, Na$^+$,K$^+$-ATPase, Na$^+$/Ca^{2+} exchange,[6] and C kinase-dependent protein phosphorylation.[7] It is because of this lack of specificity that results of amiloride inhibition of Na$^+$ flux can only be interpreted (as inhibitory to Na$^+$/H$^+$ exchange) under the most rigorously controlled conditions. A more prudent approach is to employ amiloride as well as analogs known to be

[4] P. M. Cala, *J. Gen. Physiol.* **82,** 761 (1983).
[5] S. Grinstein, C. A. Clarke, A. Du Pre, and A. Rothstein, *J. Gen. Physiol.* **80,** 801 (1982).
[6] S. Sariban-Sohraby and D. J. Benos, *Am. J. Physiol.* **250,** C175 (1986).
[7] J. M. Besterman, W. S. May, Jr., H. Le Vine III, E. J. Cragoe, Jr., and P. Cuatrecasas, *J. Biol. Chem.* **260,** 1155 (1985).

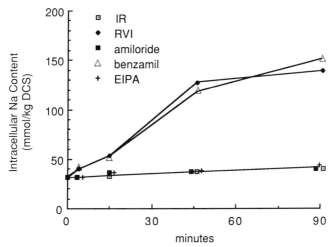

FIG. 2. Inhibition of Na^+/H^+ exchange by amiloride and some amiloride analogs. Intracellular Na content does not change over 90 min in IR (isotonic Ringer's) or in RVI (hypertonic Ringer's, 336 mOsm) in the presence of amiloride or the 5-amino-substituted analog ethylisopropylamiloride (EIPA). Net Na^+ influx seen in the RVI is similar to that seen in the presence of the analog benzamil, which specifically inhibits conductive Na^+ flux. All drug concentrations are 1 mM. (The authors thank Dr. E. Cragoe of Merck, Sharp, and Dohme for supplying amiloride and its analogs.)

specific inhibitors of Na^+ conductance or Na^+/H^+ exchange. For example, benzamil inhibits Na^+ conductance yet not Na^+/H^+ exchange (Fig. 2). In contrast, amiloride analogs with more hydrophobic groups in the 5-amino position (i.e., ethylisopropylamiloride) inhibit Na^+/H^+ exchange but not Na^+ conductance. As such, studies combining both amiloride and amiloride analogs can be more reliably interpreted than those using any of the compounds alone. It must be recognized, however, that even in the best case, secondary effects of these compounds cannot be ruled out. Consequently, studies with amiloride and its analogs must be interpreted with caution and should generally be used only as a prelude to more rigorous tests (see Appendix, Protocol B).

Kinetic Studies of Alkali Metal/H⁺ Exchange

Changes in pH as Indicators of Alkali Metal/H⁺ Exchange. Since H^+ is a substrate for Na^+/H^+ exchange, net Na^+/H^+ exchange flux will, under appropriate conditions, give rise to changes in cell and/or media pH. Changes in media and/or intracellular pH as a function of net Am flux can therefore indicate the function of an Am/H^+ exchanger (see Appendix, Protocol C,1). If the cells under investigation have a robust $Cl^-/$

HCO_3^- exchange, this anion-transport pathway will tend to "buffer" H^+ which is transported by the Am/H^+ exchange. Consequently, net Am/H^+ exchange flux will be seen as net $Am + Cl^-$ flux as depicted in Fig. 1 (see also Cala[1]). In order to evaluate the Am/H^+ stoichiometry net Cl^-/HCO_3^- must be stopped. This can be accomplished with anion-exchange inhibitors such as the disulfonic stilbenes DIDS and SITS or a host of other agents.[8] Alternately, experiments can be performed in the absence of substrate for Cl^-/HCO_3^- exchange. In room air $[HCO_3^-]$ is low (yet sufficient to support Cl^-/HCO_3^- exchange at physiological pH), and its removal can prevent Cl^-/HCO_3^- exchange without changes in volume and pH which accompany Cl^- removal. In practice, HCO_3^- removal to less than micromolar concentrations is necessary but extremely difficult.[9,10] In contrast, removal of media Cl^- by replacement with some anion that will not move on the Cl^-/HCO_3^- exchange will prevent net HCO_3^- flux through the pathway but will lead to cell shrinkage and alkalinization as the cells initially lose Cl^- in exchange for HCO_3^-. The cell shrinkage can be overcome by replacing Cl^- in medium with an appropriately reduced osmolarity, yet cell alkalinization cannot be avoided.

Systems with Net Alkali Metal/H^+ and Cl^-/HCO_3^- Exchange Flux: Tests for the Presence of Cl^-/HCO_3^- Exchange. A quick, effective means of ascertaining the presence of Cl^-/HCO_3^- exchange in a particular cell system involves measuring buffer capacity of buffer-free solutions with and without cells (see Cala[11] and Appendix; Protocol C,2). If the cells in question have a Cl^-/HCO_3^- exchanger, H^+ and OH^- equivalents added to a cell suspension have access to intracellular buffers through the anion exchange. Consequently, the cell suspension will have the buffer power of cells plus medium (Fig. 3). If the cells do not have a Cl^-/HCO_3^- exchanger or the Cl^-/HCO_3^- exchanger has been inhibited, then the buffer power of the suspension is that of the medium alone. As such, titration of poorly buffered Ringer's solution (PBR) (1) in the presence and absence of the cells to be studied and (2) in the presence and absence of Cl^-/HCO_3^- exchange inhibitors will allow an assessment of the existence of cell Cl^-/HCO_3^- exchange and the effectiveness of its inhibition. If a Cl^-/HCO_3^- exchanger is present, as inferred from increased buffer capacity in the presence of cells, then the study can be performed in media replaced by an anion such as gluconate which is not transported by the Cl^-/HCO_3^- exchanger (see Table I). In the absence of Cl^- as a substrate the Cl^-/HCO_3^- exchanger will be inhibited and the buffer capacity

[8] P. A. Knauf, *Curr. Top. Membr. Transp.* **12,** 249 (1979).
[9] M. L. Jennings, *J. Membr. Biol.* **28,** 187 (1976).
[10] M. L. Jennings, S. M. Douglas, and P. E. McAndrew, *Am. J. Physiol.,* in press.
[11] P. M. Cala, *Mol. Physiol.* **8,** 199 (1985).

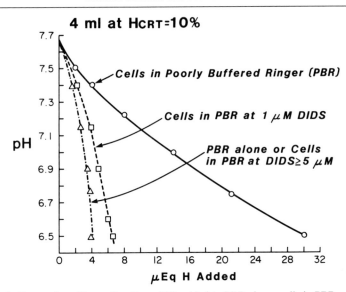

FIG. 3. Suspension pH as a function of H^+ added to PBR alone, cells in PBR, or DIDS-treated cells in PBR. DIDS concentrations were 1, 5, 10, 20, 40, 80, and 100 μM. Note that at DIDS ≥ 5 μM the buffer capacity of cells plus DIDS in PBR is indistinguishable from that of the PBR alone.

of the cell suspension should approach that of the medium alone. Similarly, this approach can be used to assess anion-transport inhibitors and the concentration of such compounds necessary to completely block Cl^-/HCO_3^- exchange.

Alkali Metal/Cl^- Cotransport vs Alkali Metal/H^+ and Cl^-/HCO_3^- Exchange. The Cl^-/HCO_3^- exchanger operating in parallel with electroneutral net Am/Cl^- cotransport will result in pH changes associated with net Am flux (Fig. 4d). In contrast, parallel Am/H^+ and Cl^-/HCO_3^- transport will tend to prevent changes in pH (Fig. 4a). That is, net Am/Cl^- cotransport can cause changes in [Cl^-] which cause the Cl^-/HCO_3^- exchange to mediate net Cl^- and HCO_3^- fluxes and produce pH changes (due to net HCO_3^- flux). In contrast, net H^+ fluxes via Am/H^+ exchange can cause changes in [HCO_3^-] which drive net Cl^-/HCO_3^- exchange and buffer H^+ associated with net Am/H^+ exchange (H^+ and HCO_3^- move in the same direction, see Cala[12]). In order to distinguish between these two possibilities studies can be performed on cells where Cl^-/HCO_3^- has been inhibited (see Appendix, Protocol C,3).

[12] P. M. Cala, *Mol. Physiol.* **4**, 33 (1983).

ALKALI METAL ION–DEPENDENT H₂O+Cl MOVEMENTS

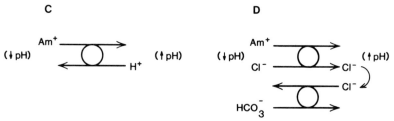

ALKALI METAL ION (Am⁺)–DEPENDENT CHANGES IN pH

FIG. 4. Diagrams depicting net alkali metal ion-dependent changes in net salt and H_2O fluxes or changes in pH. Note that both Am/H^+ exchange and Am/Cl^- cotransport pathways can produce the same net overall effect depending upon the presence or absence of a Cl^-/HCO_3^- exchanger. (A) and (B) show that if Am/H^+ exchange (A) is operating in parallel with a Cl^-/HCO_3^- exchanger, the net result is net Am/Cl^- (and H_2O) flux as is the case for an Am/Cl^- cotransporter (B). (C) and (D) illustrate that net Am/H^+ exchange (in the absence of a Cl^-/HCO_3^- exchanger) will give rise to changes in pH as will an Am/Cl^- cotransport pathway if in parallel with Cl^-/HCO_3^- exchange. Note that Am/H^+ exchange and Am/Cl^- cotransport couple to Cl^-/HCO_3^- exchange through $\Delta[HCO_3^-](\Delta pH)$ and $\Delta[Cl^-]$, respectively. Also, changes in pH due to Am/H^+ exchange or parallel Am/Cl^- cotransport Cl^-/HCO_3^- exchange are such that pH is decreased on the side of the membrane away from net Am flux and increased on the side of the membrane in the direction of net Am flux.

Thermodynamic Criteria: Force-Flow Analysis

Given that all flows (fluxes) are coupled to forces, much can be learned about the nature of the flow if the relevant forces are identified. We have used this approach in order to identify net Am/H^+ flux and indeed to distinguish between Am flux via Am/H^+ exchange, Am/Cl^- cotransport, and Am conductance.[4,11–13] Briefly, if net Am flux is by Am/H^+ exchange, then it must couple to the difference in the chemical potential differences for Am and H^+ ($\Delta\mu_{Am} - \Delta\mu_H$) as its driving force (Fig. 5). If net Am flux is via Am/Cl^- cotransport, then it must couple to the chemical potential

[13] P. M. Cala, *Fed. Proc., Fed. Am. Soc. Exp. Biol.* **44,** 2500 (1985).

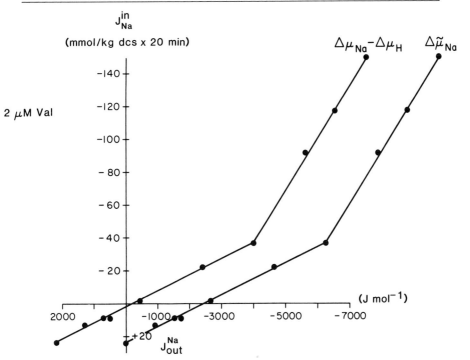

FIG. 5. Correspondence between net Na^+ uptake and the thermodynamic force for (1) Na^+/H^+ exchange ($\Delta\mu_{Na^+} - \Delta\mu_{H^+}$) and (2) Na^+ conductance ($\Delta\mu_{Na^+}$). Cells osmotically shrunken in media containing 100 μM DIDS, 2 μM valinomycin, and $[Na^+]$ varied from 6 to 150 mM. The Na^+ was varied by replacement with tetramethylammonium. (From Cala.[13])

differences for Am and Cl^- ($\Delta\mu_{Am} + \Delta\mu_{Cl^-}$). Finally, if net Am flux is conductive, then the chemical potential difference for Am and the electrical potential difference (membrane potential) are the relevant driving forces ($\Delta\mu_{Am} + zFE_m$). As the magnitudes of the above quantities change so too must the magnitude of net flux. As the sign of the quantities changes, the direction of net flux (coupled to that particular force) must change. The correlation of net Am flux with one of the above terms will permit determination of the nature of the net flux pathway based upon the nature of the forces to which it couples. Since $A\mu_{Am}$ is common to all net Am flux pathways, the discriminatory terms in the above expressions are $\Delta\mu_{H^+}$, $\Delta\mu_{Cl^-}$, and zFE_m. Consequently, attempts to distinguish between the possibilities require that the studies are performed under conditions where at least two of the terms are unequal during the course of the experiment (i.e., $-\Delta\mu_{H^+} = \Delta\mu_{Cl^-} \neq zFE_m$ or $-\Delta\mu_{H^+} \neq \Delta\mu_{Cl^-} = zFE_m$). Alternatively, using the example of the *Amphiuma* red blood cell, we had shown, based on kinetic experiments (inhibited anion exchange), that net Am flux was not associated with that of Cl^-. As such net Am flux cannot be Am/Cl^- cotransport. Since we had previously ruled out the possibility of Am/Cl^- cotransport, we needed only to be concerned with Am/H^+ exchange ($\Delta\mu_{Am} - \Delta\mu_{H^+}$) and Am conductance ($\Delta\mu_{Am} + zFE_m$) as possible flux pathways. Thus, we were able to perform studies where $-\Delta\mu_{H^+} = \Delta\mu_{Cl^-} \neq zFE_m$.

Based on these studies we were able to establish Am/H^+ exchange as the flux pathway. The most convincing criticism being the ability of H^+ ($\Delta\mu_{H^+}$) to drive net Am flux against a gradient of electrochemical potential. Using the example in Fig. 5 the net Na^+ influx (upper right gradient, $\Delta\mu_{Na^+} - \Delta\mu_{H^+}$) at all forces less than about 2800 J mol^{-1} represent net Na^+ flux driven by H^+. Note that valinomycin was included in order to establish inequality between $-\Delta\mu_{H^+}$ and zFE_m (see Appendix, Protocol D).

Appendix

Protocol A: Volume Sensitivity of Ion Flux

1. Obtain cells
2. Preincubate cells in isotonic Ringer's solution (1R)
3. Transfer cells to ouabain (added to block Na^+-K^+ pump) containing anisotonic Ringer's [see Table II for typical experimental (anisotonic) media] to activate AM flux pathways:
   a. Isotonic Ringer's (1R—controls; no activation): 1R—1R
   b. Hypotonic Ringer's (regulatory volume decrease or RVD): 1R—RVD
   c. Hypertonic Ringer's (regulatory volume increase or RVI): 1R-RVI
4. Sample cells with time[14] at time 0, 15, 30, 60, and 90 min
5. Analyze: see general methods

[14] The method described in this chapter requires ~1 min in order to obtain the first sample. While this is acceptable for *Amphiuma* red blood cells, in which volume regulatory ion and water fluxes proceed for 30–120 min, it may not provide sufficient resolution for cells whose volume regulatory ion fluxes are more rapid and of shorter duration. For such cells, variants of the described method involving rapid filtration for sampling and/or the use of low temperature to slow ion and water transport may be necessary. In addition, the sensitivity of this method requires minimum net fluxes of 10–15 mM/kg of dry cell solid (dcs)/hr. In some systems (i.e., mammalian erythrocytes), net volume and pH regulatory fluxes are <10 mmol/kg dcs/hr and it may be more appropriate to employ radioisotopes (i.e., ^{23}Na) for improved sensitivity.

TABLE II
EXPERIMENTAL MEDIUM USED FOR "TYPICAL" STUDY
INVESTIGATING EXISTENCE OF VOLUME-SENSITIVE
ION FLUXES

Experimental medium	Osmolarity composition
Isotonic (1R)	1R = 1 × isotonic
Hypotonic[a]	0.55 RVD = 0.55 × isotonic
Hypertonic[a]	1.5 RVI = 1.5 × isotonic

[a] If medium is made too dilute (hypotonic less than 0.5 the osmolarity of isotonic) or hypertonic (greater than 1.9 times the osmolarity of isotonic medium) the *Amphiuma* red blood cells will lyse. Using red blood cells it is relatively easy to determine prelytic osmolarities since surpassing the limits result in a red supernatant because of Hb release. For studies of cells which do not contain a pigment it is necessary to employ more sophisticated means of establishing the limits of prelytic osmolarity.

Interpretations

If after initial swelling or shrinkage in anisotonic ringer, the cells change ion *content* (not concentration) and H_2O volume with time and approach the control or unperturbed volume, then volume-activated ion flux pathways exist.

This experiment provides no information concerning the *mechanism* of the flux pathway. These types of experiments are discussed in subsequent sections.

Protocol B: Pharmacological Tests: Studies Employing Amiloride and Its Analogs

Identity of Na^+ Flux Pathways

1. Activate Na^+ flux (RVI) in the absence and presence of amiloride, benzamil, and a 5-amino-substituted analog
2. Measure net Na^+ flux over time

Interpretations

If amiloride inhibits net Na^+ flux, data are inconclusive.

If net Na^+ flux is inhibited by benzamil it is probably not Na^+/H^+ exchange.

If net Na^+ flux is inhibited by amiloride and 5-amino-substituted analogs, but not by benzamil, Na^+ flux is probably by Na^+/H^+ exchange.

Caution. Specificity of pharmacological probes is always open to question. These experiments will provide data consistent with a particular hypothesis yet not necessarily unique to it.

Protocol C: Kinetic Studies

1. Changes in pH as Indicators of Am^+/H^+ Exchange

Activate net Am^+ flux and

1. Measure ΔpH cells (see Freeze–Thaw method) and medium
2. Measure ΔpH cells and medium using Cl^-/HCO_3^- exchange inhibitors

Interpretations

If under conditions of net Am flux, anion-exchange inhibitors produce an increase in Am-dependent ΔpH, then net Am flux is via Am/H^+ exchange.

If Am-dependent ΔpH decreases in the presence of anion-exchange inhibition, then net Am flux is via Am/Cl^- cotransport.

Hints/Caution

If Am/Cl^- flux is via parallel Am/H^+ and Cl^-/HCO_3^- exchange, then anion-exchange inhibitors should dissociate net Am and Cl^- flux.

If Am/Cl^- flux is due to Am/Cl^- cotransport in parallel with Cl^-/HCO_3^-, exchange then net Cl^- flux should increase upon inhibition of Cl^-/HCO_3^- exchange.

The above protocol assumes that net Am flux is electroneutral. If net Am and Cl^- fluxes are conductive and there is a Cl^-/HCO_3^- exchange, then anion-exchange inhibition could produce the same effects as those expected for parallel Am/Cl^- cotransport and anion exchange.

2. Systems with Cl^-/HCO_3^- Exchange: Testing for the Presence of Cl^-/HCO_3^- Exchange

1. Titrate poorly buffered isotonic ringer (PBR)
2. Titrate cells (at least 10% hematocrit) in isotonic PBR
3. Repeat above using cells exposed to anion-exchange inhibitors or cells in Cl^--free medium

Interpretations

The difference in buffer capacity in cases 1 and 2 above reflects access of medium acid or base to intracellular buffers.

If in the absence of anion-exchange inhibitors the buffer capacity of medium alone is the same as that of cells plus medium, then the cells do not have a Cl^-/HCO_3^- exchanger.

If in the presence of anion-exchange inhibitors, buffer capacity of cells plus medium is equal to that of medium alone, then Cl^-/HCO_3^- exchange is inhibited.

Caution

Gently mix cell suspension during titration to avoid lysis.

If the cells have a robust Am/H^+ exchanger or high Am and H^+ or H^+ and Cl^- conductance, then H^+ can gain access to intracellular buffers through pathways other than Cl^-/HCO_2 exchange.

3. Am/Cl^- Cotransport vs Am/H^+ and Cl^-/HCO_3^- Exchange

Activate net Am flux and

1. measure change in media pH vs time
2. measure net Am flux vs time
3. repeat above under conditions where Cl^-/HCO_3^- exchange is inhibited

Interpretations

If in the presence of anion-exchange inhibitors, a given net Am flux produces no change in pH (or a ΔpH decrease) while in the absence of anion-exchange inhibitors net Am flux leads to changes in pH (or increased ΔpH, Fig. 4d), then net Am flux is via Am/Cl^- cotransport.

If Am-dependent ΔpH is increased by exposure of cells to anion-exchange inhibitors (Fig. 4c), then the net Am flux pathway is probably Am/H^+ exchange.

[Note: This interpretation is dependent upon having established that (1) net Am flux is electroneutral and (2) the membrane conductance to H^+ is low.]

Protocol D: Thermodynamic Studies: Force-Flow Analysis

Activate net Am flux and

1. if counter or cotransported species (i.e., H^+ or Cl^-) are at electrochemical equilibrium, alter cell E_m by altering membrane conductance to some nonequilibrium-distributed ion (use ionophore)
2. vary [Am] and measure net Am flux
3. vary the concentration of the ion to which you think Am couples ([X]) and measure net Am flux

4. calculate driving force $\Delta\mu_{Am} \pm \Delta\mu_X = RT \ln ([Am]_i/[Am]_o) \pm RT \ln ([X]_i/[X]_o)$ (+, for cotransport; −, for exchange).
5. plot net Am flux against $\Delta\mu_{Am} \pm \Delta\mu_X$ (see Fig. 5)

Interpretations

Tight correspondence between force and flow will identify the pathway.

Am flux should be able to drive the flux of X against its gradient and X should be able to drive the net flux of Am against its gradient depending on the relative signs and magnitude of $\Delta\mu_{Am}$ and $\Delta\mu_X$.

Hints/Caution

If Am is coupled to H, intra- (freeze–thaw) and extracellular (supernatant) pH must be measured.

Flux (and freeze–thaw if necessary) measurements must be made over a period:

1. of slight change in $\Delta\mu_{Am}$ and $\Delta\mu_X$
2. short relative to the time required for ions to respond to the ionophore-induced voltage and flux

[20] Preparation of Resealable Membranes Maximally Depleted of Cytosolic Components by Gel Filtration and Tryptic Digestion: Resealable White Ghosts

By Phillip G. Wood

Introduction

The human erythrocyte membrane contains a number of transport proteins that have occupied the attention of physiologists for many decades. To better understand how the various membrane-bound transport systems function and their interaction with cytoplasmic factors, detailed kinetic and biochemical studies are employed. To characterize the system kinetically, it is generally necessary to control the chemical conditions on both sides of the membrane. Toward these ends, the resealed erythrocyte ghost has been developed, where the chemical environment of both sides

METHODS IN ENZYMOLOGY, VOL. 173

of the membrane can be reestablished after lysis (for reviews, see Refs. 1 and 2).

In general, to form resealable ghosts: (1) intact cells are mixed with a hypotonic medium, allowing the cells to lyse, (2) to the lysate the desired components are added and ionic strength elevated to normal levels, and (3) the cells are resealed by incubation at an elevated temperature. Generally these operations are conducted in a beaker and a certain percentage of the cytosol remains entrapped. Thus, the red blood cells are converted into pink resealed ghosts. Intermediate washing or dialysis steps may be added, but the percentage of resealable ghosts generally decrease as a result.[3,4]

Because of this problem a continuous flow method was developed, whereby the cytosol would be able to continuously diffuse out of the lysed membranes while passing through an agarose gel filtration column.[5] The ghost membrane formed during lysis on a gel filtration column provides the limiting case where all diffusible components found in the cytoplasm are removed. The conditions in the column are extremely mild and lysed membranes are easily maintained at constant temperature and are not subjected to high centrifugal forces. Through the addition of proteolytic enzymes, both surfaces of the membrane may be within limits modified during the resealing step. Previous publications have compared various methods and characteristics of resealable white ghosts.[2,6]

General Principles

Many laboratories have participated in developing conditions that allow us to form resealed ghosts.[7–11] The overall conditions utilized for the preparation of white resealed human erythrocyte ghosts were largely adapted from the well-established and frequently used procedures of Bodemann and Passow[12] and Lepke and Passow[13] for the formation of

[1] G. Schwoch and H. Passow, *Mol. Cell. Biochem.* **2**, 197 (1973).
[2] P. G. Wood and H. Passow, *Tech. Life Sci.: Physiol.* **P1**, 1 (1981).
[3] W. J. Mawby and J. B. C. Findlay, *Biochem. J.* **172**, 605 (1978).
[4] J. T. Dodge, C. Mitchell, and D. J. Hanahan, *Arch. Biochem. Biophys.* **100**, 119 (1963).
[5] P. G. Wood, *Fed. Proc., Fed. Am. Soc. Exp. Biol.* **34**, 249 (1975).
[6] P. G. Wood, this series, Vol. 149, p. 271.
[7] T. Teorell, *J. Gen. Physiol.* **35**, 669 (1952).
[8] F. B. Straub, *Acta Physiol. Acad. Sci. Hung.* **4**, 235 (1953).
[9] G. Gardos, *Acta Physiol. Acad. Sci. Hung.* **6**, 191 (1954).
[10] J. F. Hoffman, D. C. Tosteson, and R. Whittam, *Nature (London)* **185**, 186 (1960).
[11] J. F. Hoffman, *Circulation* **26**, 1201 (1962).
[12] H. Bodemann and H. Passow, *J. Membr. Biol.* **8**, 1 (1972).
[13] S. Lepke and H. Passow, *Biochim. Biophys. Acta* **255**, 696 (1972).

pink resealed ghosts. Under these conditions more than 90% of the ghosts could be resealed for potassium when hemolysis was conducted at 0°, pH 6. At this temperature spontaneous resealing is negligible and allows one to introduce substances into the membranes. The pH at hemolysis is most important for maintaining the cation barriers in a state that can be restored later during resealing at 37°. In contrast, the anion barrier is less sensitive to pH at hemolysis.[2] In the batch method hemolysis occurs when 1 vol of a 50% cell suspension is mixed at 0° with 10 vol of a pH 3.2, 4 mM MgSO$_4$ using 1.2 mM acetic acid as a buffer. The actual concentration of acetic acid in the hemolysing medium is selected such that on mixing with cells pH 6.0 is attained in the final lysate. Prior to the readdition of salt the pH is readjusted to 7.2 and the new cytosol restored through the addition of an appropriate volume of a concentrated stock solution. The membranes are resealed by incubation at 37°. Under these dilution conditions, about 5% of the cytosol remains.

Cell Filtration

The depletion of the lysed cell of its cytosol on a gel filtration column can be accomplished in a manner analogous to that used in the desalting of protein on a column. A gel is required which has an exclusion limit greater than the largest protein that may be found in the cytoplasm. Washed cells in isotonic saline are allowed to flow into an agarose gel filtration column which has been preequilibrated with a hypotonic buffer. As illustrated in Fig. 1, the red blood cells can only flow through the void spaces between the gel beads (Fig. 1A). In the hypotonic buffer the cells burst to release their cytosol. The released contents can diffuse into the beads and void spaces (Fig. 1B). The effective path length for the released cytosol is three to four times that of the membranes. Consequently the cell membranes diffuse away from the lysate. Any remaining cytosolic components in the cells are free to diffuse into fresh buffer.

In order to deplete the cells of their cytosolic contents an agarose gel with an exclusion limit of 50 million daltons is used in this laboratory (Bio-Rad, Richmond, CA; agarose A50m, 50–100 mesh). The large mesh size of the gel allows sufficient space for the cells to pass freely through the gel bed.

A jacketed column is necessary which can maintain the column temperature between 0 and 2°. It is essential to maintain temperature in this range or the membranes will spontaneously reseal while on the column. It is equally important to avoid freezing the gel in the column which will fragment the gel beads and change the flow characteristics of the bed. A short, broad column is needed because the purpose of the column is not

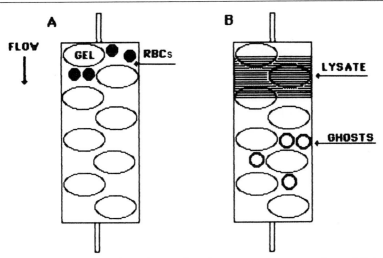

FIG. 1. The separation of membranes from lysate on an agarose column: (A) before hemolysis; (B) after hemolyis.

resolution or separation of a component that enters the gel matrix but the separation of the excluded membranes from their accompanying medium which is initially the lysate. Table I gives the nominal characteristics of columns that may be used.

The flow rate must be such that a flat velocity profile is maintained in the bed rather than an extended parabolic profile. The optimal flow rate for particular columns may be determined in a preexperiment by passing a volume of cells equivalent to that to be used in the ghost preparation through a column preequilibrated with isotonic saline. The cells should

TABLE I
NOMINAL COLUMN CHARACTERISTICS

	Column		
Column property[a]	I	II	III
Column diameter (cm)	2.5	5	10
Gel bed length (cm)	30	30	25
Gel volume (liter)	0.15	0.6	2
Estimated nominal void volume (liter)	0.04	0.2	0.5

[a] Gel: Agarose A-50m, 50–100 mesh (Bio-Rad, Richmond, CA).

flow as a flat band. Since the cells are found only in the void space, the red band should appear three to four times wider than expected for the loaded volume. In addition, when the flow rate is too high, the agarose bed becomes compressed and restricts the flow through the bed. Figure 2A illustrates the desired flow profile. In Fig. 2B the flow rate is too high; mixing of membranes and lysate is possible as the ghosts leave the column. Columns with flow adaptors (Pharmacia) are well suited for ghost preparation. However, the 10-μm nylon mesh delivered with the flow adaptors should be exchanged for a 25-μm mesh to allow free passage of the membranes. The flow adaptors must rest directly on the gel bed during flow to assure that the cells enter the gel bed and do not accumulate in a dead volume above the gel bed. All tubing leading from the column should be packed in ice to prevent premature warming of the membranes.

Hemolysis is carried out at pH 6.5 which is a compromise of the conditions used by Lepke and Passow.[13] A strong buffer with a pK_a in this range is used and PIPES (piperazine-N,N'-bis-2-ethanesulfonic acid, pK_a 6.8) has been found to be satisfactory. Constant temperature and pH must be maintained throughout the column to allow continuous diffusion of the cytosol from the membranes. A variation of a few degrees from 0° causes the membranes to relax and prevents the eventual entry and exit of proteins, although the membranes remain permeable to inorganic ions and may be resealed to normal levels later.

The column can be preequilibrated with a hemolysing solution containing $MgSO_4$ or chelators such as EDTA or EGTA. Table II gives

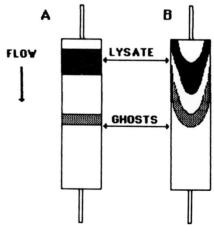

FIG. 2. Flow profiles on an agarose column: (A) plug flow; (B) extended parabolic flow.

typical compositions that may be used. As long as an acidic pH is maintained at this ionic strength (which is about equivalent to 1 : 10 hemolysis) the membranes reseal for anions and most cations, including Na^+ and choline. When cells are hemolysed above pH 7.6, the conditions become similar to those used in the preparation of Dodge ghosts.[4] White membranes are formed, but they are leaky to Na^+ and K^+.[14]

We have previously shown that the Ca^{2+}-activated K^+ channel becomes activated when the cytosol in the resealed ghosts is diluted by more than a factor of 20.[15,16] This effect seems to be caused by the dilution during hemolysis of the endogenous chelators normally present in the cytosol which help maintain low free Ca^{2+} levels. Because the Ca^{2+} concentration in unbuffered solutions is 2 μM or more, sufficient free Ca^{2+} is present to activate the channel. This activation can be prevented when either Mg^{2+} or a chelator is present at the time of hemolysis to compete with or to buffer the background Ca^{2+} levels near the channels. The intact cell maintains the internal concentration of free Ca^{2+} below 10^{-7} M through the presence of an efficient Ca^{2+} pump and a high level of endogenous chelators such as ATP (for reviews, see Refs. 17–19). Overall, efficient control of free metal ion concentrations is achieved with hemolysis solution C (see Table II, Step 1), which contains EDTA. More specific control of Ca^{2+} can be obtained through the substitution of EGTA for EDTA.

Cell Preparation

We generally use 1- to 7-day-old Red Cross blood bank cells conserved in ACD (acid–citrate–dextrose) buffer. The erythrocytes are washed three or four times in isotonic solution A (see Table II), pH 7.6 at room temperature. Due to the high buffering capacity of the cells, the pH of the supernatant after pelleting should be checked to be certain that the pH is still near 7.6. If this is not the case, washing should be continued until the pH stabilizes. As reported by Dodge[4] hemoglobin binding increases as decreases, thus it is essential to reduce hemoglobin binding as much as possible prior to hemolysis. Just prior to the addition of cells to the column, an isotonic buffer cushion is layered onto the column (see Table II, Step 2). This helps prevent the cells from premature lysis in the

[14] P. G. Wood, *Protides Biol. Fluids* **29**, 283 (1982).
[15] P. G. Wood and U. Rossleben, *Biochim. Biophys. Acta* **553**, 320 (1979).
[16] P. G. Wood, *Biochim. Biophys. Acta* **774**, 103 (1984).
[17] R. W. Meech, *Annu. Rev. Biophys. Bioeng.* **7**, 1 (1978).
[18] W. Schwarz and H. Passow, *Annu. Rev. Physiol.* **45**, 359 (1983).
[19] B. Sarkadi and G. Gardos, *in* "The Enzymes of Biological Membranes" (A. N. Martonosi, ed.), Vol. 3, p. 193. Plenum, New York, 1985.

TABLE II
PREPARATION OF COLUMN

Step[a]	Column		
	I	II	III
1. Preequilibration: elute column with solution B or C; pH 6.5 buffer (liter)	0.4	1.8	5
2. Conditioning cushion: elute column with solution A; pH 7.6 buffer (ml)	15	60	200
3. Addition of cells: load column with a 10–15% washed red cell suspension (ml)	20	75	250

[a] Solution A (erythrocyte wash medium): 146 mM NaCl, 20 mM HEPES, pH 7.6 at room temperature. Solution B (magnesium hemolysis buffer): 5 mM $MgSO_4$, 5 mM KCl, 5 mM PIPES, pH 6.5 at 0°. Solution C (EDTA hemolysis buffer): 0.1 mM EDTA, 15 mM PIPES, pH 6.5 at 0°.

tubing. After washing the cells, the pellet is mixed with sufficient wash medium to form a nominal 10% cell suspension (Step 3). Table II gives typical loading volumes that may be used.

Lysis

The cell suspension is applied to the column under gravity flow at a rate that allows plug flow through the column. Isotonic wash buffer (solution A) may be used as eluent. A volume of washed cells equivalent to ⅛ of the bed volume is applied to the column under gravity flow with a hydrostatic head of 30–40 cm. The flow rate should be as high as possible to minimize the total time of preparation. However, if the flow rate is too high, then there will be too much distortion of the flow profile and dilution of the membranes. If the later is the case, larger volumes must be collected to recover the cells initially applied. Under optimal flow conditions the hemoglobin band should appear as a flat band. The lysed membranes exit the column after the equivalent of one void volume has been eluted,

TABLE III
MEMBRANE RECOVERY

Step	Column		
	I	II	III
Maximum expected volume of ghosts (ml)	2	7.5	25
After the first appearance of membranes; collect at 0° (ml)	20	75	250
Restore isotonicity with 3 M KCl, 0° (ml)	1.06	3.98	13.25

which is about $\frac{1}{3}$–$\frac{1}{4}$ of the bed volume. The membranes are collected in a volumetric flask submerged in a ice–water bath. It should not be necessary to collect a volume much greater than that initially applied. The membranes should be opalescent white and indistinguishable from the agarose gel. After collecting the membranes, they may be quickly subdivided into aliquots with a precooled glass cylinder. The desired components may be added from precooled concentrated stock solutions. Table III gives typical values for membrane recovery and resealing.

Incorporation of Salts and Higher Molecular Weight Compounds

The introduction of salts or other components into the cytosol can be achieved through the addition of a suitable volume of a concentrated stock solution. With the addition of salt, there is a transient osmotic shrinkage of the open ghosts.[20] The osmotic outflow of water may restrict the entry of high-molecular-weight compounds. Thus, it is better to add the high-molecular-weight compounds prior to the addition of the concentrated salt solutions. Between additions incubation for 10 min at 0° is sufficient to allow equilibration of the compounds. We have incorporated albumin, carbonate dehydratase, various dextrans, and trypsin (see below) in this manner.[2,6]

Resealing

The cell suspensions may be resealed by elevating the temperature. Incubation at 37° for 45 min is usually sufficient for the resealing or an-

[20] J. Funder and J. O. Wieth, *J. Physiol.* (*London*) **262**, 679 (1976).

nealing of the fissures formed in the membranes during hemolysis. Further washings may then be conducted to establish the desired external conditions.

Partial Tryptic Digestion

In investigating the properties of the Ca^{2+}-activated K^+ channel, we found that when the white ghost membranes are loaded with low levels of trypsin and partially digested, a modification of the selectivity filter of the channel takes place.[21] From the inner surface progressive tryptic digestion seems to cause initially an increase in the rate of Ca^{2+} activated K^+ efflux in the resealed ghosts. At higher levels the selectivity filter of the Ca^{2+}-activated K^+ channel seems to become degraded, resulting in a loss of discrimination between Na^+ and K^+. At higher levels the channel becomes refractory to added Ca^{2+}. Efflux measurements were done in the presence of TLCK (sodium p-tosyl-1-lysine chloromethyl ketone) to prevent any further digestion. Since externally applied trypsin after resealing did not have an effect on the channel, we conclude that the filter was present at the inner surface. Not all transporters are affected by tryptic digestion. Lepke and Passow[22] have shown that the kinetic properties of the anion transporter are not affected by partial tryptic digestion of the inner surface in resealed pink ghosts.

In white resealable ghosts trypsin (10–100 ng/ml, 7300 units/mg) has been incorporated after the ghost membranes exit the column and prior to the addition of salt. Very mild conditions have been used for digestion. The ghosts are incubated at pH 6.5 for 2 hr at 37° in the EDTA-containing hemolysis solution C. Although digestion is conducted far from the pH optimum of the enzyme, sufficient activity is present to digest the accessible proteins. Even at pH 6.5 the spectrin bands are first to be truncated, then followed by the band 3 anion transporter protein[21] as visualized on SDS–PAGE gels.[23] While the enzyme is present on both sides of the membrane under these conditions, the digestion occurs primarily at the inner surface. Steck et al.[24] have shown that only glycophorin is degraded from the outer surface. However, glycophorin does not directly participate in any known transport process. Under these conditions the resealed ghosts appear as intact membranes under a phase-contrast light microscope with minimal amounts of vesiculation. However, at elevated pH

[21] P. G. Wood and H. Mueller, *Eur. J. Biochem.* **141,** 91 (1984).
[22] S. Lepke and H. Passow, *Biochim. Biophys. Acta* **455,** 353 (1976).
[23] U. K. Laemmli, *Nature (London)* **227,** 680 (1970).
[24] T. L. Steck, G. Fairbanks, and D. F. G. Wallach, *Biochemistry* **10,** 2617 (1971).

values or trypsin concentrations complete vesiculation of the membrane can take place.

Caution should be exercised when using proteolytic enzymes. In general for biochemical studies, it is advisable to use inhibitors such as DFP (diisopropyl fluorophosphate) or PMSF (phenylmethylsulfonyl fluoride) to stop digestion. Thereafter, membranes may be prepared or solubilized without the danger of continuing digestion. For kinetic studies this approach is not always possible since the inhibitors react not only with the incorporated enzyme but also with all the membrane proteins. Thus, the transport proteins will also be modified. Specific inhibitors such as TLCK for trypsin or TPCK (N-tosyl-1-phenylalanine chloromethyl ketone) for chymotrypsin may be of some help.

In the case of trypsin after 2 hr of incubation at resealing, digestion reaches a stable level. The protein pattern on SDS-PAGE gels does not change with longer periods of incubation. In general as long as the pH is not changed in subsequent preparations, transport measurements may be made. However, TLCK should be added to the wash medium as an extra precaution, especially when changing the pH of the medium in later washings.

Care of Column

After the recovery of the ghost membranes, the flow through the column can be reversed and the column freed from the lysate with the continued elution of two to three bed volumes of solution A. At the ionic strength of the buffer there should not be any appreciable binding of hemoglobin or other proteins to the agarose gel and the column should regain its original white color. The column can be used repeatedly. When the column is not in use, 0.02% sodium azide should be added to the wash buffer to prevent bacterial growth and the column should be stored in a cold room.

Acknowledgments

    The author thanks Miss H. Müller for her excellent technical assistance and Professor H. Passow for reading and discussing the chapter.

[21] Hemolytic Holes in Human Erythrocyte Membrane Ghosts

By MICHAEL R. LIEBER and THEODORE L. STECK

The aim of this chapter is to describe the nature of the holes induced in red cell membranes by osmotic hemolysis, how to prepare intact membranes (ghosts) with holes of a desired size, and how to measure the size and the number of holes per ghost.

The human erythrocyte is a near-perfect osmometer.[1] Because its membrane is highly permeable to water, the cell swells rapidly when placed in hypotonic media.[2] When stretched by only a few percentages, the bilayer of an osmotically swollen cell yields to hydrostatic pressure by allowing the dilation of a hole. The rupture may reach hundreds of nanometers in the few seconds during which the membrane is stressed.[3] Once the cytoplasmic contents exit, the tension abates and the hole constricts to a stable size determined by the composition of the buffer, but typically 10–20 nm in radius.[4,5]

The residual membrane plus hole is called a ghost. The ghost membrane appears to be unchanged for the most part, although bilayer molecules may diffuse from one surface of the membrane to the other across the rim of the hole.[6] The hemolytic lesion is extremely stable in size over days in ghosts stored unperturbed on ice. However, the dimensions of the hole can be increased or decreased by warming the ghosts in buffers of varied composition, from a radius of less than 1 nm to a radius of more than 1 μm (corresponding to areas of approximately 10^{-8} to 10^{-2} that of the ghost surface).[4,5] It appears that the hole dilates in response to electrostatic repulsions among the anionic groups abundant on this membrane. Hole closure may be driven by the hydrophobicity of its edge. Hole size may then reflect the balance point between these opposing forces.

Can these holes be truly resealed, i.e., abolished? We believe not. However, they can contract to a radius of less than 0.5 nm. Such small holes sieve even inorganic ions, so that the passive permeability to Na^+ or K^+ of these ghosts may be less than 10 times that of the parent red cell.[7]

[1] J. C. Freedman and J. F. Hoffman, *J. Gen Physiol.* **74**, 157 (1979).
[2] H. A. Massaldi, A. Fuchs, and C. H. Borzi, *J. Biomech.* **16**, 103 (1983).
[3] J. P. Yee and H. C. Mel, *Biorheology* **15**, 321 (1978).
[4] M. R. Lieber and T. L. Steck, *J. Biol. Chem.* **257**, 11651 (1982).
[5] M. R. Lieber and T. L. Steck, *J. Biol. Chem.* **257**, 11660 (1982).
[6] Y. Lange, A. Gough, and T. L. Steck, *J. Membr. Biol.* **69**, 113 (1982).
[7] P. J. Bjerrum, *J. Membr. Biol.* **48**, 43 (1979).

METHODS IN ENZYMOLOGY, VOL. 173

Preparation of Ghosts[8,9]

Human blood may be fresh or outdated from the blood bank. Most procedures are done on ice. The red cells are typically washed three times in a clinical centrifuge in 5 vol of 150 mM NaCl–5 mM NaP$_i$ (pH 8) by sedimenting at 2300 g_{max} for 10 min. The packed red cells are lysed by rapid and thorough suspension in 40 vol of 5 mM NaP$_i$ (pH 8). (We often add 10 μM MgSO$_4$ to standardize the variable divalent cation concentration in our deionized water.) The suspension is incubated for 5 min to allow hemoglobin to escape and is then centrifuged in 50-ml polycarbonate tubes at 15,000 rpm for 10 min in a Sorvall SS-34 rotor. The supernatant is aspirated away. The tube is then tipped and rotated on its axis so that the loosely packed ghosts slide away from a small hard button, rich in the lysosomal proteases of contaminating leukocytes, which may then be aspirated away. These ghosts contain a few percentages of their original hemoglobin and are red. Two additional washes in lysis buffer, with 5-min pauses after each resuspension, remove the remaining hemoglobin, yielding pearl-white hemoglobin-free ghosts. The hemolytic hole during this time is 10–20 nm in radius, so that small solutes or proteins can be introduced into the intracorpuscular space during a few minutes incubation.

Resealing Ghosts

While a variety of sealing protocols have been published,[7,9-14] there are only a few critical variables. (1) Sealing requires the reduction or screening of electrostatic repulsions among the fixed anionic charges on the membrane. For this reason, resealing buffers often are mildly acidic (pH 5.5 to 6.0) and contain monovalent cations exceeding 100 mM and/or millimolar Mg^{2+} or Ca^{2+} (see Figs. 1 and 2). Calcium ion is approximately 100 times more potent than monovalent cations and about 4 times more potent than Mg^{2+} in promoting hole closure. Sealing appears to be indifferent to the species of buffer anion. (2) Resealing is highly temperature dependent; it progresses negligibly on ice and rapidly at 37° (Fig. 3). (3) The ability of the membrane hole to change size is lost with time, especially at low ionic strength and at elevated temperatures, so that sealing

[8] G. Fairbanks, T. L. Steck, and D. F. H. Wallach, *Biochemistry* **10**, 2606 (1971).
[9] T. L. Steck and J. A. Kant, this series, Vol. 31, p. 172.
[10] H. Bodemann and H. Passow, *J. Membr. Biol.* **8**, 1 (1972).
[11] G. Schwoch and H. Passow, *Mol. Cell. Biochem.* **2**, 197 (1973).
[12] R. M. Johnson, *J. Membr. Biol.* **22**, 231 (1975).
[13] S. Jausel-Hüsken and B. Deuticke, *J. Membr. Biol.* **63**, 61 (1981).
[14] T. L. Steck, *Methods Membr. Biol.* **2**, 245 (1974).

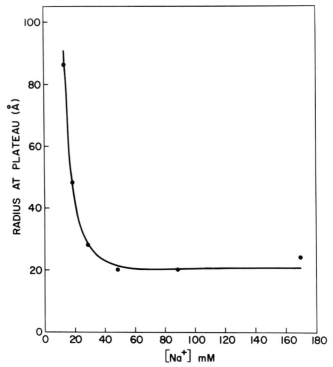

FIG. 1. The dependence of hole radius on [NaCl].[5] Ghosts were prepared in 5 mM NaP$_i$ containing 0.01 mM MgSO$_4$ at pH 8.0 and resuspended in 24 vol of 5 mM NaP$_i$ (pH 7.2) containing 0.1 mM Na$_4$EDTA and NaCl to give the values shown. The suspensions were held at 20° and the radius size of the holes estimated periodically from the fractional flotation of the ghosts (described in Fig. 6) until a stable hole size was reached. Reproduced from Lieber and Steck[5] by permission.

works best immediately after hemolysis. Indeed, ghosts reseal spontaneously, even at 0°, when red cells are lysed in 5 mM NaP$_i$ containing 1 mM Mg^{2+} or in 0.15 M NH$_4$HCO$_3$ (an isotonic salt which nevertheless rapidly permeates the red cell membrane).[9,14] It may be prudent at times to reseal ghosts without washing them, if low permeability is more important than purity. Approximately 60% of such ghosts will be impermeable to sucrose (Stokes radius, 0.52 nm). In our hands, resealing an hour after lysis by warming in isotonic saline at pH 6 reduces the hole radius to 0.9 to 1.8 nm; less than 20% of these ghosts are sealed to sucrose. Delaying resealing by a day typically yields holes of radius 1.5 to 2.0 nm.

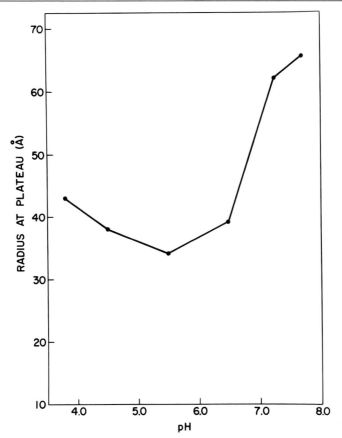

FIG. 2. The dependence of hole radius on pH.[5] Ghosts were suspended in 49 vol of sodium citrate, sodium phosphate buffers at $[Na^+] = 155$ mM at the pH values shown. The suspensions were held on ice and ghost hole size estimated as described in Fig. 1. Reproduced from Lieber and Steck[5] by permission.

Assessing Hole Size and Number

Solute Trapping[4,12]

Ghosts are equilibrated on ice with solutes of known Stokes radius and then sealed at 37° in an appropriate buffer (e.g., isotonic saline). The size of their holes can be estimated from the Stokes radius of the solutes which they trap. Such experiments do not reveal the number of holes per ghost and depend on an accurate measurement of the size of the aqueous com-

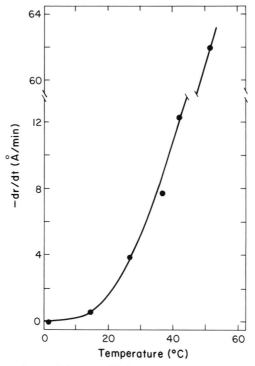

FIG. 3. The dependence of the rate of hole sealing on temperature.[5] Ghosts were suspended in 49 vol of 50 mM NaCl containing 5 mM NaP$_i$ and 0.1 mM EDTA at pH 7.2 at the temperatures shown. At timed intervals, aliquots were chilled on ice and ghost hole size estimated promptly as described in Fig. 1. Reproduced from Lieber and Steck[5] by permission.

partment of the ghosts. Furthermore, probes slightly smaller than the hole will leak out slowly, causing the apparent hole size to vary with the timing and vigor of the ghost isolation steps.

Solute Efflux Kinetics[4]

Unsealed erythrocyte ghosts are loaded with a probe by incubating the packed membranes on ice with a concentrate of the test solute for at least 10 exit half-times. The half-time for efflux can be estimated using Eq. (1). The loaded ghosts are then sealed to the extent desired by warming at 37°. These are injected into a beaker containing 30 to 60 vol of briskly stirred buffer using a syringe equilibrated to the temperature of the beaker. Other than the probe solute, the solution in the beaker must be the same as in the ghosts, and the solute load in the ghosts must not be so great as to cause excessive osmotic swelling. Aliquots are aspirated at intervals through Millipore prefilter/filter assemblies and the probe concentration measured in the filtrates. The filtration device consists of a Swinnex-25 filter holder, inverted for aspiration, and containing a Millipore AP 2502500 prefilter and a SSWPO2500 nitrocellulose filter with 3-μm pores. The prefilters are pretreated with 0.1% gelatin in buffer for a few minutes at 37° and then washed with water and dried at 50° to prevent the adsorption of probe enzymes during aspiration.[15]

Fick's first law, $dn/dt = -DA'(dC/dx)$, is used to determine A', the average apparent area of the holes in a single ghost. dn/dt is the rate of diffusion of solute molecules through the holes driven by the concentration gradient, dC/dx; D is the diffusion coefficient for the solute in the buffer. Expressed in terms of solute concentrations: $dC/dt = (DA'/\Delta x)$ $(1/V_i + 1/V_0)(C_{eq} - C_t)$. By integrating, we get

$$\ln[(C_{eq} - C_t)/(C_{eq} - C_0)] = -DA'(1/V_i + 1/V_0)t/\Delta x \qquad (1)$$

C is the concentration of the probe in the filtrate (in mol/cm^3) at zero time (0), at equilibrium (eq), and at intermediate intervals (t, in seconds). C_0 is estimated by extrapolation of the time course to $t = 0$. V is the volume of solvent in the extracorpuscular (V_0) and intracorpuscular space (V_i), both expressed as cm^3/ghost. V_i is estimated by multiplying the aqueous volume per ghost in the injected sample (i.e., sample volume/ghost number) by the factor $(C_{eq} - C_0)/C_{eq}$, which defines that fraction of the sample volume which is intracorpuscular. V_0 is calculated as the total suspension volume per ghost minus V_i. The diffusion path length, Δx, is the thickness of the membrane which appears to be 6 nm in these experiments. We used published diffusion coefficients or made reasonable estimates based on the literature and calculated A', the apparent hole area per ghost, in cm^2.

An illustration of ferritin escape kinetics is shown in Fig. 4. Note the following features: (1) The zero time intercept was unavoidably high because of the extracorpuscular volume trapped in the packed ghost suspension. (2) The internal volume per ghost, estimated as described above, was 56 μm^3. Maximal ghost volume would be approximately 150 μm^3. (3) The time course of ferritin escape was hyperbolic with no initial lag and a horizontal plateau. (4) A semilog plot of the data according to Eq. (1) was linear. The kinetics are thus compatible with a homogeneous, first order, two-compartment diffusion process. The linearity also signifies that the average hole size was constant during the experiment. (5) Using the slope of the semilog plot and a Stokes radius of 6.1 nm for ferritin, an average apparent diffusion window was calculated from Eq. (1) to be $A' = 1.76 \times$

[15] H. J. Kliman and T. L. Steck, *J. Biol. Chem.* **255**, 6314 (1980).

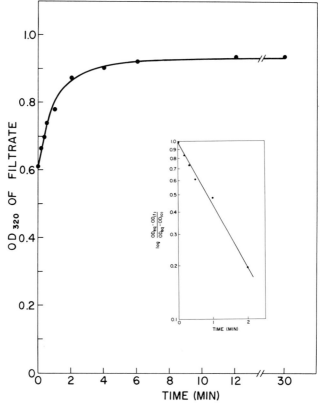

FIG. 4. Time course of ferritin escape from ghosts.[4] Packed ghosts were mixed with ferritin (2.8 μM) in the presence of 25 mM NaCl, 5 mM NaP$_i$, and 0.1 mM EDTA at pH 7.2 and allowed to equilibrate for 4 hr on ice. Three milliliters of the suspension (containing 6.3×10^9 ghosts/ml) were rapidly mixed into 47 ml of the identical buffer lacking ferritin and kept stirring briskly on ice. At intervals, aliquots were filtered and the escaped ferritin in the buffer determined spectrophotometrically. Inset, initial time course plotted according to Eq. (1). Reproduced from Lieber and Steck[4] by permission.

10^2 nm^2/ghost. This value is approximately one-millionth the membrane surface area (1.4×10^8 nm^2 ghost) and roughly 50% larger than the cross-sectional area of the probe itself (1.17×10^2 nm^2).

Control experiments[1] documented the following points: (1) There was complete equilibration of the probe with the entire ghost population. (2) The entire ghost population participated in the kinetics. (3) There was uniformity of the probe concentration within the ghosts. (4) Escape kinetics do not depend on probe charge, conformation, and chemical composition. In calculating A', we have found it justified to make the following assumptions: (1) The diffusion path can be treated as a right circular cylinder. (2) The properties of the solvent in a hole are the same as bulk solvent. (3) Unstirred layer effects and the viscous drag between solute probes and the walls of the hole can be neglected.

Ghost populations sealed for increasingly long times demonstrated progressively slower efflux kinetics. Equilibrium probe trapping experiments correlated well with these, indicating that the efflux time courses reflected the size of membrane holes.

Determination of the Number and Size of Hemolytic Holes[4]

Imagine that the total true cross-sectional area of the diffusion path per ghost is distributed among n equivalent circular holes of radius r. Then, $A = n\pi r^2$. The values determined for A' underestimate the true hole area because of sieving; i.e., the reduction in the permeability of pores to solutes of nonnegligible size. Furthermore, these values do not reveal the number of such holes per ghost. However, the true size and number of the holes can be deduced by comparing the apparent areas reported by the differential sieving of two probe solutes of unequal Stokes radius, a and b, exiting through the same holes, as follows. The total apparent areas reflected by the two probes are $A'_a = n\pi(r - a)^2$ and $A'_b = n\pi(r - b)^2$. Solving for the two unknowns of interest,

$$r = [b(A'_a)^{1/2} - a(A'_b)^{1/2}]/[(A'_a)^{1/2} - (A'_b)^{1/2}] \tag{2}$$
$$n = A'_a/[\pi(r - a)^2] = A'_b/[\pi(r - b)^2] \tag{3}$$

More generally, for a series of probes of varied radius a, Eq. (3) can be rearranged as follows[16]:

$$(A')^{1/2} = -a(n\pi)^{1/2} + r(n\pi)^{1/2} \tag{4}$$

A plot of values for $(A')^{1/2}$ as a function of the corresponding Stokes radius of the various probes, a, has a slope of $-(n\pi)^{1/2}$ and a y-intercept of $r(n\pi)^{1/2}$. Hence, $r = -(y\text{-intercept})/\text{slope}$ and $n = (\text{slope})^2/\pi$.

The method is illustrated in Fig. 5. Ghosts were loaded with both ferritin and FITC-dextran T20 and rapidly washed with ice-cold 5 mM NaP$_i$ (pH 7.2). Packed, loaded ghosts were injected into 19 vol of the same buffer and the extracorpuscular concentration of the probes determined with time. From the probe exit rates, we calculated the average apparent area per ghost for ferritin to be 1.33×10^2 nm^2 while that for

[16] M. R. Lieber, Y. Lange, R. S. Weinstein, and T. L. Steck, *J. Biol. Chem.* **259**, 9225 (1984).

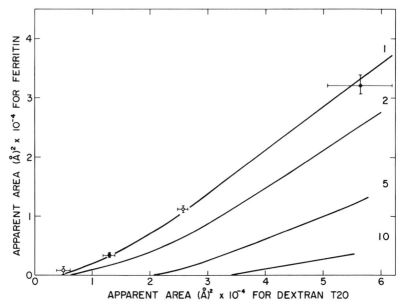

FIG. 5. Determination of ghost hole number from the differential sieving of two solutes.[4] Ghosts were prepared in two ways: abruptly in 5 mM NaP$_i$ plus 0.01 mM MgSO$_4$ as described above[4,8] (●, ■, □) or gradually, by immersing a dialysis bag containing red cells in a vast excess of this hemolysis buffer (○). The ghosts were washed in cold 5 mM NaP$_i$ (pH 7.2) lacking (●, ○) or containing (■, □) 100 mM NaCl. The same sample of packed ghosts was allowed to equilibrate with both 2.8 μM ferritin and 5.0 μM fluoresceinated dextran T20 (Pharmacia) for 3 hr on ice. A portion of the ghosts was partially sealed by incubation for 2.5 min at 30°. Aliquots of 3 ml were then mixed into 57 ml of the corresponding buffer, and the concurrent efflux of the two probes determined at intervals as in Fig. 4. Apparent hole areas and their standard deviations (error bars) were calculated for each probe according to Eq. (1) from linear semilog plots of these data. The apparent areas for the pairs of probes in each sample were plotted against one another. The solid curves in the figure are hypothetical, calculated for ghosts with the number of circular holes indicated at the right, assuming a path length, $\Delta x = 5$ nm. Reproduced from Lieber and Steck[4] by permission.

FITC-dextran T20 was 3.1×10^2 nm^2 for the same holes. Using Eqs. (2) and (3), the number of holes per ghost, n, was determined to be 1.2 and the true hole radius to be 12.1 nm. Numerous experiments with various probe combinations and ghosts which were sealed to a varied extent gave values of n from 0.5 to 1.4, signifying a single hole per ghost. Such analyses also suggested that the hole is essentially circular in profile and has a path length of 6 nm, approximately the thickness of the membrane. Note that once we can assume $n = 1$, the true area of the hole can be determined from the escape rate of a single probe solute.

Density Barrier Assay for Hole Size[4]

While accurate and rigorous, analysis of the kinetics of efflux is too laborious for routine assessment of hole size. Therefore, we developed a convenient method for measuring the porosity of ghosts by the degree to which they float or sediment when centrifuged onto barriers containing solutes of selected molecular radius. Briefly, ghosts which are permeable take up the ambient barrier solutes and therefore sediment under centrifugal force until they reach the bouyant density of the membrane (approximately 1.15 g/cm³). In contrast, impermeable ghosts retain an internal compartment the density of which is close to that of the sealing buffer.

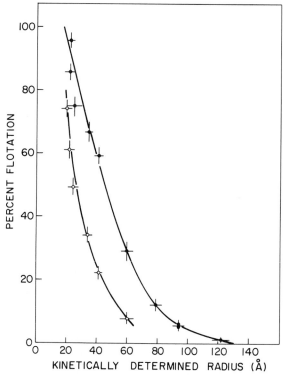

FIG. 6. Fractional flotation of ghosts on dextran barriers as a function of hole radius.[4] Aliquots of ghosts were caused to seal to various degrees. Their hole radius was estimated from the efflux rates of various probes as described above. A portion of each ghost preparation was analyzed for fractional flotation on barriers of dextran T70 (●) and dextran T10 (○) and the results plotted against hole radius. The bars indicate the standard deviations for the two parameters. Reproduced from Lieber and Steck[4] by permission.

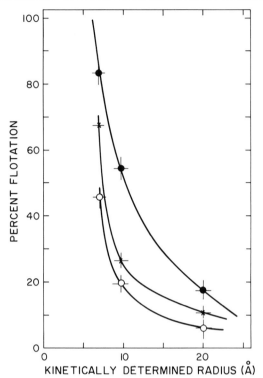

FIG. 7. Fractional flotation of ghosts on barriers of sucrose, mannitol, and CsCl as a function of hole radius.[4] Ghosts were rapidly prepared in 5 mM NaP$_i$ (pH 8.0) and aliquots promptly sealed to different degrees by incubation for 0.5, 1.5, and 50 min in 24 vol of 150 mM NaCl, 5 mM NaP$_i$, 0.1 mM EDTA, and 4.5 nM [^{3}H]sucrose at pH 6.0 and 37°. The time course of [^{3}H]sucrose efflux was then determined, and from these data hole sizes were estimated (assuming one hole per ghost). However, the semilog plots (as in Fig. 4) were not linear and irreversible trapping of 2, 22, and 39% of the label was observed in the three preparations. These findings suggest heterogeneity in the hole radius in the range of the probe size.[16] Therefore, a weighted average value for each hole radius was calculated and plotted on the abscissa. The corresponding fractional flotation of these ghosts on barriers of sucrose (●), mannitol (×), and CsCl (○) was plotted on the ordinate. The bars represent the standard deviations for the two parameters. Reproduced from Lieber and Steck[4] by permission.

Ghosts fully expanded with water would have an overall density of approximately 1.001 g/cm^3. In practice, the barrier is made to a density of approximately 1.01 g/cm^3, so that even partially shriveled ghosts will float if impermeable to the solute.

To use the fractional flotation of ghosts on barriers of various solutes as a convenient assay for hole size, the system must be calibrated as follows. Ghosts are sealed to different degrees by incubation in solutions of varied ionic strength and temperature for defined intervals. Both their flotation on several density barriers and the escape kinetics of selected solutes are determined for each sample as described above. True hole radii are calculated from the kinetic data and plotted against fractional flotation to generate standard curves for hole radii (Figs. 6 and 7).[16] The method is sensitive, employing only 1–5 μl of packed ghosts/assay. The method is precise; the average difference in flotation among 53 pairs of duplicate determinations on several kinds of barriers was 3.5%. The method is also simple, as described below.

Density barrier solutions are prepared from 10-fold concentrates in the same buffer which bathes the ghosts. The barrier solutions contain dextran T70 or T10 of density 1.01 g/cm^3 (3.07 g of dextran plus 100 g of solvent or 2.67 g of dextran/100 ml of solution), sucrose of 100 mM (1.014 g/cm^3), mannitol of 150 mM (1.010 g/cm^3), or CsCl of 50 mM (1.005 g/cm^3). These barrier concentrations and densities are optimal but are not critical, since they fall on a plateau of fractional flotation versus density. However, they should be kept constant. It is important in all cases to match the buffer in the barrier solution to that in the ghost suspension, lest the disparity in density and/or osmotic activity alter the flotation.

Typically, 0.15-ml aliquots of 1–5% ghost suspensions (0.8–4 × 10^8 ghosts/ml) are layered on 1.35 ml of the barrier solution and spun at 15,000 rpm for 12 min (for dextrans) or 18,000 rpm for 10 min (for sucrose, mannitol, and CsCl) in 3-ml polycarbonate tubes in a Sorvall SS-34 rotor. It is important to keep the time the samples spend on top of the gradient constant, since ghosts will take up some barrier solutes slowly and will then vary in their penetration of the barrier. After centrifugation, the contents of the tube are divided somewhat arbitrarily into two fractions. We prefer to collect the entire supernatant fraction (approximately 1.45 ml) by careful aspiration with a Pasteur pipet along the side of the tube opposite the pellet. The pellet is then resuspended to 1.5 ml in the sample buffer. The ghosts in each fraction are diluted in filtered saline and counted in a Coulter counter. The fraction of ghosts floating is calculated after correction for buffer blank and coincidence error, and the mean size of the hole read from the standard curve.

[22] Preparation and Properties of One-Step Inside-Out Vesicles from Red Cell Membranes

By Javier García-Sancho *and* Javier Alvarez

Introduction

Sealed inside-out vesicles are an important tool for the study of transport mechanisms with active sites oriented toward the cytoplasmic side of the plasma membrane. The preparation procedure described here is based on that originally reported by Lew and co-workers.[1-3] Hypotonically lysed red cells vesiculate spontaneously on incubation at pH 7.6 and 37° as long as the ionic strength is kept low and divalent cations are absent. The yield of everted membrane surface is typically 20–30% and transport and membrane-associated enzyme activities are preserved. The advantages of this preparation over Steck-type inside-out vesicles[4] are its simplicity (one single step), rapidity (a viable preparation may be ready in 1 hr), and the preservation of Ca^{2+}-dependent K^+ transport, which is lost[5] or altered[6] when other vesiculation methods are used.

The process of inside-out vesiculation has been recently described by Lew and co-workers.[6a] Contrary to earlier views of vesiculation by endocytosis, they have shown that lysis generates a persistent membrane edge which spontaneously curls, cuts, and splices the membrane surface to form single or concentric vesicles. The procedure for preparation of inside-out vesicles from human red cells is described below, but other nonnucleated red cells can be vesiculated similarly.[1,2,7]

Standard Vesiculation Procedure

Red cells from freshly drawn blood or from the blood bank are washed three times with an isotonic solution containing 0.1 mM EGTA or EDTA

[1] V. L. Lew, S. Muallem, and C. A. Seymour, *J. Physiol. (London)* **307,** 36P (1980).
[2] V. L. Lew and C. A. Seymour, *in* "Techniques in Lipid and Membrane Biochemistry" (T. R. Hesketh, H. L. Kornberg, J. C. Metcalfe, D. H. Northcote, C. I. Pogson, and K. F. Tipton, eds.), Vol. B415, pp. 1–3. Elsevier/North-Holland, Amsterdam, 1982.
[3] V. L. Lew, S. Muallem, and C. A. Seymour, *Nature (London)* **296,** 742 (1982).
[4] T. L. Steck and J. A. Kant, this series, Vol. 31, p. 172.
[5] S. Grinstein and A. Rothstein, *Biochim. Biophys. Acta* **508,** 236 (1978).
[6] H. Sze and A. K. Solomon, *Biochim. Biophys. Acta* **554,** 180 (1979).
[6a] V. L. Lew, A. Hockaday, C. J. Freeman, and R. M. Bookchin, *J. Cell Biol.* **116,** 1893 (1988).
[7] O. E. Ortiz and R. A. Sjodin, *J. Physiol. (London)* **354,** 287 (1984).

at pH 7.6. Cells from the blood bank, kept in acid–citrate–dextrose (ACD), tend to be acidic (pH about 6.8). It is essential to bring the pH of the cells to 7.6 before use. This can be done by titration with 300 mM Tris base. One volume of ice-cold washed packed cells (about 80% hematocrit) is lysed by vigorous mixing with 40 vol of ice-cold lysing solution containing 0.1 mM EGTA (Na$^+$, K$^+$, or Tris salt) and 2.5 mM K$^+$–HEPES, pH 7.6. EGTA can be replaced by EDTA. The presence of K$^+$ is essential to avoid inactivation of Ca^{2+}-dependent K$^+$ channels. After 2–5 min at 2°, the lysate is centrifuged 15 min at 27,000 g at 2°. This step is not essential but convenient since it allows handling of smaller volumes in the following steps. The supernatant is aspirated and the pellet resuspended at 2° in a convenient volume (usually 1–2 ml/ml of original cells) of fresh lysing solution or the same supernatant solution. Any solute added at this stage will incorporate into the vesicles.

Vesiculation is started by transferring the ghost suspension to an incubation bath at 37–38°. Vesiculation begins after 2–5 min and is complete within 30–45 min. The vesiculation process is critically temperature dependent and occurs over days in the cold. The presence of Mg^{2+} (0.1–5 mM) prevents vesiculation. On warming resealed ghosts are formed instead. Vesiculation can be conveniently followed by examining samples taken after different incubation periods under the phase-contrast microscope at a magnification of 1000. The cooling to room temperature on contact of the sample with the slide will stop vesiculation at the time of sampling. The process can also be continuously followed by using a thermostated slide at 37°.[8] The following sequence of events is usually observed during vesiculation (Fig. 1): (1) decrease of the diameter of some ghosts simultaneous with the appearance of one or several vesicles attached to the plasma membrane; (2) the number of ghosts with vesicles increases very quickly; some vesicles separate from the membrane and move freely inside the ghosts; (3) the ghost structure disappears and the vesicles are released to the medium; most of them seem bound to each other, forming clusters; (4) with time, few or no ghosts are seen and the vesicles appear smaller and individualized. For a more detailed description of the process of vesiculation, see Ref. 6a.

After incubation at 37°, disruption of clusters and some homogenization of size can be achieved by forcing the vesicle suspension through a 27-gauge needle three to five times. The same result can be achieved by vigorous vortexing of the suspension while still warm. The vesicles are then suspended in a suitable medium, the composition of which depends on the aim of the particular experiment. In most of our transport experi-

[8] V. L. Lew and C. A. Seymour, *J. Physiol. (London)* **308,** 8P (1980).

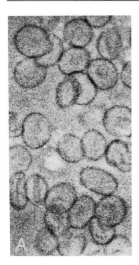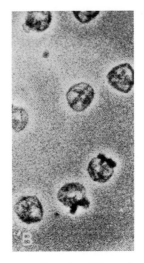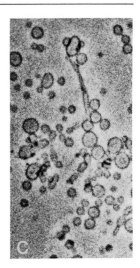

FIG. 1. Appearance of one-step inside-out vesicles under the phase contrast microscope at different vesiculation stages. From left to right: (A) ghosts before incubation at 37°; (B) beginning of vesiculation after 3 min at 37°; (C) more advanced vesiculation after 15 min at 37°.

ments the suspending medium had the following composition (mM): KCl, 18; K$^+$–HEPES, 16.5; pH 7.5 (medium F). Under these conditions the concentrations of K$^+$, Cl$^-$, and HEPES should be about the same inside and outside the vesicles. Other additions to the medium, suitable for particular experimental designs, are detailed below. The vesicles can, at this stage, be easily sedimented, if required, by centrifugation in Eppendorf tubes at 10,000 g for 5 min.

Sidedness and Enzyme Activity

The preparation is composed of unsealed membranes and sealed inside-out and right-side-out vesicles. The sealed everted membrane surface, determined by acetylcholinesterase accessibility,[4] is typically 20–30% of the total. Loss or reorientation of acetylcholinesterase during the preparation or washing of the vesicles is meaningless (<5%). If Mg^{2+} is added to the vesiculation medium no inside-out vesicles can be detected by acetylcholinesterase accessibility. Sealed right-side-out membrane surfaces, assessed by glyceraldehyde-3-phosphate dehydrogenase accessibility,[4] ranged from 0 to 10% in different experiments. Calcium-activated ATPase activity can also be demonstrated. It was found to be increased by the addition of calmodulin.[2]

Distribution of Vesicle Sizes

Observation under the phase contrast microscope or the electron microscope reveals a large heterogeneity of shape and size. Diameters range between 0.05 and 1 μm for most of the vesicles. The resealed vesicular space, measured with ^{86}Rb, ^{45}Ca-EGTA, or ^{22}Na, is usually 5–10% of that of the original cells. Using the values of the resealed vesicular space contained in inside-out vesicles and the everted membrane surface, a mean vesicular radius ($r = 3 \times$ volume/surface) of about 0.3 μm can be estimated. The apparent density of the vesicles, measured by sedimentation through Percoll gradients, is about 1.026 and uniform for the whole population. Centrifugation at 2500 g for 30 min sediments a fraction enriched in larger vesicles with mean radius of about 1.5 μm and which includes about 10% of the total acetylcholinesterase activity.

Measurements of Ion Transport

Transport through the Na$^+$ pump, the Ca^{2+} pump, or the Ca^{2+}-dependent K$^+$ channel can be conveniently studied with the aid of radioactive isotopes. For efflux experiments the tracer is loaded during vesiculation. For influx experiments it is added to the incubation medium. Medium F (see Standard Vesiculation Procedure section) with the adequate additions described below is suitable for transport experiments either at 37° or at room temperature. A multiplace magnetic stirrer can be of utility if several conditions are to be tested simultaneously.

Removal of extravesicular radioactivity at the end of the incubation period is easily obtained by passage of the samples from the vesicle suspension through ion-exchange resins.[9,10] Dowex 50-X8-100 resin is activated with sodium and packed in Pasteur pipet plugged with glass wool (about 0.8 ml of resin/pipet). Each column is washed once with 0.9 ml of elution buffer (see below) containing 10 mg/ml of bovine serum albumin (fraction V) and then twice with the elution buffer described below. The albumin treatment prevents binding of the vesicles to the resin. For each new batch of resin the pH of the eluent should be tested to be within the range of 7.2 to 7.8.

After a convenient period of incubation aliquots of 50–200 μl of the vesicles suspension are placed on top of the columns and eluted immediately with two consecutive 0.9-ml aliquots of ice-cold elution buffer. A convenient composition for the elution buffer is as follows: sucrose,

[9] O. D. Gasko, A. F. Knowles, H. G. Shertzer, E.-M. Soulinna, and E. Racker, *Anal. Biochem.* **72**, 57 (1976).
[10] J. García-Sancho, A. Sanchez, and B. Herreros, *Nature (London)* **296**, 744 (1982).

0.2 M; albumin, 2 mg/ml; Na$^+$–HEPES, 5 mM, pH 7.5. The effluent from the column, which contains only the radioactivity that was intravesicular at the time of sampling, is then counted for radioactivity. Each column can be used for at least 10 samples without meaningful spillover of extracellular radioactivity. When studying Ca^{2+} transport the presence of EGTA in the extravesicular medium should be avoided since it could draw some ^{45}Ca through the column.

Na$^+$ Pump

Na$^+$-pump activity can be evidenced either by ^{86}Rb extrusion from preloaded inside-out vesicles or by ^{22}Na uptake.[2,3] A typical example of the last procedure is shown in Fig. 2A. The experiment is best performed at an equivalent hematocrit (the amount of membrane expressed in terms of volume of the original cells) of about 50% in medium F containing 10 mM NaCl and 0.4 mM EGTA and at 37° to optimize the difference between active and passive fluxes. Sodium-22 (about 2 × 10^6 cpm/ml) is added to the medium. After a few minutes, to allow for the equilibration of leaky vesicles, MgATP is added to start the operation of the pump. The

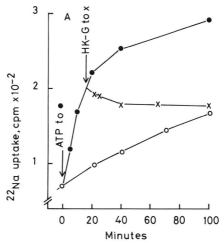

FIG. 2. ATP-dependent Na$^+$ uptake (A), ATP-dependent Ca^{2+} uptake (B), and Ca^{2+}-dependent ^{86}Rb uptake (C) by one-step inside-out vesicles. The following additions were done at the times marked with arrows: ATP: 2 mM ATP + 3 mM MgCl$_2$; HK-G: 50 U/ml hexokinase + 10 mM glucose; A23187: 0.2 μM A23187. (A) was performed at 37° and the other two at 20°. The apparent decrease of ^{45}Ca levels in (ATP + HK-G)-treated vesicles below levels in control vesicles on additon of A23187 is an artifact reflecting the osmotic shrinkage of the vesicles due to the additions. (C) was redrawn by permission from García-Sancho et al.[10] (A) and (B) correspond to unpublished experiments by the authors.

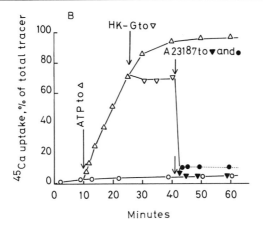

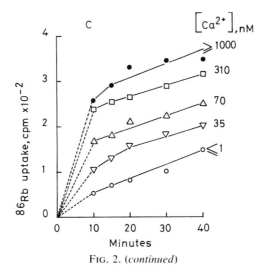

Fig. 2. (*continued*)

uptake of ^{22}Na is linear during the first 10–15 min and reaches a steady state at 60–90 min, with intravesicular concentrations well above the medium concentration. The treatment of the cells with ouabain prior to lysis or the addition of Ca^{2+} to the vesicle suspension inhibit this ATP-dependent ^{22}Na uptake.

Ca²⁺ Pump

Ca²⁺-pump activity can be evidenced by the uptake of ^{45}Ca on addition of ATP in the presence of Mg^{2+}. A typical experiment is shown in Fig. 2B.

It is convenient to wash the vesicles once with medium F prior to the transport measurements in order to remove the EGTA present in the vesiculation medium. This can be conveniently done by a 10-min centrifugation at 10,000 g in an Eppendorf-type tube. Vesicles are resuspended in medium F containing 10 mM NaCl at 20% hematocrit and ^{45}Ca is added (about 2 × 10⁶ cpm/ml). Total Ca^{2+} concentration should not exceed 20 μM, since higher concentrations might cause osmotic stress of the vesicles on accumulation inside. MgATP is added to start the operation of the pump. The experiment can be conduced either at room temperature or at 37°. In the last case the uptake of ^{45}Ca shall be 5–10 times faster. At 20° the uptake is linear during the first 10–15 min. After 60–90 min virtually all the ^{45}Ca has been taken up by the vesicles. The addition of calmodulin increases the rate of ^{45}Ca uptake,[2] especially at the lower Ca^{2+} concentrations. The degree of depletion of calmodulin should be checked carefully with each inside-out vesicles preparation since it depends largely on the time of storage of the vesicles, whether or not they have been washed, at what ionic strength, etc.

Ca^{2+}-Dependent K^+ Channels

The activity of Ca^{2+}-dependent K^+ channels can be evidenced either from the loss of preloaded ^{86}Rb or by the uptake of ^{86}Rb from the medium on addition of Ca^{2+}. A typical example of the last procedure is shown in Fig. 2C. The experiment is best performed at 10–30% equivalent hematocrit in medium F containing Ca^{2+}/EGTA buffers and ^{86}Rb (about 5 × 10⁶ cpm/ml). The start is marked by the mixing of an aliquot of concentrated vesicles suspension in EGTA-containing medium F with the incubation medium (medium F containing ^{86}Rb, Ca^{2+}, and EGTA to give the desired Ca^{2+} concentration). Temperature of incubation makes little difference in the results.

The one-step inside-out vesicles show what has been called an all-or-nothing response to Ca^{2+} (see Refs. 3, 10, and 10a). That means that, at submaximal Ca^{2+} concentrations, the channels belonging to some vesicles are active whereas those in others remain silent. Since the mean number of channels per vesicle is low,[2,10b] some of the vesicles do not take up ^{86}Rb at submaximal Ca^{2+} concentrations, so that an increasing fraction of the total vesicular space equilibrated with ^{86}Rb is observed on increasing the Ca^{2+} concentration (Fig. 2C). At 1 μM Ca^{2+} maximal activation is reached. When Mg^{2+} is present in the incubation medium the apparent sensitivity to Ca^{2+} is largely decreased.[10]

[10a] J. Alvarez, J. García-Sancho, and B. Herreros, *Biochim. Biophys. Acta* **859**, 56 (1986).
[10b] J. Alvarez and J. García-Sancho, *Biochim. Biophys. Acta* **903**, 543 (1987).

The Ca^{2+}-activated uptake of ^{86}Rb is very fast, with a half-equilibration time of less than 1 min at maximal activation (not shown in Fig. 2) and does not follow single exponentials.[10,10a,11] The rate of change of the Ca^{2+} concentration in the medium may alter the response of the channels. Calmodulin has no effect on Ca^{2+}-dependent K^+ transport in one-step inside-out vesicles.[10,11a]

Passive Permeability

Two different intravesicular compartments can be distinguished. (1) One amounts to between 5 and 30% of the total resealed vesicular space, which equilibrates quickly and at about the same rate for ^{86}Rb, ^{22}Na, and ^{45}Ca. The apparent rate constants for transport in this compartment are about 10–20 hr^{-1}. This space corresponds probably to leaky vesicles or ghosts. (2) The second is a tight compartment which equilibrates slowly with ^{86}Rb, ^{22}Na, and ^{45}Ca when the specific transport systems are not activated. The apparent rate constants are similar for the three cations, about 0.3 hr^{-1}.

Stability of One-Step Inside-Out Vesicles

Once prepared inside-out vesicles are quite stable if stored in EGTA-containing medium F at 4°. At least 50% of the ATP-dependent ^{22}Na and the ^{45}Ca uptake observed in the fresh preparation remains after 2 days of storage at 4°. The vesicular space which depends on Ca^{2+} for ^{86}Rb equilibration does not change significantly after 2 days of storage at 4°. The space insensitive for Ca^{2+} was occasionally seen to increase with storage.

During the preparatory steps both the Na^+ and Ca^{2+} pumps seem very stable, but the persistence of functional Ca^{2+}-dependent K^+ channels was extremely sensitive to experimental maneuvers. If vesiculation had not been complete or the sealed everted membrane surface was below 20%, Ca^{2+}-dependent ^{86}Rb transport was very weak or nonexistent. In about 20% of the one-step inside-out vesicles preparations the activity of Ca^{2+}-dependent K^+ channels was lost for no apparent reason.

Effects of pH on Vesiculation

The relevant pH occurs during the vesiculating incubation at 37°. The pH during lysis seems much less important and it can be adjusted before warming without much effect on the final results. If the solutions used are those described in the section on "Preparation," the pH of the lysate will

[11] J. Alvarez, J. García-Sancho, and B. Herreros, *Biochim. Biophys. Acta* **771**, 23 (1984).
[11a] J. Alvarez, J. García-Sancho and B. Herreros, *Biochim. Biophys. Acta* **860**, 25 (1986).

drop by about 0.17 pH units on warming from 2 to 37°. If the cells have been lysed without buffer the pH drop on warming could be about twice as much. If the pH is below 7.1 at 37° vesiculation is never observed. Above pH 7.3 vesiculation is always essentially complete, very few ghosts remaining at the end of the incubation period. The yield of everted membrane surface increases with pH from 7.1 to 8 (Fig. 3). An unexpectedly large yield of everted sealed membrane surface is usually obtained at pH 7.1 to 7.3, coincident with the transition pH at which vesicle formation begins. Sealed right-side-out surface is always small (0 to 10%) and does not change significantly with pH in the range of 6.8 to 8.2 When vesiculation is incomplete by the end of the incubation period, the addition of Mg^{2+} to the vesicles suspension may induce slow resealing. This could be misinterpreted for uptake.

The transport activity observed increases with the area of everted membrane surface, as measured by acetylcholinesterase latency. Some decrease of the activity of all the three transport systems studied per unit of acetylcholinesterase activity is usually observed at the more alkaline

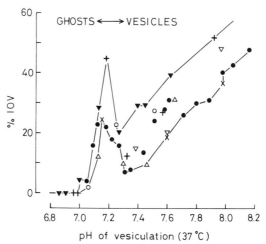

FIG. 3. Effects of pH of vesiculation on the yield of everted membrane surface, estimated from acetylcholinesterase accessibility. The results of experiments with several batches of cells lysed 1/40 either with standard lysing medium or with 0.1 mM EGTA solution are reported. Different pH values were obtained by adding to the lysate, kept at 2°, either HEPES-free acid or Tris base. The values of pH were measured at 37°. Lines have been drawn connecting either the highest or the lowest values of everted membrane surface obtained at each pH value. "Ghosts" and "vesicles" refers to the appearance under the phase-contrast microscope by the end of the incubation period. The double arrow delimits the transition pH range, where a mixed population of unvesiculated ghosts and vesicles was observed. IOV, Inside-out vesicle.

pH values. Highest pH values should be avoided when studying Ca^{2+}-dependent K^+ transport, since this condition has been reported to cause the inactivation of the channels.[3]

Acknowledgments

We thank Dr. V. L. Lew, Physiological Laboratory, Cambridge, England, for valuable comments during the preparation of this chapter. This work was supported by a grant from the Spanish DGICYT (PB 86-0312).

[23] Na+,K+-Pump Stoichiometry and Coupling in Inside-Out Vesicles from Red Blood Cell Membranes

By Rhoda Blostein and William J. Harvey

In an earlier chapter in this series, methods to measure sodium pump-mediated Na^+ and K^+ translocation and Na^+,K^+-ATPase activity in inside-out plasma membrane vesicles (IOV) derived from human red blood cells were described.[1] This vesicle preparation has a number of distinct advantages, notably its stability, low pump density, and relatively low surface : volume ratio. In this chapter we show that even though porous or unsealed membranes are present in the preparation, it is possible to assess quantitatively activities of only the well-sealed inside-out vesicles. Furthermore, vesicle heterogeneity in terms of size and pump density per vesicle is also not a problem since measurements of $Na^+ : K^+(Rb^+)$ coupling ratios indicate values close to the expected average value of 1.5 even under conditions which cause inactivation of a substantial number of pump sites.[2]

Rationale

To measure the pump stoichiometry of only those membranes which are intact, "competent" inside-out vesicles, assays of Na^+,K^+ pump and Na^+,K^+-ATPase are carried out in the absence and presence of valinomycin with K^+ (or Rb^+) added only at the cytoplasmic surface, i.e., added to the assay medium. Under these conditions, little ATP-dependent or strophanthidin-sensitive $^{22}Na^+$ uptake is expected in the absence of va-

[1] R. Blostein, this series, Vol. 156, p. 171.
[2] W. J. Harvey and R. Blostein, *J. Biol. Chem.* **261**, 1724 (1986).

METHODS IN ENZYMOLOGY, VOL. 173

linomycin in the initial phase of the reaction because little K^+ (or Rb^+) should permeate the nonleaky IOV and be available for exchange with Na^+. In the presence of valinomycin, K^+ permeates rapidly and should stimulate ATP-dependent Na^+ uptake. Similarly, the fraction of strophanthidin-sensitive Na^+,K^+-ATPase activity measured in the absence of valinomycin can be attributed to fragmented and permeable membranes and the valinomycin-stimulated hydrolysis should reflect Na^+,K^+-ATPase activity of nonleaky IOV.

Measurements of Na^+/P_i Coupling Ratio

Figure 1 depicts the results of an experiment carried out in the absence and presence of valinomycin. Vesicles were prepared and assayed as described elsewhere in this series.[1] It should be noted that the baseline uptake of $^{22}Na^+$ (ATP absent) is not altered by the addition of valinomycin to the incubation medium. The small ATP-dependent increase in net $^{22}Na^+$ influx observed in the absence of valinomycin represents, presumably, either the $Na^+/0$ uptake component of pump activity[3] and/or vesicles permeated, to a small extent, by K^+ added to the medium. The valinomycin-insensitive strophanthidin-sensitive hydrolysis is due, presumably, to the activity of permeable membranes. Thus, the ratio of the valinomycin-sensitive component of ATP-dependent $^{22}Na^+$ influx to the valinomycin-sensitive component of strophanthidin-inhibitable ATP hydrolysis is a measure of the Na^+/P_i coupling. Eleven separate experiments aimed at measuring the Na^+/P_i coupling ratio were carried out as described in Fig. 1, each at a single appropriate time point (4 min, cf. Fig. 1). A value of 2.5 $\pm$ 0.3 was obtained, which is close to the well-documented value of 2–3 for intact cells assayed with optimal concentrations of Na^+ and K^+.

Measurements of Na^+/K^+ (Rb^+) Stoichiometry

As emphasized earlier,[4] the main precaution needed for measurements of pump-mediated net Na^+/Rb^+ exchange in IOV is to ensure minimal depletion of intravesicular Rb^+. Otherwise the $Na^+ : Rb^+$ stoichiometry is overestimated due to $Na^+/0$ flux into Rb^+-free vesicles. A typical experiment is shown in Table I. Conditions chosen were similar to those generally used (see Fig. 1), except that the incubation time and temperature are reduced so that the net loss in intravesicular Rb^+ is less than about 30%.

[3] I. M. Glynn and S. J. D. Karlish, *J. Physiol.* (*London*) **256**, 465 (1976).
[4] R. Blostein, *J. Biol. Chem.* **258**, 12228 (1983).

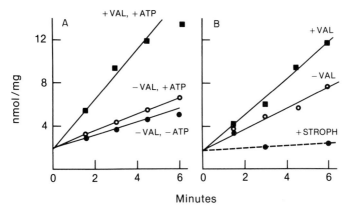

Minutes

FIG. 1. Valinomycin-stimulated Na$^+$ translocation (A) and ATP hydrolysis (B). Vesicles (4.0 mg/ml) were equilibrated with 50 mM choline chloride, 0.5 mM KCl, 0.5 mM MgSO$_4$, and 20 mM Tris-glycylglycine, pH 7.4, and then assayed at 37° in media comprising 45 mM choline chloride, 5 mM NaCl, 0.5 mM KCl, 0.5 mM MgSO$_4$, and 20 mM Tris-glycylglycine, pH 7.4. Assays were carried out in the presence or absence of 2.2 μM valinomycin as indicated. ^{22}NaCl was included in assays of ^{22}Na$^+$ uptake and [γ-^{32}P]ATP, in assays of ATP hydrolysis. Strophanthidin (0.1 mM) and ATP (0.1 mM) were added as indicated. (Data are from Harvey and Blostein.[2])

With Na$^+$ concentrations $\geq$1.8 mM, the Na$^+$: K$^+$ stoichiometry is close to the expected average value of 3 : 2. This is shown in Table I for assays at both low and relatively high ATP concentrations. Since Na$^+$/0 flux is saturated at very low levels of ATP, the contribution of Na$^+$/0 flux to the Na$^+$: Rb$^+$ stoichiometry should be negligible under the latter condition.

TABLE I
Na$^+$: Rb$^+$ STOICHIOMETRY AT LOW AND HIGH ATP CONCENTRATION[a]

ATP concentration (μM)	ATP-dependent ^{22}Na$^+$ influx (normal efflux) (a, pmol/mg/min)	ATP-dependent ^{86}Rb$^+$ efflux (normal influx) (b, pmol/mg/min)	Ratio (a/b)	95% confidence interval
2	108.7 ± 20.9	75.5 ± 7.8	1.44	(0.78–2.28)
100	1102.9 ± 53.3	729.1 ± 16.4	1.51	(1.32–1.72)

[a] Vesicles were equilibrated with 0.5 mM RbCl (or ^{86}RbCl) and cytoplasmic ^{22}NaCl (or NaCl) was 4.1 mM. Reactions were carried out for 1.5 min at 100 μM ATP and for 5 min at 2 μM ATP. Each value is the difference in the mean ± SE of two sets of four determinations (100 μM ATP) or five determinations (2 μM ATP) with and without ATP. (Data from Blostein.[4])

Comments

Hydrolytic activity of only competent sealed IOV can be assessed by adding the impermeable cardiac glycoside, ouabain. The problem with this approach is that intravesicular K^+ reduces the effectiveness of ouabain which must be added at relatively low concentrations to avoid any detergent-like effect of the glycoside. In fact, in experiments aimed at measuring the Na^+/P_i coupling ratio with ouabain added ($5 \times 10^{-5} M$), a value of only 1.7 was obtained. In studies of the coupling ratios of pump modes occurring in the absence K^+ ($Na^+/0$ flux, ATP-dependent Na^+/Na^+ exchange), addition of ouabain to inhibit the activity of membrane fragments is feasible and under these conditions $Na^+ : P_i$ coupling ratios of 2–3 are observed.[5]

The above approach to assessing the activity of competent inside-out vesicles should be generally applicable to preparations other than those derived from red blood cells provided the entry of K^+ (or Rb^+) is not limited by a diffusion potential. This should be readily circumvented by including a second ionophore such as the proton conductor carbonyl cyanide m-chlorophenylhydrazone as described for experiments with reconstituted vesicles of kidney Na^+,K^+-ATPase.[6]

[5] R. Blostein, *J. Biol. Chem.* **258**, 7948 (1983).
[6] S. J. D. Karlish and U. Pick, *J. Physiol.* (*London*) **312**, 505 (1981).

[24] Preparation of Red Cell Membrane Skeleton Proteins

By WILLIAM C. HORNE, THOMAS L. LETO, and RICHARD A. ANDERSON

The red cell cytoskeleton, or membrane skeleton, is a highly cross-linked network of proteins attached to specific integral membrane proteins. This structure contributes to cell shape, stabilizes the membrane, and imparts to the cell a remarkable elasticity when subjected to shear forces.[1] The membrane skeleton is operationally defined as the insoluble complex of proteins isolated following extraction of red cells or their membranes with nonionic detergents, typically Triton X-100.[2] SDS–PAGE analysis of such "shells" prepared under a variety of conditions

[1] W. B. Gratzer, *Biochem. J.* **198**, 1 (1981).
[2] J. Yu. D. A. Fischman, and T. L. Steck, *J. Supramol. Struct.* **1**, 233 (1973).

METHODS IN ENZYMOLOGY, VOL. 173

reveals the consistent presence of bands 1 and 2 (spectrin monomers), band 4.1, band 4.9, and band 5 (actin)[3] (nomenclature of Steck[4]).

Spectrin monomers associate laterally to form heterodimers which further assemble in a "head-to-head" fashion to form tetramers, hexamers, and higher order oligomers.[5-7] Band 4.1, filamentous actin, and spectrin form a ternary complex at the end of the spectrin molecule opposite the oligomerizaton site.[8-10] Band 4.9 mediates the lateral interaction of actin filaments.[11] In addition, other actin-binding proteins, including myosin[12] and tropomyosin,[13] have been found in erythrocytes and it is likely that other minor components of the membrane skeleton will be found. The interactions of these proteins lead to formation of the extended two-dimensional network which is attached by two linking proteins to integral membrane proteins. Band 2.1 binds to band 3[14] and to spectrin near the oligomerization site.[15] Band 4.1 links the opposite end of spectrin[8] as well as the associated actin filament to a complex of glycophorin and phosphatidylinositol 4,5-diphosphate.[16,17] Protein 4.1 also binds to the cytoplasmic domain of band 3, but the presence of spectrin and actin blocks this association.[18] Variable amounts of band 2.1 and band 3 are isolated with the insoluble shells, depending on the ionic strength of the detergent extraction buffer.[3] Glycophorins A and B are not found with the detergent-insoluble complex,[19] consistent with findings that detergents block the interaction of band 4.1 with glycophorin (see below).

Spectrin, band 2.1, and band 4.1 are easily isolated by sequential extraction of erythrocyte membranes with low ionic strength followed by

[3] M. P. Sheetz, *Biochim. Biophys. Acta* **557**, 122 (1979).
[4] T. L. Steck, *J. Cell Biol.* **62**, 1 (1974).
[5] J. S. Morrow and V. T. Marchesi, *J. Cell Biol.* **88**, 463 (1981).
[6] D. M. Shotten, B. E. Burke, and D. Branton, *J. Mol. Biol.* **131**, 303 (1979).
[7] E. Ungewickell and W. Gratzer, *Eur. J. Biochem.* **88**, 379 (1978).
[8] C. M. Cohen, J. M. Tyler, and D. Branton, *Cell* **21**, 875 (1980).
[9] V. Fowler and D. L. Taylor, *J. Cell Biol.* **85**, 361 (1980).
[10] E. Ungewickell, P. M. Bennett, R. Calvert, V. Ohanian, and W. B. Gratzer, *Nature (London)* **280**, 811 (1979).
[11] D. L. Siegel and D. Branton, *J. Cell Biol.* **100**, 775 (1985).
[12] V. M. Fowler, J. Q. Davis, and V. Bennett, *J. Cell Biol.* **100**, 47 (1985).
[13] V. M. Fowler and V. Bennett, *J. Biol. Chem.* **259**, 5978 (1984).
[14] V. Bennett and P. J. Stenbuck, *Nature (London)* **280**, 468 (1979).
[15] J. M. Tyler, B. N. Reinhardt, and D. Branton, *J. Biol. Chem.* **255**, 7034 (1980).
[16] R. A. Anderson and V. T. Marchesi, *Nature (London)* **318**, 295 (1985).
[17] R. A. Anderson and R. E. Lovrien, *Nature (London)* **307**, 655 (1984).
[18] G. R. Pasternack, R. A. Anderson, T. L. Leto, and V. T. Marchesi, *J. Biol. Chem.* **260**, 3676 (1985).
[19] T. J. Mueller and M. Morrison, *in* "Erythrocyte Membranes" (W. C. Kruckeburg, J. W. Eaton, and G. J. Brewer, ed.), Vol. 2, p. 95. Liss, New York, 1981.

high ionic strength buffers.[20,21] In addition to these extraction procedures, methods for isolating the skeleton components from Triton X-100-extracted shells have been developed. Methods for extraction and purification of spectrin, band 2.1, and band 4.1 from ghost membranes will be described, followed by a description of the preparation of band 4.9 proteins from Triton shells. We will also describe methods for labeling band 4.1 with ^{32}P, using membrane-bound kinase activities and a protein kinase which elutes from DEAE ion-exchange columns with band 4.1.

Procedures

As discussed by Siegel et al.[22] and Bennett,[23] care must be taken to prevent proteolysis of erythrocyte membrane proteins during the isolation procedure. Most nonerythroid sources of proteolytic activity (plasma and leukocytes) are removed by carefully washing the red cells. The leukocytes form a white layer (the buffy coat) on top of the packed red cells which can be removed by careful aspiration. In addition, care must be taken to inhibit erythrocyte proteases. Neutral protease activity of erythrocytes can be inhibited by calcium-chelating agents such as ethylenedinitrilotetraacetic acid (EDTA) and ethylene glycol bis(oxyethylenenitrilo)tetraacetic acid (EGTA), inhibitors of serine proteases such as diisopropyl fluorophosphate (DFP) and phenylmethylsulfonyl fluoride (PMSF) and leupeptin. Both DFP and PMSF are rapidly hydrolyzed in aqueous conditions, especially at basic pH, and should therefore be added to the buffer immediately before use. PMSF is sparingly soluble in water, and is most conveniently handled by first making a 150 mM stock solution in 2-propanol, which is then added to the buffer with rapid stirring.

Solutions

Solution A: Acid–citrate–dextrose (ACD): 75 mM trisodium citrate, 38 mM citric acid, 139 mM dextrose

Solution B: Phosphate-buffered saline (PBS): 135 mM NaCl, 5 mM sodium phosphate, 1 mM EDTA, 0.1 mM PMSF, pH 7.5

Solution C: Lysing buffer: 5 mM sodium phosphate, 1 mM EDTA, 0.1 mM DFP, pH 8.0

Solution D: 0.1 mM EDTA, 0.1 mM DFP, pH 9.3–9.5

[20] V. T. Marchesi and E. Steers, Jr., *Science* **159**, 203 (1968).
[21] J. M. Tyler, W. R. Hargreaves, and D. Branton, *Proc. Natl. Acad. Sci. U.S.A.* **76**, 5192 (1979).
[22] D. L. Siegel, S. R. Goodman, and D. Branton, *Biochim. Biophys. Acta* **598**, 517 (1980).
[23] V. Bennett, this series, Vol. 96, p. 313.

Solution E: 260 mM KCl, 40 mM NaCl, 20 mM Tris, 2 mM EDTA, 0.4 mM NaN$_3$, 0.5 mM 2-mercaptoethanol, pH 7.5

Solution F: 130 mM KCl, 20 mM NaCl, 10 mM Tris, 1 mM EDTA, 0.2 mM NaN$_3$, 0.5 mM 2-mercaptoethanol, pH 7.5

Solution G: 20 mM KCl, 5 mM sodium phosphate, 1 mM EDTA, 0.5 mM 2-mercaptoethanol, 0.1 mM PMSF, pH 7.6

Solution H: 50 mM citric acid, pH 5.0

Solution I: 130 mM KCl, 20 mM NaCl, 20 mM HEPES, pH 8.0

Solution J: 100 mM Tris–HCl, 10 mM MgCl$_2$, 0.5 mM EGTA, 0.2 mM DFP, 1 μg/ml leupeptin, 2 μM cAMP, pH 7.0

Solution K: 7.5 mM sodium phosphate, 1 mM EGTA, 0.1 mM DFP, pH 8.0

Solution L: 200 mM Tris–HCl, pH 8.3

Solution M: 20 mM KCl, 20 mM Tris, 1 mM EGTA, 0.5 mM 2-mercaptoethanol, 0.1 mM PMSF, 0.02% NaN$_3$, pH 8.3

Preparation of Erythrocyte Membranes

Blood is anticoagulated with 1 vol of solution A/7 vol of blood. Typically, two units of freshly drawn anticoagulated blood is rapidly cooled on ice. The blood is divided between four 250-ml plastic centrifuge bottles and the red cells sedimented by centrifuging for 10 min at 2500 g at 4° with a swinging bucket rotor. The plasma, along with the buffy coat, is removed by aspiration. The red cells are washed three times in cold (0–4°) solution B, each time removing any remaining white cell layer. Alternative methods for removing nonerythroid cells more completely include filtration through crystalline cellulose[24] or sedimentation of the erythrocytes through a dextran solution at 1 g[23] followed by washing in solution B. At this time the cells may be lysed and the membranes isolated, or the cells may be resuspended in solution B supplemented with 10 mM glucose, 1 μg/ml leupeptin, and 0.1 mM DFP, and stored overnight on ice.

The packed red cells are lysed by diluting 10-fold with ice-cold solution C. Several proteins, including tropomyosin,[13] require divalent cations in order to bind to the membrane. These can be isolated with the ghosts by including 2 mM MgCl$_2$ and substituting EGTA for EDTA in solution C. The membranes are isolated by centrifugation at 19,000 g for 35–40 min and washed free of cytoplasmic protein in solution C until they are milky white (four to five washes). During the later washes, the rotor is allowed to decelerate without the brake to avoid disturbing the loosely packed membrane pellet. Following each centrifugation, the supernatant is care-

---

[24] E. Beutler, C. West, and K.-G. Blume, *J. Lab. Clin. Med.* **88**, 328 (1976).

fully removed by aspiration, and the centrifuge bottle is then swirled to separate the erythrocyte membranes from a small pellet of leukocytes which adheres to the bottle. This pellet is rich in proteases and is removed by aspiration.

Purification of Spectrin from Erythrocyte Membranes

Spectrin is extracted from membranes by incubation in a low ionic strength alkaline buffer. The membranes are washed once in equal parts of solutions C and D to reduce the ionic strength. They are then resuspended to a final volume of 2.6 liters in solution D prewarmed to 37°, and incubated for 20 min at 37° with gentle swirling. After checking the suspension under a phase microscope to be sure that the "ghost" membranes have fragmented to form small inside-out vesicles (a consequence of removing spectrin and actin from the membrane), the suspension is centrifuged for 45 min as above (no brake), and the supernatant collected by aspiration. The pelleted vesicles are further processed to yield proteins 2.1 and 4.1 (see below). The supernatant is recentrifuged at 19,000 g for 1 hr to remove residual vesicles, and the spectrin is precipitated with 50% saturated ammonium sulfate (313 g dry ammonium sulfate/liter spectrin solution), with the pH maintained between 7.0 and 8.0. After addition of the ammonium sulfate, the suspension is allowed to sit overnight on ice. The following morning the precipitated protein is collected by centrifugation for 10 min at 4000 g in a swinging bucket rotor. Typically, two 250-ml centrifuge bottles are repeatedly filled with the protein suspension without removing the previously pelleted protein, thus concentrating the protein and minimizing the amount of solution transferred between bottles when the protein is resolubilized. The protein pellets are gently resuspended in solution E (a final volume of 50 ml or less) and dialyzed against two 6-liter changes of solution E. The protein solution is then centrifuged for 45 min at 113,000 g to remove insoluble material.

An alternative method for isolating spectrin is particularly useful when working with small volumes of blood. In this method, the washed erythrocyte membranes are transferred to a dialysis bag and dialyzed against several changes of solution D for 48 hr at 4°. The membranes are then removed by centrifugation twice for 1 hr at 25,000 g.

Spectrin isolated from the erythrocyte membrane by the procedures outlined above exists as a mixture of dimers, tetramers, and larger complexes, some of which contain other membrane skeleton proteins, such as actin, protein 4.1, and protein 4.9.[5-7] These species of spectrin multimers are separated from one another by gel filtration chromatography on Sepharose CL-4B. The interconversion between spectrin dimers and tetramers involves a large activation energy, and can be minimized by carrying out all procedures at low temperatures where the species are effectively trapped.[7] In addition, the relatively high ionic strength of solution E increases the dimer–dimer association constant[7] and favors the preservation of the oligomeric forms present in the sample. The dialyzed spectrin solution from two units of blood is loaded onto a 90 × 5 cm column equilibrated with solution E and chromatographed at a flow rate of 40 ml/hr. The protein elutes in three partially resolved peaks, which can be monitored by absorbance at 280 nm. Analysis of the eluant fractions by nondenaturing polyacrylamide gel electrophoresis[25] reveals that the material eluting in the excluded volume contains large oligomers of spectrin as well as complexes of spectrin, band 4.1, band 4.9, and actin. The first included protein peak is composed primarily of tetrameric spectrin, while the slowest eluting material is primarily dimeric spectrin. Protein 4.1, actin, and protein 4.9 which are not complexed with spectrin elute after dimeric spectrin. If experiments require pure preparations of dimeric or tetrameric spectrin, fractions from the tetramer to the dimer peaks are pooled, concentrated by precipitation in 50% ammonium sulfate or by vacuum dialysis in a Micro-ProDiCon concentrator (Bio-Molecular Dynamics), and then rechromatographed. Fractions are analyzed by nondenaturing and SDS–PAGE for the degree of oligomerization and the presence of proteolytic products and nonspectrin proteins (primarily actin). The desired fractions are pooled, dialyzed against solution F, and stored on ice, where they remain in the oligomeric state for 2 to 4 weeks.

Purification of Bands 4.1 and 2.1

The majority of bands 4.1 and 2.1 remains associated with the inside-out membrane vesicles (IOVs) following the extraction of spectrin and actin by low ionic strength buffer and can be subsequently extracted with high ionic strength buffers. The spectrin-depleted vesicles are washed once in cold solution D to remove as much residual spectrin as possible. This is particularly important if band 2.1 is to be isolated, since spectrin and band 2.1 are not well separated by DEAE ion-exchange chromatography. The vesicles are gently resuspended in solution C (DFP increased to 0.2 mM) with a Dounce homogenizer. Resuspending in this way increases the yield of protein 4.1 by about 30–40%. Dry KCl is added to give a final concentration of 1 M in a final volume of 200 ml. The pH is adjusted to 7.7, and the membranes are incubated for 30 min at 37°. This extraction step removes bands 2.1, 4.1, 4.9, and 6, as well as any spectrin remaining

[25] J. S. Morrow and W. B. Haigh, Jr., this series, Vol. 96, p. 298.

following low-salt extraction and wash. Selective removal of proteins 4.1, 4.9, and 6 from the membrane is accomplished by incubation on ice at pH 6.75 during the high-salt extraction. Ankyrin can then be extracted by resuspending in 1 M KCl at pH 7.7 and incubating at 37°.[26] Following centrifugation for 30 min at 113,000 g in a fixed-angle ultracentrifuge rotor, the 1 M KCl supernatant is decanted (not aspirated), dialyzed against two 6-liter changes of solution G, and loaded onto a 0.9 × 10 cm column of Whatman DE-52 cellulose equilibrated with solution G. The use of this size column maximizes the recovery of protein without compromising the quality of purification. The column is washed with 50 ml of solution G and the proteins eluted with a 500-ml linear gradient of 20–300 mM KCl in solution G. Others have used discontinuous salt gradients,[10,15] which more effectively separate protein 4.1 from proteolytic fragments of band 2.1. Although earlier methods incubate the membranes in isotonic buffer to remove band 6 prior to extracting bands 2.1 and 4.1,[20] this step is not necessary, since band 6 does not bind to the DE-52 under these conditions. Elution of band 4.1 occurs at 80–100 mM KCl whereas band 2.1 elutes at 180–220 mM KCl. Fractions are analyzed by SDS–PAGE to determine the purity of proteins 4.1 and 2.1. The appropriate fractions are pooled, concentrated by vacuum dialysis in solution F, and stored on ice. Protein 4.1 appears to be quite stable for periods of up to a month, while protein 2.1 is very susceptible to proteolysis and is not stable for much more than a week.

Protein 4.1 prepared by DEAE ion-exchange chromatography appears as a major doublet (78 and 80 kDa) with two minor doublets above and below the major bands when examined by SDS–PAGE run according to Laemmli.[27] All of these bands exhibit extensive homology as judged by two-dimensional peptide mapping.[28] The purity of this preparation is suitable for structural studies or for investigating the interactions with IOVs and other membrane skeleton components. This preparation does, however, contain contaminating kinase activities which are capable of phosphorylating most erythrocyte membrane skeleton proteins and also phosphatidylinositol 4-phosphate.[29] (Spectrin and band 2.1 prepared by the methods described here also exhibit protein kinase activity.) While the contaminating protein kinase activity is a useful way to radiolabel protein 4.1 (see below), it presents problems in studies involving the phosphorylation of isolated protein 4.1 by other protein kinases. In these

[26] W. R. Hargreaves, K. N. Giedd, A. Verkliej, and D. Branton, *J. Biol. Chem.* **255**, 11965 (1980).

[27] U. K. Laemmli, *Nature (London)* **227**, 680 (1970).

[28] T. L. Leto, unpublished observations (1983).

[29] T. Tobe, T. L. Leto, and R. A. Anderson, unpublished observations (1985).

cases, the kinase activities can be separated from protein 4.1 by gel filtration.[30]

Affinity Purification of Protein 4.1

Protein 4.1 specifically associates with the membrane through the cytoplasmic domain of the transmembrane glycoprotein glycophorin. This association requires the phospholipid phosphatidylinositol 4,5-diphosphate as an obligate cofactor. Glycophorin, when purified by the lithium diiodosalicylate/phenol method,[31] forms a water-soluble protein–phospholipid micelle[32] which contains tightly bound intrinsic phosphatidylinositol 4,5-diphosphate and binds to protein 4.1[16] The glycophorin–phospholipid micelle is sufficiently stable to be covalently coupled to Affi-Gel (Bio-Rad) and used as an affinity matrix to isolate protein 4.1.

Isolated glycophorin (15 mg/ml) is desialated with *Clostridium perfringens* neuraminidase (0.1 U/ml) in solution H for 1 hr at 37°.[17] The desialoglycophorin solution is dialyzed into solution I. Affi-Gel 10 is washed with solution I in a scintered glass funnel and combined with the desialoglycophorin. Typically 4 mg of desialoglycophorin is added per milliliter of packed Affi-Gel beads. The mixture is incubated for 24 hr at 4°, and the beads are then washed with solution F containing 1 M KCl. Under these conditions about 50% of the glycoprotein is coupled to the Affi-Gel.

To isolate protein 4.1, the 1 M KCl extract from spectrin-depleted IOVs is dialyzed into buffer F and applied to the glycophorin–Affi-Gel column (3 mg of protein/ml of packed beads). The column is then washed with solution F until the absorbance of the effluent returns to base line. Bound protein is eluted with solution F containing 1 M KCl. Analysis of the eluted fractions by SDS–PAGE shows that only protein 4.1 is bound to the immobilized glycophorin (Fig. 1).[33]

Radiolabeling of Protein 4.1

A number of methods have been used to label protein 4.1, including iodination with ^{125}I using the Bolton–Hunter method or Iodogen and, more recently, labeling with [32]P. Methods for labeling protein 4.1 with ^{125}I have been previously reported[20,34,35] and will not be further discussed

[30] T. Tobe, T. L. Leto, and V. T. Marchesi, unpublished observations (1986).
[31] H. Furthmayr and V. T. Marchesi, this series, Vol. 96, p. 268.
[32] H. Furthmayr, *in* "Biology of Carbohydrates" (V. Ginsburg, ed.), p. 123. Wiley, New York, 1981.
[33] R. A. Anderson, *in* "Membrane Skeletons and Cytoskeletal-Membrane Associations" (V. Bennett, C. M. Cohen, S. E. Lux, and J. Palek, eds.), p. 223. Liss, New York, 1986.
[34] C. M. Cohen and S. Foley, *Biochim. Biophys. Acta* **688,** 691 (1982).
[35] K. A. Shiffer and S. R. Goodman, *Proc. Natl. Acad. Sci. U.S.A.* **81,** 4404 (1984).

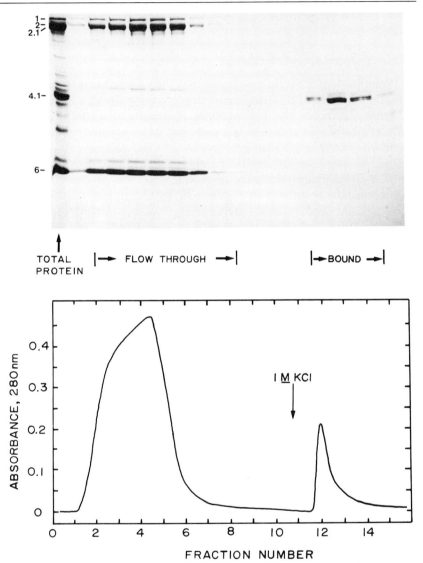

FIG. 1. Affinity purification of protein 4.1 on desialated glycophorin immobilized on Affi-Gel 10. The 1 *M* KCl extract from spectrin-depleted IOVs was dialyzed into solution F and applied to a column of immobilized glycophorin–phospholipid micelles equilibrated with solution F. After washing the column, the bound protein was eluted with solution F containing 1 *M* KCl. *Top:* SDS–PAGE of the applied protein sample (left lane) and fractions of the column effluent. *Bottom:* The absorbance profile of the column effluent. (Reprinted from Anderson[33] by permission.)

here, except to note that the covalent coupling of the Bolton–Hunter reagent to protein 4.1 blocks the binding site for glycophorin in about 40% of the labeled protein 4.1 molecules, although the remainder of the labeled protein appears functionally normal. Similar results have been noted with iodination of protein 4.1 using the Iodogen method.[36] An alternative method to label protein 4.1 is by phosphorylation of protein 4.1 using [γ-^{32}P]ATP. Two methods have been used—phosphorylaton of protein 4.1 bound to IOVs using membrane-associated cAMP-dependent protein kinase,[37] and phosphorylation of protein 4.1 in solution using the protein kinase activity which coelutes with protein 4.1 from the DEAE column. Phosphorylation of membrane-bound protein 4.1 is accomplished using spectrin-depleted IOVs at a concentration of 10 mg of protein/ml. The phosphorylation is carried out in solution J containing 50 μM ATP with 200 μCi/ml of [γ-^{32}P]-ATP. After incubating for 90 min at 22°, the ATP concentration is adjusted to 0.8 mM with unlabeled ATP, and the incubation is continued for 90 min at room temperature, or 24 hr on ice. The ^{32}P-labeled proteins 4.1 and 2.1 are extracted and purified as described above. The specific radioactivity of protein 4.1 is typically 50,000–100,000 cpm/μg. Purified protein 4.1 is labeled at a concentration of 1–2 mg of protein/ml in solution J lacking cAMP. In this case, protein 4.1 is incubated for 48 hr on ice with 50 μM ATP and 1 mCi/ml of [γ-^{32}P]ATP, the ATP concentration is adjusted to 0.2 mM, and the incubation is continued for an additional 48 hr on ice. The specific radioactivity of protein 4.1 labeled under these conditions is 200,000–500,000 cpm/μg. Protein 4.1 labeled by both methods interacts with glycophorin and spectrin–actin.

Isolation of Membrane Skeletal Proteins from Triton Shells

Methods for isolating cytoskeletal components (bands 4.1 and 4.9 and a calmodulin-binding protein) from Triton shells have been reported.[11,38,39] Such methods can be useful for obtaining these proteins for structural studies and for some functional studies. However, exposure to detergents may alter the properties of the proteins of interest. For example, the ability of protein 4.1 to associate with glycophorin–phsophatidylinositol 4,5-diphosphate is abolished in the presence of a variety of nonionic detergents.[40]

---

[36] R. A. Anderson, Ph.D. thesis. University of Minnesota, St. Paul, 1982.
[37] T. L. Leto and V. T. Marchesi, *J. Biol. Chem.* **259**, 4603 (1984).
[38] V. Ohanian and W. Gratzer, *Eur. J. Biochem.* **144**, 375 (1984).
[39] K. Gardner and V. Bennett, *J. Biol. Chem.* **261**, 1339 (1986).
[40] R. A. Anderson, unpublished observations (1981).

The procedure of Siegel and Branton[11] has been modified to isolate two erythrocyte membrane skeletal proteins with the same apparent molecular weight as band 4.9. Hemoglobin-free ghosts are incubated in 10–20 vol of buffer B for 20 min on ice to remove band 6.[41] After washing twice in solution K, the membrane pellets are pooled and resuspended in an equal volume of ice-cold solution K containing 1% Triton X-100 with the following protease inhibitors: 2 mM DFP, 1 μg/ml leupeptin, 1 μg/ml pepstatin. After incubating on ice for 30 min, the insoluble shells are collected by centrifugation at 45,000 g for 30 min and washed once with solution K containing 0.5% Triton X-100 and once with solution K with no detergent. The washed skeletons are suspended in 10 vol of warmed solution D containing the protease inhibitors used in the detergent lysis step and incubated at 37° for 45 min with gentle swirling. After centrifuging for 45 min at 113,000 g to remove any remaining insoluble material, the supernatant is brought to 20 mM Tris by the addition of 1/9 vol of solution L. The solubilized skeletal proteins are loaded at a rate of 30 ml/ hr onto a DEAE-Sephacel column (90 × 1.5 cm) equilibrated with solution M. After loading, the column is washed with 100 ml of solution M and eluted with a 500-ml linear 20–300 mM KCl gradient. The fractions are analyzed by SDS–PAGE for the presence of proteins with an apparent M_r of 48,000. Under these conditions, two peaks of protein of this molecular weight are obtained. The first (M_r ~49,000) elutes in 75–115 mM KCl, and the second (M_r ~50,500) in 180–220 mM KCl. Peptide mapping shows that the two peaks are different proteins, and that the band 4.9 seen in one-dimensional SDS–PAGE of erythrocyte membrane proteins contains both proteins.[42] In contrast to our experience with this separation, Siegel and Branton report a single peak of protein 4.9,[11] which apparently corresponds to the 49,000-Da protein.

The two proteins are pooled separately, concentrated in Amicon stirred cells with UM10 membranes, and dialyzed against 6 liters of solution F. The 48,000-Da proteins are then further purified by rate zonal centrifugation on 5–20% sucrose gradients in solution F for 24 hr at 200,000 g, which removes contaminating traces of spectrin and adducin, the calmodulin-binding dimer recently reported by Gardner and Bennett.[39] Following this step, the 48,000-Da proteins are virtually free of contaminating proteins detectable by Coomassie blue staining on SDS–PAGE (Fig. 2). The one major exception is a protein with an approximate M_r of 57,000, which is not separated from the 49,000-Da protein by ion-

[41] J. A. Kant and T. L. Steck, *J. Biol. Chem.* **248,** 8457 (1973).
[42] W. C. Horne, H. Miettinen, and V. T. Marchesi, *Biochim. Biophys. Acta* **944,** 135 (1988).

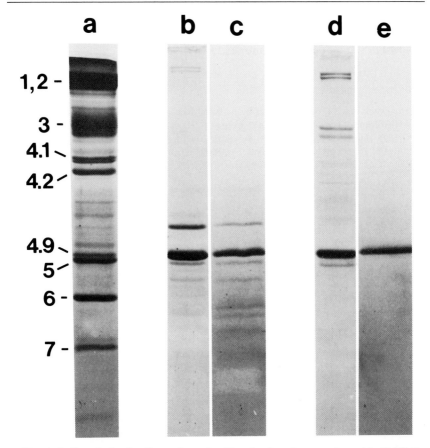

Fig. 2. Sodium dodecyl sulfate-polyacrylamide gels of erythrocyte ghosts (a); 49,000-Da protein purified by DEAE ion-exchange chromatography (b) and rate zonal centrifugation (c); and 50,500-Da protein purified by DEAE ion-exchange chromatography (d) and rate zonal centrifugation (e).

exchange chromatography, gel filtration, rate zonal centrifugation, or nondenaturing gel electrophoresis. Two-dimensional peptide mapping suggests that the 49,000-Da protein and the 57,000-Da protein are structurally related.[42]

Adducin, the calmodulin-binding heterodimer (M_r 97,000–103,000) recently described by Gardner and Bennett[39] can be obtained from the same cytoskeletal preparation used to isolate band 4.9 proteins. This protein elutes from the DEAE column in 155–205 mM KCl, just ahead of and overlapping the 50,500 Da peak. The two proteins are separated from each other, and from the spectrin present in these fractions, by rate zonal centrifugation as described above.

Acknowledgments

This work was supported in part by National Research Service Awards from the National Institutes of Health to Thomas L. Leto (No. HL06215) and Richard A. Anderson (No. GM09184), and a New Investigator Research Award, No. HL-34869, from the National Institutes of Health and a Grant-in-Aid, No. 850936, from the American Heart Association, with funds contributed in part by the AHA, Connecticut Affiliate, to William C. Horne.

[25] Reconstitution of Red Blood Cells: Cytoskeletons Coated with Phospholipid Membrane

By CHAN Y. JUNG and JOHN CUPPOLETTI

Introduction

Reconstitution of transport function of isolated transport proteins into an *in vitro* model membrane system is an essential step for the experimental elucidation of molecular and kinetic mechanisms of biological transport functions. Ideally, the model membrane systems to be used for such a reconstruction should be a lipid bilayer preparation of a well-defined lipid composition, otherwise approximating a living cell in size, geometry, and population uniformity. Most of the model membrane preparations used in the past for this purpose are unilamellar liposomes obtained by the removal of detergent from a mixed micellar solution containing lipids, protein, and detergent.[1] These liposomes are small in size, typically with a diameter less than 200 nm.[2] This gives an extremely large surface-to-volume ratio compared with cells, which makes exact quantitation of reconstituted transport activity extremely difficult for most of the biologically important transport systems.

This point is best illustrated with the carrier-mediated glucose flux in human erythrocytes. The flux is so fast that its quantitation requires considerable technical effort.[3] At 23°, the half-equilibration of D-glucose ex-

[1] E. Racker, *J. Biol. Chem.* **247**, 8198 (1972).

[2] H. G. Weder and O. Zumbuehl, *in* "Preparation of Liposomes" (G. Gregoriadis, ed.), p. 79. CRC Press, Boca Raton, Florida, 1983.

[3] C. Y. Jung, L. M. Carlson, and D. A. Whaley, *Biochim. Biophys. Acta* **241**, 613 (1971).

TABLE I
COMPARISON OF THE RATES OF GLUCOSE EQUILIBRATION IN CELLS AND IN LIPOSOME[a]

Parameters	Human erythrocytes	Liposomes
Radius (cm)	3.5×10^{-4}	10^{-5b}
Volume (cm³)	8.7×10^{-11}	4.2×10^{-15}
Surface area (cm²)	1.55×10^{-6}	1.2×10^{-9}
Volume/surface area (cm)	5.6×10^{-5}	3.3×20^{-6}
Half-equilibration time (sec)	3 (observed)	<0.18 (expected)

[a] Calculations based on 100% reconstitution.
[b] This is the maximum limit of reported values compiled in Ref. 2.

change occurs within 3 sec, and its total equilibration virtually completes within 30 sec. This fast equilibration is due not only to the high efficiency of the carrier mediation, but also to the high surface area-to-volume ratio of erythrocytes. In reconstituted systems also these two factors determine the time course of glucose equilibration. The effects of these factors on the expected rate of equilibration of glucose in an *in vitro* reconstitution in liposomes is illustrated in Table I. Assuming 100% reconstitution of carrier mediation, it is clear that with liposomes the equilibration would be completed within a second. This is much too fast to be quantified with precision using any technique presently available. One obvious compromise is to reduce the reconstitution efficiency by using lesser amount of carrier proteins to slow down the flux to a measurable speed.[4] In the case of the human erythrocyte glucose transport system, 3% of the full reconstitution would give a half-equilibration time of 6 sec in a typical liposome preparation, the lowest limit one can quantitate with precision by existing methods.

Here we describe a model system in which glutaraldehyde-crosslinked human erythrocyte ghosts are extracted with detergent, and the resulting cytoskeletal shells are used as a support on which lipid bilayers are coated. The coating is done by the reversed-phase evaporation[5] procedure, which was originally used for the preparation of liposomes. These proteoliposomes, which we term "erythrosomes,"[6] are large and stable. The tight diffusion barrier and uniform size also make them ideal for reconstitution studies of membrane transport protein.

[4] M. Kasahara and P. C. Hinkle, *J. Biol Chem.* **252,** 7384 (1977).
[5] F. Szoka, Jr., and D. Papahadjopoulos, *Proc. Natl. Acad. Sci, U.S.A.* **75,** 4194 (1978).
[6] J. Cuppoletti, E. Mayhew, C. R. Zobel, and C. Y. Jung, *Proc. Natl. Acad. Sci. U.S.A.* **78,** 2786 (1981).

Method of Erythrosome Preparation

Two key steps are involved in the preparation of erythrosomes. Cytoskeletal shells are prepared first, then they are coated with lipid bilayer. These are described separately below.

Preparation of Erythrocyte Cytoskeleton Shells

Essentially hemoglobin free, human erythrocyte ghosts are prepared by the method of Dodge *et al.*[7] Freshly drawn blood or recently outdated blood obtained from the local Red Cross blood bank may be used. The ghosts (90% hematocrit) in 20 mM sodium phosphate (pH 7.4) buffer are first cross-linked by incubating them with 100 mM glutaraldehyde for 5 min at 25°. They are then mixed with an equal volume of 5% Triton X-100 (50 mg/ml of packed ghosts) in balanced salt solution (BSS). BSS contains 125 mM NaCl, 5 mM KCl, 3.75 mM CaCl$_2$, 2.5 mM MgCl$_2$, and 10 mM Tris–HCl (pH 7.4). The mixture is kept for 30 min at room temperature. The mixture then is layered over an equal volume of 10% (w/v) sucrose in BSS (pH 7.4) and centrifuged. The cytoskeletons are readily pelleted at 4000 g in 5 min, and the supernatant and part of the sucrose cushion are removed. The pellet is mixed with an equal volume of BSS, and washed twice more as above by centrifugation with a sucrose cushion. These cross-linked and well-washed cytoskeletons are mixed with an equal volume of 5% (w/v) polyethylene glycol 6000 in BSS (pH 7.4). At this stage, the residual Triton X-100 associated with the packed cytoskeletons is approximately 0.08% (0.01 mg/mg of protein) as estimated by the use of ³H-labeled Triton X-100 as a tracer in the extraction solution. It is advisable to add 0.02% sodium azide to all solutions to prevent bacterial contamination.

Coating of Cytoskeletons with Lipids

The cytoskeletons are then coated with lipid by evaporation of ether containing phospholipid in contact with an aqueous phase containing cytoskeletons. The procedure is a modification[6] of the reversed-phase evaporation method. Triton X-100, which is insoluble in the presence of polyethylene glycol 6000 (unpublished observation), is routinely included in this procedure to precipitate residual Triton X-100.

Egg phosphatidylcholine (PtdCho) (20 µmol/ml of packed cytoskeletons) in chloroform is spread on the inside surface of a 50-ml, round-bottom glass flask, then dried to a film by rotary evaporation for 10 min at

[7] J. T. Dodge, C. Mitchell, and D. J. Hanahan, *Arch. Biochem. Biophys.* **100**, 119 (1963).

room temperature. The lipid is redissolved in 2 vol of diethyl ether. To this 1 vol of packed cytoskeletons in BSS and 1 vol of 5% polyethylene glycol 6000 in BSS (pH 7.4) are added, and the mixture is immediately subjected to rotary evaporation. Solvent boiling due to reduced pressure and rotation provides sufficient agitation to produce an emulsion, thus the sonication step of the standard reversed-phase procedure is not necessary. A 20-min rotary evaporation removes ether, and produces a gellike mixture. The gel is then dispersed in a large-volume excess (50- to 100-fold) of BSS buffer, pH 7.4, containing 5% polyethylene glycol 6000. The dispersion is then incubated for 18 to 24 hr at 35°. The lipid-coated (erythrosomes) cytoskeletons are separated from any liposomes at this point by a 5-min, 4000 g centrifugation. Lipid not associated with the erythrosomes form a floating pellet under these conditions, whereas the erythrosomes are packed at the bottom of the centrifuge tube.

For the trapping of macromolecules, the cytoskeletal shells are first incubated for 30 min at room temperature in the presence of the macromolecules at a desired concentration, and the lipid coating is carried out as described above. This procedure is found to be less satisfactory with small molecules, however, most likely because of the poor diffusion-barrier properties of the erythrosomes immediately after the lipid coating. Small molecules are best introduced into erythrosomes by equilibration at the stage of the 24-hr, 35° incubation immediately after the dispersion of the gellike mixture.

Characterization of Erythrosomes

Cytoskeletal Shells

When essentially hemoglobin free, white ghosts are treated with glutaraldehyde, and extracted with high concentrations of detergent as above, lipid-free, cross-linked cytoskeletons are obtained in good yield. Unlike cytoskeletons prepared without cross-linking,[8] these cross-linked cytoskeletons are very stable to mechanical and chemical treatments. Their morphology is not appreciably changed by centrifugation, or by treatment with 0.1 M NaOH (30 min), ionic or nonionic detergents, or organic solvents.

The cytoskeletons prepared by our procedure retain less than 4% of the native phospholipid (by Bartlett phosphorus determination). The cytoskeletons retain 20–25% of the original protein. The protein species retained in the glutaraldehyde-treated cytoskeletal shells are deduced by

[8] M. P. Sheetz, *Biochim. Biophys. Acta* **557**, 122 (1979).

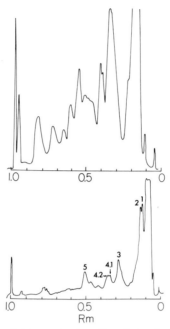

FIG. 1. Scans of Coomassie blue stained, sodium dodecyl sulfate gel electrophoresis of control ghosts (upper), and of dithiobis(succinimidyl propionate)-cross-linked, 100 mM dithiothreitol-reduced cytoskeletons (lower). Both gels represent material derived frm 2.5–5 × 10^{-8} ghosts run on 5.6% polyacrylamide gels. After reduction of the cross-linked ghosts (not illustrated), the major bands had identical mobility to that of the major bands of the untreated ghosts.

the use of dithiobis(succinimidyl propionate), a reversible amino-reactive cross-linking agent. Upon sodium dodecyl sulfate (SDS) gel electrophoresis,[9] proteins of dithiobis(succinimidyl propionate)-treated ghosts and cytoskeletons are largely excluded from the gel, indicating extensive cross-linking. Upon reduction with dithiothreitol, however, the cytoskeleton gel pattern showed bands 1, 2, 3, 4.1 + 4.2, and 5 (Fig. 1). From the observed staining intensities of these bands and their molecular weight estimates, and assuming staining intensity is proportional to mass, an approximate mole ratio of 1 : 1 : 1 : 1 : 2 was obtained for these bands.[6] An additional staining zone near the top of the gel is also observed under these conditions. The mobility of this zone may be examined with a series of lower gel concentrations.[10] Such an examination resolves the zone into approxi-

[9] G. Fairbanks, T. L. Steck, and D. F. H. Wallach, *Biochemistry* **10**, 2606 (1971).
[10] A. L. Shapiro, E. Venuela, and J. V. Maizel, *Biochem. Biophys. Res. Commun.* **28**, 815 (1967).

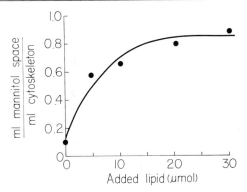

FIG. 2. Mannitol space of erythrosomes prepared from a fixed amount of cytoskeletons, as a function of the lipid in the erythrosomes. Cytoskeletons are coated with egg phosphatidylcholine (PtdCho) as described in text using an increasing amount of the lipid. Erythrosomes are equilibrated with 1 mM mannitol with tracer amounts of [^{14}C]mannitol (0.07 μCi/mol) at the 24-hr, 35°-incubation stage immediately after the lipid coating. Mannitol spaces of resulting erythrosomes are calculated from the radioactivities captured in pellets which come down by a 8000 g, 5-min centrifugation. Coated lipid of erythrosomes is measured by adding a tracer amount of [^{14}C]PtdCho in a parallel set of experiments. Abscissa values represent the amounts of the lipid associated with the erythrosomes obtained from 1 ml of cytoskeletons at the start. (Reproduced from Cuppoletti et al.[6] by permission.)

mate 550,000 and 700,000-Da bands and a minor 800,000-Da band. Similar bands have been reported[11] and identified as spectrin dimer, trimer, and tetramer, respectively. Our cytoskeletal preparations contain, however, relatively more of the 550,000- and 700,000-Da bands, but no other higher M_r species. It is possible that these aggregations may also include the peptides of bands 3, 4.1 + 4.2, and 5. The precise role of each of the proteins in lending structural stability to our cross-linked preparation is yet to be elucidated.

Lipid Coat

That the lipid coat in erythrosomes is a continuous bilayer is supported by various experimental observations: Mannitol space (ml) per erythrosome (ml) is a function of added lipids. Saturation of the mannitol space occurred between 10 and 20 μmol of phospholipid/ml of cytoskeletons (Fig. 2). This value agrees with a value of 10–20 μmol/ml of cytoskeletons required for a bilayer on structures the size of erythrosomes and with surface properties similar to the erythrocyte. The incorporation of phosphatidylcholine is quantitative under the same conditions up to 20

[11] T. H. Ji, B. J. Kiehm and R. C. Midaugh, *J. Biol. Chem.* **255**, 2990 (1980).

μmol/ml of cytoskeletons. In the presence of excess lipids, free lipids float after a 4000 g, 5-min centrifugation, and phase microscopy shows the presence of liposomes in addition to the erythrosomes.

When sealed PtdCho-coated erythromsomes are incubated with fluorescent lipid analog, such as dioctadecylindocarbocyanine, they became highly fluorescent.[6] The apparent specific association of the dye with the boundary of the erythrosome suggests that the surface of the cytoskeleton is coated with phospholipid. No structure of less than 0.5 μm is visible. When cytoskeletons alone are treated similarly, they are only slightly fluorescent. This slight fluorescence may be due to the dye binding to the cytoskeletal proteins.

When erythrosomes are formed in the presence of bovine serum ^{14}C-labeled albumin, an apparent capture volume of 0.55 ml for each ml of erythrosome is maintained after five washes. Cytoskeletons treated the same way show no radioactivity over background. If fluorescein isothiocyanate-conjugated albumin is also included, the erythrosomes become highly fluorescent.[6] Cytoskeletons treated similarly exhibited no fluorescence over the autofluorescence of the cross-linked cytoskeletons. A relatively heavy staining at the perimeter is evident. This is most likely caused by the biconcave disk shape of the erythrosome, although it is also possible that bovine serum albumin may bind to the phospholipids. SDS treatment (2%) results in the loss of fluorescent bovine serum albumin from the erythrosomes. Sucrose, mannitol, and cytosine arabinoside may also be introduced into erythromsomes with varying capture efficiencies.

Erythrosomes appear to establish their diffusion barrier gradually during the 18- to 24-hr, 35° incubation. This is indicated by the time course for the capture of small molecules, which extends 12–18 hr for the maximal trapped volume. Immediately after lipid coating, only 20% of the maximal capture is observed, indicating partial sealing. Over time, capture efficiency increases to a maximum of 85% of the packed erythrosome volume. This increased capture efficiency over time occurs only with small molecules, and possibly reflects a slow organization of the lipid into bilayer structure on the erythrosome surface.

Completely sealed erythrosomes may be used for transport studies. For example, erythrosomes are incubated for 48 hr in buffer containing 5% polyethylene glycol and L-[14C] glucose, and the time course of the efflux is followed. Such an experiment shows a typical $t_{1/2}$ for te efflux of L-glucose as 34 hr. Similarly, permeabilities of erythrosomes to other solutes may be measured. These data (Table II) indicate that erythrosomes are as tight to the diffusion of small molecules as intact cells are.

Figure 3 compares the scanning and transmission electron micrographs of ghosts, cytoskeletons, and erythrosomes. Distinct reticula are present in each of the structures. Erythrosomes have an obvious bicon-

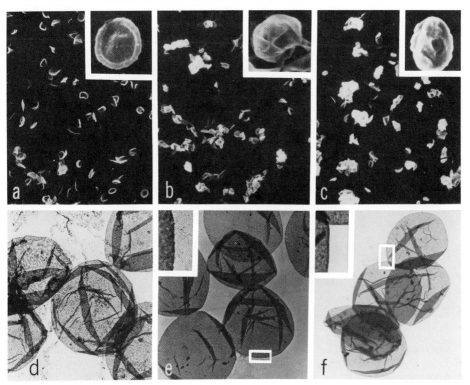

Fig. 3. Scanning (a–c) and transmission (d–f) electron micrographs of ghosts (a,d), cytoskeletons (b,e), and erythrosomes (c,f). Scanning electron micrographs are prepared by critical point drying and gold coating of structures that had adhered to polylysine-coated coverslips (×468). Insets show individual particles prepared by the same procedures after adhesion to coverslips that had not been coated with polylysine (×2340). Transmission electron micrographs are prepared by negative staining with 1% uranyl acetate (pH 4.5) after adhesion to carbon-coated grids (×2340). Insets are magnified views (×8320) of structures from representative areas.

cave shape and rough surface properties. The micrographs of erythrosomes show a structure similar to that of the ghost, with biconcave shape and rough surface. No discontinuities are evident in the structure. Transmission electron micrographs of ghosts, cytoskeletons, and erythrosomes at two different magnifications show from representative areas a granular boundary for the cytoskeletons (e), in contrast with a smooth boundary on the erythrosomes (f). The erythrosomes possess sharply defined borders, suggestive of a lipid bilayer.

When erythromsomes are analyzed by a Coulter particle counter, one typically obtains a mean size of 27 μm,[3] and the counting efficiency is as

TABLE II
DIFFUSION BARRIER PROPERTIES OF ERYTHROSOMES

Solutes	Permeability[a] $t_{1/2}$ (hr)	Capture volume[b] (liters/mol phospholipids)
Glucose	34	100
Sucrose	29	85
Sodium	170	90

[a] Each value represents a half-equilibration time calculated from a 24- to 32-hr tracer equilibrium exchange time course measured at 22°. The time course followed a single exponential function.

[b] Values are calculated from radioactivities captured in erythrosomes at their complete equilibration estimated from the observed exponential time course, and from phospholipid contents of erythrosomes using an average phospholipid molecular weight of 800.

high as 95% compared to manual counting methods. The size compares favorably with the size estimated microscopically. Cytoskeletons show a size distribution similar to erythrosomes, but a counting efficiency of less than 5%.

Concluding Remarks

The selective extraction effected by Triton X-100 treatment of erythrocytes and ghosts has been used to isolate structures depleted of integral proteins and phospholipids. The limiting structure obtained under carefully controlled conditions are essentially lipid free. Our attempts to coat these structures with lipid have been unsuccessful because they are fragile to centrifugation and to the coating procedure. Cross-linking of these structures did not improve the stability. However, one finds that cross-linking ghosts prior to the Triton X-100 extraction produces results in stable cytoskeletons, on which phospholipid can coat. These cross-linked cytoskeletons are also essentially lipid free. They are quite resistant to centrifugation damage, ionic and nonionic detergent, 0.1 M NaOH, and organic solvent. The recovery of protein is high. The observed stoichiometry between the isolated proteins suggests the existence of specific complexes in the erythrosome composed mostly of the cytoskeletal protein with some band 3.

The cytoskeletons may stabilize the phospholipid bilayer in erythrosomes as in the native membrane. The demonstration of specific interactions of the cytoskeletal proteins with phospholipids strongly suggests this possibility.[12] The time course for sealing to small molecules, as opposed to macromolecules, suggests a complex mechanism for the formation of a tight bilayer structure. The coating process may involve fusion of vesicles and interaction of micelles with the cytoskeleton. The cross-linked cytoskeletons apparently provide structural supprot for the lipid association in the erythrosome.

The lipids in the erythrosome exist in a bilayer form, coating the entire skeleton. The observed diffusion-barrier properties to small solutes support this conclusion. The solute-trapping volume of erythrosomes is calculated to be approximately 100 liters/mol of phospholipids. This is 5–400 times higher than that of liposomes (0.23–22.5 liters/mol of phospholipid).[5,13]

The utility of the erythrosome as a model system for the reconstitution of membrane transport should await protein incorporation. Nevertheless, the large internal volume and small surface-to-volume ratio would allow quantitative flux measurement for most biological functions. The tight diffusion barrier, uniformity in size, stability, and relative ease of preparation are other attributes which make erythrosomes attractive for transport studies.

[12] C. Mombers, P. W. M. Van Dijek, L. L. M. Van Deenen, J. DeGier, and A. J. Verkeij, *Biochim. Biophys. Acta* **470**, 152 (1977).
[13] H. G. Enoch and P. Strittmatter, *Proc. Natl. Acad. Sci. U.S.A.* **76**, 145 (1979).

[26] Impact of Methodology on Studies of Anion Transport: An Overview

By Aser Rothstein

The general perception is held that advances in membrane research (and biological research in general) have derived from scientific studies involving imaginative new concepts and hypotheses; that these are tested by clever experiments. The truth, in most cases, may be considerably less dramatic. I have suggested that much of our advance in membrane research has resulted from the availability of new methods and technology; that opportunism, and perhaps cleverness, in exploiting new methods has been an important factor in the advance of our knowledge[1] of membranes. I would suggest that the same premise applies to the more specific field of anion transport by the anion-exchange system of red blood cells. In order to make this point, however, it would be helpful to set the context by briefly summarizing the historical development of the membrane field, in general.

Membrane research has a relatively long history dating back to the mid-nineteenth century when Pfeffer, Naegeli, and others noted that plant protoplasts, and later marine eggs and vertebrate blood cells, placed in anisosmotic solutions would change volumes in a manner consistent with osmotic behavior. These observations led to the concept that the protoplasts were bounded by a semipermeable membrane that allowed rapid equilibration of water, but not of salts. Later, similar methods of measuring volume changes were also used to assess the "permeability" of the putative membrane to penetrating solutes (nonelectrolytes). From such simple studies Overton was able to conclude, by 1900, that the most important determinant of soute permeability was the lipid solubility of the penetrating solute molecules.[2] Furthermore, he noted that small hydrophilic molecules could permeate more rapidly than expected from their lipid solubilities. He suggested that the membrane might be a mosaic of lipid with aqueous patches. Thus he anticipated the modern concept of the lipid-mosaic model[3] by over 70 years. The next major step also came from application of size measurements taken together with lipid analysis. Gortner and Grendel in 1925[4] calculated that in the red blood cell stroma (ghost) there was just about the correct amount of lipid for a bilayer to cover the surface.

The early period of membrane research came to an end in about 1940, summarized by the publication of "The Permeability of Natural Membranes" by Davson and Danielli in 1943.[5] Rather simple procedures led to important conclusions: (1) cells were surrounded by a semipermeable (selectively permeable) membranes accounting for their osmotic behavior; (2) the basic structure of the membrane was a lipid bilayer, but possibly with aqueous patches, accounting for permeability properties to nonelectrolytes; (3) the permeation process was a physical phenomenon that could be quantified in terms of diffusion equations (derivations of the Fick

[1] A. Rothstein, *Can. J. Biochem. Cell Biol.* **62**, 1111 (1984).
[2] H. Overton, reprinted in "Biological Membrane Structures" (D. Branton and R. B. Park). Little, Brown, Boston, 1968.
[3] S. J. Singer and G. L. Nicolson, *Science* **175**, 720 (1972).
[4] E. Gortner and F. Grendel, *J. Exp. Med.* **41**, 439 (1925).
[5] A. Davson and J. F. Danielli, "The Permeability of Natural Membranes." Cambridge Univ. Press, Cambridge, England, 1943.

equation), with the limiting factor being the "resistance" of the membrane to diffusion (permeability coefficient). This was an early application of biophysical principles to biological systems; (4) the rapid penetration of undissociated (lipid-soluble) molecules and relatively slow penetration of ions could explain many phenomena related to permeation of weak electrolytes.

Techniques dependent largely on volume measurements or parameters related to volume measurement could measure relatively large net movements of water associated with solute distributions or movements. By these techniques, cell membranes, in most circumstances, behaved as though they were impermeable to ions. The cellular handling of salts and of physiological molecules such as sugars, amino acids, and nucleotides was poorly understood or explored.

In the 1940s two important technical advances led to new and important developments in membrane research: the capacity to make electrophysiological measurements using microelectrodes and, with the coming of the atomic age, techniques associated with the use of radioactive isotopes. With these methods, it became feasible to measure with precision the fluxes of ions and nonelectrolytes, to examine the electrical parameters of ion flows, and to determine the role of membrane potentials. Somewhat later, the techniques of electron microscopy with heavy metal staining, and X-ray diffraction, allowed visualization of the membrane and a detailed assessment of its basic structure. There followed a long productive period of membrane research (about 30 years) with the development of many of the current concepts of transport, including parameters relating to energetics and thermodynamics, the origins of membrane potentials and mechanisms of conduction, carrier concept-based kinetic modeling, the application of irreversible thermodynamics and the concept of coupled flows, the concept of channels, and the concept of uphill (active) transport coupled to metabolism. Membrane research proceeded at a steady pace. The number of papers published each year increased by about 8% compounded (Fig. 1).

During this period, structural studies were concerned largely with lipid components. Transport studies were concerned primarily with kinetic and thermodynamic parameters. Solubility in the lipid bilayer could account for permeation of lipid-soluble solutes, but carriers and pores or channels were "invented" to account for the fluxes of ions and hydrophilic solutes. It became increasingly evident that proteins must play an essential role. The first definitive connection of transport with a specific peptide came in 1965 with the identification of Na^+,K^+-ATPase in membrane fractions.[6]

[6] J. C. Skou, *Physiol. Rev.* **45,** 596 (1965).

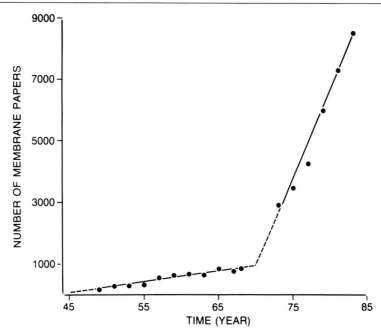

FiG. 1. The estimated annual number of published papers in the field of cell membranes.
(From Rothstein.[1])

Later it was found that small peptides could act as ionophores, allowing
rapid and specific movement of cations through bilayers,[7] and evidence
accumulated that permeability of cell membranes to many solutes could
be perturbed by sulfhydryl agents, agents that were presumed to interact
with cysteine residues of peptides.[8] Although membranes were known to
contain peptides, specific knowledge was minimal. It was not clear how
they were associated with the bilayer, how many different species were
present, and what functions they fulfilled. Knowledge of membrane proteins was largely unexplored because technologies did not exist that allowed their study.

Starting about 1970, a dramatic increase in the volume of membrane
research occurred (see Fig. 1). The sharp upward trend can be attributed
to the appearance of a number of technical and methodological innova-
tions[1]: (1) improved procedures for marking, isolating, and purifying
membranes, (2) use of detergents for "dissolving" membrane proteins,
(3) development of specific affinity probes to "mark" functional proteins

[7] B. C. Pressman, *Fed. Proc., Fed. Am. Soc. Exp. Biol.* **32,** 1698 (1973).
[8] A. Rothstein, *Curr. Top. Membr. Transp.* **1,** 1 (1970).

and functional sites, and of nonpenetrating species for determining the "sidedness" of the membrane components, (4) use of proteolytic enzymes as probes for cleaving exposed portions of membrane-bound proteins, (5) development of vesicle systems derived from particular cellular membranes, (6) reconstitution procedures for functional proteins using model membranes, (7) acrylamide gel electrophoresis for separating and identifying membrane proteins, (8) gel filtration procedures for fractionating vesicles and macromolecular membrane components, (9) development of cross-linking agents to investigate "near neighbors" among components, (10) sophisticated technologies such as microcalorimetry, nuclear magnetic resonance, electron spin resonance, infrared spectroscopy, circular dichroism, and fluorescence analysis for evaluation of the physical state of the membrane lipids and proteins, (11) the freeze-fracture technique of electron microscopy, (12) use of purified lectins to explore the sugar arrays of surface glycoproteins, (13) measurement of single-channel conductances, and (14) energy-transfer measurements to determine distances between membrane ligands.

Most of these procedures were aimed at separation, identification, and characterization of membrane proteins, and evaluation of their functional roles and their organization with respect to bilayer. The consequence was the spectacular expansion of membrane studies already noted (Fig. 1). The impact was felt not only in transport studies, but new groups of investigators were drawn into the membrane field, including virologists interested in virus–cell interactions, immunologists interested in surface antigens, endocrinologists interested in membrane receptors, cell biologists interested in growth factors and cell regulation, biochemists interested in glycoproteins and protein–lipid interactions, and others as well. In the transport field, the concept of transmembrane proteins as pathways for solute and water flow was a major conceptual advance. Numbers of transport proteins have since been identified. Their architectural arrangement with respect to the bilayer has become evident in a general way. Much, however, still remains to be done in terms of understanding the molecular nature of transport, the regulation of the transport, the genetic control of the transport proteins, etc., all very active areas of research at the present time.

The anion-exchange system of the red blood cell was one of the first to which the new technologies, outlined above, were successfully applied, leading first to the identification of a particular membrane peptide, band 3, as the mediator of anion transport,[9] and subsequently to the evaluation of its structure in the bilayer and its functional characteristics to be de-

---

[9] Z. Cabantchik and A. Rothstein, *J. Membr. Biol.* **15,** 207 (1974).

scribed in the following chapters. The physiological function is the exchange of Cl^- and HCO_3^- as part of the transfer of CO_2 from tissues to lungs. This role has been recognized for a long time. In fact, the exchange was first noted by Nasse in horse blood over 100 years ago, before the theory of electrolytic dissociation. It can, therefore, be noted that "anion exchange was discovered before anions" (Nasse measured the exchange of corresponding acids). Although the CO_2 cycle has been understood for a long time, the mechanism of the exchange was quite mysterious, and in many respects, is still mysterious. Mond,[10] in 1927, suggested that "fixed positive charges" in the membrane controlled anion permeability, but it was not until many years later that one of his students, Passow, amplified and extended the concept into a comprehensive "fixed charge" theory to account for the effects of pH and ion concentrations on anion and cation permeabilities.[11] These studies were based on evaluation of transport kinetics and they stimulated interest of several laboratories (including my own) in anion transport. They led to attempts to chemically modify the "fixed" positive charged groups of the membrane and to so perturb the transport. The "agents" found to inhibit, such as fluorodinitrobenzene (FDNB), were capable of interacting with free amino groups of peptides. Thus, their inhibitory effects were consistent with their potential neutralization of "fixed positive charges." Unfortunately, the available agents had serious drawbacks. They were not very specific in a chemical sense; they were not specific in a functional sense, influencing cation and sugar transport as well as anion transport; they were of relatively low potency, requiring high concentrations; and they penetrated rapidly so they interacted with both sides of the membrane and with the cytoplasmic contents as well. A major improvement in the methodology of chemical modification resulted from the introduction of disulfonic stilbenes as inhibitors. One of these compounds, 4-acetamido-4-isothiocyano-2,2'-stilbenedisulfonic acid (SITS), had been used earlier as a fluorescent marker for plasma membranes.[12] It turned out to be a highly potent, highly specific, impermeant inhibitor of anion transport.[13] Assessment of a series of disulfonic stilbenes[14] indicated that they might provide considerable information concerning the anion transport system: (1) they were relatively impermeant so they inhibited by binding to membrane sites exposed to the outside. Later studies were to demonstrate that they were competitive

[10] R. Mond, *Pfluegers Arch.* **217,** 618 (1927).
[11] H. Passow, *Prog. Biophys. Mol. Biol.* **19,** 425 (1969).
[12] H. Maddy, *Biochim. Biophys. Acta* **88,** 390 (1964).
[13] P. A. Knauf and A. Rothstein, *J. Gen. Physiol.* **58,** 190 (1971).
[14] Z. Cabantchik and A. Rothstein, *J. Membr. Biol.* **10,** 311 (1972).

inhibitors, so that the external site could be presumed to be the anion-binding site of the transporter[15]; (2) the potency of various analogs varied widely so that structure–function relationships allowed conclusions about the configuration of the binding site. The most potent analogs had affinities in the fractional micromolar range; (3) their action was specific for anion transport with no effects on other membrane functions; (4) inhibition occurred with analogs with or without covalent binding groups. The noncovalent analogs (reversible inhibitors) have proved useful in evaluating inhibition kinetics and the covalent analogs (irreversible inhibitors) for "marking" sites within the transport peptide; and (5) on binding their fluorescence was substantially enhanced, a property that later became useful in evaluating distances between sites by energy transfer technology.

From the properties enumerated above, it is not surprising that the disulfonic stilbenes proved to be highly useful agents. They soon provided the means for identification of the anion-transport protein. The particular agent used for this purpose was a newly synthesized analog, 4,4′-diisothiocyano-2,2′-stilbenedisulfonic acid (DIDS). It was an irreversible inhibitor, the most potent and useful of the covalent analogs.[14] Chemical modification with DIDS was used in conjunction with two other "new" procedures, the use of detergents for "solubilizing" the membranes, and SDS-acrylamide gel electrophoresis for separating the membrane peptides into molecular weight classes.[16] The results are illustrated in Fig. 2. A tritiated form of DIDS was synthesized for the purpose.[9] After separation of the membrane peptides, the agent was highly localized in band 3 in an amount that was proportional to the degree of inhibition. DIDS was also found to specifically protect band 3 against the binding of FDNB.[17] Later, a variety of other methods confirmed the conclusion that band 3 was the transport protein.[18,19]

The identification of the anion transport protein in prior years had not been feasible because appropriate methods were not available. For example, we had tried to label the transporter with less specific chemical agents, to dissolve the proteins in alcohol mixtures, and to separate by various chromatographic procedures. It was a frustrating experience and a waste of time.

[15] Y. Shami, A. Rothstein, and P. A. Knauf, *Biochim. Biophys. Acta* **508,** 357 (1978).
[16] G. Fairbank, T. L. Steck, and D. F. H. Wallach, *Biochemistry* **10,** 2606 (1971).
[17] H. Passow, L. Fasold, B. Schuhmann, and S. Lepke, *in* "Biomembranes: Structure and Function" (G. Gardos and I. Szasz, eds.), p. 197. North-Holland, Amsterdam, 1975.
[18] Z. Cabantchik, P. Knauf, and A. Rothstein, *Biochim. Biophys. Acta* **515,** 239 (1978).
[19] P. A. Knauf, *Curr. Top. Membr. Transp.* **12,** 251 (1979).

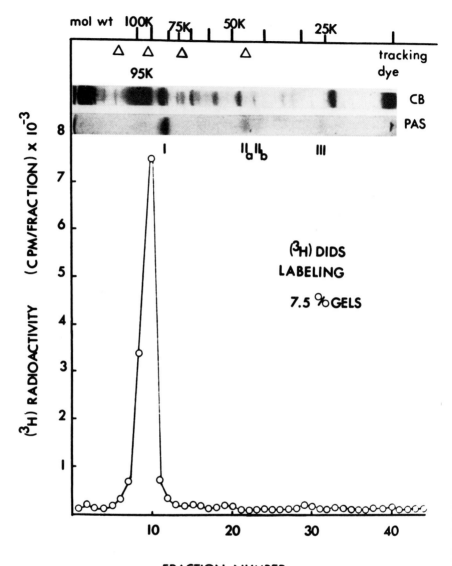

Fig. 2. The labeling of band 3 protein of red blood cells by DIDS. The CB-stained band marked 95K is band 3. (From Cabantchik and Rothstein.[9])

Once identification was made, progress was rapid. Considerable information was already available concerning band 3.[20] Within a few years (by 1976), many essential features of its role in transport could be summarized, at least in a general way[21]: (1) Band 3, as a transmembrane protein, provides the pathway for anion permeation; (2) it is asymmetrically arranged in the bilayer with a specific anion-binding site directly involved in transport; (3) the binding site is accessible from the outside to the impermeant agent, DIDS; (4) although transportable anions compete for the binding site, some degree of specificity is evident; (5) the anion-binding site is positively charged; and (6) the flux involves a one-for-one anion exchange, reflecting a conformational change in a small segment of band 3, whereby the anion-binding site is alternatively exposed inside and outside (later known as the ping-pong model).

At the time, much of the information was preliminary and the model for the arrangement of band 3 in the bilayer and its operation in transport was primitive and speculative, but the essential ideas have proved to be correct. In the interim anion exchange became one of the most studied transport systems, with hundreds of papers published in the last 10 years. As many as 36 reviews or quasi reviews (symposium articles) have appeared. Almost all of the methodologies listed earlier have been applied. The tentative general conclusions of the past have been extended and amplified, and new concepts have evolved. The models of band 3 structure and of transport mechanisms have become highly sophisticated. Two main streams of research have evolved, first an extension of the kinetic modeling of transport and second, attempts to understand band 3 structure, and architectural relationship to the bilayer. Ultimately the goal is to merge these two sets of information into a molecular understanding of how band 3 mediates transport. Although considerable progress has been made much remains to be done. Recently the application of molecular biology techniques had led to success in identifying band 3 messenger and in sequencing the peptide.[22] This approach opens new dimensions in terms of structural analysis.

The chapters to follow are written by individuals who have contributed extensively and importantly to the field. They have not only extended and stretched our concepts of the nature of the anion-exchange process and the role of band 3, but in so doing have also extended our knowledge of transport and of membrane proteins in a more general

[20] T. L. Steck, *J. Cell Biol.* **62**, 1 (1974).
[21] A. Rothstein, Z. I. Cabantchik, and P. A. Knauf, *Fed. Proc., Fed. Am. Soc. Exp. Biol.* **35**, 3 (1976).
[22] R. R. Kopito and H. F. Lodish, *Nature (London)* **316**, 234 (1985).

sense, for band 3 has served as a model for the study of transport proteins. It is also clear from each of the chapters that technology has played a key role in providing the means for experimentation leading to new kinds of information and to conceptual development.

Acknowledgment

The work on anion exchange done in my laboratory has been supported by The Medical Research Council (Canada).

[27] Isolation, Reconstitution, and Assessment of Transmembrane Orientation of the Anion-Exchange Protein

By Z. I. Cabantchik

Introduction

An essential step in the identification of transport systems is the isolation and reconstitution of the relevant membrane proteins by means which conserve the function.[1] For systems which have no intrinsic marker such as a reporter group or an associated enzymatic activity, identification can be carried out by specific labeling of the putative transporter in the intact cell or isolated membrane.[2] This is a desirable property since it facilitates tracing of the proteins throughout all the steps of purification while the protein resides in the membrane (negative purification) or is solubilized in detergent micellar structures (positive purification). In some systems binding to a specific antibody or to a reversibly binding inhibitor can be used to trace the protein even in mild detergent suspensions.[3] The alternative is to assess the protein functionally at the end of all the procedures; this encompasses both purification steps and reincorporation into membranous structures amenable to flux measurements (closed vesicles, black lipid membranes, or even living cells which have a low profile of the property in question).[1] A most ingenious approach for identifying transporters relies on the transport function as a means to fractionate reconstituted proteoliposomes, thus accomplishing isolation and

[1] Z. I. Cabantchik and A. Darmon, *CRC Crit. Rev. Biochem.* **3**, 123 (1986).
[2] Z. I. Cabantchik, P. A. Knauf, and A. Rothstein, *Biochim. Biophys. Acta* **515**, 239 (1978).
[3] A. Darmon, S. BarNoy, H. Ginsburg, and Z. I. Cabantchik, *Biochim. Biophys. Acta* **817**, 238 (1983).

reconstitution in a combined single step.[4] In general, functional reconstitution can also provide the means for assessing the modulatory role of other membranous constituents and exogenous effectors, as well as for dissecting out molecular events underlying the transport mechanism. As we shall demonstrate here, most of these strategies have been successfully applied to the anion-exchange protein (AEP) of the human erythrocyte membrane.

Anion-Exchange Protein

Anion-exchange protein, commonly referred to as band 3 (B3) on the basis of its electrophoretic mobility on SDS–polyacrylamide gels (SDS–PAGE),[5] is the major intrinsic membrane polypeptide of the human red cell membrane, comprising about 20% of the membrane protein mass, or 6% of the total membrane dry mass.[5] The relative abundance of this protein in red cells (1×10^6 copies/cell) explains the high degree of specialization of that membrane for the CBE (chloride-bicarbonate exchange) function. Both CBE and AEP-like polypeptides are present in most mammalian cells, apparently in smaller amounts.[6,7] The gene coding for the protein in chicken red cells[8] and mouse red cells[9] have been cloned and that for the mouse was also sequenced.[9] AEP from human red cells appears on SDS–PAGE as a broad band of M_r 95 K, a property which arise most probably from the presence of varying amounts of glycosidic moieties in the oligosaccharide linked to what is regarded as the predominant polypeptide entity in the 90K–100K region of the gel (Fig. 1). The carbohydrate (M_r $3-8 \times 10^3$) is of a complex type bound exofacially to an Asn-660 residue (in murine,[9] and probably also in human[10,11]). It carries a Man-Man-Man-GlcNAc-GlcNAc sequence at the core, the dissaccharide Gal-GlcNAc as a variably repeating unit with branching points at C-6 of some Gal residues,[12] fucose residues both in the periphery as well as in the core, and some sialic acid in the periphery.[13] B3 polypeptides are the

[4] S. M. Goldin and V. Rhoden, *J. Biol. Chem.* **253**, 2575 (1978).

[5] T. L. Steck, *J. Cell Biol.* **62**, 1 (1974).

[6] M. M. B. Kay, C. M. Tracey, J. R. Goodman, J. C. Cone, and P. S. Bassel, *Proc. Natl. Acad. Sci. U.S.A.* **80**, 6882 (1983).

[7] E. K. Hoffmann, *Biochim. Biophys. Acta* **64**, 132 (1986).

[8] J. V. Cox, R. T. Moon, and E. Lazarides, *J. Cell Biol.* **100**, 1548 (1985).

[9] R. R. Kopito and H. F. Lodish, *Nature (London)* **316**, 234 (1985).

[10] M. Fukuda, M. N. Fukuda, and S. I. Hakomori, *J. Biol. Chem.* **254**, 3700 (1979).

[11] C. J. Brock, M. J. A. Tanner, and C. Kempf, *Biochem. J.* **213**, 577 (1983).

[12] T. Tsuji, T. Irimura, and T. Ozawa, *Biochem. J.* **187**, 677 (1980).

[13] R. A. Childs, T. Feizi, M. Fukuda, and S. I. Hakomori, *Biochem. J.* **173**, 333 (1978).

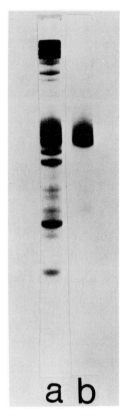

a b

FIG. 1. SDS–PAGE of red cell membranes (IC) and B3 protein isolated by detergent solubilization, cation chromatography, and affinity purification on an organomercurial-Sepharose column (see text). The amounts of protein loaded on the gels were 75 μg for IC and 15 μg for B3. The electrophoresis was according to U. K. Laemmli,[13a] the acrylamide was 7.5%.

carriers of Ii antigens, which have been regarded as the precursors of blood group ABH antigens.[13] In umbilical cord blood the antigen activity is of the Oi type, whereas in adult blood it is of the Ii type. The antigenic change associated with development has been attributed to branching of the single oligosaccharide tree in B3.[10] From the time of appearance in the plasma membrane of the cell and until the red cell is removed from circulation, the B3 polypeptides retain the basic structural and functional features. However, a minor fraction of the polypeptides cannot apparently

[13a] U. K. Laemmli, *Nature* (*London*) **277**, 680 (1970).

escape *in situ* degradation to yield a 62K fragment. This product was postulated to serve as the senescent label which is recognized by circulating autoimmune antibodies and signals removal of the senescent (or damaged) cell from circulation.[14,15]

The protein resides in the membrane and in detergent micelles apparently as a dimer, although higher oligomeric structures cannot be rejected with presently available techniques.[16] The Nt domain of B3 (40K–45K) is entirely endofacial, followed by an amphiphilic domain (50K) composed of a 15K fragment which contains the binding site for disulfonic stilbenes (DS),[17] the most potent inhibitor of CBE[18,19] and a 35K domain which contains the Ct, the carbohydrate tree and the sites for chemical modification of a series of inhibitors.[20–22] These two fragments display properties of hydrophobic proteins[22]; they contain segments which transverse the membrane several times (5–12),[22,23] two or more of which carry the anion-transport sites.[24]

Identification of CBE with the B3 AEP has been accomplished with the aid of high-affinity probes based on the DS core,[18] the most prominent of which is 4,4′-diisothiocyano-2,2′-stilbenedisulfonic acid (DIDS) or its reduced analog H_2DIDS. Inhibition correlated stoichiometrically with the specific labeling of the 95K polypeptides.[25] Similar studies conducted with other probes led to virtually the same conclusions.[19,23,26,27] Regarding the involvement of a subfraction of the 95K polypeptides in other functions such as glucose transport,[28] water transport,[29] or receptor function for

[14] M. B. B. Kay, K. Sorensen, P. Wong, and P. Bolton, *Mol. Cell. Biochem.* **49,** 49 (1982).
[15] M. B. B. Kay, S. Goodman, K. Sorensen, C. Whitefield, P. Wong, L. Zaki, and V. Rudloff, *Proc. Natl. Acad. Sci. U.S.A.* **80,** 1631 (1983).
[16] M. L. Jennings, *J. Membr. Biol.* **80,** 105 (1984).
[17] M. Ramjeesingh, A. Gaarn, and A. Rothstein, *Biochim. Biophys. Acta* **641,** 173 (1981).
[18] Z. I. Cabantchik and A. Rothstein, *J. Membr. Biol.* **10,** 311 (1972).
[19] M. Barzilay, S. Ship, and Z. I. Cabantchik, *Membr. Biochem.* **2,** 227 (1979).
[20] Z. I. Cabantchik, *in* "Structure and Function of Membrane Proteins" (E. Quagliariello and F. Palmieri, eds.), p. 271. Elsevier, Amsterdam, 1983.
[21] J. O. Wieth, O. S. Andersen, P. H. Bjerrum, and C. L. Borders, *Philos. Trans. R. Soc. London, Ser. B* **299,** 383 (1982).
[22] A. Rothstein, Z. I. Cabantchik, M. Balshin, and R. L. Juliano, *Biochem. Biophys. Res. Commun.* **64,** 144 (1975).
[23] I. G. Macara and L. C. Cantley, *in* "Cell Membranes: Methods and Reviews" (E. Elson, W. Frazier, and L. Glaser, eds.), Vol. 1, p. 41. Plenum Press, New York, 1982.
[24] S. Grinstein, S. Ship, and A. Rothstein, *Biochim. Biophys. Acta* **507,** 294 (1978).
[25] Z. I. Cabantchik and A. Rothstein, *J. Membr. Biol.* **15,** 207 (1974).
[26] H. Passow, *Rev. Physiol. Biochem. Pharmacol.* **103,** 62 (1986).
[27] P. A. Knauf, *Curr. Top. Membr. Transp.* **12,** 249 (1980).
[28] M. N. Jones and J. K. Nickson, *Biochim. Biophys. Acta* **650,** 1 (1981).
[29] P. A. Brown, M. B. Feinstein, and R. I. Sha'afi, *Nature (London)* **254,** 523 (1975).

invading *Plasmodium falciparum*,[30] these findings seem to be in dispute, demanding therefore further clarification.

Isolation

Membranes

All the isolation procedures start from membranes isolated from human blood (fresh or outdated). These are usually obtained from red cells (white cell free) by a series of hypotonic shocks and centrifugation. Starting from 300 ml whole blood one can obtain about 100 ml of red blood cells (packed at 10,000 g for 5 min). These are lysed with 5 mM NaPO$_4$ buffer, pH 8, at 1 : 10 (v : v) (5° for 10 min) followed by centrifugation (10,000 g for 10 min at 5°). The supernatants are carefully removed, the ghosts (red) resuspended in the same buffer containing 2 mM EDTA, and the above step repeated until the ghosts look white, that is with no visible trace of trapped or adsorbed hemoglobin. Approximately 64 ml packed ghosts (4 mg/ml protein) is obtained by these procedures, although the yield may vary by 30% among experimenters. These ghosts or membranes can be used within 2–3 days of preparation provided they are kept at 5° in the presence of 0.02% NaN$_3$ after having been treated with 1 mM phenylmethylsulfonyl fluoride (stock solution prepared in dimethylsulfoxide).

In order to facilitate tracing the B3 polypeptides it is recommended to label a small amount of red cells (5 ml of a 20% cell suspension in phosphate-buffered saline, pH 7.4, or any other buffer lacking primary amines) with [^{3}H]H$_2$DIDS (1 mCi/ml for 30 min at 37°), followed by washing with buffer containing 1% bovine serum albumin and buffer alone. The labeled cells are added to the unlabeled prior to the hemolysis step. The isolated membranes are washed with the same buffer used for washing the cells and centrifuged as shown above (this in order to dissociate and remove glycolytic enzymes and other extrinsic proteins associated with membrane).[5]

For many purposes it is convenient to start the purification procedure with membranes devoid of the major sialic acid-rich fragment of glycophorin. This can be accomplished without affecting B3 by digesting the cells (20% suspension) with trypsin (TPCK treated) at 1 mg/ml for 1 hr at 37°. Caution must be exercised to eliminate traces of trypsin which remain adsorbed on the membranes even after repeated washings with buffer. This can be overcome by washing the cells in the presence of albumin (1%)[25] or by treating the cells with tosyl-L-lysylchloromethyl ketone

---

[30] J. M. Wolosin, H. Ginsburg, and Z. I. Cabantchik, *J. Biol. Chem.* **252**, 2419 (1977).

(TLCK) and/or antitrypsin agents. Membranes are prepared essentially as described above for untreated cells.

So as to obtain the chymotryptic fragments of 65K and 35K which remain associated with the membrane after proteolytic incision of B3, cells are treated with chymotrypsin at 2–5 mg/ml as shown for trypsin. The proteolysis can be terminated by reacting the washed cells with PMSF (1 mM for 10 min at room temperature). Similarly, in order to obtain the 15K and the 35K fragments, membranes isolated from chymotrypsin-treated cells are digested with trypsin for 3–5 hr as shown above.

"Negative Purification" of AEP

Membranes enriched in B3 (intact or dissected into the various membrane-associated fragments) can be obtained by selective extraction of the glycophorins with either Triton X-100 (0.05%), octylglucoside (0.7%), octyl-POE (0.5%), $C_{12}E_9$ (0.05%), or equivalent detergents followed by alkali extraction of virtually all remaining major polypeptides, with the exception of minor bands in the 40K–45K area.[30] Detergent extractions are carried out at 5° for 30 min with membrane suspensions containing 0.1 mg/ml protein (final concentration) in 5 mM sodium phosphate, pH 8, followed by centrifugation (30,000 g for 20 min). The pellets are reextracted once more and washed first with buffer containing 1 mg/ml albumin (bovine) and finally with buffer alone. The pellets are resuspended in a minimal amount of buffer and exposed to 0.05 N NaOH–1 mM EDTA for 30 min at 5° after which they are centrifuged (40,000 g for 45 min), the supernatants discarded, and the pellets washed with 50 mM sodium phosphate and centrifuged. So as to avoid damaging polycarbonate centrifuge tubes by the alkali, it is recommended to use heavy Pyrex or polyallomer tubes or, alternatively, it is possible to neutralize the alkali suspension prior to centrifugation by careful titration with acetic acid or HEPES containing EDTA (10 mM final concentration). The final pellet (30 mg protein) is immediately resuspended in sodium phosphate (5 mM), pH 7.2, containing NaCl (100 mM), sucrose (100 mM), and 1 mM MnCl$_2$, divided into small alliquots, and frozen at −70°. This solution is used in order to ensure optimal sealing conditions for the vesicles obtained after thawing and sonicating the suspension in a bath sonicator (30 sec at 5°). The resulting vesicles (after sonication) contain B3 polypeptides with about equal transmembrane orientation, reflecting probably two populations of vesicles, one with right-side-out (ROV) and one with inside-out orientation (IOV). These can be separated on an AE(Aminoethyl)-cellulose column in a low ionic strength medium[31] (5 mM sodium phosphate,

[31] A. Darmon, M. Zangvil, and Z. I. Cabantchik, *Biochim. Biophys. Acta* **727,** 77 (1983).

pH 7.4) and the adsorbed IOVs are eluted from the column with high ionic strength medium (150 mM NaCl, 50 mM sodium phosphate, pH 7.4). Alternatively, they can be separated on an organomercurial-Sepharose 4B column[31] (the ROVs are eluted with a 10 mM cysteine-containing solution). All the solutions used with this procedure were deoxygenized bubbling N_2 for at least 15 min.

"Positive Purification"

Solubilization. A preliminary study of solubilization and functional reconstitution of AEP was conducted by selectively solubilizing B3 or the chymotryptic fragments from membranes with Triton X-100 (1%) and reincorporating the solubilized polypeptides into liposomes (multilamellar) by selective extraction of the detergent with toluene–lecithin solutions.[22] Reconstitution from the cationic detergent cetyltrimethylammonium bromide after purification on a concanavalin-Sepharose column gave proteoliposomes (monolamellar) which were not amenable to transport studies because of their apparently high leakage to anions (i.e., sulfate).[32]

The preparation of relatively pure and functional AEP is accomplished by solubilizing either the high ionic strength-stripped or the alkali-stripped membranes with nonionic detergents such as Triton X-100 (1%), $C_{12}E_9$ (2%), octyl-POE (2%), or equivalent detergents.[1] The concentration of protein used is 1 mg/ml to which ×10 vol of detergent [in 5 mM sodium phosphate, 5 mM dithiothreitol, 1 μM butylated hydroxytoluene (BHT), pH 7.4] are added and the mixture kept for 4 hr at 5°. At the end of this period, EDTA is added (1 mM final concentration) and the mixtures are centrifuged (100,000 g for 60 min). The supernatant is concentrated 10- to 50-fold by Amicon filtration and loaded on a 5-ml AE-cellulose column preequilibrated with the same detergent-buffered solution (5°). These steps lead to removal of the 4.1, 4.2, and 4.5 bands which coextract with B3 and glycophorins in detergent solutions. The column is washed with 5 vol of the same detergent solution or with the same buffer containing an easily dialyzable detergent (e.g., octylglucoside, octyl-POE, or equivalents) at concentrations ×1.5 the nominal critical micellar concentration (cmc). The protein is eluted from the column with 10–15 ml of NaCl (150 mM), sodium phosphate (50 mM), pH 7.2, containing the appropriate detergent and BHT (protein yield 15–20 mg). For those systems which have been subjected to any form of exofacial proteolytic treatment (i.e., excision of the sialoglycopeptides from the glycophorins) or extraction of the glycophorins from the membranes with detergents (see above), the high ionic strength eluate provides a >95% pure B3 preparation which is

[32] A. H. Ross and H. M. McConnell, *J. Biol. Chem.* **253**, 4777 (1978).

virtually ready for reincorporation into a membranous structure (see below). For other preparations, it is necessary to separate B3 from the glycophorins.[31,33] This is accomplished by affinity chromatography of the above eluate on an organomercurial-Sepharose column (*p*-hydroxymercuribenzoate linked to Sepharose 4B via a diaminoethane or diaminopropane arm. Longer aliphatic arms lead to nonspecific hydrophobic binding of B3).[31] The entire eluate is loaded on a PHMB-Sepharose column (1–2 ml wet volume) over a 1- to 2-hr period at 5° and washed with 20 ml of 1% octylglucoside or octyl-POE in 10 mM HEPES, 50 mM NaCl, 1 μM BHT, pH 7.2, at room temperature. Elution of B3 is accomplished by adding 5–10 ml of the same solution containing 25 mM cysteine or dithiothreitol (protein yield 10–15 mg). In many instances it proved advantageous to include the phospholipid of choice (usually egg lecithin or soybean lecithin at 50 mg/ml) in the elution solution. However, depending on the detergent, it might be necessary to add extra detergent in order to solubilize all the phospholipid.

As said above the various steps of purification and the respective yields of B3 can be followed by measuring the radioactivity of [³H]H₂DIDS and the protein concentration (Pierce BCA reagent). Elution profiles can be obtained by following UV absorption (280 nm) with reasonably good sensitivity or fluorescence (285 nm excitation, 340 nm emission) with excellent sensitivity when using non-UV-absorbing detergents such as octylglucoside or $C_{12}E_9$.

Reconstitution. The incorporation of the purified B3 and/or their fragments back into membrane vesicles can be accomplished by various means, depending on the stage of protein purification, the detergent present in the final eluate, and the requisite transmembrane orientation of the final product. For implantation of B3 polypeptides into membranes of living cells see Loyter et al. (this series, Vol. 171 [42]). When using nondialyzable detergents (relatively high cmc) such as Triton X-100 or $C_{12}E_9$, the method of choice for reconstitution is addition of excess phospholipid (plus BHT) to the suspension during (see above) or after elution, followed by addition of SM-2 Biobeads (Bio-Rad) or Amberlite XAD (Sigma).[33–35] Detergent removal is usually followed radioactively by using labeled detergent. The most detailed study of detergent removal from B3 suspensions suggests a gradual addition of beads for optimal reconstitution.[33]

When using dialyzable detergents (relatively high cmc), removal of detergent and spontaneous formation of vesicles can also be accom-

[33] M. F. Lukacovic, M. B. Feinstein, R. I. Sha'afi, and S. Perrie, *Biochemistry* **20**, 3145 (1981).
[34] M. J. Wolosin, *Biochem. J.* **189**, 35 (1980).
[35] D. M. Lieberman and R. A. F. Reithmeier, *Biochemistry* **22**, 4028 (1983).

plished either by dialysis or by dilution of the suspension to one-third the cmc of the detergent followed by centrifugation (50,000 g for 30 min) or by gel filtration on Sepharose 4B-like columns.[1]

Transport Assay

Irrespective of the reconstitution method used, the preparation of vesicles obtained after detergent removal is heterogeneous in size and in lamellar layers and demonstrably leaky to small molecules. So as to obtain sealed vesicles, the preparation is subjected to two cycles of freezing (straight into liquid N_2) and thawing (at room temperature) followed by sonication (30 sec in a bath sonicator at 5° in an N_2 atmosphere).[36] These vesicles are amenable to transport studies using either digital sampling or continuous tracing techniques. The first depends on the availability of relatively fast and efficient methods for separating medium from vesicles. The above vesicles can be brought down by a 2-min centrifugation in an Eppendorf table centrifuge or they can be filtered through a Sephadex G-50 (coarse) column in wet form[30] or dry form (5 ml Sephadex placed in a 6-ml syringe centrifuged for 5 min at 5000 g and subjected to a second spin after addition of vesicles) or deanionized by passing through a 2-ml column of Dowex 1-X8 (100–200 mesh) (citrate form). They can also be separated by a swift filtration through Millipore filters (0.22 or 0.40 μm).[37] All these techniques which rely on digital sampling at various time intervals have been used for measuring anion transport in either the influx or the efflux mode of [35S]sulfate exchange. An improved method for continuous monitoring of transport by fluorescence (CMTF)[38,39] is based on the use of a fluorescent substrate (nitrobenzyldiazole–taurine, NBD–taurine) and anti-NBD antibodies which quench the substrate fluorescence as it egresses from vesicles. With this method it was possible to obtain reliable measures of the anion-transport capacity of B3-reconstituted proteoliposomes[31] (see also Eidelman and Cabantchik, this series, Vol. 172 [9]). The specificity of the transport property acquired by liposomes upon reconstitution with B3 is assessed by parallel experiments carried out with B3 isolated from DIDS-labeled cells (>95% inhibition effected by 0.1 mM DIDS treatment for 30 min at 37° on a 25% cell suspension). The orientation of the polypeptides with respect to the membrane is evaluated with the aid of 4,4′-dinitro-2,2′-stilbenedisulfonic acid (DNDS), a reversibly acting membrane-impermeant inhibitor[40] which at 0.2 mM blocks more

[36] M. Kasahara and P. C. Hinkle, *Proc. Natl. Acad. Sci. U.S.A.* **73**, 396 (1976).
[37] W. Kohne, C. W. M. Haest, and B. Deuticke, *Biochim. Biophys. Acta* **664**, 108 (1981).
[38] A. Darmon, O. Eidelman, and Z. I. Cabantchik, *Anal. Biochem.* **119**, 313 (1982).
[39] W. Khone, C. W. M. Haest, and B. Deuticke, *Biochim. Biophys. Acta* **730**, 139 (1983).
[40] M. Barzilay and Z. I. Cabantchik, *Membr. Biochem.* **2**, 225 (1979).

than 90% of anion transport in intact cells. When added to vesicles the compound inhibits about 50% of the flux, a fact which suggests a random orientation of B3 polypeptides in the reconstituted proteoliposomes.[41] This can be corroborated by assaying transport after inclusion of DNDS into liposomes during freezing and thawing and sonication followed by its removal from the extravesicular space and demonstrating that also from the intravesicular surface the inhibitory effect is of a partial nature, but complementary to the other.[41] A thorough evaluation of the above reconstitution techniques and digital transport assays has been presented.[41]

The transport efficiency of the reconstituted AEP can be evaluated by several techniques.[1] The most useful one is that of initial rate per unit of protein, since it can be compared directly with that of the intact system. For this procedure it is necessary to correct for the fraction of AEP which is in leaky vesicles and for nonspecific fluxes (or leaks) occurring in the system (i.e., also in apparently sealed vesicles). The first can be obtained by comparing the intravesicular volumes measured by trapping a macromolecular probe such as radioactive dextran (V_t) relative to the volume attained at equilibrium with a radioactive anion (V_s), both measured for a given amount of vesicular phospholipid (PL). The second can be obtained by measuring the inhibitor-insensitive transport component, usually between 10 and 30% of the total flux. The technique based on initial rates demands a property which is not always affordable, except when using fast techniques such as CMTF (continuous monitoring of transport by fluorescence) or very slowly transporting substrates.

Since the AEP executes exchange of anions, it is possible to evaluate rate constants (k) from either influx or efflux measurements by measuring the fractional change of labeled substrate relative to the total content of the labeled anion in the medium or in the vesicle at equilibrium (or at zero time, as the case might be). However, comparison of rate constants obtained in vesicles with that obtained in cells demands normalization of values according to the density of B3/membrane area (D) and the surface-to-volume ratio (S/V) of the membrane systems in question ($S/V = 3/r$ for a sphere of radius r). The dimensions of r for cells and vesicles are obtained by electron microscopy (negative staining) and are used for calculating the average B3 density per unit of area. The normalized rate constant is given by $k_n = k/[(3/r)D]$, where k is the rate constant corrected for the nonspecific (DNDS-insensitive) component of fluxes.[1]

The relative efficiency of the reconstituted AEP varied between 15 and 80%, depending on the method of reconstitution, the transport assay, and the method used for correcting for the various components of the flux.[3,33,34,37,41] There is no doubt that the most reliable assessment has been

---

[41] Z. I. Cabantchik and A. Loyter, *Alfred Benzon Symp.* **14,** 373 (1980).

obtained with the CMTF procedure, since it afforded calculation of either initial rates or rate constants with a high degree of accuracy.[2,38] This is because an unlimited number of data points are obtained for each flux, thus allowing direct computer data analysis and statistical evaluation (Eidelman and Cabantchik, this series, Vol. 172 [9]).

An important and often overlooked issue in the evaluation of transport efficiencies relative to the native system is the implication that a change in the energy of activation (E_a) of the reconstituted transport system has on such evaluation. For a 10 kcal/mol difference in E_a there is a 1.7-fold change in efficiency per 10° as illustrated in Fig. 2. Therefore, if the relative efficiency at 37° is 0.14, then at 25, 15, and 5° the respective efficiencies are 0.25, 0.42, and 0.72. This clearly underscores the point that the functional capacity of the reconstituted system might be relatively lower or even higher than that of the native system depending on the temperature chosen for the comparison.[41]

The change in the temperature profile shown above can probably be attributed to the change in lipid environment afforded by reconstitution. The effect of that factor on the AEP activity, particularly the effect of cholesterol, has been well documented.[38] However, in the reconstituted

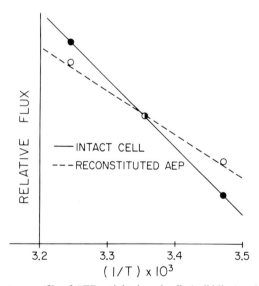

FIG. 2. Temperature profile of AEP activity in red cells (solid line) and reconstituted B3 proteoliposomes (broken line), both measured as sulfate–sulfate exchange according to Ref. 41 . The rate constants for anion exchange at various temperatures are given in values relative to the rate constant obtained at 25°. They are plotted semilogarithmically against the reciprocal of the absolute temperature (T).

state, that effect was considerably less dramatic and reproducible than in intact cells.[34,37] Optimal reconstitutions were obtained with natural mixtures of PL such as egg and soybean lecithins[1,33,37,41] and with synthetic phosphatidylcholine (ether-linked acyl chain).[34] In all these systems, however, it was not possible to discern between the effect of the lipid on the activity of the protein in the reconstituted bilayer from an effect on the reincorporation of the protein per se. From studies carried out with DNDS, it is clear that a modified susceptibility to the inhibitor is indicative of a structural change in the protein afforded by the reconstitution procedure.[39] That this effect might have been induced by the lipid is suggested by the fact that upon implantation of the solubilized AEP into membranes of living cells via reconstituted viral envelopes, full transport capacity was restored.[41]

Oriented Reconstitution

Proteoliposome Preparation. The classical method of reconstitution leads to random orientation of polypeptides unless stringent conditions are imposed on the system during the process of proteoliposome formation. Using the asymmetric properties of the protein, that is, the presence of distinct determinants at both membrane surfaces, a method was devised for immobilizing the detergent-solubilized protein at a given surface and inducing *in situ* vesiculation upon addition of lipid and removal of detergent. AEP was isolated from trypsin-treated cells and solubilized selectively in nonionic detergent as described above. For the preparation of RO B3 proteoliposomes the detergent suspension was loaded on an aminoethyl-Sepharose 4B column (2 ml), rinsed first with 10 vol of octylglucoside (1% in sodium phosphate, pH 8), subsequently with 5 vol of the same solution containing 300 mM sucrose and 10 mg/ml lecithin (egg or soybean), then with the same solution but without the lipid, and finally eluted with buffered NaCl (150 mM). This procedure leads to the formation of liposomes while the protein facing the gel matrix serves as the nucleation domain for the ensuing vesiculation. The vesicles eluted from the matrix are virtually all proteoliposomes. For the preparation of RO B3 proteoliposomes, the preparation loaded on the organomercurial-Sepharose 4B column is subjected to analogous rinsing steps except that 150 mM NaCl replaces the sucrose and the elution buffer contains 25 mM cysteine (or 2-mercaptoethanol). The eluate is dialyzed against the buffer lacking the SH compound or washed by repeated centrifugation (10,000 g for 10 min) and resuspensions.

It was recently shown[42] that B3 can be reconstituted from Triton-X-100 extracts in the presence of octylglucoside by centrifugation through

[42] V. Scheuring, K. Kollewe, W. Haase, and D. Schubert, *J. Membr. Biol.* **90,** 123 (1986).

sucrose density gradients and dialysis. The resulting proteoliposomes showed a high transport activity and retained the orientation of the native system.

Assessment of Orientation. In addition to the functional test, which consists of determining the susceptibility of the vesicle-associated transport to the inhibitory effect of the impermeant DNDS, it is possible to assess the sidedness of the DS binding site [i.e., by measuring the exofacial binding of the benzoylated and dibenzoylated derivatives of DS (BADS and DBDS)].[18,23,25]

A newly developed structural test rests on the determination of the extent of exposure of exofacial or endofacial immunodeterminants by antibody quenching of fluorescence.[43] For that purpose B3 is labeled at both domains with fluorescent tags. For the exofacial domain, cells (at 50% hematocrit in isotonic buffered saline) are treated (1 hr at 37°) with galactose oxidase (5 U/ml) followed immediately by coupling (1 hr at 37°) hydrazido or thiosemicarbazide derivatives (1 mM) of fluorescein, Lucifer yellow, or NBD (e.g., fluorescein thiosemicarbazide, Lucifer yellow itself, NBD-aspartylcarbazide). Evidently, antibodies were raised against all these groups using either keyhole limpet hemocyanin (KLH), albumin (bovine serum), or ovalbumin-coupled haptens as antigens, and their quenching efficiency independently assessed.[3,43] For the endofacial domain of B3, fluorescent reagents for SH groups have served the purpose[23] (e.g., NBD–aziridine, iodoacetamidofluorescein, etc.). The agents were reacted either with ghosts (1 mg/ml protein, 5 mM reagent for 4 hr at room temperature, pH 6.0) or with the final preparation of B3 (0.1 mg/ml protein, 0.5 mM reagent, pH 6.0 for 2 hr at room temperature). For the latter, the vesicles were dialyzed extensively as well as washed with albumin-containing solutions to remove traces of unreacted material which partitions into the lipid.

For determining the extent of exposure, excess antibody was added to the proteoliposome suspension while the fluorescence was recorded. The ratio between the maximal quenching attained with antibody before and after addition of detergent (after background subtraction) gave a measure for the degree of fluorophore exposure at a given surface.[3,43] Because of considerable scattering of fluorescence contributed by the vesicles, it is suggested to use fluorophores with relatively large Stokes shifts such as NBD or Lucifer yellow.

Acknowledgments

This work was supported in part by NIH Grants R22 AI20342 and HL 40158.

[43] S. BarNoy, A. Darmon, H. Ginsburg, and Z. I. Cabantchik, *Biochim. Biophys. Acta* **778,** 612 (1984).

[28] Proteolytic Cleavage of the Anion Transporter and Its Orientation in the Membrane

By MICHAEL J. A. TANNER

The anion-transport protein has been widely used as a model for the study of the structure and mechanism of membrane transport proteins, and it has often been used to develop and test methods for studying proteins of this type. The high abundance of this protein in the red cell membrane makes detection of the protein and most of its fragments relatively straightforward even in unfractionated membrane preparations. The determination of the complete amino acid sequence of the mouse protein[1] from the cDNA sequence was a significant advance in our potential for understanding the structure and mechanism of the protein since it provided the basis for a structural framework for interpreting past future and studies on the topography and mechanism of the protein. Kopito and Lodish[1] have suggested a model for the structure of the mouse protein which contains 12 membrane-crossing segments and is based on hydropathy analysis of the amino acid sequence. More recently the complete amino acid sequences of the human[2] and chicken[3] red cell proteins have become available as well as the sequence of the membrane domain of a closely related protein from the K562[4] cell line.

The availability of the protein sequence gives a strong impetus for carrying out detailed protein chemical studies. It is possible to build up a model of the structure of the protein using a hydropathy scan to predict the location of hydrophobic transmembrane α-helices, and so to predict the relative locations of different portions of the protein sequence. However, although the general principles of these predictive methods are almost certainly sound, their detailed validity has not been tested with any rigor for membrane transport proteins like the anion transporter. This has multiple membrane-crossing segments and is also likely to contain polar residues located within the membrane interior since the protein presumably provides polar pathways for the transport of its substrates through the bilayer. We do not know the rules deciding whether amphipathic helices containing charged and polar residues will be transmembrane or

[1] R. R. Kopito and H. F. Lodish, *Nature (London)* **316,** 234 (1985).
[2] M. J. A. Tanner, P. G. Martin, and S. High, *Biochem. J.* **256,** 703 (1988).
[3] J. V. Cox and E. Lazarides, *Mol. Cell Biol.* **8,** 1327 (1988).
[4] D. R. Demuth, L. C. Showe, M. Ballantine, A. Palumbo, P. J. Fraser, L. Cioe, G. Rovera, and P. J. Curtis, *EMBO J.* **5,** 1205 (1986).

surface seeking. Nor can we predict the way the location of a given helix is influenced by the other transmembrane helices in the structure. This can cause uncertainty in the assignment of some membrane-spanning segments. An error in this assignment can have drastic consequences on the predicted topography of a model and on the location of functional sites and possible transport mechanisms. To obtain a reliable structural model it is important to determine by experiment the topographical location of as many points in the amino acid sequence as possible. This chapter describes two approaches for doing this, proteolytic cleavage and the use of impermeant protein-labeling agents.

Proteolytic Cleavage of the Anion Transporter

A variety of proteases have been used to cleave the native protein in membranes, and their effects on the membrane-bound domain in particular will be considered here. The protein cleavage and peptide-sequencing studies have all been done on the human protein, and the recent availability of the human protein sequence makes it possible to map the sites of protease cleavage.

Trypsin Cleavage

The early work of Steck et al.[5] showed that the anion-transport protein is extremely susceptible to trypsin cleavage at an intracellular site, yielding a soluble 41-kDa fragment and a membrane-bound polydisperse 55-kDa fragment. Subsequent work[6-9] showed that the 41-kDa portion represents the N-terminal cytoplasmic domain of the protein while the 55-kDa portion is the membrane-bound domain. The latter gives a diffuse band on gel electrophoresis because of hetereogeneity in the oligosaccharide present in this region. Trypsin does not cause any significant cleavage of the protein when intact red cells are treated with the enzyme at isotonic ionic strength. However, the structure of the protein is dependent on the ionic strength of the digestion medium and the protein is cleaved by trypsin at an extracellular site when red cells are digested in an isosmotic medium of low ionic strength.[10] Cleavage occurs close to the site at which extracellular chymotrypsin (see below) and other enzymes act on the

[5] T. L. Steck, B. Ramos, and E. Strapazon, Biochemisry 15, 1154 (1975).
[6] T. L. Steck, J. J. Koziarz, M. K. Singh, G. Reddy, and H. Kohler, Biochemistry 17, 1216 (1978).
[7] R. E. Jenkins and M. J. A. Tanner, Biochem. J. 161, 139 (1977).
[8] L. K. Drickamer, J. Biol. Chem. 253, 7242 (1978).
[9] M. Fukuda, Y. Eshdat, G. Tarone, and V. T. Marchesi, J. Biol. Chem. 253, 2419 (1985).
[10] R. E. Jenkins and M. J. A. Tanner, Biochem. J. 161, 131 (1977).

protein.[10,11] The effects of trypsin on the protein when leaky membranes are digested are also markedly influenced by the ionic strength, the protein being more resistant to digestion at higher ionic strength. Digestion with high levels of the enzyme in isotonic solution also releases a small fragment located in the cytoplasm which originates from a region close to the C-terminus of the protein.[12] Jennings and co-workers[13] have established that Lys-743 is a site of intracellular cleavage by trypsin.

Chymotrypsin Cleavage

Chymotrypsin applied to intact red cells cleaves the membrane domain of the protein after Tyr-553 (see Fig. 1) about 350 amino acids from the C-terminus[1,5,6,14,15] Pronase, thermolysin, and subtilisin all cleave close to this point under similar conditions.[16,17] Extracellular trypsin at low ionic strength cleaves at Lys-562.[11] This region contains many bend-forming residues and it makes up the surface loop in the protein which is most susceptible to protease action from the outside surface of the cell.

Moderate concentrations of chymotrypsin applied to unsealed membranes also break at this point but there is an additional cleavage at a cytoplasmic site (Tyr-359) adjacent to the trypsin-susceptible site.[5,16,18,19] This is about 200 amino acids on the N-terminal side of the extracellular cleavage site. Thermolysin acting on the cytoplasmic side of the membrane cleaves after Gly-361 in the same sequence,[20] while trypsin cleaves after Lys-360 in it.[19] It is interesting that this cleavage point is in a region of high bend-forming potential but is separated from the point at which the polypeptide chain is likely to enter the membrane by a 50-amino acid segment containing many charged residues. This region is probably highly structured. The N- and C-terminal fragments produced by chymotrypsin cleavage in this way are usually described as the 17- and 35-kDa fragments on the basis of their apparent molecular weight on gel electrophoresis.

When high concentrations of chymotrypsin are used on NaOH-stripped red cell ghosts more extensive cleavage occurs and two smaller

[11] C. J. Brock, M. J. A. Tanner, and C. Kempf, *Biochem. J.* **213,** 577 (1983).
[12] D. G. Williams, R. E. Jenkins, and M. J. A. Tanner, *Biochem. J.* **181,** 477 (1979).
[13] M. L. Jennings, M. P. Anderson, and R. Monaghan, *J. Biol. Chem.* **261,** 9002 (1986).
[14] L. K. Drickamer, *J. Biol. Chem.* **251,** 5115 (1976).
[15] M. L. Jennings and J. S. Nicknish, *J. Biol. Chem.* **260,** 5472 (1985).
[16] Z. I. Cabantchik and A. Rothstein, *J. Membr. Biol.* **15,** 227 (1974).
[17] R. E. Jenkins and M. J. A. Tanner, *Biochem. J.* **147,** 393 (1975).
[18] S. Grinstein, S. Ship, and A. Rothstein, *Biochim. Biophys. Acta* **507,** 294 (1978).
[19] W. J. Mawby and J. B. C. Findlay, *Biochem. J.* **205,** 465 (1982).
[20] C. J. Brock and M. J. A. Tanner, *Biochem. J.* **235,** 899 (1986).

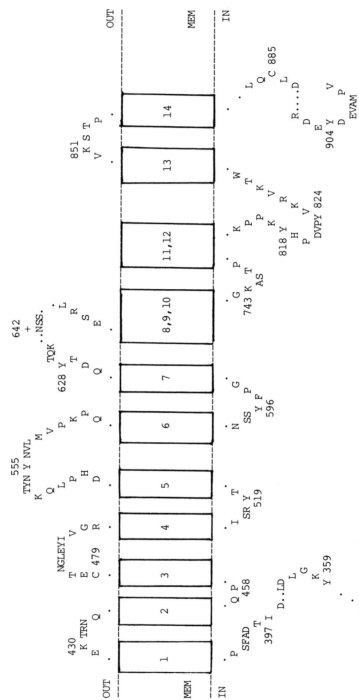

Fig. 1. A model for the topography of the human anion transporter and sites of proteolytic cleavage. The amino acid sequence of the human protein[32] is shown in the single letter code. The numbers in the blocks located within the membrane are the possible membrane-spanning segments. Sections of sequence which are adjacent but not continuous with each other are separated by dots.

fragments of apparent M_r 15K and 8K[21,22] are obtained. The 15-kDa fragment is derived from the 17-kDa N-terminal portion of the membrane domain and differs from it only at the N-terminus.[23] It probably results from cleavage closer to the membrane in the structured region discussed above. The 8-kDa fragment is derived from the 35-kDa C-terminal portion of the protein. This fragment contains two cysteine residues, which are reactive only at the cytoplasmic side of the membrane.[24] These cysteine residues probably correspond to those found at residues 843 and 885 at the extreme C-terminus of the protein. This fragment most likely originates from the action of chymotrypsin on the polar loop between residues 801 and 834, possibly at one of Tyr-818 or Tyr-824.

Cleavage by Papain and Pepsin

These two enzymes give cleavages which differ from those obtained with the above proteases. The products of extracellular papain digestion have been studied by Jennings and co-workers.[15,25] The enzyme cleaves at least three extracellular sites. A small peptide which contains the sequence from Lys-551 to Gln-564 is solubilized from the membrane. A further extracellular cleavage occurs at Gln-630, and this is associated with the inhibition of the chloride transport activity of the protein. This cleavage gives a 7-kDa fragment and a glycosylated 28-kDa portion, both of which are membrane bound and are derived from the C-terminal 35-kDa portion of the protein. Treatment with very high concentrations of papain degrades the protein so extensively that essentially only the membrane-crossing portions survive the digestion,[26] giving several fragments of about 4 kDa.

Digestion of erythrocyte ghosts by pepsin in 1 M HAc yields several membrane-bound fragments.[27] A series of overlapping fragments with identical C-termini are obtained and these have apparent M_r values of 21K, 18.5K, and 12.5K. The largest of these is very similar to the 17-kDa chymotryptic fragment and has the same C-terminus. (The differing values in apparent molecular weight are due to the use of different gel systems by different workers, and the anomalous migration of membrane-associated fragments on SDS gel electrophoresis.) Cleavage also occurs very close to the membrane at the point where the N-terminal portion of

[21] M. Ramjeesingh, S. Grinstein, and A. Rothstein, *J. Membr. Biol.* **57**, 95 (1980).
[22] M. Ramjeesingh, A. Gaarn, and A. Rothstein, *Biochim. Biophys. Acta* **729**, 150 (1983).
[23] M. Ramjeesingh and A. Rothstein, *Membr. Biochem.* **4**, 259 (1982).
[24] M. Ramjeesingh, A. Gaarn, and A. Rothstein, *J. Bioenerg. Biomembr.* **13**, 411 (1981).
[25] M. L. Jennings and M. F. Adams, *Biochemistry* **20**, 7118 (1981).
[26] J. J. Falke, K. J. Kanes, and S. I. Chan, *J. Biol. Chem.* **260**, 13294 (1985).
[27] M. J. A. Tanner, D. G. Williams, and D. Kyle, *Biochem. J.* **183**, 417 (1979).

the protein enters the membrane,[20] on the N-terminal side of residue 397. Another membrane-bound fragment, P5, is obtained from the C-terminal portion of the protein and the amino acid sequence of this has been determined.[11] It contains the 72 amino acid residues from Met-559 to Gln-630. Treatment of ghosts with very high concentrations of pepsin gives rise to several 4-kDa fragments which represent the membrane-spanning domains of the protein.[28]

Methods for Proteolytic Cleavage

There have been many publications describing methods used for proteolytic cleavage. Most of these differ only in detail, so the methods used in this laboratory are described here. The original reports describing the cleavages on which these methods are based are referred to in the discussion of the cleavages above.

Proteolysis of Red Blood Cells

Outdated human red blood cells are washed in a bench centrifuge three times with 0.15 M NaCl, taking care to remove the buffy coat from the red cell layer each time by aspiration using a water pump. The cells are finally washed in the protease digestion buffer and made up to a 20% suspension in this buffer. For digestion with chymotrypsin or subtilisin this buffer is 0.1 M sodium phosphate, pH 8.0, while digestion with thermolysin is done in 0.15 M NaCl containing 0.1 M ammonium acetate and 5 mM CaCl$_2$. Digestion with trypsin under low ionic strength conditions is done in 300 mM sucrose containing 5 mM sodium phosphate, pH 8.0. Digestion is carried out at 37° for 30 min using 0.2–0.5 mg/ml of protease. The cells are washed three times with 0.15 M NaCl containing 0.2% bovine serum albumin [or 2% (v/v) of the plasma originally present with the red cells]. After washing twice more with 0.15 M NaCl, the cells are washed with 0.1 M sodium phosphate, pH 8.0, and membranes prepared by lysis and washing in 5 mM sodium phosphate buffer, pH 8.0 at 4°.

Proteolysis of Red Cell Ghosts

Erythrocyte ghosts are prepared as described by Dodge et al.[29] using lysis into 15 mOsm sodium phosphate buffer, pH 7.4, or 5 mM sodium phosphate buffer, pH 8.0. Digestion under low ionic strength conditions is done in this buffer using 0.1–30 μg/ml protease on ice for 30–60 min. For digestion under isotonic conditions the membranes are made 0.15 M in NaCl and treated with 0.1–0.5 mg/ml of protease. The membranes are

[28] M. Ramjeesingh, A. Gaarn, and A. Rothstein, Biochim. Biophys. Acta 769, 381 (1984).
[29] J. T. Dodge, C. Mitchell, and D. Hanahan, Arch. Biochem. Biophys. 100, 119 (1963).

incubated at 0° for 20 min and then 37° for 40 min. After addition of appropriate specific enzyme inhibitor, the membrane-bound fragments are obtained by washing the membranes once with ice-cold 0.1 M NaOH containing 5 mM 2-mercaptoethanol and then twice with 5 mM sodium phosphate buffer, pH 8.0. To obtain exhaustive digestion with chymotrypsin ghosts are first digested with 0.2 mg/ml chymotrypsin as above, stripped with NaOH, and washed with 5 mM sodium phosphate buffer, pH 8.0, and then redigested using 1.5 mg/ml chymotrypsin at 37° for 30 min.

Pepsin digestion is carried out after first washing the ghosts in ice-cold 0.1 M HAc. The pellet is resuspended in 1 M HAc with the aid of a hand homogenizer and incubated at 37° for 30 min using 0.5 mg/ml pepsin. The reaction mixture is neutralized by the slow addition of an equal volume of 1 M ammonium bicarbonate and centrifuged at 30,000 g for 30 min. The membrane-bound material in the pellet is then washed twice with ice-cold 5 mM sodium phosphate, pH 8.0.

Orientation of the Protein

The topography of the protein has been studied using a variety of impermeant protein-labeling agents. The general strategy is to locate the sites in the amino acid sequence which are labeled when the reagent is accessible to each side of the membrane in turn. The folding and topography of the protein can then be deduced from the order of the labeled sites on the linear sequence of the protein. This method can only give an estimate of the minimum number of traverses of the polypeptide chain across the membrane. In order to detect all the traverses of the polypeptide chain the reagent(s) must be capable of labeling sites in every loop or tail of the protein which emerges from the membrane surface. In practice, few of the natural amino acids are susceptible to chemical modification and these will not be distributed evenly along the protein sequence, so this condition cannot be met. As a result we have quite detailed information about the topography of some regions of the protein but the folding and orientation of other regions, especially toward the C-terminus of the protein, is very poorly defined. A large number of reagents have been used to study the anion-transport protein; however, in several of these studies the site of binding of the reagent has not been defined closely enough to provide useful information on the topography of the protein. Figure 1 shows the topography of the protein based on the data currently available.

Lactoperoxidase-catalyzed radioiodination has been used to study the location of the tyrosine residues in the sequence at one or other side of the membrane.[12,27] Tyrosine-553 has been shown to be at the site of extracellular chymotryptic cleavage by many workers (see above) and this residue is radioiodinated by extracellular lactoperoxidase.[27] There is an additional extracellular tyrosine residue on the N-terminal side of this[12,27] which is likely to be Tyr-486. There are no intracellular tyrosine residues which can be radioiodinated using lactoperoxidase in the human 17-kDa chymotryptic fragment, so Tyr-519, which is predicted to be intracellular, is probably not susceptible to iodination. Tyrosine-596 is a known example of such a tyrosine residue which is refractory to labeling from either side of the membrane. This residue is located on the cytoplasmic side of the membrane. Tyr-628 can, however, be radioiodinated by lactoperoxidase and is extracellular.[11] There are several tyrosine residues present on the C-terminal side of Tyr-628, the only other radioiodinated tyrosine(s) in the C-terminal portion of the protein are intracellular.[12] Although three labeled peptides are obtained, these may be overlapping peptides containing the same tyrosine residues, since they can all be obtained from one small fragment of the protein which can be solubilized from the protein with trypsin.[12] The mobility of the peptides on peptide maps suggest that, if they are overlapping peptides they originate from a relatively basic portion of sequence. Tyrosine-784 is not labeled but this is part of a hydrophobic sequence and is probably located within the membrane. Three tyrosine residues remain, Tyr-818, Tyr-824, and Tyr-905, but the latter is located in a sequence very rich in acidic amino acids at the extreme C-terminus of the protein. This suggests that one or both of the tyrosines at positions 818 and 824 are the intracellular radioiodinated residue since they are located within a relatively basic region of sequence.

Further information on the orientation of the C-terminal region of the protein comes from studies[24,30] on the location of the cysteine groups in this region of the protein. Cysteine-843 and Cys-885 are not reactive when red cells are treated with an impermeant maleimide but are reactive when red cell ghosts are reacted with N-ethylmaleimide or when inside-out red cell vesicles are exposed to the impermeant maleimide. Cysteine-885 is present within a polar region of sequence at the extreme C-terminus of the molecule and its cytoplasmic location has been confirmed by immunochemical experiments which have shown that the C-terminus of the protein is located inside the red cell.[31,32] Cysteine-843 is part of a relatively hydrophobic sequence which is predicted to lie in the membrane interior and might be expected to react with the membrane-permeable reagent N-ethylmaleimide but not the impermeant maleimide. A further sulfhydryl is

[30] A. Rao, J. Biol. Chem. 254, 3503 (1979).
[31] D. M. Lieberman and R. A. F. Reithmeier, J. Biol. Chem. 263, 10022 (1988).
[32] S. D. Wainwright, M. J. A. Tanner, G. E. M. Martin, J. E. Yendle, and C. Holmes, Biochem. J. 258, 211 (1989).

present at residue 479 and lies close to or at the extracellular surface. This residue is unreactive with N-ethylmaleimide, but reacts on treatment of intact red cells with pCMBS[33] or with eosin-maleimide,[34] suggesting that it does indeed have an extracellular location.

Several lysine residues in the protein have been shown to be extracellular. Lysine-430 is labeled by reductive methylation of red blood cells.[35] Two other extracellular lysine residues in the 17-kDa chymotryptic fragment are also labeled under these conditions.[35,36] One of these is Lys-551, located next to the extracellular chymotryptic cleavage site. The other lysine residue is the same as that which reacts with H_2DIDS at a neutral pH. This residue appears to be Lys-539.[3] A further extracellular lysine residue which is probably involved in transport is located in the region C-terminal to Gln-638. This lysine is the site of cross-linking in the 35-kDa fragment by treatment with H_2DIDS at alkaline pH[15,37] values and is also labeled by reductive methylation.[36] Further studies[13] have shown that this lysine is located between the trypsin cleavage site at Lys-743 and a cysteine residue which can be cleaved by S-cyanylation located 5–8 kDa from the C-terminus of the protein which was assigned to be Cys-843. It was suggested that the H_2DIDS cross-linking site is one of the four lysine residues located in clusters between positions 814 and 829. However, the position of the cysteine cleavage was based on the change in apparent molecular weight of a larger fragment on SDS-gel electrophoresis after removal of the extremely acidic C-terminal portion of the protein by the cleavage. The removal of this highly negatively charged portion of the protein may give rise to errors which would result in the overestimation of the molecular weight change after the cleavage. It is possible that the S-cyanylation cleavage was actually at Cys-886. If this were the case Lys-851 could also be a potential site of cross-linking by H_2DIDS.

Pyridoxal phosphate also labels a lysine in the 35-kDa fragment (which may be extracellular) and inhibits transport.[38] Lysine-851 has been shown to be labeled by the reagent.[39] It seems likely that pyridoxal phosphate and H_2DIDS both react with the same lysine involved in transport, suggesting that Lys-851 is accessible from the extracellular surface. It is of

[33] A. K. Solomon, B. Chasan, J. A. Dix, M. F. Lukacovic, M. R. Toon, and A. S. Verkman, *Ann. N.Y. Acad. Sci.* **414,** 97 (1983).
[34] I. G. Macara, S. Kuo, and L. C. Cantley, *J. Biol. Chem.* **258,** 1785 (1983).
[35] M. L. Jennings and J. S. Nicknish, *Biochemistry* **23,** 6432 (1984).
[36] M. L. Jennings, *J. Biol. Chem.* **257,** 7554 (1982).
[37] M. L. Jennings and H. Passow, *Biochim. Biophys. Acta* **554,** 498 (1979).
[38] H. Nanri, N. Hamasaki, and S. Minakami, *J. Biol. Chem.* **258,** 5985 (1983).
[39] Y. Kawano, K. Okubo, F. Tokunaga, T. Miyata, S. Iwanaga, and N. Hamasaki, *J. Biol. Chem.* **263,** 8232 (1988).

interest to note that Lys-851 is replaced by a methionine residue in the homologous protein present in the K562 cell line.[4] Studies on anion transport in K562 cells suggest that this transporter has a significantly lower affinity for DIDS than the red cell transporter.[40]

It is likely that the polypeptide in the chymotryptic 17-kDa fragment crosses the membrane five times. As well as the extracellular site of chymotryptic cleavage, the loop containing Lys-430 and the loop containing Cys-479 are also extracellular. There is, at present, no direct evidence to establish the cytoplasmic location of the loop connecting the predicted transmembrane segments 2 and 3 (around Pro-458), or that connecting segments 4 and 5 (around Tyr-519). The region around Tyr-628 is also extracellular and is adjacent to the site of N-glycosylation at Asn-642. The lack of N-glycosylation at Asn-593 is consistent with the cytoplasmic location of the loop between membrane-crossing regions 6 and 7.[11] The C-terminal side of the N-glycosylation site is less well characterized. Figure 1 summarizes the information discussed earlier on those regions in this portion of the molecule for which there is any evidence about topological location. The figure suggests there may be 14 membrane-spanning regions and some of these will be quite short (for example, regions 13 and 14). However, the exact number and location of these membrane traverses is still very uncertain, especially on the C-terminal side of N-glycosylation site where the direct evidence on the topography of the protein is far from complete. Much more investigation will be needed before it is possible to build a reliable model of the topography of the protein.

[40] S. Dissing, R. Hoffman, M. J. Murnane, and J. F. Hoffman, *Am. J. Physiol.* **247**, C53 (1984).

[29] Functional Asymmetry of the Anion-Exchange Protein, Capnophorin: Effects on Substrate and Inhibitor Binding

By Philip A. Knauf and Jesper Brahm

Introduction

Various lines of evidence strongly suggest that the human erythrocyte anion-exchange system involves a ping-pong lock–carrier mechanism, in which the transport protein, known as band 3 or capnophorin, can exist in

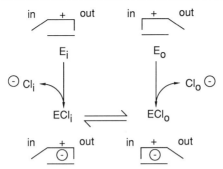

FIG. 1. Ping-pong lock–carrier model for anion exchange. The anion-transport protein can exist in a form E_i with the transport site (indicated by the + charge) accessible to the cytoplasmic compartment (in), or in a form E_o with the transport site open to the external medium (out). Conversion from one form to the other is greatly facilitated when a suitable substrate ion, such as Cl^-, is bound to the transport site to form the corresponding ECl_i or ECl_o complex. Note that in this model the transport site itself does not move across the membrane, but rather its access to the cytoplasm or the external medium is changed by the opening and closing of gates in series with the transport site.

two different forms.[1-7] In the E_o form, the transport site faces the external medium (Fig. 1), while in the E_i form it faces the cytoplasm (inside). Since the rate of net anion transport across the membrane is at least 10,000 times slower than the rate of exchange of an isotope such as ^{36}Cl,[1,8,9] the system must carry out a very tightly coupled exchange of anions, so that for each anion which leaves the cell, another anion enters. This feature suggests that the change in conformation from E_i to E_o or vice versa can only take place if a suitable anion, such as Cl^-, is bound to the transport site to form the corresponding ECl_i and ECl_o forms.

    Although the change from ECl_i to ECl_o or vice versa actually represents a change in protein conformation, which probably does not involve physical displacement of the transport site from one side of the membrane to the other, it is convenient to depict this model in the kinetically equiva-

[1] P. A. Knauf, in "Membrane Transport Disorders" (T. E. Andreoli, S. G. Schultz, J. F. Hoffman, and D. D. Fanestil, eds.), 2nd Ed., p. 191. Plenum, New York, 1986.

[2] D. Jay and L. Cantley, Annu. Rev. Biochem. 55, 511 (1986).

[3] H. Passow, Rev. Physiol. Biochem. Pharmacol. 103, 61 (1986).

[4] J. O. Wieth and J. Brahm, in "The Kidney: Physiology and Pathophysiology" (D. W. Seldin and G. Giebisch, eds.), p. 49. Raven, New York, 1985.

[5] W. Furuya, T. Tarshis, F.-Y. Law, and P. A. Knauf, J. Gen. Physiol. 83, 657 (1984).

[6] O. Fröhlich, J. Membr. Biol. 65, 111 (1982).

[7] R. B. Gunn and O. Fröhlich, J. Gen. Physiol. 74, 351 (1979).

[8] M. J. Hunter, J. Physiol. (London) 268, 35 (1977).

[9] P. A. Knauf, G. F. Fuhrmann, S. Rothstein, and A. Rothstein, J. Gen. Physiol. 69, 363 (1977).

A   B

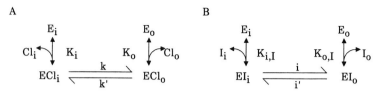

FIG. 2. (A) Kinetic representation of anion transport with Cl^- as substrate. K_i and K_o are the dissociation constants for binding of Cl^- to E_i and E_o, respectively, while k is the rate constant for the conversion from ECl_i to ECl_o and k' is the rate constant for the reverse conformationl change from ECl_o to ECl_i. (B) Corresponding representation of the transport system with I^- as substrate.

lent[10,11] form of a diffusible carrier, as is done in Fig. 2A. Such a model, in which the transport protein alternates between different conformational states, with each change resulting in the transport of a single substrate ion, may apply to other transport systems in addition to that for red cell anion exchange. It has the interesting feature that it predicts rather special effects of changes in substrate concentration on the affinity of the system for other substrates and for inhibitors. This chapter discusses some of the properties of this model, and some of the methods which may be used to determine the orientation of the transport sites and the nature of the interactions of substrates and inhibitors with the different conformations of the transport protein. Specifically, the topics to be covered include (1) determination of the asymmetry of unloaded transport sites, (2) determination of the affinities of slowly transported substrates for E_o and E_i, and (3) determination of the affinities of noncompetitive inhibitors for E_o and E_i, with and without substrate bound to the transport site. In addition to being of interest from a kinetic standpoint, such information is necessary in order to specify the conformation of the transport protein under a given set of circumstances, and therefore provides the basis for biochemical or physical studies of the change in protein conformation involved in the transport process. For other aspects of this transport system, including a discussion of the structure of capnophorin, the reader is referred to several excellent recent reviews,[1–4,12,13] as well as to several other chapters in this volume.

[10] C. S. Patlak, *Bull. Math. Biophys.* **19**, 209 (1957).
[11] R. B. Gunn, *in* "Membrane Transport Processes" (J. F. Hoffman, ed.), p. 61. Raven, New York, 1978.
[12] O. Fröhlich and R. B. Gunn, *Biochim. Biophys. Acta* **864**, 169 (1986).
[13] M. L. Jennings, *in* "The Red Cell Membrane: A Model for Solute Transport" (G. Tunnicliff and B. U. Raess, eds.). Humana, Clifton, New Jersey, 1989.

Asymmetry of Unloaded Transport Sites

One can begin by asking how many of the unloaded transport sites are facing each side of the membrane (in the E_i or E_o form) under a given set of conditions. For the human erythrocyte anion-exchange system, NMR studies with ^{35}Cl and ^{37}Cl[14] suggest that the rate-limiting steps in the transport process are the changes from the ECl_i to ECl_o conformation or vice versa, and that the anion-binding reactions can be assumed to be at equilibrium. This assumption greatly simplifies the mathematics, and will be used throughout this chapter. For a more general treatment, the reader is referred to Fröhlich and Gunn.[12] Since the exchange system is very tightly coupled, if Cl^- is the only substrate anion present the influx of Cl^- is equal to the efflux.[5] The ratio of unloaded transport sites, E_o and E_i, is therefore given by[15]

$$E_o/E_i = (K_o k/K_i k')(Cl_i/Cl_o) \tag{1}$$

if the rate constants for the conformational changes, k, and k', and the Cl^- dissociation constants, K_i and K_o, are defined as shown in Fig. 2A. From this equation it is clear that the asymmetry in distribution of unloaded transport sites will depend both on the difference in Cl^- concentrations at the two sides of the membrane (extrinsic asymmetry) and on the intrinsic asymmetry in the affinities and rate constants which characterize the transport system.

Extrinsic Asymmetry

From Eq. (1) it is evident that changes in the Cl^- gradient (Cl_i/Cl_o) will affect the distribution of unloaded sites. Even if the system were perfectly symmetric intrinsically, with $K_i = K_o$ and $k = k'$, there would be more unloaded transport sites facing the inside than facing the outside under physiologic circumstances,[4] where there is a lower Cl^- concentration inside the cell than in the outside medium.[16] The Cl^- gradient can also have a strong effect on the potency of inhibitors which interact preferentially with one of the forms of capnophorin. For example, external competitive inhibitors, which can only bind to E_o, become much more potent inhibitors when there is an outwardly directed Cl^- gradient ($Cl_i > Cl_o$), because this increases the fraction of transport sites in the E_o form.[5,6] This provides an example of one way in which the orientation of the transport

[14] J. J. Falke, K. J. Kanes, and S. I. Chan, *J. Biol. Chem.* **260**, 9545 (1985).
[15] P. A. Knauf, F.-Y. Law, T. Tarshis, and W. Furuya, *J. Gen. Physiol.* **83**, 683 (1984).
[16] J. Brahm, *J. Gen. Physiol.* **70**, 283 (1977).

sites may be artificially altered, namely by using Cl^- gradients to recruit unloaded sites to the side with the lower Cl^- concentration.

Intrinsic Asymmetry

From the standpoint of learning about the transport mechanism itself, the more important kind of asymmetry is that due to the intrinsic properties of the system, that is, to the constants k, k', K_i, and K_o. This can be conveniently expressed in terms of an asymmetry factor, A, which is defined as the ratio E_o/E_i under conditions where $Cl_i = Cl_o$. From Eq. (1), it is evident that

$$A = kK_o/k'K_i \qquad (2)$$

so $E_o/E_i = ACl_i/Cl_o$. The first step in determining the asymmetry of the system therefore is to determine the value of A.

Since A reflects an intrinsic asymmetry of *unloaded* transport sites, it does not in any way depend on the nature of the transported substrate. Thus, A should not change if we substitute, e.g., iodide for chloride at both sides of the membrane (Fig. 2B). For any substrate, an equation for A can be written like Eq. (2), except that the rate constants and dissociation constants for Cl^- are replaced by those which characterize the interactions of the other substrate ion with the system.

Determination of Unloaded Transport Site Asymmetry, A

Many different methods have been devised for measuring A (see Table I), beginning with the original work of Gunn and Fröhlich,[7] in which they showed that the Cl^- concentrations required to half-saturate the transport system were different at the inside and outside of the cell, providing the first indication that the unloaded sites are asymmetrically distributed. In practical terms, perhaps the most straightforward way to measure the asymmetry is simply to measure $K_{1/2,o}$, the concentration of external Cl^- (Cl_o) which half-saturates the system with a fixed internal Cl^- concentration (Cl_i), and then to measure $K_{1/2}$, the Cl^- concentration which gives half-saturation with $Cl_i = Cl_o$. The value of A can then be determined as follows:

$$A = [K_{1/2,o}(1 + K_{1/2}/Cl_i)]/(K_{1/2} - K_{1/2,o}) \qquad (3)$$

where Cl_i is the internal Cl^- concentration at which $K_{1/2,o}$ was measured. This method has the advantage that it is comparatively easy to determine $K_{1/2}$, either by loading ghosts with different Cl^- concentrations or else by

TABLE I

MEASUREMENTS OF A BY VARIOUS METHODS

	$K_{1/2, o}^{max}$ [a]	$K_{1/2, i}^{max}$ [b]	A	$K_{1/2}$ (predicted) [c]	$K_{1/2}$ (measured) [d]	Reference
	3.9 ± 0.6[e]	65.0 ± 18[f]	0.060	68.9	65.0 ± 5[g]	7
	3.9 ± 0.6[e]	57.0 ± 13[h]	0.068	60.9	65.0 ± 5[g]	7
	3.9 ± 0.6[e]	64.0 ± 30[i]	0.061	67.9	65.0 ± 5[g]	7
	3.9 ± 0.6[e]	61.1[j]	0.064		65.0 ± 5[g]	7, 16
	3.9 ± 1.6[k]	35.1[j]	0.111		39.0 ± 4[l]	6
	6.2 ± 0.3[m]	32.8[j]	0.189		39.0 ± 4[l]	6
			0.166[n]			6
	4.7[o]	14.5[p]	0.324	19.2	21.0 ± 3[q]	19
	2.3[r]	17.0[r]	0.137	19.3	21.0 ± 3[q]	19
	1.5[s]	5.0[s]	0.294	6.5	21.0 ± 3[q]	19
Mean:	3.8 (6)[t]	39.1 (9)[t]	0.147 (10)[t]		47.5 (4)[t]	
SD:	1.5	22.1	0.092		18.6	

[a] Concentration of external Cl^- which gives half-maximal Cl^- flux with maximal (saturating) concentrations of internal Cl^-.

[b] Concentration of internal Cl^- which gives half-maximal Cl^- flux with maximal (saturating) concentrations of external Cl^-.

[c] Predicted Cl^- concentration giving half-maximal Cl^- flux when Cl_i is kept equal to Cl_o. This value is calculated from the equation $K_{1/2} = K_{1/2, i}^{max} + K_{1/2, o}^{max}$, which holds for any ping-pong system.[12]

[d] Experimentally measured value of $K_{1/2}$.

[e] Estimated from data presented in Table I in Ref. 7.

[f] Cl_i which gives half-maximal value of $K_{1/2, o}$ (Table I in Ref. 7).

[g] From Ref. 17. Reference 24 gives a similar value (67 mM).

[h] Cl_i which gives half-maximal value of the Cl^- flux at saturating concentrations of Cl_o.[7]

[i] Cl_i which gives half-maximal Cl^- flux with $Cl_o = 23$ mM (saturating).[7]

[j] Calculated from $K_{1/2, o}^{max}$ and $K_{1/2}$, using equation shown in footnote c above.

[k] From extrapolation of $K_{1/2, o}$ values with high Cl_i and with different concentrations of 4,4'-dinitrostilbene 2,2'-disulfonate (DNDS) to zero DNDS.[6]

[l] From effect of Cl on the binding affinity of DNDS with $Cl_i = Cl_o$.[6]

[m] From effect of Cl_o on the binding affinity of DNDS with high Cl_i.[6]

[n] From the apparent dissociation constants for DNDS binding, extrapolated to zero Cl_o with high Cl_i, or extrapolated to zero Cl^- from measurements with $Cl_i = Cl_o$,[6] calculated as described in Ref. 15.

[o] Obtained by extrapolation of $K_{1/2, o}$ measured at different Cl_i values to infinite Cl_i.[19]

[p] Obtained by extrapolation of $K_{1/2, i}$ measured at different Cl_o values to infinite Cl_o.[19]

[q] From Ref. 19.

[r] From the concentrations of Cl_o and Cl_i, respectively, which give half-maximal Cl^- flux with saturating concentrations of Cl^- on the opposite side of the membrane, as in footnote h above, but from Ref. 19.

[s] From the concentrations of Cl_o and Cl_i which give half-maximal values for the Cl^- half-saturation constants at the opposite side of the membrane, $K_{1/2, i}$ and $K_{1/2, o}$, respectively, as in footnote f above, but from Ref. 19.

[t] Averages and standard deviations calculated from (n) observations which are at least partially independent. Because not all values are completely independent and because of differences in experimental techniques, these numbers are only presented to give a rough guide as to the variability in the parameters observed so far. For other determinations of A, see Ref. 15.

loading intact cells using the nystatin technique.[7,17–20] Similarly, it is easy to keep Cl_i constant and vary Cl_o, using sucrose and/or citrate to replace Cl^-. Thus, A can be determined from only two saturation curves, both of which can be obtained easily as described below.

Determination of $K_{1/2}$

Loading of Cells with Different Cl^- Concentrations Using Nystatin. Fresh blood is obtained with heparin as anticoagulant. The cells are washed two times with 150 mM NaCl, 20 mM $NaPO_4$, pH 6 and the upper white cell layer is removed by aspiration. They are then washed three times in 150 KH (150 mM KCl, 20 mM HEPES, 24 mM sucrose, pH 7.0 at room temperature), and resuspended at 50% hematocrit in 150 KH. The pH is chosen so as to make $Cl_i = Cl_o$ at the temperature of the flux measurement, in this case 0°. Unless otherwise indicated, all buffers are set to the same pH value by titrating with KOH (sodium salts and NaOH may also be used). The buffer concentration is high to prevent small changes in pH, which greatly affect Cl_i/Cl_o at low Cl^- concentrations.[21] Cells are kept on ice for all procedures, except as indicated.

To prepare cells with different Cl^- concentratons (from 10 to 150 mM), where Cl_i is designated X, 6 ml of 50% hematocrit cells is put into a 50-ml centrifuge tube, the tube is weighed on a top-loading balance, and the weight recorded. The cells are then washed once with X + 20 KH-Hi (X + 20 mM KCl, 20 mM HEPES, Y + 24 mM sucrose), where Y, the sucrose concentration supplement, is equal to $(126 - X)$ mM, or to zero if Y is negative. The cells are then resuspended in 30 ml of the same X + 20 KH-Hi buffer, but with 30 μg/ml nystatin (Calbiochem). Nystatin is freshly made up as a 5 mg/ml solution in methanol which is centrifuged to remove any insoluble material and which is then added to the treatment buffer (180 μl to 30 ml). The cells are incubated on ice for 10 min with at least occasional mixing, then centrifuged at low speed ($\sim$900 g) for 3 min. They are treated twice more with the same nystatin concentration in an identical manner, but in X KH-Lo buffer (X mM KCl, 20 mM HEPES, 24 mM sucrose). The cells are then washed five times at 21° in the X KH-Lo buffer *without* nystatin. They may be stored in this buffer overnight. They are washed at least once more with ice-cold X KH-Lo on the day of the flux measurement, after which the cells are brought to 50% hematocrit

[17] P. C. Brazy and R. B. Gunn, *J. Gen. Physiol.* **68**, 583 (1976).
[18] A. Cass and M. Dalmark, *Nature* (*London*), *New Biol.* **244**, 47 (1973).
[19] M. Hautmann and K. F. Schnell, *Pfluegers Arch. Gesamte Physiol.* **405**, 193 (1985).
[20] P. A. Knauf and J. Brahm, *Biophys. J.* **49**, 579a (Abstr.) (1986).
[21] M. Dalmark, *J. Physiol.* (*London*) **250**, 39 (1975).

by adding X KH-Lo buffer to the original weight of tube plus 50% cells. This technique produces cells with a slightly smaller than normal volume (cell water/total cell volume is about 0.65–0.69, compared with 0.70 in fresh cells), but with a Cl_i/Cl_o ratio between 0.98 and 1.1.

Preparation of Resealed Ghosts with Different Cl_i Values. For experiments at 0°, a Cl_i/Cl_o ratio near 1 can be achieved at neutral pH in intact cells, but at higher temperatures the pH for which $Cl_i = Cl_o$ decreases to pH 6.5, where the transport system is partially inhibited.[15] For this reason, and for experiments in which it is desired to replace intracellular Cl^- with a nontransported anion, it is useful to change the Cl_i by preparing resealed red cell ghosts.

Although different procedures have been described, we use the method originally described by Schwoch and Passow,[22] and later modified by Schnell (K. Schnell, personal communication), to prepare "normal" resealed ghosts. The protocols for making resealed ghosts with high or low ionic concentrations follow the same standard scheme with some slight modifications.

Essentially the preparation of ghosts consists of three steps: (1) hemolysing the red cells at 0°, (2) resealing the open ghosts at 0°C, and (3) completion of the resealing procedure at 38°. It is important for a successful preparation of uniform resealed ghosts with intact transport properties that the temperatures are as described above. This applies in particular to steps (1) and (2), where we keep the temperature within ±0.1°. During these two steps a moderate continuous stirring must be carried out, whereas occasional stirring suffices during the resealing at 38°.

Hemolysing the red cells: The heparinized blood sample is centrifuged and the buffy coat of white cells is aspirated. Next the cells are washed once in an unbuffered 165 mM KCl solution and resuspended to a hematocrit of 50% (v/v). The cells are hemolysed for 5 min at 0° by addition of a 10 times larger volume of 2 mM $MgSO_4$ and 3.8 mM acetic acid. If necessary, pH is adjusted to pH 5.9–6.2 by titrating the solution with 2 M tris.

Resealing ghosts: Resealing at 0° for 10 min is initiated by addition of a resealing solution made of 1.99 M KCl and 25 mM tris. The volume added is 1/10 of the volume of the hemolysing solution. Hence, the final chloride concentration is about 165 mM (1/12 × 1.99 M). During resealing pH is adjusted if necessary by means of 2 M tris to keep the pH around 7–7.5. After 10 min the solution is transferred to a water bath at 38° for 45 min to accomplish the resealing process. Finally, the ghosts are washed in an appropriate medium (e.g., 165 mM KCl and 2 mM KH_2PO_4) until the supernatant is clear. Centrifugation above 40,000 g is necessary to isolate

[22] G. Schwoch and H. Passow, *Mol. Cell. Biochem.* **2**, 197 (1973).

the pink-colored ghosts, whose content of hemoglobin is reduced from about 33% (w/v) in the intact red cells to <2% (w/v).

High chloride-loaded ghosts: Ghosts with an increased chloride concentration above 165 mM Cl$^-$ can be made either by simply increasing the chloride concentration of the resealing medium, increasing the volume of the "standard" resealing medium (1.99 M KCl and 25 mM tris), or a combination of these two procedures. It should be noted, however, that the maximum KCl solubility is about 3 M at 0° and thereby sets an upper limit for the increase in Cl$^-$ concentration of the resealing medium.

Low chloride-loaded ghosts: The chloride in the red cells (ca. 78 mmol/liter cell water under physiological conditions with an extracellular chloride concentration of 110 mM; ca. 100 mmol/liter cell water in red cells washed in 150 mM chloride solutions) and in the external medium of the sample used for ghost preparation contribute to the final chloride concentration of the resealed ghosts. It also appears necessary for a proper resealing of the ghost membranes that the resealing solution contain some chloride. We use a resealing solution of 42 mM KCl, 25 mM tris, and 240 mM sucrose to make ghosts 16 mmol Cl$^-$/liter cell water, which is the lowest chloride concentration obtainable by means of this procedure. Hence an internal chloride concentration between 16 and 165 mmol/liter cell water can be reached by adjusting the concentration of chloride (and sucrose to keep isosmolarity) in the resealing solution between 42 mM and 1.99 M. A certain interplay with the volumes is also possible here.

Whichever method with or without modifications is used for preparing resealed ghosts for transport measurements, it is crucially important that the cell volume be determined. In that respect one cannot use the cell content of hemoglobin, which is fairly constant in intact red cells, therefore, commonly used as the reference for a certain number of cells having certain membrane area (see below). By the procedures described above the dilution of hemoglobin inescapably varies too much for the residual hemoglobin content per cell to be a reliable measure of cell membrane area. Hence the Coulter counter should be used to determine cell number. The cell volume (e.g., in cubic centimeters, see below) is calculated from the determination of the cell number and the cytocrit of the packed ghosts (corrected for trapped extracellular fluid between the ghosts after centrifugation, approximately 8%).

Rate Coefficient, Internal Cl$^-$ Concentration, and Cl$^-$ Flux. Cells are loaded with ^{36}Cl, and the rate coefficient for unidirectional Cl$^-$ efflux (k, sec^{-1}) is measured by the methods described elsewhere in this volume (chapters [3] or [9]). The unidirectional Cl$^-$ efflux J (mol × cm^{-2} × sec^{-1}), is calculated from

$$J = k \times V/A \times Cl_i \times 10^{-6} = P_{app} \times Cl_i \times 10^{-6} \qquad (4)$$

where Cl_i (millimoles/liter cell water), the cellular chloride concentration, is determined by ^{36}Cl equilibration as described in chapter [3], in this volume. P_{app} (cm/sec) is the apparent permeability coefficient, which is related to the rate coefficient by V/A, the ratio of cell water volume to cell membrane area. The membrane area is considered to be constant, and we use a value of 1.42×10^{-6} cm^2/cell, which is a mean of several studies (see Ref. 23; J. Brahm, unpublished results). The cell water volume may, on the other hand, vary considerably. For example, though the cell volume of a ghost and an erythrocyte are equal, the cell water volume of the ghost is about 30% larger than that of an erythrocyte, because of the different content of hemoglobin. Cell water volume also depends on pH, temperature, constituents of the suspending solution, and variations in the preparation of ghosts, which emphasizes that a fixed value for the cell water volume should not be used.

Other units of flux commonly used are millimoles/(kg cell solids $\times$ sec) and millimoles/(3×10^{13} cells $\times$ sec). Particularly for work with intact cells at 0°, fluxes are often expressed in millimoles/kilogram cell solids-minute obtained by multiplying the rate constant for Cl^- exchange, k, by the intracellular Cl^- content in millimoles/kilogram cell solids, determined as described in chapter [3] in this volume. The interrelation between the different flux units is that 1 kg of cell solids equals 3×10^{13} "normal" red cells with a total membrane area of 4.4×10^7 cm^2.

The half-saturation concentration for Cl^- can be determined by a non-linear least-squares fit to the data for J as a function of Cl_i (it is better to use Cl_i if Cl_i and Cl_o are slightly different, since in red blood cells the apparent affinity for Cl^- is lower at the inside, so the flux under these conditions is primarily a function of Cl_i). $K_{1/2}$ may also be determined from the x-intercept of plots of $1/J$ versus $1/Cl_i$ or from the slope of plots of J/Cl_i versus Cl_i.

Measurement of Cl^- Flux as a Function of Cl_o

Blood is obtained and cells are prepared as described above. Fresh cells can be used directly, loading them with ^{36}Cl in 150 KH, or else nystatin can be used to alter Cl_i if desired. For fresh cells, flux media with different external Cl^- concentrations (e.g., from 1 to 20 mM) can easily be prepared by mixing HCS buffer (20 mM HEPES, 25 mM tripotassium

^{23} J. Brahm, *J. Gen. Physiol.* **79**, 791 (1982).

citrate, 224 mM sucrose) with 150 KH buffer, such that the volume fraction of 150 KH is $X/150$, where X is the desired Cl^- concentration, and the rest of the volume is made up with HCS. This is based on the idea that 200 mM sucrose and 25 mM tripotassium citrate provide an isotonic replacement for 150 mM Cl^-, while maintaining constant ionic strength.[7] For careful work at low Cl^- concentrations, it is preferable to bubble the solutions with nitrogen to remove traces of CO_2 and HCO_3^-.

Fluxes are measured as described above. The only complication in this case is that, if the external Cl^- concentration is very low, the amount of Cl^- in the cell compartment may constitute a substantial fraction of the total chloride in the cell suspension. In this case the sizes of the intracellular and extracellular Cl^- compartments must be taken into account in determining the rate constants and fluxes as described in chapter [3] of this volume.

Examples and Comparison with Other Methods

A version of this method has recently been used to determine the asymmetry of the unloaded sites at 38°,[20] where A has a value of 0.034 (that is, $E_i/E_o = 29$). One can also test this method by calculating A from earlier data, such as those of Gunn and Fröhlich.[7] Table II presents calculations of A and $1/A$ from the data for $K_{1/2,o}$ from Table I of Gunn and Fröhlich[7] and for $K_{1/2}$ from Brazy and Gunn.[17] It can be seen that, except for the data at the lowest internal Cl^- concentrations in ghosts, the results give very consistent values for A, indicating about a 15-fold asymmetry in favor of inward-facing unloaded sites (E_i).

Comparison of Table II with Table I shows that the results from this method are in general fairly similar to those obtained by other methods. All indicate an asymmetry in favor of E_i, but some recent measurements, notably those of Hautmann and Schnell,[19] indicate that the asymmetry is less than would be expected from other data sets. Experiments with bicarbonate as substrate (P. Gasbjerg and J. Brahm, unpublished data) give an A value of 0.2. Since for a ping-pong model the unloaded site asymmetry is independent of the substrate present, this would also suggest a smaller, 5-fold asymmetry in favor of inward-facing sites.

Possible Problems

The disagreement among methods may reflect several problems in making such measurements: First, since internal Cl^- at high concentrations inhibits Cl^- exchange,[24,25] there is a tendency to underestimate V_{max}.

---

[24] M. Dalmark, *J. Gen. Physiol.* **67**, 223 (1976).
[25] P. A. Knauf and N. A. Mann, *Am. J. Physiol.* **251**, C1 (1986).

TABLE II
CALCULATIONS OF A AND $1/A$

Cl_i	$K_{1/2, o}$	A^a	$1/A$
5.5	0.47^b	0.093	10.7
13.9	1.25^b	0.111	9.0
19	0.86^b	0.059	16.9
47.7	1.25^b	0.046	21.6
51.5	1.19^b	0.042	23.7
51.8	1.58^b	0.056	17.8
90	1.77^b	0.048	20.7
108	2.30^c	0.059	17.0
110	2.20^c	0.056	17.9
133	3.20^c	0.077	13.0
152	3.00^c	0.069	14.5
Mean:		0.065	16.6
SD:		0.020	4.3
SEM:		0.006	1.4

[a] Calculated from $K_{1/2}$, Cl_i and $K_{1/2, o}$ as described in text, using a $K_{1/2}$ value of 65 mM[17] and $K_{1/2, o}$ values from Ref. 7. The reciprocal of the mean A value is 15.3. This differs from the mean value of $1/A$ (16.6) because taking the reciprocal skews the distribution of measured values.
[b] Resealed ghosts.
[c] Intact red blood cells.

This leads in turn to an underestimate of $K_{1/2}$. NMR measurements of Cl^- binding with $Cl_i = Cl_o$, which do not seem to be affected by this Cl^- inhibitory site,[26] give values for $K_{1/2}$ ranging from 60 to 90 mM, at the upper end of the range obtained by flux measurement techniques. Second, anions used to replace Cl^- may not be inert, but may compete with Cl^- for binding to the transport site. For example, citrate does compete with Cl^- to a slight degree (K_i for external citrate $= 125 \pm 27$ mM[27]), and this problem is worse at lower pH, where more of the citrate is in the 2− form. Third, ionic strength may affect the affinities measured. It has been reported that the system is relatively insensitive to changes in ionic strength,[28] but this has not been tested under all circumstances used in measurements of A. Measurement of $K_{1/2}$ in intact cells by the technique

[26] J. J. Falke, R. J. Pace, and S. I. Chan, *J. Biol. Chem.* **259**, 6472 (1984).
[27] K. F. Schnell, E. Besl, and A. Manz, *Pfluegers Arch. Gesamte Physiol.* **375**, 87 (1978).
[28] R. B. Gunn, M. Dalmark, D. C. Tosteson, and J. O. Wieth, *J. Gen. Physiol.* **61**, 185 (1973).

described above involves substantial changes in ionic strength as Cl^- is varied. A possible way of avoiding this problem would be to substitute gluconate for internal Cl^-, since internal gluconate has a very low affinity for the transport site ($K_i = 720$ mM) and since gluconate can be loaded into cells[25] in the presence of high (75 μg/ml) nystatin concentrations, with long incubation times (1 hr in each solution). Fourth, since $E_m = E_{Cl}$, when Cl_o is altered to measure $K_{1/2,o}$ the membrane potential changes, and this might affect the affinities of E_o and E_i for substrate anions or the V_{max} for transport.[29] Data available at present suggest that these effects are small or nonexistent for human capnophorin.[5,7,15,30,31] but further experiments are needed to establish this conclusively.

Asymmetry in Substrate Affinity

Since the two conformations of the transport protein, E_i and E_o, must differ in structure, they will not necessarily have the same affinities for substrate, that is, K_i and K_o may differ from each other. To obtain information about the orientation of substrate-loaded transport sites, as well as about the total number of transport sites (unloaded and loaded with substrate) facing each side of the membrane, it is necessary to know the values of K_i and K_o.

Unfortunately, at present there is no obvious way to measure K_i and K_o for a rapidly transported substrate such as Cl^-. Any manipulation of Cl^- concentration at one side of the membrane leads to a rearrangement of transport site orientations in response to the new conditions. Thus, the concentration of Cl^- required to half-saturate the system at, e.g., the outside of the cell does not simply equal K_o, but rather is a complex function of all the dissociation constants and rate constants in Fig. 2A.

Measurement of K_i and K_o for Slowly Transported Substrates

For substrate anions which are transported much more slowly than Cl^-, such as sulfate or iodide, the affinities of the ions for E_i and E_o can be determined by treating the ions as nontransported competitive inhibitors. Using this method, Milanick and Gunn[32] were able to work out the scheme for binding of protons and sulfate to E_o. In general, if a substrate such as iodide (I^-) is transported much more slowly than Cl^-, it will have no effect on the distribution of Cl^--loaded transport sites, ECl_i and ECl_o.

[29] R. Grygorczyk, W. Schwarz, and H. Passow, *J. Membr. Biol.* **99**, 127 (1987).
[30] O. Fröhlich, C. Leibson, and R. B. Gunn, *J. Gen. Physiol.* **81**, 127 (1983).
[31] J. O. Wieth, J. Brahm, and J. Funder, *Ann. N.Y. Acad. Sci.* **341**, 394 (1980).
[32] M. A. Milanick and R. B. Gunn, *J. Gen. Physiol.* **79**, 87 (1982).

[For example, with iodide present only on the outside, if Cl^- influx is designated $J_i(Cl)$, Cl^- efflux $J_o(Cl)$, and I^- influx $J_i(I)$, then $J_o(Cl) = J_i(Cl) + J_i(I)$. If $J_i(I) \ll J_i(Cl)$, then it is clear that $J_o(Cl) \cong J_i(Cl)$, so the unidirectional Cl^- fluxes are almost exactly equal, just as if iodide were not present.]

To measure $K_{o,I}$ (Fig. 2B), the dissociation constant for binding of I^- to E_o, one measures the Cl^- flux with one value of external Cl^- and various concentrations of external I^-, and then plots the results on a Dixon plot of 1/flux versus I_o. If one then performs the same kind of experiment with the same Cl_i, but with a different external Cl^- concentration, the two Dixon plot lines should intersect at a point whose x value is equal to $-K_{o,I}$ (Fig. 3A). For this technique to work it is only necessary that there be no complications from modifier sites,[7,24,25] which can be

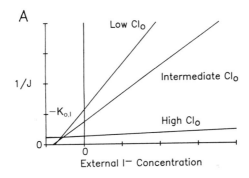

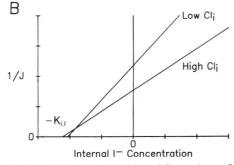

FIG. 3. (A) Determination of $K_{o,1}$. The inhibition of Cl^- exchange flux (J) by external I^- is plotted on a Dixon plot (1/flux versus external I^- concentration). Lines for different external Cl^- concentrations (with constant Cl_i) intersect near a common point, whose x value is equal to the negative of the dissociation constant for external I^- binding to E_o, $K_{o,1}$. (B) Determination of $K_{i,1}$. Dixon plot lines for internal I^- inhibition of Cl^- exchange with constant Cl_o and different values of Cl_i intersect at a point whose x value is equal to the negative of $K_{i,1}$, the dissociation constant for internal I^- binding to E_i.

avoided by choosing low concentrations of substrate and inhibitor, and that the external Cl^- concentrations be sufficiently different so that the control fluxes (in the absence of I_o) are different. In practice, it is useful to choose concentrations so that the control fluxes differ by at least a factor of 2. For example, fresh cells are prepared and loaded with ^{36}Cl as described above. Cl^- efflux is measured at external I^- concentrations from 5 to 20 mM, with either 130 or 2 mM external Cl^-. Flux media are prepared from stock solutions of 150 KH, 150 KIH (150 KH with KI instead of KCl), and HCS, where the volume fraction of 150 KH is equal to the desired Cl^- concentration divided by 150, the volume fraction of 150 KIH is equal to the desired I^- concentration divided by 150, and the rest of the volume is made up with HCS. Fluxes are measured in these media and plotted as described above. Experiments of this sort[33-35] indicate that $K_{o,\,I}$ is about 3 mM.

It is obvious that a similar technique can be used to determine $K_{i,\,I}$, the dissociation constant for binding of I^- to E_i. In this case, it is necessary to keep Cl_o constant and to measure the inhibitory effect of internal I^- at at least two different internal Cl^- concentrations. The Dixon plots for the two different Cl_i concentrations should intersect at a point whose x value is equal to $-K_{i,\,I}$ (Fig. 3B). In practice, it is necessary to prepare nystatin-treated cells with different Cl_i values, say 10 and 20 mM, and with various I_i values, say 0 to 20 mM. This is done by first preparing nystatin-treated cells with a Cl^- concentration equal to the sum of the desired $Cl^- + I^-$ concentration and then exchanging the excess Cl^- for I^- by incubating the cells in a medium with the desired Cl^-/I^- ratio.

In practice, to prepare cells with 10 or 20 mM Cl^- and 0, 5, 10, 15, or 20 mM I^-, cells are washed for nystatin treatment as described above. For each treatment, 6 ml of 50% (v/v) cells is then washed with 40 KH-Hi (40 mM KCl, 20 mM HEPES, 130 mM sucrose; all solutions are pH 7.0 at room temperature). The cells are treated as described above with 30 ml of 40 KH-Hi containing 30 μg/ml nystatin and then two more times with nystatin in a buffer consisting of $(X + Y - 10)/30$ (where X is the desired intracellular Cl^- concentration in millimolar units and Y is the desired I^- concentration) volume fraction of 40 mM KCl, 20 mM HEPES, 24 mM sucrose (40 KH-Lo), with the rest of the volume made up by 10 KH-Lo. After nystatin treatment, the cells are washed five times with solutions consisting of $X/50$ volume fraction of 50 KH-Lo, $Y/50$ volume fraction of

[33] M. A. Milanick and R. B. Gunn, *Am. J. Physiol.* **250**, C955 (1986).
[34] P. A. Knauf, N. Mann, J. Brahm, and P. Bjerrum, *Fed. Proc., Fed. Am. Soc. Exp. Biol.* (*Abstr.*) **45**, 1005 (1986).
[35] P. A. Knauf, J. Brahm, P. Bjerrum, and N. Mann, *in* "Proceedings of the Eighth School on Biophysics of Membrane Transport" (J. Kuczera and S. Przestalski, eds.), Vol. 1, p. 157. Agric. Univ., Wrocław, Poland, 1986.

50 KIH-Lo (50 mM KI, 20 mM HEPES, 24 mM sucrose), with the rest of the volume consisting of HS (20 mM HEPES, 24 mM sucrose). After each resuspension the cells are incubated for 5 min at 37° to permit Cl^-–I^- exchange. The cells are stored overnight in these solutions, then washed once more and loaded with ^{36}Cl in the same solutions. Chloride efflux is measured in 10 mM Cl^- buffers, consisting of $(X + Y - 10)/30$ volume fraction 10 mM KCl, 20 mM HEPES, 64 mM sucrose, 5 mM tripotassium citrate, with the rest of the volume made up by 10 KH-Lo. This protocol maintains constant external Cl^- of 10 mM, with internal Cl^- of either 10 or 20 mM and various I^- concentrations.

A practical difficulty which arises when using this method is that the Cl_i/Cl_o ratio is more variable and more strongly affected by I^- binding to intracellular proteins in the cells with low internal Cl^-, leading to a greater scatter in the flux determinations under these conditions. These problems are exacerbated if the value of $K_{i,I}$ is large, since the lines for the two different Cl^- concentrations must be extrapolated a considerable distance to obtain an intersection point. To avoid these problems, one can determine $K_{i,I}$ from the ID_{50} value for internal iodide at a fixed, slightly higher intracellular Cl^- concentration, since

$$K_{i,I} = ID_{50}/[1 + (Cl_i/K_{1/2})(1 + A) + ACl_i/Cl_o] \tag{5}$$

This assumes, of course, that one has reliable values of $K_{1/2}$ for Cl^- and of A. Data from both techniques give values of $K_{i,I} > 20$ mM,[34] in comparison to 3 mM for $K_{o,I}$, demonstrating considerable differences in I^- affinity between E_o and E_i.

The foregoing analysis depends critically on the assumptions that the I^- or other substrate is transported slowly relative to Cl^- and that all of the inhibition by I^- is competitive. The first assumption can be tested by measuring the flux of the slow substrate, using, for example, ^{125}I, and comparing it to the Cl^- flux under the same circumstances. In fact, the I^- flux only needs to be small enough so that the Cl^- influx $\simeq Cl^-$ efflux (see above). The competitive nature of the inhibition can be tested by plotting the Dixon plot slope against the reciprocal of the Cl^- concentration on the side at which Cl^- is varied (e.g., in the case of the inhibition of Cl^- flux by I_o, one plots the Dixon plot slope against $1/Cl_o$). For competitive inhibition, this plot should give a straight line passing through the origin. The deviation of the y intercept from the origin gives an estimate of the extent of noncompetitive inhibition.

Another possible problem is that the membrane potential is allowed to vary when the Cl_i/Cl_o ratio is changed. Since the membrane potential has been shown to alter the Cl^- exchange flux through mouse band 3, possibly due to changes in transport site orientation of the band 3 protein,[29] such membrane potential effects could lead to errors in the determination of

$K_{i, I}$ and $K_{o, I}$. Preliminary data from our laboratory, using valinomycin to alter the membrane potential, indicate that such effects for human capnophorin are small or nonexistent in the case of iodide (L. J. Spinelli, D. Restrepo, and P. A. Knauf, unpublished data), but they might be important for other substrates.

Once $K_{i, I}$ and $K_{o, I}$ are known, the ratio of the rate constants for the change from EI_i to EI_o and vice versa, i and i' (Fig. 2B), can be calculated from the form of Eq. (2) with I^- as substrate:

$$A = iK_{o, I}/i'K_{i, I} \qquad (6)$$

so

$$EI_o/EI_i = i/i' = AK_{i, I}/K_{o, I} \qquad (7)$$

Thus, with a slow anion (such as I^-) as substrate, it is possible to determine what fraction of capnophorin is in each of the forms E_i, E_o, EI_i, and EI_o. Explicitly, the equations for each form are

$$EI_i = 1/\{1 + (K_{i, I}/I_i) + (AK_{o, I}/K_{i, I})[1 + (K_{o, I}/I_o)]\} \qquad (8)$$
$$EI_o = (AK_{o, I}/K_{i, I})EI_i \qquad (9)$$
$$E_i = (K_{i, I}/I_i)EI_i \qquad (10)$$
$$E_o = (AK_{o, I}^2/K_{i, I}I_o)EI_i \qquad (11)$$

These equations can be easily programmed by using, for example, a spreadsheet so that the fractions of band 3 in each form under a variety of circumstances can be calculated. From this sort of analysis, it is clear that by changing internal and external concentrations of a slow substrate anion, such as I^- (with no Cl^- present), it is possible to make defined changes in the orientation of the transport protein. Such experiments require the presence of the slow substrate at both sides of the membrane, whereas all of the experiments to determine the system parameters were performed with the slow substrate present only at one side of the membrane, so one must further assume that there are no transmembrane interactions between, e.g., iodide inside and iodide outside. Also, the concentrations of the substrate must be kept in a range where noncompetitive interactions with the system are minimized, since such interactions could affect the orientation of the transport protein, as shown in the next section.

Affinities of Noncompetitive Inhibitors for Various Forms of Capnophorin

In theory, noncompetitive inhibitors can bind to all forms of capnophorin shown in Fig. 2, since binding of substrate and inhibitor are not

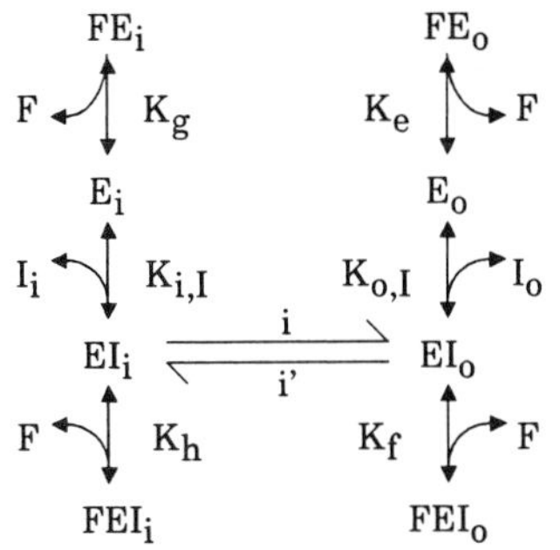

FIG. 4. Interaction of noncompetitive inhibitor, F, with capnophorin. The anion transport system is shown with I^- as substrate, as in Fig. 2B. The inhibitor F can bind to all of the forms of capnophorin, but with different dissociation constants, designated K_e, K_f, K_g, and K_h. Note that the binding site for F is located at only one side of the membrane. For example, the binding site for F may be accessible only from the external medium, but external F can bind to its site even when the transport site faces inward.

mutually exclusive. Changes in the protein conformation may affect inhibitor binding, however, so the affinities of inhibitor for the different forms of capnophorin may be different. The dissociation constants for binding of an inhibitor, designated F, to E_o, EX_o, E_i, and EX_i (where X is substrate) may be arbitrarily designated K_e, K_f, K_g, and K_h (Fig. 4). Such differences may provide important information concerning the conformational changes involved in substrate binding and/or translocation of the anion-binding site. Inhibitors may also be helpful in experimentally altering the conformation of the system, since an inhibitor will tend to recruit the transport protein toward the form for which it has the highest affinity.

Affinities for Unloaded Forms, E_i and E_o

The constants K_e and K_g can be easily determined by using the same methods as were used for determining the binding affinity of slowly transported substrates, such as iodide, in the previous section. The only difference is that usually the inhibitors work at such low concentrations that one need not compensate for their osmotic contribution to the solutions, as was the case with I^-. Thus, if one keeps Cl_i constant and measures the inhibition by an inhibitor, F, at two or more different external Cl^- concentrations, the intersection point of the Dixon plots (1/flux versus F) for different Cl_o values has an x value equal to $-K_e$, the dissociation constant for binding of F to E_o (Fig. 5A). Similarly, if one keeps Cl_o constant, the intersection point for Dixon plots of F inhibition at different Cl_i values has an x value equal to $-K_g$, the constant for the binding of F to E_i (Fig. 5C). It should be noted that these techniques work only if the inhibitor has a

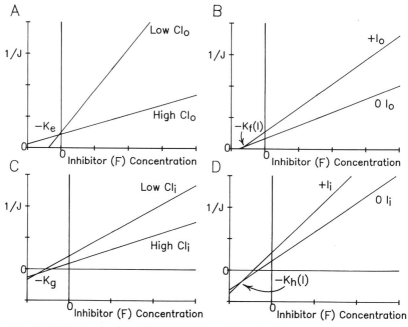

FIG. 5. (A) Determination of K_e for binding of F to E_o. Cl_i is kept constant, and the x value of the intersection point of Dixon plot lines for inhibition of Cl^- exchange by F with different external Cl^- concentrations is equal to the negative of K_e. (B) Determination of K_f for binding to F to EI_o. Cl_i and Cl_o are constant, and the intersection point for lines with different external I^- concentrations gives $K_f(I)$. (C) Determination of K_g for binding of F to E_i. The x value of the point at which Dixon plot lines with the same Cl_o and different Cl_i values intersect gives the negative of K_g. (D) Determination of K_h for binding of F to EI_i. With Cl_i and Cl_o constant, the x value of the intersection point of Dixon plot lines with different internal I^- concentrations gives the negative of $K_h(I)$.

single inhibition site, since otherwise the Dixon plots will be curved and therefore useless. It does not matter, however, whether or not the inhibitor penetrates the membrane nor whether its effects are extracellular or intracellular, as long as there is a single site of action.

Affinities for Substrate-Loaded Forms, EX_i and EX_o

In approaching this problem, the same difficulties arise as when attempting to determine substrate affinities. That is, for a rapidly transported substrate such as Cl^-, it is not yet possible to get precise information because of difficulties in defining the fraction of sites in the ECl_i and ECl_o forms. It is possible, however, to determine the affinity of the inhibitor, F, for the forms of capnophorin loaded with a slowly transported substrate such as iodide or sulfate. Such information may provide some indication of the effects of substrate binding, from which the effects of chloride binding may be roughly estimated.

For example, to determine the affinity of EI_o for the inhibitor, one keeps both Cl_i and Cl_o constant and measures the inhibitory effect of F on Cl^- exchange at different values of the external, slowly transported substrate, I_o. The Dixon plots (1/flux versus F) at different I_o values intersect at a point whose x value is $-K_f$, the dissociation constant for binding of F to EI_o (Fig. 5B). Similarly, if one keeps Cl_i and Cl_o constant and varies internal I^-, the Dixon plots for F inhibition at different I_i values intersect at a point with x coordinate equal to $-K_h$ (Fig. 5D).

Several variations of this procedure are possible. For example, instead of using the Dixon plot (1/flux versus F), one can plot the data as flux/F versus flux (an inhibitor analog of the Eadie–Hofstee plot for substrates). At the intersection point, the x coordinate divided by the y coordinate gives the negative of whichever dissociation constant is being determined. In this method, as in the case of the substrate affinities discussed in the preceding section, membrane potential changes may in theory affect the results, but no substantial effects have yet been observed (L. J. Spinelli, J. Brescia, and P. A. Knauf, unpublished data).

Using these techniques, it has been possible to determine that several inhibitors, including N-(4-isothiocyano-2-nitro)-2-aminoethyl sulfonate (NIP–taurine),[36] flufenamic acid,[37] and niflumic acid[38] bind with greater affinity to E_o than to E_i, indicating that the change in conformation of the transport site affects other regions of the protein besides the substrate site itself. In the case of the latter two inhibitors, binding of iodide decreases the affinity for the inhibitor. Surprisingly, binding of internal I^- to E_i has a greater effect than does binding of external I^- to E_o, despite the fact that the inhibitor binding site probably faces the outside.[37,39] This provides evidence for a substantial conformational change involved in binding of substrate, sufficient to produce a transmembrane effect.

If one knows the dissociation constants K_e through K_h, the effect of inhibitors on the orientation of the transport protein can be assessed by constructing a spreadsheet incorporating the terms for inhibitor binding as well as substrate binding. That is, in addition to the term for E_o there will be a term for FE_o which is simply $E_o(F/K_e)$, as well as corresponding

[36] P. A. Knauf, N. A. Mann, J. E. Kalwas, L. J. Spinelli, and M. Ramjeesingh, *Am. J. Physiol.* **253,** C652 (1987).

[37] P. A. Knauf, L. J. Spinelli, and N. A. Mann, *Fed. Proc., Fed. Am. Soc. Exp. Biol.* (*Abstr.*) **46,** 534 (1987).

[38] P. A. Knauf, N. A. Mann, and L. J. Spinelli, *Biophys. J.* **51,** 566a (Abstr.) (1987).

[39] J. L. Cousin and R. Motais, *J. Membr. Biol.* **47,** 125 (1979).

terms for other forms of the transport protein (Fig. 4). Thus, the orientation of the protein under any given set of circumstances can be calculated. For Cl^-, where the values of K_f and K_h are unknown and where K_o and K_i are also uncertain, calculations can be done assuming various values for K_o and K_i and using K_f and K_h values ranging from those expected if substrate binding has no effect (in which case, $K_e = K_f$ and $K_g = K_h$) to those expected if Cl^- has as large an effect as I^- on the affinity for F. Such calculations can predict the inhibitory potency (ID_{50}) under different conditions and, from the fit of the experimental data to such predictions, one can obtain an estimate of the values of K_o and K_i.[40]

Conclusions

This chapter presents some fairly straightforward methods for determining the asymmetry of the unloaded transport sites, E_i and E_o, the affinities of E_i and E_o for slowly transported substrates, and the affinity of unloaded and substrate-loaded forms of band 3 for noncompetitive inhibitors. The methods are specifically designed for work with the anion-transport protein, capnophorin, but may also be adapted to other ping-pong transport systems with similar characteristics.

The ultimate purpose of such methods is to learn more about the native conformational distribution of capnophorin and to devise ways for experimentally altering its conformation. Use of these methods has already revealed that the distribution of unloaded transport sites is highly asymmetric, favoring inward-facing sites[1,3,7,12,20]; that with I^- as substrate,[34] but probably not with Cl^-,[40] this asymmetry is primarily due to differences in the substrate affinities of E_i and E_o; and that binding of substrates and changes in transport site orientation can have large effects on the binding of noncompetitive inhibitors.[36–40]

Despite the success of such experiments, there are still many unresolved questions. One of these centers around the possible effects of membrane potential. Since substantial changes in membrane potential do not cause major changes in human erythrocyte anion transport,[5,7,15,30,31] it does not seem likely that membrane potential causes a shift in orientation of the transport sites, such as that which has been proposed to explain the effect of membrane potential on the transport of Cl^- by mouse band 3.[29] If, however, the sites of substrate or inhibitor binding lie in an area where the potential is affected by the transmembrane field, changes in membrane potential may alter their dissociation constants. Although preliminary experiments have not detected large effects of potential, this result might be

[40] P. A. Knauf and L. J. Spinelli, *J. Gen. Physiol.* **90**, 24a (Abstr.) (1987).

due to compensation of two or more effects, so further investigation is necessary to more clearly delineate the effects of membrane potential in this system. The second principal problem concerns the inability to determine the affinities of E_i and E_o for the rapidly transported substrates of primary biological interest, Cl^- and HCO_3^-. Although it is possible to make educated estimates of these parameters, further development of methodology is necessary before they can be specified precisely, so that the native orientation of capnophorin in Cl^- or HCO_3^- media can be determined.

In the future, the methods described here may be useful for exploring in a more precise fashion the effects of chemical and genetic modifications of capnophorin on the individual steps in the transport cycle. Also, methods for orienting the system toward one conformation or another should be useful for probing the conformational changes which occur during substrate binding or transport, by using proteolytic enzymes, fluorescent probes, and other techniques to monitor the accompanying changes in protein structure.

Acknowledgments

The authors would like to thank Mrs. Elizabeth Garrand and Ms. Laurie J. Spinelli for preparing the illustrations and Dr. Poul Bjerrum for helpful discussions. This work has been supported by Public Health Service Grant DK27495 and by a Fogarty Senior International Fellowship TW00975.

[30] Measurement of Erythroid Band 3 Protein-Mediated Anion Transport in mRNA-Injected Oocytes of Xenopus laevis

By R. Grygorczyk, P. Hanke-Baier, W. Schwarz, and H. Passow

Introduction

Oocytes of *Xenopus laevis* have been used for several years as an expression system for microinjected mRNA from many different sources. It was possible to demonstrate the successful biosynthesis of a large variety of protein species, such as hemoglobin and zein (for a review, see Ref. 1). More recently, the oocytes became quite popular as expression systems for membrane proteins, notably ion-conducting channels. After microinjection of suitable mRNA preparations many membrane proteins

---

[1] H. Soreq, *CRC Crit. Rev. Biochem.* **18,** 199 (1985).

appear in the plasma membranes of *Xenopus* oocytes in a functional state (for a review, see Ref. 2). The successful expression of ion-conducting channel proteins has been demonstrated with electrophysiological methods, involving most commonly the measurement of ion-specific currents under voltage clamp and their modification by highly selective inhibitors or activators of the respective channels. Recently it has been shown in our laboratory that the band 3 protein from mouse red cells can also be expressed in the oocytes.[3] This transport protein mediates a rapid anion exchange which does not contribute to the electrical conductance of the cell membrane. As a consequence the successful expression and insertion of the band 3 protein into the oocyte membrane cannot be demonstrated by the electrophysiological methods mentioned above.

Figure 1 shows the currently most widely accepted reaction scheme for the band 3 protein-mediated, "electrically silent" anion exchange: it follows ping-pong kinetics. This involves anion binding at one membrane surface, subsequent conformational changes of the complex between anion and transport protein, and the final establishment of a conformation which allows the release of the bound anion at the other surface of the membrane. A return of the band 3 protein to the original conformation is not possible as long as there is no anion bound. One can state, therefore, that anion binding catalyzes the conformational change of the band 3 protein from inward to outward facing and vice versa, and that the bound anion is transported during the catalytic process.

Thus the band 3 protein mediates a rapid exchange of chloride which can easily be measured with radioactive chloride, but cannot accomplish a net flow of anions which could be measured as a contribution to the electrical conductance of the cell membrane. Nevertheless there seems to exist a band 3 protein-mediated component of anion flux which contributes to the conductance of the membrane (for reviews, see Refs. 4 and 5). In the red cell this flux component is very much smaller than the exchange flux and in the oocytes it is not easily, if at all, measurable. Hence for measuring the anion transport activity of the band 3 protein in the oocyte membrane, one needs to measure anion fluxes by means of radioisotopes. The present chapter deals with the methods required (1) to obtain the mRNA, (2) to microinject the mRNA, (3) to incubate for expression, (4) to

[2] G. Yellen, *Trends Neurosci.* **7**, 457 (1984).
[3] M. Morgan, P. Hanke, R. Grygorczyk, A. Tintschl, H. Fasold, and H. Passow, *EMBO J.* **4**, 11927 (1985).
[4] P. Knauf, *in* "Physiology of Membrane Disorder" (T. Andreoli, J. F. Hoffman, D. D. Fanenstil, S. G. Schultz, eds.), pp. 191–220. Plenum, New York, 1986.
[5] H. Passow, *Rev. Physiol. Biochem. Pharmacol.* **103**, 61 (1986).

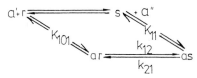

FIG. 1. Ping-pong kinetics of band 3 protein-mediated anion exchange. At surface ', the substrate a reacts with the r conformer of the transport protein to form the complex ar. The complex undergoes a conformational change into as, which enables the anion to be released at surface ". The unoccupied conformer s is capable of combining with another anion a for the journey in the opposite direction. The formation of the substrate–protein complex is assumed to be fast compared with the rates of conformational transitions ar $\rightleftharpoons$ as, which are designated by k_{12} and k_{21}. For this reason, substrate binding is described by the mass law constants K_{101} and K_{11}. The rate constants for the interconversions r $\rightarrow$ s and vice versa are assumed to be zero: there is no slippage and hence neither net flow nor electrical conductance. Designations as in Passow.[5]

demonstrate the biosynthesis of the protein, and (5) to measure anion fluxes across the plasma membrane of the oocyte.

The list includes the demonstration of the successful expression of the protein in the oocytes. This is desirable since in case of no measurable transport activity, one should ascertain whether this is due to lack of expression of the protein or whether the expressed protein is not inserted into the membrane in a functional state. A convenient method for the demonstration of band 3 protein expression is immunoprecipitation.

Procedures

Sources of mRNA

Intact mammalian red cells contain little if any mRNA and little, if any, protein biosynthesis occurs. Reticulocytes do contain mRNA, but mostly this is mRNA encoding proteins of molecular weights below the range of the band 3 protein ($\approx$97K). Thus, the mRNA for microinjection into *Xenopus* oocytes should be obtained from early red cell precursors in erythropoietic tissues. Nevertheless, even in these tissues the fraction of mRNA encoding the band 3 protein is rather small and does not seem to suffice for the expression in the oocytes of numbers of band 3 molecules sufficient to accomplish transport at easily measurable rates.

For example, microinjection of mRNA preparations from human bone marrow (from hip joints of orthopedic patients) did not lead to the appearance of band 3 protein-mediated anion fluxes in the oocytes. In order to obtain the necessary amounts of band 3-encoding mRNA, it seems necessary to use artificially stimulated erythropietic tissue. A most convenient source of such tissue is the spleen of adult mouse.[6] In the normal animal, the organ is of minor significance for erythropoiesis. However, in mice with anemia induced by repeated bleedings, hemolysis, or administration of phenylhydrazine, the production of erythropoietins is augmented. The spleen now assumes an important role in erythropoiesis. The most convenient method for the stimulation of erythroid development in the spleen consists of phenylhydrazine treatment. According to Sabban et al.,[7] female BALB/c mice receive three intraperitoneal injections of 0.8% neutralized phenylhydrazine solution (0.1 ml/20 g weight) at 0, 16, and 24 hr. The spleens are removed 60 hr after the last injection and mRNA is prepared as described by Chirgwin et al.[8] with modifications taken from the work of Braell and Lodish.[9] For details of the method, see Ref. 3. Recently mouse band 3 cDNA has become available.[10,11] Hence mRNA transcripts from band 3 cDNA will soon become available. Such transcripts will constitute an important source of mRNA for band 3 protein expression and will open up an avenue for the expression of artificially modified band 3 protein in the oocytes.

Preparation of Oocytes and Microinjection

Pieces of the ovaries are removed under anesthesia from females of *Xenopus laevis* (for details, see Ref. 12). Subsequently they are suspended in Barth's solution (in millimoles/liter: 88 NaCl, 1.0 KCl, 2.4 NaHCO$_3$, 0.82 MgSO$_4$, 0.33 Ca(NO$_3$)$_2$, 0.41 CaCl$_2$, 0.08 penicillin, 0.03 streptomycin, 5.0 HEPES, pH 7.6) containing about 1.2 to 2 U/ml collagenase (Serva, Heidelberg) and incubated for 8–16 hr at $\approx$18–20°. After subsequent washings in Ca^{2+}-free Barth's medium, prophase arrested, full-grown oocytes (stage VI),[13] are collected in normal (i.e., Ca^{2+}-containing) Barth's medium.

For microinjection the oocytes are placed in a Lucite chamber, the bottom of which contains several grooves. The oocytes are placed into

---

[6] H. W. Dickermann, T.-C. Cheng, H. H. Kzazian, Jr., and J. L. Spivak, *Arch. Biochem. Biophys.* **177**, 1 (1976).

[7] E. Sabban, V. Marchesi, M. Adesnik, and D. D. Sabatini, *J. Cell Biol.* **91**, 637 (1981).

[8] J. M. Chirgwin, A. E. Przybyla, R. J. MacDonald, and W. J. Rutter, *Biochemistry* **18**, 5294 (1979).

[9] W. A. Braell, and H. F. Lodish, *J. Biol. Chem.* **256**, 11337 (1981).

[10] R. D. Demuth, L. C. Showe, M. Ballantine, A. Palumbo, P. J. Fraser, L. Cioe, G. Rovera, and P. J. Curtis, *EMBO J.* **5**, 1205 (1986).

[11] R. R. Kopito and H. F. Lodish, *Nature (London)* **316**, 234 (1985).

[12] A. Colman, in "Transcription and Translation—A Practical Approach" (B. D. Hames and S. J. Higgins, eds.), pp. 49–69. IRL Press, Oxford, England, 1984.

[13] J. N. Dumont, *J. Morphol.* **136**, 153 (1972).

these grooves where they are fixed and yet easily accessible for microinjection. The injection pipets are pulled from glass capillaries (1 mm outer diameter) to give long, narrow shanks. After pulling, the tip of the micropipet is broken off with watchmaker forceps in order to obtain openings of about 3–8 μm. The long shank of the pipets allows one to shorten the tip in case it is blocked during injection or sucking of the mRNA. After filling with oil, the pipet is mounted on a motor-driven micropump (Bachofer, Reutlingen, FRG). A desired amount of mRNA (3–10 μl) is prepared for microinjection by filtering it by centrifugation through polycarbonate filter (pore size 0.2 μm) fixed between the opening and the cap of a 1-ml Eppendorf centrifugation tube. In order to avoid evaporation, the resultant drop of mRNA is immersed in oil. About 3 μl of mRNA is sucked into the pipet which is sufficient to inject up to 60 oocytes. During injection each oocyte receives within 30 sec approximately 50 nl water containing about 25 ng of mRNA. Control oocytes, if needed, are either mock injected without RNA or not injected.

Incubation for Translation

The microinjected oocytes are incubated for time periods from at least 16 hr up to 15 days at 18–20° in Barth's medium. For incubation the oocytes are placed on a nylon net immersed in a beaker containing a large volume of Barth's medium. This improves the survival of the oocytes during longer incubation periods. The survival is further increased if the temperature of incubation is reduced to 10–12° after the first 24 hr of incubation at 18–20°. Control experiments have shown that appearance of functional band 3 protein in the plasma membrane increases up to about 7 days after microinjection of the mRNA and that the plateau level reached is maintained up to at least the fourteenth day after injection (Fig. 2). The plateau level is higher in oocytes obtained in the months April–November than in oocytes obtained in winter.

Immunoprecipitation

For immunoprecipitation microinjected oocytes are incubated 16–36 hr in Barth's solution containing 3 μCi [^{35}S]methionine (Amersham)/ oocyte/10 μl Barth's solution at 19°. Homogenization and immunoprecipitation were carried out by combining the methods of Colman[14] and Anderson and Blobel[15] as described by Morgan et al.[3]: oocytes are homogenized in 10 μl/oocyte of homogenization buffer containing 50 mM Tris, 50 mM

[14] A. Colman, *in* "Transcription and Translation—A Practical Approach" (B. D. Hames and S. J. Higgins, eds.), pp. 271–302. IRL Press, Oxford, England, 1984.
[15] D. J. Anderson and G. Blobel, this series, Vol. 96, p. 111.

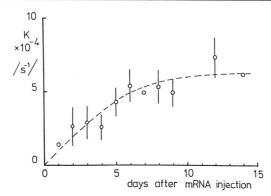

FIG. 2. Rate constant of band 3-mediated anion efflux from *Xenopus* oocytes as a function of incubation time after microinjection of anemic mouse spleen mRNA. Bars indicate ±SD ($n = 2$–6). Experiments were done on oocytes from five different females. (According to Grygorczyk et al.[18])

NaCl, 1% SDS, 0.1 mM phenylmethylsulfonyl fluoride (PMSF), pH 7.2 at 0°. After centrifugation (2 min, Eppendorf centrifuge) 50-μl aliquots of the extract of five oocytes were mixed with 2 μl 25% SDS and heated at 100° for 4 min. Then the samples are diluted with 200 μl dilution buffer containing 1.25% Triton X-100, 190 mM NaCl, 60 mM Tris, 6 mM ethylenediaminetetraacetic acid (EDTA), 0.1 mM PMSF, pH 7.4 followed by an overnight incubation at 4° with antibody. Control samples include microinjected oocytes reacted with preimmune serum, or noninjected oocytes reacted with specific antibodies. The resulting immunocomplex is bound to protein A–Sepharose (Sigma, 30 μl of a 1:1 suspension of protein A–Sepharose in dilution buffer). After end-over-end mixing for 4 hr, the beads are washed several times. The bound protein is eluted with sodium dodecyl sulfate-polyacrylamide gel electrophoresis (SDS–PAGE) sample buffer containing 1% SDS and 0.5% 2-mercapethoethanol. After heating at 100° for 4 min, the samples were applied to a SDS–PAGE[16] using a 3% stacking gel and an 8–20% gradient separating gel.

Measuring Fluxes

Basically it is possible to measure either influx or efflux of anions. Influx measurements are preferable if one needs high sensitivity for the detection of the ion movements. Efflux measurements are preferable if one wishes to obtain continuous records of the transport process or to study dose dependencies of inhibition on a single oocyte.

---

[16] U. K. Laemmli, *Nature (London)* **227,** 680 (1970).

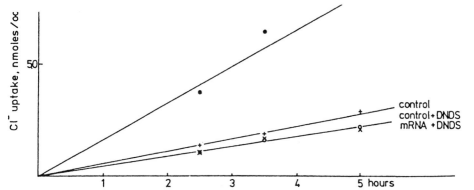

FIG. 3. Cl⁻ influx into *Xenopus* oocytes. "Control" and "control + DNDS" refer to untreated oocytes, in Barth's medium without and with 0.4 mmol/liter DNDS, respectively. "mRNA" and "mRNA + DNDS" refer to mRNA-injected oocytes in Barth's medium without or with 0.4 mmol/liter DNDS. (According to Morgan et al.[3])

Influx. The mRNA-injected oocytes are placed at the end of the "expression period" into a solution which contains the radioactively labeled anion species. After suitable time intervals (see Fig. 3), individual oocytes are collected by aspiration into a Pasteur pipet (with broad opening) and transferred into nonradioactive solution.

This medium should contain an inhibitor of the band 3 protein-mediated anion transport, such as 4,4′-dinitrostilbene-2,2′-disulfonate (DNDS), 4,4′-diisothiocyanostilbene-2,2′-disulfonate H₂DIDS, phloretin, etc. The oocytes are now washed three times in this medium by cycles of sedimentation, removal of supernatant, and resuspension in radioactivity-free solution. Finally the individual oocytes are transferred by means of a Pasteur pipet into individual scintillation vials and counts are taken. Since at the start of the experiment the oocytes were devoid of radioactivity, the sensitivity of the influx measurements is only limited by the sensitivity of the radioactivity determinations. After counting, the averages of the measurements obtained in the individual oocytes and the standard deviations are calculated. Figure 3 shows that, in oocytes which had not been microinjected with mRNA, an influx component can be seen which cannot be inhibited by DNDS or other specific inhibitors of band 3 protein-mediated anion transport, indicating the existence of an endogenous chloride transport system. The figure further shows that in oocytes which had been microinjected with mouse spleen mRNA, chloride influx is enhanced far above the level of the endogenous chloride transport. The mRNA-induced Cl⁻ influx is abolished by DNDS, indicating that the band 3 protein, the synthesis of which had been demonstrated by immunopreci-

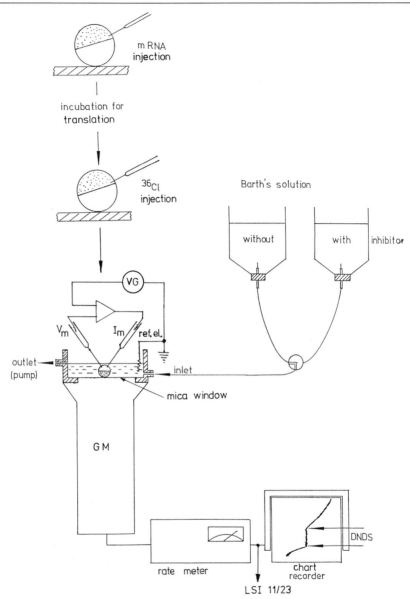

FIG. 4. Experimental arrangement for the measurement of $^{36}Cl^-$ efflux from *Xenopus* oocytes. GM, Geiger-Müller tube; V_m, I_m, voltage measuring and current delivering microelectrodes respectively; ref. el., reference electrode; VG, voltage generator; LSI 11/23, computer.

pitation, is inserted into the plasma membrane of the oocyte in a functional state.[3]

Efflux. After microinjection of mRNA and incubation for expression in Barth's solution, a single oocyte is now microinjected with $^{36}Cl^-$ ($\approx$5 nCi in about 50 nl of 200 mmol/liter NaCl) and placed into a hair loop on top of the mica window of a GM tube (GZ 7/5 Berthold, Wildbad, FRG), with its animal hemisphere facing the surface of the window (Fig. 4). This window serves as the bottom of a flow chamber which is constantly perfused with Barth's solution to maintain the $^{36}Cl^-$-containing oocyte in a nonradioactive environment. The pulse rate of the radioactive emission is continuously recorded using a suitable rate meter (e.g., LB 2040, Berthold, Wildbad, FRG). In control oocytes into which no mRNA had been injected, little if any efflux can be observed (Fig. 5A).

However, in successfully mRNA-injected oocytes, $^{36}Cl^-$ efflux can easily be observed (Fig. 5B). To demonstrate that the efflux is in fact mediated by band 3 protein, one can switch from perfusion with Barth's solution to perfusion with Barth's containing DNDS or some other reversible inhibitor of band 3 protein-mediated anion transport. This should lead to a strong inhibition which disappears after returning to perfusion with inhibitor-free Barth's solution (Fig. 5B). The chairlike record obtained in an intact mRNA-injected oocyte contrasts with the record obtained in damaged oocytes (Fig. 5C). In the latter the efflux cannot be inhibited by DNDS or other inhibitors of band 3 protein-mediated anion exchange, such as flufenamate.

The inhibition by DNDS can be used as a control for the performance of an oocyte during the flux measurements. Using this standard the protocol for the determination of the action of inhibitors on anion transport is the following (Fig. 6). The $^{36}Cl^-$ release from the mRNA-containing oocyte is followed for 10 to 20 min. The inhibitor-free perfusion medium is then changed to a medium containing the inhibitor at the desired concentration and the perfusion is continued for another 10–20 min. Subsequently follows another 10- to 20-min period of perfusion with Barth's solution containing DNDS at a maximally inhibitory concentration. The rate of efflux at maximum inhibition is used as a measure of the $^{36}Cl^-$ efflux via transport systems other than band 3 protein or via leaks, and deducted from the rate of efflux measured in the absence and the presence of the inhibitor to be studied (Fig. 6). By this method, it is possible to determine concentration–efficiency curves in single oocytes. Figure 7 represents an example, demonstrating the curve for DNDS. The half-maximal ($K_{1/2}$) effect is obtained at about 1–2 μmol/liter, which is close to the value observed in mouse red blood cells

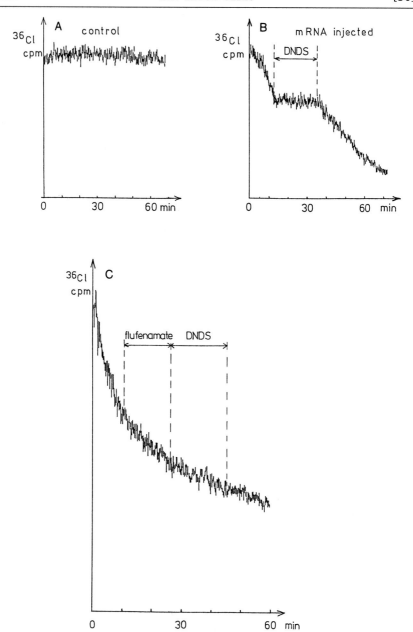

FIG. 5. Records of ³⁶Cl⁻ efflux from *Xenopus* oocytes. (A) Control oocyte. (B) Oocyte after microinjection of anemic mouse spleen mRNA. Efflux can be reversibly inhibited by 0.4 mmol/liter DNDS. (C) Leaky oocyte. Efflux can neither be inhibited by DNDS (0.4 mmol/liter) nor flufenamate (0.1 mmol/liter) in perfusate.

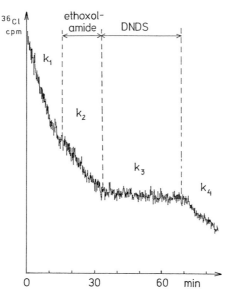

FIG. 6. Protocol for quantitative determination of the action of an inhibitor (1 μmol/liter ethoxolamide) on band 3 protein-mediated chloride efflux from *Xenopus* oocyte. To estimate inhibition, it is necessary to calculate rate constants k from the time course of the radioactivity decrease in the oocyte. For many purposes it suffices to approximate the efflux during the various time periods, in the absence or presence of the inhibitors by hand-drawn lines and calculating $k = \ln(y_2/y_1)(t_2 - t_1)^{-1}$ where y_1 and y_2 represent radioactivity in the oocyte at times t_1 and t_2, respectively. Respective k values (in sec^{-1} 10^{-4}) are $k_1 = 7.02$, $k_2 = 4.55$, $k_3 = 0.27$, $k_4 = 6.58$. (See also legend to Fig. 8.)

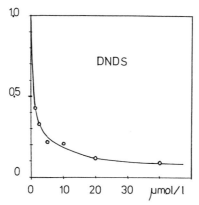

FIG. 7. ^{36}Cl$^-$ efflux (ordinate) as a function of DNDS concentration (abcissa) in the perfusate. The efflux, expressed as a fraction of the efflux in the absence of DNDS, is determined as shown in Fig. 6, except that in place of the ethoxolamide various concentrations of DNDS are used. The rate constants are corrected for residual efflux at 400 μmol/liter DNDS, as explained in the legend to Fig. 6. The half-maximal effect is reached at 1–2 μmol/liter DNDS and thus is close to the value observed with mouse red cells.[17]

at $0°$.[17] An exact comparison was not yet possible since the Cl^- transport in the mouse red cell is too fast to be measurable at $20°$, the temperature at which the $K_{1/2}$ value in the oocytes was determined. Nevertheless, it is clear that the apparent $K_{1/2}$ value in the oocytes is of the expected order of magnitude.

In the intact red cells of mouse and man, DNDS exerts its inhibitory action only if applied at the outer membrane surface. There is no inhibition if this nonpenetrating agent is incorporated into mouse or human red blood cell ghosts. Similarly, microinjection of sufficient DNDS into the oocytes to obtain a final concentration of 0.4 mmol/liter in the cytosol does not lead to an inhibition of the band 3 protein-mediated Cl^- efflux. It has, however, the interesting effect that the semilogarithmic representation of the efflux data follows a straight line for a much longer period of time than without microinjected DNDS (Fig. 8). Apparently, DNDS blocks the access of Cl^- to certain, as yet unidentified internal compartments and facilitates the even distribution of the microinjected $^{36}Cl^-$ in the cytoplasm, which is a relatively slow process and takes about 5–10 min. Unequal distribution of the microinjected $^{36}Cl^-$ requires that the orientation of the oocyte (vegetative–animal) relative to the window of the GM tube does not change during perfusion. One should be aware of this fact, especially when one changes from perfusion with one solution to perfusion with another.

Measuring Fluxes under the Voltage Clamp. Although the band 3 protein-mediated Cl^- flux is "electrically silent," it may still depend on membrane potential. This is due to the fact that (1) anion binding to conformers with inward- or outward-directed transfer site and/or (2) the equilibrium between the conformers with inward- and outward-directed transfer site may be functions of electrical field strength. To study such effects it is useful to measure $^{36}Cl^-$ efflux under voltage clamp. Since red blood cells are too small for such measurements, it is useful to employ mRNA-injected oocytes. For this purpose, the oocyte is placed in a perfusion chamber and injected with 50 nl of $^{36}Cl^-$ plus DNDS. The membrane potential is then clamped to a desired level by conventional, two-microelectrodes technique (see Fig. 4).

The counting rate determined with the GM tube is read into a computer every 30 sec and stored on disk. The $^{36}Cl^-$ efflux is followed after clamping the membrane potential at different levels for time periods of about 10 to 40 min. At the end of each time period at a given potential, DNDS is added to the perfusate and inhibition of efflux is observed, to ensure that the oocyte shows no unspecific leak for Cl^-. For each mem-

[17] P. Hanke-Baier, M. Raida, and H. Passow, *Biochim. Biophys. Acta* **940,** 136 (1987).

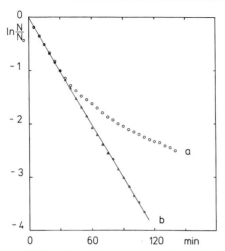

FIG. 8. Time course of $^{36}Cl^-$ release from mouse erythroid band 3 protein-containing oocytes. mRNA-injected oocytes received at the end of the expression period (○) $^{36}Cl^-$ or (△) $^{36}Cl^-$ plus sufficient DNDS to give a final DNDS concentration inside the oocyte of 400 μmol/liter. Ordinate: log radioactivity at time t divided by radioactivity at time zero. In view of the observations depicted in this figure, for all quantitative studies of $^{36}Cl^-$ efflux, the $^{36}Cl^-$ was microinjected together with sufficient DNDS to give the final concentration of 0.4 mmol/liter in the cytosol. This was done in the experiments represented in Figs. 6 and 7. (According to Grygorczyk *et al.*[18])

brane potential the rate constant k for $^{36}Cl^-$ efflux is determined off-line by fitting the exponential $N = N_0 \exp(-kt)$ to the data points by a least-squares procedure. A plot of the rate constants derived from flux measurements over the voltage range from -10 to -100 mV shows an increase by about 50% (see Ref. 18).

Discussion

When mRNA populations from whole cell extracts are expressed in the oocytes, it is necessary to verify that the transport process under study is in fact mediated by the transport protein the properties of which one wishes to study. In different types of somatic cells different variants of the band 3 protein may be expressed[10] and in the course of the development of erythroid cells nontransporting forms of band 3 protein may be replaced by transporting forms. In the case of the mouse band 3 protein as expressed in the oocytes, the susceptibility of the transport system to various inhibitors of mouse band 3 protein agrees well with the corre-

[18] R. Grygorczyk, W. Schwarz, and H. Passow, *J. Membr. Biol.* **99,** 127 (1987).

sponding data in the mouse red cell.[17] This functional similarity is complemented by the fact that band 3 proteins expressed in the oocyte are susceptible to immunoprecipitation with anti-mouse band 3 antibodies.[3] Thus we can feel reasonably sure that the Cl^- transport that appears after mRNA injection is indeed mediated by mouse erythroid band 3 protein.

A determination of the density of mouse band 3 molecules in the oocyte membrane has not yet been possible. Based on a comparison of Cl^- fluxes in oocytes and mouse red cells, assuming equal turnover numbers in both types of cells, it has been suggested that the density is quite small, perhaps 10^{-2} μm^{-2}. This number should be compared with 7000 μm^{-2} in the red blood cell. It is unclear whether this enormous difference is due to the small amount of microinjected band 3 mRNA or to a deficiency of the translocation and insertion of the final product into the plasma membrane. Experiments involving microinjection of large amounts of pure band 3 cRNA will yield the missing information in the near future.

Acknowledgment

We thank P. Eckard for critical reading of the manuscript.

[31] Chemical Modification of the Anion-Transport System with Phenylglyoxal

By Poul J. Bjerrum

Dicarbonyl reagents such as phenylglyoxal, butanedione, and cyclohexanedione react in alkaline solution with arginyl residues in proteins. These reagents have played a decisive role in the identification of arginyl residues as essential binding sites for negatively charged substrates and cofactors in numerous enzymes.[1,2] Patthy and Thész[3] explain the selectivity by suggesting that the pK of arginyl residues at anion-binding sites is lower than that of other arginine residues in the proteins. The rate of reaction, which preferentially depends on the deprotonated arginine, is therefore increased.

[1] J. F. Riordan, K. D. McElvany, and C. L. Borders, Jr., *Science* **195**, 884 (1977).
[2] J. F. Riordan, *Mol. Cell. Biochem.* **12**, 3915 (1979).
[3] L. Patthy and J. Thész, *Eur. J. Biochem.* **105**, 387 (1980).

We have previously shown that chloride exchange in human erythrocytes is a function of the protonation of one or more exofacial titratable groups having an apparent pK of 12.[4] The groups are likely to be arginine residues, due to the high pK value. By chemical modification of these groups it was shown that phenylglyoxal can be used for characterization and identification of essential arginyl residues in the anion-transport system of red cells.[5,6] A similar conclusion was reached by Zaki,[7,8] who has examined phenylglyoxal inhibition of sulfate exchange in red cells. It is the purpose of this chapter to describe methods by which essential arginines in the anion-transport protein can be modified and to discuss conclusions about their localization.

Reaction of Arginyl Residues with Phenylglyoxal

Phenylglyoxal (PG) is an α,α'-dicarbonyl compound which in alkaline solution reacts preferentially with the guanidino moieties of proteins. It has been shown that the reagent reacts with the nonionized form of arginine[5,9] and that reaction is only slow with α- and ε-amino groups.[9,10] The chemistry of the phenylglyoxal arginine reaction is still not completely understood. It was suggested by Takahashi[10] that phenylglyoxal reacts with the guanidino group of arginine with 2:1 stoichiometry. The proposed reaction scheme is demonstrated in Fig. 1. The first reaction, which is reversible, is assumed to be rate limiting, forming an initial 1:1 complex. This complex may then react rapidly with a second equivalent of phenylglyoxal to form a more stable 2:1 adduct (adduct **I**, Fig. 1). This product is found to be stable in acidic media but decomposes slowly in alkaline solution.[10]

Most of the reports on the uses of phenylglyoxal to modify essential arginine residues at enzyme active sites have either substantiated or assumed that phenylglyoxalation occurs with 2:1 stoichiometry.[10] However, in a few investigations, a 1:1 stoichiometry has been reported.[11-13] This result was obtained either in the presence of borate buffers [12] or at

[4] J. O. Wieth and P. J. Bjerrum, *J. Gen. Physiol.* **79**, 253 (1982).
[5] J. O. Wieth, P. J. Bjerrum, and C. L. Borders, Jr., *J. Gen. Physiol.* **79**, 283 (1982).
[6] P. J. Bjerrum, J. O. Wieth, and C. L. Borders, Jr., *J. Gen. Physiol.* **81**, 453 (1983).
[7] L. Zaki, *Protides Biol. Fluids* **29**, 279 (1982).
[8] L. Zaki and T. Julien, *Biochem. Biophys. Acta* **818**, 325 (1985).
[9] S. T. Cheung and M. L. Fonda, *Biochem. Biophys. Res. Commun.* **90**, 940 (1979).
[10] K. Takahashi, *J. Biol. Chem.* **243**, 6171 (1968).
[11] C. L. Borders, Jr. and J. F. Riordan, *Biochemistry* **14**, 4699 (1975).
[12] M. M. Werber, M. Moldovan, and M. Sokolovsky, *Eur. J. Biochem.* **53**, 207 (1975).
[13] B. Vandenbunder, M. Dreyfus, O. Bertrand, M. J. Dognin, L. Sibilli, and H. Buc, *Biochemistry* **20**, 2354 (1981).

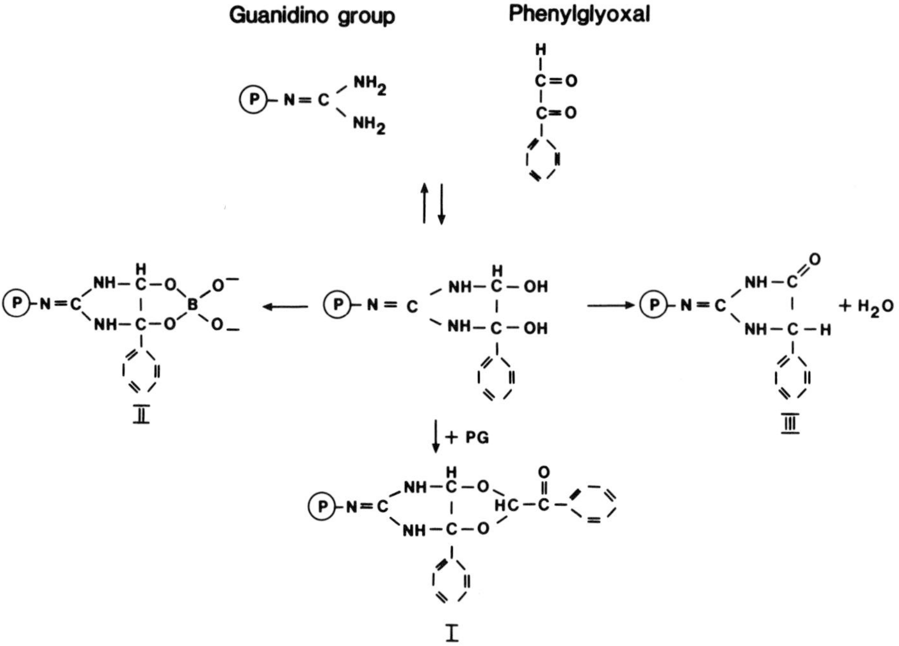

FIG. 1. Scheme for reaction of guanidine free base of an arginyl residue with phenylglyoxal. Phenylglyoxal is proposed[10] to form a 2:1 adduct with arginine, product **I**. A 1:1
adduct stabilized with borate has been suggested[11] as product **II**. Product **III** shows another
stable 1:1 adduct in accordance with Riordan.[14] (See text for details.)

low phenylglyoxal concentrations.[11] Borate can presumably complex the
cis-diol of the 1:1 primary complex, thus preventing the condensation of
another phenylglyoxal molecule (adduct **II**, Fig. 1). The stabilizing effect
of borate is known from modification with both 2,3-butanedione[14] and 1,2-
cyclohexanedione.[15] Riordan[14] has suggested that the primary 1:1 butanedione–arginine adduct apparently can be stabilized by a pinacol-type rearrangement. The suggestion is based on the observation that the 1:1
butanedione–arginine product dissociates slowly when the reagent is removed 1 hr after the reaction, whereas the 1:1 adduct apparently rear-
ranges if the reaction is allowed to continue for 20 hr, since the product is
now stable even in slightly alkaline solution. Moreover, the rearranged
product no longer reacts with borate.[14] The analogous adduct for the
phenylglyoxal–arginine complex is shown as product **III** in Fig. 1. An
analog adduct has also been observed when cyclohexanedione reacts with

[14] J. F. Riordan, *Biochemistry* **12**, 3915 (1973).
[15] L. Patthy and F. L. Smith, *J. Biol. Chem.* **250**, 557 (1975).

arginine under strong alkaline conditions,[16] and the results shown later in Tables I and II and Fig. 5 in this chapter may indicate that the same reaction also takes place when phenylglyoxal inactivates anion transport. It is therefore not known whether the phenylglyoxal reaction of essential arginines in the red cell anion-transport protein proceeds with 2 : 1 or 1 : 1 stoichiometry. Further chemical work on the reaction is clearly required to fully characterize the chemistry of the phenylglyoxal arginine complex formation in the anion-transport protein.

Chemical Reagents

Phenylglyoxal was purchased from EGA-Chemie, Steinheim-Albuch, Federal Republic of Germany, and DNDS (4,4'-dinitrostilbene-2,2'-disulfonic acid) from ICN K & K Laboratories, Inc., Plainview, New York, or from Pfaltz & Bauer, Inc., Stamford, Connecticut. DIDS (4,4'-diisothiocyanostilbene-2,2'-disulfonic acid) was prepared according to the method of Funder *et al.*[17] The buffers CHES [2-(*N*-cyclohexylamino)ethanesulfonic acid], pK_a 9.5 at 20°, and CAPS [3-(*N*-cyclohexylamino)-1-propanesulfonic acid], pK_a 10.4 at 20°, and papain from *Papaya carica* were from Calbiochem-Behring Corp., Switzerland. TES [*N*-tris(hydroxymethyl)methyl-2-aminoethanesulfonic acid], pK_a 7.5 at 25°, albumin from bovine pancreas, PMSF (phenylmethylsulfonyl fluoride), and nonactin from *Streptomyces griseus* were from Sigma Chemical Co., St. Louis, Missouri.

[^{14}C]Phenylglyoxal ([^{14}C]PG), specific activity 25–35 mCi/mmol, was obtained from GEA, Gif-Sur-Yvette, France, and was kept at −20° in methanol under a nitrogen atmosphere. K^{36}Cl, specific activity 7.2 mCi/g Cl, was purchased from The Isotope Laboratory of the Experimental Research Station at Risø, Denmark.

Exofacial Modification of Resealed Ghosts with Phenylglyoxal

Resealed human erythrocyte ghosts containing 165 mM KCl, 2 mM Tris, and 0.2–0.5 mM EDTA were prepared as described by Funder and Wieth[18] with the modification described by Bjerrum *et al.*[6] Treatment of ghosts with phenylglyoxal was carried out at different extracellular KCl concentrations by using mixtures of 165 mM KCl and 25 mM potassium citrate plus 200 mM surcose buffered with either CAPS or CHES (2.5–5 mM). Phenylglyoxal, typically 10–20 mM, was dissolved in the thermo-

[16] K. Toi, E. Bynum, E. Norris, and H. A. Itano, *J. Biol. Chem.* **242**, 1036 (1967).
[17] J. Funder, D. C. Tosteson, and J. O. Wieth, *J. Gen. Physiol.* **71**, 721 (1978).
[18] J. Funder and J. O. Wieth, *J. Physiol. (London)* **262**, 679 (1976).

statted medium and titrated to the pH of the experiment with 1 M KOH immediately prior to the addition of the packed ghosts.

Modification procedure A was used to determine phenylglyoxal binding (of [^{14}C]PG) to resealed ghosts or intact cells from the exofacial side of the membrane. Phenylglyoxal permeates the red blood cell membrane rapidly,[19] with a membrane permeability of 10^{-3} cm · sec^{-1} at 25°. In order to modify only the exofacially located arginyl residues it was therefore necessary to perform the labeling under conditions of neutral intracellular pH and alkaline extracellular pH. Fortunately, the erythrocyte membrane is not very permeable to protons,[4] and is even less so during phenylglyoxal inactivation, due to reversible inhibition of the anion-transport system by phenylglyoxal; a pH gradient can therefore be obtained for a couple of minutes even at 38°.[5,6] Packed resealed ghosts (44,000 g, 15 min, 0–4°), cytocrit 85–90% (determined from the [^{3}H]inulin space of the centrifuged sample), were treated with phenylglyoxal (10–30 mM, specific activity 0.03–0.06 mCi/mmol) in 3–4 vol of an alkaline, well-buffered electrolyte medium.[5,6] Final phenylglyoxal concentration, electrolyte composition, temperature, and pH are stated in the text or in the individual legends to the figures and tables. Samples (2–5 ml) were withdrawn from the well-stirred reaction suspension at appropriate intervals and the modification was interrupted by mixing the alkaline cell suspension with 4 vol of an ice-cold acidic buffer solution containing 165 mM KCl, 4 mM KH$_2$PO$_4$, and 2 mM EDTA. The acidity of this solution was adjusted to give a pH of 7.2–7.4 at 0° after addition of reaction suspension.

In *modification procedure B*, the ghosts were exposed to phenylglyoxal from the exofacial side with a higher specific activity (0.5–3 mCi/ mmol) when labeling was performed for subsequent polyacrylamide gel electrophoresis, and only one sample was processed at a time.[6] Packed ghosts (0.5 ml) were mixed with 1.5 ml of the alkaline medium typically containing 20 mM [^{14}C]phenylglyoxal. The chemical reaction was stopped after appropriate intervals (2–30 sec) by diluting the suspension with 8 ml of the above-mentioned ice-cold acidic buffer solution.

Intact red blood cells can also be reacted with PG from the exofacial side (by procedure A and B) as long as a high PG concentration, which prevents hemolysis, is used ($\geq$10 mM).

In order to remove membrane-adsorbed phenylglyoxal in the binding experiments, it was necessary to wash the ghosts in an ice-cold albumin-containing (0.5%) electrolyte solution and store the membranes in the same solution overnight (20 hr at 0°) before further incubation for 30 min

---

[19] J. O. Wieth, P. J. Bjerrum, J. Brahm, and O. S. Andersen, *Tokai J. Clin. Exp. Med.* **7**, 91 (1982).

at 38° in an albumin-free medium.[6] With this procedure it is possible to remove all noncovalently bound phenylglyoxal. The degree of inactivation of chloride transport in a given sample was found to be unaffected by the removal of adsorbed phenylglyoxal. After a 20-hr incubation at 0° and neutral pH, no loss of inhibition was seen even after further incubation in a slightly alkaline medium at 38°. This contrasted with the situation when ghosts were incubated at 38° in a phenylglyoxal-free medium immediately after phenylglyoxalation, as can be seen from Tables I and II and Fig. 5.

Determination of Inhibition and Membrane Binding of PG

Determination of chloride exchange flux was performed as described by Dalmark and Wieth.[20] Membrane binding of phenylglyoxal was calculated from the specific activity of the reagent and from the determination of the ^{14}C activity of the extensively washed ghost membranes.[6] The number of membranes in a given sample was determined both by counting with a Coulter counter model DN and by protein determination using a modified method of Lowry.[6,21] By correlating protein analysis to cell counting, an average content of 5.25×10^{-10} mg protein/ghost (SEM 0.04×10^{-10}, $n = 265$) was found.

Extracellular Enzymatic Treatment

Treatment of ghosts with extracellular chymotrypsin was performed as described by Bjerrum et al.[6] Treatment with papain was performed on chymotrypsin-treated cells. The cells were incubated in a medium containing 150 mM KCl, 15 mM phosphate, and 2 mM DTT (pH 7.3, 38°). The papain cleavage was performed at a concentration of 1 mg/ml ghost suspension (hematocrit 30%) for 1 hr. The reaction was stopped by adding 1 vol 10 mM iodoacetic acid in buffered saline (pH 7.3) and the ghosts were washed three times in flux medium (pH 7.3), containing 1 mM PMSF and 0.5% albumin, before determination of efflux and binding.

SDS–Polyacrylamide Gel Electrophoresis

The binding of [^{14}C]phenylglyoxal to the erythrocyte membrane proteins and protein fragments after enzymatic cleavage was examined by SDS–polyacrylamide gradient gel electrophoresis using a modification of the system introduced by Laemmli.[22] The procedure and the reagents used are described by Bjerrum et al.[6]

---

[20] M. Dalmark and J. O. Wieth, J. Physiol. (London) **224**, 583 (1972).

[21] O. H. Lowry, N. J. Rosebrough, A. L. Farr, and R. J. Randall, J. Biol. Chem. **193**, 263 (1951).

[22] U. K. Laemmli, Nature (London) **227**, 680 (1970).

Irreversible Inhibition of the Anion Transport System by Phenylglyoxal

Phenylglyoxal readily permeates the red cell membrane, the half-time of equilibration being 40 msec at 25°.[19] At neutral intracellular pH, however, only those arginyl residues that are exposed to the alkaline extracellular medium react rapidly with phenylglyoxal. The time course of an inactivation experiment with phenylglyoxal performed from the exofacial side of the membrane is shown in Fig. 2. The chloride self-exchange fluxes were determined at 0° and pH 7.3, and are expressed as a percentage of the exchange flux for an untreated control specimen. The reaction conditions are specified in the figure legend. Figure 2a shows that the exchange flux gradually approaches a residual value of about 10% of the control flux (J_0) and that the inhibition is not increased after a 60-sec exposure. The inhibitable flux, $J_0 - J_\infty$, is therefore approximately $(100 - 10) = 90\%$. As shown in Fig. 2b, the irreversible inactivation was found to be an apparent first order reaction at constant phenylglyoxal concentration. When the phenylglyoxal concentration was varied at constant pH,

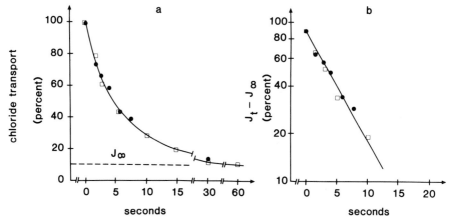

FIG. 2. Rate of irreversible inactivation of the chloride transport system (a). The fractional chloride exchange flux as a function of duration of exposure of resealed ghosts (neutral intracellular pH) to 16.7 mM phenylglyoxal in the 165 mM KCl, 3.5 mM CAPS medium (38°, pH 10.3). The phenylglyoxal reaction was interrupted by diluting, cooling, and acidifying the samples at the times indicated on the abscissa. The residual flux after extensive treatment (J_∞) was 10% of the control flux (J_0). The apparent inhibitable flux under these labeling conditions is therefore $J_0 - J_\infty$. The irreversible inactivation expressed as $\log(J_t - J_\infty)/(J_0 - J_\infty)100$ was a linear function of time at constant phenylglyoxal concentration (b), indicating apparent first order kinetics. Inactivation proceeded with a rate coefficient of 0.15 sec^{-1}, as determined by linear regression analysis of the slope of the line in (b). The two different symbols show results obtained in experiments on two different days.

the process was found to be a second order reaction.[5] The rate coefficient at 38° and pH 10.3 was 10 liters mol^{-1} sec^{-1}.

Temperature, pH, and Chloride Dependence of Irreversible Inactivation

The rate of inactivation decreased approximately 10-fold when the temperature was lowered from 38 to 25°C. This decrease corresponds to an apparent activation energy of 33 kcal mol^{-1}.[5]

Although the modified groups are important for the transport system their modification does not necessarily imply that they are functionally essential groups involved in either binding and/or translocation of the transported anions. The groups may be of importance simply as a result of unspecific or allosteric interactions with the binding site. On the other hand, if the groups are directly involved in anion binding or translocation the modification process should be sensitive to the presence of substrate anions in the medium. This requirement was fulfilled for the modified groups.[5] The second order rate coefficient for phenylglyoxal inactivation was decreased by an increase in extracellular chloride, as shown in Fig. 3.

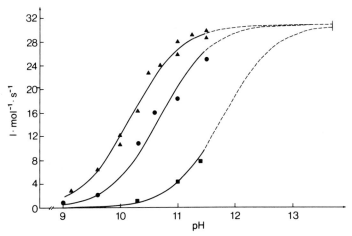

FIG. 3. The pH dependence of the rate of irreversible inactivation of the anion transport system by phenylglyoxal at extracellular chloride concentrations of 2 (▲), 8 (●), and 165 mM (■). Resealed ghosts were treated with phenylglyoxal at 25° in the appropriate sucrose–citrate and KCl media, and the rate of inactivation given as the second order rate coefficient in liters mol^{-1} · sec^{-1} was determined by following the time course of transport inhibition (cf. Fig. 2). The experimental values were fitted to simple titration curves with pK values of 10.2, 10.7, and 11.8, respectively. (From Wieth et al.[5])

The figure demonstrates the pH dependence of the rate of irreversible inactivation of the anion-transport system at three different extracellular chloride concentrations at 25°. At the lowest chloride concentration (2 mM), the rate of inactivation approaches an apparent maximal second order rate coefficient of 31 liters mol^{-1} sec^{-1} at pH values above 11. By varying the pH and the anionic composition of the reaction medium, as demonstrated in Fig. 3, it was found not only that the kinetics of transport inactivation are fully compatible with the assumption that phenylglyoxal only reacts with the deprotonated form of the functionally essential arginine, implying that the rate of reaction is reduced by a factor of approximately 10 when pH is lowered one unit at pH < 10, but also that the arginine residues apparently are more reactive at low KCl concentration due to an apparently lower pK under these conditions.[5] This conclusion is in agreement with titration studies, which show that the functionally essential transport groups, with pK around 12 (in 165 mM KCl), decrease their pK value at low extracellular chloride.[4]

The pseudo-first-order kinetics of the phenylglyoxal inactivation were also found when the anion-transport protein was modified from both sides of the membrane at neutral intra- and extracellular pH, as demonstrated by Zaki and Julien.[8] They found that reactive functional arginyl residues can be substantially protected against inactivation by substrate anions under these conditions. Their result is qualitatively similar to the result which can be observed for exofacial modification at alkaline pH values, as demonstrated in Fig. 4. Figure 4 represents a replot of data from Wieth *et al.*,[5] showing the rate of inactivation of anion transport with phenylglyoxal as a function of extracellular chloride concentrations. When the data are replotted according to the equation of Scrutton and Utter,[23] as used by Zaki and Julien,[8] a straight line is obtained which intersects with the ordinate close to the origin, indicating that phenylglyoxal is either unable to react with the chloride-loaded binding site, which is the case if the line passes through the origin, or that the reaction rate is significantly reduced, probably because the pK of the reacting guanidino group increases significantly (more than 1–2 pH units) by binding the chloride anion. The similarity in inactivation kinetics between modification from the exofacial side of the membrane at alkaline pH and modification under symmetrical conditions and neutral pH may imply that inactivation only takes place from the outside of the membrane.

Identical rates of exofacial phenylglyoxal modification were obtained at the same extracellular chloride concentration irrespective of whether the intracellular chloride concentrations were the same or different.[5] This

---

[23] M. C. Scrutton and M. F. Utter, *J. Biol. Chem.* **240**, 3714 (1965).

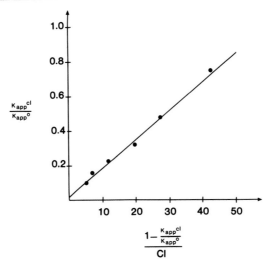

FIG. 4. The rate of phenylglyoxal inactivation obtained at various extracellular chloride concentrations (from Fig. 5 of Wieth *et al.*[5]) plotted according to the equation of Scrutton and Utter[23]:

$$K^{Cl}_{app}/K^{0}_{app} = K_2/K_1 + [(1 - K^{Cl}_{app}/K^{0}_{app})/Cl] K^{Cl}_{D}$$

where K^{0}_{app} and K^{Cl}_{app} are the apparent first order rate constants at zero chloride and at a given chloride concentration, respectively, K_1 and K_2 are the fractional order rate constants for inactivation of the chloride-unloaded and -loaded "reaction site" (time^{-1}), K^{Cl}_{D} is the dissociation constant for chloride at the "reaction site," and Cl is the chloride concentration in the reaction medium. For the equation to be valid the following reaction scheme is required,

$$\text{ECl*PG} \underset{\text{PG}}{\overset{K_2}{\longleftarrow} } \text{ECl} \overset{K^{Cl}_{D}}{\rightleftharpoons} \text{E} \underset{\text{PG}}{\overset{K_1}{\longrightarrow}} \text{E*PG}$$

where E is the chloride-unloaded reaction site and the other symbols are as already defined. Resealed ghosts (165 mM) were reacted in the buffered mixture of 165 mM KCl and the isotonic sucrose–citrate medium containing 13.5 mM phenylglyoxal (pH 10.3, 25°). The plot yields a straight line which intersects the ordinate close to the origin. From the regression analysis of the data it is impossible to conclude whether the results indicate that the reaction is significantly reduced (~50-fold, as apparently found) when the chloride ion is bound ($K_2 \ll K_1$) or whether phenylglyoxal is completely unable to react with the chloride-loaded binding site (the line going through the origin implying that $K_2 = 0$). The slope of the straight line, equal to K^{Cl}_{D}, was found to be approximately 17 mM under the alkaline conditions indicated.

result indicates that the rate of inactivation is dependent only on the extracellular chloride concentration and not on conformational changes between inward and outward configuration of the transport-binding site. This could also mean that the modified group is not directly involved in binding and/or translocation of the transported anion, but the result rather reflects the situation that phenylglyoxal, besides producing irreversible inactivation, also inactivates the transport system in a reversible manner, possibly locking the transport system in an outward-facing configuration.

Reversible Inhibition of Chloride Exchange by Phenylglyoxal

It is essential to distinguish between reversible and irreversible effects of phenylglyoxal on membrane anion transport. Phenylglyoxal also reacts as a reversible inhibitor of anion exchange with a rapidly established binding equilibrium.[5] In the report by Zaki and Julien[8] an unexpectedly higher relative increase in the rate of irreversible inactivation at neutral pH was observed when the phenylglyoxal concentration was increased at low (<4 mM) compared to higher PG concentrations. This observation was used to suggest that reversible, apparently saturating binding of PG to the transport site might precede the subsequent irreversible inactivation. This conclusion cannot be harmonized with the results obtained at more alkaline pH as discussed here for several reasons: First, when data for reversible phenylglyoxal inhibition of chloride self-exchange at pH 10 and 0° were examined in a Hill plot, a positive cooperativity with a Hill coefficient of 1.7 was obtained in contrast to the irreversible inactivation at pH 10 and 25° which Wieth et al.[5] have shown is first order in phenylglyoxal.[5] The pH-dependent reversible inhibition with phenylglyoxal is also observed at 25°.[24] Second, the rate of irreversible phenylglyoxal inactivation increases linearly with phenylglyoxal concentration at concentrations where the transport system is saturated with reversibly bound phenylglyoxal.[5] Moreover, the reversible inhibition with phenylglyoxal is rather uninfluenzed by changes in the extracellular chloride concentrations in comparison with the significant chloride dependence of the irreversible inactivation, at both neutral and alkaline pH.[24] These different observations make it unlikely that the reversible, rather chloride-insensitive phenylglyoxal binding should simply precede the very chloride-dependent irreversible inactivation. Instead, it is more likely that reversible inhibition by phenylglyoxal arrests all transporters in some extracellular "reactive" conformation. This would explain why only extracellular chloride is important for the irreversible inactivation at alkaline pH. The assumption also indicates that changes in distribution between inward and outward transport site conformations, due to changes in, e.g., the chloride gradient across the membrane, should only be able to influence irreversible PG inactivation at phenylglyoxal concentrations below those leading to

[24] P. J. Bjerrum, unpublished observations.

"full" reversible inhibition. The unexpectedly higher relative increase in PG reactivity observed, by Zaki and Julien[8] by increasing the PG concentration at very low phenylglyoxal and chloride concentrations, might thus be explained by the occurrence of fractions of highly reactive transport conformations under these conditions.

Nature of the Residual Flux

Complete inactivation of anion transport in alkaline media was usually not obtained, even after prolonged treatment with phenyglyoxal. In a long series of experiments a residual flux of 9.1% was obtained.[5] A probable explanation for this could be that the maximally modified transport system, by analogy with findings for certain enzymes,[10] operates at one-tenth of the rate of the intact system. This possibility was excluded because the residual flux of the transport system is completely inhibited after covalent reaction of the phenylglyoxal-modified membranes with a number of DIDS molecules that suffice only to inhibit 10–15% of the transport protein molecules.[5] Complete inhibition (98–99%) is obtained if phenylglyoxal-treated cells are washed free of excess phenylglyoxal at room temperature and later subjected to a second treatment. Another observation on the residual flux is that the magnitude depends on the temperature at which the cells are kept after modification. In an attempt to understand the reason for development of the residual flux and how incubation temperatures might influence it, the residual flux was examined after incubation at different temperatures.

The results presented in Table I show the reactivation of the transport system as a function of the incubation time (minutes, at 38°) shortly after preparation. The maximum obtainable flux after incubation at 38° was 30–40% of the control flux. This increase in the residual flux was found to be temperature and time dependent, but was not seen if the cells were kept at 0° for more than 20 hr before incubation (Table II and Fig. 5). The conclusion from these experiments is that the primary formation of a covalent phenylglyoxal–arginine complex is reversible to some extent immediately after reaction, and moreover that the primary complex can slowly rearrange irreversibly to another stable product as a function of time. The difference in the size of the final residual flux obtained by incubation in a PG-free medium at different temperatures probably reflects a different temperature dependence for the two opposite reactions: (1) slow irreversible stabilization of the primary formed phenylglyoxal–arginine complex, and (2) slow dissociation of the primary formed complex. In contrast to the situation at higher temperatures only the irreversible stabilizing reaction is significant when the reacted resealed ghosts are kept at 0°. The

TABLE I

REACTIVATION OF THE PHENYLGLYOXAL MODIFIED
TRANSPORT SYSTEM AT 38° AS A FUNCTION OF TIME[a]

Incubation time (min)	Chloride self-exchange (pmol cm^{-2} sec^{-1} × 10^{12})	Residual flux (%)
0.0	85.1	3.2
0.5	149.9	5.7
1.0	292.3	11.0
2.0	478.9	18.1
3.0	636.0	24.0
6.0	750.0	28.0
20.0	838.2	31.6.

[a] Isotonic resealed ghosts with neutral intracellular pH were reacted in 25 mM PG in the buffered isotonic sucrose–citrate medium at low chloride concentration (pH 10.3 and 25° for 60 sec by method A). The extracellular "maximally" modified ghosts were resuspended in 165 mM KCl, 2 mM sodium phosphate (pH 7.3, 0°) and kept strictly at 0° during transfer to nylon tubes before being incubated at 38° (1 hr after modification) for the indicated times. Immediately after incubation at 38° the samples were cooled and packed at 0° for measurement of the ^{36}Cl efflux. A mean volume of 92 μm^3 was used for calculation of the exchange efflux at pH 7.3 and 10°. The residual flux was calculated from an untreated control efflux of 2647 pmol · cm^{-2} · sec^{-1} at 10°.

TABLE II

TEMPERATURE-DEPENDENT REACTIVATION OF PHENYLGLYOXAL MODIFIED ANION
TRANSPORT SYSTEM[a]

Incubation temperature (°C)	Incubation 1 hr after PG treatment		Incubation 24 hr after PG treatment	
	Incubation (min)	Residual efflux (%)	Incubation (min)	Residual efflux (%)
0	—	3.2	—	6.3
15	60	8.0	45	6.4
25	45	15.7	45	7.4
38	20	31.6	30	8.4

[a] Chemical modification, incubation, and determination of chloride efflux as described in Table I.

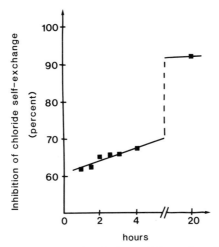

FIG. 5. The inhibition obtained by incubation at 0° for various times after the chemical modification. Resealed ghosts containing 165 mM KCl were reacted for 60 sec at low chloride concentration (8 mM) in 20 mM phenylglyoxal in the isotonic sucrose–citrate medium buffered with 5 mM CHES (pH 10.3, 25°). The reacted ghosts were stored in a PG-free medium at 0° and samples were incubated at 38° for 30 min after appropriate periods of time, before determination of the chloride exchange. The residual flux (100% − inhibition %) obtained decreases as a function of the incubation time at 0°.

irreversible reaction could be a pinacol-type rearrangement, as suggested in Fig. 1 (product **III**), implying a 1 : 1 stoichiometry, but another explanation involving rearrangement of the second phenylglyoxal molecule (in product **I**) is also possible. This would require a 2 : 1 stoichiometry. The residual flux was not significant after modification at neutral pH,[8] this is probably due to the long incubation time, allowing the slow irreversible stabilization to take place during modification.

Selectivity of Exofacial Modification

In labeling experiments[6] using [14C]phenylglyoxal, modification of the ca. 150 × 10[6] arginines present per ghost occurred rapidly when extra- and intracellular pH were both above 10 (Fig. 6). After 5 min, approximately 350 × 10[6] phenylglyoxal molecules were bound per cell, i.e., sufficient to modify all arginyl residues in the membrane proteins. Modification of only extracellular arginines, however, was achieved when ghosts with a neutral intracellular pH were reacted for a short period of time in a medium containing 165 mM KCl, 16 mM phenylglyoxal, and 5 mM CAPS (pH 10.3) (filled circles, Fig. 6). During the first 2 min, phenylglyoxal

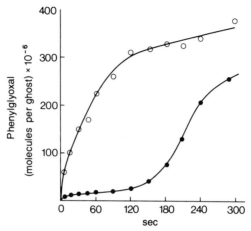

FIG. 6. Phenylglyoxal molecules bound per cell as a function of the duration of phenyl-glyoxal treatment at an extracellular pH of 10.3 at 38°; (○) leaky white ghosts, intracellular pH 10.3; (●) resealed pink ghosts, initial intracellular pH 6.4. Reaction medium: 16 mM phenylglyoxal, 165 mM KCl, 2.5 mM CAPS. The experimental conditions are described in more detail by Bjerrum *et al.*[6]

binding proceeds at a much lower rate than when the intracellular pH is alkaline, the membranes having bound only ca. 10×10^6 phenylglyoxal molecules/cell after 20 sec. After 2 min of exposure, however, the pH gradient across the membrane dissipates and the rate of phenylglyoxal incorporation increases dramatically to a value similar to that seen in ghosts without a pH gradient across the membrane. Only extracellular arginines are labeled when the intracellular pH is neutral. This is demonstrated by the fact that the intracellularly located peripheral membrane protein spectrin was not labeled when the intracellular pH was low. This is in contrast to the situation with symmetrical modification. (For comparison of the extracellular and intracellular labeling of the membrane proteins, see Figs. 8 and 13.)

To determine the apparent minimum number of reactive groups which were modified from the exofacial side of the membrane at full inhibition, the correlation between [^{14}C]phenylglyoxal binding and transport inactivation was examined. Full inhibition was obtained by binding of 9×10^6 phenylglyoxal molecules/cell when the reaction was performed at high extracellular chloride concentration (165 mM KCl; Fig. 7, open circles). This binding is equivalent to modification of 4–5×10^6 arginines/membrane, assuming a 2 : 1 phenylglyoxal/arginine stoichiometry. About 60% of the radioactive label, corresponding to 2–3×10^6 arginines/membrane, was found in the capnophorin fraction of each membrane.[6] The selectivity

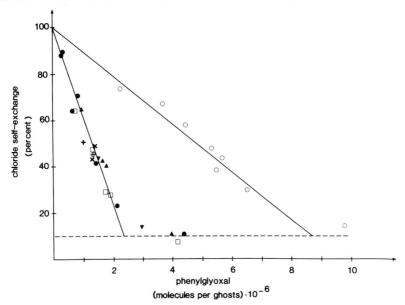

FIG. 7. Selectivity of [¹⁴C]phenylglyoxal binding as a function of extracellular chloride concentration. The ordinate indicates the chloride self-exchange in percentage of the untreated control sample. (○) Phenylglyoxal binding to resealed ghosts reacted from 1 to 40 sec with 15 mM phenylglyoxal, 165 mM KCl, 5 mM CAPS (pH 10.3, 38°). The other symbols show the phenylglyoxal binding at low chloride concentration. Reaction medium: 15 mM phenylglyoxal, 8 mM KCl, 5 mM CAPS (pH 10.3, 25°). Maximum inhibition of chloride exchange was accompanied by binding of 9×10^6 phenylglyoxal molecules/cell at high chloride concentration (165 mM). This binding was reduced to 2.5×10^6 phenylglyoxal molecules/cell when the reaction was performed at low chloride concentration (8 mM).

of modification was found to increase when inactivation was performed at low extracellular chloride concentrations. At 8 mM KCl, maximal inhibition is obtained by binding of 2.5×10^6 phenylglyoxal molecules/ghost, as demonstrated in Fig. 7.[6] This result indicates that the important groups are more reactive at low chloride concentrations, as expected if the pK of the functionally essential groups is decreased by reducing the chloride concentration.[4] Of the 2.5×10^6 [¹⁴C]phenylglyoxal molecules/cell, calculated by extrapolation to full inhibition, the major fraction was found to be located in the 10^6 capnophorin molecules present per cell (Fig. 8). This observation demonstrates that the reaction at low KCl concentration is so selective that only 1 single exofacial arginine out of the 44 arginines in capnophorin needs to be modified in order to give full inactivation of the transport molecule, assuming 2 : 1 stoichiometry.

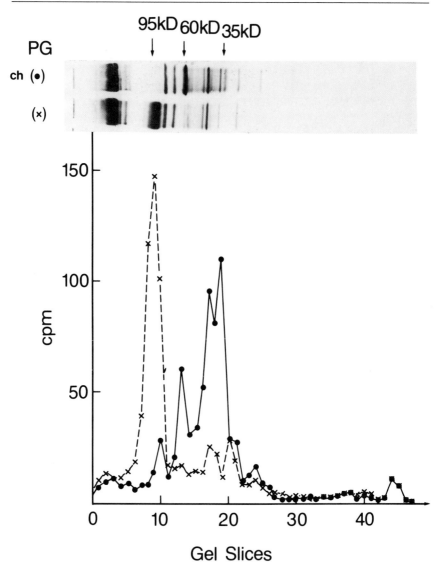

FIG. 8. Distribution of [^{14}C]phenylglyoxal in red cell membrane proteins before and after chymotrypsin cleavage of capnophorin. The selective exofacial modification (method B) was performed at low chloride concentration (8 mM) for 8.8 sec in the sucrose–citrate medium containing 20 mM phenylglyoxal, 5 mM CAPS (pH 10.3, 25°). The membrane proteins were separated on a 12–25% SDS–polyacrylamide gradient gel. The two stained gels (80 μg of membrane protein) before (×) and (●) after treatment with chymotrypsin (ch) are shown in the upper part of the figure. The counted gel contained 120 μg of membrane protein. The inhibition of chloride self-exchange in the ghosts preparation was 68.9% and the [^{14}C]PG binding was 1.8 × 10^6 PG molecules/ghost.

Location of the Functionally Essential Arginyl Residue

Capnophorin can be cleaved by extracellular chymotrypsin to produce a 35,000- and a 65,000-Da fragment (see Fig. 12). The major fraction of the [^{14}C]phenylglyoxal was found in the C-terminal 35,000-Da fragment of the protein when the labeling was performed under the most selective conditions (Fig. 8). This result indicates that the 35,000-Da segment contains at least one essential arginyl residue but may contain two, since two arginines were labeled under the less specific conditions (in 165 mM KCl). Both arginines were found in the C-terminal 35,000-Da fragment.[6] Further cleavage with extracellular papain splits the 35,000-Da segment into two new membrane-bound fragments, a smaller N-terminal fragment of 7000 Da and a bigger C-terminal segment with a molecular mass of 28,000.[25] The major fragment still contains the radioactive phenylglyoxal, as demonstrated in Fig. 9.

In an attempt to further localize the modified residue, the transport protein was degraded with chymotrypsin from both sides of the membrane (1.5 mg chymotrypsin/ml, pH 7.6, for 1 hr). By this treatment about half of the activity was lost during the first 10 min of incubation, i.e., before complete cleavage of the proteins has taken place. The rest of the activity contained in the pellet after chymotrypsin treatment was found in a diffuse, 9000-Da membrane-bound fragment (Fig. 10).[26] The localization of this fragment in the 28,000-Da C-terminal papain fragment is not known.

Modification with Phenylglyoxal from Both Sides of the Membrane

Experiments were performed to further characterize the exofacial modified arginyl residues. Phenylglyoxal modification was performed in the presence and absence, respectively, of the reversible inhibitor of anion transport DNDS, in an attempt to demonstrate protection of the exofacial binding site against phenylglyoxal inactivation.[27] DNDS, which is a competitive inhibitor,[28] should be able to protect the binding site from modification. Unfortunately, the inhibitory potency of DNDS is significantly reduced at pH values around 10, so the inactivation experiments have to be performed under mild alkaline conditions and therefore with the same pH on both sides of the membrane.

[25] M. L. Jennings, M. Adams-Lackey, and G. H. Denney, *J. Biol. Chem.* **259,** 4652 (1984).
[26] P. J. Bjerrum, *in* "Structure and Function of Membrane Proteins" (E. Quagliariello and F. Palmieri, eds.), p. 107. Elsevier, Amsterdam, 1983.
[27] J. O. Wieth and P. J. Bjerrum, *in* "Structure and Function of Membrane Proteins" (E. Quagliariello and F. Palmieri, eds.), p. 95. Elsevier, Amsterdam, 1983.
[28] O. Fröhlich, *J. Membr. Biol.* **65,** 111 (1982).

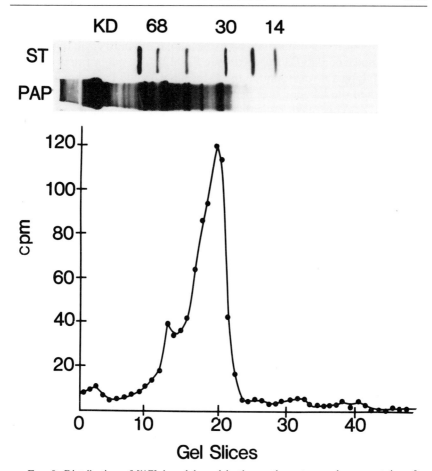

FIG. 9. Distribution of [^{14}C]phenylglyoxal in the erythrocyte membrane proteins after extracellular treatment with chymotrypsin followed by papain. The phenylglyoxal reaction and SDS–polyacrylamide gel electrophoresis were performed as described in Fig. 8. The enzymatic cleavage was carried out as described in the text. About 25% of the [^{14}C]phenylglyoxal was lost during papain treatment. The major fraction of activity was located in the C-terminal papain fragment with an apparent M_r of about 30,000 in the gel. The inhibition of chloride self-exchange after phenylglyoxal reaction was 60%. The effects of papain treatment and PG modification on the chloride self-exchange flux were found to be additive.

The resealed ghosts, containing 40 mM K_2SO_4, 4 mM KCl, and 10 mM TES (pH 8.3), were prepared by a modification of the standard procedure.[6] Human red cells were washed three times in 125 mM K_2SO_4, 10 mM Tris (pH 7.7 at 25°). The cells were allowed to equilibrate for 5–10 min between washes and were finally resuspended at 50% hematocrit. One volume of ghosts was hemolysed in 10 vol of a 4 mM $MgSO_4$, 3.8 mM

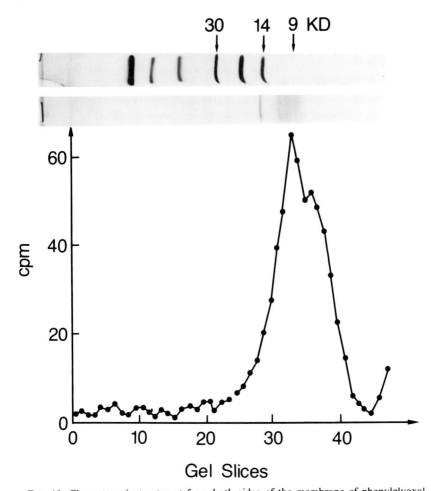

FIG. 10. Chymotrypsin treatment from both sides of the membrane of phenylglyoxal-treated erythrocytes; distribution of [^{14}C]phenylglyoxal in the membrane-bound peptide fragments. Packed, resealed isotonic ghosts (neutral intracellular pH) were selectively modified with [^{14}C]phenylglyoxal for 8.8 sec at low extracellular chloride concentration (8 mM), phenylglyoxal concentration was 20 mM, and the other conditions were as described in Fig. 8. The inhibition of chloride self-exchange was 68.9% with 1.8 × 10^6 phenylglyoxal molecules bound per ghost after the reversibly adsorbed phenylglyoxal was removed as described in the text.[6] The modified resealed ghosts (treated with extracellular chymotrypsin) were hemolysed and washed three times in 20 vol of hemolysis buffer, 10 mM sodium phosphate, 1 mM EDTA (pH 7.4, 0°), in order to remove hemoglobin. The membranes were then alkali stripped in 10 ml of 0.1 mM NaOH at 0° and washed twice in hemolysis buffer in order to remove the peripheral proteins before incubation with chymotrypsin (1.5 mg/ml) for 1 hr at 38°, pH 7.6 in the same buffer. After enzymatic cleavage the membrane was washed three times in hemolysis buffer including 0.2 mM PMSF before a new alkali stripping was performed. The loss of [^{14}C]phenylglyoxal during alkali stripping was approximately 10%. By chymotrypsin treatment from both sides of the membrane, approximately 50% of the activity was lost in the supernatant, most of it within the first 10 minutes of incubation. The electrophoresis was performed as described in Fig. 8. The recovery of ^{14}C radioactivity in the sliced gel was 99.6% of the applied activity.

acetic acid and 10 mM TES buffer, (pH 3.6–3.8). After 5 min at 0°, 1 vol of ice-cold reversal solution, 400 mM K$_2$SO$_4$, 60 mM KCl, and 6 mM EDTA, was added. The pH of the reversal solution was adjusted with KOH so that mixing with the hemolysed suspension of ghosts gave a pH of 7.2–7.3 at 0°. After a further 10 min at 0° the ghosts were resealed by incubating for 45 min at 38°. The ghosts were washed three times in efflux medium, 40 mM K$_2$SO$_4$, 5 mM KCl, and 10 mM TES (pH 8.3), and were labeled in the same medium at 25° with 25 mM phenylglyoxal (final concentration 21.4 mM PG after addition of ghosts). The pH of the medium was kept constant during reaction (pH 8.2–8.4) by addition of small amounts of 1 M KOH when pH dropped below 8.2. After reaction for the appropriate time the cells were washed in the efflux medium without phenylglyoxal at pH 7.3 and 0°, and stored overnight in the same medium containing 0.5% albumin. The chloride self-exchange and membrane binding were determined as already described.

The irreversible inactivation was essentially a pseudo-first-order process also under these symmetrical reaction conditions (in accordance with the observations by Zaki and Julien[8]) with an apparent rate coefficient of 1.3×10^{-3} sec^{-1} and a second order rate constant of 6.2×10^{-2} liters mol^{-1} sec^{-1}. This second order rate constant is about 40% lower than the expected (10.5×10^{-2} liter mol^{-1} sec^{-1}) if calculated from the rate constant for exofacial modification at 5 mM KCl (pH 10.3, 25°) (Wieth et al.; Table II[5]) and assuming a 10-fold decrease in reactivity per pH unit. The reason for this difference is likely to be the presence of 40 mM K$_2$SO$_4$ in the reaction medium but also an increased chloride protection against the phenylglyoxal reaction at pH 8.3 due to a lower K_D^{Cl} at neutral pH (~10 mM)[8] as compared to pH 10.3 (~17 mM).[5] Although sulfate apparently has no influence on inactivation at alkaline pH values,[5] Zaki and Julien[8] have shown that sulfate protects against PG inactivation at neutral pH with an apparent "half-protection" concentration of 40–60 mM. The exofacial inactivation at alkaline pH values and the reaction under symmetrical conditions at pH 8.3 is therefore apparently with the same arginyl residue(s) or at least with a group(s) which inactivate with the same rate constant and chloride dependence as expected for the exofacial group(s).

The phenylglyoxal inactivation experiments at pH 8.3 show, moreover, that the transport inactivation is very effectively abolished by 1 mM DNDS (lower curve, Fig. 11A). In the absence of DNDS, transport inactivation was 92–96% after 1 hr of incubation. However, in contrast to the considerable differences in inactivation, the binding of phenylglyoxal to the membrane appears similar (Fig. 11B). About 100 phenylglyoxal molecules/ghost were bound after 1 hr of incubation both in the presence and absence of the inhibitor. This demonstrates that DNDS can protect

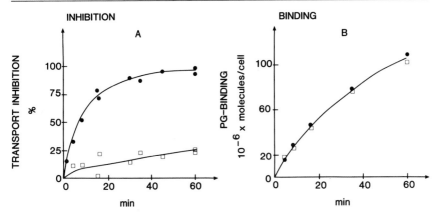

FIG. 11. Transport inactivation (A) and [14C]phenylglyoxal binding (B) of resealed human erythrocyte ghosts as a function of reaction time in 21.5 mM PG, 40 mM K_2SO_4, 5 mM KCl, 10 mM TES (pH 8.3, 25°) with (□) and without (●) 1 mM DNDS present. The resealed ghosts, prepared and reacted as described in the text, were modified under symmetrical conditions from both sides of the membrane.

against phenylglyoxal inactivation but also that about one-third of the arginines in the transport protein can be modified without serious allosteric effects on membrane anion transport. The small difference between the binding curves indicates that one, or at least only a few, arginine residue(s) are essential for anion binding or translocation. The experiment also indicates that chemical modification of critical arginines with phenylglyoxal under normal conditions apparently does not take place from the inside of the membrane, at least not when the transport sites are blocked by extracellular DNDS. We are also able to conclude from the pseudo-first-order kinetics for the reaction that if inactivation takes place from the intracellular site of the membrane it should involve only one or a few groups reacting with the same rate coefficient as the exofacial groups. In an attempt to further clarify whether arginines facing inward in the cell can be modified with phenylglyoxal, experiments on selective inactivation from the inside of the membrane were tried.

Modification from the Intracellular Side of the Membrane

Modification of internal arginines was performed under conditions of neutral extracellular pH and alkaline intracellular pH. Resealed ghosts containing 25 mM CHES, 40 mM K_2SO_4, 5 mM KCl, and 0.5 mM EDTA were prepared as described above, with the hemolysing medium containing 25 mM CHES instead of 10 mM TES. The resealed ghosts were washed three times in the efflux medium, 40 mM K$_2$SO$_4$, 25 mM CHES, and 5 mM KCl (pH 9.8, 25°). The ghosts were then packed to a hematocrit of 80–90% and titrated to pH 9.6 at 25° for 30 min after adding 10 μl of 10^{-3} M nonactin and 10 μl of 10^{-1} M NH$_4$SO$_4$/ml ghost suspension. After titration with KOH, the cells were washed once in the flux medium at pH 9.8 and 25° before packing for phenylglyoxalation. Intracellular pH (above 9.6) before phenylglyoxalation was checked by hemolysis of a small sample of packed ghosts in distilled water. One volume of ghosts was reacted in 2 vol of 30 mM PG, 25 mM TES, 40 mM K$_2$SO$_4$, and 5 mM KCl (pH 7.4 at 25°). At appropriate intervals, 2–50 sec, samples of the reaction suspension were transferred to 8 vol of ice-cold flux medium (pH 7.3). After washing the ghosts several times in the flux medium the cells were stored overnight in the same medium containing 0.5% albumin.

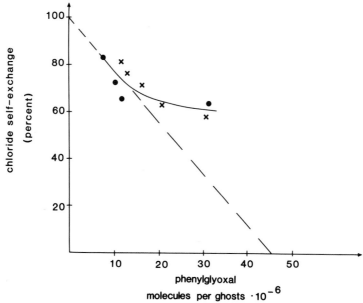

FIG. 12. Chemical modification of intracellular exposed arginine residues. The figure shows the transport inactivation as a function of [^{14}C]phenylglyoxal binding at the intracellular side of the membrane. One volume of packed resealed ghosts containing 25 mM CHES, 40 mM K$_2$SO$_4$, 5 mM KCl, and 0.5 mM EDTA, prepared and titrated to pH 9.6 (as described in the text), was reacted in 2 vol of 30 mM [^{14}C]phenylglyoxal, 25 mM TES, 40 mM K$_2$SO$_4$, 5 mM KCl (pH 7.4 at 25°). After appropriate intervals (2–50 sec) the reaction was stopped by dilution of samples with 8 vol of ice-cold flux medium (pH 7.3). The results of experiments performed on two different days are given. Maximal inactivation never exceeded 50% transport inactivation. Extrapolation of initial binding to 100% transport inactivation corresponds to binding of approximately 40 × 10^6 phenylglyoxal molecules per ghost.

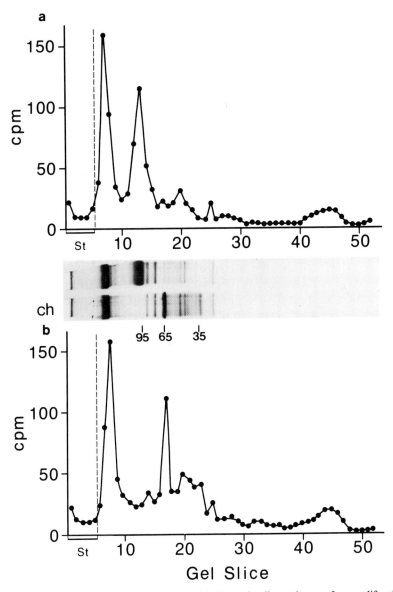

FIG. 13. Distribution of [^{14}C]phenylglyoxal in the red cell membrane after modification from the intracellular side of the membrane. The modification, which was performed for 10.4 sec under the conditions described in Fig. 12, gave a chloride self-exchange efflux of 71% and a binding of 8.2×10^6 molecules phenylglyoxal per ghost (a). After extracellular chymotrypsin treatment only the activity corresponding to band 3 was changed, with about equal amounts of [^{14}C]phenylglyoxal distributed between the 65,000- and the 35,000-Da fragments (b). The amount of applied membrane protein in the two gels was ~60 μg. The membrane proteins were separated by SDS-polyacrylamide gel electrophoresis as described in Fig. 8. The two stained gels before and after chymotrypsin treatment are shown in the middle of the figure.

The chloride self-exchange flux in the efflux medium at pH 7.2 and 0°, and the [^{14}C]PG binding of the reacted cells were determined as described.[6]

A rapid inactivation of anion transport was seen within the first 10 sec of modification but only up to about 40% inhibition (Fig. 12). The stop in inactivation was not due to a drop in pH of the intracellular fluid, because it was possible to show by hemolysing a small sample of ghosts, that the intracellular pH was still high after 1 min of exposure to phenylglyoxal. The selectivity of inactivation was not very high. Extrapolation of the initial rapid inhibition to full inhibition gave a binding of 40×10^6 phenylglyoxal molecules/ghost (Fig. 12). About one-third of the modified arginines were located in capnophorin (Fig. 13a) and these labeled molecules appear to be equally distributed between the 65,000-Da and the 35,000-Da fragments obtained after extracellular chymotrypsin treatment (Fig. 13b). It is therefore not clear whether the intracellular modification is a specific reaction which reduces the transport to only about 60% by alteration of a single group (which could be important but not essential) or whether the inactivation represents an unspecific reaction of intracellular groups which are "hidden" at neutral pH but which, at the high intracellular pH, become reactive and give some unspecific inhibition. The unspecific inhibition is apparently less pronounced at lower pH (cf Fig. 11).

Conclusion

Based on the pseudo-first-order kinetics, the chloride dependence, and selectivity of extracellular phenylglyoxal modification it is concluded that the anion-transport proteins contain at least one functionally essential arginyl residue. The residual flux seen when inactivation is performed at alkaline extracellular pH values is explained by a reaction model which indicates a two-step reaction process, the last step rendering the inactivation completely irreversible.

Modification of as much as one-third of all arginines in the transport protein under conditions where the transport binding site is protected by DNDS does not influence the anion transport significantly, showing that a large proportion of arginines can be modified without any allosteric effect on anion binding and translocation.

A maximum of only about 40% transport inactivation was obtained when modification was selectively performed from the intracellular side of the membrane at high intracellular pH and neutral extracellular pH. It is not evident from the experiment whether the inactivation represents a specific reaction with modification of one single group or whether the inactivation is the result of an unspecific reaction with different arginyl residues which influences the transport process by allosteric interactions and which only occurs at more alkaline pH values. In favor of the last hypothesis is the observation that the kinetics of inactivation are pseudo-first-order, also when ghosts are reacted from both sides of the membrane in the symmetrical situation as reported here and by Zaki and Julien.[8] Such simple reaction kinetics would not be expected if inactivation proceeded from both sides of the membrane, the intracellular modification giving, moreover, only about half-maximal inhibition. It is therefore proposed that modification of functionally essential arginyl groups under condition of high PG concentrations only takes place from the exofacial side of the membrane and that the outward-facing reactive conformation of the protein is stabilized by the reversible, relatively chloride-insensitive phenylglyoxal inhibition.

The amino acid sequence of capnophorin has recently been determined from the DNA sequence of the gene coding for the murine anion-transport protein.[29] A model of the primary structure, folded into the membrane as proposed by Kopito and Lodish[29] and later modified by Passow,[30] is shown in Fig. 14. The exofacial modified arginine should, according to this model, be either arginine residue number 664, 675, 748, 778, or 800, all of which are localized on the exofacial side of the 28,000-Da papain fragment. Of these five possible candidates only arginine-664 and arginine-675 are located in a segment which has been proved to be located on the exofacial side of the membrane.[30,31] This is in contrast to the two other arginines (748 and 778), which may be located differently. Arginine-800 is something special with respect to localization. This arginine is sitting in a long hydrophobic stretch of amino acid residues close to the middle. This sequence is long enough to transverse the lipid bilayer twice in α-helical configuration and it is therefore unpredictable to which side of the membrane this arginine should be located as shown in Fig. 14. This is the case even if it turns out that the exofaciale hydrophilic loop (amino acid residues 748–778) should be located on the intracellular side of the membrane. A part of the hydrophobic segment, amino acid residues 794–813, has recently been isolated from human capnophorin after enzymatic (pepsin) treatment followed by chemical (CNBr) degradation. This sequence is found to be completely conserved between the human and the murine protein,[32] indicating that it may have functional importance.

The possible specific chymotrypsin cleavage points in the amino acid sequence of 28,000-Da fragment were considered in order to localize the

[29] R. R. Kopito and H. F. Lodish, *Nature (London)* **316,** 234 (1985).
[30] H. Passow, *Rev. Physiol. Biochem. Pharmacol.* **103,** 62 (1986).
[31] D. Jay, *Annu. Rev. Biochem.* **55,** 511 (1986).
[32] C. J. Brock and M. J. A. Tanner, *Biochem. J.* **235,** 899 (1986).

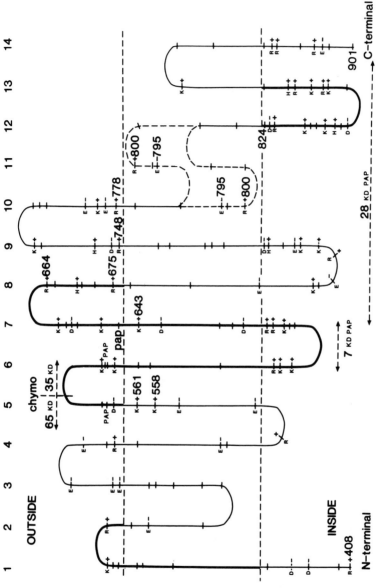

Fig. 14. Folding of the primary structure of murine capnophorin in the membrane bilayer, as proposed by Kopito and Lodish[29] and later modified by Passow.[30] The dashed line shows the two possible configurations suggested by Bjerrum.[41] The model shows the location of the charged amino acid residues in the C-terminal part of the molecule responsible for transport function. The exofacial cleavage points for chymotrypsin (chymo) and papain (pap) are shown, together with the primary binding site for covalently bound DIDS (lysine-558 or -561). The heavily drawn lines show the part of the sequence where extra- and intracellular localization has been established experimentally. R, Arg; K, Lys; H, His; D, Asp; and E, Glu. Prolines and glycines are also indicated by breaks in the sequence chain (—)

9000-Da chymotrypsin fragment in the sequence. The only sequence without obvious cleavage points and which was long enough to be a 9000-Da fragment, was the stretch from about amino acid residue number 770 to about 830, including the long hydrophobic sequence with arginine-778 and -800. This proteolytic peptide has recently been identified by partial sequence analysis (seven amino acids) as a fraction of the heterogeneous 9000-Da fragment and which begins with the N-terminal amino acid residue number 771.[33] It is therefore very likely that the functionally essential arginyl residue identified by phenylglyoxal modification is located in this segment and that the arginine could be either amino acid number 778 or 800.

The hydrophobic segment containing arginine-800 has a very special construction. It contains a total of five glycine and two proline residues, both of which types of groups are regarded as being α-helix breakers in hydrophilic proteins. These seven residues are arranged in a very orderly manner, several being spaced by five amino acid residues. Arginine-800 sits in a segment defined by two prolines, namely residues 796 and 802. Proline residues give much less rotational freedom to the peptide chain than e.g., glycine residues. They probably are of importance for the stabilization and relative position of this part of the segment and therefore also for the conformational position of arginine-800. All other bending points are glycine groups, apparently rendering the hydrophobic sequence rather flexible and thereby allowing it to undergo conformational changes.

Several molecular mechanisms appear to be able to explain the kinetics of anion transport.[19,34–37] In order to explain the ping-pong kinetics of the system, which has been demonstrated in many reports[38,39] as reviewed,[30,36,40] it is necessary to assume that the protein molecule undergoes conformational changes during translocation of the anion. The basic proposal of translocation models presented so far is that the transported anion interacts with positively charged essential groups in the transport protein changing the electric field around these groups. These changes allow a local conformational change to take place in the protein. By this

[33] H. Sigrist and K. A. Stauffer, personal communication.
[34] P. A. Knauf and F. Y. Law, Alfred Benzon Symp. 14, 448 (1980).
[35] H. Passow, L. Kampmann, H. Fosold, M. Jennings, and S. Lepke, Alfred Benzon Symp. 14, 345 (1980).
[36] I. G. Macara and L. C. Cantley, in "Cell Membranes: Methods and Reviews" (E. Elson, W. Frazier, and L. Glaser, eds.), Vol. 1, p. 41. Plenum, New York, 1983.
[37] J. Brock, M. J. A. Tanner, and C. Kempf, Biochem. J. 213, 577 (1983).
[38] R. B. Gunn and O. Fröhlich, J. Gen. Physiol. 74, 351 (1979).
[39] J. J. Falke and S. I. Chan, J. Biol. Chem. 260, 9537 (1985).
[40] R. B. Gunn and O. Fröhlich, in "Chloride Transport in Biological Membranes" (J. A. Zadunaisky, ed.), p. 33. Academic Press, New York, 1982.

mechanism the anion is exposed to the other side of the membrane. Whether this translocation involves one single step or several consecutive steps (e.g., a zipper mechanism[19]) is not known. Although no documented molecular mechanism for translocation can be presented, it appears likely that arginine-800, due to its capability of being placed at any position in the hydrophobic region of the membrane, could be one of the functionally essential groups involved in anion binding and/or translocation. The "flexible" hydrophobic segment containing arginine-800 may be surrounded by, and may interact with, a closed circle of the other α-helices forming the translocation pathway, this arginine being a mobile positive group positioned somewhere along the path. A model of this kind, in which the arginine is flexible enough to move across the membrane, has recently been proposed.[41]

Acknowledgments

I am grateful to Ingeborg and Leo Daning Foundation for Medical Research for a research fellowship, to Lise Mikkelsen, Tove Soland, and Ann Dorthe Davel for technical assistance, and to Julianne Halkier for secretarial help.

[41] P. J. Bjerrum, in "Proceedings of the Eighth School on Biophysics of Membrane Transport," Vol. 2. Agric. Univ., Wroclaw, Poland, 1986.

[32] Purification and Characterization of Band 3 Protein

By Joseph R. Casey, Debra M. Lieberman, and
Reinhart A. F. Reithmeier

Band 3 protein comprises approximately 25% by weight of the proteins in the erythrocyte membrane.[1] The purification of band 3 protein involves the preparation of red cell ghost membranes, removal of the extrinsic membrane proteins, solubilization of the intrinsic membrane proteins with detergent, and finally ion-exchange and sulfhydryl affinity chromatography. Several alternative purification procedures are also presented. The membrane domain of band 3 protein generated by mild trypsin treatment of ghost membranes can be purified using a similar protocol. Procedures for immobilizing band 3 protein and for forming matrix-bound band 3 monomers are described. The purified proteins can be characterized by measuring the binding of fluorescent inhibitors of anion exchange, by circular dichroism, and by cross-linking.

[1] G. Fairbanks, T. L. Steck, and D. F. H. Wallach, *Biochemistry* **10**, 2606 (1971).

Washing of Cells and Preparation of Ghost Membranes[2]

The first stage of band 3 protein purification is the preparation of red blood cells free of other blood cell types. This is particularly important because white blood cells contain proteases which could be released on cell lysis. Protease action can be minimized by fast work, the use of protease inhibitors, and the maintenance of the buffer temperature at 0–4° at all times.

Blood

In our laboratory recently outdated (i.e., 1 month from date of collection) red blood cell concentrate (230 ml in 63 ml of anticoagulant citrate, phosphate, dextrose, adenine) derived from 1 unit of blood (450 ml) from the blood bank is regularly used. Freshly drawn blood may also be collected, using anticoagulants such as EDTA (50 ml of 1.5% Na_2EDTA/500 ml blood), acid–citrate–dextrose [trisodium citrate dihydrate (2.2 g), citric acid monohydrate (0.8 g), dextrose (2.5 g) in 100 ml of distilled water/700 ml blood], or heparin (2250 units in 30 ml of buffered saline/500 ml blood).

Solutions

5P8: 5 mM sodium phosphate, pH 8.0

Phosphate-buffered saline: 150 mM NaCl, 5 mM sodium phosphate, pH 7.4

0.9% NaCl

Phenylmethylsulfonyl fluoride (PMSF) stock solution: 100 mM in dimethylsulfoxide or 2-propanol

Spectrin extraction buffer: 2 mM EDTA, 0.2 mM dithiothreitol, 20 μg/ml PMSF (added fresh), pH 8.0

KI extraction buffer: 1 M KI, 1 mM EDTA, 1 mM dithiothreitol, 20 μg/ml PMSF (added fresh), 7.5 mM sodium phosphate, pH 7.5

Procedure

All steps are carried out at 0–4° unless stated otherwise. To remove anticoagulants, blood additives, and white cells, red blood cells are washed three times. Blood is mixed with 5–10 vol of 0.9% NaCl or phosphate-buffered saline and centrifuged at 3000 g for 5 min (5000 rpm in a Beckman JA-20 or Sorvall SS34 rotor). After sedimentation, a layer of white cells, the buffy coat, will sit on top of the red cells. This is carefully

[2] J. T. Dodge, C. Mitchell, and D. J. Hanahan, *Arch. Biochem. Biophys.* **100,** 119 (1963).

removed along with the wash solution, by gentle aspiration. The procedure is repeated another two times.

Red blood cell ghosts are prepared by osmotic lysis of the washed red blood cells.[2] Optimum removal of hemoglobin from ghosts occurs at pH values above 7.4 at ionic strengths of between 10 and 20 mOsm.[2] The washed cells are mixed with 10 vol of ice-cold 5P8 containing 0.2 mM dithiothreitol, 20 μg/ml PMSF (added just before use), and incubated 10 min on ice. The ghosts are then sedimented for 20 min at 27,000 g (15,000 rpm in a Beckman JA-20 or Sorvall SS34 rotor) at 4°. Hemolysate is removed by aspiration but because the hemolysate is so dark in color, the ghost pellet is obscured and it is best to remove no more than one-half of the volume from the tube in this first aspiration. Washes then proceed using 5P8. After the second centifugation and buffer removal, the centrifuge tube should be rolled on its side, to expose a small pellet of white cells which should be carefully removed. Later washes may reveal white opalescent resealed ghosts at the bottom of the tube which should be left. Continue to wash the ghosts until the pellet is milky white, which usually requires four or five washes. Approximately 100 ml of ghost membranes at a protein concentration of 4 mg/ml can be prepared from 1 unit of packed red cells. Sodium dodecyl sulfate gel electrophoresis is used to characterize the ghost membrane proteins and to monitor the purification of band 3 protein (Fig. 1).

An alternative procedure for the preparation of ghost membranes from a unit of packed red blood cells involves the use of a Millipore Pellicon cassette system.[3] This system consists of a variable speed-drive pump with a 1.5 liter/min pump head, an assembled Pellicon cassette acrylic cell, and a 0.5-μm HVLP Durapore membrane cassette. Millipore silicone tubing should be used since regular Tygon tubing becomes brittle and cracks on prolonged use at 4°. Lysed red cells are pumped across the equivalent of 5 ft^2 of membrane that retains the ghost membranes but allows the hemolysate to be filtered. The membranes can be washed and concentrated using this apparatus. We find this procedure useful for the large-scale processing of multiple units of blood.

Procedure

The procedure is carred out in a cold room and all buffers are stored on ice. Cells (250 ml) are washed as described above and then poured into at least 20 vol of *ice-cold* 5P8. The lysate is allowed to flush through the system at a flow rate of 0.5 liters/min and then is concentrated to 300 ml

[3] T. L. Rosenberry, J. F. Chen, M. M. L. Lee, T. A. Moulton, and P. Onigman, *J. Biochem. Biophys. Methods* **4**, 39 (1981).

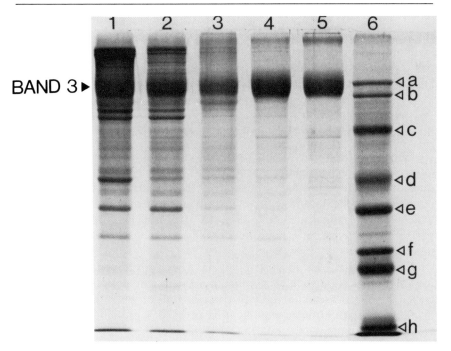

FIG. 1. Sodium dodecyl sulfate polyacrylamide gel showing various stages of a typical band 3 protein purification. 1, Erythrocyte ghost membranes; 2, EDTA-stripped ghost membranes; 3, KI-EDTA stripped ghost membranes; 4, band 3 protein (6.5 μg) purified by aminoethyl-Sepharose chromatography; 5, band 3 protein (6.5 μg) further purified by p-CMB-Sepharose sulfhydryl affinity chromatography; and 6, molecular weight markers: a, β-galactosidase (M_r 116,000); b, phosphorylase b (M_r 97,400); c, bovine serum albumin (M_r 66,000); d, ovalbumin (M_r 45,000); e, glyceraldehyde-3-phosphate dehydrogenase (M_r 36,000); f, carbonate dehydratase (M_r 29,000); g, trypsinogen (M_r 24,000); and h, trypsin inhibitor (M_r 20,100). The position of band 3 protein (M_r 105,000) is indicated by the closed arrow. Purified band 3 protein contains variable amounts of band 3 dimer, which migrates near the top of the gel, and small amounts of proteolytic fragments, visible when the gel is overloaded with respect to band 3 protein (lane 5).

using the recirculating retentate flow pattern at the same flow rate. Ice-cold 5P8 is then introduced using the constant volume wash mode and run at a flow rate of 1 liter/min until the ghosts are white. This requires about 10 liters of buffer. The entire procedure takes about 1 hr and yields approximately 400 mg of ghost membrane protein/unit of packed red cells. Scrupulous cleaning of the membrane as recommended by Millipore is accomplished by successively flushing the system with 0.9% NaCl (3 liters, or until no more hemoglobin is visible), distilled water (3 liters), 0.1 N NaOH (2 liters allowed to sit for 30 min only), and finally distilled water

(5–10 liters). The system is flushed with 2% formalin and the membrane cassette is stored in place fully wetted in 2% formalin. The membrane maintains efficient performance for at least a year. The "Millipore Pellicon Cassette System Assembly, Operation and Maintenance Instruction Manual" should be consulted for further details.

Stripping of Peripheral Proteins

Hemoglobin, glycolytic enzymes, ankyrin, and associated cytoskeletal proteins (spectrin, actin, and band 4.1) are bound to the cytoplasmic domain of band 3 protein. Removal of these extrinsic proteins leaves stripped ghost membranes containing only intrinsic membrane proteins of which band 3 constitutes 50% (Fig. 1). It is important to remove the peripheral proteins before chromatographic steps when their removal is more difficult.

Many protocols for ghost stripping have been developed, including the use of lithium diiodosalicylate, dimethylmaleic anhydride, *p*-chloromercuribenzoate, and guanidine hydrochloride,[4] all of which result in some degree of denaturation or modification of band 3 protein. While highly effective, stripping with 0.1 M NaOH[4] denatures the cytoplasmic domain of band 3 protein.[5] Similarly, 0.17 M acetic acid selectively denatures the membrane-bound domain.[5] The protocol which follows uses gentler stripping procedures to produce native band 3 protein.

Procedure

The cytoskeletal proteins, spectrin, and actin may be depleted by extraction at low ionic strength and alkaline pH (Fig. 1).[6] Ghosts are washed in 10 vol of ice-cold spectrin extraction buffer and pelleted by centrifugation at 46,000 g (19,000 rpm in a JA20 rotor) for 20 min at 4°. Poor extraction may result from the use of insufficient extraction buffer or a too-low pH. The resulting pellet is resuspended, using a syringe, in 10 vol of spectrin extraction buffer and incubated at 37° for 30 min. The ghosts are then pelleted for 30 min as above and the 37° extraction repeated. Since spectrin extraction is often incomplete, stripping may be monitored by sodium dodecyl sulfate polyacrylamide gel electrophoresis at this stage, and a further extraction at 37° should be performed if necessary. The ghosts are then washed once more with ice-cold spectrin extraction buffer.

[4] T. L. Steck and J. Yu, *J. Supramol. Struct.* **1,** 220 (1973).
[5] P. S. Low, *Biochim. Biophys. Acta* **864,** 145 (1986).
[6] V. Bennett, this series, Vol. 96, p. 313.

Extraction with high salt removes the remaining peripheral proteins (Fig. 1). Spectrin-depleted ghosts are resuspended with a syringe in at least 10 vol of KI extraction buffer and incubated for 30 min at 37°. The ghosts are then pelleted by centrifugation for 30 min at 46,000 g (19,000 rpm in a JA20 rotor). To remove residual KI, which may interfere with subsequent chromatography, stripped ghosts are resuspended in 5P8, containing 20 μg/ml PMSF and incubated at 4° for 10 min. The ghosts are centrifuged as above and washed once more in 5P8.

Alkali Extraction

A rapid, efficient alternative to the EDTA/KI procedure just described is to extract ghosts with 10 vol of ice-cold 2 mM EDTA, pH 12.[7] Immediately after dilution of the ghosts, the membranes are pelleted by centrifugation at 48,000 g (20,000 rpm in a JA-20 or SS34 rotor) for 30 min. The membranes are washed three times with 5P8.

Solubilization of Band 3 Protein

Choice of Detergent

Chromatographic purification of band 3 protein requires that the protein be in soluble form, removed from the membrane. Two considerations are important to the choice of detergent: efficiency of solubilization and retention of band 3 native structure. Ionic detergents such as sodium dodecyl sulfate, dodecyltrimethylammonium bromide, and N-dodecyl-N,N-dimethylglycine and zwitterionic detergents such as Zwittergent 3-10 and Zwittergent 3-12 are potent solubilizers of band 3, but denature the protein.[8] Bile acid detergents such as sodium cholate and 3-[(3-cholamidopropyl)dimethylammonio]-1-propyl sulfonate (CHAPS) solubilize band 3 protein poorly. These two bile acid detergents and sodium deoxycholate at concentrations of 0.5–1% eliminate inhibitor binding to band 3 protein.[8,9] Nonionic detergents are milder and a better choice. As regards the solubilization efficiency of these detergents for human band 3 protein, dodecyldimethylamine oxide (DDAO) $\gg$ octylglucoside = octaethylene glycol n-dodecyl ether ($C_{12}E_8$) = Triton X-100.[8] Octylglucoside at high concentrations ($>$40 mM) causes band 3 protein to aggregate and to have a lower affinity for inhibitors.[10] Circular dichroism,[11,12] tryptophan expo-

[7] T. L. Steck, B. Ramos, and E. Strapazon, *Biochemistry* **15**, 1154 (1976).
[8] R. Moriyama and S. Makino, *Biochim. Biophys. Acta* **832**, 135 (1985).
[9] D. M. Lieberman and R. A. F. Reithmeier, *Biochemistry* **22**, 4028 (1983).
[10] P. K. Werner and R. A. F. Reithmeier, *Biochemistry* **24**, 6375 (1985).
[11] J. Yu and T. L. Steck, *J. Biol. Chem.* **250**, 9170 (1975).
[12] K. Oikawa, D. M. Lieberman, and R. A. F. Reithmeier, *Biochemistry* **24**, 2843 (1985).

sure,[8] and stilbene disulfonate binding[8,9] have been used as measures of native structure of detergent-solubilized band 3 protein. By these criteria, $C_{12}E_8$ and DDAO give the most native band 3 protein. Although Triton X-100 has been widely used for band 3 purification,[11,13] it is not an ideal detergent. It absorbs light at 280 nm,[14] thereby interfering with spectroscopic protein determinations, it is composed of a heterogeneous population of molecules,[15] and it contains oxidizing impurities.[16,17] Detergent choice is also important with regard to dialysis or reconstitution but the detergent used for solubilization need not necessarily be the same as that used in latter stages because detergent may be exchanged by column chromatography.

Highly purified $C_{12}E_8$ may be purchased at a reasonable cost directly from the manufacturer (Nikko Chemical Co., Ltd., Tokyo, Japan) in 25-g aliquots in sealed ampoules. Fluka Chemical Corporation, Ronkonkoma, New York, and Calbiochem, a division of American Hoechst Corporation, San Diego, California, sell smaller amounts (i.e., 1 g). The detergent is stored at $-20°$ in 1-ml aliquots. The detergent is a solid at room temperature and should be melted by heating to 50°. The melted detergent is dissolved with rapid stirring in warm water or buffer solutions to prepare a 10% (v/v) stock solution. Detergent stock solutions are stored at 4° in the dark and can be used for up to 1 month. Storage under nitrogen in sealed ampoules and the addition of 1 mM EDTA or antioxidants such as 0.1 mM tertiary butylated hydroxytoluene have been recommended to prevent peroxide formation in detergent solutions.[16,17]

Solubilization of Stripped Ghosts

Procedure

Stripped ghosts (15–20 mg protein/ml) are resuspended using a syringe in an equal volume of 5P8. To this suspension are added 5 vol of 1% (v/v) $C_{12}E_8$ in 5P8. Since membrane-bound proteases may be released during solubilization, it is worthwhile to add fresh PMSF (to 20 μg/ml) to the solubilization buffer. After thorough mixing the solution should clarify. Turbidity indicates unsolubilized lipid which can be solubilized by small additions of a 10% stock detergent solution. The solution is left on ice for 20 min and then insoluble material is removed by centrifugation at

[13] M. F. Lukacovic, M. B. Feinstein, R. I. Sha'afi, and S. Perrie, *Biochemistry* **20,** 3151 (1981).
[14] R. S. Matson and S. C. Goheen, *J. Chromatogr.* **359,** 285 (1986).
[15] A. Helenius, D. R. McCaslin, E. Fries, and C. Tanford, this series, Vol. 46, p. 63.
[16] M. Lever, *Anal. Biochem.* **83,** 274 (1977).
[17] H. W. Chang and E. Bock, *Anal. Biochem.* **104,** 112 (1980).

80,000 g for 45 min (30,000 rpm in a Beckman 75 Ti rotor). The supernatant contains band 3 protein, glycophorin, band 4.5 and band 7 proteins (Fig. 1).

Selective Solubilization of Band 3 Protein

An alternative to the stripping procedures and detergent solubilization of stripped ghosts described above is selective detergent solubilization of band 3 protein from ghosts. This leaves an insoluble cytoskeletal matrix that is removed by centrifugation. A low-concentration detergent extraction may be used to deplete selectively glycophorins,[18] prior to the band 3 protein solubilization. This procedure is faster and simpler than EDTA/KI stripping but band 4.2[18] and band 6[10] tend to remain associated with band 3 protein, in contrast to using EDTA/KI stripping. The yield of band 3 protein is low as well, about 20%, because the membrane is not quantitatively solubilized.[19]

Procedure[11,19]

Ghosts are resuspended in 5 vol of 0.9% NaCl, 20 mM sodium phosphate, pH 8.0 and centrifuged at 39,000 g for 10 min (18,000 rpm in a JA20 rotor) to remove band 6 protein. If desired, sialoglycoproteins are extracted by incubation with 6 vol of 0.05% Triton X-100 or $C_{12}E_8$ in extraction buffer (0.18% NaCl, 4 mM sodium phosphate, pH 8.0).[18] Membranes are pelleted as above and washed with extraction buffer. Band 3 is then solubilized with 5 vol of 0.5% Triton X-100 or $C_{12}E_8$ in extraction buffer at 0° for 20 min. Residual membrane and cytoskeleton are pelleted by centrifugation at 40,000 g for 30 min at 4°.

Chromatography of Band 3 Protein

Once band 3 protein has been solubilized with detergent, the protein may be treated chromatographically like a classical soluble protein. Band 3 protein may be purified by anion exchange followed by sulfhydryl affinity chromatography. Both DEAE[13,19,20] and aminoethyl resins[11,21] have been used to purify band 3 protein. The cytoplasmic domain of band 3 protein contains a highly negative region which allows for binding to anion-exchange resins.[5] Band 3 protein binds tightly to DEAE resins, requiring the use of high ionic strength buffers for elution.[13] In contrast,

[18] J. M. Wolosin, *Biochem. J.* **189,** 35 (1980).
[19] G. Pappart and D. Schubert, *Biochim. Biophys. Acta* **730,** 32 (1983).
[20] A. D. M. Zangvill and Z. I. Cabantchik, *Biochim. Biophys. Acta* **727,** 77 (1983).
[21] H. Nakashima and S. Makino, *J. Biochem.* (*Tokyo*) **88,** 899 (1980).

band 3 protein can be readily eluted from aminoethyl-Sepharose by 100 mM NaCl.[9,11,21] Band 4.5 protein (glucose transporter) does not bind to these resins and glycophorin elutes before band 3 protein.[21] Aminoethyl-Sepharose is superior to aminoethylcellulose in resolving glycophorin and band 3 protein.[21] Sulfhydryl affinity chromatography is useful as a purification step, for band 3 protein concentration, and for buffer exchange. Band 3 protein is separated from the sialoglycoproteins of the red cell membrane by thiol-reactive resins because the sialoglycoproteins contain no cysteine residues. While band 3 protein is bound to the resin it is possible to exchange detergents and buffers for reconstitution or other purposes. After purification, time-dependent band 3 protein aggregation may be minimized by addition of 2-mercaptoethanol to 10 mM.[13,19] Addition of low concentrations of ionic detergents (i.e., 0.05% deoxycholate) has been reported to prevent aggregation.[19] We usually use band 3 protein preparations immediately but the solution may be stored at 4° for several days without significant proteolysis or may be quickly frozen and stored at −20° for several months.

Synthesis of Aminoethyl-Sepharose[13]

Materials and Solutions

Cyanogen bromide (Pierce or Eastman)
Sepharose CL-4B (Pharmacia)
Ethylenediamine (Fisher Scientific)
0.1 M sodium bicarbonate, pH 9.0
0.05 M NaOH
0.1 M acetic acid

Procedure

Caution: Due to the toxicity of cyanogen bromide, the following procedure[22] must be carried out in a well-ventilated fume hood. Ethylenediamine (25 ml) is added to 200 ml distilled water, adjusted to pH 9.0 with 6 N HCl, and made up to a total volume of 300 ml. To activate Sepharose, 10 g of cyanogen bromide is added to 100 ml of washed resin diluted with an equal volume of water. The reaction is carried out at room temperature and the pH is maintained at 10–11 by addition of 5 N NaOH. The reaction mixture is gently shaken for 15 min and the reaction is terminated by filtering the resin on a sintered glass funnel and washing with ice-cold water. The washing should take no more than 2 min. The activated

---

[22] S. Shaltiel and Z. Er-El, *Proc. Natl. Acad. Sci. U.S.A.* **70**, 778 (1973).

Sepharose is suspended in 2 vol of ice-cold 0.1 M sodium bicarbonate, pH 9.0, and the ethylenediamine solution (300 ml) is added immediately. The coupling reaction is allowed to proceed overnight at 4° with gentle shaking. The resin is washed successively with 1 liter each of ice-cold water, 0.1 M sodium bicarbonate, pH 9.0, 0.05 M NaOH, water, 0.1 M acetic acid, and finally water. The resin is stored at 4° in an equal volume of water containing 0.02% sodium azide.

Synthesis of p-CMB-Sepharose[13]

Materials and Solutions

p-(Chloromercuri)benzoic acid (Sigma Chemical Company), (p-CMB)
1-Ethyl-3-(3-dimethylaminopropyl)carbodiimide-HCl (Pierce)
40% Dimethylformamide
0.1 M Sodium bicarbonate, pH 8.8

p-CMB-Sepharose is synthesized since commercially available resin (Affi-Gel 501) has been reported to have a high level of nonspecific binding.[13] To 50 ml of aminoethyl-Sepharose (washed with distilled water) in 90 ml of 40% dimethylformamide is added 1.0 g (2.81 mmol) of p-(chloromercuri)benzoic acid and the pH is adjusted to 4.8. Excess p-CMB is present as a precipitate below the resin slurry. Carbodiimide (1.35 g, 7.04 mmol) is added and the suspension is stirred gently for 1 hr at room temperature, maintaining the pH at 4.8 with 1 N HCl. The reaction is allowed to proceed overnight at room temperature with gentle shaking. The resin should be decanted away from the unreacted p-CMB. The resin is filtered on a coarse sintered glass funnel and washed with 4 liters of 0.1 M sodium bicarbonate, pH 8.8, over 6–8 hr and finally washed with water. The washed gel is suspended in 90 ml of 40% dimethylformamide and the coupling procedure is repeated as described above. The resin is stored at 4° in an equal volume of water containing 0.02% sodium azide and it may be used for several months.

Aminoethyl-Sepharose Chromatography[9,11]

Procedure

Band 3 protein is purified by column chromatography using aminoethyl-Sepharose (Fig. 1). Batchwise purification gives poor resolution from contaminating proteins. Aminoethyl-Sepharose (0.2–0.5 ml resin/mg solubilized protein) equilibrated with 0.1% $C_{12}E_8$, 5P8 (column buffer) is packed into a 1-cm-diameter column. Solubilized ghost membrane protein is pumped onto the column and the column is then washed with at least one bed volume of column buffer. A 10-bed volume linear 0–0.3 M NaCl gradient in column buffer is applied to the column. Band 3 protein is eluted at 100 mM NaCl. Alternatively, the column can be eluted stepwise with 0.1% $C_{12}E_8$ in 20, 50, and 100 mM sodium phosphate, pH 8.0, band 3 protein being eluted by 50 mM sodium phosphate.

DEAE-Cellulose Procedure[13]

DEAE-cellulose resin is prepared for use by washing successively with 10 vol of 0.5 M KOH, 0.4 M potassium phosphate, pH 7.5 and 0.1% $C_{12}E_8$ in 36 mM potassium phosphate, pH 7.5. A 1-cm-diameter column is packed with about 0.3 ml resin/mg solubilized protein. The solubilized ghost protein pool prepared as above is concentrated 5-fold (to about 4–5 mg protein/ml) by ultrafiltration on Amicon PM10 apparatus and made up to 36 mM sodium phosphate, pH 7.5, by addition of phosphate concentrate. It is important to increase the ionic strength of the sample in order to reduce irreversible binding of band 3 protein to the DEAE-cellulose. After pumping onto the column, washing proceeds with 0.1% $C_{12}E_8$, 36 mM sodium phosphate, pH 7.5. Band 3 protein is then eluted with 0.1% $C_{12}E_8$, 150 mM NaCl, and 150 mM sodium phosphate, pH 7.5. Based on sodium dodecyl sulfate-polyacryamide gel electrophoresis, fractions containing band 3 protein are then pooled.

p-CMB-Sepharose Chromatography[9,13]

p-CMB-Sepharose has a high binding capacity for band 3 protein (7 mg/ml[23]) so that small resin volumes may be used for chromatography. Band 3 protein from the aminoethyl column (0.5 mg protein/ml) in 100 mM NaCl, 0.1% $C_{12}E_8$, 5P8 is pumped onto a column of p-CMB-Sepharose at about 30 ml/hr. The resin is washed with 10 bed volumes of the buffer in which the band 3 protein was applied. The column is washed with 0.1% $C_{12}E_8$ in 150 mM NaCl, 150 mM sodium phosphate, pH 7.5. It is then washed with several column volumes of 0.1% $C_{12}E_8$, 5 mM sodium phosphate, pH 7.5, or with a suitably buffered detergent-containing solution into which band 3 protein needs to be exchanged. Band 3 protein is eluted with the same buffer containing 0.1–1.0% 2-mercaptoethanol (Fig. 1). Band 3 protein can also be eluted with a freshly prepared solution containing 0.1 mM cysteine.[13]

An alternative procedure involves the use of activated thiol-Sepharose

[23] A. Boodhoo and R. A. F. Reithmeier, *J. Biol. Chem.* **259**, 785 (1984).

(Pharmacia).[24] Activated thiol-Sepharose (0.4 ml swollen resin/mg protein) in distilled water is washed on sintered glass with 15 vol of column buffer (0.3 M NaCl, 1 mM sodium EDTA, 0.1% $C_{12}E_8$, 0.1 M Tris-chloride, pH 7.2), deaerated and packed into a column. Solubilized band 3 protein is pumped onto the column and it is washed with four bed volumes of the above buffer, at 5 ml/hr. Band 3 protein is eluted with column buffer containing 50 mM L-cysteine. This procedure is inferior to the p-CMB-Sepharose procedure since band 3 protein binds slowly via disulfide exchange to thiol-Sepharose.

Lectin Chromatography

Band 3 protein bears a single N-linked oligosaccharide chain, which makes it a candidate for lectin chromatography. However, because band 3 protein is found in a membrane containing large amounts of other glycoproteins, lectin chromatography is not so successful. Band 3 protein has been purified on a concanavalin A (Con A)-Sepharose column[25,26] but only with a 20% yield from the column. The low yield is probably due to the heterogeneity of band 3 protein glycosylation since band 3 protein in the unbound fraction will not bind to the column.

Affinity Chromatography[27]

Stilbene disulfonates such as 4-acetamido-4'-isothiocyanostilbene 2,2'-disulfonate (SITS) are potent inhibitors of anion exchange in red blood cells. These inhibitors bind to a single site per band 3 monomer that is accessible from the cell exterior. Immobilization of SITS on Affi-Gel 102 provides a matrix that binds band 3 protein with high affinity.[27]

Materials and Solutions

Affi-Gel 102 (Bio-Rad)
4-Acetamido-4'-isothiocyanostilbene 2,2'-disulfonate (SITS, U.S. Biochemicals, Calbiochem)
0.1 M sodium bicarbonate, pH 8.5

Procedure

Affi-Gel 102 (1 ml) resin (15 μmol -NH$_2$/ml settled gel) is washed three times with 10 vol of distilled water and once with 10 vol of sodium bicar-

[24] A. Kahlenberg, *Anal. Biochem.* **74**, 337 (1976).
[25] J. B. C. Findlay, *J. Biol. Chem.* **249**, 4398 (1974).
[26] A. H. Ross and H. M. McConnell, *J. Biol. Chem.* **253**, 4777 (1978).
[27] S. W. Pimplikar and R. A. F. Reithmeier, *J. Biol. Chem.* **261**, 9770 (1986).

bonate buffer. The resin is suspended in an equal volume of bicarbonate buffer and 16.6 mg of SITS (30 μmol) dissolved in 2 ml of bicarbonate buffer is added. The suspension is shaken gently at 37° for 1 hr. The resin is washed three times with 10 vol of bicarbonate buffer followed by water and is stored at 4° as a 1:1 suspension in water containing 0.1% sodium azide.

Small amounts of band 3 protein can be rapidly purified using the SITS-Affi-Gel resin. All steps must be carried out quickly on ice. EDTA/ KI-stripped ghosts are solubilized with 4 vol of 1% $C_{12}E_8$ (v/v) in 228 mM sodium citrate, pH 6.5–8.0 on ice. Insoluble material is removed by centrifugation at 80,000 g for 45 min. One milliliter of the solubilized protein (1.5–2 mg protein/ml) is incubated for 15 min in a microfuge tube with 25– 100 μl of affinity resin that has been washed with the same buffer. The high ionic strength buffer promotes binding of band 3 protein to the affinity resin. After removal of the supernatant, the resin is quickly washed three times with 10 vol of 0.1% (v/v) $C_{12}E_8$ in citrate buffer. Band 3 protein is eluted immediately by shaking the resin with 1–4 vol of 1 mM BADS in 0.1% (v/v) $C_{12}E_8$ in 5P8 for 15 min. Tightly bound band 3 protein can be removed by incubation with 1% (v/v) sodium dodecyl sulfate. Large-scale purification of band 3 protein using SITS-Affi-Gel and column chromatography has produced poor yields of band 3 protein. The protein binds tightly to the resin during column chromatography and elution with 1 mM BADS is ineffective.

Immobilization of Band 3 Protein

Band 3 protein can be covalently linked to p-CMB-Sepharose or CNBr-activated Sepharose 4B. The first forms linkages via cysteine residues which are reversible; however, coupling to CNBr-activated Sepharose forms irreversible bonds via lysine residues. Immobilized band 3 protein can be used to study band 3–inhibitor interactions,[23] detergent binding,[10] band 3 protein-protein interactions, and to purify antibodies.

Matrix-Bound Band 3 Oligomers and Monomers

Band 3 protein contains five sulfhydryl groups, all but one of which can be modified by N-ethylmaleimide treatment of intact cells. The remaining sulfhydryl can be labeled with p-CMBS. Band 3 protein can be attached to p-CMB-Sepharose under conditions such that only one subunit links the band 3 oligomer to the matrix. The unattached subunits can be dissociated, leaving band 3 monomers attached to the matrix. The protein can be released from the matrix by addition of reducing reagents.

Procedure

Band 3 protein is purified by aminoethyl chromatography as described above. Band 3 protein in 0.1% $C_{12}E_8$ is applied to a 1–5 ml column (1 cm diameter) of p-CMB-Sepharose at a flow rate of 1–2 ml/hr. The effluent is monitored at 280 nm with a Pharmacia UV monitor. Band 3 protein is applied to the column until the matrix is saturated (7 mg protein/ml) as judged by a plateau in the absorbance reading (usually 20 column volumes of sample is required to reach saturation). The column is washed with 5–10 column volumes of 0.1% $C_{12}E_8$, 50 mM sodium phosphate, pH 8.0, then the same buffer containing 150 mM NaCl and finally 0.1% $C_{12}E_8$, 28.5 mM sodium citrate, pH 7.4. Band 3 protein with five of its sulfhydryl groups blocked by N-ethylmaleimide treatment of cells is still capable of binding to the p-CMB matrix. Crosslinking of band 3 protein by Cu^{2+} o-phenanthroline or DIDS prelabeling of band 3 protein in cells also does not prevent band 3 protein binding to the matrix.

Matrix-bound band 3 monomers can be generated by reducing the amount of ligand on the p-CMB-Sepharose. This is done by reducing the amount of cyanogen bromide used initially to activate the Sepharose. Normal amounts of band 3 protein are bound to the column if 2.5–250 mg of cyanogen bromide/ml resin is employed during the activation reaction. No band 3 protein can be removed from resin by sodium dodecyl sulfate if the resin was prepared using 250 mg cyanogen bromide/ml of resin. As the amount of ligand on the resin is reduced the amount of band 3 protein eluted by sodium dodecyl sulfate increases, reaching a maximum of 45% when 1.25 mg of cyanogen bromide is used per milliliter of resin. Little band 3 protein can be removed by dodecyl sulfate if band 3 protein is cross-linked to a dimer before application to the column. The band 3 protein remaining on the column after washing with 0.1% dodecyl sulfate represents band 3 monomers. The matrix-bound band 3 monomer is not capable of binding BADS. Release of the monomer from the matrix by 0.1% 2-mercaptoethanol allows the protein to redimerize and then to bind BADS.

CNBr Coupling

Procedure

(As outlined in the Pharmacia product leaflet): Band 3 protein can be immobilized on CNBr-activated Sepharose. Up to 8.5 mg band 3 protein/ ml resin can be coupled, with a coupling efficiency of greater than 80% over a broad range of band 3 protein concentrations (0.2–1 mg protein/ml of coupling suspension). The amount of band 3 protein coupled is directly proportional to the amount of band 3 protein added in the 0–5 mg band 3 protein/ml resin range.

Protocol. CNBr-activated Sepharose 4B (1 g dry weight gives about 3.5 ml swollen) is washed and swollen on sintered glass with 1 mM HCl (200 ml/g dry resin). Band 3 protein solution is made up to 0.1 M sodium bicarbonate, pH 8.3, 0.5 M NaCl, 0.1% $C_{12}E_8$, by addition of concentrated stocks. The buffer should not contain any free amino groups which will react with the cyanogen bromide-activated Sepharose (i.e., Tris buffers). The band 3 protein solution is added to the washed resin and allowed to couple with end-over-end mixing for 2 hr at room temperature or 16 hr at 4°. Spent coupling solution is removed by filtration on sintered glass. Uncoupled activated sites are blocked by incubation for 2 hr at room temperature with 1 M ethanolamine or 0.2 M glycine in 0.1 M sodium bicarbonate, pH 8.3, 0.1% $C_{12}E_8$. The resin is washed with 0.5 M NaCl, 0.1 M sodium bicarbonate, pH 8.3, 0.1% $C_{12}E_8$. Determination of the amount of protein in the spent coupling solution and washes enables one to calculate the degree of band 3 protein coupling. The amount of protein coupled to the gel can also be determined directly by hydrolyzing an aliquot of the gel (25–100 μl) in 6 N HCl at 105° for 24 hr followed by amino acid analysis.

Purification of the Membrane Domain of Band 3 Protein[12]

Materials

TPCK-treated trypsin (Sigma, Boehringer-Mannheim)
Phenylmethylsulfonyl fluoride (Sigma)
n-Dodecyloctaethylene glycol monoether ($C_{12}E_8$, Nikko Chemical Co., Tokyo, Japan)
DEAE-Sepharose CL-4B (Pharmacia)

Solutions

5 mM sodium phosphate, pH 8.0 (5P8)
2 mM EDTA, pH 12
1% (v/v) $C_{12}E_8$, 5 mM sodium phosphate, pH 8.0
0.1% (v/v) $C_{12}E_8$ in 5, 20, 50, and 100 mM sodium phosphate, pH 8.0

Procedure

Ghosts are washed with 5P8 and suspended in 1 vol of the same buffer (2–3 mg protein/ml). Ghosts are treated with TPCK-trypsin (5 μg/ml) at 0° for 1 hr. The digestion is stopped by addition of phenylmethylsulfonyl fluoride to a final concentration of 1 mM. After incubation at 0° for 15 min, the membranes are washed with 5P8. Extrinsic protein fragments are removed from the membrane by extraction with 10 vol of 2 mM EDTA, pH 12. The membranes are washed twice with 5P8 and then solubilized with 5 vol of 1% (v/v) $C_{12}E_8$ in the same buffer. The extract is applied to a 5-ml column (1 × 6.5 cm) of DEAE-Sepharose CL-6B equilibrated with 0.1% $C_{12}E_8$, 5P8. After sample application, the column is washed successively with two column volumes each of 5, 20, 50, and 100 mM sodium phosphate, pH 8.0, containing 0.1% $C_{12}E_8$. The membrane domain is eluted at 100 mM sodium phosphate and has an apparent M_r of 55,000 as assayed by SDS–polyacrylamide gel electrophoresis according to Laemmli.[28]

Assays for the Native Structure of Band 3 Protein in Solution

Once purified, it is desirable to ensure that band 3 protein has not been denatured and that it is functional. This can be accomplished through reconstitution and measurement of transport function, although this technique is time consuming and not reproducible. We propose three techniques that can be used to determine the status of a band 3 protein preparation. The first involves the measurement of inhibitor binding. The determination of a binding constant for stilbene disulfonates provides a rapid, sensitive, and convenient measure of the native state of the protein and the membrane domain. Structural information on the secondary structure of the protein can be obtained by circular dichroism. Finally, the status of the cytoplasmic domain and the oligomeric structure of band 3 protein can be probed by cross-linking.

Synthesis of Stilbene Disulfonates[29]

A number of stilbene disulfonates are commercially available (DIDS from Pierce, DNDS from Aldrich, SITS from Calbiochem or US Biochemicals). 4-Benzamido-4'-aminostilbene 2,2'-disulfonate (BADS) and 4,4'-dibenzamidostilbene 2,2'-disulfonate (DBDS) can be readily synthesized by benzoylation of 4,4'-diaminostilbene 2,2'-disulfonate (DADS).

Materials and Solutions

4,4'-Diaminostilbene 2,2'-disulfonate (Kodak)
Benzoyl chloride (Aldrich)
Silica gel (Bio-Rad Bio-Sil A, 100–200 mesh)
Column buffer [pyridine–acetic acid–water (10 : 1 : 40)]
6 N HCl

[28] U. K. Laemmli, *Nature (London)* **227**, 680 (1970).
[29] A. Kotaki, M. Naoi, and K. Yagi, *Biochim. Biophys. Acta* **229**, 547 (1971).

0.4 M Sodium bicarbonate
Methanol
Ethyl ether

Procedure

The synthesis of BADS and DBDS is as described by Kotaki *et al.*[29] 4,4'-Diaminostilbene 2,2'-disulfonate (14.8 g, 40 mmol) is dissolved in 300 ml of 0.4 M sodium bicarbonate. An equimolar amount of benzoyl chloride (5.44 g) is added dropwise with vigorous stirring at room temperature. The reaction is allowed to proceed for 3 hr at room temperature. The mixture is acidified to pH 1 with 6 N HCl and the resulting precipitate is collected on filter paper. The precipitate is dissolved in 1 liter of methanol by warming to 37°. After filtering, the solution is concentrated to saturation by rotary evaporation. The crude BADS is precipitated by addition of 500 ml ethyl ether and collected on a filter paper. The precipitate should be air dried and stored in the dark. BADS and DBDS are purified by applying 0.5 g of the air-dried mixture, dissolved in 5 ml column buffer, to a fresh silica gel column (4 × 40 cm) equilibrated with column buffer. The column is eluted by column buffer at room temperature and 5-ml fractions are collected. Three peaks absorbing at 340 nm are resolved, the first small peak being unreacted DADS, the second BADS, and the third DBDS. The purity of the fractions is tested by silica gel thin-layer chromatography using column buffer as solvent. The peak fractions are pooled, concentrated under vacuum, and recrystallized from methanol and water. BADS has a maximum absorption at 343 nm with an extinction coefficient of $\varepsilon_{343} = 34$ mM^{-1} cm^{-1} while DBDS absorbs maximally at 336 nm and has an extinction coefficient of $\varepsilon_{336} = 50$ mM^{-1} cm^{-1}.

Fluorescent Assay for Stilbene Disulfonate Binding[9]

The binding of BADS and DBDS to band 3 protein in ghost membranes or after solubilization and purification can be measured using a fluorescence assay. BADS is prefered to DBDS since the binding of the first DBDS interferes with the binding of the second DBDS to band 3 dimers. This assay provides a rapid, convenient measure of the native structure of band 3 protein. BADS inhibits anion exchange in red cells with a K_i of 1 μM. A similar binding constant is obtained by the fluorescence assay in ghosts. The binding of BADS is dependent on the ionic strength of the medium, being tighter at higher ionic strength. BADS binds to band 3 protein with normal affinity in 0.1% $C_{12}E_8$, Ammonyx LO (DDAO), and octylglucoside.[9] A lower binding affinity is observed in the presence of 1% octylglucoside[8,10] or 0.1% deoxycholate,[9] while no binding is observed for band 3 protein in 0.1% sodium dodecyl sulfate[9] or 1% deoxycholate, cholate, and N,N-dimethyl-N-dodecylglycine.[8] No binding is observed if the membranes are stripped with 0.1 N NaOH or 0.1 N acetic acid.

Procedure

Steady-state fluorescence measurements are performed on a Perkin-Elmer model MPF fluorometer with a cell compartment thermostatted to 20°. A 0.3 × 0.3 cm quartz microcuvette and adapter (Farrand Optical Co., Valhalla, NY) are used to minimize inner filter effects. The excitation slit width is set at 2.5–5.0 nm. Samples (5 μl at a protein concentration of 0.5–5 mg/ml) are diluted into 200 μl of 28.5 mM sodium citrate, pH 7.4, containing detergent if desired. Fluorescence is measured by exciting the probe at 340 nm and measuring the fluorescence at 450 nm or by exciting the protein at 280 or 295 nm and measuring the fluorescence at 450 nm by energy transfer. Excitation of BADS directly gives high background fluorescence in the presence of detergents especially above the critical micellar concentration of the detergent, due to the interaction of BADS with the detergent. Fluorescence data are corrected for dilution, self-quenching of the probe, and background fluorescence of the sample and the probe. The final BADS concentrations used should vary from 0 to 100 μM.

Circular Dichroism Spectroscopy of Solubilized Band 3 Protein

The circular dichroism spectra are recorded on a Jasco model J-550-C spectropolarimeter. A minimum of four spectral scans in the far-UV range (185–260 nm) is accumulated and averaged by a Jasco model 500N data processor. Spectra are usually taken at ambient temperature (20–23°). Cells with path lengths of 0.01 or 0.05 cm are employed. Protein concentrations (0.1–1 mg/ml) are determined by the Lowry assay in the presence of 1% sodium dodecyl sulfate. The absolute protein concentrations are also determined by amino acid analysis. The Lowry protein determinations using bovine serum albumin as a standard are multiplied by the following correction factors to obtain the absolute protein concentrations: band 3 protein, 0.98, M_r 55,000 membrane domain, 0.87; M_r 41,000 cytoplasmic domain, 1.07. The buffer concentration must be kept as low as possible if spectra are desired at wavelengths less than 200 nm. Measurements in millidegrees are converted to mean residue ellipticity [θ] in degrees cm^2 dmol^{-1} using the relationship

$$[\theta]_\lambda = \theta_\lambda \mathrm{MRW}/100dc$$

where θ_λ is the observed rotation at wavelength λ in degrees, MRW is the mean residue weight (115), d is the cell path length in decimeters, and c is the concentration of the protein in grams per milliliter.

An estimate of the fraction of α-helix (f_H) is calculated[30] from the ellipticity at 222 nm using the equation

$$f_H = ([\theta]_{222} + 2340)/-30,300$$

where $[\theta]_{222}$ is the mean residue ellipticity at 222 nm. The secondary structure of band 3 protein is calculated over the wavelength range 190–240 nm according to Chen et al.[31] using CONTIN, a FORTRAN IV program developed by the Data Analysis Group, European Molecular Biology Laboratories, Heidelberg, West Germany. Band 3 protein in 0.1% $C_{12}E_8$, 50 mM sodium phosphate, pH 8.0 has an α-helical content of 46%.[12] An earlier study using Ammonyx LO gave a helical content of 43%.[11] The helical content of band 3 protein is constant between pH 5.0 and 8.0 with decreases at pH extremes.

Crosslinking of Band 3 Oligomers[32–34]

Band 3 protein in the membrane or in the solubilized state can be crosslinked via a pair of cytoplasmic sulfhydryl groups to a covalent dimer by Cu^{2+} o-phenanthroline.[32–34] The reaction must be carried out in the absence of reducing agents. The lack of crosslinking points to a change in the structure of the cytoplasmic domain of band 3 protein and is consistant with but does not prove dissociation of band 3 oligomers. Membranes at a protein concentration of 1 mg/ml in 5 mM sodium phosphate, pH 8.0 are treated with 1 mM o-phenanthroline and 0.2 mM $CuSO_4$ at 0° for 15 min.[32,33] The reaction is stopped by addition of 1 mM EDTA and 5 mM N-ethylmaleimide and the membranes are washed with 5 mM sodium phosphate, pH 8.0. Band 3 protein in detergent solution can be crosslinked under identical conditions or by 10 μM $CuSO_4$/50 μM o-phenanthroline at room temperature for 20 min.[34] Crosslinked products may be detected by separating the reaction products in the absence of reducing agents by sodium dodecyl sulfate-polyacrylamide gel electrophoresis. The crosslink can be readily reversed by 0.1% 2-mercaptoethanol.

[30] Y. H. Chen, J. T. Yang, and H. M. Martinez, *Biochemistry* **11**, 4120 (1972).
[31] Y. H. Chen, J. T. Yang, and K. H. Chau, *Biochemistry* **13**, 3350 (1974).
[32] T. L. Steck, *J. Mol. Biol.* **66**, 295 (1972).
[33] R. A. F. Reithmeier and A. Rao, *J. Biol. Chem.* **254**, 6151 (1979).
[34] J. Yu and T. L. Steck, *J. Biol. Chem.* **250**, 9176 (1975).

[33] Protein Associations with Band 3 at Cytoplasmic Surface of Human Erythrocyte Membrane

By AHMAD WASEEM and THEODORE L. STECK

Band 3 protein is the M_r 100,000 membrane-spanning polypeptide sub-unit of the oligomeric anion-exchange protein of the erythrocyte membrane.[1-3] This glycoprotein can be subdivided with proteases *in situ* to yield an integral M_r 55,000 glycopeptide and a complementary M_r 45,000 water-soluble peptide which projects into the cytoplasmic space.[4] The latter domain is not clearly involved in anion transport but instead provides attachment sites for two kinds of peripheral membrane proteins. The first class is tightly bound, i.e., not detected in the cytosol or eluted from the membrane with physiologic saline *in vitro*. This class includes ankyrin (band 2.1), which also binds to spectrin and thus links the submembrane skeleton to the membrane proper,[5] and band 4.2, a protein of unknown function.[6] The second class is loosely bound, i.e., in rapid equilibrium between band 3 protein and the cytosol and readily eluted with physiologic saline *in vitro*. This class includes three consecutive glycolytic enzymes (phophofructokinase,[7,8] aldolase,[8,9] and glyceraldehyde-3-phosphate dehydrogenase[10,11]) as well as hemoglobin.[12-14] There is sufficient band 3 protein to bind all of these glycolytic enzymes but, of course, not all of the hemoglobin. The second class of binding reactions has thus far been observed only in human red cells.

The sites of binding of ankyrin and band 4.2 protein to band 3 protein

[1] I. G. Macara and L. C. Cantley, *Cell Membr. Methods Rev.* **1**, 47 (1983).
[2] M. L. Jennings, *Annu. Rev. Physiol.* **47**, 519 (1985).
[3] H. Passow, *Rev. Physiol. Biochem. Pharmacol.* **103**, 61 (1986).
[4] T. L. Steck, B. Ramos, and E. Strapazon, *Biochemistry* **15**, 1154 (1976).
[5] V. Bennett, *Annu. Rev. Biochem.* **54**, 273 (1985).
[6] C. Korsgren and C. M. Cohen, *J. Biol. Chem.* **261**, 5536 (1986).
[7] J. D. Jenkins, F. J. Kezdy, and T. L. Steck, *J. Biol. Chem.* **260**, 10426 (1985).
[8] J. D. Jenkins, D. P. Madden, and T. L. Steck, *J. Biol. Chem.* **259**, 9374 (1984).
[9] E. Strapazon and T. L. Steck, *Biochemistry* **16**, 2966 (1977).
[10] J. Yu and T. L. Steck, *J. Biol. Chem.* **250**, 9176 (1975).
[11] H. J. Kliman and T. L. Steck, *J. Biol. Chem.* **255**, 6314 (1980).
[12] J. M. Salhany, *J. Cell. Biochem.* **23**, 211 (1983).
[13] S. N. P. Murthy, R. K. Kaul, and H. Köhler, *Hoppe-Seyler's Z. Physiol. Chem.* **365**, 9 (1984).
[14] J. A. Walder, R. Chatterjee, T. L. Steck, P. S. Low, G. F. Musso, E. T. Kaiser, P. H. Rogers, and A. Arnone, *J. Biol. Chem.* **259**, 10238 (1984).

have not been further defined. However, all of the cytoplasmic proteins in the second class appear to compete for the very amino-terminal portion of the band 3 molecule. This region of human band 3 protein is extraordinarily acidic in composition[15] and associates with these cytoplasmic proteins through salt-reversible electrostatic interactions with their binding sites for polyanionic substrates and effectors (e.g., ATP, NAD, and 2,3-diphosphoglycerate). This region of the anion transport is also both the substrate and the binding site for a membrane-bound band 3 tyrosine kinase.[16] While the physiologic function of the binding of the second class of proteins has not been explicated, the phenomenon merits our interest for at least four reasons: first, the catalytic properties of the enzymes are characteristically altered on binding; second, the binding occurs *in vivo*[8,11]; third, there may be additional members of this class yet to be identified; and fourth, similar associations of glycolytic enzymes with membranes and cytoskeletal structures have been widely observed.[17,18] The phenomenon thus seems general and may reflect an organizational principle in cell biology.

Approaches to the study of these binding reactions is the topic of this chapter.

Preparation of Ghosts and Vesicles

Erythrocytes are usually lysed osmotically and the membranes washed in 5 mM NaP$_i$ at pH 7 to 8 to yield hemoglobin-free ghosts.[19] These ghosts are permeable to proteins by virtue of the holes induced by hemolysis, but they can readily be resealed by warming in isotonic saline.[20] Resealed ghosts do not bind specifically the proteins under discussion and provide useful control material. The unsealed ghosts can be broken down into vesicles with either a predominant right-side-out or inside-out orientation.[19,21] The inverted vesicles, like the parent unsealed ghosts, provide a convenient substrate for studies of binding to the cytoplasmic aspect of band 3 protein.

[15] R. K. Kaul, S. N. P. Murthy, A. G. Reddy, T. L. Steck, and H. Köhler, *J. Biol. Chem.* **258**, 7981 (1983).
[16] A. H. Mohamed and T. L. Steck, *J. Biol. Chem.* **261**, 2804 (1986).
[17] J. E. Wilson, *Trends Biochem. Sci.* **3**, 124 (1978).
[18] B. I. Kurganov, in "Organized Multienzyme Systems: Catalytic Properties" (G. R. Welch, ed.), p. 241. Academic Press, New York, 1985.
[19] T. L. Steck and J. A. Kant, this series, Vol. 31, p. 172.
[20] M. R. Lieber and T. L. Steck, this volume [21].
[21] T. L. Steck, *Methods Membr. Biol.* **2**, 245 (1974).

Solubilization of Protein Ligands

Proteins of the second class seem to be associated with the cytoplasmic domain of band 3 protein primarily through electrostatic interactions, in that their solubilization is promoted by elevation of ionic strength and pH. Hemoglobin is the most weakly bound and is usually entirely eluted, even at low ionic strength, unless an acidic buffer is employed. The retention of the various glycolytic enzymes during ghost preparation also varies with the pH, ionic strength, buffer anion species, and extent of washing (mass action) as expected for weak, rapidly reversible associations. For example, in the washing protocol mentioned above, the loss of phosphofructokinase is complete[7] and aldolase is extensive[9] but that of glyceraldehyde-3-phosphate dehydrogenase is small (i.e., approximately 70% of this enzyme is recovered on the well-washed ghosts).[10] Band 3 tyrosine kinase is eluted at an ionic strength greater than 0.10.[16]

Ankyrin dissociation from the membrane requires the prior removal of its second binding site: spectrin. This is usually accomplished by briefly warming standard ghosts in an alkaline buffer at very low ionic strength: typically, 15–30 min at 37° in 1 mM NaP$_i$ at pH 8.[21] The extraction of spectrin destabilizes the membrane, promoting the spontaneous generation of inside-out vesicles.[19,21] The subsequent solubilization of ankyrin from spectrin-free vesicles is usually accomplished with molar salt, elevated temperature (37°), and mild denaturants, suggesting that some degree of unfolding must accompany charge screening in its elution.[22,23] Of course, all of the peripheral proteins at the cytoplasmic surface of the membrane can be extracted *in toto* with strong denaturants[24]; however, neither they nor the residual membranes may then be fit fur further binding studies.

Purification of the Protein Ligands

Convenient methods have been described for the isolation of human erythrocyte glyceraldehyde-3-phosphate dehydrogenase,[25] aldolase,[9] phosphofructokinase,[26] and ankyrin.[22,23] Tyrosine kinase[16] and band 4.2 protein[6] have been partially purified. The overwhelming abundance of hemoglobin makes its chromatographic purification trivial.

[22] V. Bennett and P. J. Stenbuck, *J. Biol. Chem.* **255**, 6424 (1980).
[23] W. R. Hargreaves, K. N. Giedd, A. Verkliey, and D. Branton, *J. Biol. Chem.* **255**, 11965 (1980).
[24] T. L. Steck, *J. Cell Biol.* **62**, 1 (1974).
[25] J. A. Kant and T. L. Steck, *J. Biol. Chem.* **248**, 8457 (1973).
[26] N. S. Karadsheh, K. Uyeda, and R. M. Oliver, *J. Biol. Chem.* **252**, 3515 (1977).

Reassociation of the Protein Ligands with Membranes

The aforementioned proteins readily bind to stripped unsealed ghosts and inside-out vesicles (but not to sealed ghosts or right side-out vesicles) under appropriate buffer conditions. For the glycolytic enzymes, hemoglobin, and tyrosine kinase, the medium must be low in ionic strength and not too alkaline. As with their dissociation, these associations proceed very rapidly (i.e., within seconds at 37° and minutes at 0°), although time should be allowed if the ligands must traverse the single hole in the ghost (15–30 min at 0°).[20] For ankyrin, a quasi-physiologic buffer is usually employed; rebinding is poor at low ionic strength.[22,23] (Since the native association of ankyrin with the membrane is stable at very low ionic strength, one must wonder whether the reconstituted complex is entirely physiological.) Ankyrin-binding rates are relatively slow; usually 60–90 min is allowed at 0°.

Free ligand is usually separated from bound by differential centrifugation with or without a density barrier containing 5–10% sucrose. Free or bound ligand (or both) are determined and the data analyzed according to Scatchard (e.g., Refs. 9, 10, 22, and 23) or simply expressed as fractional binding.

Quantitation of binding may utilize the unique optical absorbance of the Soret bands of hemoglobin or the catalytic activity of the various enzymes. Accurate measurements of binding have also been made by quantitative densitometry of stained protein bands following gel electrophoresis.[9,10] For ankyrin, radiolabeling by enzymatic phosphorylation from [^{32}P]ATP and radioiodination with labeled Bolton–Hunter reagent have been described.[27,28] Heat-denatured ankyrin (65° for 10 min) provides a measure of the nonspecific background, which is typically less than 15% and is simply subtracted from the test values.[22,23]

The outcome of such studies suggest that each band 3 monomer can bind one copy of any of the second class of ligands in competition with the others. Although these ligands are generally tetramers, they seem to bind only a single band 3 polypeptide. Both positive and negative cooperativity have been observed for the binding of glyceraldehyde-3-phosphate dehydrogenase and aldolase to red cell membranes.[10,29] We suspect that the basis for this complex behavior is not a conformational change in band 3 protein but simply attractive interactions between copies of the former

[27] V. Bennett, this series, Vol. 96, p. 313.
[28] P. W. Lee, C. J. Soong, and M. Tao, *J. Biol. Chem.* **260**, 14958 (1985).
[29] E. Strapazon and T. L. Steck, *Biochemistry* **15**, 1421 (1976).

enzyme and repulsive interactions between copies of the latter enzyme when closely juxtaposed on the cytoplasmic tails of oligomeric band 3 proteins. (See Ref. 30 for an alternative view). The stoichiometry measured for ankyrin is less straightforward. The binding capacity of inside-out vesicles is 10–15% of the number of band 3 molecules present,[22,23] although band 3 protein isolated in solution binds ankyrin with molar stoichiometry.[31]

Association of Protein Ligands with Band 3 Protein in Solution

Band 3 protein has been isolated by ion-exchange chromatography following solubilization of the membrane in the nondenaturing detergent, Triton X-100.[10,32] It appears that band 3 protein exists both in the membrane[33] and in detergent solution[34] as a mixture of monomers, dimers, and tetramers, in an association equilibrium. Under the conditions of most solution studies, the dimeric species predominates.[10] Because of the strong propensity of band 3 protein to aggregate in Triton X-100 solutions (increasing with time, temperature, ionic strength, and the acidity of the medium),[10,32] it may be more advantageous to utilize freshly extracted, partially purified band 3 protein than to risk aggregation during complete purification. Alternatively, the cytoplasmic pole of band 3 protein may be liberated in soluble form as M_r 40,000–45,000 peptides by trypsin[4] or chymotrypsin[22] digestion of inside-out vesicles and these used in binding studies.

Free ligands can be resolved from complexes by familiar analytical methods: rate zonal centrifugation in sucrose density gradients,[9,10] polyacrylamide gel electrophoresis in the absence of denaturants,[35,36] or immunoprecipitation with antibodies to one of the reactants.[22,37] In addition, band 3 protein and proteolytic fragments of its cytoplasmic domain reversibly inhibit the catalytic activity of glyceraldehyde-3-phosphate dehy-

[30] G. E. Kelley and D. J. Winzor, *Biochim. Biophys. Acta* **778**, 67 (1984).

[31] V. Bennett, *Biochim. Biophys. Acta* **689**, 475 (1982).

[32] B. J. England, R. B. Gunn, and T. L. Steck, *Biochim. Biophys. Acta* **623**, 171 (1980).

[33] R. S. Weinstein, J. K. Khodadad, and T. L. Steck, *Alfred Benzon Symp.* **14**, 35 (1980).

[34] G. Pappert and D. Schubert, *Biochim. Biophys. Acta.* **730**, 32 (1983).

[35] D. C. Weaver, G. R. Pasternack, and V. T. Marchesi, *J. Biol. Chem.* **259**, 6170 (1984).

[36] G. R. Pasternack, R. A. Anderson, T. L. Leto, and V. T. Marchesi, *J. Biol. Chem.* **260**, 3676 (1985).

[37] V. Bennett, *Nature (London)* **281**, 597 (1979).

drogenase[38] and aldolase[9,39] but activate phosphofructokinase,[7] so that binding can be studied simply by monitoring enzyme activity.

Associations of Protein Ligands with Band 3 Protein
in the Intact Red Cell

Studies on the interaction of band 3 protein with these several proteins *in vitro* have inspired attempts to document binding *in situ*. One approach to this problem was to exploit the preferential interaction of glyceralde-hyde-3-phosphate dehydrogenase with the alkylating agent, iodoacetate.[40] Cells allowed to take up the tritiated form of this compound (presumably via the anion transporter) showed selective labeling of this enzyme and, as shown by autoradiography, the tritium was concentrated close to the membrane. Evidence for hemoglobin–band 3 protein interactions *in situ* has come from fluorescence-quenching studies.[41] It should be noted that some recent studies have led to negative conclusions concerning these band 3 protein binding reactions *in vivo* (see Ref. 42), so that the question remains open.

This laboratory has measured the extent of binding of three glycolytic enzymes *in vivo* by the rapid separation of hemolysate from membranes by filtration immediately following hemolysis.[8,11] Briefly, fresh red cells are rapidly injected into a large volume of vigorously stirred buffer at 37° containing sufficient saponin to lyse the cells completely within a small fraction of a second. Over the following few seconds, multiple samples are aspirated through Millipore prefilters plus 3.0-μm nitrocellulose filters on syringes. The filtrates are analyzed for their content of the enzyme of interest and compared to the unfiltered input. It is observed that the rate of release of the glycolytic enzymes from the lysed cells lags behind that of hemoglobin, presumably reflecting the association of the enzymes with the membrane. The rate and extent of release of the enzymes increases with the ionic strength and pH of the lysis buffer; at low ionic strength, for example, initial release is followed by slow rebinding. Such data demonstrate that both the binding and debinding processes have time constants on the order of seconds under these conditions. The initial extent of binding *in vivo* is estimated by extrapolating the elution time courses back

[38] I.-H. Tsai, S. N. P. Murthy, and T. L. Steck, *J. Biol. Chem.* **257,** 1438 (1982).
[39] S. N. P. Murthy, T. Liu, R. K. Kaul, H. Köhler, and T. L. Steck, *J. Biol. Chem.* **256,** 11203 (1981).
[40] M. Solti, F. Bartha, N. Hasz, G. Toth, F. Sirokman, and P. Friedrich, *J. Biol. Chem.* **256,** 9260 (1981).
[41] J. Eisinger, J. Flores, and J. M. Salhany, *Proc. Natl. Acad. Sci. U.S.A.* **79,** 408 (1982).
[42] G. T. Rich, J. S. Pryor, and A. P. Dawson, *Biochim. Biophys. Acta* **817,** 61 (1985).

to zero time. In the best experiments, the curves generated at several ionic strengths extrapolate to a single zero-time intercept. The convergence point for fresh, normal red cells suggests that one-half to two-thirds of the glyceraldehyde-3-phosphate dehydrogenase, aldolase, and phosphofructokinase is bound in the intact cell.[8,11] Depleting the cells of ATP and 2,3-disphosphoglycerate (which displace the enzyme from the membrane *in vitro*) increases the fraction of the enzymes associated with the membranes in this assay.

Finally, the binding of glyceraldehyde-3-phosphate dehydrogenase to the membranes of intact human erythrocytes has also been demonstrated by immunofluorescent staining.[43]

[43] A. A. Rogalski, T. L. Steck, and A. Waseem, *J. Biol. Chem.* **264,** in press.

Section II

Other Mammalian Cells: Intact Cells

[34] Transport of Glutathione, Glutathione Disulfide, and Glutathione Conjugates across the Hepatocyte Plasma Membrane

By THEO P. M. AKERBOOM and HELMUT SIES

The tripeptide glutathione (GSH) is the major thiol compound in mammalian cells and functions in a variety of cellular processes.[1-5] It is involved in detoxication by its ability to reduce hydroperoxides to the corresponding alcohols with the formation of glutathione disulfide (GSSG)[6] and by its reactivity toward electrophilic toxic compounds to form glutathione S-conjugates, which can be excreted and further metabolized via the mercapturic acid pathway.[7,8] In addition, glutathione takes part in a variety of reactions, e.g., in metabolic regulation by the formation of mixed disulfides with enzymes,[9,10] or by acting as a cofactor or substrate, e.g., in the formation of peptidoleukotrienes.[11]

Although most cells possess the capacity to synthesize glutathione from its amino acid precursors, there is an active intercellular turnover, resulting in low but significant amounts of glutathione in blood plasma.[12,13] The major organ releasing glutathione is the liver,[14] which is responsible

[1] N. S. Kosower and E. M. Kosower, *Int. Rev. Cytol.* **54,** 109 (1978).

[2] H. Sies and A. Wendel (eds.), "Functions of Glutathione in Liver and Kidney." Springer-Verlag, Berlin, 1978.

[3] A. Meister and M. E. Anderson, *Annu. Rev. Biochem.* **52,** 711 (1983).

[4] N. Kaplowitz, T. Y. Aw, and M. Ookthens, *Annu. Rev. Pharmacol. Toxicol.* **25,** 715 (1985).

[5] A. Larsson, S. Orrenius, A. Holmgren, and B. Mannervik (eds.), "Functions of Glutathione: Biochemical, Physiological, Toxicological, and Clinical Aspects." Raven, New York, 1983.

[6] L. Flohé, *Curr. Top. Cell. Regul.* **27,** 473 (1985).

[7] W. B. Jakoby, *Adv. Enzymol.* **46,** 383 (1978).

[8] H. Sies and B. Ketterer (eds.), "Glutathione Conjugation: Its Mechanism and Biological Significance," Academic Press, London, 1988.

[9] D. M. Ziegler, *Annu. Rev. Biochem.* **54,** 305 (1985).

[10] R. Brigelius, *in* "Oxidative Stress" (H. Sies, ed.), p. 243. Academic Press, London, 1985.

[11] S. Hammarström, *Annu. Rev. Biochem.* **52,** 355 (1983).

[12] D. Häberle, A. Wahlländer, and H. Sies, *FEBS Lett.* **108,** 335 (1979).

[13] A. Meister, *Trends Biochem. Sci.* **6,** 231 (1981).

[14] G. M. Bartoli and H. Sies, *FEBS Lett.* **86,** 89 (1978).

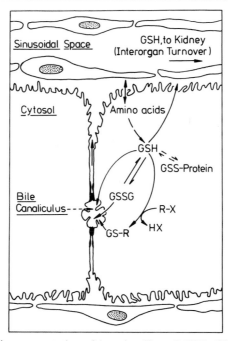

FIG. 1. Schematic representation of hepatic efflux of GSH, GSSG, and glutathione S-conjugates (thioethers). (Modified from Ref. 62).

for 90% of plasma glutathione in the rat.[15] Glutathione is mainly taken up by the kidney,[12,16] but the lung[17] and intestine[18] also show the capacity for uptake. In liver, the glutathione release corresponds to the intracellular turnover of the thiol.[14,15] A partitioning was found for the different glutathione species: GSH was mainly released into the sinusoidal space, GSSG and glutathione S-conjugates into the bile[19,20] (Fig. 1). Hepatic transport of these compounds was recently reviewed.[20a]

[15] B. H. Lauterburg, J. D. Adams, and J. R. Mitchell, *Hepatology* **4**, 586 (1984)

[16] M. E. Anderson, R. J. Bridges, and A. Meister, *Biochem. Biophys. Res. Commun.* **96**, 848 (1980).

[17] M. Berggren, J. Dawson, and P. Moldeus, *FEBS Lett.* **176**, 189 (1984).

[18] L. H. Lash, T. M. Hagen, and D. Jones, *Proc. Natl. Acad. Sci. U.S.A.* **83**, 4641 (1986).

[19] H. Sies, A. Wahlländer, and C. Waydhas, *in* "Functions of Glutathione in Liver and Kidney" (H. Sies and A. Wendel, eds.), p. 120. Springer-Verlag, Berlin, 1978.

[20] A. Wahlländer and H. Sies, *Eur. J. Biochem.* **96**, 441 (1979).

[20a] E. Petzinger, R. K. H. Kinne, and H. Sies (eds.), "Hepatic Transport of Organic Substances." Springer-Verlag, Berlin, 1989.

Assay of Glutathione in Biological Samples

Enzymatic and nonenzymatic methods for the determination of GSH, GSSG, and glutathione–protein mixed disulfides and methods of sample processing have been described previously in this series.[21–23] Newer methods and modifications to existing methods have since been described and are mentioned briefly here.

Sample processing for the assay of glutathione requires special attention since GSH is prone to autoxidation.[21] For the determination of GSSG in biological material containing appreciable amounts of GSH, the sample should be treated with a GSH-trapping agent.[24] For this purpose N-ethylmaleimide (NEM) proved to be most useful.[21,24,25] Because of interference with the assay, excess NEM must be removed. This can be achieved by chromatography[21] or by extraction with organic solvents.[25] Recently, it was suggested that the excess NEM can be selectively destroyed by alkalinization without affecting GSSG.[26]

Total glutathione (GSH + GSSG) is assayed by a kinetic method[21,25,27] in which GSH reduces 5,5'-dithiobis-(2-nitrobenzoate) (DTNB) to give the colored thiolate anion and GSSG, which is converted back to GSH by limiting amounts of glutathione reductase. Problems arising with this method may be circumvented by carrying out the assay at pH 6, in which the reduction of glutathione with DTNB is made rate-limiting instead of the glutathione reductase reaction.[28]

Glutathione and glutathione-related compounds can be determined using the HPLC technique.[29–32] Sensitive detection of GSH and GSSG without prior derivatization steps is achieved by an on-line postcolumn kinetic assay system[30] or by dual-electrode electrochemical detection.[31,32]

A simple enzymatic assay of GSH in biological samples[33] is presented here.

[21] T. P. M. Akerboom and H. Sies, this series, Vol. 77, p. 373.
[22] H. Sies and T. P. M. Akerboom, this series, Vol. 105, p. 445.
[23] M. E. Anderson, this series, Vol. 113, p. 548.
[24] S. K. Srivastava and E. Beutler, Anal. Biochem. **25,** 70 (1968).
[25] F. Tietze, Anal. Biochem. **27,** 502 (1969).
[26] P. Sacchetta, D. DiCola, and G. Federici, Anal. Biochem. **154,** 205 (1986).
[27] C. W. I. Owens and R. V. Belcher, Biochem. J. **94,** 705 (1965).
[28] P. Eyer and D. Podhradsky, Anal. Biochem. **153,** 57 (1986).
[29] D. J. Reed, J. R. Babson, P. W. Beatty, A. E. Brodie, W. W. Ellis, and D. W. Potter, Anal. Biochem. **106,** 55 (1980).
[30] A. J. Alpert and N. F. Gilbert, Anal. Biochem. **144,** 553 (1985).
[31] A. F. Stein, R. L. Dills, and C. D. Klaassen, J. Chromatogr. **381,** 258 (1986).
[32] J. P. Richie and C. A. Lang, Anal. Biochem. **163,** 9 (1987).
[33] R. Brigelius, C. Muckel, T. P. M. Akerboom, and H. Sies, Biochem. Pharmacol. **32,** 2529 (1983).

Principle

Glutathione is enzymatically converted to the 2,4-dinitrophenyl S-conjugate (DNP-SG) glutathione *S*-transferase according to the reaction

GSH + 1-chloro-2,4-dinitrobenzene

$$\rightarrow \text{2,4-dinitrophenyl-}S\text{-glutathione} + HCl$$

The formation of the glutathione conjugate can be monitored directly at 340 nm (ε_{340} = 9.6 × 10⁶ cm²/mol) or, at low levels of GSH, a dual-wavelength spectrophotometer can be used at the wavelength-pair 340–400 nm ($\varepsilon_{340-400}$ = 6.9 × 10⁶ cm²/mol) at room temperature.

Reagents

Buffer: 0.1 M potassium phosphate, pH 7.0; prepare daily from stock solutions of KH_2PO_4 (0.1 M) and K_2HPO_4 (0.1 M)

1-Chloro-2,4-dinitrobenzene (CDNB), 20 mM: Dissolve 5 mg CDNB in 1 ml ethanol

Glutathione transferase, 200 U/ml: Dialyze commerical enzyme (7 mg/ml) (Sigma, e.g., equine liver; CDNB is accepted as a universal substrate reactive with most of the isozymes) against 100 mM potassium phosphate buffer, pH 7.0, and store aliquots at −20°. The dialyzed enzyme solution is stable for several weeks

Procedure

Pipette into a cuvette 1.0 ml of buffer; sample at a volume containing 2 to 100 nmol GSH and 10 μl CDNB. Mix and record the baseline level. Start the reaction by the addition of 5 μl glutathione transferase. The reaction is completed within a few minutes.

Partitioning of Transport

Glutathione and Glutathione Disulfide

Hepatocyte polarity of transport is one determinant in the fate of glutathione synthesized in the liver cell. Transport rates of GSH, GSSG, and glutathione conjugates are different for the basolateral and canalicular domains of the plasma membrane.[19,20] Studies with perfused rat liver showed that GSH is released from the liver mainly into the sinusoidal space, with an efflux rate of about 15 nmol/(min · g) liver.[34] GSSG is found

---

[34] H. Sies, G. M. Bartoli, R. F. Burk, and C. Waydhas, *Eur. J. Biochem.* **89**, 113 (1978).

in the perfusate at low amounts[35,36] and can be accounted for by GSH autoxidation. Transport of GSH across the sinuosidal membrane is unidirectional; no uptake of GSH and GSSG is observed in the intact organ.[37]

Both GSH and GSSG are found in appreciable amounts in the bile: GSH at a low millimolar concentration, GSSG at about 0.4 mM.[35,36] Rates of biliary excretion of GSH are 1–3 nmol/(min · g) liver, for GSSG around 0.4 nmol/(min · g) liver.[35,36]

Glutathione Conjugates

For the study of glutathione conjugate transport 1-chloro-2,4-dinitrobenzene (CDNB) is a useful compound. Within the hepatocyte CDNB reacts very rapidly with GSH, as catalyzed by the glutathione transferases present in large amounts (120 U/g liver).[20] At infusion rates of CDNB up to 30 nmol/(min · g) liver, the glutathione conjugate is quantitatively excreted into the bile. At higher infusion rates, increasing amounts of the conjugate appear in the effluent perfusate. A maximum concentration of 36 mM is reached in bile.[20] Thus, biliary excretion of the glutathione conjugate is a saturable process, in which the sinuosidal transport only functions as a spillover when intracellular formation of the conjugate exceeds the biliary transport capacity.[20] Other glutathione conjugates, e.g., those formed from diethyl maleate,[38] ethacrynic acid,[39] bromosulfophthalein (BSP),[40] or menadione,[41,42] appear in the bile as well.

Hepatic uptake of glutathione conjugates has been reported for the BSP-conjugate[43,44] and for leukotriene C$_4$.[45] Uptake of BSP-glutathione (BSP-SG) is competitively inhibited by BSP, suggesting that BSP and BSP-SG are transported via a common translocator.[44] It is noteworthy that almost no uptake of dinitrophenyl-S-glutathione by the liver is observed.[20]

[35] T. P. M. Akerboom, M. Bilzer, and H. Sies, *J. Biol. Chem.* **257,** 4248 (1982).
[36] N. Kaplowitz, D. E. Eberle, J. Petrini, J. Touloukian, M. C. Corvasce, and J. Kuhlenkamp, *J. Pharmacol. Exp. Ther.* **224,** 141 (1983).
[37] R. Hahn, A. Wendel, and L. Flohé, *Biochim. Biophys. Acta* **539,** 324 (1978).
[38] J. L. Barnhart and B. Combes, *J. Pharmacol. Exp. Ther.* **206,** 614 (1978).
[39] C. D. Klaassen and T. J. Fitzgerald, *J. Pharmacol. Exp. Ther.* **191,** 548 (1974).
[40] G. Whelan and B. Combes, *J. Lab. Clin. Med.* **78,** 230 (1971).
[41] T. P. M. Akerboom, T. Bultmann, and H. Sies, *Arch. Biochem. Biophys.* **263,** 10 (1988).
[42] R. Losito, C. A. Owen, Jr., and E. V. Flock, *Biochemistry* **6,** 62 (1967).
[43] J. L. Barnhart and B. Combes, *Am. J. Physiol.* **231,** 399 (1976).
[44] L. R. Schwarz, R. Götz, and C. D. Klaassen, *Am. J. Physiol.* **239,** C118 (1980).
[45] N. Uehara, K. Ormstad, L. Örning, and S. Hammarström, *Biochim. Biophys. Acta* **732,** 69 (1983).

Glutathione monoethyl ester and related compounds are rapidly taken up by liver and other tissues. The esters are hydrolyzed within the cell, offering a means of transporting glutathione into cells.[46]

Sinusoidal Transport of Glutathione

Kinetics

In the perfused rat liver sinusoidal efflux of GSH as a function of intracellular GSH shows saturation kinetics with sigmoidal characteristics.[47] The transport rate is independent of extracellular GSH, but is inhibited by BSP-SG, probably acting from the interior of the cell. Sinusoidal efflux at normal intracellular glutathione levels is at about 80% of V_{max} (20 nmol/(min · g) liver). The K_m was estimated at 3.2 μmol/g liver.[47] Similar values were found for efflux from isolated hepatocytes.[48]

Hormonal Control

Hepatic efflux of GSH into the sinusoidal space in perfused rat liver can be modulated by hormones such as vasopressin and angiotensin II or the α-adrenergic agonists adrenaline and phenylephrine.[49] The effects of adrenaline and phenylephrine are blocked by the antagonists phentolamine and prazosin, respectively. Hormones like glucagon or dibutyryl-cyclic AMP are without effect.[49] It seems that the effects are related to calcium movements through the hepatocyte membrane. Omission of calcium from the perfusate, or perfusion with the calcium chelator EGTA, mimic the increased GSH efflux.[50,51] Electrical stimulation of the hepatic nerves around the portal vein also increases the GSH release from the perfused rat liver.

Influence of Methionine

Glutathione efflux is inhibited by methionine in isolated rat hepatocytes. The inhibition is exerted from the outside, possibly through allosteric interaction with the GSH translocator.[52] In the perfused rat liver a

[46] M. E. Anderson, F. Powrie, R. N. Puri, and A. Meister, Arch. Biochem. Biophys. 239, 538 (1985).
[47] M. Ookthens, K. Hobdy, M. C. Corvasce, T. Y. Aw, and N. Kaplowitz, J. Clin. Invest. 75, 258 (1985).
[48] T. Y. Aw, M. Ookthens, C. Ren, and N. Kaplowitz, Am. J. Physiol. 13, G236 (1986).
[49] H. Sies and P. Graf, Biochem. J. 226, 545 (1985).
[50] P. Graf and H. Sies, in "Calcium-Dependent Processes in the Liver" (C. Heilmann, ed.), p. 191. MTP Press, Lancaster, England, 1988.
[51] H. Höke, H. Krell, and E. Pfaff, Arch. Toxicol. 44, 23 (1980).
[52] T. Y. Aw, M. Ookthens, and N. Kaplowitz, Am. J. Physiol. 251, G354 (1986).

stimulation[53] has been reported. The reason for the discrepancy is not known. In our hands, addition of methionine to the perfusate leads to an increased release of thiols, but these are not identical with glutathione.[50]

Studies with Plasma Membrane Vesicles

The elaboration of techniques to isolate plasma membrane fractions of sinusoidal (basolateral) and canalicular origin[54–58] enabled kinetic studies of the transport of glutathione and glutathione-related compounds independent of cellular factors. Vesicles obtained from these membrane fragments have been reported to preserve their physiological orientation,[56,58,59] that is, the cytosolic surface inside and the luminal surface outside. Carrier-mediated transport for GSH, GSSG, and DNP-SG has been identified in both these membrane preparations. Until now only uptake studies have been described. This means that the direction of transport that was studied is reverse to the physiological direction.

In sinusoidal membrane vesicles, kinetic analysis of GSH transport revealed two transport systems, a high-affinity system with a K_m of 0.3 mM and a V_{max} of 4.2 nmol/(min · mg) and a low-affinity system with a K_m of 3.3 mM and a V_{max} of 11.2 nmol/(min · mg).[60] Transport was inhibited by GSSG and benzyl-S-glutathione. Sinusoidal transport of GSH was reported to exhibit no cation specificity,[60] but one report seems to show that transport was K^+ dependent.[61] The vesicles also transported GSSG and DNP-SG, showing mutual interference between all glutathione species, more in line with the existence of only one carrier system.

Canalicular Transport of Different Glutathione Species

Kinetics

The biliary excretion of GSH in rat liver showed no saturation kinetics, but rather a linear relation between the efflux rate and the intracellular

[53] B. H. Lauterburg, *Eur. J. Clin. Invest.* **16**, 494 (1986).
[54] J. M. M. van Amelsvoort, H. J. Sips, M. E. A. Apitule, and K. van Dam, *Biochim. Biophys. Acta* **600**, 950 (1980).
[55] M. Inoue, R. Kinne, T. Tran, and I. M. Arias, *Hepatology* **2**, 572 (1982).
[56] M. Inoue, R. Kinne, T. Tran, L. Biempica, and I. M. Arias, *J. Biol. Chem.* **258**, 5183 (1983).
[57] P. J. Meier, E. S. Sztul, A. Reuben, and J. L. Boyer, *J. Cell Biol.* **98**, 991 (1984).
[58] B. L. Blitzer and C. B. Donovan, *J. Biol. Chem.* **259**, 9295 (1984).
[59] P. J. Meier, A. St. Meier-Abt, C. Barrett, and J. L. Boyer, *J. Biol. Chem.* **259**, 10614 (1984).
[60] M. Inoue, R. Kinne, T. Tran, and I. M. Arias, *Eur. J. Biochem* **138**, 491 (1984).
[61] T. Y. Aw, M. Ookthens, J. F. Kuhlenkamp, and N. Kaplowitz, *Biochem. Biophys. Res. Commun.* **143**, 377 (1987).

GSH content at values up to 8 μmol/g liver was observed.[36] Phenobarbital pretreatment of the rats enhanced biliary efflux by about 250% without influencing sinusoidal GSH efflux.

For GSSG, a linear relationship between intracellular GSSG and biliary GSSG release was observed without saturation at intracellular values up to 200 μM, about seven times above the normal value.[35] In contrast to GSH, the biliary concentration of GSSG is higher than hepatic GSSG concentrations, with concentration gradients of up to 50. This indicates that biliary efflux of GSSG is an active transport process. Indeed, biliary GSSG release is dependent on intracellular ATP levels,[62] a phenomenon also observed in rat erythrocytes[63] and perfused heart.[64]

Biliary transport of GSSG, induced by the addition of hydroperoxides, and transport of glutathione conjugates like DNP-SG[65] and BSP-SG[66] show mutual competition, suggesting that these compounds share a common translocator. However, no inhibition of endogenous GSSG excretion was observed with these conjugates.[36,67]

Also, the transport of other cholephilic compounds like taurocholate and bilirubin was inhibited during increased GSSG transport.[68] Inhibition of taurocholate transport by GSSG was also observed with canalicular plasma membrane vesicles.[69]

Biliary GSH release is not inhibited by GSSG[34,66,68] nor by BSP-SG,[36,66,67] suggesting that GSH transport occurs via a different transport system. On the other hand, the cholephilic compounds indocyanine green, DBSP, and BSP inhibit biliary excretion of GSH and GSSG in parallel,[67] and BSP also competes with BSP-SG[43] for biliary transport. Thus the question is still open whether one multispecific translocator or different transport mechanisms exist for the transport of the different glutathione species.

In hepatocytes in primary culture, efflux of glutathione conjugates was studied by preloading the cells during prior incubation with the conjugate precursor at 10°, which prevents transport of the conjugate during its synthesis.[70] Raising the temperature of the medium to 37° started the

[62] H. Sies, in "Glutathione: Storage, Transport and Turnover in Mammals" (Y. Sakamoto, T. Higashi, and N. Tateishi, eds.), p. 63. Jpn. Sci. Soc. Press, Tokyo, 1983.
[63] S. K. Srivastava, Y. C. Awasthi, and E. Beutler, Biochem. J. 139, 289 (1974).
[64] T. Ishikawa, M. Zimmer, and H. Sies, FEBS Lett. 200, 128 (1986).
[65] T. P. M. Akerboom, M. Bilzer, and H. Sies, FEBS Lett. 140, 73 (1982).
[66] B. H. Lauterburg, C. V. Smith, H. Hughes, and J. R. Mitchell, J. Clin. Invest. 73, 124 (1984).
[67] N. Ballatori and T. W. Clarkson, Am. J. Physiol. 248, G238 (1985).
[68] T. P. M. Akerboom, M. Bilzer, and H. Sies, J. Biol. Chem. 259, 5838 (1984).
[69] J. C. Griffiths, H. Sies, P. J. Meier, and T. P. M. Akerboom, FEBS Lett. 213, 34 (1987).
[70] G. Lindwall and T. D. Boyer, J. Biol. Chem. 262, 5151 (1987).

efflux. A V_{max} of 0.15 nmol/(min · mg protein) and a K_m of 0.58 mM was measured for the transport of DNP-SG.[70] Intracellularly generated p-nitrobenzyl-S-glutathione acted as a competitive inhibitor. Efflux of DNP-SG from hepatocytes occurred against a concentration gradient, indicating that transport is energy dependent, in line with earlier findings with antimycin A.[71]

Oxidative Stress

Oxidation of intracellular glutathione is increased during increased hydroperoxide metabolism, as detected by increased levels of GSSG. Excretion of the disulfide into bile is enhanced correspondingly,[35] and may serve as an indicator of oxidative stress.[22] Indeed, a number of drugs able to generate high amounts of hydrogen peroxide within the cell, such as nitrofurantoin,[35] nifurtimox,[72] paraquat,[73] diquat,[66] quinones like menadione,[41] and aminopyrine in phenobarbital-pretreated rats,[74] induce elevated GSSG levels in bile. Further, compounds oxidizing GSH without the formation of hydrogen peroxide like diamide,[68] or thiourea,[75] phenylthiourea,[75] and methimazole[75] increase biliary GSSG efflux.

Biliary excretion of GSSG is also dependent on the oxygenation of the tissue. Under hypoxic conditions biliary GSSG is substantially lowered.[41,76]

Carbon tetrachloride, chloroform, or acetaminophen, known to cause lipid peroxidation, do not increase biliary excretion of GSSG.[66]

Possible Limitations for Assays of Glutathione in Bile

In bile GSH is subject to rapid autoxidation, which necessitates bile sampling at low temperatures[35] or in metaphosphoric acid.[77] Due to the biliary and cannula dead space, GSH autoxidation may still lead to overestimated values for GSSG, the extent depending on the biliary GSH/GSSG ratio and bile flow. Assuming a $t_{1/2}$ of GSH of 5 min,[77] a cannula volume of 6 μl,[68] a biliary dead space of 2.5 μl/g liver,[68] and a bile flow of 1.3 μl/(min · g) liver,[68] GSH autoxidation could amount to about 20%.

[71] L. R. Schwarz, K.-H. Summer, and M. Schwenk, *Eur. J. Biochem.* **94,** 617 (1979).
[72] M. Dubin, S. N. J. Moreno, E. E. Martino, R. Docampo, and A. O. M. Stoppani, *Biochem. Pharmacol.* **32,** 483 (1983).
[73] R. Brigelius and M. S. Anwer, *Res. Commun. Chem. Pathol. Pharmacol.* **31,** 493 (1981).
[74] N. Oshino and B. Chance, *Biochem. J.* **162,** 509 (1977).
[75] P. A. Krieter, D. M. Ziegler, K. E. Hill, and R. F. Burk, *Mol. Pharmacol.* **26,** 122 (1984).
[76] S. W. Cummings, K. E. Hill, R. F. Burk, and D. M. Ziegler, *Biochem. Pharmacol.* **37,** 967 (1988).
[77] D. Eberle, R. Clark, and N. Kaplowitz, *J. Biol. Chem.* **256,** 2115 (1981).

However, evidence has been presented that autoxidation within the dead space is much less.[77]

Pretreatment with AT-125, an irreversible inhibitor of γ-glutamyltransferase, resulted in increased concentrations of biliary glutathione, indicating that part of the glutathione excreted is metabolized during its passage through the biliary tract.[78,79] For studying canalicular glutathione transport in anesthetized rats, care must be taken that bile is not contaminated with pancreatic juice.[78]

Studies with Canalicular Plasma Membrane Vesicles

In canalicular membrane vesicles, GSH transport shows an apparent K_m of 0.33 mM and a V_{max} of 4.4 nmol/(min · mg), and was enhanced by a K^+-diffusion potential (vesicle inside positive).[80] Transport was inhibited by probenecid, and by the glutathione species GSSG and S-benzyl-glutathione. GSSG transport showed a K_m of 0.4 mM and a V_{max} of 1.1 nmol/(min · mg).[81] For DNP-SG transport the K_m was 1.0 mM and V_{max} was 5.1 nmol/(min · mg).[82] Also the conjugate transport proved to be electrogenic and was inhibited by GSH, GSSG, or S-benzyl-glutathione. DNP-SG conjugate transport was not inhibited by 1 mM BSP-SG.[82]

Concluding Remarks

Comparison with Other Systems

In the basolateral membrane of intestinal epithelial cells a sodium-dependent electrogenic transport of GSH is involved in intracellular detoxication systems that function in the protection against chemically induced injury.[18] Sodium-dependent uptake of GSH was also described for basolateral membrane vesicles derived from rat kidney proximal tubules.[83] Furthermore, there is a sodium-dependent uptake of the glutathione conjugate S-(1,2-dichlorovinyl)glutathione, in an electrogenic and probenecid-sensitive manner.[84] Mutual inhibition between the conjugate and GSH suggest that the two compounds are transported by the same system. It is unlikely that a similar system is operative in the basolateral membrane of the liver cell, since the liver is an organ that releases GSH; no sodium-dependent transport of GSH was found in liver cells.[60,61]

[78] W. A. Abbott and A. Meister, *Proc. Natl. Acad. Sci. U.S.A.* **83**, 1246 (1986).

[79] N. Ballatori, R. Jakob, and J. L. Boyer, *J. Biol. Chem.* **261**, 7860 (1986).

[80] M. Inoue, R. Kinne, T. Tran, and I. M. Arias, *Eur. J. Biochem.* **134**, 467 (1983).

[81] T. P. M. Akerboom, M. Inoue, H. Sies, R. Kinne, and I. M. Arias, *Eur. J. Biochem.* **141**, 211 (1984).

[82] M. Inoue, T. P. M. Akerboom, H. Sies, R. Kinne, T. Tran, and I. M. Arias, *J. Biol. Chem.* **259**, 4998 (1984).

[83] L. H. Lash and D. P. Jones, *Biochem. Biophys. Res. Commun.* **112**, 55 (1983).

[84] L. H. Lash and D. P. Jones, *Mol. Pharmacol.* **28**, 278 (1985).

Extensive studies concerning glutathione transport have been carried out with human erythrocytes. The erythrocyte exports glutathione disulfide[85] and glutathione conjugates.[86–88] GSH was not found in the medium.[85] Transport was further investigated using inside-out vesicles. GSSG transport was shown to be ATP driven, and GSH was not transported.[89] Glutathione disulfide transport showed low- and high-affinity characteristics,[89] of which the latter was inhibited by the glutathione conjugate DNP-SG.[90] The DNP-SG transport was ATP dependent.[91] However, glutathione disulfide was not able to inhibit DNP-SG transport,[91,92] indicating that different transport systems for glutathione disulfide and glutathione conjugates may exist.[92]

ATPase activities have been detected in the erythrocyte plasma membrane, which are stimulated by glutathione disulfide[93] and by glutathione conjugates.[94] Two GSSG-stimulated ATPases were separable by affinity chromatography, one with a low K_m and one with a high K_m for GSSG,[93] similar to the transport activities. It is feasible that the ATPases are involved in the transport of the different forms of glutathione. Interestingly, ATPase activities stimulated by GSSG and glutathione conjugates have also been detected in the plasma membrane of the liver.[95,96]

Significance of Glutathione Transport

The liver releases GSH mainly into the systemic blood, as part of an interorgan turnover, thereby maintaining a thiol redox balance in the plasma. The transport is stimulated upon addition of hormones like vasopressin, phenylephrine, and epinephrine. It was suggested that the hepatic glutathione transport may function in the protection against increased generation of reactive oxygen species in the blood caused by

[85] S. K. Srivastava and E. Beutler, *J. Biol. Chem.* **244**, 9 (1969).
[86] P. G. Board, *FEBS Lett.* **124**, 163 (1981).
[87] Y. C. Awasthi, G. Misra, D. K. Rassin, and S. K. Srivastava, *Br. J. Haematol.* **55**, 419 (1983).
[88] K.-G. Eckert and P. Eyer, *Biochem. Pharmacol.* **35**, 325 (1986).
[89] T. Kondo, G. L. Dale, and E. Beutler, *Proc. Natl. Acad. Sci. U.S.A.* **77**, 6359 (1980).
[90] T. Kondo, M. Murao, and N. Taniguchi, *Eur. J. Biochem.* **125**, 551 (1982).
[91] E. F. Labelle, S. V. Singh, S. K. Srivastava, and Y. C. Awasthi, *Biochem. J.* **238**, 443 (1986).
[92] E. F. Labelle, S. V. Singh, S. K. Srivastava, and Y. C. Awasthi, *Biochem. Biophys. Res. Commun.* **139**, 538 (1986).
[93] T. Kondo, Y. Kawakami, N. Taniguchi, and E. Beutler, *Proc. Natl. Acad. Sci. U.S.A.* **84**, 7373 (1987).
[94] E. F. Labelle, S. V. Singh, H. Ahmed, L. Wronski, S. K. Srivastava, and Y. C. Awasthi, *FEBS Lett.* **228**, 53 (1988).
[95] P. Nicotera, M. Moore, G. Bellomo, F. Mirabelli, and S. Orrenius, *J. Biol. Chem.* **260**, 1999 (1985).
[96] P. Nicotera, C. Baldi, S.-A. Svensson, R. Larsson, G. Bellomo, and S. Orrenius, *FEBS Lett.* **187**, 121 (1985).

inflammatory processes or during extreme physical exercise. Indeed, under such conditions GSH levels in liver and other organs were significantly decreased.[97,98] Much has to be learned about the detailed mechanisms of the processes.

It is known that glutathione conjugates may exert inhibitory effects on some enzyme activities like glutathione transferase[99] and glutathione reductase.[100] Efficient biliary disposal of these conjugates is therefore of extreme importance. This becomes especially critical during conditions of oxidative stress accompanied by increased canalicular GSSG transport. GSSG could depress transport of the conjugate and the accumulation of the conjugate could then be amplified by its inhibitory effect on GSSG reductase.

Injected leukotriene C_4 is rapidly taken up by the liver and metabolized via the mercapturic acid pathway.[101] The major product found in the bile depends on the species studied and is in the rat initially LTD_4, at a later stage N-acetyl-LTE_4 and ω-oxidized metabolites predominate. This pathway operates during various types of tissue trauma, including surgical interventions, endotoxin shock, and virus-induced hepatitis.[101]

Acknowledgment

Supported by the Deutsche Forschungsgemeinschaft, Grants Ak 8/1-1 and Si 255/8-1.

[97] P. C. Bragt and I. L. Bonta, *Agents Actions* **10**, 536 (1980).
[98] H. Lew, S. Pyke, and A. Quintanilha, *FEBS Lett.* **185**, 262 (1985).
[99] I. Jakobson, M. Warholm, and B. Mannervik, *J. Biol. Chem.* **254**, 7985 (1979).
[100] M. Bilzer, R. L. Krauth-Siegel, R. H. Schirmer, T. P. M. Akerboom, H. Sies, and G. Schulz, *Eur. J. Biochem.* **138**, 373 (1984).
[101] M. Huber and D. Keppler, *in* "Glutathione Conjugation: Its Mechanism and Biological Significance" (H. Sies and B. Ketterer, eds.), 449. Academic Press, London, 1988.

[35] Ca^{2+} Fluxes and Phosphoinositides in Hepatocytes

By P. F. BLACKMORE and J. H. EXTON

Introduction

Vasopressin, angiotensin II, α_1-adrenergic agonists (epinephrine, norepinephrine), and P_2 purinergic agonists (ATP, ADP) exert their effects on liver and certain other tissues by raising the Ca^{2+} concentration in the cytosol.[1] On the other hand, glucagon and β-adrenergic agonists exert

[1] J. H. Exton, *Adv. Cyclic Nucleotide Protein Phosphorylation Res.* **20**, 211 (1986).

their effects on liver by raising cAMP, although they also increase cytosolic Ca^{2+}. The alterations in liver cell Ca^{2+} induced by these agents can be monitored by measuring Ca^{2+} fluxes across the plasma membrane, changes in the total cellular content of Ca^{2+}, or changes in cytosolic Ca^{2+}.

Hormones increase cytosolic Ca^{2+} in their target cells by mobilizing intracellular Ca^{2+} stores and by altering the flux of Ca^{2+} across the plasma membrane. There is much evidence that the mobilization of internal Ca^{2+} is caused by inositol 1,4,5-trisphosphate generated by the breakdown of phosphatidylinositol 4,5-bisphosphate in the plasma membrane.[1,2] This breakdown occurs very rapidly in liver and other tissues[1,3] and inositol 1,4,5-P$_3$ has been demonstrated to release Ca^{2+} from endoplasmic reticulum stores in permeabilized hepatocytes and other cells.[3] Many other inositol phosphates have been identified in liver and other tissues, including inositol 1,3,4-P$_3$[4,5] and inositol 1,3,4,5-P$_4$.[5,6] Their functions are presently unknown.

Measurement of myo-Inositol 1,4,5-trisphosphate

The most commonly used method for measuring the level of inositol 1,4,5-P$_3$ in liver is to incubate hepatocytes or inject animals with myo-[2-³H]inositol for a period of time (1.5 hr or longer) to label the inositol phospholipids to isotopic equilibrium. Acid extracts of hepatocytes are then subjected to column chromatography on Dowex 1-X8. Columns are washed with increasing concentrations of ammonium formate to elute the various inositol phosphates followed by counting the radioactivity in the fractions. Although useful for routine studies, this method does not allow the separation of inositol phosphate isomers such as inositol 1,4,5-P$_3$ and inositol 1,3,4-P$_3$. For this, HPLC methods are used. Both methods are described below.

Isolation and Loading of Hepatocytes with myo-[2-³H]Inositol

Hepatocytes can be isolated by collagenase digestion[7] from rats injected intraperitoneally with 500 μCi of myo-[2-³H]inositol as described below. Cell suspensions (~40 to 50 mg wet wt/ml) are incubated in

[2] M. J. Berridge and R. F. Irvine, *Nature (London)* **312**, 315 (1984).
[3] J. R. Williamson, R. H. Cooper, S. K. Joseph, and A. P. Thomas, *Am. J. Physiol.* **248**, C203 (1985).
[4] R. F. Irvine, A. J. Letcher, D. J. Lander, and C. P. Downes, *Biochem. J.* **223**, 237 (1984).
[5] C. A. Hansen, S. Mah, and J. R. Williamson, *J. Biol. Chem.* **261**, 8100 (1986).
[6] I. R. Batty, S. R. Nahorski, and R. F. Irvine, *Biochem. J.* **232**, 211 (1985).
[7] P. F. Blackmore and J. H. Exton, this series, Vol. 109, p. 550.

Krebs–Henseleit bicarbonate buffer containing 1.5% (w/v) gelatin (Difco) and are continuously gassed with O_2/CO_2 (19 : 1). For labeling *in vitro*, hepatocytes (40-ml cell suspension in 250-ml Erlenmeyer flasks) are incubated for 90 or 120 min with 0.1 mM *myo*-[2-^{3}H]inositol (25 μCi/ml of cell suspension). The cells are then filtered through nylon mesh to remove any clumps of cells which may form. The cells are then washed once and resuspended in fresh medium without *myo*-[2-^{3}H]inositol or gelatin for use in experiments. The gelatin is omitted from the buffer at this stage since this protein is difficult to remove with trichloroacetic acid. After incubation for various periods under the experimental conditions desired, 1-ml aliquots are pipetted into tubes containing 0.2 ml of 100% (w/v) trichloroacetic acid at 0°. After centrifugation (5000 g for 10 min) the supernatant fluid is extracted six times with 2-ml aliquots of diethyl ether, saturated with water, to remove trichloroacetic acid. The last traces of diethyl ether are removed by blowing a stream of N_2 into the solutions which are then stored at $-20°$ before analysis. The neutralized extracts are subjected to ion-exchange chromatography or HPLC as follows.

Dowex 1-X8 Chromatography of Inositol Phosphates

One milliliter of 10 mM sodium tetraborate is added to the neutralized extract and then the entire solution is applied to 2 ml of Dowex 1-X8 (formate form) ion-exchange resin contained in P-5000 Pipetman disposable pipet tips (Rainin Instrument Co.). The resin is held in place by a small plug of fiberglass wool (Corning Glass Works). A small plug of fiberglass wool is also placed on top of the resin to prevent it from being disturbed by the eluting solutions of ammonium formate. The [^{3}H]inositol phosphates are eluted as follows. Free [^{3}H]inositol is removed from the column by washing with 15 ml of water, glycerophosphoinositol is removed by 2 × 5 ml of 5 mM sodium tetraborate/60 mM ammonium formate, inositol P_1 is eluted with 2 × 5 ml of 0.2 M ammonium formate/ 0.1 M formic acid, inositol P_2 is eluted with 3 × 5 ml of 0.4 M ammonium formate/0.1 M formic acid, inositol P_3 is eluted with 5 ml of 0.7 M ammonium formate/0.1 M formic acid, and inositol P_4 is eluted with 5 ml of 2.0 M ammonium formate/0.1 M formic acid.[8] To each eluate, consisting of 5 ml in 25-ml plastic scintillation vials, is added 15 ml of Beckman Ready-Solv EP and the contents are mixed and counted. Typically, inositol P_1 contained ~9000 cpm/ml cell suspension, inositol P_2 ~1200 cpm/ml cell suspension, inositol P_3 ~150 cpm/ml cells suspension, and inositol P_4 ~20

---

[8] M. J. Berridge, R. M. C. Dawson, C. P. Downes, J. P. Heslop, and R. F. Irvine, *Biochem. J.* **212**, 473 (1983).

cpm/ml cell suspension. For more detailed analyses, a linear gradient of ammonium formate (0–2.0 M) in 0.1 M formic acid can be applied to the column.

HPLC Analysis of Inositol Phosphates

Examination of the [^{3}H]inositol phosphates of labeled hepatocytes and other tissues by HPLC[5,9,10] reveals the presence of inositol bisphosphate isomers, inositol 1,3,4-P$_3$, and inositol 1,3,4,5-P$_4$. Samples (1.0 to 2.0 ml) of extracts from hepatocytes obtained as described above are applied to a Whatman Partisil 10 SAX anion-exchange column together with a RCSS silica Guard-Pak precolumn (Millipore) and the inositol phosphates are eluted as follows. HPLC is performed using a Beckman 421A controller, two 114 M solvent delivery modules, and a 165-variable-wavelength detector. The solvent for pump A is degassed, filtered H$_2$O (Millipore type RA, 1.2 μm). The solvent for pump B is degassed, Millipore-filtered 2.0 M ammonium formate buffered to pH 3.7 with orthophosphoric acid or 4.0 M ammonium formate buffered to pH 3.2 with formic acid. The column is washed with water at a flow rate of 1.5 ml/min (pump A) for 7 min. Between 7 and 13 min the eluant is increased linearly from 0 to 50% solvent B, then held at 50% solvent B until 18 min. The eluant is increased linearly from 50 to 100% solvent B between 18 and 28 min, then held at 100% solvent B until 34 min. The gradient is then decreased to 100% solvent A between 34 and 38 min. Fractions are collected at 18-sec intervals using a 2112 Redirac fraction collector (LKB) 2 min after commencement of the gradient program. The carousel used holds 100 miniscintillation vials, so that scintillant can be added directly and the contents counted in a scintillation counter. When extracts of cells are chromatographed, the eluate is passed through the wavelength detector and the absorbance monitored at 254 nm to monitor the nucleotides. The UV profile serves as a means of determining reproducibility and to verify that the column is not overloaded. If too much salt is present in samples, the UV-absorbing material and most of the ^{3}H-containing material elute very early in the gradient.

Figure 1 shows a typical HPLC profile. The location of ATP in the fractions coincides with the position of [^{3}H]inositol 1,3,4-P$_3$. The position of [^{3}H]inositol 1,4,5-P$_3$ is verified by chromatography of [^{3}H]inositol 1,4,5-P$_3$ obtained from Amersham or New England Nuclear. Inositol 1,3,4,5-P$_4$

[9] R. F. Irvine, E. E. Anggard, A. J. Letcher, and C. P. Downes, Biochem. J. 229, 505 (1985).
[10] S. J. Stewart, V. Prpic, F. S. Powers, S. B. Bocckino, F. E. Isaacks, and J. H. Exton, Proc. Natl. Acad. Sci. U.S.A. 83, 6098 (1986).

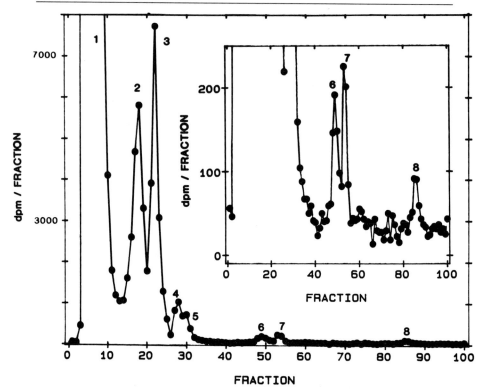

FIG. 1. HPLC analysis of hepatocyte [³H]inositol phosphates formed after 60 sec of 10^{-7} M vasopressin stimulation. Hepatocytes labeled with myo-[2-³H]inositol (American Radiolabeled Chemicals, Inc.) were incubated for 60 sec with 10^{-7} M vasopressin. A 1-ml aliquot of cells (~40 mg wet wt/ml) was deproteinized with trichloroacetic acid, extracted with diethyl ether, and then chromatographed on an HPLC column (Whatman Partisil 10 SAX anion-exchange column) as described in the text. A representative elution profile is shown. Peak identification: 1, inositol; 2, glycerophosphorylinositol; 3, inositol monophosphates; 4, inositol 1,4-P_2; 5, inositol 3,4-P_2; 6, inositol 1,3,4-P_3; 7, inositol 1,4,5-P_3; and 8, inositol 1,3,4,5-P_4. The inset is an expanded plot shown to highlight inositol 1,3,4-P_3, inositol 1,4,5-P_3, and inositol 1,3,4,5-P_4.

appears in the column profile between fractions 80 and 90. Its position can be determined by chromatography of [³H]inositol 1,3,4,5-P_4 from the above sources. For greater resolution of the inositol P_1 and inositol P_2 isomers, a linear or nonlinear gradient of ammonium phosphate can be employed with a Whatman Partisil 10 SAX column[11] or an Alltech Adsorbosphere column.[12]

[11] N. M. Dean and J. D. Moyer, *Biochem. J.* **242**, 361 (1987).
[12] T. Balla, A. J. Baukal, G. Guillemette, R. O. Morgan, and K. J. Catt, *Proc. Natl. Acad. Sci. U.S.A.* **83**, 9323 (1986).

Inositol 1,4,5-P_3 and other inositol phosphates can also be measured chemically by taking fractions from Dowex 1-X8 chromatography or HPLC and desalting and neutralizing them before conversion to *myo*-inositol by alkaline phosphatase.[13] The *myo*-inositol is then quantitated by conversion to *scyllo*-inosose by *myo*-inositol dehydrogenase, with NADH generation being measured fluorimetrically after amplification by cycling.[13,14] This method can measure picomole quantities of inositol phosphates.[13,14]

When larger amounts of inositol phosphates are present, the liberated *myo*-inositol can be measured by gas chromatography of its trimethylsilyl derivative.[15] Another method involves separation of inositol 1,4,5-P_3 by thin-layer chromatography on polyethyleneimine cellulose, extraction with NH_4OH, and acid hydrolysis of the dried extract.[16] The released phosphate is then quantitated using the Malachite Green-phosphomolybdate reaction.[16]

There are also binding assays based on intracellular receptors for inositol 1,4,5-P_3. These receptors can be purified from several tissues, but the one most commonly used for assays is from bovine adrenal cortex.[17] An assay kit based on this system which can measure 0.2–25 pmol is commercially available (Amersham).

Calcium Fluxes in Hepatocytes

Calcium fluxes in hepatocytes can be measured in three basic ways: (1) by monitoring the uptake or efflux of ⁴⁵Ca²⁺, (2) by measuring changes in total cell calcium, and (3) by monitoring changes in cytosolic free Ca²⁺. Methods (2) and (3) generally yield the most reliable quantitative information and will be described in the greatest detail.

Measurement of ⁴⁵Ca²⁺ Fluxes in Hepatocytes

Measurements of ⁴⁵Ca²⁺ uptake or efflux are frequently the easiest ways to examine hormone effects on Ca²⁺ fluxes in hepatocytes. However, the information derived may be misleading because the cells have several intracellular Ca²⁺ pools with which the isotope exchanges at different rates, and because hormones affect not only the fluxes of Ca²⁺ across the plasma membrane, but also the release and uptake of Ca²⁺ by

[13] J. A. Shayman, A. R. Morrison, and O. H. Lowry, *Anal. Biochem.* **162**, 562 (1987).
[14] L. G. MacGregor and F. M. Matschinsky, *Anal. Biochem.* **141**, 382 (1984).
[15] S. E. Rittenhouse and J. P. Sasson, *J. Biol. Chem.* **260**, 8657 (1985).
[16] R. H. Underwood, R. Greeley, E. T. Glennon, A. I. Menachery, L. M. Braley, and G. H. Williams, *Endocrinology* **123**, 211 (1988).
[17] G. Guillemette, T. Balla, A. J. Baukal, A. Spat, and K. J. Catt, *J. Biol. Chem.* **262**, 1010 (1987).

internal organelles. Thus some form of compartmental analysis is necessary to interpret the changes in $^{45}Ca^{2+}$ validly and, even then, the conclusions are limited because of the complexity of the system and the difficulty of attaining Ca^{2+} homeostasis prior to $^{45}Ca^{2+}$ uptake studies (especially after hormonal perturbation) and of attaining complete isotopic equilibrium prior to $^{45}Ca^{2+}$ efflux studies. In addition, there is usually uncertainty about the anatomical location of the various compartments.

The most complete investigations of hormonal effects on hepatocyte Ca^{2+} fluxes utilizing $^{45}Ca^{2+}$ are those of Barritt and associates[18,19] and Borle and associates.[20] The methods and analyses employed by these workers will not be described in this chapter, but any reader wishing to use $^{45}Ca^{2+}$ to study Ca^{2+} fluxes in any cell is strongly recommended to follow their procedures. As discussed at length elsewhere, simple measurements of hormonally induced changes in $^{45}Ca^{2+}$ in cells and/or incubation media are liable to give misleading information about the actual alterations in Ca^{2+} fluxes.[21,22]

Measurements of Changes in Total Hepatocyte Ca^{2+} Using Atomic Absorption Spectroscopy

Another approach to analyzing hormone effects on hepatocyte Ca^{2+} fluxes is to measure the changes in total cell Ca^{2+} using atomic absorption spectroscopy.[23,24] Isolated hepatocyte suspensions used for such studies must be equilibrated at 37° in Krebs–Henseleit bicarbonate buffer, pH 7.4, with continuous gassing ($O_2 : CO_2$, 19 : 1) for at least 10 min after isolation. This allows reaccumulation of Ca^{2+} by the cells since the collagenase digestion is carried out with perfusate containing ~50 μM Ca^{2+}. This equilibration period also allows intracellular K^+ and Na^+ to be restored to normal levels.

For time-course experiments, 10–20 ml of hepatocyte suspension is incubated with constant shaking (70–90 rpm) in 250-ml polypropylene or polycarbonate Erlenmeyer flasks with continuous gassing with $O_2 : CO_2$ (19 : 1) at a flow rate of 3–4 liters/min. For dose–response experiments, 2

[18] G. J. Barritt, J. C. Parker, and J. C. Wadsworth, *J. Physiol. (London)* **312**, 29 (1981).

[19] J. C. Parker, G. J. Barritt, and J. C. Wadsworth, *Biochem. J.* **216**, 51 (1983).

[20] R. K. Studer and A. B. Borle, *Biochim. Biophys. Acta* **762**, 302 (1983).

[21] A. B. Borle, this series, Vol. 39, p. 513.

[22] J. R. Williamson, R. H. Cooper, and J. B. Hoek, *Biochim. Biophys. Acta* **639**, 243 (1981).

[23] F. D. Assimacopoulos-Jeannet, P. F. Blackmore, and J. H. Exton, *J. Biol. Chem.* **262**, 2662 (1977).

[24] P. F. Blackmore, F. T. Brumley, J. L. Marks, and J. H. Exton, *J. Biol. Chem.* **253**, 4851 (1978).

to 3 ml of hepatocyte suspension is incubated in 25-ml polycarbonate or polypropylene Erlenmeyer flasks. Aliquots ranging between 0.5 and 1.0 ml are removed at appropriate times and layered on 10 ml of an ice-cold solution of 150 mM NaCl, 10% (w/v) sucrose, and 0.5 mM EGTA (pH 7.4) contained in 12-ml conical Pyrex test tubes. The tubes are then centrifuged for ~20 sec at ~3000 rpm to sediment the cells in either an IEC HN-SII centrifuge (Damon/International Equipment Division) with a 958 rotor or an IEC size 2 model SBV centrifuge with a 240 rotor and a mechanical foot brake.

After centrifugation, the tubes are inverted and allowed to drain for ~5 min. The insides of the tubes are then rinsed with distilled water and allowed to drain for a further 5 min, after which the insides of the tubes are wiped dry with facial tissues. The cell pellets are then dispersed into 0.5 ml of water using a vortex mixer, and 0.5 ml of 0.6 M $HClO_4$ containing 0.1% (w/v) $LaCl_3 \cdot 6H_2O$ is added to precipitate protein. The solutions are then centrifuged at 2000 g for 1 min and the Ca^{2+} in the supernatant fluid is determined by atomic absorption spectroscopy using an atomic absorption spectrophotometer (Perkin-Elmer model 603). Industrial grade acetylene is the fuel and air is used as the oxidant. Calcium standards ranging up to 150 μM are prepared in $HClO_4/LaCl_3$ diluent. Typically, basal Ca^{2+} readings are 0.03 to 0.04 absorbance units when the wet weight of the cell suspensions is ~40 mg/ml. Contamination of the cells with extracellular Ca^{2+} rarely exceeds 3–5% of cell Ca^{2+}, thus making correction for extracellular Ca^{2+} unnecessary in most experiments.

Since the effects of many hormones on net cell Ca^{2+} content are relatively small, duplicate samples should be taken and incubations carried out in duplicate or triplicate. For separation of cells and medium 7% (w/v) bovine serum albumin (Pentex) can be used instead of sucrose, and 1% (w/v) $LaCl_3 \cdot 6H_2O$ can be used instead of 0.5 mM EGTA to displace extracellular bound Ca^{2+}. The same results are obtained with either method; however, sucrose is more economical and more effectively minimizes the mixing of the incubation medium with the cell pellet.

Measurement of Hormone Effects on Free Cytosolic Ca^{2+} ($[Ca^{2+}]_i$) in Hepatocytes

An important parameter to measure in hepatocytes when analyzing hormone effects is cytosolic free Ca^{2+} ($[Ca^{2+}]_i$). By using fluorescent Ca^{2+} chelators (e.g., quin2 and fura2), which can be introduced into cells, changes in $[Ca^{2+}]_i$ can be measured.[25,26] When Ca^{2+} is bound to quin2 and

[25] R. Y. Tsien, T. Pozzan, and T. J. Rink, *J. Cell Biol.* **94**, 325 (1982).
[26] G. Grynkiewicz, M. Poenie, and R. Y. Tsien, *J. Biol. Chem.* **260**, 3440 (1985).

fura2 there are 5- and 30-fold increases in fluorescence, respectively. Quin2 and fura2 are introduced into cells as the membrane-permeant tetraacetoxymethyl and pentaacetoxymethyl esters, respectively (quin2/AM and fura2/AM). Intracellular esterases hydrolyze quin2/AM and fura2/AM to yield the membrane-impermeant free acids quin2 and fura2, which accumulate in the cytosol and thus permit the continuous measurement of $[Ca^{2+}]_i$.

Hepatocytes loaded with quin2 and fura2 can be used to measure hormone effects on $[Ca^{2+}]_i$ as follows. Hepatocytes are isolated as previously described by collagenase digestion of the liver. Cells (~50 mg wet wt/ml) are preincubated (5 ml in 25-ml polypropylene Erlenmeyer flasks) for 5–10 min at 37° before quin2/AM or fura2/AM is added. The stock solutions of quin2/AM and fura2/AM are 50 and 1 mM, respectively, in dimethylsulfoxide (Me$_2$SO). Quin2/AM (10 μl) or fura2/AM (20 μl) stock solution is added to 5 ml of cell suspension and incubated for 15 min with continuous gassing (O$_2$: CO$_2$, 19 : 1). Control cells are incubated with an equivalent amount of Me$_2$SO. After 15 min of incubation, the cells are sedimented (50 g for 1 min) and resuspended in 7 ml of fresh Krebs–Henseleit bicarbonate buffer containing 5 mg/ml gelatin and 0.5 or 2.5 mM Ca^{2+}, depending on the experimental protocol. Following a 5- to 15-min incubation with continuous gassing, 3 ml of cell suspension is added to a 12 × 50 mm borosilicate glass test tube containing a 0.9-cm-long, 0.2-cm-diameter stirring bar passing through the middle of a triangular shaped piece of polypropylene (each side ~1 cm long). Although many spectrofluorimeters specifically designed to measure certain intracellular ion changes are now commercially available (see below), a Varian model SF-330 spectrofluorometer fitted with a magnetic stirrer (Rank Brothers, Bottisham, England) and a thermostatically controlled cell housing can be employed. The cells are stirred and oxygenated by infusing O$_2$:CO$_2$ (19 : 1) through a polyethylene tube (Intramedic PE50, Clay Adams). Hormones, drugs, and agents are injected through another piece of PE50 tubing (~40 cm long) into the hepatocyte suspension, avoiding the need to open the cell housing. Routinely, 130 μl of the agent to be added is drawn up into a No. 1750 Hamilton syringe. Following injection, 30 μl of agent is added to the cells, with 100 μl still remaining in the syringe needle and PE50 tubing (dead space).

For the quin2-loaded hepatocytes, the excitation and emission wavelengths are 340 and 500 nm, respectively, with a slit width of 10 nm. At the emission wavelength of 500 nm there is a substantial contribution of pyridine nucleotide reduction when hormones are added.[27] By increasing the

[27] R. Charest, P. F. Blackmore, B. Berthon, and J. H. Exton, *J. Biol. Chem.* **258**, 8769 (1983).

emission wavelength to 520 nm, the contribution to the pyridine nucleotide fluorescence can be eliminated. However, this results in a slight loss in sensitivity. The spectrofluorometer output (millivolts) can be recorded directly on a strip chart recorder, but if averaging of several experiments is required it is better to channel the signal into an analog/digital converter and store the data in a computer. For our experiments we use a Hewlett Packard 18652A A/D converter and the computer is a Hewlett Packard 3356 laboratory automation system. The data are usually collected for 30 sec to 1 min before hormone injection and then for various times up to 30 min after hormone addition, with data in millivolts being sampled twice each second. Another advantage of using a computer is that changes in

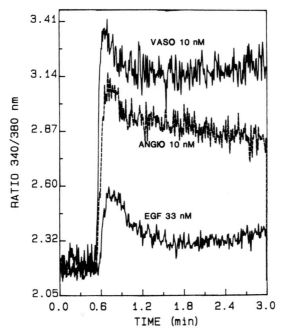

FIG. 2. Effect of vasopressin (vaso), angiotensin II (angio), and epidermal growth factor (EGF) on hepatocyte [Ca^{2+}]$_i$ using fura2. Hepatocytes were loaded with fura2 as described in the text. Agonists were added at 0.5 min and the changes in fluorescence measured at excitation wavelengths of 340 and 380 nm on two separate aliquots of cells. The data are expressed as the fluorescence values at 340 nm divided by the values at 380 nm. The traces shown are representative (in duplicate) of a single batch of hepatocytes. Autofluorescence[26,27] was subtracted in these experiments. The resting [Ca^{2+}]$_i$ calculated using Eq. (1) was 146 nM and the maximum value obtained with vasopressin was 270 nM. In another experiment (not shown) the resting level was 92 nM while the stimulated level was 293 nM. The K_d, R_{min}, R_{max}, and (S_{f2}/S_{b2}) values used were those given in Ref. 26. However, as detailed in this reference, these values should be determined for each fluorometer due to instrumental variations in sensitivity.

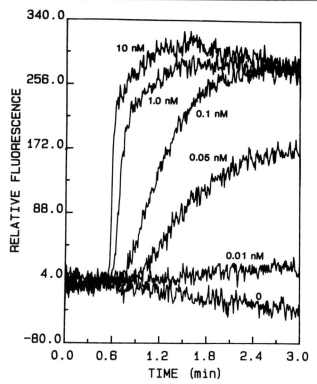

FIG. 3. Dose response of vasopressin to increase $[Ca^{2+}]_i$ in quin2-loaded hepatocytes. Hepatocytes were loaded with quin2 as described in the text. Agonists were added at 0.5 min. Each curve represents the mean of three separate traces. Each trace was normalized such that the time of hormone addition was zero fluorescence. For this to be done, a base point mean value was obtained over the 5 sec preceding the hormone injection, and this value then subtracted from all other values in the data file. Calculating the base point mean over 5 sec tends to reduce the noise level, since 10 values are averaged (slice width normally is 2/sec). For this experiment the resting $[Ca^{2+}]_i$ was 210 nM and the maximum $[Ca^{2+}]_i$ obtained with 10 nM vasopressin was 550 nM using Eq. (2). The data illustrate that the rate, magnitude, and lag (time taken before $[Ca^{2+}]_i$ increases) are dose dependent. The concentration of vasopressin for each curve is given in nanomolar units. (P. F. Blackmore, unpublished observations, 1986.)

control fluorescence (non-quin2-loaded cells) can be subtracted from the fluorescence in quin2-loaded cells.

There are now many spectrofluorometers which have been specifically designed to measure $[Ca^{2+}]_i$ in either suspensions of cells or in single cells. The software packages that come with these instruments are designed to be very user friendly. Some of the more widely used spectrofluorometers are manufactured by SPEX Industries, Inc. (Edison, NJ), Photon Technology International, Inc. (Princeton, NJ), and Tracor, Inc. (Austin, TX), although there are other excellent fluorometers available.

For the fura2-loaded hepatocytes, the two excitation wavelengths used are 340 and 380 nm, while the emission wavelength is 500 nm. If the instrument used is not capable of alternating excitation wavelengths rapidly and calculating the ratio of fluorescence intensities, then two separate runs need to be performed. To this end, the data at both wavelengths are stored in the computer which generates the ratio between the two raw data files (340 nm/380 nm). The [Ca^{2+}]$_i$ is then calculated using Eq. (1):

$$[Ca^{2+}]_i = K_d[(R - R_{min})/(R_{max} - R)](S_{f2}/S_{b2}) \quad (1)$$

The apparent K_d value used in the equation is 224 nM, R is the observed ratio (340 nm/380 nm), while R_{min} and R_{max} are the minimum and maximum ratios when the [Ca^{2+}] is zero and saturating, respectively. R_{max} can be determined in hepatocytes by adding 50 μM ionomycin to cells incubated in the presence of external Ca^{2+}. This will cause a large net uptake of Ca^{2+} into the cells, which will then saturate the dye. R_{min} can then be measured by adding 5 mM EGTA to the same cells treated with ionomycin. This will cause a slow release of intracellular Ca^{2+} and a decrease in fura2 fluorescence. The value (S_{f2}/S_{b2}) is the fluorescence of the free dye at 380 nm (S_{f2}) divided by the fluorescence of the Ca^{2+}-bound dye at 380 nm (S_{b2}). A full explanation of each of these terms is given in Grynkiewicz et al.[26] Figure 2 shows the data from a representative experiment using fura2-loaded hepatocytes stimulated with several hormones.

To quantitate the fluorescence measurements in quin2-loaded cells, the cells are lysed with Triton X-100 (10 mg/ml) after the fluorescence measurements are made and the fluorescence with saturating Ca^{2+} (at least 1.0 mM Ca^{2+}, F_{max}) and low Ca^{2+} (1.0 mM EGTA in excess of Ca^{2+} or 1.0 mM Mn^{2+}, F_{min}) are measured. The [Ca^{2+}]$_i$ is calculated using Eq. (2):

$$[Ca^{2+}]_i = K_d[(F - F_{min})/F_{max} - F)] \quad (2)$$

The apparent K_d value used in the equation is 115 nM, which was determined under ionic conditions which are similar to cytosol, namely high K$^+$ and low Na$^+$.[25] Figure 3 shows the effects of several concentrations of vasopressin on [Ca^{2+}]$_i$ in quin2-loaded hepatocytes; similar data are obtained with angiotensin II and epinephrine as agonists.[27]

[36] Magnesium Transport in Eukaryotic and Prokaryotic Cells Using Magnesium-28 Ion

By ROBERT D. GRUBBS, MARSHALL D. SNAVELY, S. PAUL HMIEL, and MICHAEL E. MAGUIRE

Introduction

The two major intracellular divalent cations are Mg^{2+} and Ca^{2+}. The latter has been extensively studied with the result that the intracellular free Ca^{2+} concentration is well accepted as a major regulator of several important functions, including stimulus–secretion coupling and muscle contraction, and as a second messenger for neurohormones.[1,2] In sharp contrast, Mg^{2+} has received relatively little attention. Available techniques for measuring intracellular free Mg^{2+} concentration are extremely limited.[3] Its transport across the plasma membrane is not completely understood, and even less is known about its transport across intracellular membranes. This paucity of information about Mg^{2+} biochemistry and physiology can be attributed in large part to two factors: a perception that techniques for measuring Mg^{2+} fluxes and intracellular concentrations are unavailable and a general feeling that intracellular Mg^{2+} is not particularly important, that it primarily serves to form MgATP or to stabilize membranes rather than playing any major intracellular regulatory role.

Within the last few years, several laboratories have provided data which speak to some of these issues. Assay techniques using the isotope ^{28}Mg for measurement of Mg^{2+} fluxes are now well established.[4–8] Further, the delineation of hormonal regulation of Mg^{2+} influx[4,5,7] and discovery that free Mg^{2+} concentration in the physiological range can be a major determinant of enzyme activity[9–16] suggests strongly that Mg^{2+} has impor-

---

[1] A. K. Campbell, "Intracellular Calcium: Its Universal Role as Regulator." Wiley, Chichester, England, 1983.

[2] W. Y. Cheung (ed.), "Calcium and Cell Function," Vols. 1–5. Academic Press, New York, 1980–1985.

[3] R. Y. Tsien, *Annu. Rev. Biophys. Bioeng.* **12**, 91 (1983).

[4] D. A. Elliott and M. A. Rizack, *J. Biol. Chem.* **249**, 3985 (1974).

[5] M. E. Maguire and J. J. Erdos, *J. Biol. Chem.* **255**, 1030 (1980).

[6] J. J. Erdos and M. E. Maguire, *Mol. Pharmacol.* **18**, 379 (1980).

[7] J. J. Erdos and M. E. Maguire, *J. Physiol. (London)* **337**, 351 (1983).

[8] J. C. Henquin, T. Tamagawa, M. Nenquin, and M. Cogneau, *Nature (London)* **301**, 73 (1983).

[9] S. J. Bird and M. E. Maguire, *J. Biol. Chem.* **253**, 8826 (1978).

[10] J. Bockaert, B. Cantau, and M. Sebben-Perez, *Mol. Pharmacol.* **26**, 180 (1984).

tant regulatory roles within the cell.[11,13,17–19] This chapter presents the methodology developed in our laboratory over the last several years for the measurement of Mg^{2+} flux and intracellular compartmentation using $^{28}Mg^{2+}$.

General Properties of $^{28}Mg^{2+}$

Source and Isotopic Properties

Magnesium-28 is the only radioactive magnesium isotope of use for biological studies since all other magnesium radioisotopes have half-lives of less than 10 min. In addition, no other cation can be substituted for Mg^{2+} at its transport site, at least in eukaryotic cells examined to date.[20] Currently, ^{28}Mg can be obtained only through the Isotope Distribution Laboratory of Brookhaven National Laboratories, which synthesizes and distributes below cost a large number of rare isotopes. Magnesium-28 can also be synthesized by cyclotron reactions, but the yield is so small as to be virtually useless unless one's laboratory is literally next door.[21]

Magnesium-28 is supplied as $MgCl_2$ in 0.9% saline at a concentration of 0.1 mCi/ml and a specific activity of >30 $\mu Ci/\mu g$ Mg. The half-life of ^{28}Mg is generally given as 21.3 hr, although values of 20.8–21.8 hr have been reported.[22] Decay of ^{28}Mg to ^{28}Al occurs initially by emission of an electron (0.46 MeV) and subsequent emission of gamma radiation (0.032, 0.40, 0.95, and 1.35 MeV). The daughter ^{28}Al decays with a half-life of 2.3 min to the stable isotope ^{28}Si, again by emission of an electron (2.86 MeV) and subsequent emission of 1.78 MeV gamma radiation. The isotope as

---

[11] S. Y. Cech, W. C. Broaddus, and M. E. Maguire, *Mol. Cell. Biochem.* **33**, 67 (1980).

[12] P. W. Flatman and V. L. Lew, *J. Physiol. (London)* **307**, 1 (1980).

[13] L. Garfinkel and D. Garfinkel, *Magnesium* **4**, 60 (1985).

[14] R. Iyengar and L. Birnbaumer, *Proc. Natl. Acad. Sci. U.S.A.* **79**, 5179 (1982).

[15] E. Racker, *in* "Chemiosmotic Proton Circuits in Biological Membranes" (V. P. Skulachev and P. C. Hinkle, eds.), pp. 377–394. Addison-Wesley, Reading, Massachusetts, 1981.

[16] M. Terasaki and H. Rubin, *Proc. Natl. Acad. Sci. U.S.A.* **82**, 7324 (1985).

[17] G. M. Walker, *Magnesium* **5**, 9 (1986).

[18] M. E. Maguire, *Trends Pharmacol. Sci.* **5**, 73 (1984).

[19] R. D. Grubbs and M. E. Maguire, *Magnesium* **6**, 113 (1987).

[20] R. D. Grubbs, C. A. Wetherill, K. Kutschke, and M. E. Maguire, *Am. J. Physiol.* **248**, C51 (1984).

[21] The University of Munich is also capable of producing ^{28}Mg upon request. Costs and shipping schedules are arranged on an order-by-order basis. Customs and shipping problems do not make this a reliable source for United States customers. Interested European users should contact one of the authors (M.E.M.) for further information.

[22] "Radiological Health Handbook," Doc. 017-011-00043-0, pp. 237–238. U.S. Public Health Service, Washington, D.C., 1970.

produced is usually free of contamination by other isotopes, although small amounts of $^{32}P/^{33}P$ (<0.05%) have been occasionally observed. Synthesis capability is 2–5 mCi/run.[23]

Because of the cost and relative unavailability of ^{28}Mg, we have rarely used it to develop assay procedures for Mg^{2+} uptake in different cell types. Our experience is that ^{86}Rb in eukaryotic cells and ^{60}Co in prokaryotic cells are excellent models. Influx and efflux protocols that work with these isotopes have, in our hands, needed little or no alteration for use with ^{28}Mg. We specifically do not use ^{45}Ca as it binds inordinately to cell surfaces, procedures for its use do not work particularly well with ^{28}Mg, and most importantly, Mg^{2+} transport properties do not resemble those of Ca^{2+} at all.

^{28}Mg Detection

Magnesium-28 can be easily detected by either liquid scintillation counting, Cerenkov counting, or gamma spectrometry. All three methods require very little sample preparation. The choice of method depends only on other isotopes that may be present and on the physical nature of the sample. Isotope can be measured in a gamma spectrometer using a wide-open window at an efficiency of 90–100%. In our laboratory, gamma counting is usually reserved for filter paper samples which are placed in the bottom of appropriate test tubes or other sample holders. The major disadvantage of gamma counting is that the background count is relatively high, since a wide-open window is necessary to detect the variety of gamma rays emitted; a major advantage is avoidance of β radiation from $^{32}P/^{33}P$ that may be present in some shipments.

Cerenkov counting in a scintillation counter can be used to detect the high-energy electron emissions from decay of ^{28}Mg and ^{28}Al and is useful primarily for small tissue samples. It is not necessary in our experience to

---

[23] Orders for ^{28}Mg should be placed through the Isotope Distribution Office, Oak Ridge National Laboratories, P.O. Box X, Oak Ridge, Tennessee 37830 [(615) 574-6984]. Department of Energy form ER-391 must be used for ordering and is obtainable from the above address. Orders will be forwarded to Brookhaven National Laboratories. Orders sent directly to Brookhaven are not accepted. The isotope is sold in nominal units of 50 μCi *at time of receipt*. It is produced every other Monday during operating periods of the linear accelerator (normally October 1 through June 1) and shipped usually by air express on Monday evenings to arrive Tuesday mornings. Potential users should specifically request shipment by commercial shippers because of the 21-hr half-life of the isotope. While passenger airlines will accept such shipments if marked for "medical" use, they often refuse to carry any radioisotopes whatsoever. Further information on ^{28}Mg can be obtained from the Isotope Distribution Office at Oak Ridge National Laboratories or from one of the authors (M.E.M.) of this chapter.

take the time to solubilize completely a few milligrams of tissue. Crude dispersion is sufficient to obtain reproducible results and saves time, an important consideration with a short-lived isotope. Sample preparation consists of tissue maceration and resuspension in 5.0 ml of 0.2 N NaOH or KOH. The major disadvantage of this method is a low efficiency of about 30%.

Most samples generated in this laboratory are cell pellets in microfuge tubes and are routinely counted in a liquid scintillation counter. The liquid above the cell pellet is aspirated and the cap of the tube cut off before placing it in a scintillation vial and adding 10 ml of scintillation fluid. Our standard scintillation fluids are 1 : 1 or 2 : 1 mixtures of toluene and Triton X-100 (purchased in 55-gallon barrels; scintillation grade is *not* necessary) with 4 g/liter diphenyloxazole (PPO). The vial is then capped, vigorously shaken, and counted. Because the electrons from decay of *both* ^{28}Mg and ^{28}Al are counted and some of the gamma emissions are also counted at low efficiency, the overall counting efficiency is approximately 200% and background is low.

Finally, it is important to correct for decay of ^{28}Mg *during* counting. Because of the short half-life, a set of 100 samples counted for 5 min each will take over 6 hr to count. Sample 100 will then have decayed by about 24% compared to sample 1. We routinely use a simple computer program to correct each sample individually for decay using the first sample counted as the zero time.[24]

Assay of Mg^{2+} Transport in Cells in Suspension Culture

^{28}Mg^{2+} Influx

The following procedure for measuring the uptake of ^{28}Mg^{2+} is useful for determination of both initial influx rates and simple accumulation of isotope. It has been used successfully with the following cell lines or types: murine S49 lymphoma, GM86 (Friend erythroleukemia), rat splenic T lymphocytes, and human T lymphocytes.

---

[24] Of interest is the observation that, in some experiments, in our haste to get a look at the data, we have counted samples within 5 min of separation of intra- and extracellular cation. Later counting of the same samples often shows an initial slight but significant *increase* in apparent counts per minute. We attribute this phenomenon to the transport of ^{28}Mg^{2+} without (presumably) the concomitant transport of the daughter isotope ^{28}Al^{3+}. Counting of a sample <5 min later does not allow sufficient time for the two isotopes to come back into equilibrium. As they do so, more ^{28}Al is detected, resulting in a slight increase in observed counts. This is apparently one of the many "joys" of working with ^{28}Mg.

Buffers

A. Incubation buffer: 140 mM NaCl (8.182 g/liter), 5.4 mM KCl (0.403 g/liter), 25 mM D-glucose (Dextrose) (4.5 g/liter), 20 mM Na–HEPES, pH 7.4 at 37° (20 ml/liter of 1 M stock) (pH adjusted with NaOH), 1.8 mM CaCl$_2$ (1.8 ml/liter of 1 M stock), 0.1 mM MgCl$_2$ or MgSO$_4$ (0.1 ml/liter of 1 M stock); osmolarity of 315–325 mOsm

B. Wash buffer: 140 mM NaCl, 5.4 mM KCl, 20 mM Na–HEPES, pH 7.4 at 4°, 1.8 mM CaCl$_2$, 2 mM MgCl$_2$; osmolarity of 300–310 mOsm

1. The desired quantity of cells (e.g., 250 ml of 1.5 × 10^6 cells/ml) is centrifuged in a tabletop centrifuge for 4 min at 200 g at room temperature. The supernatant is aspirated and each cell pellet gently resuspended in approximately 5 ml of incubation buffer at 37° by gentle trituration. The combined cell pellets are diluted to 50 ml with incubation buffer and recentrifuged. After this wash step is repeated once more, the cells are resuspended in incubation buffer to the desired volume, poured into a small wide-mouth polypropylene bottle or other suitable container, and placed in a shaking water bath (100 cpm) for 10–15 min before starting the experiment to warm the cells to 37°. This step is important in order to assume that temperature (and therefore *rate* of uptake) does not change during the experiment. It also allows time to determine the cell concentration and percentage viability.

2. The assay is initiated by the addition of ^{28}Mg^{2+} to the cells (or occasionally vice versa). With S49 cells and 0.1 mM extracellular Mg^{2+}, 3–5 × 10^5 cpm/ml ^{28}Mg^{2+} provide a good signal-to-noise ratio. The cells are quickly but vigorously swirled and a zero-time point taken in the desired number of replicates (below). The zero-time point, usually obtained within 20–30 sec of isotope addition, represents rapid binding of the isotope to the plasma membrane and a small amount of uptake. This value is subtracted from subsequent time points in calculating net uptake.

3. At the desired time, the cells are again mixed by vigorous swirling. Replicates of 1.0 ml are removed, placed in large (1.4 ml) microcentrifuge vials, and centrifuged approximately 5 sec in a Beckman Microfuge B (see Comments, below).

4. The supernatant is aspirated, leaving no overlying liquid. Using an automatic pipet, the pellet is resuspended in 1.0 ml ice-cold wash buffer by gentle trituration in the pipet tip. The tubes are recapped and again centrifuged for 5 sec. The supernatant is again aspirated and the resuspension step repeated. The cells are centrifuged a third and final time for 5–7 sec, and the supernatant carefully aspirated with particular attention to removing droplets adhering to the walls of the tube. The tubes are then

placed in scintillation vials and the next time point is ready to start. With duplicate samples (and a little practice) this routine can be done in 1.5–2 min. With triplicate samples 2.5–3 min is needed. At the end of the experiment, the caps of the microcentrifuge tubes are cut off and scintillation fluid added for counting.

5. For each tube or vessel containing cells and ^{28}Mg^{2+}, triplicate 0.1-ml aliquots are taken to determine the counts per million per unit volume in the assay. The specific activity of the ^{28}Mg as supplied is sufficiently high that the concentration of radioactive ^{28}Mg^{2+} is usually about 100 nM compared to a total Mg^{2+} concentration of usually 100 μM. Thus the actual specific activity is determined by the amount of nonradioactive Mg^{2+} added.

Silicone Oil Microcentrifugation Assay

We have also successfully used sedimentation through a silicone oil layer to obviate the need for washing steps in the above assay protocol. The procedure is identical to the above procedure until the centrifugation step. Microcentrifuge tubes are prepared beforehand and contain 100 μl of 0.25 M sucrose made 1% (v/v) with Triton X-100 overlaid with 400 μl silicone oil (General Electric Versilube F50). A 0.5-ml aliquot of cells is laid over the oil and the tube centrifuged for 30 sec in a Beckman Microfuge B. The entire aqueous supernatant and the majority of the oil layer are aspirated; the tube with the remaining oil and the entire sucrose layer are then counted as described above. The advantage of this procedure is that it is somewhat more rapid to perform than the washing protocol so that closer time points and more complicated experimental manipulations are facilitated. The disadvantages are that replication is slightly poorer, zero-time values are slightly greater, the sample size is halved, resulting in a poorer signal to noise ratio, and multiple tubes must be prepared in advance. In addition, we have been unable to purchase moderate (less than tank car loads) amounts of oil and have had to rely upon sometimes grudgingly donated bottles, usually of 1 liter, an amount that does not last very long.

Comments

Magnesium-28 ion uptake is linear over the range from 1 to 40 $\times$ 10^6 cells/ml. The best replication is usually obtained at 4 to 8 $\times$ 10^6 cells/ml. If cell viability is under 90%, the cells are not used because of increased variability. The number of washes was set at two after measuring retained intracellular ^{28}Mg^{2+} as a function of washing. About 98% of extracellular radioactivity is removed in the first wash. The second wash is primarily needed to reduce variability. Three through six washes do not remove any significant additional amount of isotope. Washing the cells at 4° gives better reproducibility than washing at 37° or at room temperature but does not alter the amount of uptake measured.

Most culture media contain 0.8–1.0 mM Mg^{2+}. The concentration is reduced in the incubation buffer to provide a relatively greater specific activity of $^{28}Mg^{2+}$. In some cell types, however, this may promote significant loss of cell Mg^{2+} over a period of 1–2 hr; for example, compare $^{28}Mg^{2+}$ accumulation and compartmentation in S49 lymphoma cells[25] versus rat adipocytes.[4]

It is important to maintain a constant osmolarity and pH in the assay. Alterations in extracellular pH markedly affect influx rate.[7,26] HEPES buffer is temperature sensitive and the pH must be measured and adjusted at the temperatures stated or appropriate corrections must be made in the pH adjustment (HEPES temperature coefficient = −0.014 pH units/°C). We have also observed that, despite care in making up stock solutions and making dilutions, the osmolarity of the final incubation and wash buffers varies sufficiently to alter uptake kinetics and cell viability, especially in long time-course experiments. We routinely measure osmolarity of each buffer and adjust to the stated osmolarity with 3 M NaCl or KCl.

Because we routinely are concerned with transport of divalent cations, our normal buffers omit phosphate to avoid divalent cation chelation and occasional precipitation of calcium phosphate [presumably due to the lowered level of Mg^{2+} (0.1 mM) in the incubation buffer]. We have, however, measured uptake in Dulbecco's modified essential medium (DMEM) in which $NaHCO_3$ has been replaced by addition of 20 mM NaCl and 20 mM Na-HEPES, pH 7.4 at 37° with or without the addition of up to 1% dialyzed horse serum. Precipitation of phosphate has not been a problem in this medium. We prefer the simplified medium to minimize interactions between test agents and medium components.

The brand of microcentrifuge is unfortunately critical. Of the various microcentrifuges available, only the Beckman model B is able to attain maximum speed rapidly (<1 sec) and stop quickly (<10 sec). The newer Beckman model E (horizontal rotor) and Fisher model 59 are acceptable. The Eppendorf, Brinkmann, and Beckman (models 11 and 12) microcentrifuges all take up to 60 sec to stop and some take several seconds to reach maximum speed. This severely limits the number of samples that can be processed within a given time period. If access to one of the

[25] R. D. Grubbs, S. D. Collins, and M. E. Maguire, *J. Biol. Chem.* **259**, 12184 (1985).
[26] R. D. Grubbs and M. E. Maguire, submitted for publication. (1989).

preferred microcentrifuges is not available, the silicone oil method described above may be an acceptable alternative.

We have attempted to use a filter rather than a microcentrifugation assay to measure $^{28}Mg^{2+}$ uptake but have been unable to find a filter type and/or washing conditions that do not cause some lysis of retained cells with consequent loss of $^{28}Mg^{2+}$. Filter types tested include glass fiber, cellulose acetate, cellulose mixed ester, cellulose nitrate, and polycarbonate, each from multiple manufacturers. This is presumably a function of cell type; a filter assay should be feasible with other cell types.

Determination of Subcellular Isotope Distribution by Digitonin Permeabilization

During our studies of Mg^{2+} transport, which were often performed in parallel with Ca^{2+} transport studies for comparison, we noticed striking differences in the amount of total cell Mg^{2+} versus Ca^{2+} that could be exchanged.[25] S49 cells contain about 4–5 nmol total Ca^{2+} but about 85 nmol $Mg^{2+}/10^7$ cells (about 1 mg protein). While essentially 100% of cell Ca^{2+} is exchangeable with extracellular $^{45}Ca^{2+}$ within 3 hr, we found that only about 2–3% of total cell Mg^{2+} can be exchanged or about 3 nmol/10^7 cells. These differences led us to adapt a cell permeabilization technique originally developed by Zuurendonk and Tager[27,28] to investigate the subcellular localization of the newly transported isotope. The technique we have used is similar to that devised by Zuurendonk and Tager; however, some revisions were necessary to optimize mitochondrial integrity and to control release of cations.[25] In practice, the procedure is an extension of the $^{28}Mg^{2+}$ influx assay described above.

The permeabilization technique is based on the highly selective interaction of digitonin and membrane cholesterol. Since plasma membrane cholesterol content is much higher than that of any subcellular membrane, it is more susceptible to the detergent action of digitonin. By carefully controlling the time of exposure and digitonin concentration, it is possible to permeabilize selectively the plasma membrane without altering the functional integrity of cell organelles such as mitochondria. When this cell treatment is combined with a rapid centrifugation step using a microcentrifuge, it is possible to separate quantitatively soluble cytoplasmic constituents such as cations and proteins from the cell particulate pellet containing all of the recognizable structures of the cell.

[27] P. F. Zuurendonk and J. M. Tager, *Biochim. Biophys. Acta* **333**, 393 (1974).
[28] P. F. Zuurendonk, M. E. Tischler, T. P. M. Akerboom, R. van der Meer, J. R. Williamson, and J. M. Tager, this series, Vol. 56, p. 207.

Buffers

In addition to the equipment and buffers required for the Mg^{2+} influx assay described above, the following additional buffers are required.

A. Intracellular buffer: 110 mM KCl (0.8195 g/100 ml), 15 mM NaCl (0.0877 g/100 ml), 50 mM K–HEPES, pH 7.15 at 4° (10 ml of 0.5 M stock/100 ml), 5 mM KPO$_4$, pH 7.0 (0.5 ml of 1 M stock/100 ml made from 1.0 M K$_2$HPO$_4$ and 1.0 M KH$_2$PO$_4$ mixed to give pH 7.0), 1 mM EGTA, pH 7.0 (adjusted with 1 M KOH; 1 ml of 100 mM stock/100 ml), 10 mM L-glutamic acid (0.1471 g/100 ml; added as free acid, will require adjustment of buffer pH with KOH), 5 mM MgCl$_2$ (0.5 ml of 1 M stock/100 ml; do not add until buffer has been diluted to volume to avoid precipitating magnesium phosphate), 2 mg/ml bovine serum albumin (fraction V), and 130 μM ADP, pH 7.0 (added just prior to use from a 100 mM stock; 130 μl/100 ml)
B. Digitonin buffer: Digitonin buffer is intracellular buffer containing *twice* the final desired digitonin concentration. For a *final* 0.003% digitonin concentration (w/v) (initial concentration 0.006%), weigh 3 mg of digitonin (Sigma, Cat. No. #D5628, not purified further) and place in a 50-ml Erlenmeyer flask. Add 50 ml of ice-cold intracellular buffer to the flask and cover with Parafilm. Sonicate for approximately 1 min, preferably in a cylindrical bath sonicator (e.g., Laboratory Supplies Company, Inc., model G112SP1G). The initially cloudy solution should now be crystal clear. If not, continue sonicating for about 10 sec *after* the solution clears. Be sure to shake droplets off sides and Parafilm so that all the digitonin is in solution. This buffer is made fresh daily.

Permeabilization

1. In order to facilitate rapid sample handling, scintillation vials should be arranged beforehand. Each row is for a single time point measured in duplicate and contains six vials in pairs. Each pair represents the duplicate samples for the time point and is composed of a *total* sample, a digitonin *supernatant,* and a digitonin (cell particulate) *pellet.*
2. The $^{28}Mg^{2+}$ influx protocol (above) is followed completely through the two washes in ice-cold wash buffer. The final pellet is then resuspended by gentle trituration in 1.0 ml intracellular buffer (*no digitonin*).
3. Immediately upon resuspension, a 0.5-ml aliquot is taken from each sample and added to a microcentrifuge tube already containing 0.5 ml ice-cold digitonin buffer. Timing of this step begins upon addition. The tubes are immediately capped, inverted twice, and replaced in the ice bath.

4. The 0.5 ml of cells remaining in each original microfuge tube is placed (tube included) in a scintillation vial. This sample constitutes the total cell-associated isotope (*total* sample).

5. After an appropriate length of time (90 sec for S49 lymphoma cells), the tubes are taken from the ice bath and are centrifuged 5–7 sec in a Beckman Microfuge B. A 0.5-ml aliquot of the supernatant is removed and placed in a scintillation vial. This sample constitutes the soluble cytosolic cell fraction (*supernatant* sample).

6. The remaining supernatant is then aspirated and the microfuge tube containing the cell particulate pellet (*pellet* sample) is placed in the third scintillation vial.

7. Aliquots (100 μl) of the original cell suspension should be taken to determine total radioisotope present.

8. For calculations, it is important to remember that, as outlined, the soluble cytosolic cell fraction taken represents 0.25 ml of the initial 1 ml of cells while each of the other samples (total cell and digitonin pellet) represent 0.5 ml of the initial 1.0 ml of cells. Calculations (and data normalization) should reflect this volume difference between samples. The easiest normalization is accomplished by multiplying the counts per minute for supernatant samples by 2 so that all three samples (total, supernatant, and pellet) reflect 0.5 ml of the initial 1.0 ml of cell suspension.

Digitonin and Cell Concentration and Exposure Time

The time course of release and the degree of release are critically dependent on both cell density and digitonin concentration. Comparison between experiments is valid only if these parameters are comparable. We routinely use a cell density of $6 \pm 0.2 \times 10^6$ cells/ml for the suspension cultured cells grown in this laboratory. This represents for most cell lines about 0.5 mg total protein. Even with such a constant cell density, the concentration of digitonin required at a given cell density to produce complete cytosolic release with minimal release of lysosomal and mitochondrial components must be determined empirically for each cell type. For example, a final concentration of 0.003% digitonin (w/v) for 90 sec works well with S49 cells whereas GM86 cells (Friend erythroleukemia) require 0.005% digitonin with 120-sec exposure for comparable release (0.001% = 8 μM digitonin). Increasing or decreasing cell density while maintaining these same digitonin and time conditions slows or speeds up release, respectively, essentially in direct proportion to the ratio of digitonin concentration to cell density.[25]

When verifying assay parameters in a new cell type, optimal conditions are determined by monitoring release of ^{86}Rb$^+$ and lactate dehydrogenase release. The time course of lactate dehydrogenase release is somewhat slower than that of $^{86}Rb^+$ at any given digitonin concentration. Release of Mg^{2+} follows an intermediate time course. Details on experimental controls and additional discussion of digitonin permeabilization have been published.[25–28]

Determination of Mg^{2+} Efflux Rates

Cell Loading and Preparation

The determination of Mg^{2+} efflux rates is somewhat simpler than measurement of influx rates since only a single centrifugation step is necessary; unfortunately, this is obviated by a somewhat more difficult preparation stage. An appropriate volume of cells is concentrated by a factor of 3 (final cell concentration of about 4×10^6/ml) by centrifugation at 200 g for 4 min and resuspended in fresh 37° culture medium containing no Mg^{2+} and 10% heat-inactivated horse serum (see Comments, below). $^{28}Mg^{2+}$ is immediately added to a final concentration of 0.2 μCi/ml, and the flask returned to the incubator for an additional 2 hr and preferably 3–4 hr. The cells are then removed and placed in an ice bath for 20 min to cool before being washed by centrifugation as described above under influx studies except that ice-cold wash buffer is used rather than incubation buffer at 37°. Obviously, adequate precautions must be taken during this step to avoid spillage and contamination since the culture medium and wash buffers will contain high amounts of radioactivity. The final cell pellet can be placed on ice for up to 5 min before initiating efflux without affecting results. We routinely measure the cell density of the cell suspension just before the last wash in order to determine the total number of cells and thus the desired final resuspension volume. Efflux is initiated by rapid resuspension of the cells to the required volume in 37° incubation buffer or other desired medium. The efflux rate is unaffected by cell densities between 1 to 20×10^6/ml with S49 cells. Aliquots of 1.0 ml are immediately centrifuged for 10 sec in a Beckman Microfuge B, and the supernatant aspirated. This set of pellets consists of a zero-time point. The entire pellet is then counted as described above. Subsequent time points are taken similarly.

Comments

It is critical that the cells are cooled before the wash step to avoid the loss of a very large proportion of the total intracellular isotope before the start of the experiment. The half-time of efflux in S49 lymphoma cells is about 50 min; therefore, a 20-min wash period at 37° or even room temperature gives ample opportunity for significant efflux to occur during washing. Control experiments in S49 cells where 1-ml cell aliquots incubated in $^{28}Mg^{2+}$ are rapidly washed either once or twice with warm buffer and then resuspended immediately in additional warm buffer give efflux data identical to that obtained by cooling the cells initially. Further, this protocol measures cpm retained by the cells rather than cpm lost into the supernatant; control experiments in which the supernatant is measured give identical efflux rates. Finally, the concentration of Mg^{2+} in the final resuspension medium is relatively unimportant for S49 cells since the mechanism of efflux has been shown not to involve Mg^{2+}–Mg^{2+} exchange.[7] However, since this may not be true in other cell types, efflux dependence on extracellular Mg^{2+} concentration should be determined.

The cost and relative availability of $^{28}Mg^{2+}$ make efflux experiments rather expensive. Thus it is critical that maximal loading of the isotope be achieved. This is accomplished primarily by increasing the specific activity during the loading phase by lowering the concentration of Mg^{2+} in the growth medium for a short period of time as described. A minimum of 2 hr of incubation is usually necessary at the specific activity noted in order to get sufficient intracellular isotope. While DME medium can be made in the laboratory, it is easier to purchase powdered medium without added Mg^{2+}. We actually purchase medium without added Mg^{2+}, Ca^{2+}, or K^+, adding whatever is necessary for the experiments planned. Flow Laboratories (McLean, VA) has consistently been the least expensive supplier. "Contaminant" Mg^{2+} (and Ca^{2+}) is about 5–10 μM in commercially supplied medium as determined by atomic absorption spectrometry. Horse serum contains about 1 mM Mg^{2+} so that the final Mg^{2+} concentration is about 0.1 mM. This increases the specific activity of the radioisotope while maintaining sufficient Mg^{2+} for viability. The disadvantage of this procedure is that acutely lowering extracellular Mg^{2+} from the normal 0.8 to even 0.2 mM causes growth arrest in S49 cells and in other cell lines.[29] Continued exposure to low Mg^{2+} concentrations for more than 48 hr leads to progressive cell death. Nonetheless, we have noted no alterations in hormonal responsiveness or in the properties of Mg^{2+} influx or efflux during exposures to 0.1 mM Mg^{2+} up to 6 hr in length versus cells incubated in medium containing 0.8 mM Mg^{2+}. Alternatively, cells can be incubated overnight with $^{28}Mg^{2+}$ in complete growth medium containing a normal Mg^{2+} concentration. While this loading procedure gives results identical to those obtained using low Mg^{2+} medium, it takes longer, re-

[29] M. E. Maguire, *Annals New York Acad. Sci.,* in press (1989).

quires more isotope, is more expensive, and generally results in fewer intracellular counts per minute.

Mg^{2+} Influx in Eukaryotic Cells Which Grow Attached to a Substratum

This procedure permits the measurement of Mg^{2+} influx in monolayer cell cultures in 6-well 35-mm plates and is well suited for determining drug dose–response effects. Individual protocols can be designed to take single or repeated samples from each well.

Equipment

The protocol described requires a thermostatically controlled "hot" box maintained at 37°. Our box is about 48 in. long × 30 in. high × 24 in. wide. Two sides consist of $\frac{1}{4}$-in. Plexiglas with hinged doors (4 × 4 and 4 × 12 in.) for access, providing good visibility and access. The remainder of the box is $\frac{3}{4}$-in. plywood. Heating is provided by a single infrared heat lamp, circulation by a small plastic fan, and thermostatic control by a Yellow Springs model 63RC controller and thermistor. Heated air is circulated through one end of the box by mounting the heat lamp and fan in opposite ends of two pieces of 90° angle flue pipe joined to form a single 180° angle unit. The wood and Plexiglas provide adequate shielding from β radiation encountered when using ^{45}Ca, ^{32}P, ^{86}Rb, and ^{28}Mg. Additional lead shielding should also be used with the latter two isotopes.

General Solutions

5% Trichloroacetic acid (TCA) made fresh daily from a 50% (w/v) stock
0.2 N NaOH
Incubation buffer and wash buffer (described above)

Initiating Influx

Cells grown in 6-well 35-mm dishes are removed from the incubator to the incubator box at 37°. The growth medium is immediately aspirated, one complete dish at a time, washed twice with 1 ml incubation buffer, and finally covered with 1 ml incubation buffer. Only three or four dishes are removed from the incubator for medium replacement at a time to avoid excessive loss of CO_2 and consequent pH changes. Fresh buffer is always pipetted against the wall of the dish to blunt the force of the stream which would otherwise dislodge cells. After the medium has been replaced in all dishes, the experiment is initiated by addition of $^{28}Mg^{2+}$ in an additional volume of (usually) 1.0 ml incubation buffer for a final volume of 2.0 ml. Other agents that are to be tested are added in additional small volumes (e.g., 10 or 20 μl) as necessary.

Incubation is terminated by aspiration of the supernatant followed by addition of 2.0 ml of ice-cold wash buffer. This is aspirated and the wash repeated once more. The final wash is also aspirated, 1.0 ml of ice-cold 5% TCA added, and the dish set aside. After all cells have been treated, the TCA is removed with a Pasteur pipet directly into scintillation vials and counted by Cerenkov counting or after additional of scintillation fluid (see above). The cell layer which remains affixed to the dish after TCA lysis is solubilized in 1.0 ml of 0.2 N NaOH for analysis of protein. Alternatively, the dishes can be covered and stored at $-20°$ for several days to allow residual $^{28}Mg^{2+}$ to decay before addition of NaOH and assay.

The major problem with this protocol is the necessity to wash the dishes repeatedly, which of course entails inevitable loss of cells as they are dislodged. Occasional loss up to and including detachment of the entire sheet of cells will occur, although little is lost after the TCA step. We have used this protocol successfully with NG108-15 (a neuroblastoma-glioma hybrid), BC3H-1 (a putative smooth muscle cell), and G8 cells (a skeletal muscle cell line).

Assay of Mg^{2+} Uptake by Prokaryotic Cells

Overview

Unlike eukaryotic cells, which must be handled very gently, most bacteria possess a rigid cell wall which affords protection against osmotic lysis and mechanical disruption. For this reason accumulation of $^{28}Mg^{2+}$ can be monitored by rapid vacuum filtration without loss of intracellular radioisotope due to cell lysis. The assay we use to detect uptake of $^{28}Mg^{2+}$ by prokaryotic cells is based on the vacuum filtration assay developed by Silver.[30] A number of modifications have been made and conditions have been optimized in order to obtain the maximum amount of data from each shipment of $^{28}Mg^{2+}$ while maintaining reproducibility. Conditions were initially optimized for uptake of $^{60}Co^{2+}$, since this ion is accumulated in *Salmonella typhimurium* via a constitutive Mg^{2+} transport system.[30-33]

[30] S. Silver and P. Bhattacharyya, this series, Vol. 32, p. 881.
[31] S. P. Hmiel, M. D. Snavely, C. G. Miller, and M. E. Maguire, *J. Bacteriol.* **168,** 1444 (1986).
[32] S. Silver, in "Bacterial Transport" (B. P. Rosen, ed.), pp. 221–324. Dekker, New York, 1978.
[33] D. L. Nelson and E. P. Kennedy, *J. Biol. Chem.* **246,** 3042 (1971).

Our uptake studies have been performed with *Salmonella typhimurium;* however, the assay should be readily applicable to other prokaryotic organisms as evidenced by previous studies of Mg^{2+} uptake using *Escherichia coli, Staphylococcus aureus,* and *Bacillus subtilis.*[32] The assay is also applicable with lower unicellular eukaryotic cells such as yeast. We have recently used it with no changes to study $^{28}Mg^{2+}$ uptake in the yeast *Schizosaccharomyces pombe.*

Cell Preparation

Cells grown overnight in minimal medium supplemented with 0.1% casamino acids and 1–10 mM $MgSO_4$ are washed once in minimal medium[33] without added $MgSO_4$ and inoculated into fresh minimal medium containing 0.1% casamino acids and the desired amount of $MgSO_4$ at an optical density[34] at 600 nm of 0.03 (usually about 1.0 ml cells/50 ml fresh medium). The specific choice of medium will depend on the species and phenotype. The cell suspension is incubated at 37° with shaking until the optical density at 600 nm reaches about 0.40 (late log phase) which usually takes 3 to 4 hr with *S. typhimurium,* but can vary considerably with different bacterial strains.

Once the appropriate optical density is reached, the cells are harvested by centrifugation at 3000 *g* for 10 min. From this point maintenance of sterility is unnecessary. The pelleted cells are washed twice with an equal volume of ice-cold, Mg^{2+}-free medium containing glucose but no amino acids. The cells are then resuspended to a cell density of 10^9 cells/ml (approximate 8-ml vol/40 ml of original exponentially growing cells) in medium containing glucose and any required amino acids. The concentration of cells is checked prior to beginning the assay. A 1 : 10 dilution of 10^9 cells/ml should give an optical density of 0.20 at 600 nm with *S. typhimurium.*

Transport Assay

Assay of ion uptake is initiated by adding 0.1 ml of ice-cold cell suspension to 0.9 ml of room temperature medium containing $^{28}Mg^{2+}$ at about 10^5 cpm/ml. Background or zero-time uptake values are determined by adding the cells to 0.9 ml of ice-cold medium containing the same amount of $^{28}Mg^{2+}$ (see below). Uptake is usually linear only for 1 to 3 min and must be determined for each strain. After the incubation period, 10 ml of

---

[34] Values given for optical density corresponding to cell number will vary with different bacterial strains and the relationship between optical density at 600 nm and cell density should be confirmed for each strain.

ice-cold medium containing 10 mM MgSO$_4$ is added to each tube, and the samples are poured over wetted cellulose acetate filters (Schleicher and Schuell #BA85, Keene, NH) under moderate vacuum (e.g., an air pump). The samples are typically poured over the filters within 30 sec after dilution with cold medium; however, no loss of cell-associated counts occurs for up to 5 min. A 12-well Millipore filtration apparatus works well; under an appropriate vacuum 10 ml of liquid takes 15–30 sec to flow through the filters, although slower drainage of even 5 min does not seem to affect the results. The filters are washed once with another 10 ml of ice-cold medium and then removed for counting, usually in a gamma counter. Magnesium ion uptake is determined by taking the difference between cell-associated counts determined at either 37° or room temperature versus 4°.

Comments

Cells are grown overnight in minimal medium to ensure that no "lag phase" occurs in cell growth upon dilution of the cells into the final growth medium. Cells grown overnight in rich medium do not react well to exposure to minimal medium and cease to grow for up to several hours. However, cells grown overnight in minimal medium and then reinoculated into minimal medium show little, if any, growth lag. We have found the addition of 0.1% casamino acids to facilitate the growth of many strains harboring mutations in Mg^{2+}-uptake systems.

Studies of ion uptake have often been performed using cells suspended in media lacking required amino acids in order to prevent cell division during the course of the experiment.[34] This presents problems due to the fact that amino acid starvation rapidly elicits the "stringent response" in most if not all bacterial species.[35] This response involves alterations in RNA synthesis and metabolic activity and results in decreased transport of a number of substances. With *S. typhimurium*, warming the cells to room temperature in the absence of required amino acids results in a decrease of nearly 50% in the Mg^{2+} influx rate within 30 min. We presume that this decrease in the rate of influx may be due to the stringent response and, while we have not rigorously tested the speed of onset, decreased rates of transport are seen within 5 min of warming the cells in the absence of amino acids. Levels of guanosine 5'-diphosphate 3'-diphosphate, one of the known mediators of the stringent response, increase substantially within seconds of the onset of amino acid starvation.[35] For this reason we include required amino acids in the reaction mixture and prevent cell growth by keeping the cells cold until the assay is started. The 10-fold dilution of the cells to start the assay appears to warm them

[35] J. A. Gallant, *Annu. Rev. Genet.* **13**, 393 (1979).

instantly and since the assay is relatively short (less than 20 min) no substantial cell division occurs. Cells stored concentrated on ice show no decrement in ion uptake for at least 2 hr.

Another problem associated with measurement of ion uptake is separation of uptake from binding to the cell exterior. In order to discriminate between uptake and binding of Mg^{2+} we compare cell-associated counts after incubation of the cells with the isotope at 37° or room temperature and at 4°. Since uptake is an active process it should not occur to any significant extent in the cold and any cell-associated counts are assumed to be bound to the cell surface. Low-affinity binding to the cell membrane would be expected to come to equilibrium very quickly even in the cold. The validity of this assumption is supported by three experimental findings: (1) there is no increase in cell-associated counts over 20 min at 4°, (2) cell-associated counts at room temperature decrease to the level of retained counts at 4° when metabolic poisons are included in the reaction mix, and (3) counts retained in the cold show only a slight nonsaturable dependence on Mg^{2+} concentration, consistent with a large number of low-affinity binding sites. Counts associated with cells at 4° also serve to provide a zero point for time-course experiments. Because both binding and transport can occur rapidly, an actual zero-time control is often difficult to obtain—a significant amount of uptake can occur in the few seconds it takes to perform the measurement.

The conditions described above for the assay of Mg^{2+} uptake have been optimized with respect to a number of parameters. First, for stopping the reaction and washing the cells, it is desirable to use medium containing a high concentration of Mg^{2+} because inclusion of unlabeled Mg^{2+} in the wash buffer improves the signal-to-noise ratio without altering the net amount of ion taken up. Addition of 10 ml of medium to stop the reaction and another 10 ml to wash the filters provides the lowest level of background while providing for the minimum accumulation of radioactive liquid. Second, polypropylene tubes are used because use of glass tubes lowers the amount of uptake significantly at low ion concentrations, presumably due to binding of Mg^{2+} to glass. Third, the type of filters used is also important. Of filters tested, cellulose based are better than glass fiber. The latter filters bind Mg^{2+} and result in an unacceptably high background. Of the cellulose based filters, the Schleicher and Schuell BA85 filters give the best results. Millipore HAWP and Gelman Metricel GN-6 filters retain 75% fewer counts than do the Schleicher and Schuell BA85 filters.

Finally, the term "Mg^{2+}-free" medium is an operational one. By atomic absorption analysis, medium with no added Mg^{2+} actually contains 15–25 μM Mg^{2+}. Given that the apparent affinity of S. typhimurium and *E. coli* Mg^{2+} transport systems is in this range, this "contaminant" amount of Mg^{2+} should alter the specific activity of the radioisotope and affect determinations of K_m and V_{max} values. However, this does not appear to be the case. Attempts to correct for "contaminant" Mg^{2+} by arbitrarily adding values between 5 and 30 μM to the total Mg^{2+} present yielded obviously impossible (indeed, negative) values for the K_m of the transport system for Mg^{2+}. We hypothesize that cells washed several times in cold, nominally Mg^{2+}-free medium become somewhat depleted of Mg^{2+}, particularly at extracellular binding sites. When they are then re-suspended at very high density in fresh nominally Mg^{2+}-free medium all of the trace Mg^{2+} in the medium rapidly associates with or is taken up into the cells to replace that lost. In practice therefore, it can be ignored in kinetic calculations.

Conclusions

   Magnesium ion transport has not been widely studied, presumably for two major reasons: a greater interest in and knowledge of the biological actions of Ca^{2+}, and the lack of a readily available and inexpensive isotope. However, measurement of Ca^{2+} fluxes with ^{45}Ca^{2+} is often complicated by excessive Ca^{2+} binding to extracellular sites and by the rapid and transient nature of the changes in Ca^{2+} flux. In contrast, in practical terms, Mg^{2+} fluxes are approximately two orders of magnitude greater than those of Ca^{2+} (although still lower than fluxes for Na$^+$ or K$^+$) and binding to extracellular sites is, in terms of assaying flux, almost nonexistent. Of more fundamental interest, alterations in the hormonal sensitivity of Mg^{2+} influx[4–8,18,19] and its intracellular compartmentation[25] suggest that major modulations of intracellular Mg^{2+} concentration may occur. Magnesium-28 ion is the only isotope usable for measuring Mg^{2+} fluxes. It has the advantages of high energy and thus easy counting, but the disadvantages of relatively high cost and short half-life. Its relatively poor availability has improved greatly in the last few years primarily because the number of users has increased. This will remain a problem until the number of users grows sufficiently to provide an economic base for a commercial supplier. The procedures described in this chapter hopefully show that measurement of Mg^{2+} fluxes is rather easy, indeed in our hands far easier than measurement of Ca^{2+} fluxes.

[37] Measurement of Amino Acid Transport by Hepatocytes in Suspension or Monolayer Culture

By MICHAEL S. KILBERG

Isolated rat hepatocytes, either in suspension or in monolayer culture, have become an important tool in the study of hepatic function.[1] Given the central role of the liver in a wide variety of metabolic interconversions, the uptake and release of nutrient molecules by hepatocytes are now recognized as important steps for potential regulation.[2,3] Consequently, solute transport by isolated rat hepatocytes continues to be a growing focus of membrane-related research. This is particularly true for the amino acids because they represent a link between protein and carbohydrate metabolism in the liver through their roles as precursors for gluconeogenesis and as substrates for oxidation by the citric acid cycle. The present chapter will describe some of the current methodology used to measure nutrient transport by isolated hepatocytes in either suspension or monolayer culture. Although amino acids are used as the substrates for this chapter, most of the procedures described could be applied to transport studies for a variety of nutrient molecules.

Liver Perfusion Apparatus

Numerous methods have been developed for the isolation of rat hepatocytes by modifications of the collagenase perfusion technique, originally described by Berry and Friend.[4] The procedure that has evolved in our laboratory is designed to be simple, efficient, and inexpensive with regard to the equipment required. From a 150-g rat, we routinely obtain $500–800 \times 10^6$ hepatocytes with a viability of 90% or greater based on Trypan Blue exclusion. One can prepare hepatocytes as well as a nonparenchymal cell fraction from the same donor animal by employing a combination of differential centrifugation and Percoll gradients.[5] We have also used this procedure for the isolation of hepatocytes and nonparen-

---

[1] R. A. Harris and N. W. Cornell (eds.). "Isolation, Characterization, and Use of Hepatocytes." Elsevier, New York, 1983.

[2] H. N. Christensen, *Biosci. Rep.* **3**, 905 (1983).

[3] P. Fafournoux, C. Demigne, C. Remesy, and A. LeCam, *Biochem. J.* **216**, 401 (1983).

[4] M. N. Berry and D. S. Friend, *J. Cell Biol.* **43**, 506 (1969).

[5] D. F. Gardner, M. S. Kilberg, M. M. Wolfe, J. E. McGuigan, and R. I. Misbin, *Am. J. Physiol.* **248**, G663 (1985).

chymal cells from other rodent species and fetal pigs[6] with similar success. An alternate method must be employed for the preparation of hepatocytes from fetal rats.[7]

Rats are anesthetized with pentobarbitol and taped, abdomen up, to a surgical table made from a $6 \times 10 \times 1/4$ in. piece of Plexiglas with 2-in. (length) legs made from 1-in.-diameter Plexiglas dowel rod. This table sits in a stainless steel tray (Fisher Scientific) that is $8 \times 12 \times 2$ in. and has a $\frac{3}{8}$-in. lip on all four edges so that it can be placed into a 12×12 in. waterbath made from $\frac{1}{4}$-in. Plexiglas. The water level of the bath is brought up to the bottom of the stainless steel tray so that some of the temperature of the bath, maintained at 37° by a submersible recirculating heater, can be transferred to the surgical tray. A 250-ml Erlenmeyer flask, containing 200 ml of perfusion buffer, is held in the waterbath by means of a ring stand and clamp so that the solution is maintained at 37°. The perfusion buffer is pumped through the liver in a single-pass mode by any commercially available peristaltic pump that can be varied in speed from 5 to 15 ml/min. The pump is placed directly beside the waterbath so that the length of tubing from the Ehrlenmeyer flask to the cannula is kept at a minimum to avoid heat loss. The cannula is made from a 21-gauge butterfly infusion set (Abbott Hospitals). The butterfly needle is cut off and the polyethylene tubing is bevelled so that the tubing itself serves as the cannula. This cannula is coupled to the pump tubing by cutting a 1-ml plastic tuberculin syringe in half and inserting the lower half into the Tygon tubing. The cannula and syringe tip are fastened by the Luer-Lok connections that have been unaltered for this purpose.

Liver Perfusion

The abdomen of the anesthetized rat is wiped with 70% ethanol to keep hair from the surgical area and to provide some degree of sterility. The surgical area can be further isolated by covering the animal with a Steri-Drape (3M Corporation). The abdominal cavity is opened and the viscera moved to the operator's right to expose the inferior vena cava, portal vein, and hepatic artery. The directions, left or right, refer to the operator. Fat and connective tissue that overlays the inferior vena cava between the liver and the left renal vein is removed by teasing the tissue with a pair of forceps. Using suture silk, a tie is placed loosely about the inferior vena cava proximal to the liver, another is placed around the vena

---

[6] P. T. K. Saunders, R. H. Renegar, T. J. Raub, G. A. Baumbach, P. H. Atkinson, F. W. Bazer, and R. M. Roberts, *J. Biol. Chem.* **260**, 3658 (1985).
[7] H. L. Leffert, K. S. Koch, T. Moran, and N. Williams, this series, Vol. 58, p. 536.

cava just above the right renal vein, and a third is placed around the left renal vein itself. The tie around the left renal vein is tightened first, followed by the one about the inferior vena cava which is most distal to the liver (i.e., the one just above the right renal vein). A small incision is made in the inferior vena cava between the two ties and the cannula is inserted so that the tie proximal to the liver can be used to secure it in place.

The perfusion pump should be operating at a slow speed (3–5 ml/min) during the insertion of the cannula. Immediately after the cannula has been secured in place, the hepatic artery and the portal vein are severed to allow the perfusate to flow out of the liver. The pump speed is increased to 10 ml/min, the thoracic cavity is opened, and the superior vena cava is clamped with a hemostat. A tissue moistened with phosphate-buffered saline (PBS, 10 mM sodium phosphate, 150 mM sodium chloride, pH 7.4) can be placed in the abdominal cavity to direct the perfusate over the edge of the Plexiglas surgical table. In this way, one can visualize the flow rate by measuring the perfusate as it drips into the stainless steel surgical tray. The abdominal cavity is then covered with a 2 × 2 in. gauze pad which has been moistened with PBS at 37°. The entire surgical tray is covered with Saran plastic wrap to help maintain humidity and temperature. Additional temperature stability is achieved by placing a 60-W light bulb approximately 6 in. away from the liver by means of a desk lamp.

The original volume (ml) of the perfusate is approximately equal to the body weight (g) of the rat. In our experience, 150- to 200-g animals produce the highest viability and best yield of hepatocytes. After one-half of the calcium-free perfusion buffer (100 ml) has been used to blanch the liver, the collagenase, suspended in 5–10 ml of perfusion buffer, is added to the 100 ml remaining in the Ehrlenmeyer flask. The perfusion is then continued at 10 ml/min until nearly all of the buffer is utilized (usually 8–10 min for a 150- to 200-g rat).

Collagenase

The calcium-free perfusion buffer consists of 25 mM sodium phosphate, pH 7.4, 3.1 mM potassium chloride, 119 mM sodium chloride, 5.5 mM glucose, 0.1% BSA (optional), and 5 mg/liter Phenol Red. The amount of collagenase used is determined for each lot of enzyme, typically 50–100 units/g body weight of the animal. Individual lots of collagenase should be tested prior to purchasing large quantities. Like others, we find that "highly purified" or "tissue-specific" preparations of collagenase are not always as suitable as those that contain significant contamination by clostripain, caseinase, and tryptic acitvity. For example, in our hands type I collagenase from Sigma (C-0130) produces hepatocyte preparations which are usually superior to those obtained with "hepatocyte-specific" enzyme preparations from the same company. Although other investigators suggest the simultaneous addition of calcium to the perfusion buffer with the collagenase, we find that added calcium is unnecessary and can cause cell aggregation.

Hepatocyte Isolation

Following the completion of the perfusion process, the liver is carefully removed and placed in a 30-ml beaker containing 10 ml of the culture medium to be used. The remaining steps are performed in a biological safety hood under sterile conditions. The cells are dispersed by forcing the tissue up and down in a 5-mm i.d. glass tube by means of a suction bulb. Using this glass tube, one can then force the suspension through a 75-μm nylon mesh which has been stretched tightly over a sterile 50-ml plastic centrifuge tube. The cells are diluted to 40 ml with ice-cold culture medium and then centrifuged for 2 min at 4° at 100 g in a refrigerated clinical centrifuge. The supernatant is either discarded or saved for the isolation of nonparenchymal cells through the use of a Percoll gradient.[5] The resulting cell pellet is resuspended and washed four more times by the same centrifugation procedure. The final cell pellet is resuspended in ice-cold culture medium to 25 ml, and then the total cell count and the percentage Trypan Blue exclusion is determined through the use of a hemocytometer. The cells are diluted with warm (37°) culture medium to the appropriate volume for either suspension or monolayer culture as described below.

Hepatocyte Culture

Acid-soluble calf skin collagen (Sigma type III, C-3511) is weighed into a autoclaved beaker containing a stirring bar and placed under a UV light for at least 1 hr. Sterile 0.5% acetic acid is used to dissolve the collagen to 1 mg/ml and the mixture is stirred at room temperature. This stock solution is stored at −20° in 5-ml aliquots. To collagen coat 24-well cluster trays, the 1 mg/ml stock is diluted 1 : 50 with sterile water and then 0.5 ml (10 μg) is added to each well. The trays are incubated at room temperature overnight and the following day the solution is aspirated from each well, leaving a collagen matrix attached to the bottom of the well. These collagen-coated trays can be stored at room temperature for several days prior to use.

For monolayer culture, the hepatocytes are diluted to 0.8×10^6 cells/ ml in 37° culture medium. These cells are plated onto the collagen-coated 24-well cluster trays (Costar) by adding 0.33 ml of the cell suspension to each well. The cells are then placed in a humidified CO_2 incubator containing an atmosphere of 5% CO_2/95% air. After approximately 1–2 hr the cells will become attached, begin spreading, and assume a cuboidal shape. Transport assays can be performed after only 30 min of plating, although some loss in cell number will be observed until the cells have completely attached and spread. The use of serum for these cultures is optional; it does increase the attachment efficiency, but it is not necessary if its use will complicate the experimental design. In fact, we have found that hepatocytes will attach and spread on the collagen-coated dishes even when suspended and cultured in a serum-free Krebs–Ringer bicarbonate buffer.[8] Interestingly, the presence of a single amino acid in the Krebs–Ringer buffer, even a nonmetabolizable analog such as 2-aminoisobutyric acid (AIB), will enhance significantly the cell attachment.[9] The reason for this amino acid-dependent effect is unclear at the present time.

Typically, the medium for suspension culture consists of Waymouth's medium or Eagle's minimal essential medium supplemented with amino acids and pyruvate as described by Seglen et al.[10] and containing 10% fetal bovine serum. Approximately 100 ml of the cell suspension (2×10^6 cells/ml) is placed in a 500-ml roller bottle and rotated at 2.5 rpm through the use of a Wheaton two-place roller bottle apparatus which is small enough to fit into a 37° humidified CO_2 incubator.

The data shown in Fig. 1 demonstrate that analogous experiments performed with either suspension or monolayer cultures of hepatocytes yield qualitatively similar results. The cells in suspension culture allow assays immediately following hepatocyte isolation, yet it is clear that within 1 hr of placing the cells in monolayer culture useful information can also be obtained. In the particular example shown, amino acid-dependent repression of glucagon-stimulated (in vivo) System A activity was monitored in the presence of inhibitors of either protein (cycloheximide) or RNA (actinomycin D) biosynthesis. The evidence for de novo synthesis of a poly(A)⁻ mRNA and a corresponding "repressor" protein has been summarized elsewhere.[11]

[8] D. S. Bracy, M. E. Handlogten, E. F. Barber, H.-P. Han, and M. S. Kilberg, J. Biol. Chem. 261, 1514 (1986).

[9] L. Weissbach and M. S. Kilberg, J. Biol. Chem. 121, 133 (1984).

[10] P. O. Seglen, A. E. Solheim, B. Grinde, P. B. Gordon, P. E. Schwarze, R. Gjessing, and A. Poli, Ann. N.Y. Acad. Sci. 349, 1 (1980).

[11] M. S. Kilberg, Trends Biochem. Sci. 11, 183 (1986).

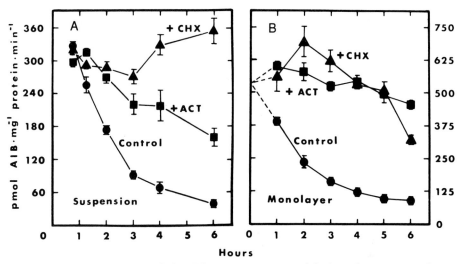

FIG. 1. Decay of glucagon-induced System A transport activity in rat hepatocytes maintained in either suspension (A) or monolayer (B) culture. Three hours prior to cell isolation the donor animal was injected with glucagon, 1 mg/100 g body weight (ip). Actinomycin D (4 μM) or cycloheximide (0.1 mM) was added to the cultures as indicated. The Na$^+$-dependent uptake of 50 μM [^{3}H]2-aminoisobutyric acid (AIB) was measured for 1 min at 37° as described in the text. The data are the averages (± SD) of three determinations. Where not shown the standard deviation bar is contained within the symbol.

Transport of Radioactively Labeled Substrates by Suspension Cell Cultures

To measure the transport of a radioactively labeled substrate by cells maintained in suspension culture, a modification of the method of McGivan et al.[12] is used. Just prior to assay, 8 ml of the suspended cells is removed from the roller bottle, separated from the culture medium by centrifugation at 100 g for 3 min, and washed twice in Na$^+$-free Krebs–Ringer phosphate buffer (KRP; 119 mM choline chloride, 5.9 mM potassium chloride, 1.2 mM magnesium sulfate, 1.2 mM potassium bicarbonate, 5.5 mM glucose, 0.5 mM calcium chloride, and 25 mM choline phosphate dibasic, pH 7.4). The choline phosphate is prepared by boiling a 10X (500 mM) solution of choline bicarbonate in the presence of 250 mM phosphoric acid for at least 1 hr to generate choline phosphate. The pH is adjusted by the addition of hydrochloric acid.

Using 2-ml microcentrifuge tubes, assay tubes are prepared by layer-

[12] J. D. McGivan, N. M. Bradford, and J. Mendes-Mourao, *FEBS Lett.* **80**, 380 (1977).

ing (in order from the bottom) 0.2 ml of 20% perchloric acid (v/v), 1 ml of
N-butyl phthalate, and 0.5 ml of either Na^+-containing (NaKRP) or Na^+-
free (CholKRP) Krebs–Ringer phosphate buffer containing the tritium-
labeled amino acid (or other labeled substrate). The Na^+-containing buffer
contains NaCl and Na_2HPO_4 instead of the corresponding choline salts.
The density of the oil layer may have to be modified through the use of
alternate oils so as to be compatible with the density of the particular cell
type under study.

A multichannel adjustable pipet is used to add 50-μl aliquots of cells
(approximately 0.7×10^6 cells) simultaneously to the top of six assay
tubes that have been equilibrated previously in a 37° water bath. After
incubation at 37° for the appropriate length of time, transport is terminated by a 30-sec centrifugation (15,000 g) in a microcentrifuge. The top
two layers (uptake buffer and oil layer) are removed by aspiration and the
residual radioactivity is removed from the side of the tube with a paper
wick. A 0.1-ml sample of the perchloric acid extract is placed into a
scintillation vial and the radioactivity determined. Following precipitation
of the total protein by the method of Bensadoun and Weinstein,[13] the
protein content of the original cell suspension is measured by a modification of the Lowry[14] procedure described below. The protein content of the
remaining 0.1 ml of perchloric acid extract for each sample can also be
used to calculate the transport velocity. The results are expressed as
picomoles of amino acid transported per milligram protein per unit time.

Transport of Radioactively Labeled Substrates by Cell Monolayers

Transport of radioactively labeled substrates by cells cultured on collagen-coated 24-well cluster trays is assayed by a modification of the
method described originally by Gazzola et al.[15] A more recent description
of this methodology has been given by Vadgama.[16] Briefly, the lids to the
24-well cluster trays are modified so that they contain either 12 × 75 mm
plastic test tubes or Beem embedding capsules (#00). These lids allow
transfer of either 2 ml (warm or cold washes) or 0.25 ml (uptake mixture)
of medium, respectively. Placing the bottom of the tray, to which the cells
are attached, upside down on either of the modified lids, and then rapidly

---

[13] A. Bensadoun and D. Weinstein, Anal. Biochem. **70**, 241 (1976).

[14] O. H. Lowry, N. J. Rosebrough, A. L. Farr, and R. J. Randall, J. Biol. Chem. **193**, 265 (1951).

[15] G. C. Gazzola, V. Dall'Asta, R. Franchi-Gazzola, and M. F. White, Anal. Biochem. **115**, 368 (1981).

[16] J. Vadgama, this series, Vol. 171, p. 133.

inverting the entire unit, transfers the solution in the tubes or capsules to all 24 wells simultaneously.

To allow a partial depletion of the intracellular amino acids and to remove extracellular sodium prior to the transport assays, 2 ml of Na^+-free Krebs–Ringer phosphate buffer (37°) is added for 10 min to each well using the lids modified with the 12 × 75 mm test tubes. After repeating this depletion procedure, the buffer is removed by inverting the tray over a dish pan and the transport assay is initiated by transferring 0.25 ml of the uptake medium to all 24 wells through the use of the lids that are modified with the Beem embedding capsules. The uptake buffers are prepared from either Na^+-free or Na^+-containing Krebs–Ringer phosphate buffer and contain the radioactively labeled amino acid. Also included here are unlabeled substrates for studies involving competitive cis inhibition. After the appropriate incubation time, the uptake mixture is discarded by rapidly inverting the cluster tray over a dishpan. The cells are rinsed four times with ice-cold Na^+-free Krebs–Ringer phosphate buffer (2 ml/well per rinse). The trays are placed upside down on absorbant paper and allowed to stand for 15 min. The cells are then solubilized by the addition to each well of 0.2 ml of 0.2% sodium dodecyl sulfate (SDS) in 0.2 N sodium hydroxide. After a 30-min incubation at room temperature, 0.1-ml aliquots of the cell extract are neutralized with 0.1 ml of 0.2 N HCl and then transferred to scintillation vials for determination of the radioactivity.

The remaining 0.1 ml in each well is used to assay the protein content by combining two previously published modifications of the Lowry procedure.[17,18] A modified copper reagent described in one of these reports[17] has several advantages over the two solutions originally described by Lowry et al.[14]: (1) it is stable for several weeks at room temperature; (2) it is stored as a single solution; and (3) including SDS, as suggested by Markwell et al.,[18] enhances protein solubilization and eliminates interference from a number of substances. To each well, 600 μl of the copper reagent (0.58 mM EDTA–copper disodium salt, 189 mM sodium carbonate, 100 mM sodium hydroxide, and 1% SDS) is added. After incubating for 10 min at room temperature, 60 μl of phenol reagent (diluted 1 : 1 with water just prior to use) is added and an additional incubation for 30 min is performed before measuring the absorbance at 500 or 750 nm. Using a multichannel adjustable pipet to add the reagents and a spectrophotometer equipped with an automated sample "sipper" system, one person can perform the protein assays for 10–12 trays in a few hours.

[17] B. Zak and J. Cohen, Clin. Chim. Acta 6, 665 (1961).
[18] M. A. K. Markwell, S. M. Has, L. L. Bieber, and N. E. Tolbert, Anal. Biochem. 87, 206 (1978).

Transport Assays for Amino Acids Not Radioactively Labeled

Monolayer cultures of hepatocytes, maintained in 100-mm-diameter collagen-coated cultured dishes (12×10^6 cells/dish), are incubated with the appropriate substrate amino acid. These incubations can be done in Na^+-containing or Na^+-free medium as described above to obtain a Na^+-dependent rate of transport when required. After the uptake incubation, the cells are rinsed three times with 20-ml portions of ice-cold PBS. The hepatocytes are removed from the dish in 5 ml of ice-cold PBS with a rubber policeman, a second 5-ml portion of PBS is used to rinse the dish, and then these two cell suspensions are combined. The cells are subjected to centrifugation at 1500 g for 10 min at 4° in a tabletop centrifuge. After discarding the supernatant, 0.3 ml of 5% sulfosalicylic acid (SSA) is added and the cell pellet resuspended with a glass rod. The precipitation is allowed to continue for 1–2 hr at room temperature and then the protein is pelleted by centrifugation in a Sorvall SM-24 rotor at 2000 g for 15 min at 4°. This SSA extract is saved and stored frozen at −20°. Just prior to analysis, the SSA extract is thawed and filtered through a disposable 0.2-μm Acrodisc filter (Gelman, Ann Arbor, MI) and then the pH of the extract is adjusted to 2.2 with 0.3 N LiOH.

An aliquot (approximately 10–20 μl) of the SSA extract is analyzed for amino acids and other primary amines by separation on an HPLC ion-exchange column followed by postcolumn derivatization by o-phthalde-hyde (OPA). The resulting derivatives can be detected and quantitated with a fluorimeter outfitted with a flow cell. For our studies, the HPLC column resin used is a Li^+-based ion-exchange column from Pickering Associates (Mountain View, CA) and the amino acids are eluted using a three-step pH gradient system as described by the manufacturer. This system is capable of separating distinctly 30–40 different primary amines. The lower limit of detection for the OPA-derivatized amino acids is approximately 100 pmol. All of the amino acids extracted from the cultured hepatocytes are present at concentrations of at least 1 nmol/12×10^6 cells (each 100-mm dish), thus providing an easily detectable amount of material as well as sufficient amounts for repeated analyses of the same extracts.

For the fluorescent detection of the amino acids by postcolumn OPA derivtization, the following stock solutions are required: (1) 400 mg of OPA is dissolved in 100 ml of methanol, filtered to remove particulate matter, and then stored at −20° in 4-ml aliquots; and (2) 45 g of boric acid is dissolved in 800 ml of water and then KOH pellets added to adjust the pH to 10.4. After the addition of 3 ml of 30% Brij to the boric acid solution, the mixture is diluted to 1 liter, filtered, and then stored at 4°.

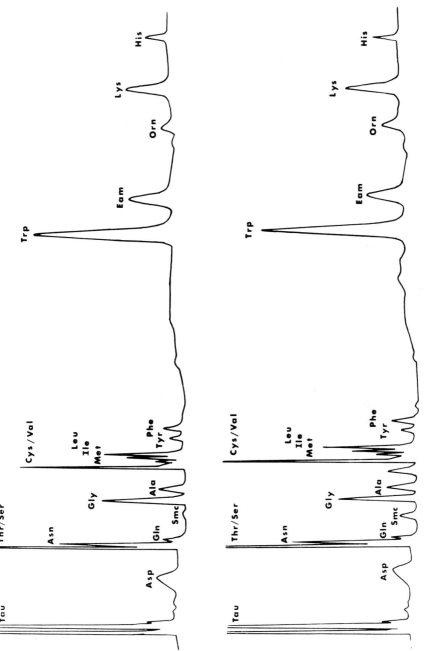

FIG. 2. Intracellular amino analysis of cultured rat hepatocytes. Cells in monolayer culture were tested for the accumulation of the unlabeled amino acid S-methyl-L-cysteine (Smc) in the presence (bottom) or absence (top) of the aminotransferase inhibitor aminooxy-acetate (AOA). The details of the assay procedure are given in the footnote to Table I.

TABLE I
INTRACELLULAR AMINO ACID ANALYSIS OF
CULTURED RAT HEPATOCYTES BY HPLC[a]

Amino acid	−AOA (mM)	+AOA (mM)
Aspartate	0.66	0.80
Threonine/serine	0.32	0.45
Asparagine	0.85	1.07
Glutamine	0.49	0.07
Glycine	0.79	0.92
Alanine	0.56	0.71
Cystine/valine	0.40	0.52
Methionine	0.23	0.35
Isoleucine	0.27	0.35
Leucine	1.59	1.99
Tyrosine	0.52	0.65
Phenylalanine	0.50	0.64
Ornithine	0.38	0.49
Lysine	0.65	0.88
Histidine	0.70	0.84
S-Methyl-L-cysteine	N.M.	0.64

[a] Isolated rat hepatocytes in primary culture were incubated for 3 hr in amino acid-free Krebs–Ringer bicarbonate (KRB) buffer. The medium was then removed and one dish of cells was incubated in fresh KRB containing 25 μM S-methyl-L-cysteine (Smc) alone, while a second dish of cells was incubated in KRB supplemented with 2 mM aminooxyacetate (AOA) and 25 μM Smc. The uptake of Smc was allowed to proceed for 60 min at 37°. The cells were then rinsed with ice-cold PBS and prepared for HPLC analysis as described in the text. The intracellular concentrations of the amino acids were calculated using an intracellular water value of 2.0 μl/mg total cell protein as measured by the method of R. F. Kletzien, M. W. Pariza, J. E. Becker, and V. R. Potter [*Anal. Biochem.* **68**, 537 (1975)]. NM, Not measurable.

Just prior to use, 4 ml of the OPA/MeOH solution is mixed with 100 ml of the boric acid solution and then 0.1 ml of 2-mercaptoethanol is added. This working stock solution is stored under nitrogen gas.

The effluent from the column is mixed with the OPA reagent by pumping the latter into a postcolumn mixing chamber at a rate equal to that of the column flow rate (0.3 ml/min). The OPA derivatization occurs in the baffled mixing chamber at room temperature after which the fluorescent derivative is detected with a fluorimeter fitted with a continuous flow cell. The data can be stored, quantitated, and analyzed further through the use of a variety of computational devices.

Figure 2 illustrates an analysis of the accumulation of S-methyl-L-cysteine, a compound not commercially available in labeled form, in the presence or absence of the aminotransferase inhibitor aminooxyacetate (AOA).[19] The hepatocytes were incubated for 3 hr in amino acid-free medium to partially deplete the intracellular amino acid pool. The quantitation of the intracellular concentration of individual amino acids in this experiment demonstrates a slower rate of catabolism in the presence of AOA for most of the amino acids measured (Table I). Indeed, S-methyl-L-cysteine does not accumulate to significant levels in the absence of the inhibitor.

Concluding Remarks

The use of isolated rat hepatocytes has become common in many investigations of hepatic function. Although a number of methodologies have been described for the isolation of these cells, we have attempted to show that their preparation and utilization for transport studies can be performed with a minimal amount of specialized equipment. Transport of small nutrient molecules has been studied at the level of whole organs, isolated cells, and reconstituted proteoliposomes. Hopefully, our understanding of transport processes will increase through the use of methods that are easily developed and that are flexible enough to allow for the study of a broad spectrum of molecules. Although the methods described here have been developed for isolated hepatocytes, the transport assays outlined can be applied, with slight modifications, to most cells that grow in monolayer or suspension culture. Likewise, these assay methods can be applied to any number of radioactively labeled substrates.

[19] J. W. Edmondson, L. Lumeng, and T.-K. Li, *Biochem. Biophys. Res. Commun.* **76,** 751 (1977).

[38] Distinguishing Amino Acid Transport Systems of a Given Cell or Tissue

By HALVOR N. CHRISTENSEN

Each enzymatic reaction is ordinarily assumed to be produced by a homogeneous population of active sites, except as the properties of one site may be modified by occupation of an adjoining one. Kinetic methods usually are applied to determine the characteristics of the reaction under the assumption that the catalytic agency is a single, homogeneous one, which with that exception may occasionally be a safe assumption. Ordinary kinetic methods are, unfortunately, quite inefficient at uncovering heterogeneity. Adequate correspondence to the Michaelis–Menten equation can be found for the sum of the action of two separate enzymes, or of an enzyme in admixture with a modified but independent form of the enzyme, each catalyzing the same reaction, in spite of considerable separations in the values of their kinetic parameters. The purpose of this chapter is not to consider the question whether significant heterogeneity is being overlooked in the enzymologic field, but rather how to recognize and resolve catalytic heterogeneity where it is more conspicuous, namely in the catalysis of membrane transport. The possibility, indeed the probability, that two or more agencies will simultaneously contribute to the transport of a single substance is now well established.

Not only does a variety of transport systems operate in parallel for a given solute, but we have come to appreciate that a membrane transport system that resembles one of the familiar systems A, ASC, or L, is not necessarily identical with that system in detail. Although the problems of designating transport systems by a brief symbolism is not my central subject here, it is difficult to review the strategy of discrimination without naming some of the examples serving that review. A glossary of the better characterized systems is shown in Table I.[1] This list includes Na^+-dependent system B, perhaps actually a family of transport systems, distinguishable from system A by its considerably wider scope among the amino acids and their analogs. The specific example of system B first uncovered occurs in murine blastocysts[2] and apparently also in the sea urchin egg and oocytes of other marine invertebrates. Superscripts $B^{0,+}$ were affixed to indicate that this system accepts not only zwitterions,

---

[1] H. N. Christensen, *J. Membr. Biol.* **84**, 97 (1985).

[2] L. J. Van Winkle, H. N. Christensen, and A. L. Campione, *J. Biol. Chem.* **260**, 12118 (1985).

including the bulky 3-amino-*endo*-bicyclo[1.2.3]octane-3-carboxylic acid, but also lysine and somewhat analogous cationic amino acids. Multiplication of known transport systems is not the only change in the list of transport systems: in at least one case, a presumably unique transport system, namely system x_A^-, turns out to represent instead the operation of system ASC[3] at low pH for anionic amino acids. The evidence that it can also operate for dibasic amino acids makes it potentially a comprehensive system.[3] In contrast, systems LI[4] and T,[5] which were for a time unrecognized and hence lumped with system L in rat hepatocytes and human erythrocytes (see Chapters 37 and 47), respectively, have been added; also a system asc[6] (note the lower case letters) suggested to represent the operation of a system ASC that had been rendered Na^+ independent by maturation of the red blood cell. Although actually a fundamental biological entity,[7] asc bears a possibly vestigial relation to Na^+ that supports the functional similarity to ASC.[8] Several allelic forms of system asc have been discriminated for the horse erythrocyte.[9] The reader might also take note of the transport of cystine by one system (last item in Table I) as a glutamate-like anion,[10,11] by another system as a lysine-like cation,[12] or by a third type of system unlike either of these.[13] On the whole the list of systems grows as study is extended, even if we disregard their slightly or considerably different expressions in various tissues and mutants.

This chapter should meet the need for an increasingly sophisticated discrimination of each catalytic contribution, i.e., by the attribution of each distinct component to a presumably homogeneous transport system. The findings of a new heterogeneity of transport mediation which necessitates the present reemphasis have no tendency to decrease the ultimate importance of recognizing the several transport systems in their noteworthy occurrences and reoccurrences in various cells and at various evolutionary and differentiative levels. We must ultimately expect these discriminations and reoccurrences either to be confirmed or denied by

[3] M. Makowske and H. N. Christensen, *J. Biol. Chem.* **257**, 14635 (1982).

[4] L. Weissbach, M. E. Handlogten, H. N. Christensen, and M. S. Kilberg, *J. Biol. Chem.* **257**, 12006 (1982).

[5] R. Rosenberg, J. D. Young, and J. C. Ellory, *Biochim. Biophys. Acta* **598**, 375 (1980).

[6] D. A. Fincham, D. K. Mason, and J. D. Young, *Biochem. J.* **227**, 13 (1985).

[7] J. V. Vadgama and H. N. Christensen, *J. Biol. Chem.* **260**, 2912 (1985).

[8] J. D. Young, D. K. Mason, and D. A. Fincham, *J. Biol. Chem.* **263**, 140 (1988).

[9] D. A. Fincham, D. K. Mason, and J. D. Young, *Biochim. Biophys. Acta* **937**, 184 (1988).

[10] S. Bannai and E. Kitamura, *J. Biol. Chem.* **256**, 5770 (1982).

[11] M. Makowske and H. N. Christensen, *J. Biol. Chem.* **257**, 5770 (1982).

[12] C. E. Dent and G. A. Rose, *Q. J. Med.* **20**, 205 (1951).

[13] R. L. Pisoni, J. G. Thoene, R. M. Lemons, and H. N. Christensen, *J. Biol. Chem.* **262**, 15011 (1987).

TABLE I

SUMMARY OF SOME KNOWN AMINO ACID TRANSPORT SYSTEMS IN TISSUES AND CELLS OF HIGHER ANIMALS[a]

Systems for dipolar amino acids	Description	Somewhat analogous system[b]
Na+-dependent systems		
System Gly	Widespread; Gly and Sar; variants known	Imino, iminoglycine, and epithelial systems
System A	Ubiquitous, serves for most dipolar aas; often repressed; variants known } Tolerance of N-Me group	
System ASC	Ubiquitous; some variation in scope; excludes N-Me aas, but often includes prolines; on protonation accepts analogous anionic aas; may accept Arg without Na+	Na+ independent asc in various red blood cells has somewhat analogous scope
System $B^{0,+}$	Widescope system, oocytes, blastocysts, probably renal and intestinal brush borders; accepts cationic and bicyclic aas despite their bulk	$b^{0,+}$, a Na+-independent analog, more sharply limited by positions of branching in preimplantation mouse blastocysts
System N	Gln, Asn, His; term so far restricted to the hepatocyte	N^m, a Gln transporter of muscle
β-System	β-Ala, Tau, 4-aminobutyrate	System specific to 4-aminobutyrate
Na+-independent systems		
System L	Ubiquitous; tolerates bulky and branched chains; high exchange property; bicyclic aas as model substrates	Separate systems L1, etc., low K_m analog develops on incubating of hepatocytes

System T	Erythrocytes, hepatocyte; favors benzenoid aas	System t in fibroblast lysosome

Na$^+$-independent systems for cationic amino acids
(systems specific each to Arg and Orn known)

System y$^+$ — Ubiquitous; fails to discriminate between Arg and Lys and homologs — Variants as in lysosomes and mouse blastocysts do not accept (homoser + Na$^+$)

Systems for anionic amino acids (capital X where Na$^+$ dependent)

System X_{AG}^- — Ubiquitous, includes neurons; Na$^+$ dependent; similarly reactive with Asp$^-$ and Glu$^-$

System X_A^- — Corrected designation ASC, operating in its protonated form; largely excludes Glu$^-$ and longer analogs

System x_G^- — Glu$^-$ and analogs, largely excluding Asp$^-$ and short analogs; Na$^+$ independent

System x_C^- — Like x_G^-, except cystine competes and exchanges with Glu$^-$; at least in some cases, locked into exchange; explains some CNS Glu$^-$ "binding"

[a] Abbreviations not defined in this table: aas, amino acids; Glu$^-$, glutamate, whereas the usual abbreviation Glu does not specify whether glutamic acid or glutamate is meant; GSH, glutathione.

[b] Note that these analogies have led to similarities in the designations, e.g., ASC and asc, B$^{0,+}$ and b$^{0,+}$, L and L1, also N and N^m. These similarities should not be taken as decisions for identity, or that one system is somehow the precursor to the other.

identification of the DNA sequences establishing their unique identities. Efficiency in the approach to that ideal calls *pro tem* for as much precision as possible in the functional recognition of differences and identity of transport systems, and subsequently of their corresponding gene products, to which the ultimate discrimination is applied. We cannot expect to ask the best questions about the mechanism or regulation of a transport unless it has met tests for homogeneity of function.

With that ultimate goal in mind, we may do well to prepare ourselves for the evolution of designations for gene products as they come to be recognized, so as to give them forms that correspond to genetic conventions, namely by the use of three successive capital letters. This objective motivates me now to a further brief excursion into the subject of the abbreviated designations for transport systems. For system ASC, no problem needs to arise, since its now familiar designation has a form appropriate for a gene product. For system Gly, a small change to GLY meets the need. For transport systems currently designated by a single capital letter, the addition of two more capitals, e.g., AT, to represent *amino acid transporter* could serve with a minimal necessary complexity. Addition of these two letters was suggested by Dale Oxender, who has helped me appreciate the emergent need. Thus a gene product characteristic of system A might be designated AAT; one characteristic of system B, as BAT; one for epithelial system G as GAT; one for system L as LAT, one for lysosomal system d, perhaps DAT, and so on. In that connection one wonders if other occurrences of any of the lysosomal systems will be found[13,14,14a]; whether each organellar occurrence is genetically unique, as is suggested by the manifestations of the congenital disease, cystinosis; or to what degree the particular context of its occurrence might cause expression to lead to different operation from that in another context. What we may perhaps expect more often than molecular identity between similar transport systems are high degrees of sequence homology between the transport-catalyzing gene products, such as that now found between the glucose transporter of the human hepatoma cell and of rat brain cells with two monosaccharide systems of *Escherichia coli*.[15]

For the gene product characteristically related to system LII, the transport designation might be LAT II, in analogy to LIV II among prokaryote gene products. To indicate for a gene product a strong preference for anionic glutamate over aspartate, a designation XAG-G⁻ might serve,

[14] R. L. Pisoni, J. G. Thoene, and H. N. Christensen, *J. Biol. Chem.* **260,** 4791 (1985).
[14a] H. N. Christensen, *Biosci. Reports* **8,** 121 (1988).
[15] M. C. J. Maiden, E. Davis, S. A. Baldwin, D. C. M. Moore, and P. J. F. Henderson, *Nature (London)* **325,** 641 (1987).

where the transport system has provisionally been designated X_G^-. The parallel forms XAT-A$^-$, XAT-C$^-$, and XAT-AG$^-$ might then serve. The letter Z, for zwitterionic, in parallel to X, for anionic, and Y, for cationic, might prove the basis for a generic designation, ZAT, thus avoiding the unfortunate implication inherent in the occasionally used adjective, neutral, namely that we are dealing with uncharged molecules. If any of the suggested designations (which for the moment are purely suggestive) has been preempted, it can of course be revised for the new application.

Some of the system designations presently under use may prove to have represented efforts to convey rather too much information when we face the needs for accommodating the designation of corresponding gene products. For example, a proposal to use lowercase letters to indicate Na^+ independency of a transport system can scarcely survive the special demands foreseen for extension of the designation of gene products, once such are identified. A designation asc might perhaps better be written ASC II for the gene product. In selecting designations for gene products, one can probably expect that such symbols, once adopted, may ultimately displace the several one-letter designations currently in use for some of the transport systems. Hence it seems of considerable practical importance that a set of designations generated as needed for the gene products preserve as much evocation of the present set of terms as is reasonably possible. Although the designation of systems and corresponding gene products is not central to my present chapter, heterogeneity in the catalysis of transport of all polar nutrients and metabolites will probably continue to be encountered. Therefore we should remain ready for strains on the set of designations used for transport systems.

The steps proposed for discriminating distinct transport systems will now be itemized. This scheme is an updating of one formalized in 1965.[16] After each item some reasons and details will be discussed.

Step I. For a given tissue or cell type, determine the time course of the uptake of each member of a group of amino acids, some of them selected because they have served in some other tissue to represent a transport system.

In some biological contexts the structure selected must be rather resistant to metabolic modification. Optically active isomers, usually the L forms of amino acids, should be selected where available, especially when the substance is to serve as a transport substrate rather than as an inhibitor. In some cases transport systems more sharply recognizable with a D-

[16] H. N. Christensen, *Fed. Proc., Fed. Am. Soc. Exp. Biol.* **25,** 850 (1966).

amino acid than with an L have, however, been described, e.g., for tryptophan as an inhibitor[5] but not a substrate,[17] or for aspartate and its analogs (for a review, see Ref. 18). Furthermore, an artificial amino acid may serve better than any natural one to represent a given transport system. Limitation of attention to the ordinary, naturally occurring amino acids, which appears to recommend itself to some investigators, may represent serious blindness to opportunity, as will be evident from the large proportion of the system-specific analogs so far identified that are artificial compounds. In short, to describe each of the natural transport systems we often need a wider range of amino acid structures than nature has given us. Our own eyes were opened to this possibility beginning in 1948[18a,19,20] when we initiated such a use of 2-aminoisobutyric acid (AIB), an innovation since followed well over 1000 times.

To permit us to proceed through the present scheme, our study of its time course should show that transport is not so fast as to prevent our observing it over an interval brief enough to approximate the initial rate, i.e., the influx minimally contaminated with subsequent efflux. A method is accordingly also needed for terminating uptake abruptly enough to measure accurately that time interval. For discrete samples of a tissue, or for cultured cells attached in a monolayer, simple physical withdrawal of the sample from the aqueous medium from which uptake is observed, or a washing of it (e.g., with a modified, ice-cold medium) may easily meet this need. For suspended cells or organelles, brief high-speed centrifugation may serve. In the latter case, separation can be enhanced where necessary by interposing a silicone oil of suitable density between the aqueous medium and the cells undergoing terminal sedimentation. Occasionally tests are made at lower temperatures, say 25°, to slow the flows. Inhibition analysis of transport can frequently be achieved even if the initial rate is scarcely accessible so that one must assume a certain amount of efflux has led to decreased net uptake; but that procedure will exaggerate the role of a slower transport process and may underestimate the role of faster processes, particularly of rapid exchange processes, an effect that can be useful for their separate description. Nevertheless, for understanding of the physiological situation, it remains important to discover the nature of the overall steady state. If the test substrate is transported by more than one route, neither route is apt to attain its own independent

[17] S. López Burillo, J. Garcia-Sancho, and B. Herreros, *Biochim. Biophys. Acta* **820,** 85 (1985).

[18] H. N. Christensen and M. Makowske, *Life Sci.* **33,** 2255 (1983).

[18a] H. N. Christensen, M. K. Cushing, and J. A. Streicher, *Arch. Biochem.* **220,** 287 (1956).

[19] H. N. Christensen, A. J. Aspen, and E. G. Rice, *J. Biol. Chem.* **220,** 287 (1956).

[20] M. W. Noall, T. R. Riggs, L. M. Walter, and H. N. Christensen, *Science* **126,** 1002 (1957).

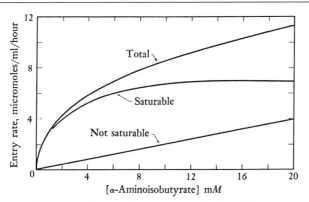

FIG. 1. Uptake by two distinct mechanisms, one of which fails to show unequivocal saturation under the conditions of the test. Deduction of a linear component (lower line, "not saturable") is presumed here to leave a rectangular hyperbola for the saturable component. Reasons for deviation of the nonsaturable component from linearity follow from the discussion of Fig. 2 (see Christensen,[56] p. 134).

steady state. Instead, entry by one process, exit by another, may dominate in various real biological situations.[21] Acceptance of this relation is decisive to understanding pathological shifts in amino acid distribution and flow, as in phenylketonuria.[22]

For selecting a group of amino acids which might be expected to identify the various transport systems of a cell, one may refer to recent summaries of the range and scope of amino acid transport systems.[1] One should not trust too rigidly, however, that this list and the list of pertinent analogs defining it are complete and applicable for every new context.

Step II. Next, investigate for the pertinent amino acids the effect of substrate concentration on initial rate. Examine the kinetic curve for segments that may correspond to a rectangular hyperbola. The presence of a second hyperbola may be suggested by the data.

Before obtaining an uncomplicated rectangular hyperbola, one may well need to deduct a component of uptake known as the nonsaturable transport. This component shows up as an apparently straight-line relation of rate to concentration, usually continuing beyond the more or less hyperbolic region to the highest reasonably accessible concentrations. The correction is illustrated in Fig. 1.[23] It is based on the apparent permeability coefficient applying to the nonsaturable component, which in the simplest case is the first order rate coefficient for uptake at the highest

[21] H. N. Christensen and M. E. Handlogten, *J. Neurotransm., Suppl.* **15**, 1 (1979).
[22] H. N. Christensen, *Biochem. J.* **236**, 929 (1986).
[23] H. Akedo and H. N. Christensen, *J. Biol. Chem.* **237**, 118 (1962).

concentrations tested. Mediated uptake may dominate so strongly over the nonsaturable that no correction need be made, as Hoare[24] concluded for leucine uptake by freshly collected human red blood cells, although we had earlier found a correction necessary where red cells from banked blood were included in the measurements.[25] If one is permitted to assume that the nonsaturable uptake includes influx only, a simple deduction will yield the mediated component. All too often the correction has thus been purely a subtraction. If, however, the interval allows substantial reexodus of accumulated solute by the nonsaturable route(s), a more complex correction needs to be made, one which takes into account whether the steady-state distribution of the solute tends toward unity[23] or toward a higher ratio (presumably true for uphill transport, or cellular uptake of a cation) or toward a lower ratio (as for an anion).[26] The correction may shift from positive to negative as the test solute concentration is increased. At low levels the nonsaturable route, even over a short interval, may allow net exodus of solute already entered by the mediated, uphill route. The declining curve in Fig. 2 shows typical behavior for the transport of an amino acid analog. As the external concentration was lowered below 0.4 mM, uptake of benzylamine by the Ehrlich ascites tumor cell, as measured by the scale on the right, already apparently became concentrative during the first minute.[26] As a result, the nonsaturable component could provide a net exodus (curve C, scale at left), as expected if the gradient was outwardly directed. This direction shifts to net entry at a 0.5 mM external test concentration. Among the three curves of Fig. 2, the deviation from linearity for the nonsaturable uptake decreases with solute concentration as the inner compartment serves better and better as a sink for the entering solute. In this chosen example, the equilibrium distribution, $[S]_i/[S]_o$, for the nonsaturable migration has been calculated for values of 0.5, 1.0, and 2.0, corresponding in theory at the transmembrane potential characteristic of the Ehrlich cell to values expected for an anion, for a molecule without net charge, and for a cation, the latter as represented by the tested benzylammonium ion at pH 7.4. If the mediated transport were fast enough to allow us to select a sufficiently short uptake interval to approximate the initial rate for nonsaturable migration, linearity would be obtained. For the slow but highly concentrative system A, the deviation from linearity is severe; but the circumstance that the mediated influx tends for that system to be nearly independent of the internal amino acid concentration allowed us to derive a relationship between the

[24] D. G. Hoare, *J. Physiol.* (*London*) **221**, 311 (1972).
[25] C. G. Winter and H. N. Christensen, *J. Biol. Chem.* **239**, 872 (1964).
[26] H. N. Christensen and M. Liang, *J. Biol. Chem.* **241**, 5552 (1966).

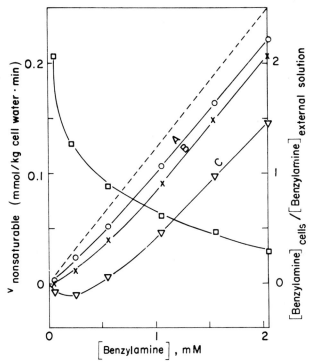

FIG. 2. The calculated net of the nonsaturable migration of benzylamine from and into Ehrlich ascites tumor cells as influenced by the steady-state position toward which the nonsaturable migration tends, and by the external test concentration of benzylamine. See text for discussion. (Reprinted from Christensen and Liang[26] by permission.)

apparent permeability coefficient and the corrections for the nonsaturable influx and efflux.[23] We later arranged this equation to show separately a multiplicative and a subtractive correction (a and b) for each relative rate (DR) observed, thus

$$v/[S] = a[(DR/n) + b]$$

where n is the equilibrium distribution ratio predicted for the nonsaturable migration.[26,27] Sets of values for a and b can be tabulated to correspond to the predicted equilibrium position of the nonsaturable migration and to the value obtained for the permeability coefficient. (Copies can be obtained on request from the author.)

For the several systems for which influx is stimulated by the height of the internal concentration of the test solute, plus the effective concentra-

[27] H. N. Christensen and M. Liang, *J. Biol. Chem.* **241**, 5542 (1966).

tion of any transport analogs endogenously present, the above equation will not, however, apply. Indeed, for such systems the internal compartment can serve so adequately as a sink for the accumulated solute that linearity of uptake may be assumed for the nonsaturable component, according to the familiar equation

$$v = [(v_{max} \times [S])/(K_m + [S])] + K_D[S]$$

or its programmed logarithmic equivalent. (Note that for system A the deviation of the nonsaturable uptake from linearity is maximal at a test concentration well below the K_m, whereas for systems that show uptake by exchange the maximal sink action of the inner phase, which allows linearity, instead is seen at low external test concentrations.)

A number of uncertainties obviously enter into the treatment of the nonsaturable component of solute transport, including (1) uncertainty as to the chemical potential of the solute in the phase into which its transport is directed. The steady-state distribution of 3-O-methylglucose[28] has been used to estimate the apparent intracellular water available as solvent for entering uncharged, hydrophilic solutes. Efforts to calculate the actual net nonsaturable uptake of an amino acid by lysosomes must await reliable measurements of the internal volume entered by a quantity of the amino acid. (2) Uncertainty as to whether any special energization may apply to this nonsaturable component. For example, the actual curvature obtained for nonsaturable benzylammonium ion uptake corresponded to an equilibrium distribution ratio of 0.5 to 0.6,[26] rather than the value of about 2 predicted for a univalent cation and from the transmembrane potential characteristic of the Ehrlich cell. This result may suggest an unexpected dominant energization of extrusion of this solute, a possibility not excluded by what is known about nonsaturable migration of amino acids and their analogs.

These two uncertainties are both pertinent to our observation[29] that α,α-dicyclopropylglycine, selected as an amino acid analog fitting none of the known transport systems, approaches in 2 hr an apparent distribution ratio of 1.45, not 1.0, between the internal and the external aqueous phases of the Ehrlich cell.

The nonsaturable transport should meet other tests that will be clear from what follows later. These tests call for failure of both influx and efflux to respond to excesses of not only the transport substrate itself, but also of any analog in the same phase (e.g., by cis inhibition) or in the

[28] R. F. Kletzien, M. W. Pariza, J. E. Becker, and V. R. Potter, *Anal. Biochem.* **68,** 537 (1975).

[29] H. N. Christensen and M. E. Handlogten, *J. Biol. Chem.* **243,** 5428 (1968) (see Fig. 14, p. 5435).

phase on the opposite side of the membrane (i.e., by trans stimulation or trans inhibition). As a historic example, mediation in the rapid passage of the chloride ion across the red blood cell membrane (by now extended to include its slower passage for various other cells) could be established as mediated only by showing inhibition by analogs having affinities much higher than that for chloride. Findings that other agents or conditions may eliminate or modify a component of transport also raise doubts as to the meaning of nonsaturability otherwise observed for it. The disappearance for a solute of either a specific component of influx or of efflux, or of trans effects on those fluxes, when a transport system is congenitally absent (e.g., in crystinosis) also points to mediation of these fluxes by the defective system, even though other tests for their mediation might so far have failed.

If inclusion in the analytical sample of a considerable extracellular phase is unavoidable in studying a tissue, for example with a tissue slice or a hemidiaphragm, the correction may be dominated by diffusion into that phase. Especially for tissues that have been subjected to damaging conditions, as for the banked red cells just mentioned, the nonsaturable uptake will include the net effect of any leakiness of the biological membrane. For compact cell isolates, the contribution of leakiness frequently is seriously exaggerated, however, when the term *diffusion* is uncritically assigned to the whole component. A classical discussion by Wilbrandt and Rosenberg[30] shows why it is a fallacy automatically to attribute to diffusion any uptake that fails to saturate under a given set of experimental conditions. Mediated transports may well become half-saturated only at inaccessibly high concentrations, so that their observed rates may bear an apparently linear relation to concentration. At least for the amino acids the nonsaturable component of uptake shows features not plausible for a major contribution by simple diffusion, such as a degree of structural specificity and a high-temperature sensitivity.[31] Structural specificity of a flow indicates for innumerable instances that pinocytosis cannot be a major component. Implausible and inconsistent porosity of the membrane to polar molecules is often presupposed by attributing nonsaturable uptake to simple diffusion. Similarly sharp limitations as to the contributions made by diffusion apply also to exodus of amino acids from cells.[29] By synthesizing amino acid analogs for which a somewhat larger proportion will be present in the most lipid-soluble form, $RCH(NH_2)COOH$, rather than the usual zwitterionic form, $RCH(NH_3^+)COO^-$, it has been possible

[30] W. Wilbrandt and T. Rosenberg, *Pharmacol. Rev.* **13,** 109 (1961) (see pp. 117, 118, and 125).

[31] H. N. Christensen and M. Liang, *Biochim. Biophys. Acta* **112,** 524 (1966).

to show that the resultant lipid solubility is still too small to contribute measurably to nonsaturable passage of free amino acids across most membranes.[32] Passage through the lipid barrier has occasionally been implied implausibly for a highly charged form, e.g., the zwitterion. The concentrations of the zwitterionic form of ordinary amino acids appear not, however, to differ sufficiently to account for the differences in the apparently passive migration of the various neutral amino acids.

What some investigators have called a second, lower affinity mediated route on the basis of an incompletely convincing curvilinearity in the continuation of a kinetic plot, other investigators have simply deducted as nonsaturable uptake. For the immediate purpose of describing the clearly mediated component, this decision may be of minor importance. When we remember possible kinetic complexities, perhaps we should, however, be careful to establish the independent character of any component of uptake that we might otherwise deduct simply to obtain the smooth rectangular hyperbola we may seek as evidence for homogeneous mediation. If we are to depend on the detailed curvature of the mediated residuum of transport as a function of concentration, we must obviously not overlook any significant curvature applying to the nonsaturable component to be subtracted. Furthermore, even though a similar mode of correction of transport kinetics may serve, whatever the functional origin of a nonsaturable component of membrane transport of small polar molecules, we should not neglect the biological importance that differences among these components may have. While nonsaturable passage may occasionally arise from artifactual or biological leaks, it appears likely often to be due instead to one or more high K_m and perhaps regulatable transport mediations that deserve closer study. Consider for example that 40% of the passage of tryptophan to the brain has been assigned to the nonsaturable category.[33] Amino acids minimally reactive with system L may also cross this barrier at physiologically important rates, even if slowly and with little evidence for saturation.

Furthermore, the standards for recognition of a true rectangular hyperbola, or of the corresponding linear segment in a linear transformation, are occasionally not high enough to justify extraction of kinetic parameters taken reliably to describe a homogeneous transport activity. In one paper kinetic parameters for a mediated transport are extracted by "linear transformation" from data already well represented by a straight-line relation between rate of uptake and concentration. It seems to me a

[32] H. N. Christensen and M. E. Handlogten, *Biochim. Biophys. Acta* **469**, 216 (1977).
[33] L. P. Miller, W. M. Pardridge, L. D. Braun, and W. H. Oldendorf, *J. Neurochem.* **45**, 1427 (1985).

Michaelis plot (v versus [S]) that is already a straight line can scarcely carry information extractable by one of the familiar linear transformations.

Returning to the correction, if we find that the nonsaturable migration indeed occurs to a substantial extent by an unidentified mediation, that mediation needs to tend toward a predictable steady state if our corrective equations are to apply precisely. Note also that each point on a kinetic plot, or on a plot of flux versus pH, needs to be corrected independently for the nonsaturable or the Na^+-independent migration applying under the conditions of that point. Furthermore, proving that a mediation is energetically passive can be even harder than proving it active. Nevertheless these corrections are the best we have for the present, and although somewhat approximate, they are often small.

The strategy of using the saturability of a catalytic process to describe the molecular recognition involved in it must be applied in the concentration range where saturability is most evident, i.e., in the vicinity of the half-saturating concentration, K_m. If transmembrane migration of the substrate persists to concentrations so high as to largely saturate the catalytic process, then tests for analog competition at such levels will largely describe a nonsaturable component, i.e., one that shows at least that single characteristic of diffusion. The result will have no validity whatever for excluding the occurrence of mediated transport at lower concentrations, whatever its physiological significance. This caution becomes especially important if the method used yields the *net* flux across the membrane, e.g., by the osmotic effect of the flux. The net mediated flux of course approaches zero as the substrate concentration chosen becomes very high relative to K_m.[30]

Step III. Beginning with a selected amino acid yielding an apparently discrete rectangular hyperbola for the plot of uptake rate versus its concentration, define the approximate range of variation in its structure that will still allow analogs to act as apparently competitive inhibitors of its migration. Select several of these inhibitors for closer study, on the basis of the behavior observed.

Step IIIA. Use these analogs to challenge the apparent homogeneity or uptake of the initially selected amino acid.

The principal concern of this chapter is how we may challenge with analog inhibition analysis the apparent homogeneity or duality of catalysis of uptake of a test solute. The difficulty we face is that a kinetic plot corresponding within reasonable standards to a rectangular hyperbola (and hence to a straight line by a linear transformation) does not establish

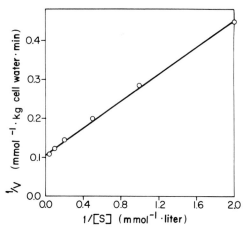

FIG. 3. Calculated total influx of a solute by two processes operating simultaneously. K_m values of 1 and 4 mM have been assumed for the two; also a V_{max} value of 5 mmol/kg cell water · min for each. See text. (Reprinted from Christensen[16] by permission.)

that the catalytic action is indeed produced by a homogeneous population of active sites. By plotting hypothetical data as in Fig. 3, the reader can easily establish that a Lineweaver–Burk plot or other linear transformation describing the summated action of two catalytic sites, one with a K_m or K_i value as much as five times the other, their V_{max} values being similar, can readily yield, as in that figure, a plot that will satisfy ordinary standards of linearity. Under these conditions the supposed value for K_m and for V_{max} obtained by the traditional graphic extrapolations would correspond to no real system. Conclusion for homogeneity of catalysis must rest on firmer evidence. Numerous cases can be cited where a transport activity that yields approximately linear plots has subsequently been shown separable into two components which also yield linear plots, i.e., hyperbolic kinetics. An example showed hyperbolic kinetics for Na^+-dependent proline uptake by fibroblasts, even though the presence of 20 mM MeAIB divided that uptake into an inhibitable component, also showing hyperbolic kinetics, and another hyperbolic component not inhibited by MeAIB.[34] The hyperbolic form of the kinetics of total Na^+-dependent uptake in that case did not, incidentally, mislead the authors into supposing that it represented a homogeneous transport system. On the basis of prior experience these two components were logically ascribed to transport systems A and ASC, and again the Eadie plots ob-

[34] S. B. Russell, J. D. Russell, and J. S. Trupin, *J. Biol. Chem.* **259,** 11464 (1984).

tained were approximately linear. We see retrospectively that per se none of the three apparently linear Eadie plots obtained in this study established transport by a single homogeneous agency.

Certain conclusions as to the meaning of the time course followed by the transmembrane accumulation of a solute, e.g., the intensity of an overshoot following addition of an external load, depend heavily on the rarely well-justified assumption that only one transport component contributes to the succession of steady states observed. Furthermore, even if a second contaminating component resists saturation, the character of the steady state toward which it tends can be critical.

One could also cite converse cases in which parameters for two supposed transport systems have unwisely been extracted directly from a plot by linear transformation, even though the deviation from linearity proves on statistical examination insufficient to establish heterogeneity. Various computer programs that yield standard erorrs for the kinetic parameters from "the double Michaelis–Menten equation" have been discussed by Duggleby.[35] Another problem arises in mathematically excluding a possible contribution by a third system. A third version of system L (beyond L1 and L2)[4] appears to be added on viral transformation of fibroblasts,[36] and to be seen when transport is tested under infinite trans conditions. Its appearance on viral transformation supports that the uniqueness indicated for the third system by inhibition analysis is probably real.

The analysis just described for Na^+-dependent proline uptake by systems A and ASC[34] was made on the basis of prior experience with the discrimination between two systems that MeAIB inhibition might provide. How do we select such a discriminatory inhibitor without that prior knowledge? One begins best by including in the test analogs that range widely in structure from the solute first selected as transport substrate, and gradually proceeding as necessary to closer analogs in order to define the full range. Mixtures of stereoisomers (e.g., b± BCH) may serve successfully for tests as transport *inhibitors*, whereas biological resolution of the isomers during a study is likely to complicate their study as transport *substrates*. Usually analogs with the same charge distribution as the substrate will be tested first, although progress has occasionally been retarded by assuming on the basis of earlier experience that a pair of amino acids presumed to be of unlike charge will not interact for transport, e.g., lysine with phenylalanine (Fig. 4); threonine with cysteine sulfinate[32]; or

[35] R. G. Duggleby, *J. Theor. Biol.* **130,** 123 (1988).
[36] S. A. Gandolfi, J. A. M. Maier, P. G. Petronini, K. P. Wheeler, and A. F. Borghetti, *Biochim. Biophys. Acta* **904,** 29 (1987).

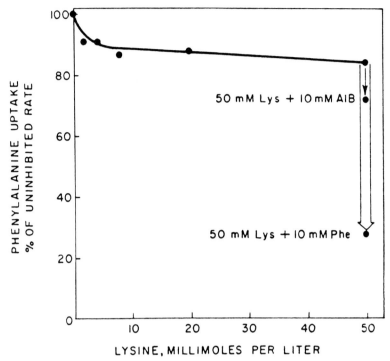

FIG. 4. A minor component of the uptake of phenylalanine by the Ehrlich ascites tumor cell appears to be readily inhibited by lysine. (Whether this component measures transport shared between the two amino acids by the Ehrlich cell would require further tests, e.g., by the ABC test of CI; also a test whether, for example, contaminating leukocytes might account for this small component.) Note then that further weak inhibition is produced by higher lysine levels, to an unattained maximum. The residual uptake of phenylalanine by the Ehrlich cell has been shown saturable and somewhat inhibitable by 2-aminoisobutyric acid (descending arrows). (Reprinted from Christensen[16] by permission.)

alanine with the anion, 2-pyrrolidone 6-carboxylate,[37] sometimes unwisely called oxoproline even though strictly it is not an amino or imino acid.[37a]

One may well begin the selection of analogs for test in the desired challenge to supposed homogeneity of a mediated transport by listing the structural features that have previously served for recognition of each of various amino acids, features which might be modified to good effect in analogs. This service might include recognition for enzyme attack (which tends to involve more specific recognition than does transport), or for cellular signaling, as well as for membrane transport. We may check for

[37] R. Ganapathy, A. Roesel, J. C. Howard, and F. H. Leibach, *J. Biol. Chem.* **258**, 2266 (1983).

[37a] H. N. Christensen, *Trends Pharmacol. Sci.* **9**, 430 (1988).

the recognition of the first charged amino and carboxyl groups of the amino acid and the distance between them. One may try the effect of N-methylation of that amino group. Replacement of the α-carboxyl group with a thiocarboxyl or sulfonate group has generally eliminated transport reactivity. We may think of the recognition of apolar mass and its distribution in space; we may try to exceed available space in one or more dimensions for some transport receptors, but not of another. For more complex amino acids, one may question whether a unique atom such as a sulfur atom in the side chain may be a key, or instead largely irrelevant to recognition. We may think of introduction of a second carboxylate group, and varying its distance from the first. Substitution of the second carboxylate group with a sulfonate has preserved recognition. One may think of recognition of a second amino group, of its distance of separation from the first one, and of its substitution by a mono-, di-, or trimethylamino group, the latter permanently cationic and hence decisively different (for the lysosomal membrane, see Ref. 38). One may guanylate the second amino group to generate an arginine analog, which may share transport with the unmodified ω-amino derivative, or may be refused by a receptor for lysine-like diamino acids, or may be recognized by a different system (for blood–brain barrier transport, see Ref. 39). The pK values of the two amino groups can be modified by changing their separation or by introducing a nearby polar substituent, with interesting consequences for sharing of a transport route among analogs. One may modify the H^+ concentration on either side of the membrane to ask which form of a dissociating group is recognized at that point. Guanylation of the distal amino group to form an arginine analog can serve informatively by elevating its pK, along with the more specific consequences already mentioned. The common recognition of histidine, glutamine, and asparagine by system N of the hepatocyte may suggest a role for recognition by H bonding to these structures (see Chapter 37).

One needs to remember that the pH range in which an amino acid becomes reactive with a corresponding transport system is not likely to be found exactly the same as the pH at which its state of charge changes on titration in free solution.[40] The free energy of the binding at the recognition site is of course a factor in the "biological titration."

Inevitably the selection of what may seem the ideal analogs for challenges to transport mediation and its homogeneity is restricted by their

[38] R. L. Pisoni, J. G. Thoene, R. M. Lemons, and H. N. Christensen, *J. Biol. Chem.* **262,** 15011 (1987) (see Tables I and III).

[39] W. H. Oldendorf, P. D. Crane, L. D. Braun, E. A. Gosschalk, and J. M. Diamond, *J. Neurochem.* **50,** 857 (1988).

[40] H. N. Christensen, *Biochim. Biophys. Acta* **779,** 255 (1984).

relative availability in relation to their promise as indicated by the collective past experience. One inevitably becomes indebted to the generosity of other persons for samples, to whom one may be able carefully to justify one's request for a minimal sample, and whom one can then keep informed of results and their significance. Ideally, the mutual dissemination of residual synthesized or isolated portions of such compounds should then serve goals common to donor and recipient scientist.

Step IIIB. Ascertain what portion of the initial rate of uptake of the test amino acid is subject to the inhibitory action of a given analog.

Ideally the inhibitory effect of an analog reaches an approximate maximum at an accessible concentration of the inhibitor. A substantial portion of the rate of uptake of the test amino acid may then remain, only slightly affected by further increases of inhibitor concentrations. An extrapolative method of deducing the maximum inhibition in the former situation was developed by Inui and Christensen,[41] not however applicable for the too-imprecise data of Fig. 4.

Step IIIC. Ascertain the complexity of the interaction, for example, by plotting the inhibitory action as in Fig. 5. The result may suggest (1) competition for an apparently homogeneous population of transport-mediating sites, or (2) competition for two apparently homogeneous populations of sites, (3) noncompetitive inhibition, or (4) complexity too great for direct analysis.

Elsewhere the interaction between a pair of analogs competing for transport has been divided into five classes, rather than the three just mentioned, disregarding possible combinations among even the five classes[42]:

    I. Complete and homogeneous (both analogs yield kinetic curves like diagram 1 in Fig. 5)
    II. Homogeneous but complete in only one direction (one curve is instead like diagram 2 in Fig. 5)
    III. Homogeneous but complete in neither direction (both curves are like diagram 2)
    IV. Complete but homogeneous in only one direction (one curve is like diagram 1, the other like diagram 3)
    V. Complete but homogeneous in neither direction (both curves somewhat like diagram 3)

[41] Y. Inui and H. N. Christensen, *J. Gen. Physiol.* **50**, 203 (1966).
[42] H. N. Christensen, *Adv. Enzymol.* **32**, 1 (1969).

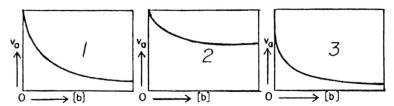

FIG. 5. Diagram to illustrate three cases (1, 2, and 3) of the slowing of the uptake of a solute by adding increasing quantities of an analog b (see text). (Reprinted from Christensen[42] by permission.)

The stubborn cases IV and V, where the heterogeneous interaction resists analysis as to whether it is based on unresolved competition or on noncompetitive inhibition, are the ones requiring further analysis. The downward-directed rectangular hyperbola shown by a saturable inhibition may terminate not with a horizontal line but with an apparently linear, gradual downward slope, showing that higher concentration of the analog produce further although weaker inhibition of residual uptake. Figure 4 shows what appears to be an extreme, real case of that kind, where the presumed K_i values lie so far apart that the uppermost K_i value can probably be neglected in evaluating the first K_i. In this figure, only a circumscribed, here minor, portion of the uptake of phenylalanine is highly sensitive to the presence of lysine at pH 7.4. The longer arrow at the right shows that most of the residual uptake of phenylalanine is still saturable. The portion inhibitable by 2-aminoisobutyric acid (AIB, shorter arrow) ordinarily corresponds approximately to Na^+-dependent uptake, for example, by system A. Figure 4 may already suggest the valuable but demanding strategy of systematically blocking transport components one by one to be described below under Item IV, to isolate presumably all the distinct components.

Replotting data corresponding to the right-hand part of Fig. 1 with a constricted abscissa scale, and thereby including higher inhibitor concentrations, may show that the apparently linear, sloping section is actually part of a second rectangular hyperbola, representing a considerably higher second K_i. The implication then to be evaluated is that uptake occurs by two distinct saturable routes. (Unless in a given membrane the affinities for the uptake of all transport substrates fall uniformly into two distinct ranges, one should be cautious in ascribing one of these to a high-affinity, the other to a low-affinity transport system, because with a different analog the affinity sequence may well be reversed. Correspondingly, diagrams that compare quantitatively the division of the uptake of various amino acids among several transport systems are apt to apply only for a single arbitrarily selected test concentration.) In a third case (diagram 3 in Fig. 5), two or more rectangular hyperbolas may overlap so much as to be resolvable only by curve-fitting methods, or more frequently not reliably resolvable at all by graphic means, given the precision ordinarily secured in measuring rates. A search may then identify a different inhibitor with heightened, perhaps excellent discrimination between the two components. Obviously, accompanying measures of variance of the data are needed for even provisional conclusions as to the nature of the heterogeneity.

Especially where one of the uptake processes half-saturates at very low external solute concentrations, one must be alert to determine whether the process may be a specific binding without a biologically significant release into a second compartment. This discrimination is beyond the scope of this chapter, although study in isolated or artifical vesicles may assist in proving the physical reality of transport. One may suspect that the physiological expectation of the investigator occasionally plays a role in this decision. For example, unexplored high-affinity fixation of aspartate and glutamate may tend to be called binding in nervous tissue but membrane transport in hepatocytes (see Ref. 18).

As another difficulty, the circumstance that uptake of a test solute shows nonparabolic kinetics that correspond approximately to the sum of two rectangular hyperbolas does not establish uptake by two independently mediated routes. The kinetic curves that mimic overlapping rectangular hyperbolas might instead correspond to one of the less frequent forms of regulatory kinetics of catalysis, namely negative cooperativity. An enzymologist would not automatically conclude that two isozymes are present just because the kinetic curve is capable of that interpretation. For membrane transport, a plausible case of this kind was presented by Glover et al.[43] to account for a perhaps spurious duality in bacterial phenylalanine and tyrosine uptake. Figure 6 presents their biphasic kinetic plots of transport of six amino acids, each of which might be interpreted to represent uptake by two mediating routes. For the last two, tyrosine and phenylalanine, the kinetics of uptake retained that form when the pH or temperature was changed, when an irreversible inhibitor was applied, or when mutants were generated with accelerated or slowed uptake of these two amino acids. Hence the authors were led to urge negative cooperativity within a single catalytic system as an explanation for the biphasic kinetics. From this case they argued for a rather general possible explanation for kinetic complexity of the sort illustrated in this

[43] G. I. Glover, S. M. D'Ambrosio, and R. A. Jensen, *Proc. Natl. Acad. Sci. U.S.A.* **72**, 814 (1975).

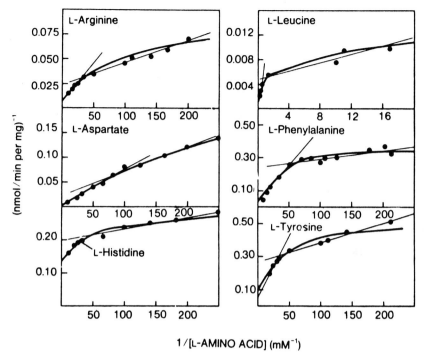

1 / [L-AMINO ACID] (mM^{-1})

FIG. 6. A case in which the Lineweaver–Burk plot for influx suggests the presence of two distinct processes, although the authors conclude that the form of the curve arises instead from negative cooperativity. See text to consider also why one cannot be sure that only two processes participate. (Reprinted from Glover *et al.*[43] by permission.)

figure. Overath and associates[44] have instead suggested that the rate of transport of a metabolite might be the sum of nonidentical contributions of the carrier in a fluid lipid domain and an ordered domain of the membrane. The existence of these possibilities are the main explanations of why our present scheme has had to be rather more detailed than one should wish.

Figure 7 illustrates a second provocative problem of this kind. Figure 7A shows MeAIB transport active by cultured MDCK cells.[45] The straight line over the tested MeAIB concentration range suggests transport limited to system A. Figure 7B and C illustrates the kinetics obtained when MeAIB transport has been either stimulated by prior amino acid deprivation (Fig. 7B) or modified by a chemical transformation of the parent cell line (Fig. 7C). In both cases the transport of MeAIB has

[44] L. Thilo, H. Träuble, and P. Overath, *Biochemistry* **16**, 1283 (1977).
[45] P. Boerner and M. H. Saier, Jr., *J. Cell. Physiol.* **122**, 308 (1985).

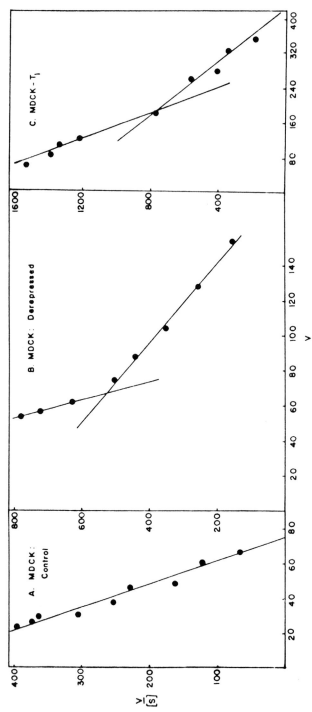

FIG. 7. Eadie–Hofstee plots of Na^+-dependent MeAIB uptake (v in pmol·min^{-1}·mg^{-1} protein) as a function of MeAIB concentration in millimolar units, in amino acid-fed (A) and in amino acid-starved (B) MDCK cells, or in a transformed variant (C) of this cell line. See text for discussion. (Reprinted from Boerner and Saier[45] by permission; this reference should be consulted for procedural details.)

attained more complex kinetics. Has a variant of system A been added, according to the usual interpretation, a system somewhat different in its sensitivity to MeAIB, or has the original system instead taken on more complex kinetics? To answer this question, one would like to know if various analogs show a difference in K_i in their action at low and at high ranges of MeAIB concentration, to see if two different transport-mediating sites are really indicated by the bent curves of Fig. 7B and C. For answering such questions for amino acid transport systems we have the advantage of access to more challenging sets of structural analogs than we have for example for the much more difficult differentiation among kinetic alternatives for anion transport by the red blood cell protein III.[46]

It is important to note that if an inhibition between two analogs is not competitive, its existence does not indicate shared transport. Confusion will arise if one overlooks that a substrate analog may inhibit transport without binding at the receptor site or without being transported as a consequence of that binding. Further, the analog might fill an alternative site completely but only slow the transport and not block it entirely, to produce partial noncompetitive inhibition and a kinetic curve resembling the upper one in Fig. 4. That result would not point to a shared route of mediated uptake. Noncompetitive inhibition of amino acid transport by analogs has been observed in a number of instances (see for example, for system ASC of the rat hepatocyte, Table I in Ref. 47). Obviously analog inhibition must be shown to arise from competition with the transport substrate to maintain the validity of interpretation of the kinetic tests described here. Even for stimulation by an alkali metal ion one should consider, along with possible cotransport, a possible allosteric action, or a response to changes produced in transmembrane potential. Conversely, a shared component of transport could lie hidden in findings of noncompetitive inhibition between two analogs.

It can be difficult to demonstrate that uptake of an analog is associated with its inhibitory action. Occasionally the uptake of an amino acid by a given route may be too small a proportion of its mediated uptake to be statistically significant. The initial search for amino acids transported by system ASC into several cells depended on resistance of their Na^+-dependent uptake to 25 mM MeAIB or N-methylalanine. The surviving rate for several amino acids (e.g., glycine) was too small to be decisive, although glycine had been seen by Vidaver *et al.*[47a] to slow alanine uptake by the pigeon red cell with a very high K_i. For the human fibroblast in culture, a

[46] H. Passow, *Rev. Physiol. Biochem. Pharmacol.* **103**, 61 (1986).
[47] M. S. Kilberg, M. E. Handlogten, and H. N. Christensen, *J. Biol. Chem.* **256**, 3304 (1981).
[47a] G. A. Vilaver, L. F. Romain, and F. Haurowitz, *Arch. Biochem.* **107**, 82 (1964).

distinct mediation of glycine uptake was observed in the presence of 10 mM MeAIB,[47b] pointing to overlap in affinities for systems Gly and A.

How then may we meet the important need to distinguish competitive inhibition for transport between two analogs from a noncompetitive inhibition action by one analog with the transport of the other? One answer is available if that transport is electrogenic: If the inhibitor supplied alone does not produce the predicted response in or to the transmembrane potential,[48,49] one may conclude that its observed uptake does not occur by the electrogenic route, and hence that its inhibition is not based on competition for a shared transport route. The measure of electrical responses for assaying electrogenic membrane transport and its inhibition by various routes appears steadily more valuable.[49a] Such studies also show that energization by K$^+$ gradients should not yet be dismissed.

The evidence that an analog is indeed reacting with the transport receptor site per se, with resultant transport, can often be strengthened by showing mutual trans stimulation between test substrate and analog. Under this phenomenon the presence of the analog accelerates the flow of the substrate across the membrane from the opposite compartment, and vice versa. The terms transstimulation, heteroexchange, and flow-driving counterflow have been applied, although not as precise synonyms. The phenomenon is believed to arise from the more rapid reorientation of the loaded carrier site than of the empty "mobile carrier," i.e., a receptor site subject to reorientation with respect to the two surfaces of the membrane. Hence features of transstimulation should tell us much about the reorientation mechanism. Failure to observe transstimulation between two prospective analogs does not, however, prove that the two do not share a common transport route, since occurrence of the phenomenon depends on which of the carrier movements is rate limiting. It was recognized early that transstimulation might be idiosyncratically missing for valine as a substrate of a transport system, given its peculiarly slow movement outward on the system L carrier.[50] As another example, a cationic amino acid sharing mediation of transport with the combination, dipolar amino acid plus sodium ion, may fail to show transstimulation, one with the other, if the movement of the carrier-bound cationic amino acid, rather than of the empty carrier site, should prove significantly rate limiting to the total exchange.[50a–52] Where the informative phenomenon of transstimulation is

[47b] R. Franchi-Gazzola, G. C. Gazzola, V. Dall'Asta, and G. G. Guidotti, *J. Biol. Chem.* **257**, 9582 (1982).
[48] S. G. Schultz and R. Zalusky, *J. Gen. Physiol.* **47**, 1043 (1964).
[49] E. M. Wright, R. E. Schell, and B. Stevens, *Biochim. Biophys. Acta* **818**, 271 (1985).
[49a] C. Bergman and J. Bergman, *J. Physiol. (London)* **336**, 197 (1986).
[50] D. L. Oxender, *J. Biol. Chem.* **240**, 2976 (1965).
[50a] E. L. Thomas and H. N. Christensen, *Biochim. Biophys. Acta* **40**, 277 (1970).
[51] E. L. Thomas and H. N. Christensen, *J. Biol. Chem.* **246**, 1682 (1971).

seen, half-maximal stimulation may occur at the K_m for migration of the stimulatory analog, although such identity appears not to be obligatory.

Conclusions for coupled exchange of ionic metabolites across a membrane should of course not be drawn simply for metabolic plausibility, nor to meet a felt need for electroneutrality. Kinetic evidence for coupled movements is essential. Just what exchanges of ionic metabolites across the inner mitochondrial membrane are coupled to each other to constitute a shuttle in the physiological context[53] must be especially vital for the conservation and regulation of energetic gradients.

Kinetic evidence for homogeneity of an interaction between two analogs is inductive and hence not absolutely complete. The inductive evidence may be reinforced by the two quantitative steps that follow:

1. (A) If the interaction between analogs A and B suggests their competition for a single population of sites, compare the value of the K_i shown by the inhibitor A with the value of K_m it shows for uptake by the same apparently homogeneous agency. (B) Reverse the roles of test amino acid and inhibitor, and check for possible apparent identity of K_m and K_i values for B. (C) If consistency is obtained, check for consistency of the K_i values of additional analogs in inhibiting the uptake of each of the first two analogs (described as the "ABC" test by Ahmed and Scholefield,[54] as elaborated by Scriver and Wilson,[55] and subsequently discussed by myself.[56] The ABC test was used[2] formally to verify the distinctiveness of system $B^{0,+}$).

I have already emphasized in Step IIIA and in Fig. 2 that an apparent hyperbolic relation between uptake rate and substrate concentration does not necessarily establish that a homogeneous population of sites is involved in the catalysis. The same applies to the relation between rate and the concentration of an inhibitory analog. A consistency between the K_m and K_i values of each member of an interactive pair of transport substrates could represent a fortuity, but this possibility becomes less and less likely as consistent K_i values are obtained for more and more dissimilar analogs. Conversely, because of the possible inclusion of an unsuspected component of transport mediation, an error can be made in concluding that two analogs do not share a mode of transport simply because an inconsistency is observed in the K_m and K_i values for the interaction between the pair.

[52] H. N. Christensen, M. E. Handlogten, and E. L. Thomas, *Proc. Natl. Acad. Sci. U.S.A.* **63**, 948 (1969).
[53] J. W. Young, E. Shrago, and H. A. Lardy, *Trends Biochem. Sci.* **12**, 423 (1987).
[54] K. Ahmed and P. G. Scholefield, *Can. J. Biochem. Physiol.* **40**, 1101 (1962).
[55] C. R. Scriver and O. H. Wilson, *Nature (London)* **202**, 92 (1964).
[56] H. N. Christensen, "Biological Transport," 2nd Ed., pp. 388–389. Benjamin, Reading, Massachusetts, 1975.

Unfortunately the kinetic precision required to show consistency between the K_m and K_i values for each of two amino acids presumed to share a transport system has not always been accessible, for example for absorption by the intact intestinal mucosa, or occasionally for uptake by lysosomes. In such instances the identification of transport systems, however plausible, must be considered particularly provisional until more precise kinetics can be obtained, perhaps by preparing membrane vesicles, or unless genetic, immunological, or other demonstration of its contribution in isolation can be obtained.

2. If the interaction suggests competition for two discrete populations of sites, each of these apparently homogeneous, one should seek to conduct the same test for correspondence of K_m and K_i values for one set of sites, or even for both sets considered one at a time.

A different inhibitor will presumably be needed to test each of these populations more rigorously for homogeneity. We must remain somewhat pessimistic about the chances for successful curve fitting to measure the K_m and K_i values for the test substrate alone to describe two modes of uptake occurring simultaneously unless the quotients V_{max}/K_m for the two populations are separated by a factor as large as 3 (recall Fig. 3).

Step IV. Unless inherent simplicity permits conclusive identification of the modes by which the test amino acid is taken up with corresponding modes by which analogs are taken up by procedure III, one should produce conditions that limit uptake (or inhibition of uptake) to each simple system so that uptake can be observed by it as nearly alone as possible. Seven examples (steps IV,A–G) follow:

Step IVA. Make the test under conditions that largely eliminate one or more routes of transport, e.g., in Na^+-free medium, at a lowered pH, in the presence of a nonanalog inhibitor, or at a substrate concentration restricting most of the uptake to one system.

For example, Inui and I[41] systematically eliminated one of the modes of methionine uptake by the Ehrlich cell by conducting the test in media in which choline replaced Na^+, and measured the transport thus lost. (See Chapter 47 for discussion of Na^+ dependence of transport.) This device, applied earlier for avian red blood cells (Fig. 2 in Ref. 57, presenting 1951 data, Ref. 58) has now served in many cases. A limited lowering of the pH may specifically protonate the transport-mediating system to eliminate or

---

[57] H. N. Christensen, in "Membranes and Ion Transport" (E. E. Bittar, ed.), Vol. 1, p. 368. Wiley (Interscience), New York, 1970.

[58] H. N. Christensen, T. R. Riggs, and N. E. Ray, J. Biol. Chem. 194, 41 (1952).

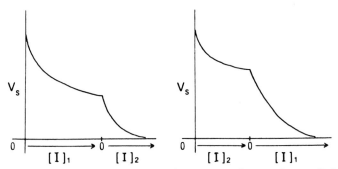

Fig. 8. Inhibition of influx of a test solute by presence of one analog, applied first, and then the superimposed presence of another, each in increasing concentration. The two have been applied in reversed order at the right. In the ideal case each has its action restricted to one component of transport. (Reprinted from Christensen[1] by permission.)

modify[3] its action, or it may change the charge of the transported molecule to modify characteristically its route of uptake.

One needs, however, to verify for each case that the presumably Na^+-dependent system is fully eliminated by the substitution of another cation or osmolite, and that the presumably Na^+-independent system remains unimpeded on that substitution. This test can scarcely make a consistently reliable discrimination by Na^+ dependence if for either one of the systems one substrate reacts only in the presence of Na^+ while the other substrate needs no Na^+ at all,[52] or for which Na^+ may even be inhibitory (Fig. 8 in Ref. 52). Choice of the osmolite substituting for Na^+ must be fitted to the situation.

Experimental elimination of a transport system may assist in discovering how it happens that two transport systems cooperate to produce the observed physiological distribution and certain distributional phenomena. For example, the contribution of system A to amino acid transport into isolated microvessels of the blood–brain barrier appears to have been eliminated experimentally by their treatment with collagenase.[59] By this means it could be shown that the transstimulation of phenylalanine flow by previously accumulated glutamine apparently requires a contribution of both system A (presumably for glutamine uptake) and system L (for the exchange).

Step IVB. Measure transport by a signal shown quite specific to one system, e.g., a decrease in transmembrane potential.

[59] P. Cardelli-Cangiano, A. Fiori, C. Cangiano, F. Barberini, P. Allegra, V. Peresempio, and R. Strom, *J. Neurochem.* **49**, 1667 (1987).

Step IVC. Measure uptake in the presence of one or more competitively inhibitory analogs shown able to eliminate uptake by every retained route but one.

Care as to the concentration of the inhibitory analog: Unless the specificity of each analog is perfect, the question is whether a "window of opportunity" is available, namely a concentration sufficiently above the K_i of the analog used to block the system in question, but not high enough to block another possibly pertinent system. One of the most common errors lies in failing to recognize that an analog sufficiently restricted to uptake by one system to serve as a model *substrate* for that system may show insufficient specificity to serve as a model *inhibitor* of that system.[60]

Suppose one were studying the uptake of glycine by an unfamiliar cell, for which system A is in fact the sole mediated route. Suppose one finds that an enormous concentration, say 20 mM, of a "system ASC substrate," e.g., cysteine or threonine, substantially inhibits glycine uptake. Does that prove that system ASC shared in glycine uptake? Even given that less than 1% of the uptake of this "ASC substrate" itself occurs by system A, nevertheless the K_m for that uptake might be 20 mM. Obviously at 20 mM the ASC substrate could eliminate about half the uptake of glycine via system A, without providing any evidence whatever for the participation of system ASC. It is therefore that one must establish at what levels the conditions of step IV,C are met in a given context.

Step IVD. Modify the structure of the analog to restrict its inhibition as nearly as possible to a single population of reactive sites, and do this for an analog serving as a substrate for each mediated route.

In an interesting recent example of one step in this approach, a suspected low-capacity, high-affinity component of α-L-alanine transport by the rabbit ileum, otherwise easily overlooked, was found to be eliminated by the presence of 80 mM β-alanine, to leave a kinetically simplified alanine influx. For this route the K_m for α-alanine influx was measured as 0.1 mM, that for β-alanine as 2 mM.[61] Incidentally, the presence of only 0.1% of α-alanine in the β-alanine preparation used would have distorted the results by almost halving L-alanine influx by this route. So high a degree of purity of the β-alanine seems plausible, however, because of the totally different origins typical for commercial preparations of these two isomers. I doubt, however, that amino acids isolated from protein hydrolysates, or their N-methylated products, would regularly meet so severe a standard.

[60] H. N. Christensen, *Trends Biochem. Sci.* **13**, 40 (1988).
[61] V. Andersen and B. G. Munck, *Biochim. Biophys. Acta* **902**, 145 (1987).

Step IVE. Identify an analog meeting the first conditions of Step IVD, but one that is itself not transported by the system it inhibits, or even better not transported at all.

D-Tryptophan is about as effective as L-tryptophan as a competitive inhibitor of transport system T in the human red cell,[62] but it has been reported not perceptibly transported by that system.[17] If the same selectivity is shown, its application to the hepatocyte should be valuable. A low-affinity transport of D-tryptophan by system L[63] or A may need to be taken into account. Compounds such as this one that can obstruct a transport system without migrating by that system have been much sought for, but rarely identified, to provide the clearest simplification desired under the present Step IV. Inhibitors so highly unlike the natural amino acid substrate (or its possible cosubstrate, Na^+) that it seems highly unlikely that they are transported by the inhibited system deserve careful investigation for a useful noncompetitive action. An example is the use of harmaline to detect an apparent Na^+-binding point on a Na^+-independent system asc.[64]

Step IVF. In parallel to Step IVD, use the modified analog instead as a transport substrate, with the result that it is taken up by a single mediated route.

(In practice of course one usually replaces one test analog with another rather than modifying its structure in steps IV,C and D; but the sequential and progressive approach to the design of the ideal system-specific model is nevertheless the essence of this strategy.)

Steps IV,C–F reach ideally the goal illustrated in Fig. 8, namely to identify an ideal pair of competitively inhibitory analogs that can be applied sequentially, the first analog to inhibit completely and homogeneously one component of the uptake of the test solute; the second analog then superimposed to inhibit similarly a second uptake component of the test solute, given that there are only two components. These two inhibitory analogs are accordingly selected as ideal for the indicated task by the tests of step II, and then applied as under step IV,C and E. The same inhibitory analogs should then serve as selective substrates according to the present step.

To show that they fully achieve their purpose, the two ideal analogs of Fig. 8 should meet one more test, namely they should serve as selective inhibitors yielding, respectively, much the same K_i values when applied in

[62] R. Rosenberg, J. D. Young, and J. C. Ellory, *Biochim. Biophys. Acta* **598,** 374 (1980).
[63] D. L. Oxender, *J. Biol. Chem.* **240,** 2976 (1965).
[64] J. D. Young, D. K. Mason, and D. A. Fincham, *J. Biol. Chem.* **263,** 140 (1987).

the reverse order. This demanding criterion arises from the circumstance that the needed concentration of the inhibitory analog might act by placing an energetic load on the cell: for example, the transmembrane potential is apt to be decreased if the transport occasions uptake of net positive charges, or other forms of energy storage such as ATP levels and Na^+ and K^+ gradients are apt to be depleted. One may then expect V_{max} to be changed, and K_m changes might also occur. The question has been raised whether a residual portion of uptake by a given transport system (say system A) could thereby be sufficiently changed in its characteristics to appear artifactually to represent uptake by a different system (thus conceivably identifying as an artifactually modified system A uptake, as some have supposed, the component called system ASC). Beginning two decades ago we had for this discrimination between systems A and ASC only the device of N-methylation to restrict transport reactivity to system A,[65,66] but no corresponding device to restrict uptake to system ASC. We emphasized during the early analytical use of inhibitory analogs[66,67] that one of these analogs in saturating a transport system might produce qualitative changes in the uninhibited residue of that transport, a possibility that still deserves consideration for new instances. More recently we have demonstrated contexts in which threonine is sufficiently specific to system ASC to allow its use in some cells not only as a model analog according to step IV,E, but also in a predetermined excess to eliminate selectively the ASC component according to step IV,C. Hence we could eliminate the two components in the order A : ASC, or in the reversed order ASC : A, by applying first one inhibitor, then superimposing the second, without substantial changes in each component when their order of application was reversed.[67,68] This success disposes, we believe, of a prior suspicion that the ASC component is really an artifactually modified portion of component A. Cysteine sulfinate applied to uptake at low pH may serve alternatively as an ASC inhibitor, an action not produced on system A under any known condition.

Suppose, however, that the ASC inhibitor were used at a concentration high enough to show significant inhibition of uptake of the same test substrate via system A in the selected context. This circumstance would not distort the partition between routes A and ASC at the selected sub-

[65] H. N. Christensen, D. L. Oxender, M. Liang, and K. A. Vatz, *J. Biol. Chem.* **240,** 3609 (1965).
[66] H. N. Christensen, M. Liang, and E. G. Archer, *J. Biol. Chem.* **242,** 5237 (1967).
[67] H. N. Christensen, *in* "Membranes and Transport" (A. N. Martonosi, ed.), Vol. 2, p. 145. Plenum, New York, 1982.
[68] M. E. Handlogten, R. Garcia-Cañero, K. T. Lancaster, and H. N. Christensen, *J. Biol. Chem.* **265,** 7905 (1981).

strate concentration when MeAIB is applied first to inhibit system A, and the ASC inhibitor applied secondly by superimposition. But a reversed order for their application would yield a distorted and erroneous partition between the two routes. Hence passage of the test of Fig. 8 necessitates use of analogs, each quite specific to one of the routes under study.

The identification of model substrates rather specific to a single transport system is important also for the test for specific protection of the transport receptor structure from an irreversible modification by a reagent applied to monitor its subsequent isolation, for example, azidophenylalanine. The authors of a recent paper[69] overlooked conversely that all five of the amino acids most effective in protecting CHO cells from thermal damage at 45° (including significantly AIB) are to some degree system A substrates (see Fig. 3 in Ref. 70). Therefore, in 1987 these authors dismissed, perhaps prematurely, their own plausible suggestion that the protective action of each amino acid was to stabilize a protein component of that transport system.

We began this chapter with several examples of the uncovering of heterogeneity in a component of transport first attributed to a single mediating system. Behind each of these several corrections lay an imperfection in the specificity of a presumed model substrate for a single transport system. Conversely, they often arose from the presence of an unsuspected additional transport system with which the model substrate also reacted.

A somewhat typical history of the presumed sufficient specificity of a model substrate for a given transport system can be given. Amino acids constructed on a bicycloheptane[71,72] or a bicyclooctane[73] ring system (BCH and BCO, Fig. 9) have appeared to be largely unreactive with systems A and ASC, and not dependent for transport reactivity on the presence of Na^+. Hence these bicyclic amino acids were for a time considered specific model substrates for system L. BCO will presumably come to replace BCH for this purpose because the former is not a racemic mixture in need of a chemical resolution, so that the possibility of a biological resolution during its experimental use is avoided with BCO; furthermore, in tested occurrences it is somewhat more reactive with

[69] M. A. Vidair and W. C. Dewey, *J. Cell. Physiol.* **131**, 267 (1987).

[70] M. A. Shotwell, D. W. Jayme, M. S. Kilberg, and D. L. Oxender, *J. Biol. Chem.* **256**, 5422 (1981).

[71] H. N. Christensen, M. E. Handlogten, I. Lam, H. S. Tager, and R. Zand, *J. Biol. Chem.* **244**, 1510 (1969).

[72] H. S. Tager and H. N. Christensen, *J. Biol. Chem.* **246**, 7572 (1971).

[73] H. N. Christensen, M. E. Handlogten, J. V. Vadgama, E. de la Cuesta, P. Ballasteros, G. G. Trigo, and C. Avendaño, *J. Med. Chem.* **26**, 1374 (1983).

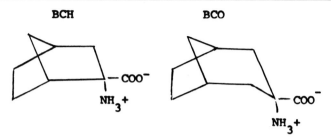

FIG. 9. Two bicyclic amino acids used as model substrates for the study of membrane transport. BCH is 2-*endo*-aminobicyclo[3.2.1]heptane-2-carboxylic acid, whereas BCO is the bicyclooctane analog, which has an additional carbon atom in the ring bearing the amino and carboxyl group, where the amino group also is endo. This change eliminates the optical asymmetry of BCH, simplifying the preparation of the analog in optically pure form, with modest gains in its specificity to system L.[73]

system L, and somewhat more specific to it. Each of these amino acid analogs has served well as a transport analog of the type called for in Fig. 8, undoubtedly because their three-dimensional bulk excludes them from several systems other than L. But what if other systems exist that will accept these bicyclic amino acids? If we suppose first that such components are also Na⁺ independent, they might unknowingly have been included in the system L measurements. Under challenges of the kind described here, such "contaminating" components have under favorable circumstances been discovered. As already mentioned, a component of the BCH-sensitive, Na⁺-independent uptake of typical substrates of system L by the human red blood cell proved disproportionately effective for the benzenoid amino acids, including tryptophan, and thus emerged system T.[5] During extended incubation of rat hepatocytes in primary culture a different variant of system L emerged. This component, provisionally designated system L1, shows an affinity preference for histidine over leucine, and an exceptionally high affinity for system L substrates in general. This trait of high affinity suggests that some other reported high-affinity occurrences of system L may also represent system L1. This feature may well account for the high affinities observed for system L transport at the blood–brain barrier and in various other cells, although a scarcely saturable transport component might be produced at this barrier by another variant of system L or for low-affinity amino acids.

In another challenge to the homogeneity of bicyclic, amino acid-inhibitable, Na⁺-independent amino acid transport, in this case in the pigeon red blood cell, a distinct Na⁺-independent mediation of transport emerged,[6] this one quite unlike system L, serving such amino acids as threonine, serine, alanine, and cysteine. This transport system appears highly similar to one seen earlier in the red cells of some sheep, and subsequently of some horses.[6] This system has earned the provisional designation system asc, through a possibly superficial resemblance to Na^+-dependent system ASC. This designation follows a provisional convention for using lower case letters for designating new, Na^+-independent systems.[74] System asc is comparatively weak in its reactivity with BCO, and slow in its approach to a steady state of uptake.

Beyond these Na^+-independent "contaminants" of system L, measured conventionally, a Na^+-dependent system unexpectedly transports, although with a lower preference, the same bicyclic amino acids into mouse blastocysts, activated from diapause, and apparently may also appear in the sea urchin egg after fertilization.[2] To represent the sharp discrimination from system A, the letter B is applied generically to refer to such systems. Because in the indicated cases this new component accepts cationic as well as neutral amino acids, we provisionally designated it $B^{0,+}$. On the basis of its wide Na^+-dependent scope, system B might well turn out to be much the same system that serves for the wide-scope, Na^+-dependent brush border uptake of amino acids by the renal and/or intestinal epithelial cell.

Figure 10 illustrates a simple version of the strategy of combining a series of steps to isolate components of transport one by one.[75] The conditions selected for eliminating each component have been identified previously by quantitative tests similar to those of Fig. 8. We see first the use of the analog MeAIB in excess to inhibit the mediated component of alanine uptake by the rat hepatocytes attributed to system A. Next we see the withholding of Na^+ from the medium, which serves to eliminate all other Na^+-dependent components, in the present case consisting largely if not entirely of system ASC. If the test substrate were, however, glycine or glutamine instead of alanine, two other transport components (added in parentheses), namely systems Gly and N, would at least for rat hepatocytes be included in the measurement by this step, along with system ASC. Still another component which could conceivably be present,[1] namely system B, also has not been excluded. As the final step of Fig. 10, another model amino acid, the one constructed on the bicycloheptane (or the bicyclooctane) ring system, serves to eliminate the component attributed to system L. Our interpretation that for the rat hepatocyte each of the three steps of Fig. 10 eliminates a single homogeneous component of uptake is provisional, although it may be quite acceptable for alanine. At

[74] S. Bannai, H. N. Christensen, J. V. Vadgama, J. C. Ellory, E. Englesberg, G. G. Guidotti, G. C. Gazzola, M. S. Kilberg, A. Lajtha, B. Sacktor, F. V. Sepúlveda, J. D. Young, D. Yudilevich, and G. Mann, *Nature* (*London*) **311,** 308 (1984).

[75] H. N. Christensen, *Perspect. Biol. Med.* 358 (1981).

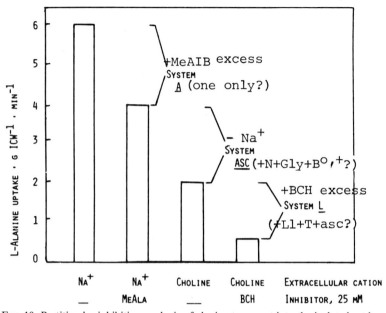

FIG. 10. Partition by inhibition analysis of alanine transport into the isolated rat hepato-cyte among discrete transport systems, adapted from Fig. 4 in Ref. 75. See text for interpre-tation. The validity of this simple scheme depends on the assumption that no significant uptake of alanine occurs by any of the six systems included in parentheses. The suggestion of their parenthetical contaminating contributions presents the unprecedented, hypothetical worst case of complexity yet supportable, conceivable for some other amino acid but not likely for alanine. Certainly if glycine were the substrate, a contribution by system Gly would need to be taken into account; if glutamine were the substrate, a contribution by system N would need to be considered. If the cell in question shows uptake by both systems ASC and $B^{0,+}$,[1] one might need to discriminate each of these in turn, first eliminating $B^{0,+}$ as the BCO-inhibitable component, and then ASC as the cysteine sulfinate-inhibitable compo-nent, in each case from the middle-bracketed interval. Similar strategies might serve specifi-cally to eliminate hypothetical Na^+-independent contributions by system L1, T, or asc. It may often be simpler to check for the presence of the less ubiquitous components by devising a single specific test, rather than arranging a comprehensive scheme. For example, we might well find we could measure a hypothetical $B^{0,+}$ component as the loss of BCO uptake occasioned by the removal of Na^+ from the external medium. Or we might measure the (Na^+-independent) asc component by the loss of threonine transport produced by N-ethylmaleimide, which does not inhibit the classical system L.[7] A favorable choice of test substrate enhances precision by proportionally making larger the difference between the two flux measurements under comparison.

the moment an aggregate of six known transport systems other than A, ASC, and L might conceivably be present (as shown in parentheses), and depending on what test substrate is used, may contaminate either of two of the three major components separated by the strategy of Fig. 10. The

Na$^+$-independent system called asc (not ASC)[7] might conceivably also contribute to alanine uptake by the hepatocyte, although a contribution by it has so far escaped notice even in fetal rat hepatocytes, even though it is present in the erythroid cells of the fetal rat liver.[76] The original scheme of Fig. 10 is sufficient for its purpose only if our choices of alanine and the rat hepatocyte have allowed that degree of simplicity in the analysis. Furthermore, the limited strategy of Fig. 10 yields the quantitative partition of the uptake of a given amino acid applying only at the selected substrate concentration. Note that measures of the inhibitory effectiveness of a series of analogous amino acids have occasionally been obtained and tabulated only for a single concentration. Tests at enough different concentrations to allow extraction of kinetic parameters manifestly provide much more satisfactory information.

Figure 11[47b] illustrates a different combination of inhibitors and transport-modifying conditions (including in this case responsiveness to amino acid starvation) for discriminating transport by three parallel systems.

Will all of the nine transport systems (A, L, ASC, and six others) referred to for the hepatocyte in the preceding paragraph ever coexist in a membrane of a single cell type? If so, we have been lucky so far in not yet encountering such complexity. May nature tend not to develop or retain together several transport systems that serve rather much the same function, so that the design of an experimental strategy serving all possible needs might prove a largely intellectual rather than a practical exercise? Are certain transport systems restricted in their occurrence, for example, to erythroid cells, to certain organelles, or to certain differentiative stages? These questions should soon be answered, at least for some specific cases. When our purpose is only to establish the existence of a new transport system and to characterize it, we have the opportunity to select the most favorable substrate, the most favorable cell, tissue, or phenotype for the purpose, perhaps one where sensitivity to regulation of the given transport component may facilitate our intent. For economy of effort, our discriminatory scheme may well be simplified as in Fig. 10 to meet the immediate needs for an actual context as indicated by prior tests.

Step IV G. Measure the transport parameters under conditions maximizing either net transport or exchange, e.g., the so-called *zero-trans* or *infinite-trans* conditions.

For example, making kinetic observations on the extent of steady-state accumulation, to the degree that these can validly be made, minimizes as already mentioned the contribution of rapid, exchanging sys-

[76] J. V. Vadgama, M. Castro, and H. N. Christensen, *J. Biol. Chem.* **262**, 13273 (1987).

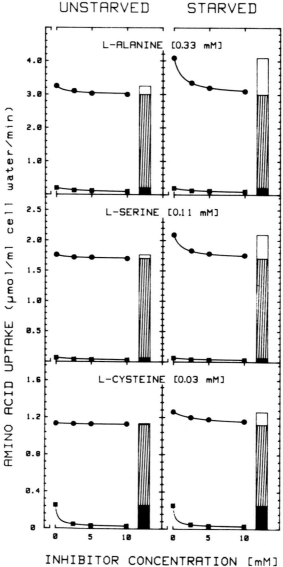

FIG. 11. Identification of transport routes of 10-sec uptake of alanine, serine, and cysteine by human fetal fibroblasts, particularly the component stimulated by 12 hr of amino acid starvation. The curves show the inhibition of the uptake of each of these three amino acids at the indicated concentrations normal for blood plasma, by the analog MeAIB in Na^+-containing medium (upper curve, ●) or BCH in Na^+-free medium (lower curve, ■), at the inhibitor concentrations indicated on the abscissa scales. The bars at the right of each panel show that the incremental uptake on amino acid starvation is Na^+ dependent and MeAIB-inhibitable, but not inhibited by BCH in Na^+-free media, and hence is attributed to system A, not to systems ASC or L. (Reprinted from Franchi-Gazzola et al.[47b] by permission; this reference should be consulted for further details.)

tems, and maximizes the contribution of slower systems producing net transport, a factor that delayed the characterization of system L until the early 1960s.[77]

At various stages we have hoped that the strategy outlined above had already yielded a model substrate identifying precisely one or another homogeneous transport-mediating system, neither transported by nor reactive with any other system under the conditions of the test. Obviously that goal is being approached by the close examination of every possibly different component of solute migration across a biological membrane. Model analogs more sharply restricted each to a distinguishable Na^+-independent component than the two bicyclic analogs may already occur among insufficiently tested synthetic analogs. This experience calls then for step V, and then the final step of this scheme.

In modifying conditions, one needs to avoid any that might intensify unstirred layer effects, which we did once presumably by unintentionally narcotizing peristaltic or villar motion, for intestinal absorption.[78]

Step V. As one begins to suppose that all components of the uptake of a given test solute have been identified, its influx may be compared at several widely spaced concentrations in the presence of the minimally adequate list of inhibitory analogs or under modified conditions to block all the identified routes. If migration still shows a dependence on concentration (beyond a small predictable one), look for a new inhibitory analog to serve to identify yet one more component of mediation, and repeat steps III and IV.

Step VI. As one then continues to suppose that all amino acid transport systems for a given cell or tissue have been identified, new analogs may be identified or designed with structural features that minimize their reaction with any of the known systems. If concentration-dependent uptake is nevertheless detected for such an analog, the process can be examined for unsuspected identity with a known system. If no identity is discovered the new analog should be carried through the foregoing scheme, in an effort to describe yet one more transport system, or the behavior predictable for the absence of mediated transport.

The dipolar amino acid, α,α-dicyclopropylglycine, was selected in this way for its minimal reactivity with the amino acid transport systems known for the Ehrlich cell. It serves so far in that cell as a model for study of the nonsaturable transport, which proved to be approximately equal

[77] D. L. Oxender and H. N. Christensen, *J. Biol. Chem.* **238,** 3686 (1963).
[78] J. A. Antonioli and H. N. Christensen, *Am. J. Physiol.* **215,** 951 (1968).

(ratio 1.45 instead of 1.00) in the two directions and only slightly greater than that measured for ordinary amino acids by extrapolation of rate of uptake to infinite concentrations.[27] This result supports the possible uniqueness of a dominant nonsaturable route of passage across the plasma membrane for dipolar amino acids.

Conclusion

Results have shown that amino acid transport for each cell or tissue (and this conclusion appears to apply equally to absorption by the intestinal or renal mucosa) is effected by a number of systems, a number smaller than and bearing no obvious relation to the number of amino acids regularly involved in our metabolism. The range of reactivity of each of these systems appears, however, defined by structural features of the amino acid, rather than biological considerations. Furthermore, competition for transport between each pair of amino acids tends to occur in two, three, or perhaps more distinct transport systems. This chapter addresses the problems of resolving that complexity and discovering all the contributing catalytic systems. In practice some of the steps visualized here become unnecessary where the interaction of a pair of amino acids for transport proves simple. Furthermore, one may not wish to explore the full range of amino acid transport for a given cell or organelle.

Lest this account seem unduly complex, we can give it a simple summary by reference again to Fig. 8: We look for two or more analogs that independently saturate each of the routes by which we may suppose a substrate passes a membrane. By suitable selection of the substrate for transport and the conditions for the test, we can try to keep minimal the number of transport components to be discriminated in such a test. We then apply the ABC test, test for transstimulation, and apply other tests for a discriminated component to assure ourselves of fully shared transport between substrate and the corresponding analog. If we fail to find a set of analogs with largely independent inhibitory action restricted each in the main to one component of the transport of the test substrate, our precision of discrimination will be low, and our conclusion for catalytic homogeneity of that component of transport may well remain more provisional than usual. The quite different biological roles indicated for these several transport systems are heralded by large differences in their regulation (for example, the recently shown roles for protein kinase C, transforming growth factors, and oncogenes in the positioning and stimulation of system A[79,80]), all of which point to the biological importance of the

[79] W. D. Dawson and J. S. Cook, *Fed. Proc., Fed. Am. Soc. Exp. Biol.* **44,** 646 (Abstr.) (1985).
[80] E. Racker, R. J. Resnick, and R. Feldman, *Proc. Natl. Acad. Sci. U.S.A.* **82,** 3535 (1985).

discrimination of the several systems. A colleague comments more optimistically that the discrimination between these transport systems has for the most part developed to the point that they have become suitable targets for identification of the gene expressing each.

In this chapter we considered how the chemical structure of substrates for transport may be manipulated to find analogs that serve best for observing operation of a single transport system alone. In pursuing this goal we need to remember that the inward and outward fluxes are often asymmetric so that the effects of such structural changes need not be identical for the two fluxes. We have long been interested in a presumably complementary role of biological changes in the structural environment of the binding site as presented first at one and then at the other surface of the membrane in producing the flux asymmetries, and hence in determining how steeply the substrate is concentrated across a membrane (see, for example, Ref. 81). Here we have then a second goal for the strategies of modifying the structure of transport substrates, namely mechanism. A series of such structures, when compared as to their handling by a single transport system, may help discover whether somewhat complementary changes may have been produced biologically as a transport receptor site is presented first at one membrane surface and then at the other.

We now have biological means of substituting one selected amino acid residue for another in the protein presumably bearing two different forms of the receptor site, and hence of modifying experimentally its environment. It has been possible in a recent example to substitute in the lactose carrier of *E. coli* the alanine residue in position 177 with valine or threonine, or the tyrosine residue in position 236 with asparagine, serine, or histidine.[82] These carrier mutants show not only modified substrate selectivity, but also modified asymmetry between the two directional fluxes, without changes, however, in the stoichiometry of H^+–galactoside cotransport.[82]

Implied in this approach is the conclusion that the directional flux asymmetry does not arise from a simple rotation of a carrier[83] or from any other mechanism which requires large movements of a protein molecule through the plane of the membrane bilayer,[84] ideas which were considered decades ago. Various leaders in the field (e.g., see Refs. 85 and 86) have nevertheless felt no need to abandon use of the term carrier, along with transporter, etc., for the molecule mediating each specific transport, a

[81] H. N. Christensen, *Adv. Protein Chem.* **15**, 239 (1960).
[82] R. J. Brooker and T. H. Wilson, *J. Biol. Chem.* **260**, 16181 (1985).
[83] A. Dutton, E. D. Rees, and S. J. Singer, *Proc. Natl. Acad. Sci. U.S.A.* **73**, 1532 (1976).
[84] J. Kyte, *J. Biol. Chem.* **249**, 3652 (1974).
[85] G. F. Baker and W. F. Widdas, *J. Physiol. (London)* **395**, 56 (1988) (see especially p. 73).
[86] B. R. Stevens, H. J. Ross, and E. M. Wright, *J. Membr. Biol.* **66**, 213 (1982).

usage which for many years has not been intended to imply that the carrier operates like a ferryboat. If there is an error in the historically important qualification of a mobile carrier, note that the defect is already present in the well-accepted term carrier. The coupling of an enzymatic reaction to the translocation event, e.g., in the phosphotransferase reaction, should not distract our attention from the unique catalytic event of translocation per se.

It is obviously important for biomedical progress to identify the transport system or systems whose regulation determines the distribution and flow of a given metabolite, as can be illustrated with a recent demonstration of the role of system N in determining hepatic glutamine flows,[87] or of the role of system L in the dual effects of phenylalanine accumulations on the distribution of amino acids in phenylketonuria.[88] These goals, which I approach elsewhere,[89] justify our greatest care in differentiating the various transport-mediating proteins.

Acknowledgments

Support is acknowledged from Grants HD01233 and DK32281 to the University of Michigan, Institute for Child Health and Human Development, National Institutes of Health. The hospitality of U.C.S.D. at the final stages is also acknowledged. The author is grateful to Jacqueline Benson for preparing the numerous typescripts through which this chapter has passed.

[87] D. Häussinger, S. Sobell, A. J. Meyer, W. Gerok, J. M. Tager, and H. Sies, *Eur. J. Biochem.* **152**, 597 (1985).
[88] H. N. Christensen, in "Amino Acids in Health and Disease" (S. Kaufman, ed.), pp. 1–17. Liss, New York, 1987.
[89] H. N. Christensen, *Physiol. Rev.,* in press (1989).

[39] Measuring Hexose Transport in Suspended Cells

By Jørgen Gliemann

Definition

Most, if not all, cells possess specific transporters for hexoses in the cell membrane which provides a barrier between the cytosol and the external milieu. These transporters utilize a concentration gradient which is normally maintained because the sugar is constantly removed from the cytosol by phosphorylation and further metabolism. If this were not the case, the sugar would achieve the same concentration on the two sides of the membrane and net transport would cease. This type of transport is called facilitated diffusion.

Techniques will be described for cells which can be obtained in suspension. This implies that with careful stirring, extracellular diffusion gradients should not represent a problem. Suspended cells may be from cell cultures [e.g., cultured human lymphocytes (IM-9), Novikoff hepatoma cells, HeLa cells, and numerous others]. They may be isolated after incubation or perfusion of tissues with collagenase (e.g., adipocytes, hepatocytes, cardiac myocytes) or they may be performed in blood and tissues (polymorphonuclear leukocytes, monocytes, macrophages, lymphocytes).

It should be noted that the amino acid sequence and structure of the glucose transporter from human hepatoma cells was reported recently[1] and this protein is highly homologous to the 55-kDa transporter (an integral membrane glycoprotein) identified in human erythrocytes.[2] It is therefore highly probable that glucose transporters, which can be studied in various cell types, are in fact very similar membrane proteins.

The glucose transporter is no doubt the most important hexose transport protein but a fructose transporter is also present in some cells. It has only been identified and partially characterized in adipocytes but it may be present in other cells as well.

The following systems will not be dealt with in this chapter: (1) cells which in addition to the equilibrating glucose transporters also possess an active transport system coupled to the flux of sodium ions down their concentration gradient, e.g., epithelial cells of the small intestine and the proximal renal tubuli; (2) tissues from which isolated cells cannot be readily prepared, e.g., skeletal muscle and (3) erythrocytes.

Principles for Measurements of Transport Rates

The basic requirement is to measure unidirectional flux of sugar from the extracellular to the intracellular compartment and vice versa. If this can be done when transport of the tracer molecule is fastest (very low substrate concentration, physiological temperature) then the system can be explored under all conditions. Most natural hexoses which are transported rapidly (D-sugars) are also metabolized rapidly inside the cell and it may therefore not be possible to separate the rate of transport from the rate of metabolism, i.e., to point out the rate-limiting step. For this reason, a nonmetabolizable sugar analog should be used, at least in the initial

[1] M. Mueckler, C. Caruso, S. A. Baldwin, M. Panico, I. Bleuch, H. R. Morris, W. J. Allard, G. E. Lienhard, and H. F. Lodish, *Science* **229,** 941 (1985).
[2] W. J. Allard and G. E. Lienhard, *J. Biol. Chem.* **260,** 8668 (1985).

studies of an unknown system. 3-*O*-methyl-D-glucose (3OMG) is an analog which has an affinity to the glucose transporter similar to that of glucose (in rat adipocytes the affinity is about twice as high[3]) and which is either not metabolized or metabolized very slowly in various cells.

From a practical point of view, the half-time of equilibration of a hexose analog such as 3OMG determines whether the unidirectional flux (i.e., initial velocity) can be measured or calculated on the basis of a simple manual technique. This is feasible when the half-time is 2–3 sec or higher. The "fastest" cells (apart from human erythrocytes) studied until now with regard to equilibration of tracer 3OMG at 37° are insulin-stimulated rat adipocytes[4] and rat hepatocytes[5] and they have half-times of that magnitude.

The half-time of equilibration is determined by the intracellular distribution volume, which is roughly equivalent to the intracellular water volume, and the permeability of the cell membrane which, in turn, is determined by the concentration of transporters in the cell membrane if we assume that their characteristics vary little among cell types. It is postulated that hexose transport in all suspended cell types except human erythrocytes can be evaluated using the techniques described below.

Transport from the extracellular buffer to the cytosol involves the following steps: (1) rapid mixing of the cell suspension with buffer containing tracer, (2) incubation, usually for a few seconds, (3) rapid mixing with a solution which prevents efflux of the labeled analog from the intracellular compartment, prevents further influx, and dilutes the extracellular tracer, and (4) separation of cells with the intracellular sugar as efficiently as possible from the extracellular buffer.

Measurement of Entry: Human Lymphocytes

Human lymphocytes of the IM-9 line are easily grown in culture, they are quite uniform in size with a mean surface area of 4.5×10^{-10} m^2, and a mean 3OMG distribution volume (approximate intracellular water volume) of 7×10^{-13} liter. Their 3OMG transport is not changed by various ligands which modulate the glucose transport system in other cell types.[6]

Materials

1. Lymphocytes, for instance 5×10^7/ml, suspended in a convenient buffer, e.g., 140 mM Na$^+$, 4.7 mM K$^+$, 2.5 mM Ca^{2+}, 1.25 mM Mg^{2+},

[3] R. R. Whitesell and J. Gliemann, *J. Biol. Chem.* **254,** 5276 (1979).
[4] J. Gliemann and W. D. Rees, *Curr. Top. Membr. Transp.* **18,** 339 (1983).
[5] Y. Okuno and J. Gliemann, *Biochim. Biophys. Acta* **862,** 329 (1986).
[6] W. D. Rees and J. Gliemann, *Biochim. Biophys. Acta* **812,** 98 (1985).

142 mM Cl$^-$, 2.5 mM H$_2$PO$_4^-$/HPO$_4^{2-}$, 1.25 mM SO$_4^{2-}$, and 10 mM HEPES

2. Round-bottom plastic tubes, about 3.5 ml (miniscintillation vials)
3. Buffer with 3-O-[^{14}C-*methyl*]glucose, for instance 2 μCi/ml. With a specific activity of 60 Ci/mol, the concentration of labeled 3OMG will be 33 μM
4. Stopping solution containing 0.3 mM phloretin and 0.1 μM HgCl$_2$. This is prepared by slowly adding 1 vol of stock solution [82 mg phloretin dissolved in 1 ml ethanol/dimethyl sulfoxide (7:3, v/v)] into 1000 vol vigorously stirring buffer or 0.9% NaCl at room temperature

Procedure

1. Place 15 μl isotope containing buffer in the bottom of a tube. It should form a vaulted droplet. Stopper and place in 37° room or water bath.

2. Squirt 50 μl suspension (37°) onto the droplet using an automatic pipe and let stand for a long time, e.g., 15 min, for determination of 3OMG distribution volume at equilibrium. Add 3 ml stopping solution, centrifuge for 1 min at about 4000 g in a rapidly accelerating bench centrifuge to pellet the cells, pour out supernatant, resuspend in 3 ml stopping solution, centrifuge again, remove supernatant, and add scintillation fluid miscible with water. A third wash may be carried out but is usually not necessary. The total aqueous volume is 65 μl and the intracellular water volume is about 1.75 μl. Thus one would expect 2.7% of the isotope to be inside the cells at equilibrium. The concentration of labeled 3OMG is about 8 μM.

3. Determine uptake at one or a few seconds using a metronome set at 120 beats/min. For 1.5 sec, squirt cells onto the isotope with one hand at beat one while the other is ready to pipette stopping solution at beat four. Place the tip with stopping solution on the side of the tube about 1 cm from the cell droplet. The "front" of the stopping solution should hit the cells at the right time. The large volume serves to dilute the extracellular tracer and should be pipetted gently to avoid splattering.

4. Determine blank values by adding about half of the stopping solution, then cells and the remaining stopping solution. With the chosen amount of cells blank values will contain about 0.12% of the total isotope, that is about 4% of the intracellular counts at equilibrium.

The coefficient of variation should be around 5% and is, after a little practice, independent of time from 1 sec and up. However, the scatter on the blank values will of course add to the scatter on the early points if the uptake is low. Therefore, initial velocities are usually best estimated from uptakes of approximately 25% of the equilibrium value.

Analysis of Uptake Curve

Figure 1 shows an example which may be the basis for calculation. However, several points should first be controlled.

1. Does the stopping solution stop efficiently? Incubate cell samples to equilibrium, add stopping solution, and let stand for up to 1-hr before cells are recovered. In this way, the half-time of efflux is determined, and it is about 30 min with the IM-9 lymphocytes. This means in practice that the stopping solution is sufficiently efficient for a few minutes. In other words, only relatively few samples (four to six) should be handled at a time.

2. Does the stopping solution stop immediately? In other words, can we be sure that there is no rapid loss of intracellular sugar analog before the slow loss detected as above? The best way is to compare the equilibrium distribution volume determined using the stopping solution with that determined without disturbing the isotopic equilibrium. The latter can be determined by transferring a volume of equilibrated cells (e.g., 100 μl) to microfuge tubes containing an oil with a density intermediate between buffer and cells[7] followed by centrifugation for 1 min. Dibutyl phthalate or a suitable silicone oil may be used. The measured distribution volume must be corrected for trapped buffer by incubation in parallel tubes (or in the same tubes if double-isotope technique is used) with a labeled extracellular marker as sucrose, methoxyinulin, or L-glucose.

3. Is 3OMG not metabolized at all? Measure the distribution space at various times which are long as compared with the half-time of transport. If the sugar analog is phosphorylated and trapped inside the cell, then the apparent distribution space will creep up and exceed the intracellular ^3H_2O space measured with the oil technique. Another simple check is to ascertain that 3OMG will wash out of the cells when buffer is added instead of stopping solution and attain the proper equilibrium calculated according to the new isotope concentration. Finally, labeled material can be extracted from cells incubated to equilibrium and identified using standard separation methods such as paper chromatography. A phosphorylation of 3OMG, although at a very slow rate, has been demonstrated in human polymorphonuclear leukocytes.[8] It should be noted that a slow

---

[7] P. Andreasen, B. Schaumburg, K. Østerlind, J. Vinten, S. Gammeltoft, and J. Gliemann, *Anal. Biochem.* **59,** 110 (1974).

[8] Y. Okuno, L. Plesner, T. R. Larsen, and J. Gliemann, *FEBS Lett.* **195,** 303 (1986).

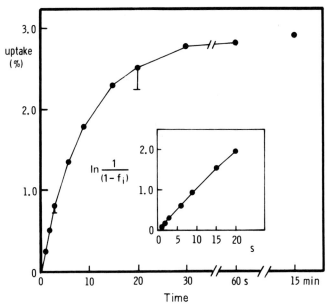

FIG. 1. Time course of exchange of 1 mM 3-O-methyl-D-glucose in IM-9 lymphocytes at 37°. Note that the substrate concentration is low as compared to the exchange K_m of 10 mM and almost the same curve would therefore be obtained when using labeled sugar analog alone. The cells (50 μl) were first equilibrated for 30 min with 1 mM unlabeled 3OMG and then squirted onto buffer (15 μl) containing both 1 mM unlabeled 3OMG and tracer [^{14}C]3OMG. Stopping solution was added at the appropriate times and intracellular radioactivity recovered (see Measurement of Entry). The ordinate shows the percentage uptake of the added tracer into 2.6×10^6 cells (final concentration 4×10^7 cells/ml). The inset shows a logarithmic transformation with the fractional filling, f_i, calculated as (cpm$_t$ − cpm$_0$)/(cpm$_\infty$ − cpm$_0$) where cpm$_0$, cpm$_t$, and cpm$_\infty$ indicate the radioactivity in the cell pellets at time zero, time t, and 15 min. Each point is the average of four replicates ± 1 SD. For calculations, see text. The rate constant $k = v/S$ is calculated by linear regression of the results shown in the inset as 0.10 sec^{-1}. The entry velocity (v) is 0.10 sec$^{-1} \cdot$ 1 mM = 0.10 m$M \cdot$ sec^{-1}. Since one cell has a distribution volume of 7×10^{-13} liter, this corresponds to 7×10^{-17} mol $\cdot$ sec$^{-1} \cdot$ cell^{-1}. Alternatively, the cells at a concentration of 4×10^7/ml initially take up about 0.3%/sec. Thus, one cell shows an initial uptake (v) of $3 \times 10^{-3} \cdot 10^{-6}$ mol/4×10^7 cells or about 7×10^{-17} mol $\cdot$ sec$^{-1} \cdot$ cell^{-1}. The permeability (P) is $k \cdot$ vol/surface ratio. Since the mean distribution volume is 7×10^{-16} m^3 and the mean surface area is 4.5×10^{-10} m^2, P can be estimated as approximately 1.5×10^{-6} m $\cdot$ 0.1 sec^{-1} = 1.5×10^{-7} m $\cdot$ sec^{-1}. The clearance per cell is 0.10 sec$^{-1} \cdot 7 \times 10^{-16}$ m^3 = 7×10^{-17} m$^3 \cdot$ sec$^{-1} \cdot$ cell^{-1} (and 1 m^3 contains 1 mol). Alternatively, 4×10^7 cells "clear" 3×10^{-9} ml/sec and one cell about 7×10^{-17} m$^3 \cdot$ sec^{-1}. (Reprinted from Rees and Gliemann[6] by permission.)

rate of 3OMG metabolism (e.g., 100 times slower than the rate of transport) does not prevent its use for measurement of influx rates. One just has to be careful in defining the distribution space of the sugar analog.

If these criteria are fulfilled and if the cells are reasonably homogeneous with respect to hexose permeability, then the equilibration curve will approximate two-compartment kinetics. The inset of Fig. 1 shows the exponential nature of the equilibration curve in lymphocytes. Similar curves, although with different half-times, have been obtained with hepatocytes, insulin-stimulated adipocytes, polymorphonuclear leucocytes, and can probably be generated using a whole variety of cells.

Calculations Based on the Exponential Curve

The rate constant (k) of 3OMG entry can be read as the slope of the curve in Fig. 1 (inset). It has the dimension $\sec^{-1}$ (ln 2 divided by the half-time) and the meaning is the fraction of the intracellular distribution space that would be filled per unit time if no back flux occurred (i.e., initially filled per unit time). It may be determined from a single point if the uptake curve is already known to be exponential. It is equivalent to the entry velocity, v (mol $\cdot$ liter^{-1} $\cdot$ sec^{-1}), divided by the substrate concentration S (mol $\cdot$ liter^{-1}).

Therefore v is obtained as kS and has the dimension $\sec^{-1} \cdot$ mol $\cdot$ liter^{-1} [intracellular 3OMG distribution space or (roughly) intracellular water space]. v is often expressed as moles per cell, i.e., mol $\cdot$ liter^{-1} multiplied by the distribution volume of one cell in liters. This expression of v may also be derived directly from the uptake curve as the initial fractional uptake multiplied by the substrate concentration and divided by the number of cells in the incubation.

The permeability (P) has the dimension m $\cdot$ sec^{-1} and is calculated as k multiplied by the distributon volume/surface ratio.

The clearance is the volume of extracellular buffer that is initially "rinsed" for isotope per unit time. It is equivalent to k multiplied by the distribution volume in a given number of cells.

The calculations are shown in the legend to Fig. 1.

Measurement of Exit

This involves loading of the cells with 3OMG until equilibrium is achieved followed by the addition of a volume of buffer and, at the appropriate time, a volume of stopping solution. It is convenient to let the total efflux volume be at least 10 times the loading volume so that the intracellular isotope at infinite time is 10% or less of the intracellular isotope at time zero. In the chosen example with IM-9 lymphocytes loading volume was 65 μl and total efflux volume should therefore be at least 650 μl. However, this volume is too large because it cannot mix instantaneously with the stopping solution. It is therefore an advantage to concentrate the cells as much as possible before efflux. Ten microliters of concentrated suspension, containing slightly less than 10 μl 3OMG distribution volume, would be appropriate and the efflux volume may be 150 μl followed by the 3 ml of stopping solution. Nevertheless, the scatter tends to be higher in efflux experiments (coefficient of variation around 10%) in part because the blank values (infinite time) are higher than in influx experiments.

The efflux curve should be the mirror image of the influx curve shown in Fig. 1 when the concentration of sugar analog is small as compared with K_m. In fact, this comparison is important for testing the technical feasibility of the experiments. The calculations are also analogous. The rate constant of exit, $k = v/S$, should be calculated as the slope of the linearized exponential efflux curve with the ordinate $(\ln(1/f_r)$, where f_r is the fraction of radioactivity remaining in the cells at a given time (analogous to inset of Fig. 1).

Hexose Transport in Adipocytes

This is the only cell type with a density lower than buffer and a technical modification is therefore required. After the addition of stopping solution add quickly (but do not splatter) 0.5 ml silicon oil with a density of 0.99 and a viscosity of 100 cS. Centrifuge for 40 sec in a rapidly accelerating centrifuge.[3] The cells should coalesce into one pellet which is easily removed, for instance, by a bent piece of pipe cleaner. Damaged cells tend to scatter around the rim of the tube.

Alternatively, add only 400 μl stopping solution, transfer immediately, and using the same pipet, 400 μl suspension to 550-μl polypropylene tubes filled with 100 μl silicone oil or dinonyl phthalate, centrifuge for 30 sec in a microfuge at about 10,000 g. Cut the tube through the oil layer using a scalpel blade mounted on a wrench.[9]

In these methods, extracellularly trapped buffer is mainly replaced by the oil and blank values are therefore low. The precision of transport data is about the same as in other cell types in spite of the very low intracellular 3OMG distribution volume of adipocytes. The method is applicable to

[9] J. E. Foley, R. Foley, and J. Gliemann, *Biochim. Biophys. Acta* **599**, 689 (1980).

human adipocytes obtained by biopsies.[10] Methods for measuring hexose transport in adipocytes have been reviewed previously.[4,11]

Substrate Concentration Dependence

General. The combination of the hexose with a finite number of transporters in the membrane leads to facilitated diffusion exhibiting saturation kinetics. This system is different from the simple model of enzyme reaction with one free enzyme because two separate binding sites are available on the transporter, one on the outside and one on the inside of the membrane. Half-saturation constants (K_m) and maximum velocities (V_{max}) will therefore have different interpretations depending on the direction of the flux and the hexose concentration on each side of the membrane. It has been shown[12] that the relationship between transport velocity (v) and substrate concentration (S) for any of the protocols described below will be given by the Michaelis–Menten equation (cf. legend to Fig. 2).

Before analysis of concentration dependence can be carried out it should be ascertained that the exponential uptake curve (Fig. 1) is almost entirely due to transfer of substrate through the specific transporters. Nonmediated ("simple") diffusion should be negligible. One way is to take advantage of the stereospecificity and the rate constant of entry of labeled L-glucose, which is usually less than 1% of that of labeled 3-O-methyl-D-glucose.[3,8] Moreover, most of the L-glucose transport is likely to be inhibited by a high concentration of unlabeled 3OMG.[3] It seems therefore to a large extent transferred through the specific transporters. If this is the case, one can safely assume that D-glucose transport is for practical purposes mediated exclusively by the specific transporters. However, 3OMG is slightly more lipophilic than glucose and its nonmediated diffusion may therefore be slightly higher. It is therefore important to show that its transport is inhibited almost completely (to less than 1%) by phloretin (cf. the stopping solution) and by a high concentration of cytochalasin B (20 μM), which we may assume is without effect on nonmediated 3OMG diffusion.[8] With this background, concentration dependence may be analyzed according to the following protocols.

Equilibrium Exchange. In this situation the total concentration of 3OMG is the same on the two sides of the membrane and measurement is initiated by establishing a concentration gradient for the labeled analog.

[10] O. Petersen and J. Gliemann, *Diabetologia* **20**, 630 (1981).
[11] J. Gliemann, in "Methods of Diabetes Research" (J. Larner and S. L. Pohl, eds.), Vol. 1, p. 105. Wiley, New York, 1984.
[12] Y. Eilam and W. D. Stein, in "Methods of Membrane Biology" (E. D. Korn, ed.), Vol. 2, p. 283. Plenum, New York, 1974.

Transport of 3OMG is the same in both directions, net transport is zero, and only flux of labeled molecules is followed as a marker for unidirectional flux. Consequently, it makes no difference whether flux of labeled molecules is measured as entry or exit. It is a good control of the system to show that the same result is obtained regardless of whether influx or efflux techniques are used.[6] Further experiments may be carried out using the easiest method.

Entry experiments are performed by first equilibrating the cells with unlabeled 3OMG at various concentrations. The time should be long enough to assure equilibration even at the highest concentrations, e.g., 30–60 min for the IM-9 lymphocytes. Then uptake is measured as explained in the legend to Fig. 1 using identical 3OMG concentrations in the isotope solution and the cell suspension. Two-compartment kinetics (i.e., exponential uptake curves) will be obtained at all substrate concentrations if it was obtained when using a very low substrate concentration, but the half-time will increase with increasing substrate concentration. The reason is that more and more transporters become occupied with unlabeled 3OMG, and equally from both sides, so that a diminishing fraction will be available for transport of the tracer.

Calculations are as those in the legend to Fig. 1. The rate constant, v/S, is calculated, from one or several points for each 3OMG concentration. K_m and V_{max} can now be calculated using one of the transformations of the Michaelis–Menten equation yielding a straight line, for instance, Hanes' transformation, [cf. Fig. 2 and Eq. (1)]. It should be possible to fit one straight line to the experimental points if there is only one class of transporters.

$$S/v = S/V_{max} + K_m/V_{max} \qquad (1)$$

Exit experiments are performed by first loading a concentrated cell suspension with labeled 3OMG at various concentrations of unlabeled sugar analog followed by washout in buffer containing the same concentration of unlabeled analog as the loading buffer (cf. Measurement of Exit). The rate constant of exit, v/S, is calculated for each 3OMG concentration and K_m and V_{max} values are calculated using a transformation of the Michaelis–Menten equation.

Zero-Trans Entry. This expression is used to describe the situation where 3OMG has a defined concentration in the buffer whereas the concentration in the intracellular compartment is zero at the start of the experiment. Transport of glucose may in some cells approach this situation under physiological conditions, namely if glucose is effectively metabolized in the cytoplasm at the given extracellular glucose concentration. Using a nonmetabolizable hexose analog the experimental problem

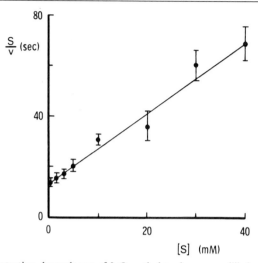

FIG. 2. Concentration dependence of 3-O-methyl-D-glucose equilibrium exchange entry in IM-9 lymphocytes at 37°. The cell suspenson was first equilibrated for 1 hr with unlabeled 3OMG at the concentrations indicated. Aliquots (50 μl) were then squirted onto 15 μl of buffer containing isotope and the same concentration of unlabeled 30MG as the loading buffer. For each concentration, the rate constant, v/S, was calculated as explained in the legend to Fig. 1. The concentration dependence is plotted according to Hanes' transformation of the Michaelis–Menten equation: $S/v = S/V_{max} + K_m/V_{max}$. The error bars represent $\pm$ 1 SE and the line is drawn by linear regression. K_m (intercept with the abscissa) is calculated as 9.9 mM. V_{max} (reciprocal slope) is calculated as 0.73 m$M \cdot \sec^{-1}$. (Reprinted from Rees and Gliemann[6] by permission.)

is the following. When 3OMG is present extracellularly at a relatively high concentration (say, well above K_m), then the entry of tracer is slower than that occurring with tracer alone. However, 3OMG which has entered the intracellular compartment is initially present there at a very low concentration and the tracer can move back to the extracellular compartment with less restriction. Therefore, the uptake curve will deviate away from an exponential curve as f_i increases, i.e., becomes more flat than the curves characterizing equilibrium exchange. There are two ways to circumvent this problem. One way is to read the initial velocity directly from the entry curve (cf. legend to Fig. 1). This can be done if the first and almost linear part can be determined with sufficient precision. The other way is to use an integrated rate replot. The purpose is to extrapolate the initial velocity to that which would be obtained with zero internal sugar analog, i.e., at time zero. The process is analogous to calculating the initial velocity from an exponential uptake curve (cf. Fig. 1).

As shown by Ginsburg and Stein,[13] the initial velocity can be obtained by plotting

$$t/C \text{ vs } -[\ln(1 - C/S_o) + C/S_o]/C \qquad (2)$$

where t is the time of uptake, C is the internal substrate concentration at that time, and S_o is the external substrate concentration. This plot yields a straight line and the intercept with the ordinate ($C = 0$) gives the reciprocal initial velocity for a given S_o (v in mM · sec^{-1}). This procedure has been used previously to obtain initial velocities in the IM-9 lymphocytes[6] and adipocytes.[4,14]

Zero-Trans Exit. According to this protocol, cells are first loaded with labeled and unlabeled 3OMG until equilibrium is achieved. A large volume of 3OMG-free buffer is then added so that flux occurs from the intracellular to the extracellular compartment with only negligible back flux. Initial velocities for various 3OMG concentrations may in principle be obtained from the first and linear parts of the exit curves but this is in practice difficult or impossible. The reason is that the scatter on the zero-time value (fully loaded cells) and the difficulty with stopping efflux into a large volume instantaneously. On the other hand, advantage may be taken from the fact that a single zero-trans exit curve with an initially high intracellular 3OMG concentration (well above K_m) contains the information needed to calculate K_m and V_{max}. This is because the intracellular concentration runs from the highest value initially and ideally zero at infinite time (with infinite dilution). Karlish et al.,[15] using erythrocytes, have provided an integrated rate equation:

$$-\ln S_t/S_0(S_0 - S_t)^{-1} = (V_{max}/K_m)(S_0 - S_t)^{-1}t - K_m^{-1} \qquad (3a)$$

where S_0 and S_t are the intracellular substrate concentrations at time zero and time t. This equation is equivalent to the Lineweaver–Burke transformation of the Michaelis–Menten equation:

$$1/S = (V_{max}/K_m)v^{-1} - K_m^{-1} \qquad (3b)$$

Therefore, $-K_m$ is obtained as the reciprocal intercept with the ordinate and V_{max}/K_m as the slope when the data are plotted according to Eq. (3a).

It is also possible to calculate the initial velocity from a zero-trans exit curve since when the left sides of Eqs. (3a) and (3b) are equal, then

[13] H. Ginsburg and W. D. Stein, *Biochim. Biophys. Acta* **382,** 353 (1975).
[14] L. P. Taylor and G. D. Holman, *Biochim. Biophys. Acta* **642,** 325 (1981).
[15] S. J. D. Karlish, W. R. Lieb, D. Ravn, and W. D. Stein, *Biochim. Biophys. Acta* **255,** 126 (1972).

$(S_0 - S_t)^{-1}t$ equals v^{-1}. These principles have been used to calculate K_m and V_{max} in the IM-9 lymphocytes.[6]

Infinite-Cis Entry. In this situation influx is carried out using one very high concentration of 3OMG in the buffer (the cis side). The experiment is easy to perform since standard influx technique is used and since uptake is very slow. It is a special case of zero-trans entry and Eq. (2) will therefore give a straight line. V_{max} is obtained as the reciprocal intercept with the ordinate. The importance of the experiment is that the substrate concentration on the inside (trans side) as a function of time changes from zero to the very high concentration used on the cis side. It therefore contains the information necessary to calculate the K_m for the inside site when a saturating concentration of substrate is present on the outside (infinite cis entry K_m). This K_m will determine the rate at which substrate at various inside concentrations will flux back to the outside buffer. It has been shown (with erythrocytes) that the intercept with the abscissa will be $-K_m/S_0^2(1 + S_0/\pi)$, where S_o is the 3OMG concentration and π is the effective osmotic concentration of the buffer[13] and K_m can thus be calculated. This approach has been used with adipocytes[14] and IM-9 lymphocytes.[6]

Infinite-Cis Exit Experiments. These are carried out by first loading the cells with a saturating substrate concentration and efflux into solutions with different substrate concentrations is followed. The specific radioactivity of the inside and outside sugar is the same. Efflux of label is slow due to the high substrate concentration on the inside. Furthermore, efflux will be linear (zero order) until the internal concentration ceases to be saturating and v is therefore quite easy to measure. v will decrease as external sugar analog increases. A plot of $1/v$ vs S gives $-K_m$ as the intercept on the abscissa.[14] The infinite-cis exit procedure provides a measure of K_m on the external side when a saturating concentration of substrate is present on the inside.

Infinite-Trans Experiments. These are also known as countertransport experiments. In the influx version, cells are first loaded with a very high concentration of 3OMG and influx is measured using external labeled sugar at different concentrations. Therefore, this is another way of obtaining a K_m for the external site when a saturating concentration is present on the inside. External labeled sugar at a low concentration on the outside will transiently accumulate on the inside and markedly exceed its intracellular distribution space at equilibrium. This phenomenon is a classical way of demonstrating the existence of a facilitated diffusion system and has been shown in polymorphonuclear leukocytes.[8] In the efflux version, cells are loaded with labeled 3OMG at various concentrations and efflux into buffer with a saturating concentration of unlabeled 3OMG is measured. When cells are equilibrated with labeled 3OMG at a low concentration and unlabeled 3OMG is added on the outside (without changing the outside isotope concentration significantly) then labeled 3OMG will transiently move against its concentration gradient, demonstrating the existence of a facilitated diffusion system.

Symmetry and Asymmetry. The hexose transport system is said to be symmetrical when K_m and V_{max} values are independent of the direction of the flux. Symmetry may be subdivided into two categories. One is that the kinetic constants are also independent of the substrate concentration in the compartment which is labeled substrate moves into (the trans side). In other words, the kinetic constants are similar according to all protocols. This seems to be the situation in rat hepatocytes at 20° [16] and in rat adipocytes at 37° [4,14] although kinetic asymmetry has also been reported in adipocytes. [17] Another category is directional symmetry with facilitation of movement of labeled substrate from one compartment to the other (cis to trans) with substrate present on the trans side. This implies that V_{max} and K_m values are higher according to equilibrium exchange than to zero-trans protocols. Such phenomenon has been reported in several cultured cell types [18]; however, the explanation may be methodological rather than mechanistic. [6]

Asymmetric transport kinetics have been reported extensively in human erythrocytes at 20° (for a review, see Ref. 19) but may be less pronounced at 37°. [20] The hexose transport system of IM-9 lymphocytes is asymmetric at 37° with three to four times higher K_m and V_{max} values for zero-trans exit and equilibrium exchange than for zero-trans entry. [6] It may well be similar to that in human erythrocytes. A clear physiological function of the asymmetry has not been pointed out.

The Fructose Transporter

This has been demonstrated in rat adipocytes. In the absence of insulin most fructose is transferred via its own transporter and a minor but significant fraction enters the cell via the glucose carrier. In the presence of insulin many more glucose transporters become operative in the

[16] J. P. Craik and K. R. I. Elliot, *Biochem. J.* **182**, 503 (1979).
[17] J. Vinten, *Biochim. Biophys. Acta* **772**, 244 (1984).
[18] P. G. W. Plagemann, R. M. Wohlheuter, J. C. Graff, J. Erbe, and P. Wilkie, *J. Biol. Chem.* **256**, 2835 (1981).
[19] W. F. Widdas, *Curr. Top. Membr. Transp.* **14**, 165 (1980).
[20] J. Brahm, *J. Physiol.* (*London*) **339**, 339 (1983).

plasma membrane[21,22] and most fructose now enters via this pathway.[23] This conclusion is in part based on uptake studies and in part in experiments with ATP-depleted cells since no nonmetabolizable fructose analog has been described. The affinity of D-glucose and 3OMG for the fructose transporter seems negligible.[23] A fructose transporter has also been identified in rat hepatocytes.[5]

Comments on Methods

The Tracer. 3OMG is available labeled with ^{14}C or ^3H. It is essential that the labeled molecule contains no trace contaminant which can be metabolized. Experience shows that labeling with ^{14}C in the methyl moiety is satisfactory in this respect. ^3H-Labeled preparations have to be rinsed frequently for impurities, for instance, by paper chromatography, but they have the advantage of being less expensive.

Other nonmetabolizable glucose analogs may be used but they have much lower affinities than 3OMG or glucose which, broadly speaking, have K_m values in the physiological glucose concentration range (3–10 mM). Thus, L-arabinose (a D-galactose derivative lacking the C-6 hydroxymethyl group) has a K_m of 50–100 mM and D-allose (a C-3 epimer of D-glucose) has a K_m of at least 200 mM. This means that saturation of a transporter requires extremely high sugar concentrations outside a practical range. However, a measure of inhibition constants of other sugars can be obtained when the reciprocal rate constant for entry of, for instance, L-arabinose is plotted against concentration of the competing sugar (cf. Fig. 2).

The Stopping. Some investigators prefer to use cytochalasin B at a high concentration (10–20 μM) because this is a more specific competitive inhibitor of hexose transport. However, it is much more expensive than phloretin, and specificity seems unimportant for a stopping solution. The small concentration of $HgCl_2$ in the stopping solution (see Measurement of Entry) is not essential to arrest the transport. It can help forming a cell pellet which is dense enough to prevent loss of cells.

The Compartments. It is evident that the cell suspension used for transport studies should ideally contain only one type of cells. The intracellular water space contributed by other cells will inevitably cause an error. For instance, the study of hexose transport in polymorphonuclear

[21] E. Karnieli, M. J. Zarnowski, P. J. Hissin, I. A. Simpson, L. B. Salans, and S. W. Cushman, *J. Biol. Chem.* **256**, 4772 (1981).

[22] K. Suzuki and T. Kono, *Proc. Natl. Acad. Sci. U.S.A.* **77**, 2542 (1980).

[23] E. Schoenle, J. Zapf, and E. R. Froesch, *Am. J. Physiol.* **237**, E325 (1979).

leukocytes[8] requires much higher cell purity than the study of glucose incorporation into glycogen or 2-deoxyglucose uptake. The reason is that hexose metabolism in erythrocytes and even lymphocytes is low as compared to the leukocytes, whereas all these cell types transport glucose, although at very different rates.

Transport of tracer 3OMG may fail to show two-compartment kinetics even when only one cell type is present. For instance, rat thymocyte suspensions consist of two populations which can be distinguished kinetically. About one-third are "active" cells which equilibrate 3OMG with a half-time of about 1 min and two-thirds are "quiescent" with an equilibration half-time of 30–50 sec.[24,25] In such cases, the subpopulations should be separated, preferably physically, or otherwise kinetically, so that the individual rate constants can be estimated. It should be remembered that the initial velocity read directly from the uptake curve (cf. legend to Fig. 1) only relates to the "active" cells.

Transport Modulation

The study of transport seems most interesting in cells with a regulation at this step. The nature of such regulations is only beginning to be understood. Insulin induces, according to most authors, a marked increase in V_{max} in adipocytes with little or no change in K_m,[3,4,14] and it causes an increase in the concentration of transporters in the plasma membrane to an extent which largely explains the V_{max} increase.[21,22] This model has been challenged since Whitesell and Abumrad[26] found the insulin effect predominantly due to a 10-fold increase in the affinity of the transporter to hexose. Nevertheless, it seems certain that transporters can move in and out of the plasma membrane, and this may be a general phenomenon in cells with regulated transport systems. Such "instability" may be the reason why details in the procedure can change transport in unexpected ways. Vigorous mechanical treatment of adipocytes has been reported to increase transport rates and it may even make a difference whether cells are pipetted into buffer or vice versa.[26] Small concentrations of glucose or pyruvate can be necessary to maintain low transport rates.[26] Low temperature has been shown to increase the transporter concentration in the plasma membrane relative to the intracellular location in adipocytes.[27]

[24] J. P. Reeves, *J. Biol. Chem.* **252**, 4876 (1977)

[25] R. R. Whitesell, L. H. Hoffman, and D. M. Regen, *J. Biol. Chem.* **252**, 3533 (1977).

[26] R. R. Whitesell and N. A. Abumrad, *J. Biol. Chem.* **260**, 2894 (1985)

[27] T. Kono, K. Suzuki, L. Dansey, F. W. Robinson, and T. L. Blevins, *J. Biol. Chem.* **256**, 6400 (1981).

The choice of buffer for incubation may be important. Thus, Tris can inhibit recycling of insulin receptors[28] to the plasma membrane and it seems conceivable that it may also interfere with cycling of transporters.

If uptake of 3OMG at a low concentration is exponential both with and without an agent which modulates transport, then it should be quite easy to measure K_m and/or V_{max} changes. Nevertheless, reports in the literature are very often conflicting and failure to analyze uptake curves may be part of the reason. If the uptake curve is exponential in one situation and not in the other, then only a fraction of the cells is modulated. If this point is not appreciated, the kinetic constants will necessarily be measured incorrectly.

A modifying condition may also change the distribution space for 3OMG (the intracellular water space). Fasting, for instance, will decrease the distribution space markedly in adipocytes. In such cases V_{max} expressed in $mM \cdot sec^{-1}$ or in $mmol \cdot cell^{-1} \cdot sec^{-1}$ will of course change disproportionally (cf. legend to Fig. 1) and the latter expression may be the most logical. It has been shown that phytohemagglutinin increases 3OMG transport in T lymphocytes.[29] However, the distribution space was increased (activation of a volume-regulating mechanism?) to the same extent as the uptake. Therefore, entry per cell was increased but not the rate constant of entry and not the initial velocity (expressed in $mM \cdot sec^{-1}$). Chemotactic peptides increase 3OMG transport per cell in polymorphonuclear leukocytes and they also cause an increase in the distribution space.[29a]

What Protocols Are the Most Important?

Investigators may wish to analyze hexose transport in a given cell suspension without going through all possible and sometimes complicated analyses. It first has to be shown that the tracer is indeed transported by a facilitated mechanism. This point may seem trivial but it was thought until recently that 3OMG was transported by nonmediated diffusion in, for instance, polymorphonuclear leukocytes.[30] In addition to the marked stereospecificity, the glucose transporter should also be nearly completely inhibited by phloretin and cytochalasin B. However, we have noted that fructose transport is not inhibited by the latter drug.[31] Residual transport

[28] P. Rennie and J. Gliemann, *Biochem. Biophys. Res. Commun.* **102**, 824 (1981).

[29] J. H. Helderman, *J. Clin. Invest.* **67**, 1636 (1981).

[29a] Y. Okuno and J. Gliemann, *Biochim. Biophys. Acta* **941**, 157 (1988).

[30] C. E. McCall, J. Schmitt, S. Cousart, J. O'Flaherty, D. Bass, and R. Wykle, *Biochem. Biophsy. Res. Commun.* **126**, 450 (1985).

[31] Y. Okuno and J. Gliemann, *Diabetologia* **30**, 426 (1987).

of a given labeled ligand in the presence of, for instance, 20 μM cytochalasin B might therefore be due to transfer via the fructose carrier rather than to nonmediated diffusion.

The analysis of the entire uptake curve with tracer at a low concentration (two-compartment kinetics?) is essential for all other protocols. Equilibrium exchange just requires repeat of the same procedure with cells preincubated for a long time with unlabeled 3OMG at different concentrations. It approaches a physiological situation in cells with a very low rate of glucose metabolism compared to glucose transport.

Zero-trans entry seems the next logical step and has also the advantage of being technically simple. The sugar transporter in most cells is the gateway for entry (the exception being hepatocytes with exit as the most important process) and it is essentially zero trans if sugar phosphorylation is very effective. The integrated rate procedure to facilitate estimation of initial velocities [Eq. (2)] is helpful and not very sensitive to bias when early time points are used. The combination of equilibrium exchange and zero-trans entry experiments should therefore give a valid picture of the function of how a particular hexose transport system is working. The kinetic analysis is not complete but in those cases where a kinetic asymmetry seems evident (most notably human erythrocytes) the major differences are actually observed between the equilibrium exchange and zero-trans entry protocols.

Integrated rate equations which require experiments with long incubation times may be dangerous to use for estimation of K_m values from entry [infinite cis, Eq. (2)] to exit [zero trans, Eq. (3)] curves because they are sensitive to bias. This may arise if the tracer moves via routes other than that investigated or if it is metabolized, even to a minor extent, with resulting accumulation of metabolic products inside the cells.

Transport as a Control Point

One of the most important problems is to determine whether the cell can phosphorylate and metabolize all glucose that enters. This is the case in adipocytes both in the absence and in the presence of insulin at very low extracellular glucose concentrations. With increasing substrate concentrations, glucose transport ceases to be rate limiting and this happens, of course, at much lower concentration in the presence than in the absence of insulin. The situation is the same when 2-deoxy-D-glucose, which is phosphorylated and not further metabolized to any major extent, is used instead of D-glucose.[9] Apparent K_m values based on 2-deoxyglucose uptakes measured after incubation for minutes will therefore be intermediate between that of the transporter and that of hexokinase(s).

2-Deoxyglucose uptake, i.e., the rate of its conversion to 2-deoxyglucose phosphate, has been used as a measure of hexose transport in several cell types but it is often not known whether transport is the rate-limiting step under the conditions used. In order to resolve this point it is necessary to measure at such early times (usually seconds) that only a small fraction (e.g., 25%) of the theoretical distribution "space" is filled. In this connection "space" means the distribution volume the sugar would have had if it was not metabolized.

The same arguments apply to measurements of the rate of glucose metabolism which may or may not equal the glucose transport rate. In adipocytes [^{14}C]glucose is converted to products which are trapped inside the cell and to $^{14}CO_2$.[32,33] The fraction converted to CO_2 is small, particularly in human adipocytes,[33] and the rate of accumulation of radioactivity in the cells (measured over several minutes) will therefore approximate the rate of transport at low substrate concentrations. Such uptake experiments are easy to do and a high precision can be obtained. They can be helpful in preliminary or "screening" experiments and in dose–response studies when it is known that transport remains rate determining both with and without the modifying agent. However, when a not previously studied condition or factor changes the rate of glucose metabolism then there is only one way to find out whether glucose transport is also changed: to measure the initial velocity.

[32] J. Gliemann, W. D. Rees, and J. E. Foley, *Biochim. Biophys. Acta* **804**, 68 (1984).
[33] A. Kashiwagi, M. A. Verso, J. Andrews, B. Vasques, G. Reaven, and J. E. Foley, *J. Clin. Invest.* **72**, 1246 (1983).

[40] Preparation and Culture of Embryonic and Neonatal Heart Muscle Cells: Modification of Transport Activity

By KARL WERDAN and ERLAND ERDMANN

Introduction

Transport processes across the cell membrane maintain cellular homeostasis. Alterations in cell metabolism require adaptive changes of these transport systems. Especially in active movements of Na^+ and K^+, these compensatory mechanisms play a dominant role: extracellular/intracellular gradients for sodium and potassium ions are maintained by active transport of Na^+ and K^+ across the cell membrane, mediated by

Na^+,K^+-ATPase (EC 3.6.1.3), the sodium pump. In the heart muscle cell, one major function of this enzyme is restoration of excitation-induced imbalance of cellular Na^+,K^+ homeostasis. In addition, it also plays an important role as pharmacological receptor for and mediator of the positive inotropic action of cardiac glycosides.[1,2]

Active Na^+,K^+ transport in the heart is subject to modification and regulation under various physiological and pathological conditions, as well as by drug therapy. These changes can be studied in detail at the cellular level in beating heart muscle cells in culture. In this chapter we shall describe the experimental techniques necessary for these investigations. The methods are outlined below.

1. Preparation and cultivation of cardiac muscle and nonmuscle cells from chicken embryos and neonatal rats. The methods given allow cultivation of heart cells from both species under identical conditions.

2. Characterization of active Na^+,K^+ transport in heart cells in culture—modification of transport activity by inhibitors and stimulators. Na^+,K^+-ATPase located in the cell membrane transports Na^+ out and K^+ into the cells, in an energy-dependent manner. The stoichiometry of this transport is $3Na^+/2K^+/1$ intracellular ATP split. Sodium pump activity is described by influx measurements of radioactively labeled $^{42}K^+$ or $^{86}Rb^+$ (used as a K^+ analog). Quantitative characterization is achieved by determination of the Michaelis constant (K_m) and the transport capacity (V) of active K^+ influx. Cardiac glycosides, e.g., ouabain, specifically inactivate Na^+,K^+-ATPase,[1] thereby inhibiting active Na^+,K^+ transport. This allows discrimination of sodium pump activity from other Na^+ and K^+ flux mechanisms across the cell membrane. The rates of active Na^+,K^+ transport depend on the beating frequency of the heart muscle cells; transport is inhibited by cardiac glycosides and stimulated by insulin.

3. Determination of the number of cardiac glycoside receptors and sodium pump molecules in heart cells in culture. Cardiac glycosides specifically bind to the Na^+,K^+-ATPase molecule,[1] the "receptor" region of the enzyme being located at the outer surface of the cell membrane. In binding experiments with [^{3}H]ouabain, the number of cardiac glycoside receptors per heart muscle cell can be quantified, being identical with the number of sodium pump molecules per cell.

4. Correlation of cardiac glycoside receptor occupation and inhibition of active Na^+,K^+ transport in cardiac muscle and nonmuscle cells from chicken embryos. Binding of a cardiac glycoside molecule to its receptor results in inactivation of the Na^+,K^+-ATPase molecule. By simultaneous

[1] E. Erdmann, *Handb. Exp. Pharmacol.* **56,** 337 (1981).
[2] C. O. Lee, *Am. J. Physiol.* **249,** C367 (1985).

measurement of [^{3}H]ouabain receptor binding and inhibition of active K^+ influx by ouabain, the stoichiometry of Na^+,K^+-ATPase inactivation and inhibition of active Na^+,K^+ transport can be evaluated. By this experimental device, a "pump reserve" can be demonstrated in chicken heart muscle cells; in chicken heart nonmuscle cells, on the contrary, a linear correlation exists between sodium pump inactivation and transport inhibition.

5. Modification of the number of sodium pump molecules of cultured chicken heart cells by cell exposure to ouabain or low K^+. Impairment of sodium pump activity either by lowering extracellular K^+ or by inactivation of part of the Na^+,K^+-ATPase molecules by ouabain induces a compensatory increase in the number of active sodium pump molecules per cell. This is demonstrated by an increase in the number of cardiac glycoside receptors per cell, as well as by a higher transport capacity of these cells. The extent of this adaptation process differs in cardiac muscle and nonmuscle cells from the same species.

6. Monitoring of beating in heart muscle cells in culture: sodium pump activity and the positive inotropic action of cardiac glycosides. A monitoring system is described for observation of beating (frequency, amplitude, and velocity of cell wall motion) in heart muscle cells in culture. The positive inotropic action of cardiac glycosides is reflected in a semiquantitative manner by an increase in amplitude and velocity of cell wall motion. This effect is due to a drug-induced inactivation of Na^+,K^+-ATPase with concomitant inhibition of active Na^+,K^+ transport and increase in intracellular Na^+.[2] Cardiac glycoside sensitivity of chicken heart muscle cells is inversely related to the number of sodium pump molecules, which is subject to regulation.

Materials and Solutions

Chemicals purchased from NEN Chemicals (Dreieich, FRG): ^{22}NaCl, carrier free; ^{86}RbCl, 0.9–4.6 mCi/mg; [^{3}H]ouabain, 14–20 Ci/mmol

Amersham-Buchler (Braunschweig, FRG): ^{42}KCl, 22 μCi/mg

Biochrom (Berlin, FRG): Collagenase "Worthington," 125–250 U/mg, CLS II; fetal calf serum; horse serum; CMRL 1415 ATM medium

Sigma Chemie (Taufkirchen, FRG): Bovine insulin, 24 IU/mg, No. 15 500; dexamethasone No. D-8893

Serva Biochimica (Heidelberg, FRG): Trypsin 1 : 250, No. 37 290; bovine serum albumin, No. 11 920; *N*-2-hydroxyethylpiperazine-*N'*-2-ethanesulfonic acid (HEPES)

Boehringer Mannheim (Mannheim, FRG): Iron-saturated transferrin, No. 652 202

All other chemicals are of analytical grade and are purchased from Merck (Darmstadt, FRG) and Boehringer Mannheim. Plastic culture flasks (25 cm^2, 175 cm^2) were from Nunclon Plastics (Roskilde, Denmark).

CMRL medium with modified K$^+$ and Ca^{2+} concentrations: K$^+$- and Ca^{2+}-free CMRL 1415 ATM medium is purchased from Biochrom and supplemented with the desired K$^+$ and Ca^{2+} concentrations

Salt solution A: NaCl (127 mM), KCl (2.7 mM), Na$_2$HPO$_4$·2H$_2$O (10.6 mM), KH$_2$PO$_4$ (2.1 mM), D-glucose (5 mM); pH 7.25

Salt solution B: NaCl (135 mM), KCl (5.4 mM), CaCl$_2$ (1.8 mM), MgCl$_2$ (1.05 mM), NaH$_2$PO$_4$ (0.36 mM), D-glucose (5 mM), HEPES (20 mM); pH adjusted with NaOH to 7.25 (room temperature)

Calcium–sorbitol solution: CaCl$_2$ (1.8 mM), sorbitol (280 mM), HEPES (3 mM); pH 7.25

Preparation and Cultivation of Cardiac Muscle and Nonmuscle Cells from Chicken Embryos and Neonatal Rats

Primary cultures of myocardial cells are prepared aseptically, either from hearts of 1- to 3-day-old rats or from 12- to 13-day-old chicken embryos, according to the method of Harary *et al.*[3]: 20–100 hearts are cut into pieces (approximately 1 mm^3) and placed in ice-cold phosphate-buffered salt solution. Thereafter, the pieces are disaggregated into single cells by repeated treatment at 37° with 10–20 ml trypsin (0.12%)–collagenase (0.03%)–salt solution A (duration 15 min each) in a 50-ml Erlenmeyer flask with a magnetic stirrer (200 rpm). The products of the first two treatment steps are discarded because they contain cell debris and mainly mesenchymal cells.[4] Subsequent supernatants (about three to five) are poured into centrifuge tubes, containing 1.0 ml precooled (4°) growth medium (CMRL 1415 ATM), according to Healy and Parker,[5] supplemented with 10% fetal calf serum, 10% horse serum, and 0.02 mg/ml gentamicin and adjusted to pH 7.40 with 1 N HCl. The cell suspensions are centrifuged (10 min, 300 g; room temperature), the supernatants discarded, and the cell pellets resuspended in fresh growth medium. The resulting cell suspension is then distributed in 175-cm^2 Nunclon plastic culture flasks (about 10^6 cells/ml, 20–30 ml/flask) and incubated at 37° in a water-saturated atmosphere for 2 hr. During this period, the majority of

---

[3] I. Harary, F. Hoover, and B. Farley, this series, Vol. 32, p. 740.

[4] B. Blondel, I. Roijen, and J. P. Cheneval, *Experientia* **27**, 356 (1971).

[5] G. M. Healy and R. C. Parker, *J. Cell Biol.* **30**, 531 (1966).

"nonmuscle" cells,[6] accounting for 30–40% of total heart cells,[4] attach to and spread out on the surface of the flasks. At the end of this incubation period, the supernatants containing the myocardial muscle cells are pooled in a beaker. Using an inverted phase-contrast microscope (Diavert, Leitz, Wetzlar, FRG), cell counts are carried out with a hemocytometer, and the volume is adjusted with additional growth medium to give the final cell concentration wanted [cell yields: about $(1-2) \times 10^6$ cells/rat heart; about $(1.5-3) \times 10^6$ cells/chicken heart]. The cell suspension is then distributed into 25-cm^2 plastic culture flasks, 5 ml suspension/flask, with cell densities of about 10^5 viable cells/cm^2. Further incubation of the cells with growth medium and daily medium changes is carried out at 37° in a water-saturated atmosphere. As CMRL 1415 ATM has a high buffer capacity, incubation of the cells in a 5% CO_2 atmosphere is unnecessary.[3]

Part of the experiments (see specific figure legends) have been carried out with heart muscle cells grown in serum-free, hormone-supplemented growth medium (adapted from Claycomb[7]): after initiation of the cell cultures, serum-supplemented growth medium (see above) is replaced after 24 hr by serum-free CMRL medium, supplemented with bovine insulin (25 μg/ml), dexamethasone (40 μg/ml), iron-saturated transferrin (25 μg/ml), and bovine serum albumin (25 μg/ml), the pH being adjusted to 7.40 (room temperature) with 1 N HCl, 290 mOsm/liter. In this serum-free medium, proliferation of nonmuscle cells is largely retarded. All studies with heart muscle cells are carried out after cultivation of the cells for 3–5 days. During this time, either a synchronously contracting monolayer of heart muscle cells has formed, or the cells are lying isolated or in small clusters of two to five beating cells, depending on the initially chosen cell density. With a seeding density of 10^5 cells/cm^2, for example, a synchronously beating cell monolayer is obtained after 3 days, with about 1.0 mg cell protein/flask. To ensure that cell cultures mainly consists of muscle cells, the percentage of beating cells was determined by phase-contrast microscopy in each preparation on the day the experiment was done. All measurements described in the text are carried out with cell preparations whose percentage of beating cells was 75% or more of the total cell population. Since every culture also contains quiescent muscle cells, the total percentage of muscle cells is even higher than the percentage of beating cells. The term nonmuscle cells (see above) refers to heart cells in culture lacking sarcomeres, mainly consisting of fibroblasts and endothelial cells.[6] In contrast to heart muscle cells, they possess a high prolifera-

[6] W. J. Marvin, R. B. Robinson, and K. Hermsmeyer, *Circ. Res.* **45**, 528 (1979).

[7] W. C. Claycomb, *Exp. Cell Res.* **131**, 231 (1980).

tion rate and rapidly attach to the culture flask after seeding (see above). These heart nonmuscle cells from neonatal rats and chicken embryos have been separated from the heart muscle cells during the cultivation procedure (see above), and have been grown separately under identical culture conditions, with the exception that horse serum is omitted from the growth medium and a higher gentamicin concentration (0.05 mg/ml) is used. Experiments with nonmuscle cells are carried out after one subcultivation (splitting ratio 1:2; detachment of the cells at 37° by 0.05% trypsin plus 0.02% EDTA in Ca^{2+}, Mg^{2+}-free salt solution).

The methods given allow cultivation of heart cells from chicken embryos and neonatal rats under identical conditions. Further protocols also include culture methods for heart cells from other species.[8-13]

Characterization of Active Na^+, K^+ Transport in Heart Cells in Culture: Modification of Transport Activity by Inhibitors and Stimulators

Sodium pump activity in cultured heart cells from chicken embryos and neonatal rats can be characterized by measurement of active K^+ flux into the cells. All determinations are carried out at 37°.

Cells grown as monolayers in 25-cm^2 plastic culture flasks (muscle cells: 0.5–1.0 mg protein/flask; nonmuscle cells 0.3–0.5 mg protein/flask) are used. Prior to the measurements, the growth medium is removed by suction, and the cells are washed three times with 5 ml each of salt solution B with lowered HEPES concentration (3 mM) at 37°. Thereafter, the cells are equilibrated for 60 min with 5 ml of salt solution B. The uptake experiment is then started by replacing this equilibration solution by 5 ml of salt solution B containing tracer amounts (3 × 10^6 cpm/5 ml) of $^{42}K^+$. Uptake is determined after 5 or 10 min by removing the radioactive assay medium and quickly washing the cells three times with 5 ml each of cold (4°) salt solution B with lowered HEPES. By these washings, extracellular $^{42}K^+$ is removed without loss of $^{42}K^+$ taken up into the cells.[14] The washed cells are then lysed in 1.3 ml NaOH (0.1 M)/EDTA (4.5 mM)

[8] T. Kazazoglou, J.-F. Renaud, B. Rossi, and M. Lazdunski, *J. Biol. Chem.* **258**, 12163 (1983).
[9] W. H. Barry, S. Biedert, D. S. Miura, and T. W. Smith, *Circ. Res.* **49**, 141 (1981).
[10] R. Kandolf, A. Canu and P. H. Hofschneider, *J. Mol. Cell. Cardiol.* **17**, 167 (1985).
[11] F. H. Kasten, *in* "Tissue Culture, Methods and Applications" (P. F. Kruse, Jr., and M. K. Patterson, Jr., eds.). Academic Press, New York, 1973.
[12] M. Lieberman, W. J. Adam, and P. N. Bullock, *Methods Cell Biol.* **21**, 187 (1980).
[13] A. L. Harvey, "The Pharmacology of Nerve and Muscle in Tissue Culture." Croom Helm, London, 1984.
[14] K. Werdan, G. Bauriedel, B. Fischer, W. Krawietz, E. Erdmann, W. Schmitz, and H. Scholz, *Biochim. Biophys. Acta* **687**, 79 (1982).

overnight at 37°. Six hundred microliters is taken for radioactivity measurement in 10 ml Unisolve (Zinsser, München, FRG) after neutralization with 1.0 M HCl, using a liquid scintillation counter. If desired, cellular K^+ content is measured by flame photometry in two 200-μl portions (see below). In this case, cells are washed three times with Ca^{2+}–sorbitol solution instead of salt solution B (see above). Two 50-μl samples are taken for protein determination according to the method of Lowry, with bovine serum albumin as standard. The variation in protein content per flask within one experiment is $\leq 7\%$.

In each series of experiments, a zero-time assay is carried out by adding radioactive assay medium to the cultures and immediately (within 10 sec) washing the cells by the standard procedure. The result of this $^{42}K^+$ influx measurement is shown in Fig. 1a for cultured rat heart muscle cells (●—●). Uptake is linear for at least 10 min. From the slope of the line, the uptake rate for K^+ can be calculated and expressed as nanomoles K^+ taken up/milligram cell protein $\times$ minutes. Extrapolating the uptake to $t = 0$ min yields the small portion of extracellular $^{42}K^+$ adhering to the cells, being identical with the inulin space.[14] Adding $10^{-3}\ M$ ouabain to the equilibration—and assay—solution reduces the subsequent $^{42}K^+$ influx to 39% (Fig. 1a, ■—■). Active $^{42}K^+$ influx (△—△) mediated by the sodium pump (Na^+,K^+-ATPase) and being inhibited by ouabain can now be calculated by subtracting the ouabain-insensitive $^{42}K^+$ influx from total influx.

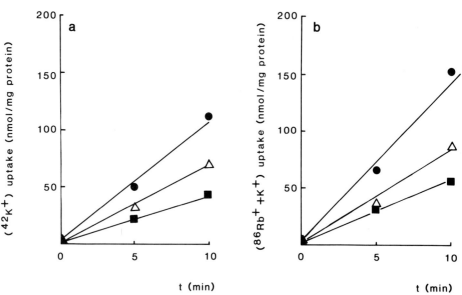

FIG. 1. Kinetics of (a) $^{42}K^+$ and (b) $^{86}Rb^+$ influx in rat heart muscle cells in culture. Values are means from closely correlating triplicates. (See text.)

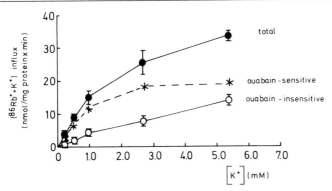

FIG. 2. K^+ dependency of ouabain-sensitive and ouabain-insensitive ($^{86}Rb^+$ + K^+) influx in chicken heart muscle cells in culture. Uptake rates are calculated from influx measurements at 10 and 60 sec. For determination of ouabain-insensitive influx, cells are preincubated for 3 hr in the presence of 10^{-4} M ouabain. $x \pm$ SD, $n = 3$. (See text.)

In the experiment of Fig. 1b, tracer amounts of $^{86}Rb^+$ instead of ^{42}K are added to the assay medium containing 5.4 mM K^+. Because Rb^+ behaves as a K^+ analog, the $^{86}Rb^+$ radioactivity taken up into the cells reflects the uptake of K^+. From the $^{86}Rb^+$ radioactivity/flask (2×10^6 cpm) and the amount of Rb^+ (2.5 nmol/flask) plus K^+ (27 μmol/flask) the influx can then be calculated as nanomoles ($^{86}Rb^+$ + K^+)/milligram protein $\times$ min.[14,15] From comparison of Fig. 1a and b it can be seen that the ($^{86}Rb^+$ + K^+)-uptake rates are nearly identical with the K^+-uptake rates as measured with $^{42}K^+$. For experimental convenience due to the longer half-life of $^{86}Rb^+$ (18.7 days) in comparison with $^{42}K^+$ (12.5 hr), all measurements of active K^+ transports described further are carried out with $^{86}Rb^+$ instead of $^{42}K^+$ as active tracer (see also Werdan et al.[14,15]).

In an identical manner as described for heart muscle cells from neonatal rats (Fig. 1), active K^+ transport can be studied in cultured heart muscle cells from chicken embryos, as well as in cardiac nonmuscle cells from these species.[16,17] Instead of salt solution B, growth medium or serum-free, hormone-supplemented CMRL medium can be taken as assay medium, yielding similar results.

Further characterization of the ouabain-sensitive and the ouabain-insensitive component of K^+ influx is shown in Fig. 2: In cultured heart muscle cells from chicken embryos, ($^{86}Rb^+$ + K^+)-influx rates have been

[15] K. Werdan, G. Bauriedel, M. Bozsik, W. Krawietz, and E. Erdmann, *Biochim. Biophys. Acta* **597**, 364 (1980).
[16] K. Werdan, B. Wagenknecht, B. Zwissler, L. Brown, W. Krawietz, and E. Erdmann, *Biochem. Pharmacol.* **33**, 55 (1984).
[17] K. Werdan, B. Wagenknecht, B. Zwissler, L. Brown, W. Krawietz, and E. Erdmann, *Biochem. Pharmacol.* **33**, 1873 (1984).

measured as described (●—●) in Fig. 1b, at different K$^+$ concentrations. Ouabain-insensitive uptake rates (○—○) due to diffusion and eventually additional transport mechanisms increase nearly linearly with increasing extracellular K$^+$ concentrations. In contrast, ouabain-sensitive influx (×––×), representing sodium pump activity, demonstrates the characteristics of a saturable transport process, with a maximal transport capacity of about 20 nmol (^{86}Rb$^+$ + K$^+$)/mg protein × min, and half-maximal transport activity (K_m) at about 0.8 mM K$^+$. Similar results are obtained with rat heart muscle cells.

Sodium pump activity of heart cells in culture can be modified by various substances: cardiac glycosides, e.g., ouabain, are specific inhibitors of Na$^+$,K$^+$-ATPase.[1] In Fig. 3, heart muscle cells and nonmuscle cells from neonatal rats and chicken embryos are incubated for 4 hr in serum-supplemented (2.5% fetal calf serum, 2.5% horse serum), HEPES-buffered (20 mM) CMRL 1415 ATM medium in the presence of different concentrations of ouabain (see abscissa to Fig. 3). To keep ouabain-insensitive (^{86}Rb$^+$ + K$^+$) influx small (see Fig. 2), the K$^+$ concentration of the

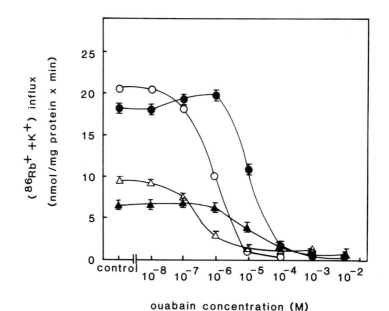

FIG. 3. Concentration-dependent inhibition of (^{86}Rb$^+$ + K$^+$)-influx rates by ouabain in cultured cardiac muscle and nonmuscle cells from chicken embryos and neonatal rats. ●—●: Rat heart muscle cells; ▲—▲: rat heart nonmuscle cells; ○—○: chicken heart muscle cells; △—△: chicken heart nonmuscle cells. Values given are means from closely correlating triplicates (●, ▲, △, ±SD) and duplicates (○).

medium is lowered to 0.75 mM. Thereafter (^{86}Rb$^+$ + K$^+$)-influx rates are measured by adding 2×10^6 cpm ^{86}Rb$^+$/flask, as described in the experiment of Fig. 1b (10-sec and 10-min measurements). In all cell types, (^{86}Rb$^+$ + K$^+$) influx is inhibited in a concentration-dependent manner, rat heart cells being less sensitive [muscle cells: EC$_{50}$ = $(1.3 \pm 0.2) \times 10^{-5} M$; nonmuscle cells: EC$_{50}$ = $(1.5 \pm 0.3) \times 10^{-5} M$] than chicken heart cells [muscle cells: EC$_{50}$ = $(5.8 \pm 0.8) \times 10^{-7} M$; nonmuscle cells: EC$_{50}$ = $(2.6 \pm 0.4) \times 10^{-7} M$; values are given as $x \pm$ SEM; n = 5–7; from Werdan et al.[16,17]]. As a consequence of inhibition of active Na$^+$,K$^+$ transport, cell Na$^+$ increases and cell K$^+$ falls (Fig. 7; see also Werdan et al.[16–18]).

(^{86}Rb$^+$ + K$^+$)-influx rates strongly depend on the beating frequency of the heart cells, the transport rate being about 2.5 as high at a beating frequency of 100/min than in quiescent rat heart muscle cells in culture.[19] In rat heart muscle cells, insulin as well as vanadate (Na$_3$VO$_4$) stimulate active (^{86}Rb$^+$ + K$^+$) influx in a concentration-dependent manner up to 100%, while the ouabain-insensitive influx of K$^+$ remains unaltered.[15,20] The consequence of this pump stimulation is a rise in cell K$^+$ and a slight fall in cell Na$^+$ (for results and further experimental details, see Werdan et al.[15]). Active Na$^+$,K$^+$ transport is completely abolished by heat shocking of the heart cells[17] (60 min at 60°). Because it is an ATP-driven transport, it is also abolished in ATP-depleted cells. This can be accomplished by incubation of rat heart muscle cells for 10 min in 0.5 mM iodoacetic acid (inhibition of glycolysis) + 0.5 μM carbonylcyanide-p-trifluoromethoxy-phenyl-hydrazone (FCCP) (uncoupling of oxidative phosphorylation). By this procedure, energy-rich phosphates of the cells (ATP, 23 nmol/mg cell protein; creatine phosphate, 11 nmol/mg cell protein) are drastically lowered (ATP, 1; creatine phosphate, 1).

Determination of the Number of Cardiac Glycoside Receptors and Sodium Pump Molecules in Heart Cells in Culture

Measurement of cardiac glycoside binding to intact heart cells in culture allows calculation of the number of Na$^+$,K$^+$-ATPase molecules per cell.[16,17] For determination of cellular [^{3}H]ouabain binding, monolayers of chicken heart muscle cells in 25-cm^2 plastic culture flasks are taken (0.5–1.0 mg cell protein/flask). Prior to the measurements, the growth medium is removed by suction, and the cells are washed twice with 5 ml each of

[18] K. Werdan, C. Reithmann, and E. Erdmann, Klin. Wochenschr. 63, 1253 (1985).
[19] G. Bauriedel, Thesis. University of Munich, Munich, Federal Republic of Germany, 1983.
[20] K. Werdan, G. Bauriedel, M. Bozsik, W. Krawietz, and E. Erdmann, Basic Res. Cardiol. 75, 466 (1980).

salt solution B with lowered HEPES concentration (3 mM) at 37°. Thereafter, the cells are equilibrated for 120 min at 37° with 5 ml of serum-supplemented (2.5% fetal calf serum, 2.5% horse serum), HEPES-buffered (20 mM) CMRL 1415 ATM medium (pH 7.45) with lowered (0.75 mM) K$^+$ ("assay medium"). Then, [^{3}H]ouabain (3.3 × 10^6 cpm; added as 100-μl portion with thorough mixing) is given to each flask, yielding a final ouabain concentration of 6.1 × 10^{-8} M. After different incubation periods, cell-bound [^{3}H]ouabain is determined, as described for ^{86}Rb$^+$ in the preceding chapter, and the values ([^{3}H]ouabain bound/mg protein) are plotted versus incubation time (Fig. 4, "association," ○—○). Binding is almost completed within 30 min; thereafter, only a small increase in cell-bound ouabain is observed. Using the same amount of [^{3}H]ouabain radioactivity, but high concentrations of unlabeled ouabain (10^{-4} M), little cell-bound radioactivity is measured (Fig. 4, "association," △—△), the amount being almost independent from the ouabain concentration within the range of 10^{-4}—10^{-2} M (experiments not shown). Another set of chicken heart muscle cells in 25-cm^2 culture flasks is incubated with [^{3}H]ouabain (6.1 × 10^{-8} and 10^{-4} M, respectively, 3 × 10^6 cpm/flask), as described above, for 120 min at 37°. Then, the non-cell-bound [^{3}H]ouabain is quickly removed by washing the cells five times in ouabain-free salt solution B with lowered (0.75 mM) K$^+$, and dissociation of cell-bound [^{3}H]ouabain is observed for the next 30 min (Fig. 4, "dissociation,"

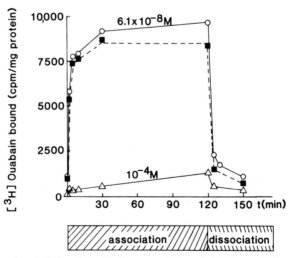

Fig. 4. Kinetics of [^{3}H]ouabain binding to and dissociation from chicken heart muscle cells in culture. ■– –■: Specific [^{3}H]ouabain binding at 6.1 × 10^{-8} M ouabain. Values given are means from closely correlating duplicates. (See text.)

○—○). Dissociation occurs rapidly; within 5 min, cells lose most of the [³H]ouabain originally bound. By a variety of controls it has been proved[16] that [³H]ouabain bound at high ouabain concentrations ($\geq 10^{-4}$ M) in incubation medium represents nonspecific ouabain binding to the cells, while specific [³H]ouabain binding to the cardiac glycoside receptor (Fig. 4, ■– –■) is obtained by subtracting this nonspecific binding (△—△) from total binding (○—○) at the very ouabain concentration chosen. Equilibrium of specific, receptor-mediated [³H]ouabain binding at 37° is achieved within 30 min, being stable throughout the whole incubation period of 120 min in the experiment of Fig. 4, and in other experiments (not shown) for at least 8 hr. As K^+ increases the dissociation constant of receptor binding of ouabain to these cells,[16,17] binding experiments are normally carried out at lowered K^+ (e.g., 0.75 mM), to achieve almost maximal binding without irreversible damage of the cells.[16,17] However, these binding experiments can also be done at physiological K^+ concentrations.[18] In a similar manner as described in Fig. 4 for chicken heart muscle cells, [³H]ouabain receptor binding can also be studied in cardiac nonmuscle cells from chicken embryos[16] as well as in heart muscle and nonmuscle cells from neonatal rats.[17] However, in rat heart cells, the more complex binding kinetics, due to the presence of at least two classes of ouabain-binding sites (see below), has to be taken into account.[17]

The concentration dependence of specific, receptor-mediated [³H]ouabain binding to chicken and rat heart muscle cells in culture is shown in Fig. 5a. For measurement, cell monolayers in 25-cm² culture flasks (0.9 and 1.0 mg protein/flask, respectively) are incubated for 4 hr at 37° in "assay medium" (see above, [K^+] = 0.75 mM; 5 ml/flask) in the presence of various concentrations of [³H]ouabain (see abscissa to Fig. 5a). Measurement of specific [³H]ouabain binding is carried out as described in the experiment of Fig. 4. Cell-bound radioactivity at 10^{-3} M ouabain is taken as unspecific binding, being 8% in chicken cells and 17% in rat cells of maximal [³H]ouabain counts bound. In Fig. 5b, the binding data of Fig. 5a are plotted according to the method of Scatchard,[21] to determine number and affinity of the ouabain-binding sites: the cell-bound ouabain/incubation volume (flask) is measured as described, the amount of free, non-cell-bound ouabain/incubation volume is given by the amount of total ouabain/flask minus cell-bound ouabain/flask. In case of chicken heart muscle cells, Scatchard plot analysis yields a straight line, being indicative of a single class of binding sites; the slope of the line represents the dissociation constant (K_D) of ouabain binding; the point of intersection with the ordinate gives the maximal binding capacity, being identical with

[21] G. Scatchard, *Ann. N.Y. Acad. Sci.* **51**, 660 (1949).

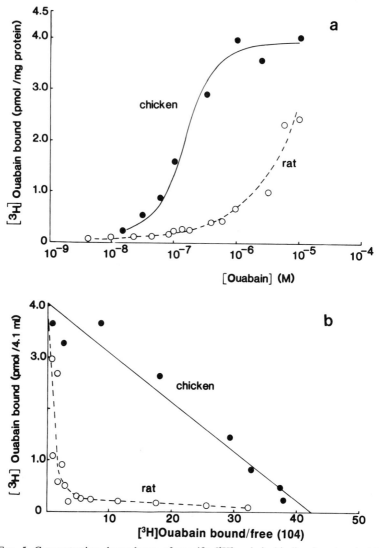

FIG. 5. Concentration dependence of specific [^{3}H]ouabain binding in rat and chicken heart muscle cells in culture. (a) [^{3}H]Ouabain radioactivity/flask: 3.7×10^6 cpm ([ouabain] $\geq 6 \times 10^{-8} M$). At ouabain concentrations $<6 \times 10^{-8} M$, the [^{3}H]ouabain solution (100 μl/flask) is diluted to give the ouabain concentration chosen (see abscissa). (b) Scatchard plot analysis[21] of binding data presented in (a). Chicken cells: linear regression analysis yields the following results: Binding capacity $(B) = 4.3$ pmol/mg protein; dissociation constant $(K_D) = 2.2 \times 10^{-7} M$; $r = -0.98$. Rat cells: analysis according to Weidemann et al.[22] for two classes of binding sites yields the following values: $B_1 = 0.2$ pmol/mg protein; $K_{D1} = 1.6 \times 10^{-8} M$; $B_2 = 4.6$ pmol/mg protein; $K_{D2} = 8.9 \times 10^{-6} M$.

TABLE I

CARDIAC GLYCOSIDE RECEPTORS IN RAT AND CHICKEN HEART CELLS IN CULTURE[a]

Cell type	Capacity		Dissociation constant (K_D) of ouabain binding (M)
	pmol/mg cell protein	Sites/cell	
Chicken heart muscle	2.6 ± 0.3	9×10^5	$(1.5 \pm 0.2) \times 10^{-7}$
Chicken heart nonmuscle	2.1 ± 0.1	3×10^5	$(1.9 \pm 0.2) \times 10^{-7}$
Rat heart muscle			
High-affinity sites	0.2 ± 0.1	8×10^4	$(3.2 \pm 2.0) \times 10^{-8}$
Low-affinity sites	2.6 ± 0.6	10^6	$(7.1 \pm 3.0) \times 10^{-6}$
Rat heart nonmuscle			
High-affinity sites	~ 0.1		$\sim 10^{-8}$
Low-affinity sites	~ 1		$\sim 10^{-6}$

[a] Data have been obtained from experiments as described in Fig. 5. Values are means ± SEM ($N = 4$–8). For calculation of the number of sites/cell, the cell protein content has been determined as (10^6 cells/mg protein): chicken heart muscle cells, 1.8; chicken heart nonmuscle cells, 4.2; rat heart muscle cells, 1.5.

the amount of ouabain-binding sites/cell protein of incubation volume (see legend to Fig. 5). With rat heart muscle cells, no straight, but a curved line is found (Fig. 5b). Applying the computer program of Weidemann et al.,[22] these binding data are compatible with the presence of two classes of ouabain-binding sites in rat heart muscle cells: a high-affinity, low-capacity and a low-affinity, high-capacity binding site (see legend to Fig. 5b). Similar data—one class of sites in chicken and two classes of sites in rat— are obtained with heart nonmuscle cells.[16,17] To determine the number of ouabain-binding sites/cell, the protein content/cell must be measured. This can be done by detaching the cells from the plastic with 0.5% trypsin plus 0.02% EDTA in Ca^{2+},Mg^{2+}-free salt solution (5–10 min, 37°), centrifuging the cell suspension for 10 min at 300 g at room temperature and washing the cell pellet twice in salt solution B. After renewed suspending, the number of cells/suspension volume is determined with a hemocytometer under the phase-contrast microscope and the cell protein content/ suspension volume is measured according to the method of Lowry. The following data are obtained by this method (10^6 cells/mg protein): chicken heart muscle cells, 1.8; chicken heart nonmuscle cells, 4.2; rat heart muscle cells, 1.5; rat heart nonmuscle cells, 2.3. With these ratios, the number of ouabain-binding sites/cell can now be calculated (Table I). The ouabain-binding sites in rat and chicken heart cells have been classified as

[22] H. J. Weidemann, H. Erdelt, and M. Klingenberg, *Eur. J. Biochem.* **16**, 313 (1970).

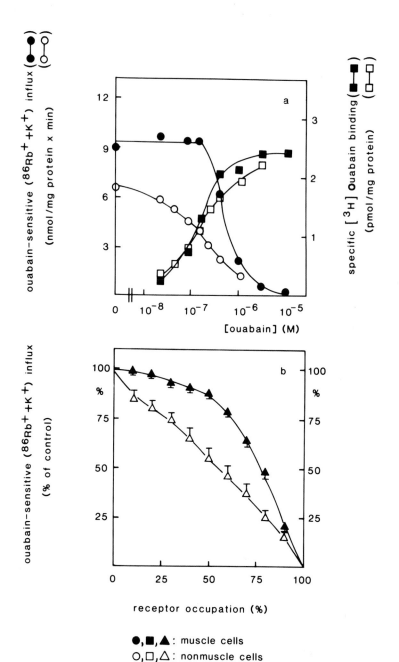

FIG. 6. [³H]Ouabain binding and inhibition of active (⁸⁶Rb⁺ + K⁺) influx in chicken heart muscle (closed symbols) and nonmuscle (open symbols) cells in culture. (a) The concentration-dependent, specific [³H]ouabain binding (■, □) and concentration-dependent inhibition of active (⁸⁶Rb⁺ + K⁺) influx (●, ○) by ouabain is given; 2.45 × 10⁶ cpm (⁸⁶Rb⁺) and (0.96–

cardiac glycoside receptors.[16–18,23] The presence of a single, saturable class of cardiac glycoside receptors in chicken heart muscle cells has been confirmed by other groups,[24,25] while Lazdunski and co-workers[8] have found two classes of receptors in these cells. The reason for this discrepancy remains to be established.

Correlation of Cardiac Glycoside Receptor Occupation and Inhibition
Active Na^+,K^+ Transport in Cardiac Muscle and Nonmuscle Cells
from Chicken Embryos

To correlate cardiac glycoside receptor occupation and active K^+ influx in heart cells, monolayers of cardiac muscle and nonmuscle cells from chicken embryos in 25-cm^2 culture flasks (1.16 and 0.25 mg protein/flask) are incubated at 37° for 4 hr in "assay medium" at $[K^+]$ = 0.75 mM (5 ml/flask) in the presence of various ouabain concentrations (see abscissa to Fig. 6a). In one set of experiments using unlabeled ouabain, active K^+ influx is measured thereafter with tracer amounts of $^{86}Rb^+$ in just the same manner as described in the experiment of Fig. 3. In the other set, using [^3H]ouabain, concentration-dependent, specific ouabain binding to the cells is determined with the method described for the experiment of Fig. 5; [^3H]ouabain binding and ($^{86}Rb^+$ + K^+) influx in the presence of 10^{-3} M ouabain is assumed to be nonspecific. The results are given in Fig. 6a. From the binding data, the dissociation constant (K_D) for specific ouabain binding to the cardiac glycoside receptor can be calculated as 2.0×10^{-7} M for heart muscle cells and 1.4×10^{-7} M for heart nonmuscle cells; the binding capacity, equivalent to the amount of cardiac glycoside

[23] H. J. Berger, K. Werdan, and E. Erdmann, in "The Na^+,K^+-Pump Part B: Cellular Aspects" (J. C. Skou, J. G. Nørby, A. B. Maunsbach, and M. Esmann, eds.), p. 345. Alan R. Liss, New York, 1988.

[24] D. Kim, W. H. Barry, and T. W. Smith, J. Pharmacol. Exp. Ther. **231**, 326 (1984).

[25] A. Lobaugh and M. Lieberman, Biophys. J. **47**, 149a (Abstr.) (1985).

3.83) $\times 10^6$ cpm [^3H]ouabain per flask. For calculation of ouabain-sensitive ($^{86}Rb^+$ + K^+) influx, the ouabain-insensitive influx of ($^{86}Rb^+$ + K^+) at 10^{-3} M ouabain (0.35 nmol/mg protein × min in muscle cells; 1.21 nmol/mg protein × min in nonmuscle cells) is subtracted from total influx rates for every ouabain concentration chosen. Unspecific [^3H]ouabain binding is determined in the presence of 10^{-3} M ouabain. Scatchard plot analysis (not shown) of [^3H]ouabain binding data of Fig. 6a yields the following results: muscle cells, $K_D = 2.0 \times 10^{-7}$ M, B = 2.7 pmol/mg protein; nonmuscle cells: $K_D = 1.4 \times 10^{-7}$ M, B = 2.3 pmol/mg protein. Values are means from closely correlating triplicates. (b) The correlation is given between cardiac glycoside receptor occupation (abscissa) and inhibition of active ($^{86}Rb^+$ + K^+) influx (ordinate). Values (x ± SEM) are based on the results of nine (muscle cells, ▲) and five (nonmuscle cells, △) experiments, respectively, as presented in (a).

receptors (R_{total}), amounts to 2.7 and 2.3 pmol/mg cell protein, respectively (Scatchard plot analysis of binding data of Fig. 6a). As binding of ouabain to its receptor follows the law of mass action,[1,16] the percentage of cardiac glycoside receptors occupied by ouabain (RO) can be calculated for every ouabain concentration chosen, according to the following formula: $K_D = [O_{free}][R_{free}]/[RO]$, where $[R_{free}]$ = concentration of free cardiac glycoside receptors at a given ouabain concentration = $[R_{total}]$ − [RO], and $[O_{free}]$ = concentration of free ouabain in incubation volume = concentration of total ouabain in incubation volume − concentration of cell-bound ouabain ([RO]). The calculation is based on the assumption that one cardiac glycoside receptor molecule binds one molecule of ouabain. From concentration effect curves, the values for ouabain-sensitive ($^{86}Rb^+$ + K^+)-influx rates are obtained for the ouabain concentrations producing 10, 20, 30, . . . 90% of cardiac glycoside receptor occupation. This correlation of cardiac glycoside receptor occupancy (abscissa) and inhibition of active ($^{86}Rb^+$ + K^+) influx (ordinate) is given in Fig. 6b: In heart nonmuscle cells (open symbols), a linear correlation exists between receptor occupation and transport inhibition. In case of heart muscle cells (closed symbols), however, up to 40–50% of all cardiac glycoside receptors can bind ouabain with only little inhibition of active ($^{86}Rb^+$ + K^+) transport; only higher receptor occupancies (>50%) lead to a stronger transport inhibition. Therefore, cardiac muscle cells, at least from chicken embryos, are characterized by a "sodium pump reserve,"[26] which is lacking in cardiac nonmuscle cells from chicken embryos as well as in human erythrocytes.[16,27,28] This cell-specific difference in coupling of receptor occupancy and transport inhibition has to be borne in mind, when comparing equal receptor occupancies in different cell types from one species, even in the same organ.

In cultured heart cells from chicken embryos, inhibition of active Na^+,K^+ transport is due to occupation of a single class of cardiac glycoside receptors (Fig. 6; see also Werdan et al.[16,18,27]). In cultured rat heart muscle cells, however, only ouabain binding to the low-affinity receptor is followed by inhibition of the sodium pump, while binding to the high-affinity receptor does not result in any measurable inhibition of active ($^{86}Rb^+$ + K^+) influx under the experimental conditions described.[17]

[26] T. Akera and T. M. Brody, *Trends Pharmacol. Sci.* **6**, 156 (1985).
[27] K. Werdan, B. Zwissler, B. Wagenknecht, W. Krawietz, and E. Erdmann, *Biochem. Pharmacol.* **32**, 757 (1983).
[28] L. Brown, K. Werdan, and E. Erdmann, in "Cardiac Glycosides 1785–1985 Biochemistry–Pharmacology–Clinical Relevance" (E. Erdmann, K. Greeff, and J. C. Skou, eds.), p. 49. Steinkopff-Verlag, Darmstadt, Federal Republic of Germany Springer-Verlag, New York, 1986.

Modification of the Number of Sodium Pump Molecules of Cultured
Chicken Heart Cells by Cell Exposure to Ouabain and Low K^+

In this section experimental procedures are given to modify the number of sodium pump molecules in cultured chicken heart cells.

In the experiment of Fig. 7, heart muscle cells from chicken embryos are cultured for 2 days in serum-supplemented CMRL medium (5.0 ml/flask). Thereafter, the medium is replaced by serum-free, hormone-supplemented, HEPES-buffered CMRL medium ($[K^+]$ = 3.5 mM) containing 3×10^{-6} M ouabain. In the presence of this concentration, 88% of all sodium pump molecules are occupied by ouabain, as can be calculated from the dissociation constant (K_D) for receptor binding of ouabain at $[K^+]$ = 3.5 mM (K_D = 4.2 $\times$ 10^{-7} M; see Werdan et al.[18]). At different time intervals, cell Na^+, cell K^+, ($^{86}Rb^+$ + K^+)-influx rates, and [^3H]ouabain binding are measured.

Measurement of cell Na^+ (exchangeable cellular pool of Na^+): Four hours prior to determination, tracer amounts of $^{22}Na^+$ (8×10^6 cpm in a volume of 50 μl) are added to each flask. Within this time period, $^{22}Na^+$ equilibrates with intracellular Na^+. Thereafter, cells are washed with ice-cold Ca^{2+}–sorbitol solution, and cellular $^{22}Na^+$ is measured as described above for $^{86}Rb^+$. From the amount of cellular $^{22}Na^+$ radioactivity, the cellular amount of exchangeable Na^+ can be calculated, which is given in nanomoles per milligram cell protein in the ordinate of Fig. 7.

Measurement of cell K^+: After removing extracellular K^+ by cell washings with Ca^{2+}–sorbitol solution (see above), this is done by lysis of the cells in 1.3 ml 0.1 N NaOH/4.5 mM EDTA/flask (see above) and determination of K^+ in a 200-μl extract by flame photometry.

Measurement of ($^{86}Rb^+$ + K^+)-influx rates: Ouabain-sensitive ($^{86}Rb^+$ + K^+) influx is determined as described above (1.5 $\times$ 10^6 cpm $^{86}Rb^+$/flask; incubation period 10 min), uptake in the presence of 10^{-4} M ouabain (incubation period 60 min; "ouabain insensitive") being subtracted from total influx.

Measurement of specific [^3H]ouabain binding to the cells: At the times indicated in the abscissa of Fig. 7, cell-bound ouabain is removed by thorough washing of the cells in ouabain-free salt solution B (five washings, 5 ml each, 37°). Thereafter, the cells are further incubated for 2 hr at 37° in ouabain-free, serum-free CMRL medium (5 ml/flask). Cell-bound ouabain (85–95%) can be removed by this treatment.[29] Subsequently,

[29] K. Werdan, C. Reithmann, W. Krawietz, and E. Erdmann, *Biochem. Pharmacol.* **33**, 2337 (1984).

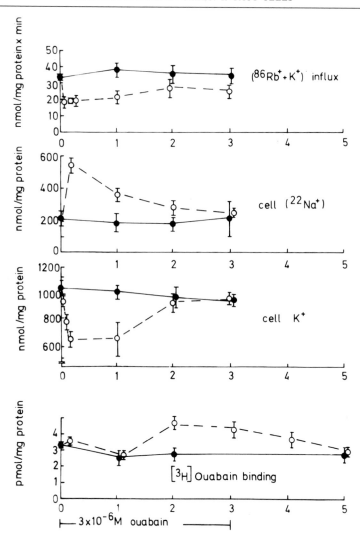

incubation period (days)

FIG. 7. Chronic ouabain exposure (3×10^{-6} M) of chicken heart muscle cells. Effect on (^{86}Rb$^+$ + K$^+$) influx, cell Na$^+$, cell K$^+$, and specific [^{3}H]ouabain binding. Cells were cultured for up to 3 days in serum-free medium, in the absence (●—●) or presence (○--○) of 3×10^{-6} M ouabain. Values are given as $x \pm$ SD, $n = 3$.

specific [^{3}H]ouabain binding to the cells under equilibrium conditions is determined as described above (4 × 10^{-7} M [^{3}H]ouabain, 2 × 10^6 cpm/ flask; incubation period 2 hr; [K$^+$] = 0.75 mM, 37°). Specific [^{3}H]ouabain binding is calculated by subtracting of unspecific binding in the presence of 10^{-4} M ouabain from total binding.

Within the first few hours of ouabain exposure of the cells, active (^{86}Rb$^+$ + K$^+$)-influx rate is depressed, cell Na$^+$ increases, and cell K$^+$ falls, due to ouabain-induced inhibition of the sodium pump (Fig. 7). However, during the further course of the experiment, (^{86}Rb$^+$ + K$^+$) influx nearly completely recovers, cell Na$^+$ falls, and cell K$^+$ rises again, the cellular cation contents being within the normal range within 2–3 days of ouabain exposure. In parallel, the capacity of the cells to bind [^{3}H]oua-bain (4 × 10^{-7} M) to the cardiac glycoside receptors increases, after a lag phase of 24 hr, reaching a new steady state after further 24 hr ([^{3}H]oua-bain binding is measured after removing the unlabeled ouabain bound to the cells during the culture period, see above). This increase in specific [^{3}H]ouabain binding is solely due to an increase in the number of cardiac glycoside receptors per cell during chronic ouabain exposure, while the affinity of the receptor for ouabain, as measured by Scatchard plot analy-sis according to Fig. 4, remains unchanged.[18] As, in addition, the protein/ cell ratio remains unaltered during chronic ouabain exposure,[18] the rise in specific [^{3}H]ouabain binding reflects an equivalent increase in the number of sodium pump molecules per cell. This process is reversible within 48 hr after removing ouabain from the culture medium (Fig. 7), it depends on protein synthesis,[18] and it is restricted to the cardiac glycoside receptor/ sodium pump system, while the number of β-adrenergic receptors re-mains unchanged.[18] By this mechanism, the heart cell, as many other cell types (for a discussion, see Werdan et al.[18]), tries to restore the impaired Na$^+$,K$^+$ homeostasis, the driving force being probably the rise in intracel-lular Na$^+$.[30]

This adaptive increase in the number of cardiac glycoside receptors and sodium pump molecules can be studied in a quantitative manner (Fig. 8): chicken heart muscle cells are cultured for 2 days in serum-supple-mented CMRL medium (5.0 ml/flask). Then, the medium is replaced by serum-free, hormone-supplemented, HEPES-buffered CMRL medium ([K$^+$] = 3.5 mM), with different concentrations of ouabain (see abscissa of Fig. 8), and cell culture is continued for three further days. Thereafter, cell-bound ouabain is removed by washing of the cells, and specific [^{3}H]ouabain binding (4 × 10^{-7} M) and ouabain-sensitive (^{86}Rb$^+$ + K$^+$)-influx rates in the absence of ouabain are determined (Fig. 8), with the

[30] D. Kim and T. W. Smith, Am. J. Physiol. **250**, C32 (1986).

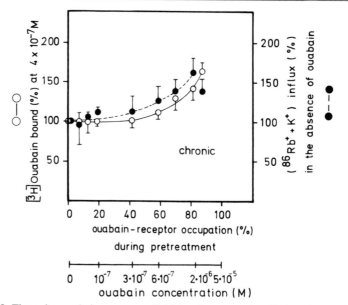

FIG. 8. Three-day ouabain exposure of chicken heart muscle cells in culture. Effect on specific [³H]ouabain-binding and on ouabain-sensitive (^{86}Rb$^+$ + K$^+$)-influx rates. Chicken heart muscle cells are cultured for 3 days at various concentrations of ouabain (abscissa). After removal of cell-bound ouabain, specific [³H]ouabain binding at 4×10^{-7} M and active (^{86}Rb$^+$ + K$^+$)-influx rates at [K$^+$] = 3.5 mM are measured, with the techniques described in the experiment of Fig. 7. Values are given as $\bar{x} \pm$ SD, $n = 3$. For calculation of cardiac glycoside receptor occupancy during the culture period, concentration-dependent [³H]ouabain binding at [K$^+$] = 3.5 mM is measured with another set of chicken heart muscle cells within the same experiment, according to the experiment of Fig. 4. From the dissociation constant $K_D = 4.3 \times 10^{-7}$ M of specific [³H]ouabain binding, receptor occupancy can be calculated for every ouabain concentration chosen during the culture period. This correlation is shown in the abscissa. For further explanation see text to Fig. 6.

methods given above: If more than 40% of all cardiac glycoside receptors are occupied by ouabain during the culture period, an increase in both [specific [³H]ouabain binding as well as ouabain-sensitive (^{86}Rb$^+$ + K$^+$) influx] is found, the effects being the more marked as more receptors become occupied during ouabain exposure (Fig. 8). These results unequivocally demonstrate an adaptive increase in functioning sodium pump molecules during chronic ouabain exposure. The transport properties of these additional sodium pump molecules are unaltered.[18] However, this compensatory mechanism only comes into play when the "sodium pump reserve" becomes exhausted (occupation of more than 40% of all cardiac glycoside receptors and thereby inactivation of more than 40% of all sodium pump molecules; compare Fig. 6). As long as this "sodium pump reserve" works, there is no need for the heart muscle cell to increase its number of sodium pump molecules.[18]

Lowering extracellular K^+ can also reduce sodium pump activity of cultured chicken heart cells. The compensatory mechanisms to overcome this reduction have been studied in detail with the experimental techniques given in this chapter[31,32]: (1) enhancement of the activity of the single sodium pump molecule, observable within minutes after low K^+ exposure, due to the rise in intracellular Na^+ ("sodium pump reserve"), and (2) induction of additional sodium pump molecules during "chronic" (≥ 24 hr) exposure to low K^+. The degree of this latter adaptation mechanism, however, seems to be strongly cell specific: in the case of cardiac muscle cells, only very low external K^+ concentrations (about 1 mM) give rise to additional sodium pump molecules. In case of cardiac nonmuscle cells from the same species, the number of sodium pump molecules already increases by a moderate lowering of external K^+ from 5.5 to 3.0 mM.[31]

Monitoring of Beating in Heart Muscle Cells in Culture: Sodium Pump Activity and Positive Inotropic Action of Cardiac Glycoside

The monitoring system (Fig. 9) is based on the description given by Kaumann and co-workers.[33] The beating cell in the perfusion chamber (see below) is observed through an inverted phase-contrast microscope (Diavert, Leitz, Wetzlar, FRG) at 300-fold magnification; it is recorded with a TV camera (Grundig FA 70 B; FRG), displayed on the screen of a monitor (Grundig BG 23T; FRG), and simultaneously fed into a video recorder (Sony U matic, VO-5630, Sony Corporation). Up to 25 beating cells of the monolayer sheet can be observed in this way under identical conditions during the experiment. The phase-contrast microscope as well as the TV camera are located in a thermostatically maintained incubation chamber, the temperature within normally kept at 37°. Though the heart muscle cells are firmly attached to the culture flask, contraction of the cells during beating causes changes in light intensity around the cell wall, the dark cell body contrasting with the light surroundings. These light intensity changes are recorded by means of a photocell (BPY 61 Siemens,

[31] K. Werdan, G. Schneider, W. Krawietz, and E. Erdmann, *Biochem. Pharmacol.* **33**, 1161 (1984).
[32] K. Werdan, C. Reithmann, G. Schneider, and E. Erdmann, *in* "The Sodium Pump 4th International Conference on Na,K-ATPase" (I. Glynn and C. Ellory, eds.), p. 679. Company of Biologists, Cambridge, England, 1985.
[33] A. J. Kaumann, R. Wittmann, L. Birnbaumer, and B. H. Hoppe, *Naunyn-Schmiedeberg's Arch. Pharmacol.* **296**, 217 (1977).

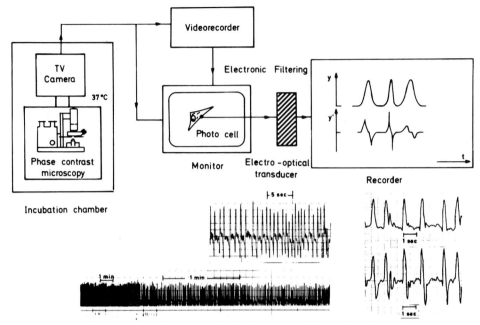

FIG. 9. Monitoring system for beating myocardial cells in culture.

FRG) stuck to the monitor in the surroundings of the cell wall. The output signal of the photocell is amplified, electronically filtered, and monitored on a conventional recorder.[34] The velocity (dy/dt) of cell motion (y), contraction as well as relaxation velocity, can be determined by a differentiating preamplifier,[34] as shown in Fig. 9.

As perfusion chambers, the 25-cm² plastic culture flasks (Fig. 10) in which the cells have been grown are used. In the neck of the flask, a rubber stopper which is perforated by two steel electrodes for external cell pacing and by two syringe needles for superfusion (long needle) and for aspiration of the medium (short needle) is inserted. Electrodes and needles are bent as shown in Fig. 10, the electrodes being placed in the medium just above the cell monolayer, and the tips of both needles dipping into the medium. The system is further fixed in the neck of the flask by the cut-out original plastic screw plug. Cells are continuously superfused with serum-free, HEPES (20 mM)-buffered CMRL medium (normally 3.8 ml/min; incubation volume 3.0–4.0 ml; pH 7.4; 37°) with modified Ca^{2+} (0.6–3.6 mM).

[34] A. J. Sinclair, H. A. Miller, and D. C. Harrison, *J. Appl. Physiol.* **29**, 747 (1970).

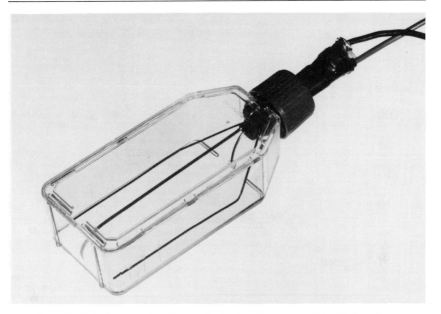

FIG. 10. Perfusion chamber for monitoring beating myocardial cells in culture.

The solution is transported in silicone tubes with the help of a roller pump (ISMATECH mp25GJ-4, Zürich, Switzerland); the aspirated medium is discarded. If desired, cells can be electrically driven by the steel electrodes connected to a Grass SD9 stimulator (Grass Instruments, Quincy, MA), pacing rate 90–150/min, 10–100 V, pulse duration about 5 msec.

Using this device, pulsation of single cells can be continuously monitored under stable conditions over an observation period of about 60 min (Fig. 9), and complete medium exchange is achieved within 2–3 min.

In Fig. 11, the effect of increasing concentrations of ouabain on contraction velocity (dy/dt) of chicken heart muscle cells in culture is presented, as measured with the method given above: Cells (seeding density $1.5 \times 10^5/cm^2$) have been cultured for 2 days in a 25-cm^2 plastic culture flask, yielding a synchronously beating cell monolayer. After washing of the cells, superfusion of the electrically driven (120 beats/min) cells is started with serum-free CMRL 1415 medium, buffered with 20 mM HEPES, as described above ([K$^+$] = 3.5 mM; [Ca^{2+}] = 0.6 mM; 37°). After an equilibration period of 60 min, monitoring is started. The effect of ouabain is determined in a cumulative manner by superfusing the cells

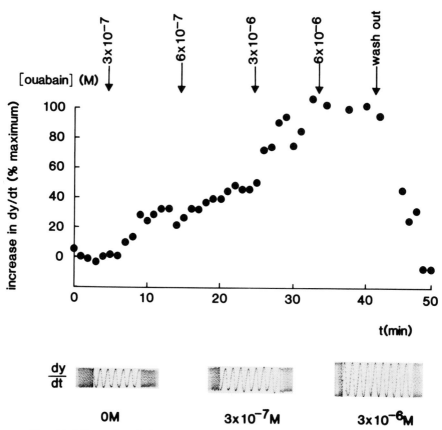

FIG. 11. Effect of ouabain on contraction velocity of chicken heart muscle cells in culture. Contraction velocity (*dy/dt*) of a single cell is monitored.

with medium, supplemented with increasing concentrations of ouabain. The maximum of the observed effect—an increase in contraction velocity—is regularly achieved within 5–7 min. In the presence of ouabain concentrations $>3 \times 10^{-6}$ M (electrically driven cells) and $>6 \times 10^{-7}$ M (spontaneously beating cells) toxic effects such as arrhythmias become visible. At the end of the experiment, the cells are superfused with ouabain-free medium, thereby reaching the control value of *dy/dt* within 7 min (Fig. 11). During every min of the observation period, *dy/dt* of five events is averaged in arbitrary units, and the ouabain-induced increase in *dy/dt* is given in the ordinate of Fig. 11 as a percentage of maximum increase. A similar increase in pulsation amplitude (not shown) parallels the increase in contraction velocity, both being equivalent to a positive inotropic effect (for references, see Werdan *et al.*[18]). From experiments as presented in Fig. 11, concentration–response curves can be constructed; an example is shown in Fig. 12 (●—●; see also Werdan *et al.*[17,18] and Erdmann *et al.*[23]). To demonstrate a positive inotropic effect of a substance, the experiments as described in Figs. 11 and 12 have to be carried out at submaximal concentrations of Ca^{2+} in the superfusion medium (EC_{50} and EC_{90} values for Ca^{2+}-induced increase in dy/dt: chicken heart muscle cells, 1.4 and 2.9 mM; rat heart muscle cells, 0.9 and 1.6 mM). To compare the positive inotropic effect of a substance tested with the maximal effect achieved by Ca^{2+}, cells are superfused within the last 5 min of the experiment with high Ca^{2+} medium (chicken heart muscle cells, 3.6 mM; rat heart muscle cells, 2.4 mM).

This experimental setting as described can be used to study the relationship between the number of cellular sodium pump molecules and the positive inotropic effect of cardiac glycosides in chicken heart muscle cells in culture. To modify the number of cellular sodium pump molecules, chicken heart muscle cells are cultured for 3 days either in medium supplemented with ouabain (3×10^{-7} and 3×10^{-6} M, respectively), or in medium with lowered K^+ (1.0 mM). At the end of the culture period, cell-bound ouabain is removed by washing of the ouabain-cultured cells. Thereafter, specific [³H]ouabain binding (4×10^{-7} M) is determined in these groups (for experimental details see Modification of the Number of

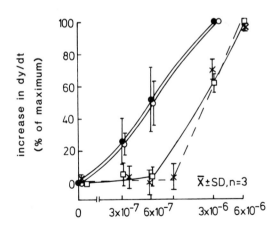

FIG. 12. Concentration–response curves of ouabain on contraction velocity in chicken heart muscle cells in culture. Control cells (●—●); cells grown for 3 days in the presence of 3×10^{-7} M ouabain (○—○), 3×10^{-6} M ouabain (□—□), or at lowered (1.0 mM) K^+ (×– –×). In every group, three cells have been monitored.

Sodium Pump Molecules of Cultured Chicken Heart Muscle Cells by Cell Exposure to Ouabain and Low K$^+$). From the binding data obtained (control: 2.3 ± 0.3 pmol/mg protein; 3 × 10^{-7} M ouabain: 2.3 ± 0.2; 3 × 10^{-6} M ouabain: 4.1 ± 0.4; [K$^+$] = 1.0 mM: 4.5 ± 0.2; $\bar{x}$ ± SD, n = 3), a 78% increase in the number of cardiac glycoside receptors/sodium pump molecules per cell by the toxic ouabain concentration (3 × 10^{-6} M) and a 96% increase by 1.0 mM K$^+$ is evident, while the positive inotropic, nontoxic ouabain concentration of 3 × 10^{-7} M is without any effect on specific [^{3}H]ouabain binding. According to the experiment of Fig. 11, the effect of ouabain on contraction velocity (dy/dt) is now tested in these groups under identical experimental conditions ([K$^+$] = 3.5 mM, [Ca^{2+}] = 0.9 mM) and the results are given in Fig. 12. With cells grown in 3 × 10^{-7} M ouabain (○—○), the concentration response curve is identical with the one of the control cells (●—●). In the cell groups with raised numbers of cardiac glycoside receptors/sodium pump molecules per cell [grown in 3 × 10^{-6} M ouabain (□—□) or 1.0 mM K$^+$ (×--×)], the cardiac glycoside effect clearly becomes attenuated, the concentration effect curves being shifted to higher ouabain concentrations (Fig. 12). Therefore, an increase in the number of cardiac glycoside receptors/sodium pump molecules per heart muscle cell attenuates the pharmacological action of cardiac glycosides.

Comments

Sodium pump activity can be determined in intact cells either by measurement of active K$^+$ influx (Figs. 1 and 2) or active Na$^+$ efflux; in the latter case, heart muscle cells have to be loaded with radioactive Na$^+$ prior to the measurement.[35] Transport activity of the sodium pump depends on the concentration of extracellular K$^+$ (Fig. 2), as well as on intracellular Na$^+$ and ATP. Under physiological conditions, intracellular Na$^+$ activity (about 10 mM) represents the rate-limiting step of sodium pump activity. For measurement of maximal transport rate (pump capacity), intracellular Na$^+$ has therefore to be increased by Na$^+$ loading of the cells.[24,26] The experimental procedures described in this chapter for determination of active K$^+$ influx can also be applied to quantification of active Na$^+$ efflux and sodium pump capacity. Binding experiments with [^{3}H]ouabain allow calculation of the number of sodium pump molecules by heart muscle cells. The method given has been thoroughly tested with respect to specific and nonspecific binding, association and dissociation kinetics,

[35] D. McCall, *Am. J. Physiol.* **236**, C87 (1979).

and internalization of the radiolabeled compounds used.[16,17] From the results it is clearly evident that specific ouabain binding in cultured heart cells reflects binding to the cardiac glycoside receptor, being part of the Na^+,K^+-ATPase molecule.[36] The stoichiometry of one molecule of ouabain bound per one molecule of Na^+,K^+-ATPase has been confirmed recently[25] by comparison of [^{3}H]ouabain binding with binding of antibodies directed to Na^+,K^+-ATPase in cultured chicken heart muscle cells.

By correlating cardiac glycoside receptor occupation with inhibition of active K^+ influx (Fig. 6), the sodium pump reserve can be characterized. Major differences exist in various cell types, e.g., heart muscle and nonmuscle cells from chicken embryos, with respect to the extent of this pump reserve. An increase in intracellular Na^+ as primary event is the most attractive hypothesis to explain this compensatory mechanism. However, other possibilities should also be taken into account which have been discussed in detail.[24,26]

The number of sodium pump molecules in heart muscle cells is subject to regulation. This is demonstrated for chicken heart muscle cells with respect to chronic ouabain exposure (Figs. 7 and 8). The method given can also be applied to sodium pump regulation by extracellular K^+,[31,32,37] extracellular Na^+,[24] and by hormones, e.g., triiodothyronine.[38] Also in rat heart muscle cells in culture, the number of sodium pump molecules, as measured by [^{3}H]ouabain binding, increases after cell exposure to triiodothyronine or to low K^+. Under these conditions, high and low affinities of cardiac glycoside receptors rise in a parallel fashion up to 200% of control.[23] Again, considerable differences are found with respect to the extent of sodium pump regulation in various cell types, e.g., chicken heart muscle and nonmuscle cells.[31]

In this chapter, a simple device is given to measure contractility in cultured heart cells (Fig. 9). Similar and more complex systems are found elsewhere.[33,34,39-41] Inhibition of the sodium pump by cardiac glycosides results in a positive inotropic effect (Fig. 11). Modification of the number of sodium pump molecules therefore alters the pharmacodynamics of these drugs (Fig. 12). The clinical relevance of the finding with respect to

[36] M. Heller, in "Heart Cells in Culture" (A. Pinson, ed.). CRC Press, Boca Raton, Florida, 1987.

[37] D. Kim, W. H. Barry, and T. W. Smith, Circ. Res. **55**, 39 (1984).

[38] D. Kim and T. W. Smith, J. Clin. Invest. **74**, 1481 (1984).

[39] G. Fayet, F. Couraud, F. Miranda, and S. Lissitzky, Eur. J. Pharmacol. **27**, 165 (1974).

[40] B. Koidl, H. A. Tritthart, and S. Erkinger, J. Mol. Cell. Cardiol. **12**, 165 (1980).

[41] J. Harary, G. Wallace, and G. Bristol, Cytometry **3**, 367 (1983).

tolerance development during drug therapy is discussed in detail elsewhere.[18]

Acknowledgments

The authors are grateful to H. J. Berger, I. Bohn, C. Reithmann, and A. Thomschke for valuable discussion and preparation of the manuscript. H. Schöffmann, H. Obwexer (Institute of Clinical Chemistry, University of Munich), W. Müller (EL MED, Augsburg, FRG), I. Bohn, and C. Reithmann were engaged in establishing the monitoring system and the perfusion chamber. This work has been supported by grants from Wilhelm Sander-Stiftung (78.014) and DFG (ER 65/4).

[41] Isolation of Calcium-Tolerant Atrial and Ventricular Myocytes from Adult Rat Heart

By MARY BETH DE YOUNG, BARTOLOMEO GIANNATTASIO, and ANTONIO SCARPA

The ability to isolate living cells from adult heart tissue has been a major prerequisite for cellular studies of myocardial functions such as intracellular Ca^{2+} homeostasis, cell contractility, receptor properties, intracellular signaling mechanisms, excitation–contraction coupling, transport, and electrophysiology. The advantage of achieving a primary cell digest is that the cells are fully differentiated and morphologically similar to cells in intact heart,[1] but lack interstitial tissue and contaminating cell types which can complicate measurements in whole tissues. Rat heart is often used as an experimental system because of its convenient size and the extensive literature available on this system. Although a wide variety of myocyte preparation methods have been reported,[2] most techniques for isolation of intact heart cells involve perfusing the heart with collagenase in the absence of Ca^{2+} (which is important for separation of the basement and plasma membranes[1]) followed by a gradual reintroduction of calcium to the isolated cells. The calcium tolerance of the cells is a major criterion of their viability and can be determined, for the most part, morphologically. Calcium-tolerant cells are rod shaped in the presence of 1 mM or more extracellular Ca^{2+}, with clear striations. Intolerant cells are intact, but hypercontracted into vesiculated spheres. A third population of periodically contracting cells can also be identified: rapidly contracting

---

[1] J. W. Dow, N. G. L. Harding, and T. Powell, *Cardiovasc. Res.* **15**, 483 (1981).
[2] B. Farmer, M. Mancina, E. S. Williams, and A. M. Watanabe, *Life Sci.* **33**, 1 (1983).

cells usually become distorted over time, but slowly contracting cells (beating several times per minute) may be normal.[3] Other criteria of viability include maintenance of a high ATP:ADP ratio, contraction in response to cell depolarization, and the presence of surface proteins for cell attachment, hormonal responses, and channel activity.

A basic cell preparation for both atrial and ventricular myocytes has been developed using a combination of existing procedures, commercially available medium, and a simplified Langendorf perfusion system. Novel aspects of the procedure include reintroduction of Ca^{2+} to the heart before complete cell dissociation and a rapid Percoll purification step to remove dead (Trypan Blue permeant) cells. The cells are morphologically normal, have a high ATP:ADP ratio, demonstrate Ca^{2+} channel activity, have releasable sarcoplasmic reticulum Ca^{2+} stores, respond to norepinephrine and extracellular ATP, contract and show increased intracellular Ca^{2+} concentration on depolarization with K$^+$, and can be cultured on embryonic feeder cells. The cells are metabolically viable and suitable for a variety of experiments for up to 6 hr following isolation.

Cell Isolation Procedure

Rationale

The procedure for isolation of cardiac myocytes is designed around two requirements: (1) the need to prevent tissue hypoxia during the cell isolation process and (2) avoidance of the "calcium paradox." The first problem is solved by maintaining the heart on an artificial circulation system which provides an appropriate balance of salts, vitamins, amino acids, glucose, and dissolved oxygen at 37°. Medium enters the heart through a plastic cannula placed in the aorta and exits through the vena cava. This retrograde perfusion maintains the left ventricle to a greater extent than the right ventricle, suggesting that ventricular myocytes obtained are enriched in left ventricular cells. Experiments attempting to isolate cells from chunks of heart incubated in the perfusion medium have been less successful.

The "calcium paradox" is a major complication to preparation of calcium-tolerant heart cells. As mentioned before, a Ca^{2+}-free medium is required to separate the cells; however, reintroduction of physiological Ca^{2+} concentrations can cause irreversible cell damage. Removal of me-

[3] M. C. Capogrossi, A. A. Kort, H. A. Spurgeon, and E. G. Lakatta, *J. Gen. Physiol.* **88,** 589 (1986).

dium Ca^{2+} from intact heart for 5 min followed by reintroduction of 1 mM $CaCl_2$ results in marked ultrastructural damage and sarcolemmal disruption.[4] We have observed changes in heart color from red to yellow/brown and formation of a fluid-filled pocket on the side of the heart with premature readdition of Ca^{2+}. Thus, $CaCl_2$ is generally reintroduced gradually to intact cells late in the preparation (1–2 hr after Ca^{2+} removal), which increases the likelihood of depletion of cellular Ca^{2+} stores and alteration of Ca^{2+} homeostatic mechanisms. Others have observed that gradual reintroduction of Ca^{2+} to intact heart following a period of collagenase digestion in Ca^{2+}-free medium does not cause the paradox, reducing the time in a Ca^{2+}-free medium to 20 min.[5] This has been confirmed and maintained as part of this procedure to produce cells with the least disruption of Ca^{2+} homeostasis.

Materials

Type I collagenase can be obtained from the Worthington Biochemical Corp. (Freehold, NJ). Lot numbers 46C9069, 67047M, and 67155M have been used successfully. (Refer to Common Difficulties for the selection procedure.) BSA fraction V, pH 7.00, is purchased from ICN Immunobiologicals (Lisle, IL). MEM (Joklik modified) tissue culture medium is from Sigma (St. Louis, MO). Percoll is obtained from Pharmacia (Piscataway, NJ) and Fura2AM is from Molecular Probes (Eugene, OR). All other materials are of reagent grade.

Reagents

Basic salt solution (BSS)
Ca^{2+}-Joklik medium, pH 7.20
Ca^{2+}-free Joklik medium, pH 7.20
Ca^{2+}-Joklik medium with 1% BSA, pH 7.40
2.5 $\times$ Ca^{2+}-Joklik with 1% BSA, pH 7.00
60% Percoll solution, pH 7.40

Stock solutions of frozen Joklik media and refrigerated BSS are maintained. BSS consists of 130 mM NaCl, 3 mM KCl, 1.2 mM KH_2PO_4, 1.00 mM $MgSO_4$, 1.25 mM $CaCl_2$, and 10 mM HEPES, pH 7.20. Joklik medium is prepared from Joklik MEM powder enriched with 10 mM Na-HEPES, pH 7.20. KCl (10 mM) is added to Ca^{2+}-free solutions, and 1.25 mM $CaCl_2$ is added to make Ca^{2+}-Joklik. Ca^{2+}-Joklik supplemented with

[4] L. E. Alto, and N. S. Dhalla, *Am. J. Physiol.* **237**, H713 (1979).
[5] R. A. Altschuld, L. M. Gamelin, R. E. Kelley, M. B. Lambert, L. E. Apel, and G. P. Brierley, *J. Biol. Chem.* **262**, 13527 (1987).

1% BSA (w/v), pH 7.40, is used for cell incubations following the isolation procedure. A concentrated ($2.5\times$) Ca^{2+}-Joklik with BSA medium is required to prepare a 60% Percoll solution of normal osmolarity. Forty milliliters of $2.5\times$ Ca^{2+}-Joklik with BSA (pH 7.00) is mixed with 60 ml of Percoll on the day of the experiment to yield 100 ml of a 60% Percoll solution with a pH approximating 7.40. All solutions are made with deionized distilled water, and medium that has been frozen is filtered before use through a Nalgene apparatus equipped with a 0.2-μm filter.

Preparation

Careful planning and preparation is important to ensure that unnecessary delays do not occur after removal of the heart which could increase the risk of cellular hypoxia. The sequence and timing of steps in the procedure and the materials/solutions needed are shown in Table I.

Method

Dissection (Step 1). A 250- to 350-g male Sprague-Dawley rat is injected ip with 0.1 ml 50 mg/ml Nembutal/100 g body weight and, within minutes, becomes insensitive to sound and touch. The animal is then rapidly decapitated and bled, which provides a clearer field for dissection and reduces blood clots. An incision is made across the abdomen below the diaphragm, and the chest is opened like a box by cutting up both sides of the rib cage and across the diaphragm, pushing the "lid" of the chest up to fully expose the heart and lungs. The heart and lungs are lifted out and cut away from the body with scissors closely following the spine, preserving the descending aorta.

The heart and lungs are immersed in cold (0°) BSS containing 10 mM glucose and quickly rinsed free of blood. The heart is kept in this medium during the cannulation procedure as much as possible to slow its metabolism. The aorta is located by lifting the thymus, which appears like a fat pad over the heart. The aorta is held between thumb and forefinger and the thymus, lungs, and other extraneous tissue are stripped away with the other hand. These portions are cut away from the heart with scissors, being careful not to pull at or cut close to the heart, leaving the aorta clearly in sight. The aortic arch is then cut to the desired length, and a piece of plastic tubing with a widened end, attached to a blunted 16- or 18-gauge needle, is inserted into the aortic opening. The end of the cannula should be at least 3 mm from the heart. The aorta is tied to the cannula with 2-0 surgical silk (Ethicon Inc., Somerville, NJ), and several milliliters of room-temperature Ca^{2+}-Joklik are pushed into the heart with a syringe. The base of the cannula is then transferred to a Luer-lock fitting

TABLE I
OVERVIEW OF THE MYOCYTE PREPARATION PROCEDURE

Step	Time (min)
1. Heart removal and cannulation of aorta	-3 to -5
Needed: 300 ml chilled BSS with 10 mM glucose, a filled 1-ml syringe with the cannula at the end, lengths of 2–0 silk, a dissection tray, scissors and forceps, Nembutal, 250–350 g rat	
2. Begin perfusion of the heart with oxygenated Ca^{2+}-Joklik (pH 7.20, 37°)	0
Needed: Perfusion apparatus (see Fig. 1), 100 ml Ca^{2+}-Joklik, 100% O_2 tank, water baths to warm the beakers (37°) and the outside of the condenser (39°)	
3. Begin perfusion with Ca^{2+}-free Joklik (100% O_2, pH 7.20, 37°)	$+5$
Needed: A second perfusion apparatus, 55 ml Ca^{2+}-free Joklik (with 10 mM KCl)	
4. Add 90 U/ml collagenase and 1 mg/ml BSA	$+10$
Needed: 22 mg of 207 U/mg collagenase type I and 50 mg of BSA fraction V dissolved in 5 ml Ca^{2+}-free Joklik (which replaces 5 ml drained off during the heart transfer)	
5. Add $CaCl_2$ to a final concentration of 0.25 mM	$+30$
Needed: 62.5 μl 0.20 M $CaCl_2$	
6. Add 0.25 mM $CaCl_2$	$+35$
Needed: 62.5 μl 0.20 M $CaCl_2$	
7. Add 0.50 mM $CaCl_2$	$+40$
Needed: 125 μl 0.20 M $CaCl_2$	
8. Add 0.25 mM $CaCl_2$ for a final concentration of 1.25 mM	$+45$
Needed: 62.5 μl 0.20 M $CaCl_2$	
9. Remove heart from the perfusion apparatus	$+47$–54
10. Separate ventricular and atrial tissue, cutting the heart into small pieces	
Needed: 5 mg 207 U/ml collagenase dissolved in 10 ml of Ca^{2+}-Joklik with 1% BSA, pH 7.20, scissors, forceps, a Petrie dish	
11. Shake the pieces in oxygenated Ca^{2+}-Joklik medium at 37° with 100 U/ml collagenase and 1% BSA for 5 min	
Needed: A shaking water bath and a closed 100-ml plastic jar with a connection for oxygen flow	

TABLE I (*continued*)

Step	Time (min)
12. Triturate the suspension gently three or four times and filter the suspension through a 250-μm Nitex screen	
Needed: A plastic 10-ml pipet and a 4 × 4 in. piece of 250-μm mesh Nitex screen	
13. Wash the cells twice in Ca²⁺-Joklik with 1% BSA (centrifugation at 100 g, 1 min)	
Needed: Several 15-ml polypropylene test tubes and 30 ml Ca²⁺-Joklik with 1% BSA	
14. Purify the cells over a layer of 45% Percoll solution (centrifugation at 500 g, 2 min) in a swinging bucket centrifuge	
Needed: 5 ml of 60% Percoll solution mixed with 1.6–1.8 ml Ca²⁺-Joklik with 1% BSA in a 15-ml test tube and 5 ml of cells at a concentration of 6 × 10⁵ cells/ml in Ca²⁺-Joklik with 1% BSA	
15. Wash cells 2× in Ca²⁺-Joklik with 1% BSA (centrifugation at 100 g, 1 min)	
16. Store the cells at room temperature in wide-mouth plastic beaker at a concentration of 2 × 10⁵ cells/ml in Ca²⁺-Joklik with 1% BSA	~6 hr

at the base of a reflux condenser (which is part of the perfusion system). The time between the first incision and the beginning of perfusion ranges between 2 and 4 min. A significant increase in blood clots and damaged heart areas is observed when the dissection and cannulation procedure lasts more than 6 min.

Perfusion (Steps 2–8). The heart is perfused with both Ca²⁺-Joklik and Ca²⁺-free Joklik medium during the digestion procedure, requiring two separate but similar closed circulatory systems. This is achieved by duplicating the beakers, tubing, and reflux condensers of the system (see Fig. 1), and pumping the medium with a Cole–Parmer Masterflex peristaltic pump (No. 7553-30) equipped with two pump heads. The beakers of oxygenated Joklik medium are kept at 37° in a water bath. Tygon tubing originates at the beakers, goes through the pump heads, and is connected to the tops of the condensers. The medium drips into a reservoir of medium in the condenser, and then flows through the heart (attached to the bottom of the condenser) at a rate of 10 ml/min. The medium leaving the heart is collected with funnels into the original beakers. The pH of the solutions is continuously monitored during the procedure with an on-line electrode attached to a commercial pH meter (Corning, NY).

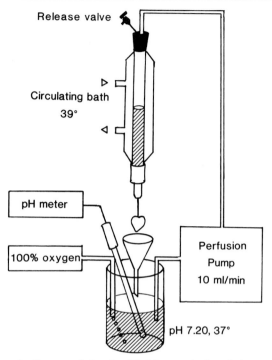

FIG. 1. Schematic diagram of the simplified Langendorf perfusion apparatus used to maintain the heart.

The heart is perfused with Ca^{2+}-Joklik for 5 min following dissection to allow the heart to recover, and to allow estimation of any damage to the heart. A healthy heart will beat vigorously, ejecting any remaining blood on its own. (This is collected before recirculation begins.) A heart that is likely to yield healthy myocytes will beat regularly, with no signs of swelling, color changes from red to brown or yellow, large reductions in medium pH, or cloudiness in the perfusate.

After 5 min, the heart is transferred to the second (Ca^{2+}-free) perfusion system. The first few milliliters of medium are collected to prevent Ca^{2+} contamination of the solution, which might interfere with the cell dissociation, and perfusion is continued for another 5 min. The heart should be completely slack and relaxed at this point: a swollen or rigid aspect indicates heart damage. Collagenase (90 U/ml) and BSA (1 mg/ml) are then added. The heart swells soon after, but the color should not be significantly changed. After a total of 30 min of perfusion (20 min with collagenase) 0.25 mM CaCl$_2$ is added back to the heart. The Ca^{2+} concentration of the medium is then increased at 5-min intervals to final concentrations of 0.50, 1.00, and 1.25 mM $CaCl_2$. The typical signs of a "calcium-paradoxed" heart (a yellow/brownish color of the heart and areas of fluid accumulation beneath the surface) are not present if $CaCl_2$ is added at this phase, although these changes can be observed with earlier additions.

The heart digestion is usually completed shortly after the reintroduction of calcium. Excessive digestion of the heart increases the yield, but results in loss of some receptor responses as well as poor longevity of the cells, so the heart is removed before it is completely slack although it is distinctly softened.

Cell Dissociation (Steps 10–13). After the heart is removed from the perfusion system, the atria and ventricles are carefully dissected to ensure that the cell types will not be mixed. Ordinarily, the ventricles digest more rapidly than the atria so that optimal conditions for isolation of each cell type are not obtained in the same preparation. The dissected tissue is then cut in quarters, teased into strands, and placed in a Ca^{2+}-Joklik solution with 1% BSA and 100 U/ml collagenase, pH 7.20. This is shaken at 37° for 5 min in a closed container with oxygen introduced through a tubing connection in the top. The resultant suspension is triturated by slowly drawing the fluid into a 10-ml plastic pipet four times, and is then filtered through a 250-μm Nitex screen (Tetko Inc., Elmsford, NY). The cells are then spun down at 100 g in a swinging bucket centrifuge and washed several times with Ca^{2+}-Joklik with 1% BSA (pH 7.40).

Cell Purification (Steps 14–15). The living and dead cells are separated by density over a Percoll step gradient. A clean separation (leaving dead cells at the interface and viable cells in a pellet at the bottom of the tube) requires a precise composition of the Percoll layer. To simplify this, a 60% stock solution is prepared and then diluted to 42–44% in the test tube, allowing optimization of the separation on a daily basis. Ordinarily, 5 ml of 60% Percoll solution is mixed with 1.7 ml of Ca^{2+}-Joklik with BSA (to form a layer of 45% Percoll) for optimal results. Occasionally, however, clumps of viable cells will be visible above the pellet or the pellet will show too many Trypan Blue-staining cells when examined under the microscope. (A normoosmolar Trypan Blue solution is mixed with the cells to a final Trypan Blue concentration of 0.001%.) If either of these problems exist, the density of the Percoll layer can be easily adjusted by increasing or decreasing the dilution of the 60% Percoll stock solution.

After layering the cells on top of the Percoll solution, the tube is centrifuged at 500 g for 2 min in a tabletop swinging bucket centrifuge at room temperature (Fisher, Pittsburgh, PA). The dead cells are removed from the medium/Percoll solution interface, and the Percoll layer is carefully removed from the centrifugation tube, leaving the intact cells in a loose pellet at the bottom. The pellet is washed several times at 100 g with Ca^{2+}-Joklik with BSA, and stored at room temperature in the same medium.

Common Difficulties

The collagenase used for the digestion is important and must be chosen carefully to produce the largest quantity of stable cells appropriate for a given application. Often, several lot numbers must be tested to find a lot that gives a high yield of calcium-tolerant cells which are stable over a long period of time. In addition, the cells should be tested in experiments for a given application to verify, for example, that necessary receptors or surface proteins are intact. After an appropriate lot of collagenase has been found, the most frequent cause of a failed myocyte preparation is a damaged or hypoxic heart. Rough handling of the heart, improper cannulation (i.e., entry of the cannula into or very close to the heart valves), inadequate oxygenation, introduction of an air bubble or foreign particles to the heart, or the presence of blood clots will rapidly be obvious. The perfusion apparatus must be cleaned thoroughly between preparations to prevent contaminants. In addition, the heart is sensitive to pH, and will show gradual damage at pH values above 7.40. In order to consistently achieve accurate pH readings during experiments despite the presence of proteins released by the heart and collagenase, a Corning combination pH electrode is used (model 476531, Corning Glass Works, Corning, NY). The junction is changed every 5 or 6 experiments to prevent inaccurate readings due to clogging. A Hach One combination pH electrode has also been used.

Assessment of Cellular Viability

Morphology. The viability of the cells can be determined in a number of ways. The first is by visual examination, as mentioned. The morphologies of healthy rat ventricular and atrial myocytes are shown in Fig. 2A and B. The striations of both types of cells are clearly visible, and the length : width ratios of the cells are greater than 5, which is a sign that the cells are not hypoxic.[1] A damaged cell is shown in the atrial preparation to demonstrate the marked difference in morphology observed with hypercontraction or cell death. The ventricular cell shown has the dimensions 125×25 μm and the healthy atrial cell is 75×10 μm. Although cell sizes are somewhat variable, atrial cells seem to be generally shorter and more narrow, which is consistent with observations made of guinea pig atrial

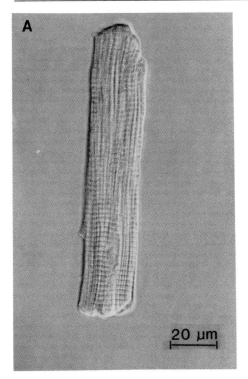

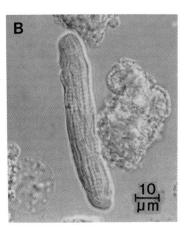

FIG. 2. Micrographs of (A) ventricular and (B) atrial myocytes isolated from adult rat heart. Photographs were taken with a Nikon camera on a Zeiss IM 35 inverted microscope equipped with Nomarski optics.

and ventricular cells.[6] The atrial cells are not Percoll purified because so few cells (about 10^4) are obtained; however, the ventricular cells are purified to give an average distribution of 73 ± 6% rod-shaped, 9 ± 4% distorted, and 17 ± 8% Trypan Blue-permeant cells (mean ± SD, $n = 16$). Thirty-two ± 11% of the rod-shaped cells ($n = 9$) show infrequent spontaneous contractions. The cell yield is 2.0 ± 0.5 × 10^6 rod-shaped cells/ heart (mean ± SD, $n = 12$) by a hemacytometer cell count and 80% of these are stable at room temperature for up to 5 hr following purification. Although the yield (approximately 13.3 mg protein/heart) is a low percentage (about 7%) of the total heart weight, the quality of the cells is very high and the number of cells obtained is sufficient for measurements of intracellular Ca^{2+} concentrations and nucleotides, as well as for electro-

[6] R. Mitra and M. Morad, *Am. J. Physiol.* **249,** H1056 (1985).

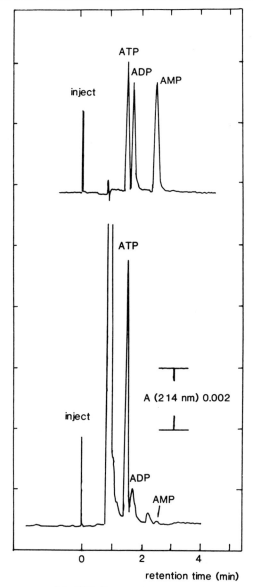

FIG. 3. A chromatogram of the HPLC separation of adenine nucleotides. A 10-μl sample of 11 μM ATP, ADP, and AMP is shown above, and 10 μl of a ventricular cellular extract (see text) is shown below.

TABLE II
NUCLEOTIDE MEASUREMENTS IN VENTRICULAR MYOCYTES[a]

Medium	Nucleotide content (nmol/mg protein)			ATP : ADP ratio
	ATP	ADP	AMP	
Ca^{2+}-Joklik with 1% BSA	56 ± 16	5.5 ± 2.0	1.2 ± 1.0	10.0 ± 0.6
BSS with 10 mM NaHCO$_3$, 10 mM glucose, 1% BSA	48 ± 9	6.5 ± 1.2	1.8 ± 1.0	6.3 ± 0.6

[a] Measurements were made by HPLC as indicated in the text. Results are expressed as the mean ± SE, $n = 4$. ATP content is 300–400 nmol/10^6 cells.

physiology and cell culture. Higher yields have been reported by other investigators[2] but their preparations were not fully characterized in terms of cell longevity, metabolic activity, and receptor responses.

Adenine Nucleotide Measurements. Both the ATP : ADP ratios and the intracellular content of ATP are important indicators of cell viability. A chromatogram of ATP, ADP, and AMP measured by HPLC as described previously[7] is shown in Fig. 3. Nucleotide standards are shown above, and cellular ATP, ADP, and AMP are shown below. The cellular nucleotides are extracted with 1 mM perchloric acid/2 mM EDTA added to a pellet of 1.5×10^5 cells following a 5-min incubation at 37°. The PCA is neutralized with KOH, and proteins and potassium perchlorate are removed by centrifugation in an Eppendorf microcentrifuge at 8500 g. The nucleotides are then separated on a Waters HPLC system over a 5-μm reversed-phase C$_{18}$ column. The mobile phase used is 50 mM ammonium phosphate (pH 6.50). Typical values are given in Table II. The ATP/ADP ratio in Ca^{2+}-Joklik medium with 1% BSA (which, compared to BSS, has an additional 10 mM NaH$_2$PO$_4$ as well as vitamins and amino acids) is higher than what is seen in BSS with 1% BSA and 10 mM NaHCO$_3$ and is also greater than the value of 6.30 reported elsewhere in a similar medium.[5] The ATP content values are also higher than reported elsewhere both in units of nanomoles/milligram protein[5,8–10] and nanomoles/10^6 cells.[5] [Protein measurements were made with a Bio-Rad protein assay (Bio-Rad, Richmond, CA)].

[7] G. R. Dubyak and A. Scarpa, *Biochemistry* **22,** 3531 (1983).
[8] J. Y. Cheung, I. G. Thompson, and J. V. Bonventre, *Am. J. Physiol.* **243,** C184 (1982).
[9] K. H. McDonough and J. J. Spitzer, *Proc. Soc. Exp. Biol. Med.* **173,** 519 (1983).
[10] R. A. Haworth, D. R. Hunter, H. A. Berkoff, and R. L. Moss, *Circ. Res.* **52,** 342 (1983).

Fura2 Measurements of Cytosolic [Ca²⁺]. The intracellular Ca^{2+} dye fura2 can be used to assess the ability of the cells to respond to appropriate stimuli. Fura2 measurements were made as previously described.[11] Cells were loaded with 2 μM fura2AM for 40 min at 37° in Ca^{2+}-Joklik medium with BSA and washed repeatedly in the same medium. For measurements, 2×10^4 cells was resuspended in 2 ml BSS (pH 7.40) with 10 mM added $NaHCO_3$, and the fluorescence of the suspension was monitored at 37° with constant stirring. Both the excitation (340 nm) and emission (510 nm) wavelengths were established with interference filters (Omega Optical, Brattleboro, VT).

Fura2 traces qualitatively demonstrating intracellular Ca^{2+} concentration changes in ventricular cells are given in Fig. 4. The cells have an extracellular ATP receptor, and its effects on cellular Ca^{2+} homeostasis are potentiated by norepinephrine as described.[11] This indicates that the receptors, G proteins, second messengers, and channels required for these responses are functional. The cells also retain a cytosolic Ca^{2+} response to K^+ depolarization, and associated cell contraction has been observed under a Nikon AFX-IIA microscope with a temperature-regulated (37°) stage. The cells maintain intracellular Ca^{2+} stores which are necessary for maximal Ca^{2+} responses and can be released by caffeine or the Ca^{2+} ionophore ionomycin. The stores are depleted by addition of 10 mM caffeine or a 30-min pretreatment with 1 μM ryanodine, demonstrating that they are sarcoplasmic reticulum in origin.[12] There is some question of the magnitude of Ca^{2+} changes that occur as the calculated values [94 ± 29 nM resting and 209 ± 28 with K^+ depolarization (mean ± SD, n = 9)] seem somewhat low. This may be a characteristic of the cell preparation as, although 94 ± 6% of the cells (253 cells from 4 experiments were observed) contracted in response to KCl, only 30 ± 9% of the cells demonstrated sustained contraction. The others demonstrated a transient beating response which lasted on the order of 15 sec. Because of the intermittance, duration, and lack of synchronicity of the response, it is possible that these cells do not make a significant contribution to the Ca^{2+} fluorescence signal. Similar low [Ca²⁺] responses have been observed with 35 mM KCl in other preparations.[13,14] This may indicate depletion of intracellular stores in a portion of the cells (i.e., the cells which are not slowly spontaneously contractile). Other concerns about the quantitation of fura2 signals are that the dynamic range of fura2 is somewhat blunted

[11] M. B. De Young and A. Scarpa, *FEBS Lett.* **223,** 53 (1987).
[12] M. B. De Young and A. Scarpa, submitted for publication.
[13] J. Y. Cheung, J. M. Constantine, and J. V. Bonventre, *Am. J. Physiol.* **252,** C163 (1987).
[14] T. Powell, P. E. R. Tatham, and V. W. Twist, *Biochem. Biophys. Res. Commun.* **122,** 1012 (1984).

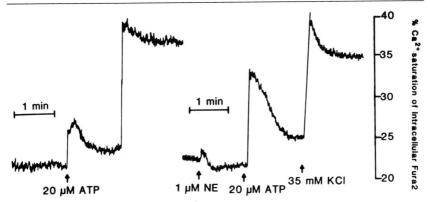

FIG. 4. A fluorescence tracing of fura2-loaded ventricular myocytes exposed to ATP and KCl with and without pretreatment with norepinephrine (NE). Measurements were made with a University of Pennsylvania Biomedical Instrumentation Group fluorimeter.

within myocytes[15] and that a portion of the dye may be compartmented within organelles. [Use of Triton X-100 following cell lysis with digitonin decreases the minimal fluorescence by another 29 ± 3% (mean ± SE, $n = 5$).]

Other Diagnostic Techniques. Besides being suitable for measurements of receptor activation, intracellular Ca^{2+} stores, and contraction, the cells were assessed for their utility in cell culture and electrophysiology. The cells were cultured on irradiated fibroblasts[16] and underwent the structural disorganization and redifferentiation characteristic of myocytes cultured under certain conditions.[17] The reconstruction phase indicates that the cells are capable of protein synthesis and reorganization of the contractile apparatus in addition to retaining attachment factors. Some electrophysiological experiments determined the presence of membrane currents and Ca^{2+} channels (data not shown) although occasionally additional collagenase treatment was needed to make the cell surface cleaner for pipet attachment.

Summary

A basic myocyte preparation method has been developed which uses standard media and a simplified perfusion system to obtain cells which are calcium tolerant, purified, stable for hours, and present in sufficient yield

[15] W. G. Wier, M. B. Cannell, J. R. Berlin, E. Marban, and W. J. Lederer, *Science* **235,** 325 (1987).

[16] J. G. Rheinwald and H. Green, *Cell* **6,** 331 (1975).

[17] R. L. Moses and W. C. Claycomb, *J. Ultrastruct. Res.* **81,** 358 (1982).

for many types of studies. The cells have been used for Ca^{2+} measurements examining the roles of ATP and norepinephrine in Ca^{2+} homeostasis[11] as well as to examine regulation of intracellular Ca^{2+} stores.[12] Further studies have shown that the cells have a high ATP : ADP ratio and can be used for electrophysiological studies and cell culture.

Acknowledgments

We would like to thank Dr. Ruth Altchuld and Dr. Gerald Brierley for their assistance in the early phases of mastering the myocyte preparation, and Linda Castell for her construction of the perfusion system. We also thank Francesco De Cobelli for his help with the ATP and protein measurements. Dr. Richard Eckert and Dr. David Van Wagoner provided their expertise by demonstrating that the myocytes could be cultured and were suitable for electrophysiology, respectively. This work was supported by NIH Grants HL 18708 and HL 07502.

[42] Measurement of Na+ Pump in Isolated Cells

By LEIGH H. ENGLISH and JOHN T. SCHULZ

Introduction

As the importance of the Na^+ pump has grown with developments in our understanding of cell biology, so also the measurement of the Na^+ pump in isolated cells has become easy and routine. Several methods are described here that have been commonly used to provide accurate analysis of the activity of the Na^+ pump in isolated cells. Choice of the appropriate method will depend on knowledge of the particular cell and the form of data needed for a particular application; however, in general all methods described below have been used frequently without serious drawbacks.

Almost all methods for measuring the Na^+ pump in isolated cells depend on the assumption that the cardiac drug ouabain binds specifically to the Na^+ pump and inhibits only this enzyme. Ouabain is a sterol glycoside that was first reported in 1785 by Withering to assist in the treatment of dropsy.[1] Since Schatzmann[2] demonstrated that cardiac glycosides inhibited active transport of Na^+ and K^+ in erythrocytes no one has ob-

[1] W. Withering, "An Account of the Foxglove and Some of Its Medical Uses with Practical Remarks on Dropsy and Other Diseases." Sweney, Birmingham, England, 1785.
[2] H. J. Schatzmann, *Helv. Physiol. Pharmacol. Acta* **11**, 346 (1953).

served any other primary activity of ouabain. Joiner and Lauf[3] later demonstrated that ouabain binds to specific receptors associated with Na$^+$ and K$^+$ transport. Although the specific sequence of the binding site on the pump has not been identified, ouabain apparently binds to the catalytic α subunit of the pump on the outer surface of the cell. A photoaffinity-labeled ouabain derivative has also been observed to bind to a small 12-kDa polypeptide[4] associated with the Na$^+$ pump. Binding of ouabain enhances the affinity of the active site aspartyl residue for PO$_4^{3-}$. By stabilizing this intermediate, the turnover of the enzyme is inhibited. The binding of ouabain to the Na$^+$ pump in a 1 : 1 ratio provides an excellent tool for determining the number of pumps on the cell plasma membrane. Also, ouabain inhibition of the Na$^+$ and K$^+$ transport catalyzed by the Na$^+$ pump provides the basis for rapid analysis of the activity of the Na$^+$ pump.

The first section in this chapter describes methods for the analysis of the number of Na$^+$ pumps in cultured cells by evaluating ouabain binding. The second section describes the measurement of the number of Na$^+$ pumps by so-called "back-door" phosphorylation, in which the active site of the Na$^+$ pump is labeled with inorganic phosphate. In the third section several methods of assaying of ouabain-inhibitable ^{86}Rb$^+$ influx are described as an indicator of the Na$^+$ pump activity. Finally, in the fourth section, the measurement of Na$^+$ efflux is described. This method for measuring the Na$^+$ pump can be particularly useful for cells that survive in low Na$^+$ environments.

The methods in this chapter describe the analysis of the Na$^+$,K$^+$ pump in isolated cells; therefore, statements made about the Na$^+$ pump in the parent tissue should be restricted unless care is taken to preserve conditions representative of that parent tissue. Many epithelial cells have enzymes distributed in a polar fashion in the tissue but when these cells are isolated this condition is lost. Also, enzymatic assays may be enhanced by the presence or absence of basal lamina or contact with adjacent cells of either the same or different types. In the tissue, the growth condition may also provide hormones or other factors in either the serum or lumen of particular organs that may alter the activity or number of pumps expressed in the isolated cells. The level of Na$^+$ pump in some isolated cells is known to vary with the tissue culture growth conditions, such as osmotic pressure, saline composition, density of the cell population, and position of the cells in the cell cycle. Also the physical act of harvesting cells that normally grow attached in tissue culture can enhance the intra-

---

[3] C. H. Joiner and P. K. Lauf, *J. Physiol. (London)* **283**, 155 (1978).
[4] B. Forbush, J. H. Kaplan, and J. F. Hoffman, *Biochemistry* **17**, 3667 (1978).

cellular Na^+ concentration, which will enhance the activity of the Na^+ pump. Cells display a wide sensitivity to ouabain and ouabain resistance expressed in some cells may be due to enzymes that are otherwise nearly identical to the Na^+ pump in function but lack ouabain sensitivity. Analysis of this condition is not covered in this chapter but is described elsewhere.[5,6] Therefore, isolated cells provide excellent material for rapid and easy analysis of the Na^+ pump; however, it may be important to consider alterations in cellular physiology created by the isolation process.

Binding of Cardiac Glycosides

The cardiotonic steroids and their glycoside derivatives are potent and specific inhibitors of Na^+,K^+-ATPase in vitro[7,8] and of sodium and potassium pumping activity in vivo.[9] The cardiac glycosides, notably ouabain, associate with the sodium pump[10] at rates comparable to their steroid analogs, but they come off the pump much more slowly. First order dissociation rate constants for the bovine brain $Na,^+K^+$-ATPase–glycoside complex formed in the presence of saturating Mg^{2+} and inorganic phosphate at $25°$ is in the range of about 0.2 to 0.5 hr^{-1} in vitro[11]; that is, the $t_{1/2}$ for glycoside dissociation is around 3 hr. Similar dissociation rates at $37°$ are observed when ouabain is bound to primate fibroblast membranes.[12]

Because they associate rapidly and dissociate slowly and because they bind to the extracellular side of the sodium pump,[13,14] the cardiac glycosides are suitable affinity ligands for determination of sodium pump stoichiometry on the cell surface of many mammalian cells. There are, however, at least four complicating factors which should be kept in mind

[5] L. H. English, J. Epstein, L. Cantley, D. Housman, and R. Levenson, *J. Biol. Chem.* **260,** 1114 (1985).
[6] L. H. English, B. White, and L. C. Cantley, in "New Insights into Cells and Membrane Transport Processes" (G. Post and S. T. Crooke, eds.). Plenum, New York, 1986.
[7] R. L. Post, C. R. Merrit, C. R. Kinsolving, and C. D. Albright, *J. Biol. Chem.* **235,** 1796 (1960).
[8] J. C. Skou, *Biochim. Biophys. Acta* **42,** 6 (1960).
[9] J. F. Hoffman, *Am. J. Med.* **41,** 666 (1966).
[10] Throughout this discussion, the terms "sodium pump" and "ouabain receptor" will be used interchangeably.
[11] A. Yoda and S. Yoda, *Mol. Pharmacol.* **10,** 494 (1974).
[12] From our unpublished results in plasma membranes prepared from African green monkey CV-1 fibroblasts and their ouabain-resistant transfectants.
[13] J. F. Hoffman, *Am. J. Med.* **41,** 666 (1966).
[14] J. R. Perrone and R. Blostein, *Biochim. Biophys. Acta* **291,** 680 (1973).

when attempting to quantify cell surface sodium pumps by ouabain binding:

1. There are at least two isozymes of the sodium pump[15] which differ in their sensitivity to ouabain. In the rat two isozymes have been identified, α and $\alpha(+)$. Rat kidney expresses only α which when purified has a K_i for ouabain inhibition of Na$^+$,K$^+$-ATPase activity of 3.2×10^{-5} M ouabain. Rat axolemma expresses only $\alpha(+)$, which is half-maximally inhibited at a much lower ouabain concentration than is α (1×10^{-7} M).[16] Some rat cell types express both α and $\alpha(+)$. Among other systems that have been studied, the various morphological divisions of the rabbit nephron express sodium pumps of different ouabain sensitivities.[17] Titration of cells which express more than one ouabain affinity will appear not to saturate unless the highest concentration of ouabain used is >10 times the K_i of the lowest affinity receptor (see below).

2. In general, the α isozyme of rodents is much less sensitive to ouabain than is the α isozyme of other species [e.g., Na$^+$,K$^+$-ATPase(α) of dog kidney, bovine brain, and primate fibroblasts]. Practically, precise quantitation of low-affinity ($K_d > 10^{-6}$ or so) ouabain receptors is very difficult because at the ouabain concentrations needed to saturate the receptor nonspecific binding is very high.[18]

3. External K$^+$ inhibits ouabain association with the sodium pump, while internal Na$^+$ promotes ouabain binding.[19,20] Hence, ouabain-binding assays should be done in isotonic buffered saline or in K$^+$-free medium.

4. Cells internalize bound ouabain. The half-time ($t_{1/2}$) for internalization in HeLa cells is approximately 5 hr at 37°.[21] The practical consequence of internalization is that the incubation time in a ouabain-binding assay must be much shorter than the $t_{1/2}$ for internalization. Because the binding time must be relatively short, equilibrium binding is not obtained at less than saturating ouabain concentrations.[22]

[15] K. Sweadner, J. Biol. Chem. **254**, 6060 (1979).

[16] K. Sweadner, J. Biol. Chem. **260**, 11508 (1985).

[17] A. Doucet and C. Barlet, J. Biol. Chem. **261**, 993 (1986).

[18] D. Charlemagne, J.-M. Maixent, M. Preteseille, and L. Lelievre, J. Biol. Chem. **261**, 185 (1986).

[19] E. van Zoelen, C. Mummery, J. Boonstra, P. van der Saag, and S. de Laat, J. Cell. Biochem. **21**, 77 (1983).

[20] J. Mills, A. Macknight, J. Jarrell, J. Dayer, and D. Ausiello, J. Cell Biol. **88**, 637 (1981).

[21] L. Pollack, E. Tate, and J. Cook., Am J. Physiol. **241**, C173 (1981).

[22] When the concentration of ouabain is equal to the K_d for ouabain binding, the rate of association is equal to the rate of dissociation. The time to approach equilibrium under these conditions is several times the $t_{1/2}$ for dissociation. In purified CV-1 cell plasma membranes bound in the presence of Mg^{2+} and phosphate the $t_{1/2}$ for ouabain dissociation

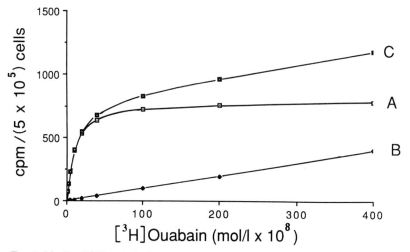

FIG. 1. Idealized [³H]ouabain titration of cells possessing one class of sodium pumps with an apparent K_d for ouabain of about 10^{-7} M. (A) Specific binding; (B) nonspecific binding; (C) total binding.

The results of a [³H]ouabain titration of whole cells depend on whether the cells express one or more than one form of the sodium pump. Figure 1 is an idealized representation of ouabain binding to cells which express only one class of ouabain receptors. The cells are titrated with increasing concentrations of [³H]ouabain of constant specific radioactivity in buffered saline at 37°. The binding of ouabain to its receptor on the cell surface is a specific process and hence saturable. Theoretically, for a single class of receptors on the cell surface, a plot of ouabain bound per cell versus ouabain concentration should give a quasi-hyperbolic plot that asymptotically approaches the maximum number of binding sites per cell as [³H]ouabain concentration increases (since binding is rarely at equilibrium at subsaturating concentrations of [³H]ouabain, the curve is not a true hyperbola). Curve A of Fig. 1 is an illustration of such a plot. The concentration of [³H]ouabain at which binding is half-maximal is the *apparent* dissociation constant for a given set of experimental conditions (again, equilibrium binding must be obtained at all ouabain concentrations in order to determine the *true* equilibrium dissociation constant). A plot

is several hours at 37° (our unpublished results). Hence, equilibration times for ouabain binding to the sodium pump on these cells is many hours at subsaturating ouabain concentrations. Binding for such a long period of time would be complicated by internalization, as well as by loss of viability and rupture of cells bound in isotonic saline.

such as curve A is rarely obtained experimentally because ouabain binds nonspecifically as well as specifically to the cells. The extent of nonspecific binding is dependent on the conditions of the binding and isolation procedures; but for a given set of conditions nonspecific binding usually increases linearly with ouabain concentration. Curve B of Fig. 1 represents nonspecific binding. In practice these types of data are obtained by adding [^{3}H]ouabain at various concentrations to cells which have already been exposed to supersaturating concentrations of unlabeled ouabain. The experimental titration curve which one obtains upon titrating cells with ouabain is the sum of specific and nonspecific binding (if binding is conducted for short times so that internalization is not a problem). Curve C of Fig. 1, the sum of curves A and B, represents data which would typically be obtained upon [^{3}H]ouabain titration of whole cells which had not been previously saturated with unlabeled ouabain. Note that for any single concentration of [^{3}H]ouabain greater than that required for saturation of specific binding, the number of binding sites per cell is simply the difference between total binding (curve C) and nonspecific binding (curve B). More importantly, extrapolation of the linear portion of curve C intercepts the ordinate at the value corresponding to the saturation number of binding sites.

The data obtained in titrations of cells which have two classes of ouabain receptors of different ouabain affinities are somewhat more complex. Figure 2 is an idealized representation of such a titration. The cells have a minor class of binding sites with an apparent K_d of $<10^{-6}$ M and a major class with an apparent K_d of about 3×10^{-5} M [similar to what one might expect to find in a rat cell which expressed mostly the α isozyme of the sodium pump and some $\alpha(+)$]. Several aspects of the figure are noteworthy: (1) extrapolation of the total binding curve segment between 10^{-4} and 2×10^{-4} M to the ordinate may yield a saturation stoichiometry which is the sum of high- and low-affinity receptors, but it does not reveal that there is more than one class of receptors. (2) Specific binding (curve A: total, nonspecific) will appear to be unsaturable if the titration is conducted only to a maximum of 1×10^{-4} M [^{3}H]ouabain.[23] (3) At concentrations of [^{3}H]ouabain sufficient to saturate the high-affinity receptors, the low-affinity receptors may already be binding ouabain, making determination of high-affinity receptor stoichiometry difficult.

Since it is impossible to predict *a priori* whether a previously uninves-

---

[23] Figure 2 has been drawn in a manner indicating that saturation of very low affinity receptors can be determined. Often, at higher ouabain concentrations, nonspecific binding is substantial and erratic enough to preclude precise definition of saturation stoichiometry.

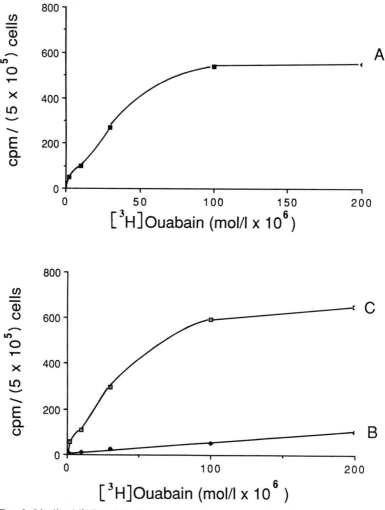

FIG. 2. Idealized [³H]ouabain titration of cells possessing two classes of sodium pump: one with low affinity for ouabain ($K_d = 3 \times 10^{-5}$) and one with high affinity ($K_d = 5 \times 10^{-7}$). (A) Specific binding; (B) nonspecific binding; (C) total binding.

tigated cell type expresses more than one class of ouabain receptor,[24] a titration is in order. If the titration reveals only one class of saturable sites (as idealized in Fig. 1) then routine determinations of sodium pumps on

[24] Determination of ouabain inhibition of ⁸⁶Rb⁺ uptake may reveal classes of pumps with different K_i values for ouabain; however, it is conceivable that a very low turnover number could make one class of pumps relatively silent in an assay of transport inhibition.

the cell surface can be conducted at a saturating concentration of [^{3}H]ouabain at which nonspecific binding is minimal. If the titration reveals more than one class of sites, then the relative contributions of each class to total [^{3}H]ouabain binding must be carefully scrutinized before reporting the stoichiometry of receptors.

Method 1[25]: Ouabain Binding to Whole Cells in Suspension

This method can be used for cells grown in suspension or for adherent cells which readily form single-cell suspensions when scraped from a dish in the presence of buffered isotonic saline.

Materials

TBS: 150 mM NaCl, 10 mM Tris–HCl, pH 7.4

TBS, 2 mM ouabain: This solution will be used both to suspend the cells used for determination of nonspecific binding and to prepare [^{3}H]ouabain dilutions for assay of nonspecific binding. Dissolve 14.58 mg of unlabeled ouabain/10 ml of TBS. Scale up as necessary

Stock [^{3}H]ouabain: Uniformly tritiated ouabain can be purchased from a number of suppliers. We obtain it from New England Nuclear at a specific radioactivity of 20 Ci mmol^{-1} suspended to 1 mCi ml^{-1} in a solution of 9 : 1 ethanol : benzene. Labeled ouabain should be diluted with unlabeled carrier to approximately 250 to 500 Ci mol^{-1}. For example, place 50 μCi of [^{3}H]ouabain (20 Ci mmol^{-1}, 1 mCi ml^{-1}) in a 1.5-ml microcentrifuge tube. Evaporate to dryness under nitrogen. Add 0.5 ml of TBS/2 mM ouabain. Vortex the solution and then sonicate for a few minutes in a bath sonicator. The molar contribution of the isotopically labeled ouabain (2.5%) can be ignored in subsequent calculations. For titrations serially dilute the 500 Ci mol^{-1} stock in TBS

Cells: We conduct assays on CV-1 monkey fibroblast transfectants grown to confluence in 10-cm round dishes in the presence of alpha medium (Grand Island Biological Co., Grand Island, NY) supplemented with 13% fetal bovine serum. The cells are approximately 7.5 × 10^4 cm^{-1} at confluence. For a ouabain titration, figure on about 5 × 10^5 cells/determination and three determinations each for total and nonspecific binding at each [^{3}H]ouabain concentration (that is about 3 to 4 × 10^6 cells per given [^{3}H]ouabain concentration)

Oil/TBS: In order to isolate cells from isotope and reduce nonspecific binding, they will be centrifuged through a layer of 1 : 1 (v/v) silicone

---

[25] Abbreviations used are as follows: TBS, Tris-buffered saline; Ci, curie; SDS, sodium dodecyl sulfate; EDTA, ethylenediaminetetraacetic acid.

oil: dinonyl phthalate. After incubation in [³H]ouabain and subsequent cooling on ice for 10 min, each experimental point (0.5 ml of cells, 1×10^6 cells ml^{-1}) will be transferred to a 1.5-ml microcentrifuge tube containing 0.3 ml of 1:1 silicone oil: dinonyl phthalate overlayed with 0.5 ml of cold TBS. Before beginning the experiment, prepare the oil solution. Place 0.3 ml of oil in each of as many tubes as necessary. Gently overlay the oil in the bottom of each tube with 0.5 ml of TBS. (Caution: if the aqueous layer is added too forcefully it will form an aqueous bubble under a layer of oil. If this occurs simply centrifuge the tubes for a few seconds in the microcentrifuge.) After preparing the oil/aqueous two-phase system in each tube, place the tubes on ice

Procedure

Cell Suspension. On the day of the experiment check the cells for confluence. Next, separate the plates into two groups. One set of plates will be resuspended in TBS and used to determine total [³H]ouabain binding, the other set will be used to determine nonspecific binding. Gently wash the confluent monolayers in each set of plates twice with 10 ml of TBS. Remove each wash by aspiration. After washing suspend the cells in sufficient solution to give a final concentration of approximately 1 to 2 $\times$ 10^6 cells ml^{-1}: to each plate designated for total binding determination add the appropriate volume of TBS, scrape the cells from the bottom of the plate with a rubber policeman, remove the suspension with a pipet, and pool in a 15-ml conical tube. Do likewise for each plate designated for nonspecific binding, substituting TBS/2 mM ouabain in place of TBS. Following suspension, vortex the cells and remove an aliquot for determination of cell number and viability. Count the cells in the presence of Trypan Blue on a hemocytometer. CV-1 transfectants are about 95% impermeant to Trypan Blue after the suspension procedure and will remain so for about 1 hr at room temperature. After 30 min at 37° in TBS, viability falls to about 90%. Hence incubations for binding determinations are kept to 20 min at 37°. If the cells to be assayed are already growing in suspension, remove them from medium by centrifugation in a clinical centrifuge; wash by resuspension in TBS. Centrifuge again. Repeat the wash, and resuspend to a final concentration of 1×10^6 cells ml^{-1} in either TBS or TBS/2 mM ouabain.

Ouabain Binding. Plan to conduct the titration by assaying binding at each [³H]ouabain concentration in three tubes of cells for total binding and three tubes for nonspecific binding. For each determination, place

0.45 ml of cell suspension (5×10^5 cells) in a 1.5-ml Eppendorf microcentrifuge tube. Add 0.05 ml of [^{3}H]ouabain, vortex the resulting solution, and place it in a 37° shaking water bath. Allow binding to proceed for 20 min with intermittent vortexing. After 20 min, place all the tubes on ice. After they have cooled, transfer the contents of each to a tube containing ice-cold oil/TBS. Once all the solutions are transferred, centrifuge each for 2 min in a microcentrifuge. The cells should form a tight pellet beneath the oil in the bottom of the tube. Carefully aspirate the aqueous and oil layer off of each pellet. Invert the tube to allow any remaining oil to drain away. With a razor blade, slice off the bottom tip of each tube and place it in a 5-ml glass scintillation vial. Add 0.4 ml of 1% SDS, and vortex occasionally for 1 hr. Finally, add 4 ml of aqueous compatible scintillation cocktail. Cap the vials, vortex each extensively, place in 20-ml carrier vials, and count on a scintillation counter. Empirically determine the specific radioactivity of [^{3}H]ouabain by adding a measured volume of stock [^{3}H]ouabain to 0.4 ml of 1% SDS in a 5-ml scintillation vial. Add scintillant and count as for the experimental samples.

Specific binding is calculated by subtracting nonspecific from total binding. The stoichiometry of [^{3}H]ouabain binding per cell can then be determined using the cell count and the empirically determined specific radioactivity of the isotope.

Comments

Incubation Time. We chose 20 min for three reasons. First, cell death is minimal after 20 min in TBS at 37°. Second, we assume 20 min to be much less than the $t_{1/2}$ for ouabain internalization. Third, we assumed that the pseudo-first order association rate constant for ouabain is about 0.64 μM^{-1} min^{-1} [26] at 25° and that the first order dissociation rate constant for the ouabain–Na$^+$,K$^+$-ATPase complex is approximately 0.22 hr^{-1} [27] at 25°.[28] Given these parameters at 25°, the $t_{1/2}$ for approach to equilibrium should be less than 1 min at 37° and 1 μM [^{3}H]ouabain.

Centrifugation through Oil. This is the most effective method we have found to reduce nonspecific binding and enhance the precision of replicate determinations. In our hands, alternative methods of separating the cells from isotope, such as rapid filtration and simple centrifugation with wash-

---

[26] A. Yoda, S. Yoda, and A. M. Sarrif, *Mol. Pharmacol.* **9**, 766 (1973).

[27] A. Yoda, *Mol. Pharmacol.* **9**, 51 (1973).

[28] This gives a K_d of ~6 nM. This is appropriate for primate cells, but note again that the kinetic parameters, as well as the K_d, will vary among different isozymes of the sodium pump and among pumps from different species.

ing, yielded much higher nonspecific binding and greatly increase the scatter in the data. However, others have had excellent success using a Whatman GFC glass fiber filter to separate cells from the isotope.[29]

Method 2: Quantitation of Ouabain Receptors in Cells that Form Tightly Adherent Monolayers

Unlike their ouabain-resistant transfectants, CV-1 African Green Monkey fibroblasts form very tightly adherent monolayers in culture. Since they are more difficult to bring into single cell suspension, it is more practical to assay ouabain binding to these cells in nondisrupted monolayers.

Materials

Materials required are the same as for method 1 with the following modifications:

1. No oil/TBS is required
2. Cells should be grown in multiwell plates which will yield confluent monolayers of approximately 1 to 2×10^6 cells/well
3. Plan on at least 0.5 ml of [^{3}H]ouabain solution at a given concentration per 9 cm^2 of a given well
4. Stock [^{3}H]ouabain should be diluted in TBS for total binding determinations and in TBS/2 mM ouabain for nonspecific determinations
5. 0.05% Trypsin/0.02% EDTA in normal saline (Irvine Scientific, Santa Ana, CA) will be used to suspend cells for counting

Procedure

Cells. Seed four six-well plates (diameter of 33 mm; surface area of 8.5 cm^2 with sufficient cells to give an initial density of about 1/10 confluence; for CV-1 cells, which have a confluent density of about 2.5×10^5 cells cm^{-2}, seed to give 2.5×10^4 cells cm^{-2}). Three wells each will be used to determine total binding at each of four [^{3}H]ouabain concentrations (e.g., 10^{-5}, 10^{-6}, 10^{-7}, and 10^{-8} M). Likewise, three wells each will be used to determine nonspecific binding at each [^{3}H]ouabain concentration.

Determination of Cell Number. Parallel plates should be seeded with cells and grown to confluence exactly as their experimental counterparts. They will be washed and incubated in TBS exactly as the experimental wells except that ouabain will be omitted from the incubation. At the end of the binding experiment, these wells should be suspended by trypsinization to get an estimate of cell number per well.

[29] P. Svoboda, J. Svartengren, M. Snochowski, J. Houstek, and B. Canon, *Eur. J. Biochem.* **203** (1979).

TBS or TBS/2 mM Ouabain Preincubation. After the cells have reached confluence (about 3 days for CV-1 cells grown at 37° in 5% CO_2), aspirate the medium off the monolayers. Wash each monolayer twice with 4 ml room-temperature TBS, being careful to add the TBS at the side of the well so as not to disrupt the monolayers. After washing add 0.5 ml of TBS to each well which will be used to determine total ouabain binding and to each well which will be used to determine cell number. To monolayers which will be used to determine nonspecific binding add 0.5 ml of TBS/2 mM ouabain. Place the dishes in a 37° incubator for 15 min. This step is designed to saturate ouabain receptors on the cells which will be used for nonspecific binding.

[^{3}H]Ouabain Binding. After the preincubation remove the solution from each well by aspiration. Add 0.5 ml of a given [^{3}H]ouabain solution to each well which will be used for total binding determination. To wells which will be used for cell counting, add 0.5 ml TBS. To wells designated for nonspecific binding add 0.5 ml of [^{3}H]ouabain in TBS/2 mM ouabain. Incubate the plates at 37° for 15 min. At the end of the incubation aspirate the solution from each well. Wash gently with ice-cold TBS (two 4-ml washes). Solubilize the cells used for determination of total and nonspecific [^{3}H]ouabain binding by placing in each well 1 ml of 0.15 N NaOH, 1% (w/v) SDS, and shaking overnight. Remove the solubilized contents into 20-ml scintillation vials, add aqueous-compatible scintillant, and count. Add 1 ml of trypsin/EDTA solution to each monolayer of cells which will be used to determine cell number. After the trypsin has removed the cells from the dish, suspend the cells in each well and count on a hemocytometer.

Comments

Cell Number. If care is taken to seed each well of the experiment with the same number of cells, then the described method will reliably give the cell number per well and [^{3}H]ouabain binding per cell can be fairly precisely determined. Alternatively, protein concentration can be determined on a portion of the cell lysate from each experimental monolayer. If protein is to be determined, solubilization conditions should be chosen so that none of the constituents of the solubilization solution interferes with the chosen method of protein determination.

Ouabain Access to the Cell Surface. We assume that in cells which do not form tight junctions [^{3}H]ouabain has unrestricted access to the bottom and the top of a monolayer.[30]

---

[30] B. Kennedy and J. Lever, *J. Cell. Physiol.* **121,** 51 (1984).

Measurement of the Number of Na^+ Pumps by
Phosphorylation of the Active Site with Inorganic
Phosphate: "Backdoor Phosphorylation." [31-33]

This method for measuring the Na^+ pump can be easily applied and is particularly useful when the affinity for ouabain of a particular cell is known to be quite low. This method relies on the physiological response of the Na^+,K^+-ATPase to ouabain rather than the number of ouabain molecules attached to the Na^+ pump. Therefore this method will not overestimate the number of Na^+ pumps because of binding to nonfunctional Na^+ pumps or due to trapped or sequestered ouabain.

Materials

$^{32}PO_4^{3-}$; assay buffer (5 mM $MgCl_2$, 100 μM PO_4^{3-}, 200 mM imidazole–HCl, pH 7.4); plasma membranes prepared from the isolated cells; Millex-GS 0.22-μm filter (Millipore, Bedford, MA); bovine serum albumin (20 mg/ml); 25% trichloroacetic acid (TCA) solution; 5% TCA, 0.1 M H_3PO_4; 0.15 M KH_2PO_4, pH 2.0; gel sample buffer (variable, depending on gel system)

Procedure. Membranes (60–100 μg from isolated cells prepared by one of several plasma membrane isolation procedures)[34-36] are suspended in 50 μl of 5 mM $MgCl_2$, 100 μM PO_4^{3-} and 200 mM imidazole–HCl, pH 7.4 buffer with or without ouabain of various concentrations. The amount of Na^+,K^+-ATPase in the membranes isolated from cells can vary widely. Therefore, it may be necessary to alter this method by either increasing the amount of membrane protein or increasing the amount of $^{32}PO_4^{3-}$ used in the assay. The phosphate in this buffer should be added as H_3PO_4 to avoid adding either K^+ or Na^+ to this assay. If present, Na^+ would drive the reaction into the Na^+-dependent conformation of the enzyme, which is unable to support phosphorylation by inorganic phosphate. Similarly, K^+ induces a conformational change in the ATPase, destabilizing the aspartyl phosphate intermediate. Therefore, both of these ions should be excluded from the assay. $^{32}PO_4^{3-}$ must be filtered through a Millex or comparable filter to remove radioactive contaminations probably in the form of polyphosphate. $^{32}PO_4^{3-}$ (10–100 μCi) is added to each sample tube. After 40 min at room temperature the reaction is terminated by the

[31] R. L. Post, G. Toda, and F. N. Rogers, *J. Biol. Chem.* **250,** 691 (1975).
[32] M. D. Resh, *J. Biol. Chem.* **257,** 11946 (1982).
[33] L. H. English and L. C. Cantley, *J. Biol. Chem.* **261,** 1170 (1986).
[34] D. M. Brunette and J. E. Till, *J. Membr. Biol.* **55,** 215 (1971).
[35] B. S. Jacobson, J. Cronin, and D. Branton, *Biochim. Biophys. Acta* **506,** 81 (1878).
[36] G. K. Chacko, F. V. Barnola, R. Villegas, and D. E. Goldman, *Biochim. Biophys. Acta* **373,** 308 (1976).

addition of 100 μl 25% TCA and the precipitate is pelleted out of the suspension. Under these conditions only the Na⁺,K⁺-ATPase is able to accept inorganic phosphate at the active site. Ouabain will stabilize this phosphoenzyme to different degrees, depending on its concentration and the affinity of the Na⁺ pump for the drug. The amount of phosphorylated Na⁺ pump can be determined by either assaying the precipitated protein by solubilizing the pellet in 10% sodium dodecyl phosphate followed by liquid scintillation counting or by visualizing the 100-kDa phosphorylated catalytic subunit of the Na⁺,K⁺-ATPase on an autoradiograph of an SDS-polyacrylamide gel run under acid conditions according to the method of Avruch and Fairbanks.[37] Alternatively, the acid gel system of Amory *et al.*[38] can be used. Whatever the gel system, it is important to keep the enzyme under acid conditions. The aspartyl phosphate bond is quite unstable at high pH. If autoradiographic visualization is required, centrifuge the suspension of precipitated protein for 2 min in a tabletop microfuge (10,000 g). Wash the pellet 3× with 1.0 ml 5% TCA, 0.1 M H₃PO₄ at 4°, and rapidly rinse once with 0.3 ml 0.15 M KH₂PO₄, pH 2.0. Finally, resuspend the pellet in 200 μl 0.25 M sucrose and the sample gel buffer for the acid gel system of choice.

Measurement of ⁸⁶Rb⁺ Uptake into Isolated Cells

Method 1: Measurement of ⁸⁶Rb⁺ Uptake into Isolated Cells[39,40]

Materials

⁸⁶Rb⁺ (New England Nuclear, MA); microfuge; appropriate physiological saline solution (the exact formula will depend on the cells under investigation); ouabain (Sigma, St. Louis, MO); dinonyl phthalate (ICN); silicone oil (Aldrich); 0.25% Trypan Blue solution (Gibco, Grand Island, NY); and 1.5-ml disposable test tubes

Procedure. If cells are not growing in suspension, cells are suspended in a small volume of growth medium supplemented with 25 mM HEPES. Other buffers may be more desirable, depending on the cells under investigation. If the cells are loosely attached, then they can be suspended by agitating the tissue culture flasks gently or gently rapping the bottom of a plate of cells. Many cells once removed from the host tissue attach to the

[37] J. Avruch and G. Fairbanks, *Proc. Natl. Acad. Sci. U.S.A.* **69**, 1216 (1972).
[38] A. Amory, F. Foury, and A. Goffeau, *J. Biol. Chem.* **255**, 9353 (1980).
[39] L. H. English, I. G. Macara, and L. C. Cantley, *J. Cell Biol.* **87**, 1299 (1983).
[40] R. L. Smith, I. G. Macara, R. Levenson, D. Housman, and L. Cantley, *J. Biol. Chem.* **257**, 773 (1982).

surface of the flask or dish and for this procedure must be brought into suspension by gently scraping the dish or flask. For an alternative to this suspension method, see method 3. The final cell suspension should be approximately 5×10^6 cells/ml. The cell growth medium supplemented with additional buffer should stabilize the pH of the assay medium. Alternatively, the cells can be suspended in a buffered physiological saline solution the composition of which is derived from the growth medium. A fraction of the cell suspension should be counted for cell viability, which can be determined quickly by Trypan Blue exclusion. Typically, Trypan Blue is added to a one-tenth dilution of the cell suspension and the number of Trypan-excluding cells is determined by hemocytometer.

Suspended cells are apportioned into separate tubes for each treatment. The simplest experiment to measure Na^+ pump requires a control tube and one for ouabain treatment. The volume of this suspension depends on the number of points needed in a timed assay of $^{86}Rb^+$ uptake. For most mammalian fibroblasts the uptake of $^{86}Rb^+$ reaches equilibrium within 40 min. Red blood cells, however, have a very slow rate of $^{86}Rb^+$ uptake and require several hours to reach equilibrium. Trial and error may therefore be necessary to determine the best time course for the particular cells under consideration. We typically use five data points to determine an uptake curve. Each data point requires 0.1–0.2 ml of suspended cells at a concentration of 10^5–10^6 cells/0.2 ml. Each curve will therefore require about 1.0 ml of suspended cells. Easy removal of the last aliquot is enhanced by having an additional 0.1 ml of volume in the suspension. Ouabain is then added to the treatment tube to a final concentration of 1 mM from a concentrated solution of 10 mM ouabain in buffered saline solution. An equal volume of saline without ouabain is added to the control tube. Ouabain (1 mM) is used because the EC_{50} for ouabain is typically in the micromolar range for most cells. This excess in ouabain concentration virtually ensures inhibition of ouabain-resistant Na^+ pumps on the cell surface. Ouabain is added to the treatment tube approximately 1 min before initiating the assay. Typically the cells are not preincubated for extended times in ouabain. Prolonged preincubation can result in the induction of alternative K^+-transport systems that can mask the real value of ouabain-inhibited $^{86}Rb^+$ uptake. The experiment is initiated by the addition of 3–5 μCi $^{86}Rb^+$ to each of the cell suspensions. The stock solution of $^{86}Rb^+$ is usually at relatively high specific activity (1–5 mCi/0.1 ml), therefore it is better to dilute the isotope into assay buffer before adding it to the cell suspensions. Also, attention must be given to the pH of the $^{86}Rb^+$ because it is usually prepared as $^{86}RbCl$–HCl and requires that the pH be adjusted before the assay. Typically, $^{86}Rb^+$ (10 μCi) of the stock $^{86}RbCl$ solution is usually added as a small volume (100 μl) of assay buffer and the pH adjusted to neutral prior to initiating the assay. Before initiating the assay with the addition of ^{86}Rb$^+$ to the cell suspensions, prepare the 1.5-ml disposable microfuge tubes. An individual tube is required for each time point in this assay. To the bottom of each tube add 300 μl of a 1 : 1 mixture of silicone oil : dinonyl phthalate. Label each tube according to the treatment and time point and set up tubes in chronological order according to their position in the protocol.

Initiate the assay by adding the isotope to the two treatments approximately 15 sec apart. Because one is adding two aliquots of the same volume to two separate tubes, e.g., two separate treatments, care should be taken to add a reproducible volume (usually between 30 and 50 μl). Thirty seconds before the time of the first time point add 800 μl of ice-cold assay buffer and place it on top of the 300 μl of oil mixture previously placed in the labeled disposable tubes. Repeat this procedure for the ouabain treatment. At the time of the first time point remove 200 μl of the suspended cells and place this into the cold assay buffer in the disposable tube. Repeat this for the treated cell. The cold will slow down and in most cases stop the ^{86}Rb$^+$ uptake. Once the two tubes are ready insert them into the microfuge and spin them for 30 sec. Remove the tubes and invert them allowing the buffer and oil to drain from the pelleted cells. These inverted tubes can usually sit until the time course is completed; however, care must be taken to ensure that the cell pellet remains in the bottom of the tube. Most isolated cells remain clumped in the tip of the tube. After the time course, excise the tips of the tubes and place each in a scintillation vial. Solubilize the pellet in detergent (100 μl of 10% SDS) and assay by liquid scintillation. Red blood cells which do not adhere to each other rapidly begin to slide down the edge of the tube following centrifugation. For red blood cells and other cells that do not adhere in a clump, it has proved useful to quickly freeze the tip of the microfuge tube in a dry ice–acetone bath immediately after removing the tube from the microfuge. The tip is then excised with a razor blade, transferred to a scintillation vial, solubilized in 10% SDS, and counted by liquid scintillation. If this is impractical consider method 4 for cells of this sort.

The values obtained by this procedure give one the counts associated with the pellet of cells. This includes the counts associated with ^{86}Rb$^+$ inside the cell and counts from ^{86}Rb$^+$ associated with the tightly bound water surrounding the cell. This nonspecifically bound ^{86}Rb$^+$ can be assessed and subtracted from the values obtained in the time course. To do so, an aliquot of the same cell suspension used in the assay, approximately 700 μl, is placed on ice and allowed to cool for 15 min. To this suspension add ^{86}Rb$^+$ to the same specific activity used in the assay. Because the cells are cold there should not be any time-dependent transport of $^{86}Rb^+$ under these conditions. Remove 200-μl aliquots of the cold cell suspension and place them on 800 μl ice-cold assay buffer on top of 300 μl dinonyl phthalate/silicone oil. Centrifuge these tubes for 30 sec and invert them to allow the oil to drain. Excise the tips after 10 min, solubilize them in 100 μl of 10% SDS, and count them by liquid scintillation. The value obtained by this procedure is the number of counts associated with the bound water or otherwise nontransported $^{86}Rb^+$ and should be subtracted from the values obtained during the transport experiment. The numbers can then be plotted as either counts per minute versus time, or converted to a more meaningful number, nanomoles of $^{86}Rb^+$-labeled K^+ transported per cell number versus time. In this conversion one uses the concentration of K^+ in the assay buffer and the specific activity of the $^{86}Rb^+$ to estimate the number of moles of $^{86}Rb^+$-labeled K^+ actually transported. For this conversion, the specific activity of $^{86}Rb^+$ must be determined for each of the treatments by liquid scintillation counting of a small aliquot (5 μl) of cell-free assay buffer from each of the treatment tubes. Correction for any difference in isotope specific activity can then be made when calculating the rate of ion transport. This conversion assumes a $1:1$ correspondence between the $^{86}Rb^+$ label and K^+. In instances where this correspondence has been assayed, it has been found to be close to $1:1$. Therefore, while $^{86}Rb^+$ transport may not represent the exact amount of K^+ transported it can give a close approximation and provide for comparisons of the amount of ouabain-sensitive $^{86}Rb^+$ uptake. Of course, if exact numbers are needed, $^{86}Rb^+$ can be replaced with $^{42}K^+$ in this assay or KCl can be completely replaced with RbCl in the assay medium.

Method 2: Measuring $^{86}Rb^+$ Uptake in Attached Cells

Materials

Cells adhered to culture dishes (35-mm culture dishes or microwell plates); aspirator for removing radioactive medium from the culture dish; and $^{86}RbCl$

Procedure. If it is necessary to ensure that the cells under consideration are kept attached during the assay, the $^{86}Rb^+$ uptake can be followed in small dishes or microwells. In this assay a single dish or well is used to assay a single time point in a $^{86}Rb^+$ uptake time course. The growth medium is removed and the assay is initiated by the addition of $^{86}Rb^+$-labeled assay medium to the cells. At the desired intervals, the labeled medium is removed by aspiration and the cells are washed with ice-cold label-free assay buffer which is also quickly removed by aspiration. The cells are then solubilized in detergent (usually 10% SDS) and a sample is counted by liquid scintillation. Another sample of the assay dish or well is saved for a protein assay. Similar to the procedure in method 2, nontransported ^{86}Rb$^+$ should be calculated and subtracted from the timed uptake points. This is done by cooling the dish of cells on ice for 15 min prior to addition of labeled assay buffer. The amount of ^{86}Rb$^+$ associated with the cells should be independent of time at this temperature. The labeled assay medium is then removed and the cells are washed as before, solubilized, and counted by liquid scintillation. This procedure requires careful addition of the washing and assay buffers so as not to disturb the attached cells. This can be done by adding the assay buffer or wash to the side of the dish or well to avoid blowing attached cells off of the dish. Care must also be taken to accurately measure the level of protein in the dish so that transport rates can be calculated, taking into account the best estimate of the number of cells still in the dish at the time of the harvest. Some cells will inevitably float away during this assay. Data can be expressed as counts per minute per milligram of cell protein or as nanomoles of Rb-labeled K$^+$ per milligram of cell protein. Generally activity expressed per cell number or per well is an inaccurate measurement in this assay, although with extreme care reproducible numbers are obtained.

Method 3: Measurement of ^{86}Rb$^+$ Uptake into Cells That Do Not Adhere to Each Other[41,42]

Materials

Dowex 50-X8-100; 6-in. Pasteur pipets; glass wool; rack for holding pipets; pipet bulbs made into receptacles by removing the upper one-fourth of the bulb; 100 mg/ml bovine serum albumin, 10 mM Tris, pH. 7.4; 0.25 M sucrose, 10 mM Tris, pH 7.4 (washing buffer)

Dowex Column Preparation

Dowex is prepared by washing the resin with several volumes of deionized H$_2$O followed by addition of solid Tris base in the ratio of 2 g Tris/1 g Dowex. This mixture is gently stirred so as not to destroy the resin. After 1 hr, the resin is again stirred and the fines removed by decanting. The resin is then washed with deionized H$_2$O in a Büchner funnel until the pH is 7.4. The resin is then packed into columns.

Preparation of Columns

Single strands of glass wool (four or five) are packed in the bottom of 6-in. pipets and the pipets are submerged in 10 mM Tris, pH 7.4. These

[41] O. D. Gasko, A. F. Knowles, H. G. Shertzer, E. M. Seuolinna, and E. Racker, *Anal. Biochem.* **72,** 57 (1976).
[42] P. Harikumar and J. P. Reeves, *J. Biol. Chem.* **258,** 10403 (1983).

columns can then be easily filled with Dowex by lifting individual tubes out of the buffer while adding the resin. Gravity and the flow of buffer through the column will allow for rapid filling. Fill to within 1/2 in. of the top of the column.

Assay Preparation

Prepare the columns for assaying cells by suspending them on a support and affix a receptacle (pipet bulbs with the end cut out) to the top of each column to hold the volume of the wash. Wash each column with 100 μl of 100 mg/ml bovine serum albumin followed by 2 ml 0.25 M sucrose, 10 mM Tris, pH 7.4.

Procedure. Ouabain-inhibitable $^{86}Rb^+$ uptake in cells such as red cells that do not adhere together following centrifugation can be assayed by passing cells over a small cation-exchange column. In this assay the cells are prepared as in method 2, but instead of stopping the transport with ice-cold saline and centrifugation through oil, the transport is stopped by loading the aliquot of cells on top of a cation-exchange column and eluting the cells with isotonic buffered sucrose. Using one column for each data point, after the appropriate interval on a time course, add a 100- to 200-μl volume of cells directly into the resin. Allow 2 sec for the cells to enter the column and add 100–200 μl of ice-cold 0.25 M sucrose, 10 mM Tris, pH 7.4. After this volume has entered the column add an additional 2 ml of this washing buffer to flush out the cells. The intact cells come out of the column with the washing buffer.

Measurement of $^{22}Na^+$ Efflux[43]

Method 1

Under most circumstances the measurement of ouabain-inhibitable $^{22}Na^+$ efflux is not easy or as accurate a measure of the Na^+ pump as $^{86}Rb^+$ uptake in isolated cells. The intracellular concentration of Na^+ is quite low, typically about 5–20 mM, while the extracellular Na^+ is much higher, about 150 mM. Under these conditions observing Na^+ efflux from cells loaded with $^{22}Na^+$ is quite difficult. The cells must be labeled in medium or buffer containing $^{22}Na^+$ and since the labeled Na^+ volume is quite small compared with the extracellular unlabeled Na^+, the background level can obscure the measurement. Second, the amount on nonspecifically bound Na^+ will be quite high.

[43] L. H. English and L. C. Cantley, *J. Cell. Physiol.* **121,** 125–132 (1984).

Materials

Physiological saline; Na$^+$-free loading buffer (physiological saline in which NaCl is replaced with choline chloride or isotonic sucrose); ^{22}NaCl; microfuge; silicone oil/dinonyl phthalate mixture as in method 1 of the third section

Procedure. Typically, cells are loaded with ^{22}Na$^+$ in high Na$^+$-buffered saline. After 1 to 4 hr at 37°, depending on the cell type, the cells are washed in excess choline–Ringer's saline and pelleted by low-speed centrifugation. The cells are then resuspended in Na$^+$-free assay buffer in which NaCl is replaced with choline chloride or isotonic sucrose and apportioned into different tubes with and without ouabain. The rate of ^{22}Na$^+$ efflux is followed by removing aliquots of the suspended cells and centrifuging them through oil as described in method 1 of the third section.

Conclusion

The methods described above have proved effective for the identification and characterization of the Na$^+$ pump in several isolated cells of both vertebrate and invertebrate origin. The authors are aware that other methods exist for measurement of the Na$^+$ pump in isolated cells and certainly new methods are continually being developed; however, the methods described above have been used by the authors and therefore we feel confident in recommending their use.

[43] Measurement of Na$^+$–K$^+$ Pump in Muscle

By RAYMOND A. SJODIN

Like other cell types that maintain a low internal sodium ion concentration and a high internal potassium ion concentration relative to the respective external concentrations of these ions, muscle cells contain a metabolically driven Na$^+$–K$^+$ pump.[1-3] The energy source for the muscle Na$^+$–K$^+$ pump is ATP, as in other cell types.[4] As skeletal muscle fibers

[1] R. B. Dean, *Biol. Symp.* **3,** 331 (1941).
[2] H. B. Steinbach, *J. Biol. Chem.* **133,** 695 (1940).
[3] E. J. Conway and D. Hingerty, *Biochem. J.* **42,** 372 (1948).
[4] M. Dydynska and E. J. Harris, *J. Physiol. (London)* **182,** 92 (1966).

maintain a resting membrane potential with an internal negativity of some -90 mV[5] and a large membrane conductance to potassium ions, the electrical driving force can exert large effects on potassium ion movements that have to be taken into account in measurements of the Na^+–K^+ pumping rate in muscle fibers.

Measurements of Na^+–K^+ pump activity in muscle fibers can be made following several approaches. The ability of muscle fibers to produce a net extrusion of Na^+ against an electrochemical gradient and a net accumulation of K^+ can be determined by following net movements of these cations analytically using the technique of flame photometry.[6,7] The rate of Na^+ pumping in the outward direction can be measured by means of radioactive isotopes, for Na^+ with appropriate corrections of the Na^+ efflux for nonpumped sources of efflux.[8–10] Likewise, the rate of K^+ pumping in the inward direction can be determined using radioactive K^+ as tracers and either applying corrections for nonpumped K^+ influx or blocking the nonpumped K^+ influx with appropriate blocking agents.[11] By combining measurements of Na^+ pumping and K^+ -pumping rates, it is also possible to determine the stoichiometric ratio of the number of Na^+ pumped outward per K^+ pumped inward.[11,12] The foreign cations rubidium and cesium are pumped inward over the K^+-specific part of the Na^+–K^+ pump in muscle fibers and measurements of Rb^+ and Cs^+ influxes have been made to determine Na^+–K^+ pump activity.[13,14] Finally, as the Na^+–K^+ pump in muscle is electrogenic (net electrical charge is transported by the Na^+–K^+ pump), changes in muscle fiber membrane potential have been used to assess Na^+–K^+ pump activity in muscle fibers.[13,15,16] Procedures for all of these techniques for measurement of Na^+–K^+ pump activity in muscle fibers will be summarized. In most cases the methods are described as applied to frog skeletal muscle but they are sufficiently general for the most part to be easily modified for other types of muscle.

[5] R. H. Adrian, *J. Physiol.* (*London*) **133**, 631 (1956).
[6] J. E. Desmedt, *J. Physiol.* (*London*) **121**, 191 (1953).
[7] H. B. Steinbach, *J. Gen. Physiol.* **44**, 1131 (1961).
[8] R. D. Keynes and R. C. Swan, *J. Physiol.* (*London*) **147**, 591 (1959).
[9] R. D. Keynes and R. A. Stenhardt, *J. Physiol.* (*London*) **198**, 581 (1968).
[10] R. A. Sjodin and L. A. Beauge, *J. Gen. Physiol.* **52**, 389 (1968).
[11] R. A. Sjodin and O. Ortiz, *J. Gen. Physiol.* **66**, 269 (1975).
[12] R. A. Sjodin, *J. Gen. Physiol.* **57**, 164 (1971).
[13] R. H. Adrian and C. L. Slayman, *J. Physiol.* (*London*) **184**, 970 (1966).
[14] L. A. Beauge and R. A. Sjodin, *J. Physiol.* (*London*) **194**, 105 (1968).
[15] R. P. Kernan, *Nature* (*London*) **193**, 986 (1962).
[16] A. S. Frumento, *Science* **147**, 1442 (1965).

Net Na$^+$ Extrusion and K$^+$ Accumulation

In this type of experiment muscles are made rich in Na$^+$ content by storage in the cold (2–6°) in a K$^+$-free solution. As external K$^+$ is an activator of the Na$^+$–K$^+$ pump and the pump is very sensitive to temperature, both of these conditions in concert decrease the rate of the Na$^+$–K$^+$ pump to levels at which Na$^+$ are gained by the fibers and an approximately equivalent amount of K$^+$ is lost from the fibers. After the period of Na$^+$ enrichment, such Na$^+$-loaded fibers extrude the gained Na$^+$ and reaccumulate the lost K$^+$ on elevation of the temperature to normal (20° for frog muscle) and restoration of an adequate value of $[K^+]_o$.[2,6,7,13] As the Na$^+$ extrusion that occurs is against an electrochemical gradient, the net changes in cation content occurring during recovery from the Na$^+$-rich state reflect activity of the Na$^+$–K$^+$ pump.

Handling of Muscles. To assure uniform results, it is best to mount all muscles on platinum frames. The procedure for handling of muscles also applies to the radioactive tracer methodology to be described subsequently. Sartorius muscles from the frog are carefully dissected free from the animal and are tied with thread at each end at the tendon. The muscles are then carefully checked under a dissecting microscope for damaged fibers due to dissection or other pathology, usually evidenced by whitish spots or opaque cloudy areas. If any pathology or damage is seen in either of the pair of muscles, both muscles are discarded as experiments are performed on muscle pairs using one of the pair as a control. Satisfactory muscles are then tied to platinum frames. The platinum frames are made by using a 15-cm length of platinum wire stiff enough to support small tensions without bending (18 to 20 gauge). The 5-cm portion at one end is bent to form a C shape with a gap of 3.5 to 4 cm with small loops at the ends of the C to facilitate the typing of threads. In the absence of available platinum wire, glass or Teflon rods can be used as a substitute. It is not wise to attempt to substitute other metals for platinum because of possible toxicity. When tied to the support frames, the sartorius muscles should be under slight tension by stretching to about 1.10 times the resting length. It is also convenient to have the pelvic end of the muscle tied at the end of the frame (bottom) and the insertion end tied to the handle end of the C frame.

Solutions

Standard Ringer's solution for dissection: NaCl, 115 mM; KCl, 2.5 mM; CaCl$_2$, 2 mM; Tris buffer, 1 mM

Sodium-loading solution: NaCl, 115 mM; CaCl$_2$, 2 mM; Tris buffer, 1 mM

Recovery solution: NaCl, 110 mM; KCl, 5 mM; CaCl$_2$, 2 mM; Tris buffer, 1 mM

Analysis-preparation wash: MgCl$_2$, 86 mM; CaCl$_2$, 2 mM; Tris buffer, 1 mM

For all solutions, pH is adjusted to 7.35 by neutralization of the Tris with HCl.

Sodium-Loading Procedure. Pairs of muscles tied to platinum frames are placed in 100- to 500-cm^3 beakers filled with Na$^+$-loading solution and equipped with magnetic stirrers. Prior to being placed in the beakers, the muscles should be passed through two rinses of Na$^+$-loading solution of about 2 min each to remove extracellular K$^+$ from the muscles. The muscles in Na$^+$-loading solution are then placed in a cold room at approximately 4°. The muscles are kept in the stirred vessels at 4° for at least 16 hr to sufficiently elevate internal [Na$^+$].

Recovery Procedure. The muscle pairs are then removed from the cold room. One pair member of each pair of muscles from the same animal is then immediately prepared for analysis. The other pair members are then placed in recovery solution for 1 to 2 hr at 20° and subsequently prepared for analysis.

Preparation for Analysis. Muscles on their frames are removed from their respective solutions and passed through a series of test tubes containing analysis-preparation wash solution, enough to cover the muscles. The total wash period is 20 min in five tubes according to the following time schedule: 1, 2, 2, 5, and 10 min. After removal from the final wash solution, muscles are cut at the ties from the wire frames and carefully blotted on filter paper to remove excess solution. The muscles are then placed in 5-cm^3 platinum crucibles previously weighed (crucibles with lids are preferable). The crucibles containing wet muscles are weighed again and subsequently placed in an oven at 95° for 30 min. The crucibles with dried muscles are then reweighed. The muscles in crucibles are next ashed in a furnace at 550° for 12–15 hr. A drop of HNO$_3$ is added to the ash samples and the ash dissolved in 5 ml of highly pure deionized water.

Cation Analysis. The dissolved ash samples are analyzed for Na$^+$ and K$^+$ by flame photometry, Na$^+$ at 589 and K$^+$ at 768 nm. Cation concentrations can be expressed per unit of wet weight of the muscles or per unit of intracellular water content. Details can be found in work by Sjodin and Beauge.[17]

Interpretation of Data. The data from a typical experiment with five pairs of sartorius muscles are tabulated in Table I. The data on the recov-

---

[17] R. A. Sjodin and L. A. Beauge, *J. Gen. Physiol.* **61**, 222 (1973).

TABLE I

Na$^+$ AND K$^+$ CONCENTRATION CHANGES DURING RECOVERY FROM Na$^+$ ENRICHMENT^a

Muscle	[Na$^+$]$_i$	[K$^+$]$_i$	Δ[Na$^+$]$_i$	Δ[K$^+$]$_i$
Na$^+$-loaded control	41.9 ± 4.4	48.0 ± 4.7	—	—
Two-hour recovery	13.3 ± 1.6	77.3 ± 3.4	−28.6 ± 4.6	+29.3 ± 3.4
Fresh muscles	6.6 ± 1.8	86.8 ± 7.2	—	—

^{a} Data in millimoles per kilogram wet weight.

ery experiment are from Sjodin and Ortiz[11] and the data on fresh muscles just removed from the animal are from Sjodin and Beauge.[17] The data show the elevation in [Na$^+$]$_i$ and the decline in [K$^+$]$_i$ that occurs on storing muscles in a K$^+$-free medium in the cold (Na$^+$-loaded vs fresh muscles) and the Na$^+$ extrusion and K$^+$ accumulation that occurs when the temperature is elevated to 20° and K$^+$ are included in the medium (Na$^+$-loaded muscles vs muscles after a 2-hr recovery). As recovering sartorius muscle fibers have an internal membrane potential of around −90 mV,[14] the Na$^+$ extrusion is clearly against an electrochemical gradient and represents activity of the Na$^+$–K$^+$ pump. Due to the large muscle membrane conductance to K$^+$ and the electrogenic action of the Na$^+$–K$^+$ pump, it is not clear how much K$^+$ reaccumulation is directly by the K$^+$ part of the Na$^+$–K$^+$ pump and how much is movement driven by the extra internal negativity generated by the pump. Other experimental approaches to be discussed must be used to clarify this uncertainty. It is noteworthy that ouabain, a well-known inhibitor of the Na$^+$–K$^+$ pump, prevents Na$^+$ extrusion and K$^+$ accumulation from occurring in a recovery experiment.[18]

Variations and Uses of Recovery Experiments. The kinds of experimental measurements just illustrated can be used in a variety of applications to test the effects of inhibitors, activators, and modulators on the Na$^+$–K$^+$ pump. In addition to the action of ouabain just mentioned, the influence of the inhibitor azide on the Na$^+$–K$^+$ pump in muscle has been determined.[19] Recovery experiments have also been useful in investigating the ability of monovalent cations other than K$^+$ to activate and be transported by the Na$^+$–K$^+$ pump. Rubidium and cesium ions can substitute for potassium ions to activate the Na$^+$–K$^+$ pump.[13,14] Due to the low membrane conductance to these foreign cations, proportionately more of

[18] J. A. Johnson, *Am. J. Physiol.* **187**, 328 (1956).
[19] L. A. Beauge and R. A. Sjodin, *J. Physiol. (London)* **263**, 383 (1976).

their total movement across the membrane is directly via the pump. Other cations studied by this method that have a K^+-like action on the Na^+–K^+ pump are lithium ions[20] and ammonium ions.[21]

Measurements of the Na^+–K^+ Pump Rate with Radioactive Tracers

Measurement of the Na^+ Pump Rate. Sodium efflux from the sartorious muscle of the frog can readily be measured using sodium-24 or -22 ions by simply loading muscles with the radioactive isotope of Na^+ and collecting the radioactivity lost from the muscles to nonradioactive solutions as a function of time.[9,12,22] The problem that arises, however, is that not all of the Na^+ efflux from muscle cells is due to the Na^+ pump. A large fraction of the Na^+ efflux occurring in skeletal muscle fibers is due to a Na^+-for-Na^+ exchange called "exchange diffusion."[8,9,20] There are two techniques available for not including exchange diffusion in the measurement of the Na^+-pumped flux in muscle fibers. One method is to measure the Na^+ efflux into Na^+-free solutions where external Na^+ has been replaced with either lithium or choline.[9,20,22] The other technique is to determine the external K^+-sensitive and strophanthidin (or ouabain)-sensitive component of Na^+ efflux.[12] The exchange diffusion component of Na^+ efflux is insensitive to external K^+ and to strophanthidin or ouabain whereas the Na^+ pump is both stimulated by external K^+ and inhibited by external strophanthidin or ouabain. There is a small component of Na^+ : Na^+ exchange that also occurs via the Na^+ pump but its presence is only evident in K^+-free solutions.[9,12] This component of Na^+ efflux is also inhibited by strophanthidin or ouabain and should be included as Na^+ pump activity whenever it occurs.

Detailed Procedure for Measurement of Na^+-Pumped Efflux. For better resolution, Na^+-enriched muscles should be used in the measurement of pumped Na^+ efflux. The magnitude of the pumped Na^+ efflux becomes amplified in such muscles and K^+ sensitivity is more readily determined. The rate constant for loss of radioactive Na^+ from Na^+-loaded muscles attains a value dependent on the external K^+ concentration and remains at a steady value for long periods of time. Muscles should be dissected and handled as indicated previously. The procedure for Na^+ enrichment should be followed as before. The K^+-free Na^+-enrichment or loading solution should be radioactively labeled with either $^{24}Na^+$ or $^{22}Na^+$ at an activity of approximately 50 μCi for each 10 ml of solution. Muscles

[20] L. A. Beauge and R. A. Sjodin, *J. Gen. Physiol.* **52,** 408 (1968).

[21] L. A. Beauge and O. Ortiz, *J. Exp. Zool.* **174,** 309 (1970).

[22] L. J. Mullins and A. S. Frumento, *J. Gen. Physiol.* **46,** 629 (1963).

should remain in the cold Na$^+$-loading solution overnight. During this period, their Na$^+$ content becomes elevated to desirable levels and they become uniformly labeled with radioactive Na$^+$ ions.

Prior to the measurement of Na$^+$ efflux muscles are warmed to room temperature in the K$^+$-free solution containing radioactive Na$^+$. A rack of test tubes is prepared which contain the solutions into which Na$^+$ efflux is to be measured. The tubes are usually and conveniently the same tubes used for scintillation counting wells. The platinum support wires are bent so that they fit into these tubes and so that about 5 ml of solution in each tube adequately covers each muscle. It is a convenient time scale to pass each muscle into each tube for a 5-min period. The first 40 min (eight tubes) of efflux should be into a K$^+$-free medium. This solution can have the same composition as sodium-loading solution. As it is often desired to change the external Na$^+$ concentration during efflux, any fraction of the Na$^+$ in the efflux solutions can be replaced with an osmotic equivalent of either lithium, Tris, or choline. Also, it is often desired to change the external K$^+$ concentration during efflux measurement. The standard Ringer's solution can be used when $[K^+]_o = 2.5$ mM. When it is desired to elevate the value of $[K^+]_o$ above this value, the value of $[Na^+]_o$ is decreased by the amount the K$^+$ concentration exceeds a value of 2.5 mM. By such maneuvers, any desired value of either $[K^+]$ or $[Na^+]$ in the efflux solution can readily be achieved.

Muscles are removed from the ^{22}Na$^+$ or ^{24}Na$^+$ labeling solution and rinsed for a 10-sec interval in a nonradioactive solution having otherwise the same composition as the Na$^+$-loading solution. This rinse solution is discarded. Muscles are then transferred sequentially for 5-min periods through the entire series of efflux tubes for the desired efflux program. The first eight periods should be in a K$^+$-free solution. The next series of tubes should contain K$^+$ at the desired concentration. Values of $[K^+]_o$ from 5 to 10 mM give good and adequate stimulation of the Na$^+$ pump rate. Usually a series of six tubes (30 min of efflux) provides an adequate period of time to measure the rate constant for radioactive Na$^+$ loss in any given solution composition. At the termination of efflux measurement, each muscle should follow the procedure detailed previously and entitled Preparation for Analysis. An aliquot of the dissolved ash solution for each muscle should be taken for assay of radioactivity remaining in the muscle at the termination of efflux. The remaining ash solution can be used for cation analysis by flame photometry.

To obtain rate constants for radioactive sodium loss from muscles, all radioactivity collected from each muscle is assayed in a scintillation counter and is appropriately background and decay corrected. All counts, beginning with the final radioactivity remaining in the muscle, are then serially back-added to obtain the amount of radioactivity in the muscle at each time interval during efflux measurement. The radioactivity counts in each muscle are then plotted versus time semilogarithmically so that rate constants can be obtained from the slopes of linear portions of the plots. The results of the procedures just described are illustrated in Fig. 1 for data from frog sartorius muscle. The data are for Na^+ efflux in solutions with $[Na^+]_o = 110$ mM throughout. Often a small change in solution osmotic pressure when KCl is added to the solution is tolerated so that the value of $[Na^+]_o$ does not change during the measurement. As the rate of

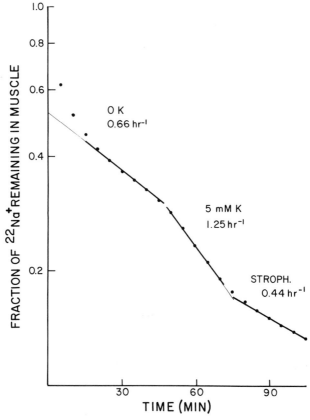

FIG. 1. The rate of loss of $^{22}Na^+$ from sartorius muscles enriched with sodium is plotted semilogarithmically versus the time in contact with different external solutions. The external Na^+ concentration is 110 mM in all solutions used. The external potassium ion concentration is indicated for the different phases of efflux. The rate constant for each phase is also indicated. The final phase of efflux is in the presence of 5 mM K^+ and 10^{-5} M strophanthidin (STROPH.).

operation of the sodium pump is sensitive to the value of $[Na^+]_o$, this is an advantage that offsets the small change in osmotic pressure. The rate constants for $^{22}Na^+$ loss from the muscle are indicated in Fig. 1 for the different phases of Na^+ efflux. The rate constant increases by 0.59 hr^{-1} units when 5 mM K$^+$ is added to the medium. This increment in Na^+ rate of loss is taken as the Na^+ pump activity for these conditions. The addition of the Na^+ pump-inhibitor strophanthidin decreases the rate constant for loss of radioactive ions to 0.44 hr^{-1}, which is close to the K$^+$-free value. As rate constants in whole-muscle experiments tend to decrease with time due to greater loss from the faster equilibrating fibers, the final rate constant cannot be said to differ statistically from the initial rate constant. The experiment illustrated by Fig. 1 thus shows activation of the Na^+ pump by external K$^+$ and inhibition of the pump by strophanthidin.

The rate constant is a convenient measure of the rate of Na^+ turnover by the pump. The rate of Na^+ extrusion by the pump, however, involves the product of rate constant and internal Na^+ concentration. In experiments where net Na^+ extrusion occurs, the internal Na^+ concentration changes with time. A combination of the net extrusion measurements described previously and the tracer methods just described is required to assess Na^+ extrusion rate. The initial value of $[Na^+]_i$ is known as complete equilibration of tracer with the muscle occurs during overnight storage and the soak solution-specific activity is known. The increment in rate constant occurring when external K$^+$ is added leads to net Na^+ extrusion and the product of (rate constant increase) (initial $[Na^+]_i$) provides a means to correct the initial value of $[Na^+]_i$ to obtain the instantaneous value of $[Na^+]_i$ at any time during efflux. This method gives results that are correct to within experimental errors. For the data in Fig. 1, the initial value of $[Na^+]_i$ is 53 mM. The initial rate of Na^+ extrusion when external K$^+$ is added is thus 0.59 hr^{-1} × 53 mM, or 31.3 mM hr^{-1}. The rate constant difference of 0.59 hr^{-1} is equivalent to $1.64 × 10^{-4}$ sec^{-1}. The value of net Na^+ efflux via the pump can be obtained by multiplying the rate constant in sec^{-1} units by Na^+ concentration and by the average volume-to-surface area ratio for the fibers. For sartorius muscle fibers this ratio is $1.9 × 10^{-3}$ cm. Thus Na^+ net efflux via the pump has an initial value of $(1.64 × 10^{-4}$ sec$^{-1})(1.9 × 10^{-3}$ cm$)[53 × 10^{-6}$ (mol/cm^3)$] = 16.5 ×$ 10^{-12} mol/(cm^2 · sec) or 16.5 pmol/(cm^2 · sec).

The data in Fig. 2 were obtained in a similar manner when all of the external Na^+ in the solutions was replaced by an osmotic equivalent of the buffer Tris neutralized to a pH of 7.35. Thus the value of $[Na^+]_o$ remains nominally zero during all of the efflux period. For this special case there can be no Na^+ influx and the instantaneous value of $[Na^+]_i$ is

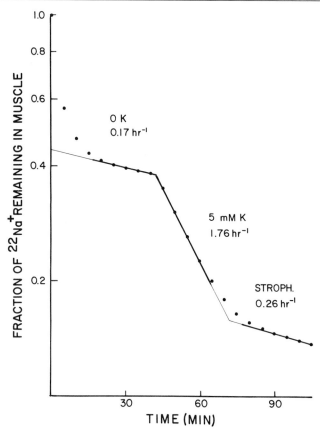

FIG. 2. The rate of loss of $^{22}Na^+$ from sartorius muscles enriched with sodium is plotted semilogarithmically versus the time in contact with different external solutions. In all of the solutions, all of the Na^+ has been replaced with an osmotic equivalent of Tris buffer neutralized to pH 7.35. As in Fig. 1, the external K^+ concentrations used and the rate constants obtained are indicated. The final phase of efflux occurs in the presence of 5 mM K^+ and 10^{-5} M strophanthidin.

always the initial value times the fraction read from the graph at each time by taking the intercept of the first straight line on the graph with the ordinate as 100% of the initial intrafiber Na^+ concentration. The initial fast fraction of ^{22}Na loss is due to the extracellular radioactive Na^+ and a small fast component of Na^+ bound to the fibers. Such plots are convenient because the value of $[Na^+]_i$ at each instant during efflux is known. The data in Fig. 2 also show that the increment in rate constant when K^+ is added to the medium is greater in the absence than in the presence of external Na^+. This is due to the fact that external Na^+ have an inhibitory

TABLE II

RATE CONSTANTS FOR ^{22}Na$^+$ LOSS FROM
MUSCLES AT DIFFERENT VALUES OF $[K^+]_o$ AND
$[Na^+]_o$

$[K^+]_o$ (mM)	$[Na^+]_o$ (mM)		
	0	60	120
	hr^{-1}		
0	0.25	0.48	0.56
2.5	1.38	0.91	0.71
5	1.60	1.07	0.92
10	2.15	1.57	1.15

effect on the Na$^+$ pump in skeletal muscle fibers.[12] The data in Table II indicate values of the rate constant measured for Na$^+$ efflux for various values of $[K^+]_o$ and $[Na^+]_o$. The data illustrate the activation of the Na$^+$ pump by external K$^+$ ions and inhibition by external Na$^+$.

Procedure for Measurement of K$^+$ Influx via Na$^+$–K$^+$ Pump Pathway. Potassium influx is readily measured in frog sartorius muscle fibers.[23] The difficulty in measuring the K$^+$ influx component due to the Na$^+$–K$^+$ pump is one of resolution. The fraction of K$^+$ influx that is via the Na$^+$–K$^+$ pump is small in normal muscle fibers. The fraction of K$^+$ influx due to the pump becomes increased in sodium-loaded muscle fibers.[10] The reason for this behavior is that most of the K$^+$ influx in normal muscle fibers is electrodiffusive in nature and occurs through specific K$^+$ channels. By increasing the internal Na$^+$ concentration, the Na$^+$–K$^+$ interchange via the pump becomes proportionally increased with respect to the electrodiffusive K$^+$ influx. The Na$^+$–K$^+$ pump in muscle fibers is electrogenic, however, meaning that its operation has a hyperpolarizing effect on the muscle fiber membrane potential.[15,16] As increased rate of operation of the Na$^+$–K$^+$ pump leads to an increased internal electrical negativity within the fibers, the electrodiffusive K$^+$ influx through K$^+$ channels becomes increased whenever the Na$^+$–K$^+$ pump is caused to operate at a greater rate. Thus the strophanthidin- or ouabain-sensitive K$^+$ influx increases in Na$^+$-loaded fibers but not all of the increase is via a direct coupling to the Na$^+$ pump. Thus ouabain-sensitive K$^+$ influx in muscle cells is not synonymous with pumped K$^+$ influx. The same is true for any cell with an electrogenic Na$^+$–K$^+$ pump and an appreciable membrane conductance to K$^+$. This fact is often overlooked and the practice of defining pumped K$^+$ influx as the ouabain-sensitive K$^+$ influx component is widespread. It

[23] R. A. Sjodin and E. G. Henderson, *J. Gen. Physiol.* **47**, 605 (1964).

should be recognized that this practice is erroneous and can lead to an overestimate of the K^+ influx directly coupled to the Na^+-K^+ pump. Though corrections for the enhanced electrodiffusive K^+ influx due to the pump can be applied,[11,13] the procedures are beyond the scope of those presented here. The only direct method for an unambiguous measurement of the K^+ influx directly coupled to the pump is to block the K^+ channels with K^+-blocking agents. A method has been described in which external barium ions are used to block the K^+ conductance.[11] Almost all of the residual K^+ influx is directly coupled to the Na^+-K^+ pump under these conditions. This method is now outlined.

Freshly dissected frog sartorius muscles are handled as in the previously described procedures. When barium ions are used to block the K^+ conductance it is not necessary to load muscles with extra sodium to measure pumped K^+ influx accurately. A radioactively labeled $^{42}K^+$ Ringer's solution is prepared by adding an aliquot of a stock solution of ^{42}KCl neutralized to a pH of 7.35. The stock solution contains approximately 1 mCi of $^{42}K^+$/ml of solution. A 1-mCi shipment of $^{42}K^+$ is enough for 2 days of experimentation given the decay rate of $^{42}K^+$. The concentration of $^{42}K^+$ in the uptake solution should be about 2 μCi/ml of Ringer's solution. Uptake is conveniently performed in 30-ml Erlenmyer flasks. About 50λ of the fresh $^{42}K^+$ stock solution suffices to label about 30 ml of Ringer's solution. Recovery Ringer's solution (see solutions previously described) should be used with the addition of 5 mM $BaCl_2$. Two radioactively labeled batches of the barium Ringer's solution should be prepared, one containing 10^{-5} M ouabain and one without the ouabain. Uptake of $^{42}K^+$ by the muscles should be measured in a water bath at 20° if room temperature is not adequately controlled. A rinse flask containing 5 mM Ba^{2+} recovery solution without $^{42}K^+$ should be prepared. Counting of $^{42}K^+$ in the muscles is performed by placing the muscles mounted on the platinum frames into plastic counting tubes which fit the well of a gamma counter.

It is convenient to run a pair of sartorius muscles from the same animal simultaneously, one in the presence and one in the absence of ouabain. Muscles are placed in the $^{42}K^+$-labeled Ringer's solution for 10-min intervals for the first hour of uptake. After the first hour of $^{42}K^+$ uptake, 30-min uptake intervals suffice. After each time interval of uptake, muscles are lowered via the platinum frames into the unlabeled rinse solution for 10 sec. The bottom of the frame is then blotted on a clean piece of filter paper and the muscle then placed in the counting tube. If the muscle is tight enough on the frame and is well centered, the muscle will not touch the sides of the counting tube and will not become damaged by the procedure. A 30-sec to 1-min counting period should provide adequate accuracy of counts. In a 10-min uptake interval, a few thousand counts per minute of

^{42}K$^+$ enter the muscle. After counting, the muscles are returned to their respective solutions for the next uptake interval.

At the end of the desired total period of uptake, muscles are prepared for analysis (see previously described procedure) after the final radioactive count determination. An aliquot of the final ash solution used for cation analysis is taken for radioassay of the counts of ^{42}K$^+$ in the muscle at the termination of the uptake period. The final counts in the muscle and an aliquot of the ^{42}K$^+$-labeled uptake solution are radioassayed in the well of a gamma counter under conditions of identical geometry and self-absorption, usually 5 ml of total solution in the counting tubes. In this way, the specific activity of ^{42}K$^+$ in the uptake solution (counts per minute per micromole ^{42}K$^+$) can be determined. Dividing the final total counts per minute in the muscle by the specific activity of ^{42}K$^+$ in the uptake solution under the conditions of identical counting efficiency gives the micromole amount of K$^+$ that entered the muscle during the entire uptake period. The counts per minute of ^{42}K$^+$ uptake obtained in the whole-muscle geometry during the uptake intervals can then be used to obtain the kinetics of the ^{42}K$^+$ entry. The initial counts thus provide the kinetics while the final counting procedures provide the calibration of the method. The typical results of the entire procedure are illustrated in Fig. 3, where counts per minute of ^{42}K$^+$ entry are plotted versus time for a pair of sartorius muscles from the same animal. The straight lines through the data points intersect at the ordinate axis at a point corresponding to the ^{42}K$^+$ contained in the extracellular space which equilibrates rapidly. The number of micromoles of K$^+$ represented by the final uptake point is accurately known so that the uptake rates for K$^+$ can be obtained from the slopes of the uptake lines. For this purpose, the zero for the uptake lines is shifted to the intersection point of the lines with the ordinate axis. This effectively corrects the total counts in the muscle for the extracellular counts which are not desired in the computation of uptake rate. Applying this procedure to the data in Fig. 3 gives values of 3.74 μmol/(g · hr) for the control value of K$^+$ influx and 0.86 μmol/(g · hr) for the uptake rate in the presence of ouabain. The electrodiffusive K$^+$ influx can be regarded as effectively blocked by barium ions so that the difference in the two rates, the ouabain-sensitive K$^+$ influx, can be regarded as representing the rate of K$^+$-inward pumping. For the data in Fig. 3, the ouabain-sensitive K$^+$ uptake rate or rate of K$^+$ pumping is 2.88 μmol/(g · hr) where g represents gram wet weight of whole muscle.

Stoichiometric Ratio of the Na$^+$–K$^+$ Pump in Muscle Fibers

Measurement by Radioactive Tracer Methods. The stoichiometric ratio of the Na$^+$–K$^+$ pump is the ratio of the number of Na$^+$ transported by

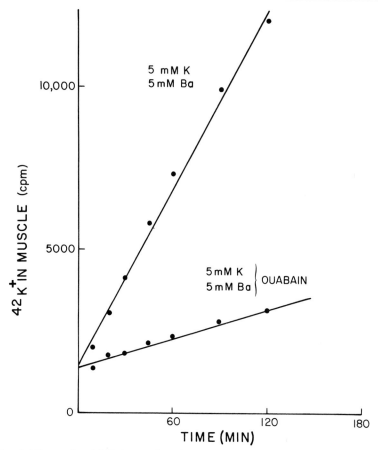

Fig. 3. The uptake of $^{42}K^+$ by a pair of fresh sartorius muscles from the same animal is plotted versus time. The Ringer's solution used contained 110 mM Na$^+$, 5 mM K$^+$, and 5 mM Ba^{2+} for both muscles. The uptake for one of the muscles took place in the presence of 10^{-5} M ouabain (bottom line).

the pump to the number of K$^+$ transported by the pump. The rate of transport of both cations by the pump can be measured by the radioactive tracer methods previously described. Thus no new methods need be described. There are two precautions to be taken in determining the stoichiometric ratio of the Na$^+$–K$^+$ pump. As both Na$^+$- and K$^+$-pumped fluxes are very dependent on the value of [Na$^+$]$_i$, the stoichiometric ratio must be obtained for some definite value of [Na$^+$]$_i$. Also, care must be taken not to include any K$^+$ influx that is due to K$^+$ being pulled into the fibers by the additional internal negativity created by electrogenic action of the Na$^+$–K$^+$ pump. This means either correcting the total K$^+$ influx for the electrodiffusive component or blocking this component with a K$^+$ channel-blocking agent as described for the case where Ba^{2+} are used.

An example calculation will be made for fresh muscles with a normal Na$^+$ content. Using Ba^{2+}, the pumped K$^+$ influx from tracer methods is 2.88 μmol/(g·hr) from the data in Fig. 3. If the same method for measuring the rate constant for ^{22}Na$^+$ loss from muscles that was described earlier is applied to fresh muscles not previously loaded with additional Na$^+$, rate constants can be obtained in a K$^+$-free medium, in a 5 mM K$^+$ solution, and in a 5 mM K$^+$ solution with added strophanthidin or ouabain. Such measurements yield a K$^+$-sensitive and ouabain-sensitive component of the rate constant equal to 0.47 hr^{-1}.[10] If this value of rate constant increment is applied to the normal Na$^+$ content of fresh muscle (6.55 μmol/g),[17] the rate of Na$^+$ pumping in fresh muscle is 3.08 μmol/ (g·hr). For fresh muscle, the stoichiometric ratio of the Na$^+$–K$^+$ pump is thus 3.08/2.88 = 1.1 to the nearest tenth. Radioactive tracer data thus predict a small degree of electrogenic action of the Na$^+$–K$^+$ pump in fresh muscle fibers, as is observed.

When the sodium contents of muscles become elevated, as in the example for Fig. 1, the K$^+$- and ouabain-sensitive component of the rate constant for radioactive Na$^+$ loss from muscles increases in value (0.59 hr^{-1} in the data for Fig. 1). The average rate constant increment measured for a large number of muscles is 0.75 hr^{-1}.[10,12] To obtain the stoichiometric ratio for such muscles requires measurement of the ouabain-sensitive K$^+$ influx in Na$^+$-loaded muscles in the presence of barium ions.[11] For an example calculation, the ouabain-sensitive K$^+$ influx in a barium solution is 10 μmol/(g·hr) at [Na$^+$]$_i$ = 20 μmol/g muscle. To get the comparable pumped Na$^+$ efflux, one multiplies the Na$^+$ content of 20 μmol/g by the Na$^+$ rate constant increment of 0.75 hr^{-1} to get a rate of Na$^+$ pumping of 15 μmol/(g·hr). For [Na$^+$]$_i$ = 20 μmol/g, the stoichiometric ratio of the Na$^+$–K$^+$ pump is thus equal to 15/10 = 1.5. Similar measurements can be made at any value of [Na$^+$]$_i$ by simply knowing the appropriate values of the rate constants for both Na$^+$ loss and K$^+$ uptake. At elevated values of the internal Na$^+$ concentration, radioactive tracer methods predict greater electrogenic action of the Na$^+$–K$^+$ pump as is observed.

Measurement by Analytical Methods and Use of Foreign Cations. In the section headed Net Na$^+$ Extrusion and K$^+$ Accumulation, methods were described in which Na$^+$–K$^+$ pump activity could be determined by measuring changes in Na$^+$ and K$^+$ content of muscles during recovery from Na$^+$ enrichment. The Na$^+$ extrusion and K$^+$ accumulation that occur over an interval of time are determined by flame photometry. If all of the Na$^+$ extrusion and K$^+$ accumulation that occurs is directly via the

Na^+-K^+ pump route, such measurements can also provide information about pump stoichiometry. The problem encountered in Na^+-loaded muscles is that electrogenic action of the Na^+-K^+ pump normally creates additional electrical negativity inside the fibers which pulls K^+ ions into the fibers by movement through K^+ channels. In this case, the net changes in cation contents do not provide direct information about pump stoichiometry. When barium ions are used to block K^+ channels as in the radioactive isotopic measurements, changes in net cation contents for Na^+ and K^+ do provide information about pump stoichiometry. Such measurements have been made and indicate a stoichiometry of 1.1 Na^+ transported per K^+ in the presence of barium ions.[11]

The stoichiometry of the Na^+-K^+ pump in muscle can thus vary with the particular conditions under which the measurement was made. Foreign cations that are not present under physiological conditions can also activate the Na^+-K^+ pump in a manner similar to the activation by external K^+. In such cases the foreign cation is said to exert a K^+-like action on the pump. The Na^+-K^+ pump in such circumstances is transformed to a Na^+-M pump, where M represents the foreign cation. If the membrane permeability to the foreign cation is low, almost all of the entry of that cation into the muscle fibers will occur via the pump. Net changes in cation content of the muscles measured by the previously described methods can, in such cases, provide estimates of pump stoichiometry. Some foreign cations to which the muscle fiber membrane has a low permeability are Li^+, Cs^+, and Rb^+. All of these cations have been found to activate the Na^+-K^+ pump in muscle fibers in a K^+-like fashion. Since muscle fiber membrane permeability to these cations is low, especially for Li^+ and Cs^+ ions, pump stoichiometry can be estimated from the net cation movements produced in Na^+-enriched muscles by the presence of the foreign cation.[13,14,20,24] All of the stoichiometries of number of Na^+ extruded per foreign cation pumped inwardly are close to 1.

Measurement of Na^+-K^+ Pump Activity in Muscle by Electrical Methods

The ability of the Na^+-K^+ pump in muscle to contribute to the resting membrane potential of fibers has already been alluded to in previous sections. This property of the Na^+-K^+ pump is termed electrogenic, which means that, due to the observed stoichiometry of the pump, Na^+ and K^+ are transported by the pump at unequal rates to create an electrical current across the membrane. The electrical current is in the outward direction across the membrane and is carried by the Na^+ moving out-

[24] L. A. Beauge and O. Ortiz, *J. Physiol. (London)* **226,** 675 (1972).

wardly uncompensated by K$^+$ moving inwardly via the pump. The pump current can be measured under voltage clamp conditions and so be used as a measure of Na$^+$–K$^+$ pump activity.[25] A knowledge of pump current and pump stoichiometry permits calculation of both Na$^+$ and K$^+$ transport rates. In fibers that are not voltage clamped, the pump current does not appear as such since it is normally balanced by the sum of all the electrodiffusive net ionic fluxes across the membrane. Under these conditions, however, the pump can contribute to the membrane potential and the pump contribution can be measured in various ways.[15,16]

Following the method of Frumento,[16] sodium-enriched muscles are prepared by overnight immersion in cold K$^+$-free Ringer's solution as described in a previous section. After the period of Na$^+$ enrichment, muscles are transferred to standard Ringer's solution (see Solutions in a previous section) and kept at a temperature of 3°. Membrane potentials are then measured with microelectrodes filled with 3 M KCl,[26,27] using an amplifier with high input impedance and other standard bioelectric recording equipment.[28,29] It is important that the bath in which the muscle is placed for measurement of membrane potential be capable of permitting the ambient temperature of the muscle fibers to be rapidly changed from 3 to 25°C. It is usually sufficient to measure the membrane potentials of from 6 to 10 surface fibers of the sartorius muscle and to average these values for each determination. After measuring membrane potentials at 3°, the temperature is rapidly changed to 25° and membrane potentials are again measured at intervals of 3 to 4 min for a total period of about 25 min. It is convenient to perform the measurements on paired sartorius muscles from the same animal. One of the muscles of the pair is not warmed to 25° and is prepared for analysis (see previous sections) immediately after the membrane potential determination at 3°. The analysis of this control muscle provides initial values for the Na$^+$ and K$^+$ contents. After the membrane potentials are determined for the warmed muscle fibers, the warmed muscle is also prepared for analysis. Cation analyses for this muscle provide the final values for Na$^+$ and K$^+$ content.

Analysis of Membrane Potential Data. The main finding in the experiment just described is that the muscle fiber membrane hyperpolarizes (fiber becomes more negative inside) when the fibers are warmed from 3 to 25°. The hyperpolarization begins immediately after warming, rises to a maximum in about 5 min, and then declines to somewhat lower values. The peak hyperpolarization observed is 10 to 15 mV on the average and

[25] P. DeWeer and R. F. Rakowski, *Nature (London)* **309**, 450 (1984).
[26] J. Graham and R. W. Gerard, *J. Cell. Comp. Physiol.* **29**, 99 (1946).
[27] G. Ling and R. W. Gerard, *J. Cell. Comp. Physiol.* **34**, 382 (1949).
[28] R. H. Adrian, *J. Physiol. (London)* **133**, 631C (1956).
[29] R. H. Adrian, *J. Physiol. (London)* **151**, 154 (1960).

values of over 40 mV have been measured. The theoretical interpretation of the observations is that the pump works at a negligibly slow rate at 3° while warming to 25° rapidly brings the pump to full activity so that the electrogenic effect on the fiber membrane potential is observed. The analytical measurements of Na^+ and K^+ contents in these experiments permit the calculation of Na^+ and K^+ concentrations and Nernst or equilibrium potentials for the cations. The theory for the muscle fiber membrane potential states that it may not become more negative than the equilibrium potential for K^+, E_K, unless the Na^+–K^+ pump is included in the theory for the membrane potential and the pump produces a net extrusion of Na^+. In the experiments just described, the membrane potential becomes more negative than E_K and a net extrusion of Na^+ occurs, supporting the idea that the Na^+–K^+ pump produces the observed hyperpolarization. In the theory for the resting membrane potential, the potential is a direct function of the absolute temperature. Even when corrected for the change in absolute temperature occurring when the bath temperature is changed from 3 to 25°, a pump-dependent hyperpolarization of the magnitude described takes place.

The electrical data obtained in these experiments can be combined with the data obtained on net Na^+ and K^+ content changes measured during recovery from the Na^+-enriched state. If the experiments are performed as described, both sets of data will be available. When the corrected value of the electrical potential change is plotted against the amount of Na^+ extruded, an excellent correlation is obtained.[16] Muscles that extrude the most Na^+ have the highest hyperpolarizations.

Ouabain-Sensitive Membrane Hyperpolarization. A variation of the electrical measurements just described is to utilize the fact that the Na^+–K^+ pump is inhibited by ouabain. The experiments are performed as in the previously described methodology except that all membrane potentials are measured at room temperature. One set of measurements is made in the absence of ouabain and one set is made in the presence of ouabain, usually on a paired muscle from the same animal. The membrane potentials during Na^+ extrusion in the absence of ouabain are more negative than in the muscles treated with ouabain where no Na^+ extrusion occurs.[13,14] The ouabain-sensitive part of the membrane potential is taken as a measure of Na^+–K^+ pump activity.

Measurement of Pump Site Density by Ouabain Binding and
    Determination of Na^+–K^+ Pump Turnover Rate

Ouabain binds to the Na^+–K^+-ATPase of the Na^+–K^+ pump at the same concentrations used for inhibition of the pump. The binding is measured by means of the uptake of radioactively labeled cardioactive steroids which inhibit the pump, usually tritiated digoxin or ouabain.[30,31] The binding is one digoxin (or ouabain) molecule per pump enzyme molecule, which permits determination of the number of pump sites from the number of digoxin molecules bound.[30] Usually a binding curve is obtained by measuring binding in the concentration range 10^{-8} to 10^{-5} M digoxin or ouabain. The binding of digoxin or ouabain to the pump is termed specific binding and total binding must be corrected for nonspecific binding. The specific binding requires the presence of ATP and Mg^{2+} and is maximal only in the presence of Na^+ in isolated ATPase studies. Also, nonradioactive digoxin or ouabain molecules compete with radioactive molecules for occupancy of pump sites whereas nonspecific binding is noncompetitive. Therefore, a large excess of unlabeled digoxin or ouabain will inhibit essentially all of the specific binding of labeled molecules, leaving the nonspecific binding largely unaffected. This fact is used to measure the specific ouabain binding in frog sartorius muscles. Though ouabain binds very rapidly to $Na^+–K^+$-ATPase, diffusion delays to the pump sites in whole-muscle tissue render labeling by [³H]ouabain a slow process, requiring about 50 min to assure equilibration.[31] Ringer's solution is labeled with [³H]ouabain and sartorius muscles are placed in contact with this solution for 50 min. Following the period of uptake, each muscle is successively transferred through six washes of 10 min each in tubes containing inhibitor-free Ringer's solution at 0°. During the 1-hr wash period in the cold, all extracellular radioactivity is lost and essentially none of the specifically bound ouabain is lost. After the 0° wash, muscles are digested in 0.2 ml NCS and counted in a toluene mixture with a liquid scintillation counter. This procedure is followed as a function of ouabain concentration. The uptake at concentrations greater than 10^{-6} M is assumed to be due to a linear nonspecific binding which is subtracted from the total binding to obtain the number of ouabain molecules specifically bound to the pump. Application of the method yields a value of 1600 pump sites/μm^2 of membrane[31] using a surface : weight ratio of 0.415 cm^2/mg for the frog sartorius muscle.[8]

Calculation of Na⁺–K⁺ Pump Turnover Rate. As the number of pump sites per unit area and the rate of Na^+ transport per unit area are both quantities that have been measured, the number of Na^+ pumped per second at each single pump site can be calculated. When sartorius muscles are loaded to about 50 mM internal Na^+ concentration and the value of $[K^+]_o$ is elevated to 10 mM, the $Na^+–K^+$ pump creates a Na^+ flux of about

[30] H. Matsui and A. Schwartz, *Biochim. Biophys. Acta* **151,** 655 (1968).
[31] D. Erlij and S. Grinstein, *J. Physiol. (London)* **259,** 13 (1976).

30 pmol/(cm^2 · sec). Converting this flux to number of ions transported per square centimeter per second and dividing by the number of pump sites per square centimeter gives a value of 112 ions per second pumped at each pump site. The maximal rate of operation of the pump is the transport of about 70 pmol of Na$^+$/cm^2 · sec. The maximal rate of turnover of the Na$^+$–K$^+$ pump in muscle is therefore around 260 ions per pump site per second.

[44] Measurement of Transport versus Metabolism in Cultured Cells

By ROBERT M. WOHLHUETER and PETER G. W. PLAGEMANN

Introduction

The relationship between the transport of nutrient molecules across the cell membrane of animal cells and the metabolism of those molecules within the cells has both physiological and methodological significance. In the physiological sense, permeation of the membrane is properly regarded as the first step in the utilization of a nutrient. Accordingly, the control strength exerted at the permeation step over the rate of utilization of a nutrient is as appropriate a subject for biochemical investigation as are the control strengths of subsequent enzymatic steps. Beyond the question of control strengths of essentially independent processes is the question of a physical association between transporter and enzymes which could result in metabolic channeling of the imported molecule. Although bona fide translocation systems have not been established for nutrient uptake in animal cells, there have been observations which suggest that exogenous substrates are used preferentially to endogenous. In general, these important questions concerning the kinetic and physical relationship of transport and metabolic systems have been neglected.

At the methodological level, an ill defined relationship between transport and metabolism has often been a pitfall to the study of transport per se. Rigorous elucidation of the specificity, kinetic characteristics, temperature dependence, and concentrative capacity of a given transport system may well be impossible when the permeant used to assay the transport system is subject to metabolic conversion.

Compounding the methodological difficulties is the rapidity of some transport systems: facilitated diffusion of hexoses,[1,2] nucleosides,[3] and

[1] J. C. Graff, R. M. Wohlhueter, and P. G. W. Plagemann, *J. Cell. Physiol.* **96,** 171 (1978).
[2] J. Vinten, J. Gliemann, and K. Osterlind, *J. Biol. Chem.* **251,** 794 (1976).

METHODS IN ENZYMOLOGY, VOL. 173

nucleobases[4] has been observed in several cell types to proceed with half-times of 10 sec or less. It is a truism that an initial, constant rate of permeant uptake reflects its rate of transport regardless of subsequent metabolism. However, to establish that a measured rate of uptake is indeed an initial rate is no mean task, particularly in the familiar case of rapid permeation by facilitated diffusion followed by phosphorylation of the permeant. In this case, as the older literature on nucleoside uptake so amply illustrates,[5] two phenomena make it especially difficult to discern a true, initial rate of uptake. (1) The accumulation of isotope in phosphorylated metabolites impermeable to the membrane is so great in relation to the accumulation of intracellular, free permeant that the time course of metabolite accumulation appears to extrapolate through zero time. And (2) even the briefest of washes may suffice to leach any free permeant from the cell with the result that the time course of metabolite accumulation in fact extrapolates through zero time.

With these precautionary notes in mind, one can sketch out the attributes of an ideal experimental protocol for measuring transport in intact cells. The method should be capable of monitoring uptake as a function of time with a temporal resolution of at least 1 sec and beginning at zero time. It should quench the transport process virtually instantaneously and avoid any washing of the cells under conditions where permeant that has traversed the membrane might flow out again. It should be sufficiently sensitive to detect intracellular concentrations of permeant well below the extracellular concentration of permeant. Furthermore, if the aim of the research is to clarify the relationship between rates of transport and metabolism, the method should quench enzymatic activities simultaneously with transport and present the cell contents in a form amenable to the fractionation of metabolites.

A number of techniques have been devised which achieve some of these goals, and which may be completely satisfactory with a given cell/permeant combination. They can be summarized in terms of a few principles: inhibitor-quench, in which transport is stopped by exposure of the cells to a potent, fast-acting inhibitor; segregation into a nonaqueous phase, usually by centrifugation into an oil or by rapid filtration; indirect measurement, which relies on some optical signal generated upon movement of permeant into or out of the cell; metabolic inertness, which avoids the complications of metabolic conversion of permeant. In applica-

---

[3] R. M. Wohlhueter, R. Marz, and P. G. W. Plagemann, *Biochim. Biophys. Acta* **553,** 262 (1979).

[4] P. G. W. Plagemann and R. M. Wohlhueter, *Biochim. Biophys. Acta* **688,** 505 (1979).

[5] P. G. W. Plagemann and D. P. Richey, *Biochim. Biophys. Acta* **334,** 263 (1974).

tion, these principles are not mutually exclusive, and have often been combined to good effect. Nor are they inherently limited to a particular temporal resolution, although very rapid sampling usually requires special gadgetry.

Choice of a metabolically inert permeant has proved invaluable to the study of amino acid[6] and hexose transport.[7] Prerequisite to its use is a demonstration that the permeant enters by a single route and is a reasonable analog to the physiological permeant it stands in for. It is, of course, not suited to a comparison of transport and metabolic flux.

Use of an inhibitor to quench transport presupposes the existence of such an inhibition for the transport system under study, and requires the demonstration that it acts "instantaneously" relative to the time scale appropriate to that system. The method has been elegantly applied in a continuous flow system to the measurement of nucleoside transport at subsecond intervals.[8] An inhibitor of transport will, in general, not halt metabolism; thus its use does not ensure against the complications due to metabolic conversion nor lend itself to studies aimed at a comparison of transport and metabolic flux.

Indirect measurement of influx is exemplified by the early studies of Sen and Widdas[9] on hexose transport. Influx of glucose into suspended cells is accompanied by a change in light scattering from the cells due to solvent drag. The scattered light can be monitored continuously, without any mechanical or chemical intrusion into the system. The method has low sensitivity in that it detects only osmotically appreciable transfers of substrate. A more sensitive approach employs fluorescent substrates whose emission is significantly different between the extracellular and intracellular milieu.[10] It is limited mostly by its requirement for specially designed substrates which are inherently nonphysiological.

Centrifugation into nonaqueous media has long been used for the segregation of cells and organelles from an aqueous phase, usually for purposes of metabolite analysis.[11,12] Over the past 10 years we have applied this principle extensively to studies of nucleoside, nucleobase, and hexose transport in various types of animal cells. During that time we have developed a set of experimental protocols which possess most of the

[6] H. N. Christensen, in "Biological Transport," 2nd Ed. Benjamin, Reading, Massachusetts, 1975.
[7] D. M. Regen and H. E. Morgan, Biochim. Biophys. Acta 79, 151 (1964).
[8] A. R. P. Paterson, E. R. Harley, and C. E. Cass, Biochem. J. 224, 1001 (1984).
[9] A. K. Sen and W. F. Widdas, J. Physiol (London) 160, 392 (1962).
[10] O. Eidelman and Z. I. Cabantchik, Anal. Biochem. 106, 335 (1980).
[11] W. C. Werkheiser and W. Bartley, Biochem. J. 66, 79 (1957).
[12] J. M. Oliver and A. R. P. Paterson, Can. J. Biochem. 49, 262 (1971).

attributes discussed above. These protocols should be of general utility to any cell/permeant combination, provided the cells can be put in suspension and the permeant is available with isotopic label. They rely upon rapid mixing of cell suspension and permeant solution by means of a dual syringe device, followed by rapid segregation of the cells into an oil phase, or, optionally, through an oil phase into an underlying aqueous acid. Used with cognate pairs of metabolically inert and active substrates this methodology permits characterization of a given transport system per se and its role in relation to subsequent events in the metabolism of the substrate.

In this chapter we focus on methodological details; we have reviewed elsewhere[13-15] many of the results obtained by these methods.

Equipment and Procedures

Dual-Syringe Device. The task of rapidly mixing precise proportions of cell suspension and permeant solution we accomplish with a manually operated dual syringe, pictured in Fig. 1 and diagrammed in Fig. 2. The ratchet grooves built into the guide rods permit 12 repetitive deliveries of equal volumes of mixture with a single loading of the syringes. The spacing of these notches determines, for syringes of given cross section, the total volume delivered.

Two rods with different spacing can be easily exchanged by flipping the pusher plate 180°; others can be installed to accommodate different experimental protocols. The spacing required to approximate any desired volume can be calculated for any pair of syringes, but for optimum accuracy it is prudent to choose syringes of such dimension that the 12-ratchet intervals occupy as much of barrel length as possible. In any case the actual delivery of the syringes should be measured gravimetrically by expressing water into tared vials.

The dimensions of the V-groves given in the diagram are appropriate for a 6-ml disposable syringe (1/2-in. plunger diameter, Monoject, St. Louis, MO), to contain cell suspension, and a 1-ml disposable syringe (3/16-in. plunger diameter), to contain permeant solution. Such syringes, with the 0.137-in. ratchet spacing indicated in the diagram, provide about $448 + 61$ μl/interval. For additional versatility, a set of adapters can be machined to center smaller syringes in a maximally sized V.

[13] P. G. W. Plagemann and R. M. Wohlhueter, *Curr. Top. Membr. Res.* **14,** 225 (1980).
[14] R. M. Wohlhueter and P. G. W. Plagemann, *Int. Rev. Cytol.* **64,** 171 (1980).
[15] P. G. W. Plagemann, R. M. Wohlhueter, and C. Woffenden, *Biochim. Biophys. Acta* **947,** 405 (1988).

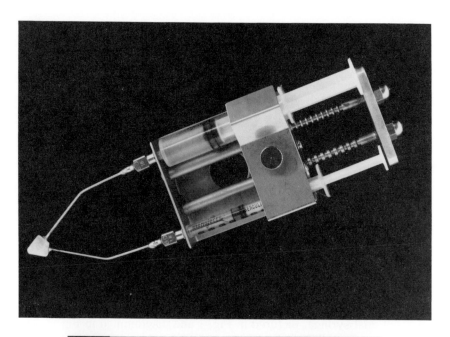

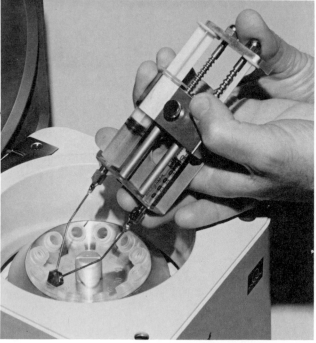

FIG. 1. Photographs of the dual-syringe device (top) and operator's hold on it (bottom).

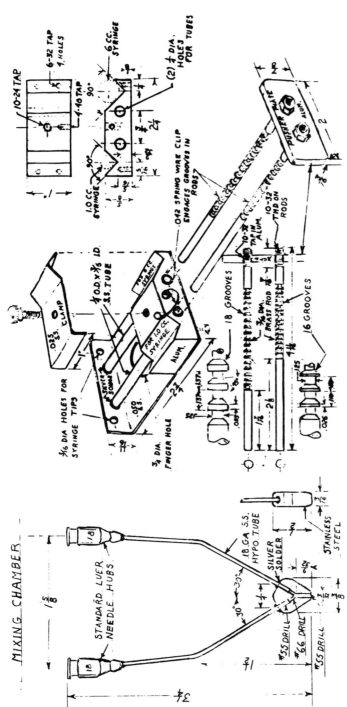

Fig. 2. Machinist's drawing of dual-syringe device. Dimensions are appropriate for use with 6- and 1-ml disposable syringes and show ratchet spacings of 0.137 and 0.110 in.

719

To advance the ratchet one notch, the operator depresses the spring momentarily with the left thumb, then pushes the plunger bar with the right thumb, allowing the spring to snap into the next notch. With practice the operation can be performed at 1-sec intervals.

The mixing chamber is conveniently made by silver soldering two cannulae, terminated with Luer adapters, onto the two arms of a Y-shaped boring in stainless steel. In operation the tip should be touched to the side of the vial into which the mixed sample is delivered in order to prevent retention of a drop of sample on the tip.

Centrifuge, Tubes, and Oil Phase. The Eppendorf model 5414 micro-centrifuge provides a 12-place, fixed-angle rotor which is accelerated to 12,000 g in 5 sec. Our observations (see below) indicate that cultured cells of typical dimension and density are effectively out of contact with aqueous medium 2 sec after starting the centrifuge.

The standard 1.5 ml-conical, snap-cap tubes, available from several sources, may be used, but for closely spaced measurements they must have their caps clipped off, since there is not time enough to close them. With uncapped tubes no more than about 600 μl total fluid volume can be added to a tube without loss upon rotation. A slight advantage can be gained with Sarstedt screw-top tubes (#72.692), because their treaded collar extends the length of the tube, and they can be purchased without caps.

The primary requirement for a suitable oil phase is that it be intermediate in density between that of cells, typically 1.07 kg/liter, as measured in sucrose gradients, and the aqueous cell suspension, about 1.00 kg/liter. Other considerations include its viscosity, chemical inertness, and ability to wet the centrifuge tube. Mixtures of Dow-Corning #550 silicone fluid (density = 1.07 kg/liter) and mineral oil (density = 0.844 kg/liter) are satisfactory in all these regards. For example, an 84 : 16 (w/w) solution of 550 fluid and mineral oil gives a density of 1.034 kg/liter (at 24°), which is appropriate for most of the cells we have used. Use of an oil mixture facilitates small adjustments in density if a particular cell type should demand it. Indeed, our empirical approach to a new cell type is to make a series of oil mixtures of closely spaced composition, and to evaluate by eye whether the cells sediment cleanly through the oil and whether the oil–water interface is sharp and stable. The potential for versatility is neatly illustrated by the studies of Vinten et al.,[2] who measure the rapid transport of hexoses into adipocytes by centrifuging the bouyant cells upward into silicone oil less dense than water.

If total uptake of permeant into the cell is to be measured, that is, if metabolic activity is irrelevant to the experimental design, then a small volume of oil, 50–100 μl, in the bottom of the centrifuge tube is sufficient and leaves more room for cell suspension. If the experimental design requires metabolic quenching, then the oil should be underlaid with 50–100 μl of an aqueous acid solution of high density: 0.5 M trichloroacetic or perchloric acid in 10% (w/v) sucrose or 15% (w/v) glycerol. In this case we find it advisable to increase the volume of oil to 150–200 μl as a hedge against cell suspension contacting the acid phase as the sample is first squirted into the tube. The practicable sample volume is accordingly reduced, if tubes are to be centrifuged without caps.

Dispensing Samples. Cells are suspended in an appropriate medium at a density calculated to deliver around 10 μl of cell volume per sampling. Thus, for the syringe and ratchet described above, and for cells of 1 μl/million, about 2 × 10^7 cells/ml will do; 6 ml of such a suspension is needed for a single, 12-point time course. The suspension is loaded into the syringe via flexible tubing slipped over a ground-off, Luer-lock needle which replaces the mixing chamber cannula. The plunger is drawn back to the bar, which should be positioned a little bit before the first notch of the ratchet. Inclusion of a small air bubble is useful to help stir the suspension during waiting periods.

Cell suspensions may, of course, be pretreated as desired. Of special interest is the protocol for equilibrium exchange,[16] in which the cells are preincubated with nonlabeled permeant. Concentrations of nonlabeled and labeled permeant are chosen so that, after mutual dilution by mixing, they are equal.

Labeled permeant is loaded into the second syringe from a stock solution of concentration calculated to give the desired final concentration upon dilution. The concentration of radioactivity is adjusted to be reasonably detectable at 1/10 equilibrium with cell water. As an example on the syringes described above, which involve a 8.3-fold dilution of permeant, and a cell density chosen to deliver 10 μl/sample: a stock solution at 10 μCi/ml would result in about 2700 dpm in intracellular water when permeant was still at an intracellular concentration only 1/10 of the extracellular concentration. For studies involving a series of permeant concentrations, we find it convenient to utilize a constant concentration of radioactivity while increasing the chemical concentration of permeant.

Tubes containing oil are loaded into the centrifuge and spun for a couple of seconds to consolidate the phase(s); this is particularly important if an acidic underlay is present. Samples are dispensed beginning at rotor position 12 and proceeding in a countdown fashion. The centrifuge is started immediately after dispensing the last sample; 5 sec is more than

---

[16] W. F. Stein, "Transport and Diffusion across Cell Membranes." Academic Press, Orlando, 1986.

adequate to sediment the cells. The cover interlock switch can be exploited to start the centrifuge by closing the cover with the timer running.

Timing is best done auditorily, especially for very short intervals; an electric metronome provides the tempo. Two to four beats per sample may help the operator coordinate deliveries. If the experimental design requires sampling over an entire time course to attainment of equilibration of permeant across the membrane, it is advisable to stagger intervals such that early time points are more closely spaced, while the latter ones are lengthened progressively.

Our favorite stagger, because of its mnemonic ease, is to deliver each sample at the number of beats equal to its position in the rotor. This generates a time series of 0, 1, 3, 6, 10 . . . beats, whereby one beat can take on any value the metronome and the operator can handle. If sampling schedules result in periods of waiting, the syringes should be rocked to keep the cells in uniform suspension. For very long times (greater than a couple of minutes), the dual syringe loses its advantage. For such times, it becomes easier to mix cells and permeant in the same proportion as delivered by the syringes, incubate them with agitation, and manually pipette samples to tubes containing oil at the desired times. We have frequently combined data gathered by means of the dual syringe at short times with that gathered by pipetting at longer times into a single time course.

Controlling the ambient temperature of solutions and apparatus is the only practicable form of temperature control available to our technique. It is most convenient, of course, to conduct the experiments at the bench in a stably climatized laboratory. We have been able to encompass the whole range of temperatures from about 2 to 40°, however, by enclosing the entire apparatus in a modified glove box thermostatted by circulating coolant from a bath through a radiator mounted in the box.[3]

Intra- and Extracellular Water Space. It is essential to the method to measure the extracellular water that is carried along with the cells through the oil phase, because total uptake of isotope into the cells requires correction for the radioactivity attributable to extracellular permeant. It is desirable to measure the intracellular water, because that permits the judgment of whether permeant or a metabolic derivative thereof is concentrated beyond that predicted for equilibrium between intracellular and extracellular water.

We routinely measure these quantities simultaneously with ^3H_2O (NEN, Boston, MA) and [*carboxyl*-^{14}C]carboxylinulin (NEN, Boston, MA). A solution of about 5×10^6 dpm of inulin and 10^7 dpm of H_2O/ml of cell suspension medium is prepared. This solution is mixed with cell suspension in the same proportion as permeant solution is, allowed to equilibrate (1 min is ample), and an aliquot equivalent to the total syringe delivery is spun through oil. Triplicate determinants at the beginning and end of an experiment are made. Thence, the cell pellets are worked up in the same way as other samples, counted in a scintillation spectrometer with a dual-isotope program, and compared to the counts per million of ^3H and ^{14}C counted under identical conditions with a measured volume of the original solution roughly equal to the volume of the cell pellet. Total pellet H_2O space minus inulin space is taken as intracellular water.

With several types of cultured cells, we have found the extracellular water accompanying cells through the oil to be between 10 and 20% of total pellet water, and constant for any given batch of cells. Furthermore, both total and extracellular water space vary linearly with the number of cells sedimented, and both extrapolate through zero.[17]

Sample Workup. Preparation of samples for scintillation counting will need to be tailored in some degree to experimental design. Supernatant fluid may be aspirated to waste, or saved for analysis. After its removal, the upper part of the tube is rinsed carefully with water, which is then aspirated to waste, along with most of the oil.

In the absence of an aqueous underlay, the whole cell pellet is taken for counting. Practically any means of dispersing cell contents homogeneously into a scintillant solution should suffice; tissue solubilizer, trichloroacetic acid (0.5 M), or Triton X-100 (1%) work about equally well. To simplify quantitative transfer, we have generally put the entire sample tube into the scintillation vial, capped the vial, and shaken it vigorously. More recently, we have adopted a different approach which is easier and more economical. Ready-Solv HP scintillation cocktail (Beckman Instruments, Fullerton, CA) has sufficient solubilizing power so that 0.5 ml can be added directly to the cell pellet, homogenized with a Teflon pestle machined to match the centrifuge tube (Kontes, Vineland, NJ), capped, and counted by placing it upright in a standard scintillation vial.

When an aqueous underlay is present sample workup will usually involve neutralization of the acid in preparation for chromatographic fractionation of the metabolic derivatives of the permeant. A method which lends itself well to the situation is that of Khym[18]: After rinsing the upper part of the tube and removing most of the oil, dilute buffer (e.g., 50 mM Tris, pH 8) is added to increase the volume of the acid phase to, say, 300 μl. The cell pellet is broken up with a small pestle, and 800 μl of a 0.5 M solution of trioctylamine (Sigma Chemical Co., St. Louis, MO) in Freon

[17] R. M. Wohlhueter, R. Marz, J. C. Graff, and P. G. W. Plagemann, *Methods Cell Biol.* **20,** 211 (1978).
[18] J. Y. Khym, *Clin. Chem.* **21,** 1245 (1975).

113 (1,1,2-trichlorotrifluoroethane; Matheson, East Rutherford, NJ) is added. The tube is shaken vigorously, centrifuged to separate the phases, and the neutralized (upper) aqueous phase removed for further analysis. Oil residues are dissolved in the Freon phase.

For permeants such as glucose, nucleosides, and nucleobases, where metabolism involve formation of phosphate esters, it may suffice the experimenter's needs to separate only phosphorylated and nonphosphorylated forms of the permeant. A variety of batch separations have been employed by us and others, including precipitation as barium[19] or lanthanum[20] salts and binding to anion exchangers such as polyethyleneimine[21] or Dowex 1.[22] Even such a simple fractionation scheme can serve to monitor simultaneously the kinetics of appearance of free, isotopically labeled permeant within the cell and of its conversion to phosphorylated intermediates, and, thus, to give valuable information on the relationship between the two processes.

Metabolism is not always so straightforward, however. An example of a more demanding permeant is adenosine, which is phosphorylated, but also rapidly deaminated in many types of cells to inosine, which is, in turn, phosphorolyzed to hypoxanthine and ribose 1-phosphate. Isotopic label in the adenine moiety soon appears in extracellular inosine and hypoxanthine, as well as in intracellular adenine nucleotides. In such a case it is prudent to analyze samples of the supernatant fluid chromatographically. An illustration is given below.

Metabolically Inert Systems. A rigorous characterization of a transport system does not end with a determination of specificity and influx kinetics. Questions of concentrative capacity, (a)symmetry with respect to direction of flux, and the relative resistivities to movement of substrate-loaded and empty transporter are important to an understanding of physiology as well as mechanism. The available experimental protocols (cf. Stein[16]) which address these questions work only in situations where the permeant employed is not subject to metabolism. A classical means to this end is the preparation of plasma membrane vesicles devoid of metabolic activities. There are several treatises on these techniques.[23,24]

The use of nonphysiological analogs of natural substrates has been of great utility in the study of several transport systems. α-Methylaminoiso-

[19] R. F. Kletzien and J. F. Perdue, *J. Biol. Chem.* **249,** 3366 (1974).

[20] B. Bakay, M. A. Telfer, and W. L. Nyhan, *Biochem. Med.* **3,** 230 (1969).

[21] R. M. Wohlhueter, R. Marz, J. C. Graff, and P. G. W. Plagemann, *J. Cell. Physiol.* **89,** 605 (1976).

[22] J. Katz, P. A. Wals, S. Golden, and R. Rognstad, *Eur. J. Biochem.* **60,** 91 (1975).

[23] D. F. H. Wallach and R. Schmidt-Ullrich, *Methods Cell Biol.* **15,** 235 (1977).

[24] T. L. Steck and J. A. Kant, this series, Vol. 31, p. 172.

butyric acid and 2-aminonorbornyl-2-carboxylic acid are prominent examples from amino acid transport[6]; 3-O-methyl-D-glucose[7] and 6-deoxy-D-glucose[25] are metabolically inert substrates for mammalian glucose transporters. 5'-Deoxyadenosine has been used successfully in some cells as a metabolically inert nucleoside analog,[26] but is rapidly phosphorolyzed in several cell types.[27] Rapid mixing and sampling with the dual-syringe methodology described here complements the use of these inert permeants, especially where their transport is vary rapid.

Three other experimental techniques should be considered in the design of a metabolically inert system. One is the relative ease of obtaining mutant cells lacking an "activating" enzyme for nonessential nutrients.[28] In the case of nucleosides and bases, kinase- or phosphoribosyltransferase-deficient variants may be selected by exposure to toxic analogs or toxic concentrations of the natural compound. For clones of these variant cells the natural transport substrate is metabolically inert, permitting an analysis of transport without interference from metabolism as well as a comparison to the metabolically competent parent line.

A second technique uses specific enzyme inhibitors. The deamination of adenosine, for instance, can be completely inhibited with low concentrations of deoxycoformycin,[29] thereby eliminating a rapid and ramified pathway of adenosine metabolism. In fact, cells selected for the absence of adenosine kinase and treated with deoxycoformycin are almost completely inactive toward adenosine (although slow conversion of adenosine to adenine may persist), and serve as an ideal system for studying adenosine transport.[30]

A third technique, generally applicable to phosphorylated permeants where enzyme-deficient mutants are not readily available, is to inhibit substrate level and oxidative phosphorylation. Brief treatment of several cultured cell lines with 5 mM iodoacetamide plus 5 mM KCN fully depletes their pools of ATP and phosphoribosylpyrophosphate, and thereby prevents the phosphorylation of nucleosides and hexoses and the phosphoribosylation of nucleobases.[31] The normal operation of several systems of facilitated diffusion may be observed in such cells. This approach, though useful, demands careful controls on cell membrane integrity, on the effectiveness of pool depletion, and on the insensitivity of the transport system itself to iodoacetate.

[25] A. H. Romano and N. D. Connell, *J. Cell. Physiol.* **111,** 77 (1982).
[26] D. Kessel, *J. Biol. Chem.* **253,** 400 (1978).
[27] P. G. W. Plagemann and R. M. Wohlhueter, *Biochem. Pharmacol.* **32,** 1433 (1983).
[28] L. H. Thompson and R. M. Baker, *Methods Cell Biol.* **6,** 209 (1973).
[29] I. H. Fox and W. N. Kelley, *Annu. Rev. Biochem.* **47,** 655 (1978).
[30] P. G. W. Plagemann and R. M. Wohlhueter, *J. Cell. Physiol.* **116,** 247 (1983).
[31] P. G. W. Plagemann, R. Marz, and J. Erbe, *J. Cell. Physiol.* **89,** 1 (1976).

TABLE I
ISOTOPICALLY LABELED PERMEANTS[a] WHOSE TRANSPORT HAS
BEEN MEASURED BY THE DUAL-SYRINGE TECHNIQUE[b]

Nucleosides	Nucleobases	Hexoses
Adenosine	Adenine	2-Deoxyglucose
Arabinosylcytidine	8-Azaguanine	6-Deoxyglucose
Cytidine	Cytosine	Galactose
2'-Deoxyadenosine	5-Fluorouracil	Glucose
5'-Deoxyadenosine	Hypoxanthine	3-O-Methylglucose
Deoxycytidine	6-Mercaptopurine	
Inosine	Orotic acid	
Uridine	6-Thioguanine	
Thymidine	Uracil	
Tubercidin		

[a] Available from commercial sources either as ^{3}H or ^{14}C: Moravek Biochemicals, Brea, California; NEN, Boston, Massachusetts; Research Products International, Mount Prospect, Illinois; Amersham, Arlington Heights, Illinois.
[b] See Refs. 13–15 for citations to original literature.

Integrated Rate Equations. In most of our studies on the transport of nucleosides, nucleobases, and hexoses we have analyzed our data in terms of integrated rate equations derived for a simple, carrier-mediated, facilitated diffusion system.[16] Our reliance on integrated rate treatment is primarily pragmatic: even with 1- to 2-sec sampling intervals commencing after a 2-sec sedimentation delay, and at 25°, we usually cannot collect sufficient data to define an initial, linear rate (see examples below). We are thus obliged to use curvilinear extrapolation to zero time. The integrated rate equations for simple transporters of various symmetries are formally similar to the curve for an exponential approach to equilibrium

TABLE II
CELL TYPES IN WHICH TRANSPORT HAS BEEN MEASURED BY
DUAL-SYRINGE TECHNIQUE[a]

Rat	Mouse	Human	Hamster	Dog
Novikoff hepatoma	L929 cells	Erythrocytes	CHO cells	Thymocytes
Lymphocytes	L1210 leukemia	HeLa		Leukocytes
Erythrocytes	P388 leukemia			
Hepatocytes	S49 lymphoma			

[a] See Refs. 13–15 for citations to original literature, except for rat hepatocytes, which is unpublished work.

TABLE III
HALF-TIMES OF TRANSMEMBRANE EQUILIBRIUM OF SELECTED TRANSPORT SYSTEMS[a]

Permeant	Cells	Temperature (°C)	$t_{1/2}$ (sec)	Ref.
3-O-Methyl-D-glucose	Novikoff hepatoma	25	35	1
	HeLa	25	12	1
	Rat adipocytes[b]	37	2	2
	Rat hepatocytes	22	7	36
Adenosine	Novikoff hepatoma	25	5	30
	L5178Y lymphoma	22	1	8
	CHO[c]	25	3	30
Thymidine	Novikoff hepatoma	25	6	3
	Human erythrocytes	25	3	33
Hypoxanthine	Novikoff hepatoma	25	12	4

[a] Half-times ($t_{1/2} = \ln 2/k$) are calculated for concentrations of permeant in the first-order range, assuming an exponential approach to equilibrium, either from initial rates at low concentration ($k = $ rate/[P]) or from Michaelis–Menten parameters ($k = V_{max}/Km$).
[b] Treated with insulin.
[c] Deficient in adenosine kinase and treated with deoxycoformycin.

(but deviate significantly from it when [S] is significant relative to K_m). To us, they seem the most appropriate curve for extrapolation.

Integrated rate treatment is also of theoretical value. We find, for example, that the kinetics of nucleoside transport in several cultured cell lines is fitted well by the simple, symmetrical carrier model, and consider these findings strong evidence for the validity of that model. Integrated rate analysis of isotopic flux in multistep, metabolic networks becomes quickly intractible with increasing complexity. For simplified treatments of transport in tandem with a single, Michaelian trapping reaction, see Heichal et al.[32] and Wohlhueter and Plagemann.[14]

Results and Discussion

The dual-syringe methodology described here has been applied by us to investigations of transport of hexoses, nucleosides, and nucleobases in a wide variety of mammalian cell types, both cultured and dispersed from tissues. Table I lists some of the permeants employed in these studies; Table II lists some types of cells used. A comprehensive discussion of the

[32] O. Heichal, D. Ish-Shalom, R. Koren, and W. D. Stein, Biochim. Biophys. Acta 551, 169 (1979).
[33] P. G. W. Plagemann, R. M. Wohlhueter, and J. Erbe, J. Biol. Chem. 257, 12069 (1982).

results obtained is not germaine to this chapter. They have been reviewed in references.[13–15] We present here only a few results illustrative of the methodology.

Rapidity of Transport. Table III summarizes some of the more rapid transport systems we and others have observed by way of emphasizing the time scales which may be necessary to measure transport rates accurately. The half-time for transmembrane equilibration (where concentration of permeant is low relative to the Michaelis–Menten constant) provides a succinct measure of rapidity of these transport systems. Note that for these first order situations, the rate of permeant entry will have diminished to 50% of initial by half-time.

Figure 3 shows time courses for uridine entry into erythrocytes and cultured hepatoma cells at a concentration in the first order range.[34,35] The figure serves to illustrate several features of the methodology. The location of the origin of these plots is, in essence, a measured quantity: The y axis is offset by the amount of radioactivity in the cell pellet which is attributed to the extracellular space in the pellet as measured by [^{14}C]inulin, usually a correction of 10 to 15%. The x axis is offset by 2 sec to adjust for the effective delay in centrifugally removing cells from contact with the aqueous medium. This delay we have estimated by fitting exponential curves to several such sets of data, but without offsetting the x axis, nor constraining the curve to pass through the origin. The x intercepts of such curves cluster around -2 sec, and justify our routinely adding 2 sec to nominal sample times. These data attest to the necessity of curvilinear extrapolations to initial velocities, and suggest that the experimentally determined offsets to the axes are reasonable. They also demonstrate the advantage of measuring intracellular water space as an aid to interpretation of results. The curves seem to be approaching asymptotically an intracellular concentration expected for a nonconcentrative transport system in which substrate radioactivity is not being trapped intracellularly as impermeable metabolites.

The behavior of an actively metabolizing system is illustrated in Fig. 4. Here uptake of radioactivity from exogenous uridine proceeds nearly linearly to levels well above that expected for transmembrane equilibrium. The nonphosphorylating control (Fig. 4A) resembles the uptake curve in Fig. 3, and suggests that the augmented uptake in metabolizing cells is due to impermeable metabolites being trapped intracellularly. This suggestion is confirmed in Fig. 4B, which divides total, intracellular radio-

[34] P. G. W. Plagemann, R. Marz, and R. M. Wohlhueter, *J. Cell. Physiol.* **97**, 49 (1978).

[35] P. G. W. Plagemann and R. M. Wohlhueter, *in* "Regulatory Function of Adenosine" (R. M. Berne, T. W. Rall, and R. Rubio, eds.), p. 179. Nijhoff, The Hague, The Netherlands, 1985.

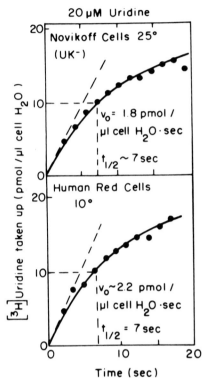

FIG. 3. Examples of uptake of uridine into Novikoff rat hepatoma cells devoid of uridine kinase and into human erythrocytes. The dual-syringe device was operated at 2-sec intervals at an ambient temperature of 25 or 10°. Uridine (20 μM) contained 240 cpm of tritium/μl. Cells were centrifuged into oil and processed as described in text, using 0.5 M trichloroacetic acid to release radioactivity from the cells. In the case of erythrocytes the trichloroacetic acid extract was centrifuged to remove precipitated hemoglobin and an aliquot taken for scintillation counting; with Novikoff cells the entire extract was counted. The solid line shows the exponential curve best fitting the data; the broken line through the origin is the velocity at zero-time calculated for the exponential curve. Note that 2 sec has been added to the nominal sampling times, and that the radioactivity in the total cell pellet has been corrected for that in the extracellular space as measured by [^{14}C]inulin. For Novikoff cells intracellular and extracellular water per pellet was 20.5 and 2.0 μl, respectively; for erythrocytes, 32.2 and 4.0 μl, respectively. Data are from Refs. 34 and 35.

activity into different metabolite categories, thus demonstrating that it is the accumulation of uracil nucleotides which sustains the overall rate of uptake of radioactivity. Figure 4C shows that these rates of nucleotide accumulation are a Michaelian process. That process, taken in context, is clearly intracellular phosphorylation of uridine (the K_m for uridine transport as measured in either uridine kinase-negative cells or ATP-depleted

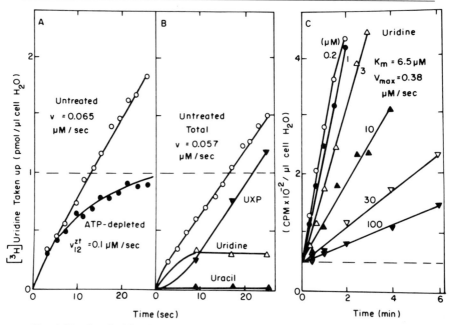

FIG. 4. Uptake of uridine into wild-type Novikoff hepatoma cells at 25°. (A) Time courses of appearance of tritium from 1 μM exogenous uridine in total cell pellets of metabolizing and ATP-depleted cells, corrected for extracellular radioactivity. (B) Time courses of appearance of tritium into various uridine metabolites as determined by paper chromatography of the acid-soluble extracts of cells which had been centrifuged through oil into 0.5 M trichloroacetic acid in 10% sucrose. (C) Samples collected at 30 sec to 6 min by pipetting cell-permeant mixtures into mirocentrifuge tubes containing oil and centrifuging. Exogenous uridine was at various concentrations, as indicated on the individual curves, but with constant concentration of radioactivity (570 cpm/μl). The straight lines drawn through the data points were replotted against exogenous concentration and the Michaelis–Menten equation fitted to them; the best fitting Km and V_{max} are indicated. In all panels, the broken, horizontal lines indicate the concentration in cell water which is equal to extracellular concentration of permeant. The data of (A) and (B) are from Ref. 34; those of (C) are from Ref. 35.

cells is about 250 μM; V_{max} about 26 μM/sec[13,15]). Unfortunately, such kinetics of "uptake" have frequently been construed as kinetics of transport.[5,13] If, in fact, a given transport process is rate limiting for permeant metabolism, then equating uptake kinetics with transport kinetics may be valid and useful, but the burden of proof lies with the investigator, and will most likely require a demonstration of transport kinetics in the absence of metabolism.

[36] J. D. Craik and K. R. Elliot, *Biochem. J.* **182**, 503 (1979).

Another situation in which monitoring the net movement of radioactivity into cells without regard to metabolic conversions can lead to misinterpretation is illustrated in Fig. 5. Here the complication is not just rapidity of permeant transport, but the fact that some metabolites may themselves be permeable and exit the cell rapidly. The rate of adenosine uptake in CHO cells (Fig. 5A) appears to extrapolate through time zero and to remain linear for some 30 min. Fractionation of radioactivity in the supernatant medium (Fig. 5B) reveals that adenosine had entered the cells much more rapidly than appears from total uptake and is depleted from the medium by 30 min. A kinetic characterization of adenosine transport

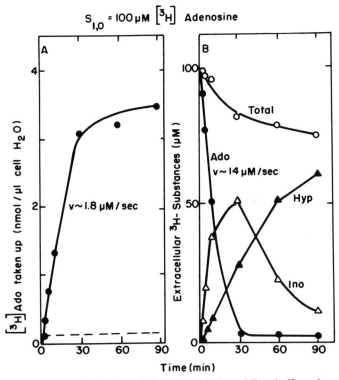

FIG. 5. Uptake of adenosine into wild-type CHO cells at 25° and efflux of metabolites. [³H]Adenosine (100 μM, 8 cpm/pmol) was mixed with cells suspended at 4.8×10^6/ml; at the indicated times 0.5-ml aliquots were centrifuged through oil. In (A), the total radioactivity in the cell pellet, expressed as adenosine (Ado) equivalents, is shown. In (B), the results of paper chromatographic fractionation of radioactivity in the supernatant medium are shown. The indicated velocities were estimated graphically. The broken, horizontal line in (A) represents a concentration of 100 μM in cell water. Data are from Ref. 35.

measured in CHO cells devoid of adenosine kinase and with adenosine deaminase blocked by treatment with deoxycoformycin gave $K_m = 89 \ \mu M$ and $V_{max} = 24$ pmol/μl cell water/sec.[30,35] The transport rate expected for 100 μM adenosine is thus about 13 pmol/μl cell water/sec, comparable to that estimated from the rate of disappearance of adenosine from the medium.

The lessons to be drawn from these examples are two. Measurement of rates of transport in intact cells with permeants subject to metabolism is fraught with difficulties. And once that transport system has been characterized in the absence of metabolic interferences, valuable physiological information is to be had by monitoring isotope movements across the membrane and into various metabolite pools.

Acknowledgments

The authors wish to thank their former and present colleagues who contributed to the development of these methods, particularly Dr. Richard Marz, Dr. Jon C. Graff, and Mr. John Erbe. Ms. Laurie Erickson was responsible for much of the technical work, and Ms. Yvonne Guptill for the preparation of the manuscript. The research was supported by Grants CA33082 (to R.M.W.) and GM24468 (to P.G.W.P) from the USPHS.

[45] Purification and Reconstitution of the Phosphate Transporter from Rat Liver Mitochondria

By Ronald S. Kaplan, Raymond D. Pratt,
and Peter L. Pedersen

Introduction

The transport of inorganic phosphate (P$_i$) across the mitochondrial inner membrane into the matrix compartment is essential for oxidative phosphorylation. This transport is catalyzed by at least two transport systems.[1-4] One system, the P$_i$ carrier, catalyzes an NEM-sensitive high-magnitude P$_i$/H$^+$ symport. The other system, the P$_i$/dicarboxylate carrier, catalyzes an exchange of P$_i$ and/or dicarboxylates across the inner membrane. The latter transporter is insensitive to NEM but can be inhibited by

---

[1] P. L. Pedersen and J. P. Wehrle, in "Membranes and Transport" (A. Martonosi, ed.), Vol. 1, pp. 645–663. Plenum, New York, 1982.
[2] K. F. LaNoue and A. C. Schoolwerth, Annu. Rev. Biochem. 48, 871 (1979).
[3] A. Fonyo, Pharmacol. Ther. 7, 627 (1979).
[4] J. Bryla, Pharmacol. Ther. 10, 351 (1980).

other sulfhydryl reagents. Both transporters have been extensively characterized in intact mitochondria with respect to their kinetics, substrate
specificities, and inhibitor sensitivities.[5–10] Such investigations have indicated that the P_i transporter is responsible for supplying most of the P_i
required for oxidative phosphorylation.

In order to understand the functioning of the mitochondrial phosphate
transporter at the molecular level, recent efforts[11–17] have focused on
isolating this carrier and reconstituting its transport activity in phospholipid vesicles. In the present chapter, we describe a procedure that we
have developed[12] for the purification of the reconstitutively active P_i/H^+
symporter from rat liver mitoplasts.

Methods

Stock Solutions of Buffers and Inhibitors

1. Buffer A: 20 mM KP_i, 40 mM KCl, 2 mM EDTA, pH 7.2
2. Buffer B: 10 mM KP_i, 20 mM KCl, 1 mM EDTA, pH 7.2
3. Buffer C: Buffer B + 1% Triton X-114 + 1 mg cardiolipin/ml
4. Buffer D: 10 mM KP_i, 50 mM KCl, 20 mM HEPES,[18] 1 mM EDTA,
   pH 6.6
5. Buffer E: 50 mM KCl, 20 mM HEPES, 1 mM EDTA, pH 7.0
6. N-Ethylmaleimide (Sigma): 270 mM dissolved in water. Stored at
   room temperature (shielded from light) for up to several weeks

[5] H. Freitag and B. Kadenbach, *Eur. J. Biochem.* **83**, 53 (1978).
[6] F. Palmieri, G. Prezioso, E. Quagliariello, and M. Klingenberg, *Eur. J. Biochem.* **22**, 66 (1971).
[7] W. A. Coty and P. L. Pedersen, *J. Biol. Chem.* **249**, 2593 (1974).
[8] W. A. Coty and P. L. Pedersen, *Mol. Cell. Biochem.* **9**, 109 (1975).
[9] R. S. Kaplan and P. L. Pedersen, *Biochem. J.* **212**, 279 (1983).
[10] M. Crompton, F. Palmieri, M. Capano, and E. Quagliariello, *Biochem. J.* **142**, 127 (1974).
[11] J. P. Wehrle and P. L. Pedersen, *Arch. Biochem. Biophys.* **223**, 477 (1983).
[12] R. S. Kaplan, R. D. Pratt, and P. L. Pedersen, *J. Biol. Chem.* **261**, 12767 (1986).
[13] H. Wohlrab, *J. Biol. Chem.* **255**, 8170 (1980).
[14] H. V. J. Kolbe, D. Costello, A. Wong, R. C. Lu, and H. Wohlrab, *J. Biol. Chem.* **259**, 9115 (1984).
[15] F. Bisaccia and F. Palmieri, *Biochim. Biophys. Acta* **766**, 386 (1984).
[16] P. Mende, H. V. J. Kolbe, B. Kadenbach, I. Stipani, and F. Palmieri, *Eur. J. Biochem.* **128**, 91 (1982).
[17] M. Muller, D. Cheneval, and E. Carafoli, *Eur. J. Biochem.* **140**, 447 (1984).
[18] Abbreviations used are: HEPES, 4-(2-hydroxyethyl)-1-piperazineethanesulfonic acid; NEM, N-ethylmaleimide; pCMB, p-chloromercuribenzoic acid; SDS, sodium dodecyl sulfate; SDS–PAGE, sodium dodecyl sulfate–polyacrylamide gel electrophoresis; dansyl, 5-dimethylaminonaphthalene-1-sulfonyl.

7. Asolectin (soybean phosphatidylcholine type IV-S, Sigma): 200 mg/ml in chloroform; stored under nitrogen, at $-20°$, in the dark
8. Cardiolipin (sodium salt, from bovine heart, in ethanol, Sigma): Prior to use, the ethanol is evaporated (rotary evaporator) and the cardiolipin is then dissolved in chloroform and stored under nitrogen, at $-20°$, in the dark. Immediately prior to use, the chloroform is removed from the cardiolipin utilizing the procedure described below for asolectin. The cardiolipin is then dispersed in a given buffer via sonication in a bath sonifier for several minutes
9. Ortho [^{32}P]phosphate (Amersham; carrier free in dilute HCl): Stored at room temperature, shielded appropriately

Purification of the P_i/H^+ Symporter

Hypotonic Shock Treatment of Mitoplasts. Frozen rat liver mitoplasts are employed as the starting material. Mitoplasts (approximately 300 mg of protein) are thawed and then mixed with an equal volume of buffer A. All subsequent steps are performed at $0-4°$. The mitoplast suspension is then diluted to 30 mg protein/ml with buffer B, placed on ice for 5 min, and then further diluted to 3.6 mg protein/ml with buffer B. Following this hypotonic shock treatment, the suspension is centrifuged at 20,000 g (average) for 10 min. The pellets are then resuspended with buffer B to the same volume as in the earlier 5-min incubation.

Transporter Extraction and Hydroxylapatite Chromatography. The phosphate transporter is extracted with Triton X-114 in the presence of added cardiolipin as follows. The diluted suspension is mixed with an equal volume of a solution that consists of buffer B plus 6% (v/v) Triton X-114 plus 6 mg dispersed cardiolipin/ml. Following a 20-min incubation, the suspension is centrifuged at 130,000 g (average) for 35 min (including acceleration time). The supernatant (i.e., Triton X-114-solubilized mitoplasts) is removed and applied to hydroxylapatite (HA; BioGel HTP, Bio-Rad) columns. These columns are prepared by placing approximately 0.5 g of dry HA into each 5.75-in. Pasteur pipet which has been fitted with a glass wool plug. The pipet columns are then packed by tapping them lightly against the counter top for approximately 30 sec. The high-speed supernatant (500 μl) is then added to each column. The columns are subsequently eluted with buffer B plus 1% (v/v) Triton X-114 plus 2 mg cardiolipin/ml. The first 1 ml of flow-through from each column is collected.

Chromatography on DEAE-Sepharose CL-6B. A column containing DEAE-Sepharose CL-6B (Pharmacia) is prepared as follows. The resin is first washed with buffer B plus 1% Triton X-114. The DEAE-Sepharose

CL-6B is then placed into a column (Pharmacia K 9/15) and packed (the final volume after packing is approximately 9 ml) and subsequently washed again with buffer B plus 1% Triton X-114. The resin is then washed a final time with 7 ml of buffer C. Following the addition of the combined HA eluates (1.5–3 mg protein), the column is washed with 12 ml of buffer C. The functional P_i transporter is then eluted with a linear 0– 0.3 M NaCl gradient (50 ml) in buffer C. The gradient-eluted material is then pooled and subsequently added to a column containing Affi-Gel 501 as described below.

Chromatography on Affi-Gel 501. A column of Affi-Gel 501 (Bio-Rad) is constructed utilizing approximately 3 ml of resin in a 3-ml syringe (B-D 5570) which has been fitted with a glass wool plug. The resin is initially washed and packed utilizing deionized water and then equilibrated with 7.5 ml of buffer C. Following the addition of the DEAE eluate (i.e., 0.3– 0.6 mg protein; DEAE salt-eluted fraction) the resin is washed with 14 ml of buffer C. One of two different procedures, depending on the capacity of a given batch of Affi-Gel 501, is then utilized to elute the active P_i transporter. *Method I* is employed with low-capacity Affi-Gel 501 (i.e., 4–5 μmol Hg/ml) and involves elution with a linear 0–24 mM gradient of 2-mercaptoethanol in buffer C (48 ml). The reconstitutively active P_i transporter appears in the eluate after the first several milliliters. *Method II* is carried out with high-capacity Affi-Gel 501 (i.e., 8–9 μmol Hg/ml). It is important to note that with Method II, prior to the washing and packing steps, the high-capacity Affi-Gel 501 resin is diluted with an equal volume of BioGel A-5m (100–200 mesh, Bio-Rad). Elution involves the addition of a two-step 2-mercaptoethanol gradient. The first step consists of the addition of 1.5 mM 2-mercaptoethanol in buffer C (17 ml). The active P_i transporter is then eluted with 30 mM mercaptoethanol in buffer C. Typically, the first 1–1.5 ml of eluate following the 30 mM 2-mercaptoethanol addition is devoid of transport activity. The transporter then elutes in approximately the next 5 ml. Utilizing either methods I or II, eluted fractions are incorporated into liposomes and assayed for transport activity as described below. The purified P_i transporter fractions are stable for at least several days when stored frozen (in liquid nitrogen) in liposomes.

Preparation of Liposomes

Asolectin (233 mg) is transferred into a 30-ml Corex tube and dried under a stream of nitrogen. The dried lipid is redissolved in diethyl ether, and the solvent is removed as above. The dried lipid is evacuated for at least 30 min (while protected from light) in order to remove remaining traces of solvent. Following the addition of 2.1 ml of buffer D, the tube is flushed with nitrogen, sealed, and vortexed. The lipid is then dispersed in a bath sonicator until the mix appears transparent. This step typically requires 15–30 min.

Preparation of Proteoliposomes

An aliquot of freshly isolated protein (0.25–0.35 ml; typical protein ranges are 0.5–0.9 mg of hypotonically shocked Triton X-114-solubilized mitoplasts; 15–40 μg of HA eluate; 2–4 μg of DEAE eluate; 2–20 μg of Affi-Gel 501 eluate) is mixed with 0.53 ml of liposomes. The mixture is then vortexed and rapidly frozen in liquid nitrogen. Immediately prior to assay, the sample is thawed in a room-temperature water bath for 10 min. The thawed suspension is then sonicated with a probe sonicator (Biosonik; Bronwill Scientific, Rochester, NY) for a total of 19 sec (i.e., two 7-sec pulses and one 5-sec pulse, applied over a 30-sec time interval) at room temperature, employing the microtip and a probe intensity setting of 20. The proteoliposomes are subsequently diluted by the addition of 125 μl of buffer D, pH 7.0, allowed to temper for at least 5 min at room temperature, and then assayed for transport activity.

Measurement of NEM-Sensitive Phosphate/Phosphate Exchange

Transport reactions are conducted at room temperature in the presence of 10 mM phosphate. These incubations are carried out behind a Plexiglas shield (greater than 1 cm thick) and two pairs of disposable gloves are worn by the investigator in order to minimize exposure to radiation. Additionally, a survey meter (Ludlum Measurements, Inc.) with an appropriate detector is utilized in order to rapidly detect any radioactive spills or contamination. Experimental reactions are initiated by the addition of 20 μl of carrier-free $^{32}P_i$ [3–6 × 10^5 Bq (becquerel); final specific radioactivity in the reaction mix = 0.9–1.9 × 10^5 Bq/μmol] to 320 μl of proteoliposomes. The transport incubations are typically carried out for 60 sec and are then quenched by the addition of 20 μl of 270 mM NEM (15 mM final concentration). Controls are incubated with the same amount of NEM for 60 sec prior to the transport-triggering addition of $^{32}P_i$. After a given transport incubation is quenched, a 30-μl aliquot of the reaction mixture is removed and placed into 255 μl of ice-cold buffer D, pH 7.0. Aliquots (80 μl) of the diluted reaction mixture are then placed onto Sephadex columns and the intraliposomal $^{32}P_i$ is quantified as described below. The NEM-sensitive phosphate/phosphate exchange reaction is then calculated by subtracting the control value from the experimental value. Finally, it should be noted that this calculation requires a determination of the specific radioactivity (i.e., counts per minute per micromole P_i) of the P_i in the reaction mix. This value is determined by measuring the counts per minute (via Cerenkov counting) of a known volume of the undiluted reaction mix and then dividing this value by the calculated amount of chemical P_i in this volume.

Direct Determination of Initial Phosphate Transport Rates and the Amount of Protein That Associates with Pelletable Proteoliposomes

The initial rate of phosphate transport, normalized to the amount of protein that actually associates with pelletable proteoliposomes, can be measured as follows. After the freeze-thaw sonication steps, the entire contents of three tubes of proteoliposomes are combined, diluted with ice-cold buffer D, pH 7.0 (4.0–5.3 ml), and centrifuged at 135,000 g (average) for 35 min. The resulting pellet is then resuspended in 1 ml of buffer D (room temperature, pH 7.0). Aliquots (200 μl) of the suspension are assayed for NEM-sensitive P_i/P_i exchange utilizing the following procedure. After the addition of 20 μl of water, transport is triggered by the addition of 30 μl of $^{32}P_i$ (final specific radioactivity in the reaction mix is approximately 2.2–4.4 × 10^5 Bq/μmol). At various time intervals (i.e., 20, 40, and 60 sec), 60-μl aliquots are removed and placed into 3.53 μl of 270 mM NEM (final NEM concentration = 15 mM). The quenched reaction mix is then diluted with approximately 8 vol of ice-cold buffer D, pH 7.0. Aliquots (80 μl) are added to Sephadex columns and processed as described below, except that the eluates (and subsequent washes) from four tubes are added to a single scintillation vial (i.e., one time point). Controls are preincubated with 20 μl of 165 mM NEM (final NEM concentration during the preincubation = 15 mM) instead of water, for 60 sec prior to the $^{32}P_i$ addition. At various times following the addition of isotope (30 μl) to control incubations, 60-μl aliquots are removed and placed into 3.53 μl of 45.6 mM NEM (final NEM concentration = 15 mM). The control reactions are then diluted and processed identically to the experimental reactions. Protein is directly measured in the remainder of the proteoliposomal suspension by the method of Kaplan and Pedersen.[19] Blanks for the protein assay contain liposomes, processed as described above, except that buffer C, containing 30 mM 2-mercaptoethanol, is added instead of protein. The NEM-sensitive transport rate is then normalized to the amount of protein measured in the proteoliposomal suspension. It is important to note that the protein which pellets with the liposomes represents the sum of the incorporated protein plus any nonspecifically adsorbed protein. Since only the former can catalyze NEM-sensitive P_i/P_i

[19] R. S. Kaplan and P. L. Pedersen, *Anal. Biochem.* **150**, 97 (1985).

exchange with the internal liposomal volume, the specific transport activity value determined by this method represents a minimum estimate. This estimate is likely to be considerably more accurate than the value obtained if one assumes that all the added protein does in fact associate with the liposomes.

Determination of Intraliposomal $^{32}P_i$

Following all transport reactions, intraliposomal $^{32}P_i$ is separated from the extraliposomal label via centrifugation through Sephadex columns and is then quantified as follows. Sephadex columns are prepared via a method based on that of Penefsky.[20] Disposable 1-ml tuberculin syringes (Plastipak; B-D 5602) are fitted with polyethylene filters (4-mm diameter, 35-μm porosity, 1/16-in. thickness, Bel Art Products, Pequannock, NJ). Sephadex G-50 medium, which has been swollen in buffer E, is poured into the syringes via Pasteur pipet. The syringes are filled to the top and the buffer is allowed to drain. The syringes are then placed into Pyrex test tubes (13 × 100 mm) and centrifuged at approximately 2100 rpm for 90 sec (excluding acceleration and deceleration time; IEC model HN-SII centrifuge; IEC rotor 958). Following centrifugation, the syringes are placed into clean test tubes.

After a given transport incubation, 80-μl aliquots of the diluted and quenched reaction mixtures are placed onto the Sephadex. These columns are then centrifuged again at approximately 2100 rpm for 90 sec. The eluates are transferred into scintillation vials (two eluates per vial) and each test tube is rinsed with 2 ml of buffer B plus 3% (v/v) Triton X-100. The washes are also transferred into the same vials. Radioactivity is then determined in a liquid scintillation counter via Cerenkov counting through an open window. Employing this procedure, intraliposomal $^{32}P_i$ passes through the columns and is then quantified, whereas the external label is retained on the columns.

Covalent Modification of the Phosphate Transporter with NEM

The purified phosphate transporter can be labeled with NEM as follows. The Affi-Gel 501 eluate (10–19 μg protein) is incubated with 40 mM NEM for 5 min on ice. The reaction mix is then diluted with 7–9 vol of acetone and allowed to remain on ice for at least an additional 5 min. The incubation mix is then centrifuged in an Eppendorf centrifuge (model

[20] H. S. Penefsky, this series, Vol. 56, p. 527.

5412) for 2 min. The resulting pellets are then washed via the addition of 150 μl of water followed by 6 vol of acetone per pellet. After an additional 5-min incubation on ice, the mixtures are recentrifuged as above. The final pellets are prepared for application to polyacrylamide gels[12] and then combined.

Miscellaneous Procedures

Mitochondria and mitoplasts are prepared as described by Pedersen et al.[21] employing the modifications that have been previously discussed.[22] Protein (in the absence of liposomes) is determined by the Lowry procedure[23] modified as described previously.[22] SDS–polyacrylamide gel electrophoresis and sample preparation are carried out as previously detailed.[12,22]

Properties of the Purified Phosphate Transporter

Employing the procedures described above, highly active phosphate transporter can be purified from rat liver mitochondria in either a two-subunit (method I) or a single-subunit (method II) form. Starting with 300 mg of initial mitoplast protein, methods I and II yield approximately 100–200 and 50–100 μg of purified P_i transport protein, respectively. The SDS-polyacrylamide gradient gel electrophoretic profile of sequential steps in the P_i carrier purification scheme (method I, low-capacity Affi-Gel 501) is depicted in Fig. 1. The final purified material (lane 5) consists primarily of two protein bands with molecular masses of approximately 33 and 35 kDa. As shown in Table I, the purified material displays an NEM-sensitive P_i/P_i exchange activity of 8.07 μmol/min/mg protein which is enhanced 161-fold relative to the initial Triton X-114-solubilized mitoplasts. The time course for the $^{32}P_i$ transport reaction into proteoliposomes, which are formed utilizing the purified transporter, is depicted in Fig. 2. The transport proceeds linearly for approximately 15 sec and reaches a plateau after 180–300 sec. Analysis of the data assuming an exponential approach to isotopic equilibrium (Fig. 2B) yields a first order rate constant of 0.85 min^{-1} and a $t_{1/2}$ of 49 sec.

[21] P. L. Pedersen, J. W. Greenawalt, B. Reynafarje, J. Hullihen, G. L. Decker, J. W. Soper, and E. Bustamante, *Methods Cell Biol.* **20**, 411 (1978).

[22] R. S. Kaplan and P. L. Pedersen, *J. Biol. Chem.* **260**, 10293 (1985).

[23] O. H. Lowry, N. J. Rosebrough, A. L. Farr, and R. J. Randall, *J. Biol. Chem.* **193**, 265 (1951).

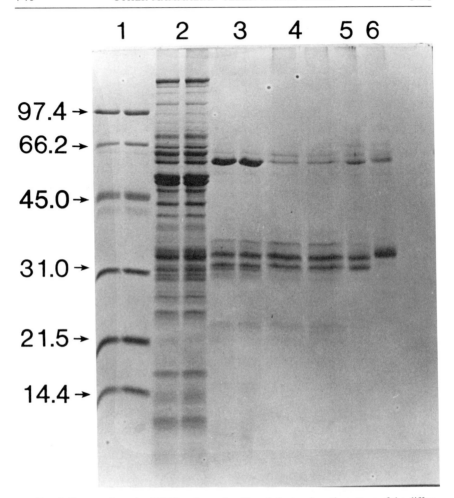

FIG. 1. Coomassie-stained SDS–polyacrylamide gel electrophoretic pattern of the different stages of purification of the active phosphate transporter from rat liver mitoplasts. Proteins were run in a 4.5% polyacrylamide stacking gel followed by a highly resolving 14–20% gradient gel. Lane 1, 2 μg of each Bio-Rad SDS–PAGE low-molecular-weight standard protein: phosphorylase b (97,400), bovine serum albumin (66,200), ovalbumin (45,000), carbonate dehydratase (31,000), soybean trypsin inhibitor (21,500), lysozyme (14,400). Lane 2, 40 μg of the hypotonically shocked, Triton X-114-extracted mitoplasts. Lane 3, 15 μg of the hydroxylapatite eluate. Lane 4, 12 μg of the DEAE eluate. Lane 5, 19 μg of the Affi-Gel 501 eluate (method I). Lane 6, 19 μg of the NEM-labeled Affi-Gel 501 eluate. (From Kaplan et $al.$,[12] by permission of the publisher.)

TABLE I

PURIFICATION OF THE PHOSPHATE TRANSPORTER FROM RAT LIVER MITOPLASTS[a]

Fraction	NEM-sensitive P_i/P_i exchange (μmol/min/mg protein)	Specific activity enhancement (-fold)
Triton X-114 solubilized mitoplasts	0.05 (8)	—
Hydroxylapatite eluate	2.00 (8)	40
DEAE eluate	4.99 (8)	100
Affi-Gel 501 (methods I + II) eluate	8.07 (12)	161

[a] Frozen rat liver mitoplasts (approximately 300 mg of protein) were employed as starting material. Purified protein fractions were incorporated into phospholipid vesicles in the presence of 10 mM KP$_i$. Transport reactions were carried out for 60 sec. Other conditions were as described above. Values in parentheses represent the total number of preparations assayed. (From Kaplan *et al.*,[12] by permission of the publisher.)

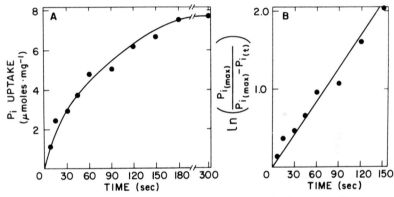

FIG. 2. Time dependence of the NEM sensitive P_i uptake into proteoliposomes. (A) Direct plot of P_i uptake versus time. The Affi-Gel 501 eluate (method I) was incorporated into phospholipid vesicles in the presence of 10 mM P$_i$. An aliquot (20 μl) of either water (experimental reaction) or 165 mM NEM (control reaction) was added to 200 μl of proteoliposomes. Sixty seconds later, carrier-free ^{32}P$_i$ (20 μl) was added. Successive 20-μl aliquots were removed at various times from a given incubation and placed into 15 mM NEM (final concentration). The quenched reactions were then diluted with ice-cold buffer D (pH 7.0, 170 μl), and intraliposomal ^{32}P$_i$ was determined as described above. All values were corrected for NEM-insensitive uptake of label. Each data point represents a mean of five to seven incubations. (B) Logarithmic plot of saturation of exchange versus time. $P_{i\,(max)}$ is the maximum extent of P_i uptake. $P_{i\,(t)}$ is the extent of uptake at time t. The rate of the phosphate exchange is calculated as the product of the first order rate constant (0.85 min^{-1}) and the $P_{i(max)}$ value (7.69 μmol of P$_i$/mg of protein, observed at 300 sec). Under the conditions of this assay, the rate was 6.54 μmol of Pi/min/mg of protein. (From Kaplan *et al.*,[12] by permission of the publisher.)

The SDS–polyacrylamide gradient gel electrophoretic profile of the single-subunit form of the purified phosphate transporter, which is prepared utilizing method II (high-capacity Affi-Gel 501), is depicted in Fig. 3, lane 2. This material consists mainly of the 33-kDa protein. Most importantly, a reconstituted specific transport activity of 8 μmol/min/mg protein is observed with purified P_i transporter prepared by either methods I or II. This finding indicates that both the 33- and the 35-kDa protein species appear to be functionally competent and catalyze P_i transport to similar extents.

With the purified transporter one can also carry out direct measurements of both the initial rate of the NEM-sensitive P_i/P_i exchange as well as the amount of protein that actually associates with the proteoliposomes (i.e., protein that pellets with the proteoliposomes during ultracentrifugation). We have performed these measurements with the purified transporter (prepared via method II) and determined a specific transport activity of 20.8 μmol/min/mg protein. This value is enhanced severalfold over the values typically obtained in the absence of initial rate and direct protein association measurements.

Effect of Covalent Labeling Agents on the Purified Phosphate Transporter

Incubation of the purified P_i transporter with NEM, a sulfhydryl reagent which inhibits phosphate transport, causes a marked decrease in the electrophoretic mobility of the 33-kDa band. This is observed with purified P_i carrier prepared either via method I (Fig. 1, lanes 5 and 6) or method II (Fig. 3, lanes 2 and 3). Moreover, with the method I preparation, the 33- and 35-kDa bands migrate as a single 35-kDa species. Finally, the transport activity of the reconstituted purified P_i carrier is nearly completely abolished by other sulfhydryl reagents such as mersalyl (15 mM) and pCMB (4 mM).

Other types of covalent labeling agents[24] also inhibit the activity of the purified transporter following its reconstitution into phospholipid vesicles. Thus, reagents selective for tyrosine (N-acetylimidazole, 15 mM), histidine (diethyl pyrocarbonate, 16 mM), lysine and the amino terminus of a protein (dansyl chloride, 3 mM) cause a nearly complete inhibition (i.e., >95%) of transport. Moreover, the guanidine-selective reagent phenylglyoxal (5 mM) is somewhat inhibitory (i.e., 64%).

---

[24] For a review, see R. L. Lundblad and C. M. Noyes, "Chemical Reagents for Protein Modification," Vols. 1 and 2. CRC Press, Boca Raton, Florida, 1984.

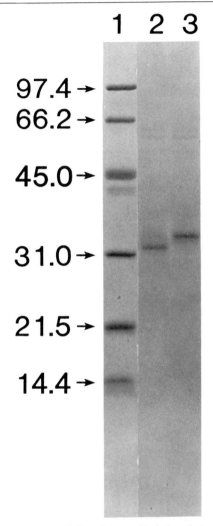

FIG. 3. Effect of NEM on the mobility of the purified phosphate transporter (method II) in an SDS–polyacrylamide gradient gel. Lane 1, 3 μg of each Bio-Rad SDS–PAGE low-molecular-weight standard. Lane 2, 10 μg of the Affi-Gel 501 purified (method II) P_i carrier (preincubated with water rather than NEM). Lane 3, 10 μg of the Affi-Gel 501 eluate which was preincubated with 40 mM NEM. SDS–polyacrylamide gel was as described for Fig. 1. (From Kaplan et al.,[12] by permission of the publisher.)

Comparison with Other Preparations of the Phosphate Transporter

It is useful to compare the properties of our phosphate transporter preparation from rat liver mitochondria with preparations from other tissues that have been obtained by different laboratories. Thus, Kolbe *et al.*[14,27] have obtained highly purified P_i carrier from bovine heart mitochondria utilizing chromatography on sequential HA columns. These chromatographic steps were carried out first in the absence and then in the presence of SDS and urea. SDS–PAGE of their final fraction indicated that it consisted of two protein bands, each with a molecular mass of approximately 34 kDa. As we observe with the purified rat liver P_i carrier (method I, Fig. 1, lanes 5 and 6), alkylation of their preparation with NEM also caused the two protein bands to comigrate. Reconstitution experiments indicated that their final preparation displayed some transport activity. However, the magnitude of their specific activity was substantially decreased when compared with their less purified material (i.e., the initial HA eluate)[14,25] presumably due to the presence of urea and SDS. Bisaccia and Palmieri[15] have purified the phosphate transporter from porcine heart mitochondria, utilizing hydroxylapatite chromatography in the presence of exogenous cardiolipin, as the primary purification step. Their preparation is approximately 90% pure and runs as a single band on SDS–PAGE with an apparent molecular mass of 33 kDa. Upon incorporation into phospholipid vesicles, this fraction catalyzed P_i transport with a specific activity of 26.0 μmol/min/mg protein.

Final Comment

An important point which remains unresolved concerns the nature of the two protein bands that are revealed by SDS–PAGE of the purified phosphate transporter obtained from either rat liver (method I) or beef heart mitochondria. The question arises as to whether these bands reflect (1) intraprotein sulfhydryl–disulfide interactions, which may restrict protein unfolding and may continue to occur during SDS–PAGE,[14] even though the samples have been pretreated under reducing conditions prior to electrophoresis and, in the case of the beef heart preparation, the electrophoresis was performed in the presence of mercaptoacetic acid; or (2) two distinct but very homologous proteins that differ in primary sequence and/or posttranslational modification and comprise the functional phosphate transport system.[12,14] This controversy will be resolved only

[25] H. Wohlrab, H. V. J. Kolbe, and A. Collins, this series, Vol. 125, p. 697.

after these two proteins have been separated, reconstituted, and the primary sequence of each determined.

Acknowledgment

This work was supported by National Science Foundation Grant DMB-8606759 to P.L.P.

[46] Measurement of Vacuolar pH and Cytoplasmic Calcium in Living Cells Using Fluorescence Microscopy

By FREDERICK R. MAXFIELD

Introduction

As described in other chapters in this series, many methods have been developed for the measurement of ion transport processes in cells and organelles. The method described in this chapter, quantitative fluorescence microscopy using indicator dyes, is particularly useful for making local measurements of ion concentrations within living cells with good spatial and temporal resolution. It is ideally suited for measuring dynamic processes or changes taking place asynchronously in a population of cells. Current limitations of the method include some difficulties in interpretation of data and the relatively high cost of instrumentation. During the past decade there have been dramatic improvements in microscopy instrumentation and in the development of improved indicator dyes. These improvements are leading to the widespread use of quantitative fluorescence microscopy for studying cell physiology. In this chapter, application of these methods to measure free Ca^{2+} in the cytoplasm and pH within endocytic and lysosomal compartments is described. With appropriate indicator dyes, the same methods can be used for measuring other ions.

Principles of Measurement

The fluorescence properties of indicator dyes are altered by binding of metal ions or by titration of ionizable groups. Fluorescein derivatives are the most widely used dyes for measurement of pH within cellular compartments. The pH dependence of the excitation profile for the fluorescein isothiocyanate derivative of a protein is shown in Fig. 1A. The fluorescence is strongly pH dependent between pH 5.0 and 7.0. A very useful

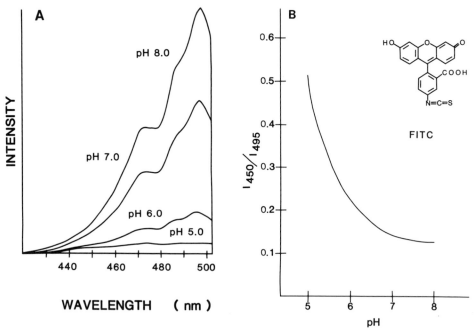

FIG. 1. pH dependence of fluorescein fluorescence. (A) The serum protein α_2-macroglobulin was labeled using fluorescein isothiocyanate. The protein was dissolved in buffers at the indicated pH values, and fluorescence excitation profiles were obtained using a spectrofluorometer. The emission was monitored at 520 nm. (B) Fluorescein-labeled α_2-macroglobulin was dissolved in buffers between pH 4.0 and pH 8.0. Fluorescence intensities of solutions were measured with 450- and 490-nm excitation using a microscope spectrophotometer, and I_{450}/I_{490} at each pH was determined. Inset: fluorescein isothiocyanate (FITC).

parameter is the ratio of fluorescence intensities with excitation at 450 and 490 nm (I_{450}/I_{490}). This ratio, which is dependent on pH but independent of dye concentration, is shown in Fig. 1B. Intensity ratio values are particularly useful in fluorescence microscopy measurements since the amount of dye in the sample volume is generally unknown.

Several fluorescent indicators for the measurement of cytoplasmic [Ca^{2+}] have been synthesized by Tsien and collaborators.[1,2] Quin 2 and, more recently, fura 2 have been used in many laboratories. The excitation spectra for quin 2 and fura 2 are shown in Fig. 2. Both indicators are sensitive to [Ca^{2+}] in the range 0–1 μM. As with fluorescein measurement

[1] G. Grynkiewicz, M. Poenie, and R. Y. Tsien, J. Biol. Chem. 260, 3440 (1985).
[2] R. Y. Tsien, T. Pozzan, and T. J. Rink, J. Cell Biol. 94, 325 (1982).

of pH, a fluorescence intensity ratio can be used for measuring $[Ca^{2+}]$. Figure 2C shows the Ca^{2+}-dependent I_{340}/I_{380} ratio for fura2.

Fura2 has several significant advantages over quin2.[1] These include markedly higher fluorescent intensity (especially in the Ca^{2+}-free form), resistance to photobleaching, and better selectivity for Ca^{2+} vs Mg^{2+}. Since the dissociation constant of Ca^{2+} bound to fura2 is somewhat higher than for quin2, the range of fura2 sensitivity includes higher Ca^{2+} concentrations. The increased fluorescence brightness allows one to use lower indicator concentrations, thereby reducing the perturbation of the cells.

Ideally, fluorescent indicator dyes would be sensitive to only one environmental factor (e.g., pH or $[Ca^{2+}]$). In fact, fluorescence properties of all indicator dyes are sensitive to a variety of environmental influences,

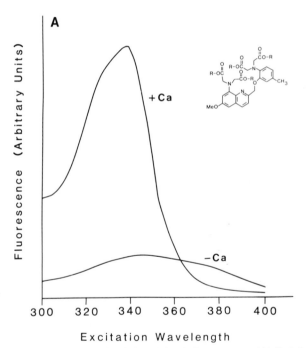

Excitation Wavelength

Fig. 2. Calcium dependence of quin2 and fura2 fluorescence. (A) Quin2 (2 μM) was dissolved in buffers containing no calcium (2 mM EGTA) or saturating levels of calcium ($[Ca^{2+}] = 100$ μM). Fluorescence excitation spectra were obtained with the emission at 500 nm using a spectrofluorometer. Inset: Quin2. (B) Same as in (A) except fura2 (500 nM) was used and the emission wavelength was 510 nm. Inset: fura2. (C) Fura2 (500 nM) was dissolved in buffers with $[Ca^{2+}]$ from 10 nM to 10 μM (see text for description of buffers). Fluorescence intensities of the solutions were measured with 340- and 380-nm excitation using a microscope spectrophotometer, and the I_{340}/I_{380} value at each $[Ca^{2+}]$ was determined.

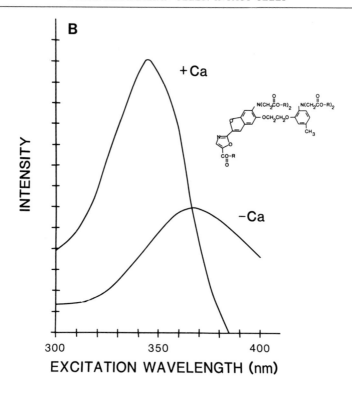

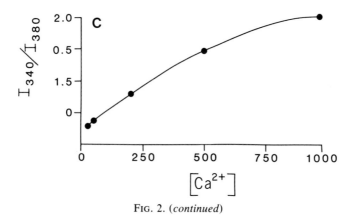

FIG. 2. (*continued*)

and this must be kept in mind when interpreting fluorescence spectra of indicator dyes in cells. Effects such as changes in solvent polarity or viscosity, adsorption to proteins or other cellular components, interference from other ions, the presence of quenching agents, or covalent modification of the indicator within the cells can all interfere with the accurate measurement of a specific ion in living cells. These are not merely theoretical considerations. Examples of each of these types of interference have been encountered in the measurement of pH or $[Ca^{2+}]$ in cells. As discussed in later sections of this chapter, methods have been developed for reducing the possibility for misinterpretations due to interference from these types of effects. However, it is impossible to anticipate all of the possible cellular effects on fluorescence, and no simple set of protocols can guarantee that the fluorescent indicators are behaving as desired.

The basis of measurement using indicator dyes is the equilibrium binding of the appropriate cation to the dye. In a thermodynamically defined solution one could use the dyes to measure thermodynamic activities of the cations, using Debye–Hückel corrections to obtain the individual ionic activity coefficients. However, within cells the microenvironment of the dye is not well-enough defined to justify this thermodynamic interpretation. The quantity measured is empirically defined by the test equilibrium (e.g., $Ca + fura2 \rightleftharpoons Ca:fura2$). This equilibrium measures the concentration of Ca^{2+} effectively available for binding to the dye (i.e., free Ca^{2+}). The Ca^{2+} concentration in the cells is determined by comparison with the dye fluorescence in calibration solutions where $[Ca^{2+}]$ is known. It should be noted that the measurements of ionic concentrations or pH within cells will only be accurate on an absolute scale to the extent that the calibration buffers duplicate all of the other important factors (other ions, polarity, etc.) within the cell.

Instrumentation

A system for quantitative fluorescence microscopy is illustrated in Fig. 3. The system can be used for microspectrofluorometry using the photomultiplier and for digital image analysis using the video camera and image processor. It has proved very useful to have both capabilities on a single microscope. An inverted microscope, as shown in Fig. 3, has several advantages. Cells are grown on a coverslip which forms the lower surface of a tissue culture dish. With this arrangement the cells are accessible for rapid changes of the culture medium or for addition of agents to the culture medium during microscopic observation.

There are several important features of the microscope spectrophotometer shown in Fig. 3. The electronic shutter on the fluorescence illumi-

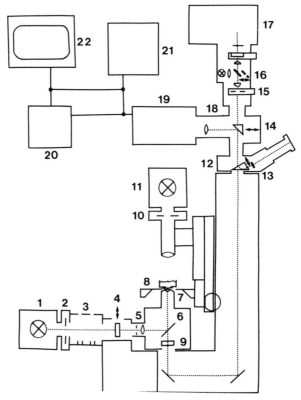

FIG. 3. A fluorescence microscope equipped for quantitative fluorescence microscopy. The diagram is based on the Leitz Diavert microscope with a Kinetek (Yonkers, NY) photometry system in use in the author's laboratory. Similar systems are available from other manufacturers. The use of an inverted microscope allows easy access to the sample chamber. Important components are indicated by numbers in the diagram. (1) Fluorescence illuminator with glass (visible excitation) or fused silica (UV excitation) collecting lens. (2) Electronically controlled shutter. (3) Slots for neutral density filters. (4) Exchangeable excitation bandpass filter. (5) Adjustable field diaphragm for limiting area of illumination. (6) Dichroic beam splitter which reflects excitation wavelength and transmits fluorescence emission. (7) Objective (fluorite for 340-nm excitation). (8) Cells growing on coverslip which forms lower surface of culture dish. (9) Emission filter. (10) Electronically controlled shutter. (11) Bright-field illuminator. (12) Exchangeable prism for reflecting light to eyepiece. (13) Electronically controlled shutter. (14) Movable partial reflecting mirror. (15) Diaphragm for limiting region of photometry measurement. (16) Lamp and removable mirror for adjusting measurement diaphragm. (17) Photomultiplier tube and housing. (18) Eyepiece for projecting image onto video camera. (19) Silicon intensified target video camera. (20) Video tape or disk recorder. (21) Image analysis system. (22) Video monitor.

nator is essential for regulating the time of exposure to high-intensity light. The shutter on the eyepiece reduces the entry of stray light, and the shutter on the halogen illuminator is used to block the transmitted light path during fluoresence measurements. Electronic controllers for the coordinated opening and closing of these shutters are available with commercially available microscope spectrophotometry systems.

The excitation wavelength is changed by placing appropriate filters in the light path. Filters can be changed manually or using motorized filter changers. It is advisable to avoid moving the dichroic beam splitter which reflects incident light into the microscope objective. Slight changes in the orientation of the dichroic mirror can lead to significant changes in the pattern of fluorescence illumination, and it is often difficult to achieve adequate alignment of two dichroic mirrors. Thus, the excitation filters should be mounted separate from the dichroic beam splitter. In the past few years computer-controlled microscope photometry systems have become available from several microscope manufacturers and instrumentation suppliers. The system in the author's laboratory (Kinetek, Yonkers, NY) controls all of the microscope shutters, changes excitation wavelengths via a six-position motorized filter wheel, records photometer output, and plots output values. Two measurements per second can be made with this arrangement. For higher time resolution, rapidly rotating filter wheels or dual illumination systems[3] can be used. It is possible to use a monochromator in place of the excitation filters. The advantage of this is great flexibility in excitation wavelength. However, filters generally have a higher light throughput, and they are much less expensive if only a few wavelengths are to be used. Narrow-bandpass filters covering the ultraviolet and visible spectrum are available from several suppliers.

Emitted fluorescent light is collected by the objective and passed through the dichroic beam splitter and one or more filters chosen to match the emission profile of the fluorophore. Complete illumination and emission systems for fluorescein and fura2 are listed in Table I.

The area of illumination is limited by a diaphragm placed at an optical image plane. The position of this diaphragm and the size of the opening can be adjusted during the alignment of the microscope. Restricting the illumination to the structure of interest (e.g., a single cell or a region within the cell) reduces the effect of stray light, and it prevents bleaching of nearby regions prior to measurement of their fluorescence. A second diaphragm, on the emission light path, is used to restrict the region of measurement. This diaphragm is adjustable in size and is centered on the optical axis of the microscope. In some commerical microscope photome-

---

[3] B. A. Kruskal and F. R. Maxfield, *J. Cell Biol.* **105**, 2685 (1987).

TABLE I
OPTICAL SYSTEMS FOR pH AND [Ca^{2+}] MEASUREMENTS

Dye	Excitation filter[a] (nm)	Dichroic beam splitter[b] (nm)	Emission filter[a] (nm)	Comments
Fluorescein	450 (10) 490 (10)	510	515 (longpass) 525 (20)	525-nm emission bandpass filter can be used to reduce autofluorescence, if necessary; 75-W xenon lamp is the light source
Fura2	340 (10) 380 (10)	400	430 (longpass) 510 (40)	340-nm excitation light is absorbed by glass. All optical components must be of fused silica or quartz. Nikon UV-fluor fluorite objectives have adequate transmission at 340 nm; 510-nm emission filter can reduce autofluorescence; 100-W mercury or 75-W xenon lamp is the light source; 340-nm filter can be replaced by a 350-nm filter with some loss of intensity

[a] The numbers given are the center transmission wavelength with the bandwidth at 50% transmission in parentheses.
[b] The beam splitter reflects light below the indicated wavelength and transmits light above that wavelength.

try systems (e.g., Leitz MPV), a small lamp in the photometer head is used to project the image of this diaphragm onto the sample plane. Careful alignment of the diaphragms and matching their size to the region of interest can reduce stray light significantly.

For whole-cell intensity recordings, the photometer can be operated at room temperature, and extremely sensitive photomultipliers are usually not required. The output from the photomultiplier can be recorded in analog or digital form. It is convenient to integrate the signal for the entire time the shutter is open, with the average intensity recorded at the end of the interval.

Image Analysis

The photomultiplier system can provide accurate intensity readings over a wide range of fluorescence intensities. One can obtain spatial information using a scanning stage, but it is often preferable to use a video image analysis system. As shown in Fig. 3, the video camera is mounted on the fluorescence microscope so that an image is focused on its faceplate. We have found it convenient to use a beam-splitting prism which reflects 90% of the light to the video camera while transmitting 10% of the light to the photomultiplier. This allows simultaneous acquisition of intensity and video recordings.

A schematic diagram of an image-processing system is shown in Fig. 4. An important consideration in the choice of video components is the intensity response of the devices. Photomultipliers have a large dynamic range—they will produce a linear response to incident light over several orders of magnitude of intensity variation. For video cameras the useful range is usually less than a factor of 100. If the signal falls outside this range, either the light intensity must be altered, or the operating parameters of the video camera must be changed. For measurement of intravesicular pH or cytoplasmic $[Ca^{2+}]$, a silicon-intensified target (SIT) video camera provides an adequate level of sensitivity and resolution. For quantitative work, it is absolutely required that the camera black level and gain be manually adjustable with all automatic regulation suppressed. It is also helpful if the intensity response curve of the camera is linear. (A nonlinear curve can be corrected during later processings steps, if necessary.)

Image intensifiers coupled with video cameras can be used as very sensitive detectors with favorable characteristics for quantitative image analysis.[4] Alternatively, charge coupled devices (CCDs) have excellent sensitivity, dynamic range, and signal-to-noise properties, especially when cooled.[5] The principal limitations of CCD-based image acquisition have been the speed of read-out and the difficulty of rapidly storing large numbers of high-resolution images. It is likely that both of these problems will be eliminated soon.

Output from a video camera can either be digitized directly or recorded for later analysis. If a video recorder is used, it is again important to make sure that no automatic signal modulators are used. (Many recorders automatically adjust the video signal to an "optimal" level, and this type of adjustment must be suppressed.) If data are recorded on video tape, it is often necessary to pass the signal through a time base corrector before digitization. The time base corrector adjusts for any timing errors which may come from slight variations in the video tape speed during recording or playback.

The signal from the video camera or tape recorder is then digitized and stored in memory. Flash digitizers work at video rates; an entire image

[4] K. R. Spring and R. J. Lowy, *Methods Cell Biol.* **29**, 269 (1989).
[5] R. S. Aikens, D. A. Agard, and J. W. Sedat, *Methods Cell Biol.* **29**, 291 (1989).

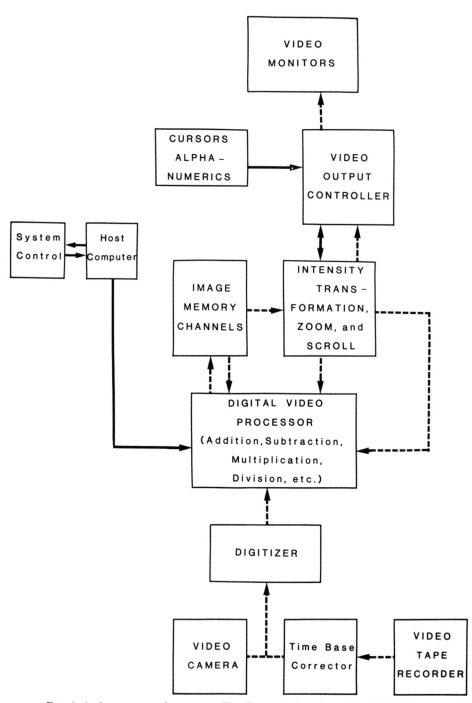

FIG. 4. An image-processing system. The diagram is based on a Gould IP8500 image-processing system with a Micro VAX II host computer which is in use in the author's

can be digitized and stored in 1/30 sec. Typically, the images are stored as a 512 × 512 array of picture elements (pixels) with 1 byte (256 gray levels) of intensity information per pixel. It is usually necessary to use signal averaging to reduce the noise level. On most image analyzers sequential video frames can be averaged together, and this is very helpful for low light level fluorescence images.

Beyond digitizing and storing the video data, a variety of functions can be carried out by the image processor. Many of these functions could also be carried out on the host computer but at a slower speed. Frequently used functions include adding or subtracting constant intensity values and the pointwise addition, subtraction, multiplication, or division of images. Intensity transformations are used for several image-processing operations. In an intensity transformation each intensity value in the input image is converted to a new intensity value in the output image. The output intensity values are contained in an array with 256 elements, and with appropriate hardware an intensity transformation of a 512 × 512 image can be accomplished in 1/30 sec. Intensity transformations can be used to correct for nonlinear response in video components, to enhance contrast, or to perform other types of image enhancement. Other functions useful for image analysis include generation of intensity histograms and display of intensity profiles. Useful books on image processing have been written by Castleman,[6] Rosenfeld and Kak,[7] and Inoué.[8]

Measurement of Endosomal and Lysosomal pH

The pH of various vacuolar compartments can be measured by delivering a fluorescein derivative to the appropriate compartment and then measuring the fluorescence intensity at two excitation wavelengths (e.g.,

[6] K. R. Castleman, "Digital Image Processing." Prentice-Hall, Englewood Cliffs, New Jersey, 1979.
[7] A. Rosenfeld and A. Kak, "Digital Picture Processing," Vols. 1 and 2. Academic Press, New York, 1982.
[8] S. Inoué, "Video Microscopy." Plenum, New York, 1986.

laboratory. Similar systems are available from other manufacturers. All image-processing steps are initiated by instructions from the host computer. Video output from a camera or video recorder is digitized at video rates by the digitizer. The digitized images (512 × 512 pixels, 1 byte/pixel) can be stored directly in image memory channels or they can be processed (e.g., averaged to reduce noise levels) using the digital video processor. The contents of image memory channels can be viewed on video monitors. Many mathematical operations on the images in memory are carried out with the intensity transformation unit and the digital video processor, which allow very fast parallel processing of image data. Some complex operations are carried out by the host computer. Digitized images can be stored on disks and magnetic tapes associated with the host computer.

450 and 490 nm). Excitation at 490 nm is used rather than 495 nm, which corresponds to the excitation maximum, to avoid problems with reflected light passing through the emission filters. The method for delivering the probe to the appropriate compartment will depend on the cell function being investigated. By appropriately choosing the fluorescent probe, the temperature, and the time of incubation, it is often possible to have most of the fluorescein in a single type of compartment such as lysosomes or nonlysosomal multivesicular endosomes.[9] In other cases, spatial separation of organelles can also help to limit the measurement to a single type of compartment (e.g., tubular structures near the Golgi complex).[10] It is usually necessary to use several biochemical and morphological methods to establish the nature of the compartment containing a fluorescent probe. For example, polypeptide ligands can be double labeled with fluorescein and ferritin so that the localization can be determined using both light and electron microscopy. Similarly, the presence of certain enzymes (e.g., acid phosphatase) within the compartment can be determined by cytochemical staining procedures. Radiolabeled ligands can also be used to follow entry of ligands into lysosomes or release back into the culture medium. Lysosomes are usually labeled by long incubations with fluorescein dextran (1–20 hr) followed by a chase period (1–3 hr). The dextran accumulates in the lysosomes without being degraded.[11]

Photometry Measurements

Once the probe has been delivered to an endocytic organelle, the fluorescence intensities are measured. The pH is determined from I_{450}/I_{490} after comparison with values from an appropriate calibration curve (see below). Usually, some additional problems must be overcome before a useful pH value can be obtained. The two major problems which are frequently encountered are intrinsic cellular fluorescence (autofluorescence) and the delivery of the probe to more than one type of compartment (e.g., probe remaining at the cell surface). If even a small amount of probe is in a neutral or high pH environment, the higher fluorescence brightness of the fluorescein at high pH (Fig. 1) will unequally weight this component in a measurement of average pH.

In many cases the autofluorescence can be accounted for simply by measuring the fluorescence intensity of unlabeled cells and subtracting the average intensity of an unlabeled cell from the intensity of a labeled cell. This correction is adequate if the autofluorescence is low compared to the fluorescein fluorescence and if the autofluorescence intensities are

[9] D. J. Yamashiro and F. R. Maxfield, *J. Cell Biol.* **105**, 2723 (1987).
[10] D. J. Yamashiro, B. Tycko, S. R. Fluss, and F. R. Maxfield, *Cell* **37**, 789 (1984).
[11] S. Ohkuma and B. Poole, *Proc. Natl. Acad. Sci. U.S.A.* **75**, 3327 (1978).

uniform within a population of cells. As an alternative, when the auto-fluorescence is high and variable, the autofluorescence of a field of cells (or a single cell) can be measured. The cells are then incubated with the fluorescein-labeled probe on the microscope stage. After the incubation, extracellular ligand is removed by repeated rinses and aspiration of the medium. All operations are performed without moving the cells. After the incubation and rinsing, I_{450} and I_{490} are again measured on the same field. By this procedure an exact correction for autofluorescence can be made. The disadvantage is that the rate of data acquisition is very slow since only one field can be measured per incubation.

Fluorescence from fluorescein remaining at the cell surface is a difficult problem for whole cell (photometer) measurements. A "null point" method has been developed which corrects for this problem.[12] The extra-cellular medium is set at a test pH using a nonpermeant buffer, and fluorescein fluorescence is measured with 490 nm excitation. The pH of all endocytic compartments is then rapidly equilibrated to the test pH by addition of ammonium acetate and methylamine (both 20 mM). If the fluorescence intensity increases upon addition of the weak acid and weak base, the average pH of the endosomes was below the test pH. The intensity will fall if the endosomes were more alkaline than the test pH. The pH where no change is observed corresponds to the average endoso-mal pH.

Image Analysis Measurements

Digital image analysis can yield all of the information available from whole-cell photometry readings with the potential for additional spatial resolution. If one simply takes the total fluorescence intensity in the entire video field, or a defind region within the field, then the fluorescence intensity ratio values should be the same as those obtained with a photometer. (In fact, the proportionality between the photometer intensities and digitized video intensities should be checked carefully to be sure that the entire video system is responding linearly to intensity.) Digital image analysis is extremely useful since it allows the measurement of fluorescence intensities from individual vesicles or regions of the cell in which vesicles are highly concentrated. In several cases, it has been possible to distinguish two types of vesicles within a cell or to analyze only the vesicular fluorescein fluorescence against a background of autofluorescence.[9]

In order to analyze the fluorescence from individual vesicles or groups of vesicles, a method for detecting such vesicles within the image must be developed. Unfortunately, there is no uniquely best method for doing

[12] D. J. Yamashiro and F. R. Maxfield, *J. Cell Biol.* **105**, 2713 (1987).

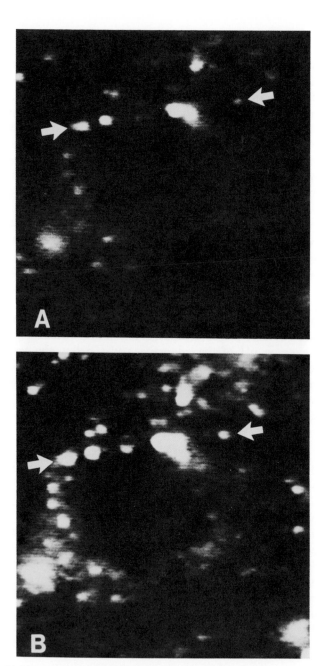

FIG. 5. Digitized images of cells for intravesicular pH measurements. Chinese hamster ovary cells were incubated with fluorescein-labeled dextran for 30 min. Fluorescence images were recorded with excitation at (A) 450 nm and (B) 490 nm. The recorded images were digitized and eight sequential frames were averaged. Intensity profiles along the 10-μm line between the arrowheads are shown in Fig. 6.

this. We have developed a procedure which has proved useful for measuring pH in several types of endocytic compartments. In several test cases, the pH values were confirmed by whole-cell photometry readings.

Figure 5 shows a fluorescence micrograph of a cell containing fluorescein in endocytic vesicles. The intensity profile along a line is shown in Fig. 6. It can be seen that the fluorescence within a vesicle is superimposed on a slowly varying (diffuse) background of fluorescence. The width of a vesicle (at half-maximum intensity) is approximately 6 pixels, corresponding to 1 μm in this case. It should be noted that apparent size in the video image is not necessarily the same as the actual size of the object. The optical properties of the microscope, the properties of the camera, the tape recorders, and other electronic devices on the data path can all lead to blurring or spreading the image of an object.

After a digitized image has been obtained, the measurement procedure is divided into two steps. First, the background fluorescence is estimated. This is the fluorescence expected at the same location in the cell if the vesicle were not present. We determine this from fluorescence intensities near the vesicle. A background image is formed such that each pixel's intensity is the median intensity of a 32 × 32 pixel square around that point. Since the square is much larger than the vesicle, the median (fiftieth percentile intensity) should not be affected by the presence or absence of one or two vesicles within the square. The size of the square should be adjusted depending on the properties of the image. The square should be small enough so that the background represents local variations in intensity but large enough so that fluorescence from vesicles will not affect the median value. The background image is then subtracted from the original image, yielding a background-corrected image. The dashed line in Fig. 6 shows the background intensities calculated along a line in Fig. 5.

A threshold value is now used to identify vesicles in the background-corrected image from which the I_{450}/I_{490} ratio will be determined. All pixels below the threshold are assigned an intensity value of zero, and all pixels above the threshold are unchanged. This threshold must be chosen to include most of the vesicles while excluding the remaining nonvesicular fluorescence. The absolute intensities of individual vesicles are greatly altered by variations in the threshold value, but ratio values are not significantly affected over a wide range of threshold values. The threshold is usually determined using the 490-nm image, and all of the fluorescent spots above the threshold are identified. The fluorescence intensity of exactly the same pixels are determined in the 450-nm image, and the I_{450}/I_{490} ratio is calculated for each spot. Assuming that all parts of a spot have the same pH, the measured I_{450}/I_{490} ratio should not be altered by selecting the brightest pixels with a threshold. On the other hand, the signal-to-noise ratio is most favorable in regions of high intensity. The background-

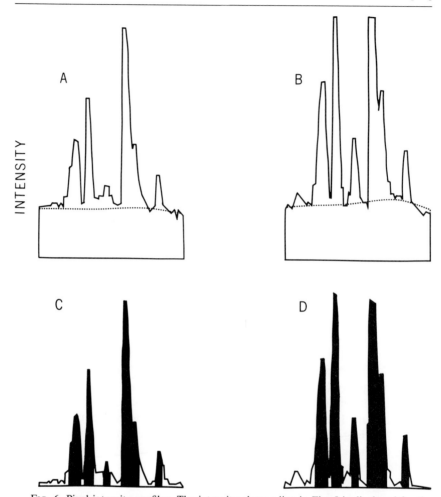

FIG. 6. Pixel intensity profiles. The intensity along a line in Fig. 5 is displayed for the images with (A) 450-nm and (B) 490-nm excitation. The image-processing steps for pH measurement in vesicles are illustrated. The solid lines show the intensity of individual pixels along the line. The dashed lines are the local background values calculated separately for each image. The local background for a pixel is defined in this case as the median intensity in a 32×32 pixel (6×6 μm) region around the pixel. This local background is subtracted from the original image (negative values are set to zero), and the intensities in the difference images are shown in (C) and (D). A threshold intensity is determined for the 490-nm image, and all intensities above the threshold are indicated by the shaded part of the curve. The intensities of the same pixels are shaded in the 450-nm image. I_{490} is determined for all of the pixels which form a contiguous region above the threshold intensity, and I_{450} is determined for the same region. The pH of the vesicle is then determined from the I_{450}/I_{490} value.

corrected intensities of pixels above the threshold intensity are illustrated in Fig. 6 as the filled parts of the curve.

Once the fluorescence intensity values of individual spots have been determined, the I_{450}/I_{490} ratio for each spot is calculated. The ratio is converted to a pH value by reference to a calibration curve. The pH value obtained is the internal pH of a single vesicle or the collection of vesicles which form a contiguous spot in the 490-nm image after background correction and application of a threshold.

This procedure has been used successfully to measure the pH of individual endocytic vesicles and lysosomes in cells with a high level of autofluorescence.[9] Since the autofluorescence was diffuse (as is often the case), it could be subtracted out automatically during the background correction.

It is very useful to be able to measure the pH of lysosomes or endocytic vesicles individually. In a whole-cell measurement, the pH values will be strongly biased toward higher pH values as a result of the reduction in fluorescein fluorescence at acidic pH values. For example, if 20% of the fluorescein were in a pH 7.4 environment and 80% in a pH 4.6 environment, the "average" pH measured for the whole cell would be 6.0.

Calibration Curves

Since the fluorescence from indicator dyes can be affected by several influences other than the ion of interest it is desirable to obtain calibration curves for the fluorophore when it is in an environment very close to that found in living cells. This calibration curve can be compared to one obtained with the fluorophore free in solution to determine if any local environmental effects are altering the fluorescein behavior.

Calibration buffers are prepared to cover the pH range from 4.5 to 7.5 in increments of 0.5 pH units. These are supplemented with monensin (10 μM) and methylamine (10 mM) to collapse any pH gradients across membranes. Cells are incubated with the fluorescein probe exactly as they would be for a measurement on live cells. At the time the measurements would have been made, the cells are fixed with 2% formaldehyde in phosphate-buffered saline for 5 min at room temperature. This fixes the fluorescent probe in the compartment from which measurements were made in living cells. The cells are then placed sequentially in the calibration buffers, and I_{450}/I_{490} measurements are made on the fixed cells using either photometry or image analysis as described above for living cells. The calibration curve is obtained by plotting I_{450}/I_{490} vs pH (Fig. 1B). The same curve can be generated for the fluorescent protein in solution, using all of the same microscope optical components as were used for the measurements on cells. The fluorescent solution can be placed in a hemocytometer chamber (with the coverslip facing the objective) or placed in the well of a coverslip bottom dish.

In most cases, the calibration curves obtained in solution have agreed with the curves obtained from the cells within about 0.2 pH units over the range from pH 5.0 to pH 7.0. This is in large part a result of the specificity of fluorescein as an indicator of pH. It is not significantly affected by factors such as ionic strength, protein concentration, or other ions in the range normally encountered in mammalian cells.[11,13] However, some exceptions have been observed, and these indicate the importance of obtaining calibration curves in the cells. In some phagocytic cells, the generation of oxidants may alter the fluorescence properties of fluorescein through covalent modification of the dye.[14] Measurements with fluorescein-labeled proteins bound to the surface of a fresh water amoeba also gave values inconsistent with the bulk pH of the solution.[15] In this case the low ionic strength may have allowed Donnan effects, arising from fixed charges on the membrane, to alter the local pH as sensed by the fluorescein. These effects would be substantially diminished by the ionic strength of physiological salts in mammalian cells.

In-cell calibrations provide a reasonable indication that the fluorescein is behaving as expected in the intracellular environment. In addition, the agreement between solution calibration curves and the pH dependence of I_{450}/I_{490} as obtained by image processing helps to validate the image-processing steps described in the previous section. If there were systematic errors in the image processing protocol, the curves of I_{450}/I_{490} vs pH obtained by image analysis would not match the curves from photometry measurements of solutions.

Measurement of Cytoplasmic Free Ca^{2+}

Much of the instrumentation used to measure pH within endocytic compartments can be used with minor modification to measure cytoplasmic [Ca^{2+}] using the fluorescent indicator fura2. As with fluorescein, the measurements are based on the ratio of fluorescence intensities with two

[13] J. M. Heiple and D. L. Taylor, in "Intracellular pH: Its Measurement, Regulation, and Utilization in Cellular Functions," p. 22. Liss, New York, 1982.
[14] J. K. Hurst, J. M. Albrich, T. R. Green, H. Rosen, and S. Klebanoff, J. Biol. Chem. 259, 4812 (1984).
[15] P. L. McNeil, L. Tanasugarn, J. B. Meigs, and D. L. Taylor, J. Cell Biol. 97, 692 (1983).

excitation wavelengths. In the case of fura2, the I_{340}/I_{380} ratio is strongly dependent on $[Ca^{2+}]$. Since many microscope objectives do not transmit 340-nm light, excitation at 350 nm can be used as an alternative with some loss of intensity.

Loading the Dye into the Cytoplasm

Since fura2 and quin2 are highly charged molecules, they do not freely cross cell membranes. In many cases, the cells can be loaded using membrane-permeant esters of the dyes which are cleaved to the tetracarboxylate indicator form by cytoplasmic esterases.[2] Acetoxymethyl esters of both indicators (fura2AM and quin2AM) are commercially available. Cells are incubated with fura2AM (approximately 1 μM) or quin2AM (20–50 μM) for 10–30 min to allow entry of the indicator. The extracellular ester is removed by repeated rinsing, and the cells are incubated for several minutes to allow complete cleavage of the ester. This hydrolysis can be monitored using the fluorescence spectra of cell extracts and comparing spectra of them with the ester and acid forms at low and high $[Ca^{2+}]$.

Several precautions must be used when loading the cells with the esters. Fura2AM is very poorly soluble in aqueous buffers. When diluted from DMSO stock solutions into aqueous buffer, fura2AM is only partially dissolved even at concentrations as low as 1 μM. Solubility can be improved by diluting the fura2AM into buffer containing 5–10 mg/ml fatty acid-free albumin. Undissolved fura2AM is removed by centrifugation (10,000 g for 5 min). A nonionic dispersing agent, Pluronic F-127 (Molecular Probes, Junction City, OR), can be used to increase the solubility of fura2AM in aqueous cell loading medium. The following solutions are mixed sequentially: 4 μl of fura2AM (5 mM) in DMSO, 1 μl of Pluronic F-127 (25% w/v) in DMSO, 30 μl of calf serum, and 1 ml of loading medium (e.g., HEPES-buffered saline). If a suspension containing undissolved particles is added to cells, the particles can become attached to the surface of the cells or engulfed by the cells. This creates regions of very high fluorescence arising from undissolved fura2AM. This problem is much less serious with quin2AM, but microcentrifugation is still recommended to remove some undissolved fluorescent particles.

Many cell types can be loaded successfully using the acetoxymethyl esters. However, in other cell types the dye is not trapped efficiently in the cytoplasm. In some cell types very little fluorescence becomes cell associated, perhaps as a result of low cytoplasmic esterase activity.

Alternate procedures have been developed for introducing fura2 into the cytoplasm. These include direct microinjection through micropipets into the cytoplasm[16] and transient permeabilization of cells by ATP.[17]

Even after introduction of the dye into the cytoplasm, the distribution must be monitored carefully. In some cell types, the fura2 fluorescence is disproportionately high in the nucleus, suggesting that the dye is concentrated there by an unknown mechanism. A potentially serious problem is that in many cases the fura2 concentration in the cytoplasm decreases with time. This effect appears to be reduced at lower incubation temperatures. This dye movement has characteristics similar to organic anion transport systems, and in macrophages it has been shown that fura2 loss can be blocked by probenecid, a drug that inhibits organic anion transport.[18] The fura2 may leave the cell by crossing the plasma membrane, or it may become concentrated in organelles. For example, when PtK2 rat kangaroo kidney epithelial cells were incubated with fura2AM at 37°, fluorescent staining was observed in a punctate pattern throughout the cell. Homogenization and centrifugation of these cells showed that almost half of the fura2 in the cells sedimented with cellular organelles.[17] This dye redistribution would not be detected by whole-cell intensity recordings. In summary, the loading of fura2 into the cell is not a routine procedure, and care must be taken to be sure that the fluorescence arises from cytoplasmic fura2. Although it has many disadvantages, quin2 seems to be more predictable than fura2 in terms of its distribution in the cell when loaded as the acetoxymethyl ester.[2]

The calcium indicator dyes are themselves calcium chelators, and they are generally present at concentrations which are significant compared to the normal Ca^{2+} buffering capacity of the cell. Cytoplasmic concentrations of quin2 are often in the $0.5-1$ mM range, and it has been shown that this concentration can significantly diminish transient increases in cytoplasmic $[Ca^{2+}]$. Since fura2 gives approximately 30-fold higher fluorescence than quin2, much lower concentrations of fura2 can be used (approximately $20-100$ μM). This should reduce the effect of the dye on overall $[Ca^{2+}]_i$ changes. However, the dye is still present in large excess over $[Ca^{2+}]_i$, and this may have a significant effect on localized, transient increases in free Ca^{2+}. This effect can be examined by measuring the

[16] M. Poenie, J. Aldeton, R. Y. Tsien, and R. A. Steinhardt, *Nature (London)* **315**, 147 (1985).

[17] R. R. Ratan, M. L. Shelanski, and F. R. Maxfield, *Proc. Natl. Acad. Sci. U.S.A.* **83**, 5136 (1986).

[18] T. H. Steinberg, A. S. Newman, J. A. Swanson, and S. C. Silverstein, *J. Cell Biol.* **105**, 2695 (1987).

magnitude and duration of $[Ca^{2+}]$ transients with different intracellular concentrations of fura2. Illumination of cells with high intensity visible or UV light can be toxic, and shutters should be used to minimize the duration of exposure. Neutral density filters should be used to attenuate the light intensity to the minimum value that provides adequate signal-to-noise. It is necessary, therefore, to demonstrate that loading the cells with the dye and making the fluorescence measurements does not affect the cells or the process being investigated.

Once the dye has been loaded into the cytoplasm, $[Ca^{2+}]_i$ is measured from the ratio of fluorescence at two excitation wavelengths. For fura2, the I_{340}/I_{380} ratio is usually used.[1] In both cases, the fluorescence ratio is converted to $[Ca^{2+}]$ by comparison with an appropriate calibration curve (see below). The major advantage of the fluorescence ratio method is that the ratio is independent of the amount of indicator dye in the measurement area. Thus, the ratio is independent of cell thickness and dye concentration.

Whole-Cell Photometry Measurement

Cells are grown on a coverslip which forms the lower surface of an incubation chamber. The excitation and measurement areas are aligned and adjusted to match the size of a cell. Autofluorescence is measured using both 340- and 380-nm excitation. For sparse cultures it is useful to measure the intensity in regions with no cells to determine the relative contributions of autofluorescence and scattered (or reflected) light. In cells loaded with fura2, the autofluorescence is often 10% or less of the fura2 fluorescence. I_{340} and I_{380} are measured on the cells loaded with fura2. If necessary, the background or autofluorescence values are subtracted from each measurement before calculating the I_{340}/I_{380} value. The duration of measurements is usually limited to less than 1 sec to minimize exposure of the cells to light, especially when repeated measurements of the same cell are made.

Agents thought to raise $[Ca^{2+}]_i$ can be added directly to the culture medium. The open dish on an inverted microscope allows very fast addition or exchange of incubation medium. If the dish is held rigidly in place, repeated medium changes can be made without detectable motion of the cells. Repeated measurements of I_{340} and I_{380} can be made following the stimulus to determine the time course of changes in $[Ca^{2+}]_i$. The time between measurement of I_{340} and I_{380} should be as short as possible so that the ratio represents essentially a single time point.

Image Analysis for Local Measurement of [Ca²⁺]ᵢ

Cells are loaded with fura2 and placed on the microscope as discussed in the preceding sections. The video camera is set to manual control (all automatic signal regulation disabled). The gain and black levels are adjusted to provide a usable video signal. No parts of a fluorescent cell should appear black on the video monitor, and none should have saturating levels of brightness. Quantitatively, the intensities of all pixels should be within the linear range of the video camera. This can be checked by digitizing images of the same field with a series of neutral density filters used to attenuate the incident light. The pixel intensities should be proportional to the incident light intensity. If the brightest pixels are too high in intensity, they may show nonlinear response; in that case, the camera gain or the illumination intensity should be reduced. If the brightest pixels are too low in video output intensity, the full gray scale range of the digitizer may not be used. This can lead to some loss of intensity information. The linearity should be checked with the brightest and least bright images expected for a series of experiments (e.g., with $[Ca^{2+}]_i$ set high or low using ionophores with high or low extracellular $[Ca^{2+}]$). If necessary, corrections for nonlinearity should be made using intensity transformations as described in the instrumentation section.

Once the camera controls have been set, the images with 340- and 380-nm excitation are digitized directly or recorded for later digitization. The advantage of recording is that images of rapidly changing cells can be stored for later analysis. Repeated 340- and 380-nm images are obtained throughout the time course of the experiment. For image analysis it is especially important that the pair of 340- and 380-nm images are obtained within a very short time interval since even slight changes in cell shape between measurements can make large changes in the I_{340}/I_{380} ratio. For example if a region of the cell becomes thicker between measurement of I_{340} and I_{380}, the I_{380} value would be increased, leading to an erroneous conclusion that $[Ca^{2+}]$ was low in that region. This type of effect can be especially pronounced at the edges of motile cells. One control for this type of artifact is to reverse the order of measurement. If the ratio image depends on whether the 340- or 380-nm image was obtained first, then it is likely that motion artifacts are present.

In addition to the images of cells, images of regions without cells are obtained. These images are used to correct for camera black level and light coming from noncellular sources. It is important that these background images be obtained under exactly the conditions used to obtain images of fluorescent cells. Images of cells without the fluorescent dye should also be obtained to verify that autofluorescence is an acceptably low fraction of total fluorescence.

The background images at each wavelength are subtracted pointwise from the images of the fluorescent cells. Regions of the image corresponding to cells are determined using a threshold intensity value in either the 340- or 380-nm image. Local [Ca^{2+}]$_i$ can be determined from the I_{340}/I_{380} ratio at each point in the image. A ratio image can be formed by dividing the 340-nm image by the 380-nm image. It is useful to change the ratio image into an image where intensity or color is proportional to [Ca^{2+}]. This can be accomplished using an intensity transformation table which converts ratio values into [Ca^{2+}] using the calibration curve. The [Ca^{2+}] image is a useful form of display for analyzing local [Ca^{2+}] gradients within a cell (Fig. 7). Intensity profiles through the cell can be used to display gradients graphically (Fig. 8).

Changes in [Ca^{2+}]$_i$ for the whole cell can also be obtained from the images. It is important to remember that the total (background corrected) intensity of the cell in the 340-nm image ($I_{340, \text{total}}$) should be divided by the total intensity of the same pixels in the 380-nm image ($I_{380, \text{total}}$). The average I_{340}/I_{380} value obtained from the ratio image, $(I_{340}/I_{380})_{av}$, is not an appropriate parameter to use since regions of low and high fluorescence (e.g., thin and thick regions of the cell) contribute equally to this average. In general, $(I_{340, \text{total}}/I_{380, \text{total}})$ is not equal to $(I_{340}/I_{380})_{av}$.

The practical limits of spatial and temporal resolution have not been determined. Video cameras require 1/30 sec per frame, and it is often necessary to average four or more frames to reduce noise levels. Thus, differences in Ca^{2+} can be detected with about 1-sec resolution using standard video cameras.

As shown in Fig. 8, we have been able to detect significant [Ca^{2+}] gradients over distances as short as 2 μm. It is possible that steeper gradients could be observed in thinner cells. Spatial resoltuion is limited by the usual constraints of light microscopy. In principle, resolution of about 0.3 μm can be achieved. However, this would only be achievable in an extremely thin part of the cell where the entire thickness of cytoplasm would be in focus. Images of I_{340}/I_{380} or [Ca^{2+}] are two-dimensional representations of three-dimensional cells. Contributions from out-of-focus planes are a major limitation in accurate measurement of [Ca^{2+}] gradients within the cell.

Calibration of [Ca^{2+}]$_i$ Measurement

In order to relate the fluorescence ratios to free [Ca^{2+}], a calibration curve is obtained. A series of dye-containing solutions of varying [Ca^{2+}] is prepared and the fluorescence ratio (at excitation wavelengths λ_1 and λ_2) of each solution is measured in the experimental instrument. The relation-

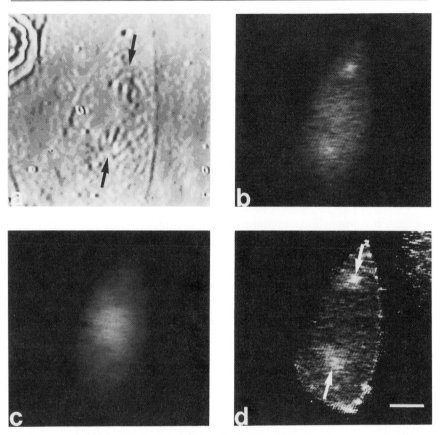

FIG. 7. Local measurement of cytoplasmic [Ca^{2+}]. A PtK2 rat kangaroo cell (a) was loaded with approximately 100 μM fura2 and observed during anaphase (Ref. 16). Digitized fluorescence images of the cell with (b) 340-nm excitation and (c) 380-nm excitation were obtained. Images of fields without cells were recorded under the same conditions and these background images were subtracted from the images of the fluorescent cells. An I_{340}/I_{380} image was obtained by pointwise division of the images. The I_{340}/I_{380} values were converted to [Ca^{2+}] using a calibration curve as described in the text, and the image of [Ca^{2+}] is displayed in (d). Bar, 5 μm.

ship between [Ca^{2+}] and the fluorescence ratio is expressed as follows[1]:

$$[Ca^{2+}] = [K_D (R_{exp} - R_0)F_{f, \lambda_2}]/[(R_{sat} - R_{exp})F_{b, \lambda_2}] \tag{1}$$

where K_D is the calcium-dye dissociation constant, R_{exp} is the fluorescence ratio from experiment, R_{sat} is the ratio of fluorescence intensities at

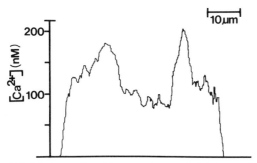

FIG. 8. Profile of [Ca²⁺] along a line. The cytoplasmic [Ca²⁺] along a line through the cell in Fig. 7 is displayed.

saturating calcium, R_0 is the fluorescence ratio at zero calcium, F_{f, λ_2} is the fluorescence intensity of free dye (no Ca²⁺) at λ_2 and F_{b, λ_2} is the fluorescence intensity of calcium-bound dye at λ_2. For fura2, λ_1 and λ_2 are generally chosen as 340 and 380 nm, respectively.

The results of a calibration curve can be used to obtain K_D in the appropriate calibration buffer (i.e., one which approximates the intracellular environment). R_0, R_{sat}, F_{f, λ_2} and F_{b, λ_2} are obtained from measurements on solutions with no free Ca²⁺ (less than 5 nM) or high free Ca²⁺ (>100 μM).

The source of calcium for preparing the calibration solutions is an equimolar CaEGTA stock at approximately 0.5 M. Equimolar amounts of the Ca²⁺ and EGTA stocks are determined by a pH titration procedure.[19] Stock solutions of nominally 1 M CaCl₂ and EGTA are prepared. One hundred-fold dilutions are made of each into 50 mM MOPS, pH 8.0. The dilute EGTA is titrated to a pH minimum with the dilute CaCl₂. (Protons are released from the EGTA as calcium binds). After the titration, the original (nominally 1 M) stocks are mixed in the same proportions which yielded the pH minimum. This produces an equimolar CaEGTA solution. It is then necessary to know the absolute concentration of the equimolar stock. This can be determined if the concentration of either the calcium or EGTA stocks is known accurately. The [EGTA] can be determined by titration against a Ca²⁺ standard solution using a calcium-selective electrode,[20] or the total calcium in the equimolar stock can be determined by atomic absorption spectrometry or standard quantitative analysis meth-

[19] O. G. Moisescu and H. Pusch, *Pfluegers Arch. Eur. J. Physiol.* **355,** R122 (1975).
[20] D. M. Bers, *Am. J. Physiol.* **242,** C404 (1982).

ods (e.g., precipitation as the oxalate and determination of the oxalate by permanganate titration).

A calibration buffer similar to that used by Tsien and colleagues[2] for quin2 and fura2 contains 115 mM KCl, 10 mM NaCl, 10 mM MOPS, 2.066 mM EGTA, 1.066 mM MgCl$_2$, adjusted to pH 7.05 with KOH. This buffer provides a free EGTA concentration of 2 mM and [Mg^{2+}]$_{free}$ = 1 mM. Fura2 (500 nM) or quin2 (2 μM) is added to a sufficient volume of buffer to make up all the desired calibration solutions. Aliquots are then removed for the solutions at each [Ca^{2+}] desired, and CaEGTA from the equimolar stock is added in an amount determined as follows:

Volume of CaEGTA added =

$$V_{CaEGTA} = ([EGTA][Ca^{2+}]_{free}\ V_{tot})/(K_{D,\ CaEGTA}[CaEGTA]_{stock}) \quad (2)$$

Where [Ca^{2+}]$_{free}$ is the final concentration of free Ca^{2+} in the calibration solution, V_{tot} is the final volume of the calibration solution after addition of CaEGTA, $K_{D,\ CaEGTA}$ is the apparent dissociation constant for CaEGTA under the conditions of the measurement, and [CaEGTA]$_{stock}$ is the concentration of the CaEGTA stock solution as determined experimentally. The $K_{D,\ CaEGTA}$ must be corrected for temperature, pH, and ionic strength.[21] For this calibration buffer at 20°, $K_{D,\ CaEGTA}$ is 306 nM. Equation (2) is obtained from the mass action law as follows:

$$[EGTA][Ca^{2+}]_{free} = K_{D,\ CaEGTA}\ [CaEGTA] \quad (3)$$

From the total concentration of EGTA and Mg^{2+} along with the dissociation constant of MgEGTA[21], [EGTA] = 2 mM. [EGTA] does not change significantly upon addition of CaEGTA. Since all of the Ca is added as CaEGTA,

$$V_{tot}[Ca^{2+}]_{tot} = V_{CaEGTA}[CaEGTA]_{stock} \quad (4)$$

Where [Ca^{2+}]$_{tot}$ = [Ca^{2+}]$_{free}$ + [CaEGTA].
Solving for V_{CaEGTA}, we get

$$V_{CaEGTA} = [Ca^{2+}]_{tot}\ V_{tot}/[CaEGTA]_{stock} \quad (5)$$

In the final solution, [Ca^{2+}]$_{tot}$ ≃ [CaEGTA] (i.e., there is little free Ca^{2+}). Then using Eq. (3) to replace [Ca^{2+}]$_{tot}$ we obtain Eq. (2).

21 A. E. Martell and R. M. Smith, "Critical Stability Constants," Vol. 1. Plenum, New York, 1974.

For saturating calcium, add $CaCl_2$ from a 1 M stock to a final concentration of 3 mM in the calibration buffer. The pH of each solution must be readjusted to pH 7.05 ± 0.01. Fluorescence of each solution is then measured at each excitation wavelength in the experimental instrument.

It is important to use deionized water which is uncontaminated with calcium. Glass serves as a Ca^{2+} reservoir, so solutions should not be exposed to glass. Previously used plasticware should be rinsed with boiling 10 mM Na_2EDTA and then thoroughly with deionized water. The use of Ca^{2+}-chelating resins such as Chelex is not recommended, since the resins may release free chelator into solution. The affinity of EGTA for Ca^{2+} depends on pH, so pH must be adjusted accurately. Calibration solutions are usually stable for 1–2 months at 4°. Ratio values are not absolute, but depend on the instrument. Thus, calibration data must be obtained on the same instrument in exactly the same configuration as used for obtaining experimental data.

Acknowledgments

I am grateful to P. Marks, B. Kruskal, D. Yamashiro, B. Tycko, R. Ratan, C. Keith, and M. L. Shelanski for many discussions of these methods. Financial support for NIH Grants DK27083 and GM34770 is acknowledged.

[47] Transport in Mouse Ascites Tumor Cells: Symport of Na+ with Amino Acids

By A. ALAN EDDY and E. R. JOHNSON

Introduction

Mouse ascites tumor cells, especially the so-called Ehrlich–Lettre line, have long served as a model system for studying the mechanism of Na^+-dependent solute transport in mammalian cells.[1] The cells are grown in the mouse peritoneal cavity for 7–10 days, a yield of 1 ml of packed cell volume per mouse typically being obtained. After washing and resuspension in a standard Ringer's solution, the tumor cells accumulate various common amino acids extensively, either during respiration of endogenous substrates or during glycolysis. For instance, 2-aminoisobutyrate can be concentrated 30 to 50-fold from a 0.1 mM solution in 30 min. Similar

---

[1] H. N. Christensen, *Biochim. Biophys. Acta* **779**, 255 (1984).

gradients of methionine or glycine can be formed with little concomitant metabolism of the amino acid. The accepted explanation is that the process is driven by the electrochemical gradient of Na^+ acting across the plasma membrane.[1,2] In contrast to the amino acids, common monosaccharides, including glucose, are absorbed by facilitated diffusion. Electrolyte transport in the tumor cells has also been studied widely in relation to the sodium pump, volume regulation, and the origin of the membrane potential.[3,4]

Materials and Methods

Ringer's Solutions. The standard Ringer's solution contains 155 mM Na^+, 8 mM K^+, 131 mM Cl^-, 16 mM orthophosphate, and 1.2 mM $MgSO_4$, at pH 7.4. Sodium ion or K^+ may be replaced by choline isotonically. Such solutions are prepared daily by mixing 0.77 M solutions of choline chloride, NaCl, KCl, or $MgSO_4$ with 0.1 M Na_2HPO_4 (or K_2HPO_4) at pH 7.4 and distilled water. These concentrated stock solutions are stored at 0° for up to 1 month.

Cell Preparations. After collection and washing, the tumor cells (1 vol) are incubated for 30 min at 37° in the standard Ringer's solution (25 vol). It is often useful to filter the warm cell suspension through a nylon coffee strainer at that stage. The cells are collected in the bottom of a graduated centrifuge tube (1500 g, 1 min). Contaminating erythrocytes are removed with the supernatant solution and from the top of the cellular pellet. The volume of packed cells is noted. After washing twice, using 25 vol of warm Ringer's solution each time, the packed cells can be kept at 4°. They should be used for experiments within 40 min.

The foregoing procedures substantially deplete the cellular endogenous amino acid pool, the components of which are known to exchange with extracellular amino acids. Further depletion of that pool can be achieved by subjecting the tumor cells to a hypotonic shock.[5]

Depletion of Cellular ATP Content. In the absence of glucose, addition of 2 mM NaCN to the tumor cell suspension, either with or without 10 mM 2-deoxyglucose, lowers the cellular ATP content at 37° by at least 95% during 3 min.[5] Exposure of the preparation to deoxyglucose in these

[2] A. A. Eddy, *in* "Amino Acid Transport in Animal Cells" (D. L. Yudilevich and C. A. R. Boyd, eds.), pp. 47–86. Manchester Univ. Press, Manchester, England, 1987.

[3] E. K. Hoffmann, *Biochim. Biophys. Acta* **864**, 1 (1986).

[4] E. Gstrein, M. Paulmichl, and F. Lang, *Pfluegers Arch.* **408**, 432 (1987).

[5] M. Morville, M. Reid, and A. A. Eddy, *Biochem. J.* **134**, 11 (1973).

circumstances for more than 30–45 min is deleterious. To avoid loss of cyanide the cell suspensions are shaken at 37° in stoppered flasks and washed in Ringer's solution containing cyanide.

Manipulation of Cellular [Na$^+$] and Cellular [K$^+$]. In order to lower cellular Na$^+$ when ATP is depleted, the tumor cells (1 vol) are suspended in 25 vol of Ringer's solution containing 163 mM K$^+$ and 2 mM NaCN. Similarly, soaking the preparation in a Ringer's solution containing 163 mM Na$^+$ leads, in the presence of 2 mM NaCN, to depletion of cellular K$^+$. When respiration is not impaired, the action of the sodium pump can be restricted by suspending the tumor cells at 37° in a Ringer's solution containing 163 mM Na$^+$ without added K$^+$ and including 0.3 mM ouabain. Cellular [Na$^+$] increases to at least 100 mM in 30–50 min and cellular [K$^+$] decreases to 60 mM or less.

Incubation with Amino Acid. As cell preparations depleted of ATP tend to accumulate Na$^+$ rapidly from the standard Ringer's solution, it is often convenient to resuspend the cell pellet as follows. At time zero the selected amino acid (0.1–20 mM), labeled with ^{14}C (0.03–0.3 μCi ml^{-1}) and dissolved in an appropriate Ringer's solution (10 ml), is added at 37° to the cellular pellet (20–60 mg dry wt) which is brought into suspension promptly by means of a glass rod. The cell suspension is shaken (1–2 Hz) and sampled (1–5 ml) at intervals up to 60 min. Whereas preparations depleted of ATP are studied in suspensions containing up to 6 mg cells (dry wt) ml^{-1}, those in which respiration or glycolysis is proceeding are used at 2 mg ml^{-1} to avoid changes in oxygen tension or pH.

The sampling procedures used are governed by the need to assay the respective extracellular and cellular contents of water, Na$^+$, K$^+$ and amino acid. For that purpose, a sample containing 20–50 mg of cellular dry wt is mixed with [^{3}H]inulin (1 μCi) and chilled to 0°. The tumor cells are separated from the medium by centrifugation (2 × 10^3 g, 1 min), the supernatant solution (M) being retained. The pellet of cells is weighed and extracted with 10 ml of 0.01 M HNO$_3$ for 2 hr. After separation of cellular debris the extract is assayed for ^{14}C, ^{3}H, Na$^+$, and K$^+$. The corresponding assays for the medium M are performed and the respective cellular and extracellular concentrations of amino acid, Na$^+$, and K$^+$ computed.

It should be noted that relatively small (1–2 mg) samples of the tumor cells can be separated from the suspending medium by centrifugation through 1-bromododecane.[6] Such samples are suitable for analysis of Na$^+$ and K$^+$ by atomic absorption spectroscopy but their ionic content is too small for analysis by flame emission spectroscopy.

[6] E. Johnson and A. A. Eddy, *Biochem. J.* **226,** 773 (1985).

Dependence of $[A]_i/[A]_o$ on $[Na^+]_o/[Na^+]_i$ in Tumor Cell Preparations Depleted of ATP

The above procedures are used to show how glycine uptake and efflux parallels the flow of Na^+ ions across the cell membrane when the sodium pump is not functioning owing to depletion of cellular ATP. The tumor cells are first depleted of both ATP and Na^+, by keeping them for 30 min at 37° in a K^+-based Ringer's solution lacking Na^+. In phase 1 of the experiment they are then suspended in the presence of 2.5 mM glycine in a cyanide-Ringer's solution containing 15 mM Na^+, a concentration similar to the initial cellular concentration of 25 mM Na^+. Under these conditions the cellular glycine concentration increases to about 80% of the extracellular concentration in 15 min. For phase 2 of the experiment the tumor cells are transferred at 23 min to a Ringer's solution containing 150 mM Na^+ and 1 mM glycine. During the next 15 min Na^+ flows into the cells and glycine is concentrated there 3 to 4-fold. Phase 3 starts at 54 min when cellular $[Na^+]$ is about 100 mM. The tumor cells are then transferred to a Ringer's solution of similar composition to the one used in phase I. Glycine now flows out of the tumor cells with Na^+. These events together with the behavior of the controls are illustrated in Fig. 1.

Inspection of lines C and D of Fig. 1 shows that the accumulation of glycine reached a maximum value that subsequently declined. This peak represented the transient maximum glycine gradient that could be sustained by the transient sodium gradient prevailing at that time. An extensive series of observations, like those illustrated in Fig. 1, showed that the transient steady state of glycine, as defined above, varied systematically with the sodium gradient according to the relationship[7]

$$\log([A]_i/[A]_o) = (1.257 \pm 0.068)\log([Na^+]_o/[Na^+]_i) + 0.068 \pm 0.023$$

Refinements of the above experiment include the demonstration that by elevating the membrane potential in the presence of valinomycin, the maximum amino acid gradient formed without the intervention of ATP is similar in magnitude to that formed during energy metabolism.[8]

Demonstration of Symport of Na^+ with Methionine

The tumor cells initially absorb glycine or methionine from a 10 mM solution at about 25 nmol min^{-1} mg^{-1} of cells (dry wt). They absorb Na^+ at an initial rate of about 50 nmol min^{-1} mg^{-1} when the sodium pump is inhibited either by depletion of ATP or in the presence of ouabain. Our

[7] A. A. Eddy, *Biochem. J.* **108,** 489 (1968).
[8] M. Reid, L. E. Gibb, and A. A. Eddy, *Biochem. J.* **140,** 383 (1974).

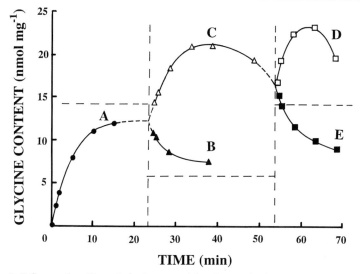

FIG. 1. Inflow and outflow of glycine caused by changes in the extracellular concentrations of Na^+ or glycine in preparations depleted of ATP. The vertical interrupted lines show the times at which the tumor cells were transferred to a new suspension medium. The horizontal interrupted lines show the cellular glycine content (nmol mg^{-1}) at which the cellular and extracellular glycine concentrations were equal (observations of M. C. Hogg). The experiment was divided into three phases:

	Extracellular concentration (mM)		Approximate cellular	Time
	[Na]	[Gly]	[Na] (mM)	(min)
Phase 1				
Line A (●)	15	2.5	25	0
Phase 2				
Line B (▲)	15	1	25	23
Line C (△)	150	1		
Phase 3				
Line D (□)	150	2.5	100	54
Line E (■)	15	2.5		

unpublished work indicates that ^{22}Na present in the standard Ringer's solution exchanges with cellular Na^+ with a half-time of about 0.4–0.5 min so that isotopic equilibrium is almost achieved in 2 min. This basal rate of exchange is not lowered by 0.5 mM amiloride or 1 mM furosemide. After such a preliminary incubation with ^{22}Na the amino acid is added, the further absorption of ^{22}Na serving as a measure of the cellular content of Na$^+$. A control lacks added amino acid. In the recommended procedure ouabain is present to prevent expulsion of cellular Na$^+$, together with valinomycin to minimize changes in the basal flow of Na$^+$ due to membrane depolarization by the amino acid.

Collect, preincubate, and wash the tumor cells (4 ml packed volume) as outlined above. Suspend them in the standard Ringer's solution (6.5 ml final vol) at 0°. Next, respective portions of this suspension are incubated (1) with ^{22}Na$^+$ in the presence of 10 mM methionine, (2) with ^{22}Na$^+$, and (3) with 10 mM methionine labeled with ^{14}C. The procedure is as follows: (1) Add 1 ml of the above cell suspension to 3.55 ml standard Ringer's solution at 37° and incubate for 1 min. (2) Add in turn 0.2 ml of 7.5 mM ouabain dissolved in the Ringer's solution, 10 μl of ethanol containing 50 μg of valinomycin and, finally, 2.5 μCi ^{22}Na. Incubate at 37° for 2 min. (3) Add 0.25 ml of 200 mM methionine dissolved in the standard Ringer's solution and containing ^{22}Na at the same specific activity as the cell suspension. Sample at nine intervals up to 2 min. For this purpose each sample (0.2 ml) is added to 0.8 ml of the standard Ringer's solution containing 0.3 mM ouabain at 0°. This stopping solution has been layered beforehand over 0.5 ml of 1-bromododecane at 0° in a microfuge tube. The

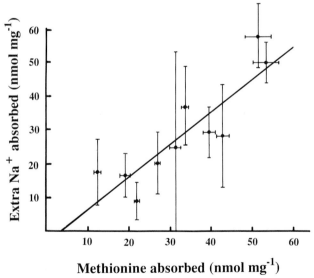

FIG. 2. Amount of extra Na$^+$ absorbed as a function of the amount of methionine absorbed in the same time. The results of four experiments were pooled. A linear regression analysis corresponded to a slope of 0.963 ± 0.191 equivalents of Na$^+$ per methionine equivalent and a coefficient of correlation $r = 0.872$.

tube is then subjected to centrifugation (20,000 g, 1 min) in an Eppendorf microfuge centrifuge. The supernatant solution and cell pellet are subsequently collected and analyzed for ^{22}Na. Further similar assays are performed in which, on the one hand, the methionine is labeled with ^{14}C (1 μCi) and ^{22}Na is omitted and, on the other hand, the amino acid is omitted and ^{22}Na is present. Figure 2 shows typical results obtained in four experiments. The scatter of the observations leads to considerable uncertainty in the estimate of the stoichiometry (n) of the coupling process ($n = 0.963 \pm 0.191$).

A recent study of Na$^+$ absorption in the presence of isoleucine also led to an estimate of the stoichiometry that was near 1, the symport stoichiometry being inferred from observations made during incubations of 30-min duration. The mathematical procedure employed took into account the considerable flow of isoleucine entering the tumor cells without Na$^+$ ions outside the symport.[9]

[9] A. A. Eddy, P. Hopkins, and E. R. Johnson, *Biochem. J.* **251**, 111 (1988).

[48] Measurements of Cytoplasmic pH and Cellular Volume for Detection of Na$^+$/H$^+$ Exchange in Lymphocytes

By S. Grinstein, S. Cohen,* J. D. Goetz-Smith, and S. J. Dixon

The Na$^+$/H$^+$ antiport is an amiloride-sensitive, electroneutral ion-exchange system present in the plasma membranes of most, if not all, mammalian cells. In nonepithelial cells the antiport is virtually quiescent at physiological cytoplasmic pH (pH$_i$), but it is markedly activated when the cytoplasm becomes acidic, suggesting a role in the regulation of pH$_i$. It has also been implicated in the regulation of cellular volume and in the initiation of proliferation by growth factors.[1] The activity of the antiport can be detected as a transmembrane flux of Na$^+$ (measurable either as the unidirectional flux of radioactive Na$^+$ or by the resulting changes in net cellular Na$^+$ or osmotic content), or by the intra- and/or extracellular pH changes that accompany transmembrane H$^+$ (equivalent) flux.

Two detection methods applicable to isolated cell suspensions are described below. One is based on the measurement of the osmotic cellular swelling which, under certain circumstances, accompanies Na$^+$/H$^+$ exchange. The other involves the direct monitoring of pH$_i$ by fluorescence

* Deceased.
[1] R. L. Mahnensmith and P. S. Aronson, *Circ. Res.* **56**, 773 (1985).

spectroscopy. Because the antiport is maximally active at low (subnormal) pH_i, procedures for acid loading the cells are also described. These methods have been implemented in our laboratory for the study of Na^+/H^+ exchange in rat thymic lymphocytes (thymocytes). A method for the isolation of these cells is also presented below.

Isolation of Rat Thymic Lymphocytes

Principle

The rat thymus is a convenient source of large numbers of relatively homogeneous cells that are viable in suspension for many hours. This organ is composed largely (>90%) of immature T lymphocytes which can be readily dispersed into a single cell suspension by mechanical means, avoiding the use of enzymes or divalent cation chelators.

Reagents

NaCl medium: 140 mM NaCl, 1 mM KCl, 1 mM CaCl$_2$, 1 mM MgCl$_2$, 10 mM glucose, 20 mM N-2-hydroxyethylpiperazine-N'-2-ethane-sulfonic acid (HEPES)-Na, pH 7.3

HEPES-buffered RPMI: Nominally HCO$_3^-$-free solution RPMI 1640 (Gibco) with 20 mM HEPES, pH 7.3

Procedure

One male Wistar rat weighing 150–200 g is sacrificed by asphyxiation with CO_2. An incision through the skin and superficial fascia is made with scissors along the midline. The thymus is exposed by cutting the ribs along the sternum, taking care not to puncture the heart or great vessels. The thymus is then removed using fine forceps, and kept in 5 ml NaCl medium at room temperature. Contaminating blood is removed by rinsing in NaCl medium. The thymus is then cut into small (1–2 mm^3) fragments with fine scissors, diluted with an additional 30 ml of NaCl medium at room temperature, and transferred to a 40-ml Dounce-type homogenizer fitted with the loose (B) pestle. The cells are then dissociated by 20–30 gentle strokes. To remove large cell aggregates and connective tissue, the resulting suspension is filtered through six layers of surgical gauze and the filtrate is then centrifuged at room temperature at 200 g_{max} for 5 min. The supernatant is discarded and the pellet washed by resuspension in 20 ml NaCl medium and resedimentation as before. The final pellet is resuspended in the desired volume of HEPES-buffered RPMI.

Comments. An average of 6–9 × 10^8 cells is obtained from each thymus. The viability of the cells, estimated by Trypan Blue exclusion, is normally ≥95%. If the cells are maintained at room temperature in HEPES-buffered RPMI at concentrations ≤5 × 10^7/ml, the viability remains essentially constant for up to 8 hr and the cells do not aggregate. To prevent cell settling, the cell suspension can be gently rotated, avoiding foaming.

Detection of Na$^+$/H$^+$ Exchange-Induced Cell Swelling by Electronic Sizing

Principle

This method detects the swelling that accompanies activation of Na$^+$/H$^+$ exchange in cells incubated with the Na$^+$ salt of weak organic acids (e.g., propionate).[2] As shown in Fig. 1a, the organic anion is in equilibrium with the protonated form (step 1), which can rapidly penetrate through the membrane (step 2). Intracellular dissociation (step 3) brings about cytoplasmic acidification and consequently activation of the antiporter, which will exchange extracellular Na$^+$ for internal H$^+$ (step 4). The continued presence of the weak acid, together with the sustained operation of the antiport, result in intracellular accumulation of Na$^+$ and of the organic anion. This is accompanied by uptake of osmotically obliged water, manifested as cell swelling. Therefore, measurement of the rate of cell swelling can provide an indirect estimate of the rate of Na$^+$/H$^+$ countertransport.

To ensure that swelling is indeed due to Na$^+$/H$^+$ exchange, the inhibitor amiloride can be added to the sodium propionate medium or a parallel experiment can be performed in Na$^+$-free potassium propionate medium. Both conditions will prevent swelling associated with Na$^+$/H$^+$ exchange. That cells suspended in the presence of amiloride or in potassium propionate are being acidified by propionate and are still capable of swelling can be demonstrated using exogenous alkali cation/H$^+$ exchangers. Monensin, an ionophore which catalyzes Na$^+$/H$^+$ exchange, will bypass the inhibitory effect of amiloride and induce marked swelling of cells in sodium propionate medium. Similarly, the K$^+$/H$^+$ exchanger nigericin will induce swelling of cells in potassium propionate.[2]

Fast, accurate, and reproducible measurements of the volume of suspended cells can be made by electronic sizing. The instrument most widely used for this purpose is the Coulter counter, though other instruments are commercially available. The operation of the Coulter counter is based on differences in the electrical conductivity between a cell and the

---

[2] S. Grinstein, J. D. Goetz, W. Furuya, A. Rothstein, and E. W. Gelfand, *Am. J. Physiol.* **247**, C293 (1984).

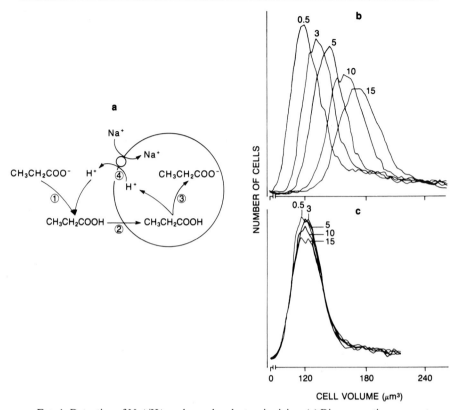

FIG. 1. Detection of Na^+/H^+ exchange by electronic sizing. (a) Diagrammatic representation of the sequence of events leading to cell swelling in isotonic sodium propionate medium: (1) Protonation of extracellular propionate; (2) permeation of undissociated propionic acid; (3) intracellular dissociation of propionic acid, generating propionate and lowering pH_i; (4) activation of Na_o^+/H_i^+ exchange, with intracellular accumulation of sodium propionate. (b and c) Volume changes of rat thymocytes suspended in sodium propionate medium with (c) or without (b) 100 μM amiloride. The curves are size distributions of cells at increasing times (in minutes) after resuspension in sodium propionate. Typical Coulter Channelyzer curves are illustrated. From S. Grinstein, S. Cohen, J. D. Goetz, and A. Rothstein [Fed. Proc. Fed. Am. Soc. Exp. Biol. **44**, 2508 (1985)].

suspending medium. The cell suspension is drawn by a vacuum pump through an aperture, through which a constant electric current is applied. As a cell passes through the aperture, the electrical resistance increases and a voltage pulse is generated, the magnitude of which is proportional to cell volume. The absolute volume of the cell can be calculated if the system is calibrated with standard particles of known size, taking into account the shape, deformability, and conductivity of both the cells and the standards.[3]

The volume distribution of a population of cells can be determined by connecting the Coulter counter to a pulse-height analyzer such as the C1000 Coulter Channelyzer. This instrument allocates the pulses according to their size into serially arranged memory locations ("channels"). At the end of each run the memory can be displayed visually on an oscilloscope screen or printer. A typical set of graphs is illustrated in Fig. 1b, which shows the progressive cell swelling observed when thymocytes are suspended in sodium-propionate-containing medium.

Reagents

Sodium propionate medium: 140 mM sodium propionate, 1 mM KCl, 1 mM CaCl$_2$, 1 mM MgCl$_2$, 10 mM glucose, 20 mM HEPES, pH 6.7 at 37°

Potassium propionate medium: 140 mM potassium propionate, 1 mM KCl, 1 mM CaCl$_2$, 1 mM MgCl$_2$, 10 mM glucose, 20 mM HEPES, pH 6.7 at 37°

Amiloride hydrochloride: 10 mM (in water)

Monensin: 5 mM (in ethanol)

Nigericin: 1 mM (in ethanol)

Procedure

When using a particular Coulter counter adapted with a specific aperture, the optimal settings to analyze a particular cell type must be determined at the outset. To measure rat thymocytes in the Coulter model ZM equipped with the 100-μm aperture, the current is set at 1 mA and the attenuation at 8, with a 20 kΩ matching. The Channelyzer base channel threshold is set at 5.

To follow the time course of swelling induced by Na$^+$/H$^+$ exchange, 1–2 × 10^6 cells is suspended in 20 ml sodium propionate medium at 37°; 500-μl samples are counted at increasing time intervals and the resulting size-vs-frequency graphs plotted as illustrated in Fig. 1b and c. The specificity of the swelling can be tested adding amiloride ($\geq$100 μM) to the sodium propionate medium or using potassium propionate medium. If the volume of the cells fails to increase under these conditions, their ability to swell can be confirmed using monensin (5 μM) or nigericin (1–5 μM).

From the volume measurements, the activity of the antiport can be quantified as the rate of amiloride-sensitive Na$^+$ uptake, if it is assumed

[3] G. B. Segel, G. Cokelet and M. A. Lichtman, *Blood* **57**, 894 (1981).

that swelling is isosmotic and that the volume gain arises exclusively from the uptake of sodium propionate and accompanying water.[2]

Comments. The electronic sizing method is highly reproducible: the coefficient of variation of 14 determinations of the initial median volume was only 0.02. Similarly, the rate of sodium propionate-induced swelling is consistent: the coefficient of variation of the initial rate of swelling, measured over a period of 3 min, was 0.11.

To obtain reproducible results, care must be taken to remove all the debris and large cell aggregates before counting. Blockade of the aperture is the only significant problem occasionally encountered during these measurements.

The pH of the propionate medium (6.7) was selected to increase the fraction of undissociated acid (to improve acid loading), while minimizing competitive inhibition of Na^+ uptake through the antiport by extracellular H^+.[2]

Detection of Na^+/H^+ Exchange by Measurement of Intracellular pH (pH_i)

pH_i Determination

Principle. The activity of the Na^+/H^+ antiport can be detected by monitoring the resulting changes in pH_i. A convenient and reproducible method for the measurement of pH_i in small suspended cells involves measuring the fluorescence of cells loaded with carboxylated fluorescein derivatives such as 2',7'-bis(carboxyethyl)-5,6-carboxyfluorescein (BCECF). As shown in Fig. 2, the fluorescence intensity of this compound is exquisitely pH dependent, increasing at more alkaline levels. The apparent pK_a of the fluorescence change is 6.97,[4] making this dye suitable for the study of pH_i in the physiological range. Because BCECF contains four dissociable carboxylate groups, it permeates biological membranes very poorly. However, the intracellular compartment can be loaded with the fluorescent probe by synthesizing it *in situ*. This can be accomplished by incubating the cells with a permeant precursor which is converted to BCECF by cytoplasmic enzymes. The acetoxymethyl ester of BCECF (BCECF-AM) is a suitable precursor, inasmuch as it is highly lipid soluble and readily cleaved by intracellular esterases.

Unlike BCECF, BCECF-AM is virtually not fluorescent (Fig. 2A and B). This provides a simple means of monitoring the course of the loading procedure. When a cell suspension is incubated with BCECF-AM, the

[4] T. J. Rink, R. Y. Tsien, and T. Pozzan, *J. Cell. Biol.* **95,** 189 (1982).

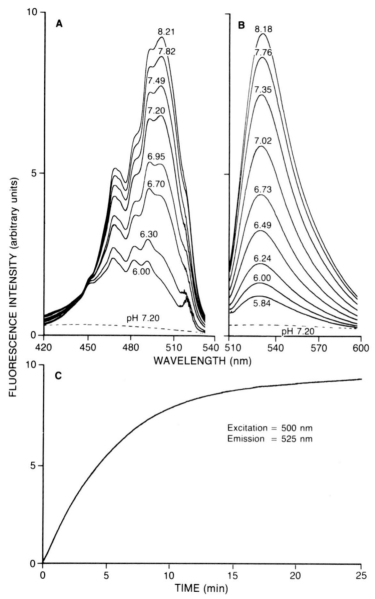

FIG. 2. Excitation (A) and emission (B) spectra of BCECF (solid lines) and an equimolar concentration of BCECF-AM (broken lines), at the indicated pH values. Excitation spectra were recorded with emission at 525 nm. Emission spectra with excitation at 500 nm. The spectra have not been corrected for the lamp or photomultiplier characteristics. (C) Time course of deesterification of BCECF-AM by thymocytes. A suspension of 3×10^7 cells in 1 ml of NaCl medium was added to a fluorescence cuvette and maintained at 37°. At time zero, 3 μg/ml BCECF-AM was added and the appearance of fluorescence was monitored at the indicated wavelengths.

conversion to the free acid can be followed by the appearance of fluorescence at 525 nm, using an excitation wavelength of 500 nm (Fig. 2C). In thymocytes, loading is complete within 30 min. The trapped dye is not concentrated in organelles but rather appears to be largely in the cytoplasmic compartment, as suggested by direct microscopic observation and by the finding that, with increasing digitonin concentrations BCECF is released from the cells before lactic dehydrogenase, a large cytoplasmic marker.

The fluorescence intensity of a suspension of BCECF-loaded cells is a function of their pH_i. The most direct method for calibration of fluorescence vs pH_i involves releasing the probe into the medium by lysing the cells and recording the fluorescence intensity at known values of pH. If the dye behaves differently in the medium than it did inside the cells, a correction must be introduced. The correction factor can be estimated by releasing the dye from the cells under conditions where the intra- and extracellular pH is identical. In this case, any fluorescence change observed when the cells are lysed must be attributed to interaction of the probe with cellular components. Equilibration of intra- and extracellular pH can be obtained using the K^+/H^+ exchange ionophore nigericin, which sets $[K^+]_i/[K^+]_o = [H^+]_i/[H^+]_o$. Therefore if the cells are suspended in media with a K^+ concentration similar to that of the cytoplasm, pH_i will follow pH_o. In practice, cells are suspended in K^+-rich solution, nigericin is added, and, after the fluorescence intensity has stabilized, detergent is added to lyse the cell. The ensuing fractional fluorescence change, which is relatively constant at varying pH, is taken as the correction factor.

If the starting pH values of a series of samples are identical, the need for individual (internal) calibration using detergent lysis is obviated. In this case a single external calibration can be applied to a number of samples. For this purpose, an equivalent sample is suspended in K^+-rich medium, nigericin is added, and pH_o (and therefore pH_i) is varied in steps, while recording the fluorescence intensity. Because in all samples the starting fluorescence and the quenching of the dye by cellular components are similar, the calibration curve obtained with nigericin is directly applicable to all the samples. It is important to note that, since quenching of the dye can vary with intracellular concentration, the nigericin/K^+ calibration applies only to cells loaded under identical conditions.

Reagents

BCECF-AM: 1 mg/ml (in dimethyl sulfoxide). Can be stored at $-20°$ for several months (Molecular Probes, Junction City, OR)
HEPES-buffered RPMI (see above)

K$^+$ solution: 141 mM KCl, 1 mM CaCl$_2$, 1 mM MgCl$_2$, 10 mM glucose, 20 mM HEPES-Na, pH 7.3 at 37°

Nigericin: 1 mM (in ethanol)

Triton X-100: 5% (v/v)

Procedure. The following procedure describes the conditions routinely used in our laboratory for rat thymocytes. Cells (3–10 × 10^7/ml) in HEPES-buffered RPMI are incubated for 30 min at 37° in the presence of 1–3 μg/ml BCECF-AM. The cells are then sedimented at 200 g_{max} for 5 min at room temperature, washed once with fresh HEPES-buffered RPMI and resuspended in this medium at the desired concentration (usually 5–10 × 10^7/ml). Aliquots of the loaded cell suspension are then transferred to a fluorescence cuvette containing the desired medium, to a final concentration of 3–5 × 10^6 cells/ml. Fluorescence is measured under continuous magnetic stirring in a fluorescence spectrophotometer with excitation and emission wavelengths of 500 and 525 nm, respectively, using 5-nm slits. A continuous recording of fluorescence (pH$_i$) can be obtained connecting the fluorimeter output to a Y (ordinate axis) vs time plotter.

If internal calibration is required, the cells are lysed with Triton X-100 (0.05%, v/v final), and the pH of the medium is changed stepwise by addition of small volumes of concentrated acid [e.g., 1 M 2-(N-morpholino)ethanesulfonic acid] or base (e.g., 1 M Tris), while monitoring fluorescence and measuring pH by inserting a small combination probe into the cuvette. The data are plotted in a linear graph of pH vs fluorescence intensity (in arbitrary units). To calculate the pH of the sample, the fluorescence recorded prior to lysis is multiplied times the correction factor, followed by interpolation in the calibration graph. The correction factor is estimated by suspending a sample in K$^+$ solution, adding 2–5 μM nigericin and, after a steady reading is obtained, lysing the cells with 0.05% Triton X-100. The correction factor is the ratio of the fluoresence observed before and after addition of detergent.

For external calibration a sample is suspended in K$^+$ solution and nigericin (2–5 μM) is added. The pH of the medium is then changed stepwise with acid or base and the pH and fluorescence are recorded after each addition as described above. In the absence of detergent, equilibration of pH$_i$ with pH$_o$ is not instantaneous, requiring typically 1–3 min after each pH step. The pH$_i$ of the test sample can then be directly interpolated in the resulting calibration curve.

Comments. In the case of thymic lymphocytes, loading with BCECF is very uniform. Within a population of cells, the fluorescence intensity is rather homogeneous (main graph, Fig. 3) with a small coefficient of variation, which can be partly accounted for by variation in cell size. Only

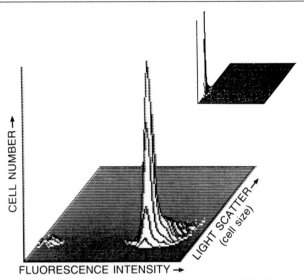

FIG. 3. Fluorescence and size distribution of a population of BCECF-loaded thymocytes. Multiparameter histogram of the green fluorescence intensity (BCECF fluorescence; x axis) and forward angle light scatter (a measure of cell size: z axis) of a typical preparation of BCECF-loaded rat thymocytes, obtained with the EPICS V flow cytometer. Inset: Multiparameter histogram of cells not loaded with the dye. Notice that cell autofluorescence is negligible.

about 5% of the cells are not labeled, which is similar to the fraction of Trypan Blue-permeable (nonviable) cells.

The efficiency of loading is very high. Most (>70%) of the BCECF-AM added to the medium is converted intracellularly to the free acid. For cells loaded as described above with 1–3 μg/ml BCECF-AM, the intracellular concentration of the acid can reach 5–15 μg/10^8 cells (Fig. 4), equivalent to an intracellular concentration of 84–252 μM. This concentration, which is sufficient for accurate pH_i determinations, has no effect on cellular viability or on the ability of cells to grow in culture. Moreover, the dissociable groups in the dye do not contribute significantly to the intrinsic buffering capacity of the cells, estimated at 25 mmol · liter cells^{-1} pH^{-1}.[5] This is in contrast to some fluorescent probes for intracellular Ca^{2+} determination, which markedly increase the total Ca^{2+}-buffering power of the cells.

The main problem encountered in some cell types is that of rapid leakage of BCECF. While in thymocytes loss of dye at room temperature

5 S. Grinstein, S. Cohen, and A. Rothstein, *J. Gen. Physiol.* **83**, 341 (1984).

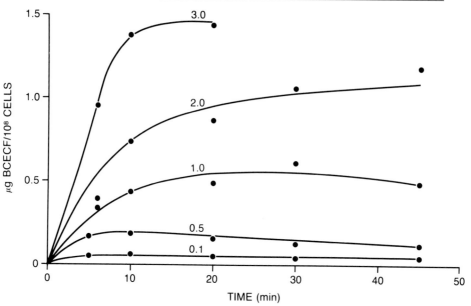

FIG. 4. Time and concentration dependence of BCECF loading. Rat thymocytes (10^8/ml) were incubated at 37° for varying times with the concentration of BCECF-AM (in micrograms per milliliter) indicated on the curves. The amount of BCECF generated was determined by sedimenting the cells and releasing the probe with 0.05% Triton X-100. The concentration of BCECF was determined fluorimetrically by comparison with BCECF standards containing an equivalent concentration of detergent.

is less than 7% hr^{-1}, other cell types are much more permeable. In extreme cases, such as some cultured erythroleukemic cells, their marked leakiness precludes the use of BCECF and analogous probes. For any particular cell type, the leakage rate can be easily determined by sedimenting aliquots of loaded cells at increasing time intervals and determining the fluorescence of the pellet and/or supernatant.

Acid-Loading and Fluorimetric Determination of Na$^+$/H$^+$ Exchange

Principle

In many cells the Na$^+$/H$^+$ antiport is nearly inactive at normal pH$_i$, but it is markedly activated when the cytoplasm is acidified. This section describes a convenient method to acid load suspended cells which is compatible with the fluorimetric determination of pH$_i$. The method is based on the use of nigericin, a carboxylic ionophore that catalyzes the electroneutral (tightly coupled) exchange of K$^+$ for H$^+$, but does not transport large organic cations. Addition of nigericin to cells suspended in media devoid of Na^+ and K^+ will result in the exchange of cytoplasmic K^+, the main intracellular cation, for extracellular H^+, the only transportable cation. The resulting intracellular acidification can be terminated by removal of the ionophore, which is accomplished by addition of a scavenger such as defatted serum albumin. If the course of the acidification is monitored with BCECF (e.g., Fig. 5), the extent of acid loading can be varied at will by adding albumin at the appropriate time.[5]

Once the cells have been acid loaded, Na^+/H^+ exchange can be initiated by introducing Na^+ to the medium. Activation of the antiport results in a clear cytoplasmic alkalinization, which can be monitored fluorimetrically (Fig. 5). The specificity of the alkalinization can then be tested by performing similar experiments in the presence of inhibitory doses of amiloride or by substituting nontransported cations such as Rb^+, choline$^+$, or N-methyl-D-glucamine$^+$ for extracellular Na^+.

Reagents

NMG medium: 140 mM N-methyl-D-glucamine chloride, 1 mM KCl, 1 mM CaCl$_2$, 1 mM MgCl$_2$, 10 mM glucose, and 20 mM HEPES-Tris, pH 7.3
Nigericin: 1 mM (in ethanol)

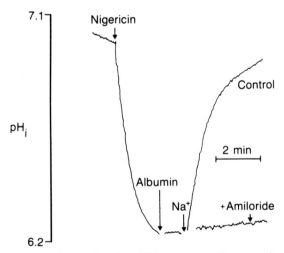

FIG. 5. Acid-loading and activation of Na^+/H^+ exchange. pH_i was monitored fluorimetrically in BCECF-loaded thymocytes suspended in NMG$^+$ medium. Where indicated, nigericin (0.4 μM) was added to initiate acid loading. Next, albumin (5 mg/ml, final) was added to scavenge the ionophore and terminate acid loading. Na^+/H^+ exchange was then activated by addition of 50 mM NaCl with (lower trace) or without (upper trace) 100 μM amiloride.

Essentially fatty acid-free bovine serum albumin: 100 mg/ml in NMG medium

NaCl medium (see above)

Concentrated NaCl stock: 2 M in water

Procedure. The conditions routinely used in our laboratory for rat thymocytes are described. BCECF-loaded cells, suspended in NMG medium at a concentration of 3–5 × 10^6 cells/ml, are transferred to a fluorescence cuvette and their pH$_i$ monitored as described above. To initiate acid loading, nigericin (0.1–0.5 μM) is then added to the suspension. Higher concentrations of nigericin are used when rapid and more pronounced acid loading is required. Lower concentrations are recommended when a thorough removal of the ionophore is desired. The course of the acidification can then be followed by monitoring the resulting decrease in BCECF fluorescence. When the desired pH$_i$ is approached, acid loading is terminated by addition of 5 mg/ml (final) of defatted serum albumin, which scavenges the ionophore.

After acid loading has been completed, Na$^+$/H$^+$ exchange can be activated by introducing Na$^+$ to the medium. The simplest procedure involves direct addition to the cuvette of the desired amount of NaCl from a concentrated (2 M) stock. In this case, however, the tonicity of the medium is increased and the cells shrink. This problem may be circumvented by the following procedure: after acid loading the cells are rapidly sedimented, resuspended in a small (25–50 μl) volume of fresh NMG medium, and this suspension is transferred into a fluorescence cuvette containing NaCl medium or a mixture of NMG and NaCl media containing the desired concentration of Na$^+$. This procedure ensures that isosmolarity is maintained throughout the measurement and further removes any residual nigericin from the membranes. The specificity of the Na$^+$-induced alkalinization can be tested with amiloride, which does not interfere with the acid-loading procedure, or by cation substitution experiments.

Comments. The rate of the Na$^+$/H$^+$ exchange can be quantified precisely as the product of the rate of Na$^+$-induced change of pH$_i$ (in pH units min^{-1}) times the buffering power of the cell (in moles · liter cells^{-1} · pH unit^{-1}). The former is estimated directly from the calibrated fluorescence recording. The buffering power can be measured by a number of procedures, including direct titration of cell lysates or by measuring ΔpH$_i$ while pulsing with weak electrolytes.[6]

The method for acid loading is very reproducible. Under the conditions described, thymocytes consistently attain a minimum pH$_i$ of 6.2–6.4. We have also used it successfully with other types of lymphoid cells,

[6] A. Roos and W. F. Boron, *Physiol. Rev.* **61**, 296 (1981).

neutrophils, and Chinese hamster ovary cells. Similarly, the activation of the antiport is reproducible and applicable to other cell types.

The nigericin method for acid loading is comparatively slow, requiring several minutes for completion. If more rapid acid loading is required, two other methods can be used: (1) pulsing the cells with weak acids, which enter the cell in the protonated form (see above) and (2) preloading the cells with NH_4^+, followed by resuspension in NH_4^+-free solution. In the latter case NH_3, which is formed by the deprotonation of NH_4^+, rapidly leaves the cell, acidifying the cytoplasm.[6]

The Na^+/H^+ exchanger can also be activated without prior acid loading. In some cell types, including thymocytes, the antiport is activated by phorbol esters, diacylglycerol, or growth-promoting factors.[7] In addition, activation of exchange without prior acidification is also obtained by osmotically shrinking the cells. In these cases, the resulting changes in pH_i can be detected by the fluorescence procedure, but the associated volume changes are too small to be reliably detected by electronic sizing. For fluorescence monitoring, the cells are suspended in isotonic NaCl medium and, after a baseline pH_i has been defined, the activators are added or the osmolarity is increased by adding a small volume of concentrated NaCl or N-methyl-D-glucamine chloride.[8]

[7] S. Grinstein and A. Rothstein, *J. Membr. Biol.* **90**, 1 (1986).
[8] S. Grinstein, J. D. Goetz, S. Cohen, and A. Rothstein, *J. Cell Biol.* **101**, 269 (1985).

Author Index

Numbers in parentheses are footnote reference numbers and indicate that an author's work is referred to although the name is not cited in the text.

A

Abbott, R. E., 328
Abbott, W. A., 531, 532(78)
Aberlin, M. E., 175
Abumrad, N. A., 631
Adam, M., 131, 156(52)
Adam, W. J., 639
Adams, J. D., 523, 524(15)
Adams, M. F., 427
Adams-Lackey, M., 293, 300, 301(9), 308(9), 311(9), 312(9), 329(9), 483
Adelberg, E. A., 280
Adesnik, M., 456
Adorante, J. S., 291, 332
Adragna, N., 82, 184, 281
Adrian, R. H., 696, 706(13), 711
Ahmed, H., 533
Ahmed, K., 601
Akedo, H., 583, 584(23)
Akera, T., 650, 660(26)
Akerboom, T. P. M., 524, 525, 526, 527, 529(35), 530, 531(22, 35, 41, 68), 532, 534, 553, 556(28)
Akerman, K. E. O., 94
Al-Saleh, E. A., 127, 128(29)
Albrich, J. M., 762
Albright, C. D., 678
Aldeton, J., 764
Alegra, P., 603
Alifimof, J. K., 184
Allard, W. J., 617
Aloj, S. M., 94
Alpert, A. J., 525
Altamirano, A. A., 293
Altendorf, K., 94
Altman, P. L., 199

Alto, L. E., 664
Alvarez, J., 374, 375
Ambesi-Impiombato, F. S., 94
Ammann, D., 267
Amory, A., 689
Amsler, K., 280
Andersen, B. L., 301, 312(13)
Andersen, O. S., 413, 470, 472(19), 493(19), 494(19)
Andersen, R., 215
Andersen, V., 604
Anderson, D. J., 457
Anderson, H. M., 224
Anderson, J. M., 328
Anderson, M. E., 523, 524, 527
Anderson, M. P., 425, 431(13)
Anderson, R. A., 381, 386, 387, 389, 517
Anderson, S., 334
Andreasen, P., 620
Andreoli, T. E., 288
Andrews, J., 634
Anggard, E. E., 537
Ansay, M., 251
Antonioli, J. A., 126, 131(21, 22), 613
Anwer, M. S., 531
Apel, L. E., 664, 673(5)
Apitule, M. E. A., 528
Aranibar, N., 276
Archer, E. G., 606
Archibald, A. L., 251
Arias, I. M., 528, 529, 532
Armbrustmacher, P., 312
Arnone, A., 513
Aronson, P. S., 293, 777
Ash, O., 184
Ashley, D. L., 200
Aspen, A. J., 582

Assimacopoulos-Jeannet, F. D., 540
Atkins, G. L., 139
Atkinson, P. H., 565
Atschuld, R. A., 664, 673(5)
Aubby, D. S., 232
Aubert, L., 302
Augustin, H. W., 233
Ausielo, D., 679
Auvil, J., 233
Avendaño, C., 607, 608(73)
Avruch, J., 689
Aw, T. Y., 523, 527, 528, 529, 532(61)
Awasthi, Y. C., 530, 533

B

Baar, S., 130
Babson, J. R., 525
Bakay, B., 724
Bakeeva, L. E., 94
Baker, G. F., 241, 243, 245(39), 615
Baker, R. M., 725
Bala, T., 538, 539
Baldi, C., 533
Baldwin, S. A., 580, 617
Ballantine, M., 423, 432(4), 456
Ballasteros, P., 607, 608(73)
Ballatori, N., 530, 531
Balshin, M., 413, 416(22)
Bang, O., 239
Bannai, S., 125, 577, 609
Barber, E. F., 568
Barberini, F., 603
Bareford, D., 130
Barker, J., 80
Barlet, C., 679
Barnhart, J. L., 527
Barnola, F. V., 688
BarNoy, S., 410, 419(3), 422
Baroin, A., 293
Barritt, G. J., 540
Barry, W. H., 639, 649, 660(24), 661
Bartha, F., 518
Bartley, W., 716
Bartoli, G. M., 523, 524(14), 526, 530(34)
Barton, T. C., 194, 200
Barzilay, M., 413, 418
Basketter, D. A., 241
Bass, D., 632

Bassel, P. S., 411
Battocletti, J. H., 200
Batty, I. R., 535
Baukal, A. J., 538, 539
Baumbach, G. A., 565
Bauriedel, G., 639, 640(14), 641, 643
Bazer, F. W., 565
Beatty, P. W., 525
Beauge, L. A., 293, 696, 698, 699, 700, 705(10), 710
Becker, J. E., 574, 586
Belcher, R. V., 525
Bellomo, G., 533
Belt, J. A., 301, 329(25)
Bender, D. A., 126
Benga, G., 200
Bennett, P. M., 381
Bennett, V., 6, 381, 389, 391(39), 498, 513, 515, 516, 517
Benos, D. J., 295, 336
Bensadoun, A., 570
Beredsen, W., 231
Berendsen, W., 230
Berger, H. J., 649
Berggren, M., 523
Bergmann, W. L., 223
Berkoff, H. A., 673
Berkowitz, L. R., 282
Berlin, J. R., 675
Berne, R. M., 251
Berridge, M. J., 535, 536
Berry, C. A., 222
Berry, M. N., 564
Bers, D. M., 770
Berthon, B., 542, 545(27)
Bertles, J. F., 37
Bertrand, O., 467
Besi, E., 305, 443
Bessis, M., 3, 12(1), 36, 40(1), 42, 43, 44(9), 45, 48, 51
Besterman, J. M., 336
Beutler, E., 50, 383, 525, 530, 532, 533
Bevers, E. M., 231
Beyer, E., 300, 301(3), 307(3, 7), 308(7), 309(3, 7), 310(3, 7), 311(7), 312(3, 7), 326(3, 7), 329(3, 7)
Bhattacharyya, P., 559
Bieber, L. L., 571
Biedert, S., 639
Biempica, L., 529

O

P

Subject Index

A

F

Facilitated diffusion, 616, 725
  integrated rate equations for, 726–727
  rapidity of, 714–715
Fast-flow techniques, to measure sugar
  fluxes, in RBC, 247–248
Fast transport technique, for RBC, 160–
  175
  effect of temperature, 173–175
FDNB. *See* Fluorodinitrobenzene
Ferritin, escape from ghosts, 360–363
Fibroblasts
  CV-1 African Green Monkey, tightly
    adherent monolayers of, assay of
    ouabain receptors in, 686–687
  in suspension, ouabain binding assay,
    683–686
Filter technique, for efflux studies, 256–
  257
Filtration, of water, across membranes,
  192
Flow tube technique, for studying rapid
  transport, 161–175
Flufenamate, inhibitor of band 3 protein-
  mediated anion exchange, 461–462
Fluorescein
  excitation profile, pH dependence of,
    745
  measurement of pH and intracellular
    calcium in living cells using, optical
    system for, 751–752
Fluorescence microscopy, for measurement of vacuolar pH and cytoplasmic
  calcium in living cells, 745–771
  calcium measurement protocol, 762–771
  calibration of calcium measurement,
    769–771
  image analysis, 752–755
    for local measurement, 766–769
  indicator dyes, 746–749
  instrumentation, 749–752
  loading dye into cytoplasm, 763–765
  optical systems for, 751–752
  pH measurement protocol, 755–762
    calibration curves, 761–762
    by image analysis, 757–761
    by photometry, 756–757
  principles of, 745–746
  whole-cell photometry, 765

Fluorescent chromophore, to measure
  water transport in cells and vesicles,
  222
Fluorescent indicator dyes
  environmental influences on, 747–749
  measurement using, principle of, 749
Fluorodinitrobenzene, 406
  inhibition of monocarboxylate transport
    in RBC, 325
Fructose transport, 617, 629–630, 632–633
Fura2, 105
  fluorescence
    calcium dependence of, 747
    environmental influences on, 747–749
  in hepatocytes, for analysis of hormone
    effects on cytosolic free Ca^{2+}, 541–
    545
  loading into cytoplasm, 763–765
  in measurement of intracellular calcium,
    746–752, 762–771
    advantages over quin2, 747
    in atrial and ventricular myocytes,
      674–675
  measurement of pH and intracellular
    calcium in living cells using, optical
    system for, 751–752
Furosemide, inhibition of cation–anion
  cotransport, in RBC, 282

G

G8 cells, attached to substratum, assay of
  Mg^{2+} influx, 558–559
Gene products, designations for, 580–581
Ghost membranes
  characteristics of, 496
  preparation of, 496–498
Ghosts
  cell water volume of, 441
  chemical modification of critical argi-
    nines with phenylglyoxal
    from both sides of membrane, 483–487
    from inside of membrane, 487–490
  digestion of, by pepsin, 427–428
  extracellular chymotrypsin treatment,
    471
  fractional flotation of, on dextran barri-
    ers, as function of hole radius, 365–
    367
  fragmentation of, 46–47, 50